Highly Efficient OLEDs

Highly Efficient OLEDs

Materials Based on Thermally Activated Delayed Fluorescence

Edited by Hartmut Yersin

Editor

Hartmut Yersin
University of Regensburg, Germany
Universitätsstr. 31
93053 Regensburg
Germany

All books published by **Wiley-VCH** are carefully produced. Nevertheless, authors, editors, and publisher do not warrant the information contained in these books, including this book, to be free of errors. Readers are advised to keep in mind that statements, data, illustrations, procedural details or other items may inadvertently be inaccurate.

Library of Congress Card No.: applied for

British Library Cataloguing-in-Publication Data
A catalogue record for this book is available from the British Library.

Bibliographic information published by the Deutsche Nationalbibliothek
The Deutsche Nationalbibliothek lists this publication in the Deutsche Nationalbibliografie; detailed bibliographic data are available on the Internet at <http://dnb.d-nb.de>.

© 2019 Wiley-VCH Verlag GmbH & Co. KGaA, Boschstr. 12, 69469 Weinheim, Germany

All rights reserved (including those of translation into other languages). No part of this book may be reproduced in any form – by photoprinting, microfilm, or any other means – nor transmitted or translated into a machine language without written permission from the publishers. Registered names, trademarks, etc. used in this book, even when not specifically marked as such, are not to be considered unprotected by law.

Print ISBN: 978-3-527-33900-6
ePDF ISBN: 978-3-527-69173-9
ePub ISBN: 978-3-527-69175-3
oBook ISBN: 978-3-527-69172-2

Cover Design Formgeber, Mannheim, Germany
Typesetting SPi Global, Chennai, India
Printing and Binding C.O.S. Printers Pte Ltd, Singapore

Printed on acid-free paper

10 9 8 7 6 5 4 3 2 1

Contents

Preface *xv*

1 TADF Material Design: Photophysical Background and Case Studies Focusing on Cu(I) and Ag(I) Complexes *1*
Hartmut Yersin, Rafał Czerwieniec, Marsel Z. Shafikov, and Alfiya F. Suleymanova

1.1 Introduction *1*
1.2 TADF, Molecular Parameters, and Diversity of Materials *4*
1.2.1 TADF and Phosphorescence *6*
1.2.2 Minimizing $\Delta E(S_1-T_1)$ *7*
1.2.3 Importance of $k_r(S_1-S_0)$ *7*
1.3 Case Study: TADF of a Cu(I) Complex with Large $\Delta E(S_1-T_1)$ *15*
1.3.1 DFT and TD-DFT Calculations *16*
1.3.2 Flattening Distortions and Nonradiative Decay *16*
1.3.3 TADF Properties *18*
1.3.4 Radiative $S_1 \rightarrow S_0$ Rate, Absorption, and Strickler–Berg Relation *20*
1.4 Case Study: TADF of a Cu(I) Complex with Small $\Delta E(S_1-T_1)$ *22*
1.4.1 DFT and TD-DFT Calculations *22*
1.4.2 Emission Spectra and Quantum Yields *23*
1.4.3 The Triplet State T_1 and Spin–Orbit Coupling *23*
1.4.4 Temperature Dependence of the Emission Decay Time and TADF *28*
1.5 Energy Separation $\Delta E(S_1-T_1)$ and $S_1 \rightarrow S_0$ Fluorescence Rate *30*
1.5.1 Experimental Correlation Between $\Delta E(S_1-T_1)$ and $k_r(S_1 \rightarrow S_0)$ for Cu(I) Compounds *31*
1.5.2 Quantum Mechanical Considerations *32*
1.6 Design Strategies for Highly Efficient Ag(I)-Based TADF Compounds *34*
1.6.1 Ag(phen)(P_2-nCB): A First Step to Achieve TADF *34*
1.6.2 Emission Quenching in Ag(phen)(P_2-nCB) *36*
1.6.3 Sterical Hindrance. Tuning of the Emission Quantum Yield up to 100% *38*
1.6.4 Detailed Characterization of Ag(dbp)(P_2-nCB) *40*
1.7 Conclusion and Future Perspectives *45*
Acknowledgments *46*
References *46*

2   Highly Emissive d^{10} Metal Complexes as TADF Emitters with Versatile Structures and Photophysical Properties *61*
Koichi Nozaki and Munetaka Iwamura
2.1   Introduction *61*
2.2   Phosphorescence and TADF Mechanisms *62*
2.3   Structure-Dependent Photophysical Properties of Four-Coordinate [Cu(N^N)$_2$] Complexes *64*
2.4   Flattening Distortion Dynamics of the MLCT Excited State *76*
2.5   Green and Blue Emitters: [Cu(N^N)(P^P)] and [Cu(N^N)(P^X)] *77*
2.6   Three-Coordinate Cu(I) Complexes *79*
2.7   Dinuclear Cu(I) Complexes *80*
2.8   Ag(I), Au(I), Pt(0), and Pd(0) Complexes *84*
2.9   Summary *85*
         References *86*

3   Luminescent Dinuclear Copper(I) Complexes with Short Intramolecular Cu–Cu Distances *93*
Akira Tsuboyama
3.1   Introduction *93*
3.2   Overview of Luminescent Dinuclear Copper(I) Complexes *94*
3.2.1   Structure *94*
3.2.2   Luminescence Properties *99*
3.3   Structural and Photophysical Studies of the Dinuclear Copper(I) Complexes: [Cu(μ-C∧N)]$_2$ (C∧N = 2-(bis(trimethylsilyl)methyl) pyridine Derivatives) *100*
3.3.1   Outline *100*
3.3.2   X-ray Crystallographic Study *101*
3.3.3   Photophysical Properties *102*
3.3.3.1 Absorption Spectrum *102*
3.3.3.2 DFT Calculation *103*
3.3.3.3 Emission Properties *104*
3.3.3.4 Emission Decay Kinetic Analysis *105*
3.3.4   OLED Device *110*
3.3.5   Experimental *111*
3.3.5.1 Synthesis *111*
3.3.5.2 Measurement, Calculation, and Device *111*
3.3.5.3 X-ray Structure Analysis *112*
3.3.5.4 DFT Calculation *112*
3.3.5.5 OLED Device *112*
3.4   Conclusion *112*
         Acknowledgment *113*
         References *114*

4	**Molecular Design and Synthesis of Metal Complexes as Emitters for TADF-Type OLEDs** *119*
	Masahisa Osawa and Mikio Hoshino
4.1	Introduction *119*
4.2	Cu(I) Complexes for OLEDs *122*
4.2.1	Energy Levels of Molecular Orbitals in Tetrahedral Geometries *122*
4.2.2	Ligand Variation *123*
4.3	Mononuclear Cu(I) Complexes for OLEDs *126*
4.3.1	Bis(diimine) Type *131*
4.3.2	[Cu(NN)(PP)]$^+$ Complexes with phen or bipy Derivatives as Ligands *131*
4.3.3	[Cu(NN)(PP)]$^+$ Complexes with NN Ligands Other Than phen or bipy Derivatives *134*
4.3.4	Tetrahedral Cu(I) Complexes with the LUMO on the PP Ligand *142*
4.3.5	Charge-Neutral Three-Coordinate Cu(I) Complexes *146*
4.4	Dinuclear Cu(I) Complexes for OLEDs *155*
4.4.1	Dinuclear Cu(I) Complexes Possessing {Cu$_2$(μ-X)$_2$} Cores *155*
4.4.2	Other Dinuclear Cu(I) Complexes *157*
4.5	Another Group of Metal Complexes Exhibiting TADF *157*
4.6	Conclusion *160*
	Acknowledgments *160*
	Appendix *161*
4.A.1	Schematic Structures of 1–86 *161*
4.A.2	Abbreviations and Molecular Structures of Materials for OLEDs *168*
	References *171*

5	**Ionic [Cu(NN)(PP)]$^+$ TAD9727 F Complexes with Pyridine-based Diimine Chelating Ligands and Their Use in OLEDs** *177*
	Rongmin Yu and Can-Zhong Lu
5.1	Introduction *177*
5.2	The Influence of Molecular and Electronic Structure on Emissive Properties of Cu(I) Complexes *178*
5.3	Heteroleptic Diimine/Diphosphine [Cu(NN)(PP)]$^+$ Complexes with Pyridine-Based Ligand *181*
5.3.1	[Cu(NN)(PP)]$^+$ Complexes with 2,2′-bipyridyl-based Ligands *181*
5.3.1.1	[Cu(NN)(PP)]$^+$ Complexes with 2-(2′-pyridyl)benzimidazole and 2-(2′-pyridyl)imidazole-based Ligands *182*
5.3.2	[Cu(NN)(PP)]$^+$ Complexes with 5-(2-pyridyl)tetrazole-based Ligands *185*
5.3.3	[Cu(NN)(PP)]$^+$ Complexes with 3-(2′-pyridyl)-1,2,4-triazole-based Ligands *187*
5.3.4	[Cu(NN)(PP)] Complexes with 2-(2-pyridyl)-pyrrolide-based Ligands *188*

5.3.5	[Cu(NN)(PP)]$^+$ Complexes with 1-(2-pyridyl)-pyrazole-based Ligands *189*	
5.3.6	[Cu(NN)(PP)]$^+$ Complexes with Carbazolyl-modified 1-(2-pyridyl)-pyrazole-based Ligands *191*	
5.3.7	[Cu(NN)(PP)]$^+$ Complexes with 1-phenyl-3-(2-pyridyl)pyrazole-based Ligands *192*	
5.3.8	[Cu(NN)(PP)]$^+$ Complexes with 3-phenyl-5-(2-pyridyl)-1H-1,2,4-triazole-based Ligands *193*	
5.4	Conclusion and Perspective *194*	
	References *195*	

6 Efficiency Enhancement of Organic Light-Emitting Diodes Exhibiting Delayed Fluorescence and Nonisotropic Emitter Orientation *199*
Tobias D. Schmidt and Wolfgang Brütting

6.1 Introduction *199*
6.2 OLED Basics *200*
6.2.1 Working Principle *200*
6.2.2 Electroluminescence Quantum Efficiency *202*
6.2.3 Delayed Fluorescence *203*
6.2.4 Nonisotropic Emitter Orientation *204*
6.2.5 Optical Modeling *205*
6.3 Comprehensive Efficiency Analysis of OLEDs *206*
6.4 Case Studies *209*
6.4.1 Treating the OLED as a Black Box *209*
6.4.2 Highly Efficient Thermally Activated Delayed Fluorescence Device *214*
6.4.3 Low Efficiency Roll-Off Triplet–Triplet Annihilation Device *218*
6.5 Conclusion *222*
    Acknowledgments *223*
    References *223*

7 TADF Kinetics and Data Analysis in Photoluminescence and in Electroluminescence *229*
Tiago Palmeira and Mário N. Berberan-Santos

7.1 TADF Kinetics *229*
7.1.1 Introduction *229*
7.1.2 Excitation Types *231*
7.1.3 Photoexcitation *232*
7.1.3.1 Rate Equations *232*
7.1.3.2 Fluorescence and Phosphorescence Decays *232*
7.1.3.3 Steady-state Fluorescence and Phosphorescence Intensities *233*
7.1.3.4 Excited-state Cycles *235*
7.1.3.5 TADF Onset Temperature *238*
7.1.3.6 Conditions for Efficient TADF *239*

7.1.4	Electrical Excitation	*240*
7.1.4.1	Steady State	*240*
7.1.4.2	Conditions for Efficient Electroluminescence	*241*
7.1.5	More Complex Schemes	*244*
7.2	TADF Data Analysis	*245*
7.2.1	Introduction	*245*
7.2.2	Steady-state Data	*245*
7.2.2.1	Delayed Fluorescence and Phosphorescence Intensities as a Function of Temperature: Rosenberg–Parker Method	*245*
7.2.2.2	Prompt and Delayed Fluorescence Intensities as a Function of Temperature	*245*
7.2.2.3	Delayed Fluorescence Intensity as a Function of Temperature	*249*
7.2.3	Decay Data	*249*
7.2.4	Combined Steady-state and Decay Data	*250*
7.2.4.1	Linear Relation Between Delayed Fluorescence Lifetime and Intensity Ratio	*250*
7.2.4.2	Linearized Relation for the Determination of ΔE_{ST}	*250*
7.3	Conclusion	*252*
	Acknowledgment	*252*
	References	*252*

8 Intersystem Crossing Processes in TADF Emitters *257*

Christel M. Marian, Jelena Föller, Martin Kleinschmidt, and Mihajlo Etinski

8.1	Introduction	*257*
8.1.1	Electroluminescent Emitters	*257*
8.1.2	Thermally Activated Delayed Fluorescence	*258*
8.2	Intersystem Crossing Rate Constants	*259*
8.2.1	Condon Approximation	*260*
8.2.1.1	Electronic Spin–Orbit Coupling Matrix Elements	*261*
8.2.1.2	Overlap of Vibrational Wave Functions	*262*
8.2.2	Beyond the Condon Approximation	*263*
8.2.3	Computation of ISC and rISC Rate Constants	*264*
8.2.3.1	Classical Approach	*265*
8.2.3.2	Statical Approaches	*265*
8.2.3.3	Dynamical Approaches	*265*
8.3	Excitation Energies and Radiative Rate Constants	*266*
8.3.1	Time-Dependent Density Functional Theory	*266*
8.3.2	DFT-Based Multireference Configuration Interaction	*267*
8.3.3	Fluorescence and Phosphorescence Rates	*268*
8.4	Case Studies	*269*
8.4.1	Copper(I) Complexes	*269*
8.4.1.1	Three-Coordinated Cu(I)–NHC–Phenanthroline Complex	*270*
8.4.1.2	Four-coordinated Cu(I)–bis-Phenanthroline Complexes	*275*
8.4.2	Metal-Free TADF Emitters	*277*
8.4.2.1	1,2,3,5-Tetrakis(carbazol-9-yl)-4,6-dicyanobenzene (4CzIPN)	*279*

8.4.2.2	Mechanism of the Triplet-to-Singlet Upconversion in the Assistant Dopants ACRXTN and ACRSA *282*	
8.5	Outlook and Concluding Remarks *285*	
	References *286*	

9 The Role of Vibronic Coupling for Intersystem Crossing and Reverse Intersystem Crossing Rates in TADF Molecules *297*
Thomas J. Penfold and Jamie Gibson

9.1	Introduction *297*	
9.1.1	Background to Delayed Fluorescence *300*	
9.1.2	The Mechanism of rISC *302*	
9.2	Beyond a Static Description *303*	
9.2.1	Obtaining the Potential Energy Surfaces *304*	
9.2.1.1	Vibronic Coupling Model Hamiltonian *306*	
9.2.2	Solving for the Motion of the Nuclei *309*	
9.2.2.1	Multiconfigurational Time-Dependent Hartree Approach *310*	
9.2.2.2	Density Matrix Formalism of MCTDH: ρMCTDH *311*	
9.3	Case Studies *312*	
9.3.1	Ultrafast Dynamics of a Cu(I)–phenanthroline Complex *313*	
9.3.2	The Contribution of Vibronic Coupling to the rISC of PTZ-DBTO2 *316*	
9.4	Conclusions and Outlook *322*	
	References *323*	

10 Exciplex: Its Nature and Application to OLEDs *331*
Hwang-Beom Kim, Dongwook Kim, and Jang-Joo Kim

10.1	Introduction *331*	
10.2	Formation and Electronic Structures of Exciplexes *332*	
10.3	Optical Properties of Exciplexes *336*	
10.3.1	Photoluminescence of Exciplexes *336*	
10.3.2	Absorption Spectra of Exciplexes *338*	
10.4	Decay Processes of the Exciplex in Solution *339*	
10.4.1	Fluorescence Rate Constant for the Exciplex State *340*	
10.4.2	Contact Radical Ion Pair (CRIP) Versus Solvent-separated Radical Ion Pair (SSRIP) *342*	
10.4.3	Charge Separation Versus Charge Recombination *343*	
10.4.4	Intersystem Crossing (ISC) in the Exciplex *345*	
10.5	Exciplexes in Organic Solid Films *346*	
10.5.1	Prompt Versus Delayed Fluorescence *347*	
10.5.2	Spectral Shift as a Function of Time *350*	
10.6	OLEDs Using Exciplexes *353*	
10.6.1	Exciplexes as Emitters *353*	
10.6.2	Exciplexes as Sensitizers *356*	
10.7	Summary and Outlook *360*	
	Appendix *360*	
10.A.1	Small Molecular Pairs of Donors and Acceptors Forming Exciplexes *360*	

10.A.2	Small Molecules with Electron-donating Moieties Forming Exciplexes *360*	
10.A.3	Small Molecules with Electron-accepting Moieties Forming Exciplexes *365*	
10.A.4	Small Molecules with Electron-donating and Electron-accepting Moieties Forming Exciplexes *368*	
	References *370*	

11 Thermally Activated Delayed Fluorescence Materials Based on Donor–Acceptor Molecular Systems *377*
Ye Tao, Runfeng Chen, Huanhuan Li, Chao Zheng, and Wei Huang

11.1	Introduction *377*	
11.2	TADF OLEDs *380*	
11.2.1	Device Structures and Operation Mechanisms of TADF OLED *380*	
11.2.2	TADF Molecules as Emitters for OLEDs *382*	
11.2.3	TADF Molecules as Host Materials and Sensitizers for OLEDs *382*	
11.2.4	Host-free TADF OLEDs *383*	
11.3	Basic Considerations in Molecular Design of TADF Molecules *384*	
11.3.1	Design Principles of Donor–Acceptor Molecular Systems for TADF Emission *384*	
11.3.2	Control of Singlet–Triplet Energy Splitting (ΔE_{ST}) *386*	
11.3.3	Modulation of Luminescent Efficiency of TADF Emission *389*	
11.4	Typical Donor–Acceptor Molecular Systems with High TADF Performance *391*	
11.4.1	Cyano-based TADF Molecules *391*	
11.4.2	Nitrogen Heterocycle-based TADF Molecules *396*	
11.4.3	Diphenyl Sulfoxide-based TADF Molecules *405*	
11.4.4	X-bridged Diphenyl Sulfoxide-based TADF Molecules *407*	
11.4.5	Diphenyl Ketone-based TADF Molecules *408*	
11.4.6	X-bridged Diphenyl Ketone TADF Molecules *410*	
11.5	Organoboron-based TADF Molecules *411*	
11.6	TADF Polymers *412*	
11.7	Intermolecular D–A System for TADF Emission *413*	
11.8	Summary and Outlook *417*	
	References *417*	

12 Photophysics of Thermally Activated Delayed Fluorescence *425*
Andrew Monkman

12.1	Introduction *425*	
12.2	Comments on the Techniques Used in Our Studies *428*	
12.3	Basic Absorption and Emission Properties *428*	
12.4	Phosphorescence and Triplet State Measurements *438*	
12.5	Characteristics of the Delayed Fluorescence *440*	
12.5.1	Time-resolved Emission in Solution *440*	
12.5.2	Time-resolved Emission in Solid State *446*	

12.5.3	Kinetics of the ^{1}CT Prompt State	*449*
12.6	Understanding Which Excited States are Involved	*450*
12.7	Excited-state Properties	*452*
12.8	Dynamical Processes	*455*
12.9	Emitter–host Interactions	*457*
12.10	Energy Diagram for TADF	*459*
12.11	Final Comments	*459*
	Acknowledgments	*461*
	References	*461*

13 Thioxanthone (TX) Derivatives and Their Application in Organic Light-emitting Diodes *465*
Xiaofang Wei, Ying Wang, and Pengfei Wang

13.1	Organic Light-emitting Diodes	*465*
13.2	Pure Organic TADF Materials in OLEDs	*467*
13.3	TX Derivatives for OLED	*468*
13.3.1	High Efficient OLEDs Based on TX-based TADF Materials	*468*
13.3.1.1	Design and Characterization of TX-based TADF Emitters	*468*
13.3.1.2	Nondoped OLEDs Based on TADF Emitters with Quantum Well Structure	*481*
13.3.1.3	White OLEDs Based on Blue Fluorescent Emitter and Yellow TX-based TADF Emitters	*486*
13.3.2	TADF Host for Phosphorescent Emitters	*490*
13.4	Concluding Remarks and Outlook	*495*
	Acknowledgments	*496*
	References	*496*

14 Solution-Processed TADF Materials and Devices Based on Organic Emitters *501*
Nidhi Sharma, Michael Yin Wong, Ifor D.W. Samuel, and Eli Zysman-Colman

14.1	Introduction	*501*
14.1.1	Solution-Processed Blue TADF Materials and Devices	*504*
14.1.2	Solution-Processed Green TADF Materials and Devices	*512*
14.1.3	Solution-Processed Yellow-to-Red TADF Materials and Devices	*523*
14.1.4	Comparison of State-of-the-Art Solution-Processed OLEDs to Vacuum-Deposited Counterparts	*526*
14.1.5	Solution-Processed TADF Polymers and Dendrimers	*527*
14.2	Summary and Outlook	*537*
	References	*538*

15 Status and Next Steps of TADF Technology: An Industrial Perspective *543*
Alhama Arjona-Esteban and Daniel Volz

15.1	What Does the Market Want?	*543*
15.1.1	The Emitter Materials: Heart of the OLED	*544*
15.1.2	Processing Aspects	*547*

15.1.3 Sustainability Aspects *549*
15.1.3.1 Availability Issues *549*
15.1.3.2 Recycling Considerations *550*
15.1.4 Realization of Efficient and Stable Blue OLEDs *550*
15.1.4.1 The Blue Gap *550*
15.1.4.2 Key Performance Indicators *551*
15.2 Mastering Blue OLEDs with TADF Technology *552*
15.2.1 Current Status of Blue TADF Technology: Academia *552*
15.2.2 Current Status of Blue TADF Technology: Industry *554*
15.3 An Alternative Approach: TADF Emitters as (Co) Hosts *559*
15.3.1 General Remarks *559*
15.3.2 First Attempts of Using TADF as Hosts *561*
15.3.3 Discussion of Various Concepts *562*
15.3.3.1 TADF as Host for Other TADF Emitters *562*
15.3.3.2 TADF as Host for Fluorescent Materials *563*
15.3.3.3 TADF as Host for Phosphorescent Emitters *564*
15.4 Outlook: What to Expect from TADF Technology in the Future *566*
References *567*

Index *573*

Preface

Organic light emitting diodes (OLEDs) are already commercially applied for smartphone and TV displays. And OLED lighting comes in the focus of interests. Currently in commercial applications, the process of OLED light emission is mainly based on iridium metal complexes that are, however, expensive since the iridium metal is very rare. The Ir(III) center has the advantage of inducing high spin–orbit coupling that is important for harvesting all generated singlet (25%) and triplet (75%) excitons in the lowest triplet state and for transforming the excitation energy with almost 100% internal efficiency into light. This *triplet harvesting mechanism* was already proposed about 20 years ago. However, a disadvantage of Ir-based light emitters is not only given by the high costs of Ir, in particular, if mass production for OLED illumination is aspired, but its use is also problematic for blue light emission due to stability problems. As a consequence, an alternative OLED light emitting mechanism has been proposed at the University of Regensburg already in 2006. It allows one to exploit the effect of thermally activated delayed fluorescence (TADF) in OLEDs for harvesting all generated triplet and singlet excitons through emission from the lowest excited singlet state. This leads to the *singlet harvesting mechanism*. The great advantage of this mechanism is given by the fact that high TADF and device efficiency can be achieved with low-cost Cu or Ag complexes or purely organic molecules, even for the blue spectral range. Meanwhile, the development of this new technology is characterized by an exceptional interdisciplinary research in the fields of chemistry, physics, and material sciences.

In this volume, leading scientists present comprehensive reviews, which provide insight into TADF properties of organometallic and purely organic emitters, the mechanisms of electroluminescence, the development of new emitter and host materials, and device structures. The different contributions are written in a style that enables researchers from related fields and industrial laboratories as well as graduate students to follow the highly informative presentations. I am convinced that this book demonstrates the attractiveness and the great potential of TADF compounds and triggers further studies towards a better understanding of optoelectronic properties and mechanisms. These studies will not only open larger-scale application of OLED displays and lighting systems, but will also stimulate future progress in organic electronics. For instance, it is attractive to advance a new OLED generation based on *direct* singlet harvesting, using emitter molecules with almost zero-engery gap between the lowest excited triplet and singlet state, or to develop electrically pumped OLED lasers.

Regensburg, Germany *Hartmut Yersin*
July 2018

1

TADF Material Design: Photophysical Background and Case Studies Focusing on Cu(I) and Ag(I) Complexes[a]

Hartmut Yersin[1,b]*, Rafał Czerwieniec*[1]*, Marsel Z. Shafikov*[1,2]*, and Alfiya F. Suleymanova*[1]

[1] *University of Regensburg, Department of Chemistry, Institute of Physical Chemistry, Universitätsstr. 31, Regensburg, D-93053, Germany*
[2] *Ural Federal University, Department of Technology of Organic Synthesis, Institute of Chemical Technology, Mira str. 19, Ekaterinburg, 620002, Russia*

1.1 Introduction

Basic research of photophysical and chemical properties of organo-transition metal compounds was strongly activated by their potential commercial use. This became particularly apparent for classes of compounds that may be applied as emitters in organic light-emitting diodes (OLEDs) [1–19] or in light emitting electrochemical cells (LEEC) [7, 20–28]. These scientific investigations led to a much deeper understanding of the photophysical principles and of the compound's properties resulting in the development of an enormous number of new materials in part with drastically improved properties for OLED applications [4, 9, 29–52]. Improvements were also stimulated in the fields of related functional materials based on metal complexes for sensing of oxygen or temperature [53–59] or for photocatalysis [60–66].

For luminescent materials to be applied in OLEDs, it is essential that all excitons generated in the emission layer are harvested and converted into photons. Since the statistic ratio of the formed excitons is 1 singlet to 3 triplets [67, 68], special mechanisms that allow to harvest all of them are required, as the two types of excitons show different relaxation properties [67]. Already about 20 years ago, it was discovered that third-row transition metal complexes, especially those with Ir(III), Pt(II), or Os(II) metal centers, are well suited for such harvesting processes, since the metal centers can induce efficient spin–orbit coupling (SOC) [69–77] between the lowest triplet state T_1 and higher-lying singlet states S_n (with $n > 1$) [1, 9, 11, 12, 67, 69–89]. As a consequence, fast intersystem crossing (ISC) to the lowest triplet state of several tens of femtosecond [88, 90] can occur, and relatively high radiative phosphorescent rates from the T_1 state to the electronic

---

[a] Previously published in ChemPhysChem, 2017, 18, 3508–3535 with permission by Wiley-VCH.
[b] Author for correspondence: ORCID ID: http://orcid.org/0000-0003-3216-1370.

Highly Efficient OLEDs: Materials Based on Thermally Activated Delayed Fluorescence,
First Edition. Edited by Hartmut Yersin.
© 2019 Wiley-VCH Verlag GmbH & Co. KGaA. Published 2019 by Wiley-VCH Verlag GmbH & Co. KGaA.

ground state S_0 are induced. These latter rates can become as high as $\approx 10^6$ s^{-1} [70, 89, 91]. Therefore, these phosphorescent compounds are frequently denoted as triplet emitters. As a consequence, when applied in OLEDs, these materials can harvest all singlet and triplet excitons in the lowest excited triplet state. Accordingly, the corresponding mechanism is denoted as *triplet harvesting effect* [69, 78]. Indeed, using, for example, Ir(ppy)$_3$ (with ppy = 2-phenylpyridinate), OLEDs with almost 100% internal quantum efficiency could be produced [11, 81].

However, these triplet emitter complexes require high-cost rare metals, and this may become a limiting factor, when OLED lighting goes into mass production [70, 92, 93]. Therefore, an alternative harvesting mechanism that may work with low-cost materials has been proposed more than one decade ago [94]. This mechanism is based on the molecular effect of thermally activated delayed fluorescence (TADF), according to which also all excitons generated in the emission layer may be harvested. In this situation, however, the emission does not stem from the lowest excited triplet state, but from the thermally activated singlet state S_1. Hence, this mechanism has been denoted as *singlet harvesting mechanism* [9, 30, 31, 37, 40, 56, 70, 92, 94–97]. Accordingly, the luminescent materials do not need to contain high SOC-inducing metal centers (high-cost materials), since the (thermally activated) singlet state usually carries sufficient allowedness with respect to the transition to the singlet ground state (spin-allowed transition). Therefore, an efficient path for photon generation becomes available.

Obviously, thermal activation, from the lowest triplet state T_1 to the higher-lying singlet state S_1, requires a relatively small energy separation $\Delta E(S_1-T_1)$ between these states. For example, at ambient temperature ($T = 300$ K), a thermal energy of $k_B T \approx 210$ cm^{-1} (26 meV) is available (k_B = Boltzmann constant). Hence, as a rule of thumb, efficient thermal activation with fast up-ISC or reverse intersystem crossing (rISC) is not expected to occur for $\Delta E(S_1-T_1)$ distinctly above 10^3 cm^{-1} (≈ 130 meV). Indeed, such energy separations can be realized with environmentally friendly and low-cost Cu(I) [4, 9, 27, 30–35, 37–56, 70, 92, 93, 95–105] and Ag(I) [104, 106–109] complexes as well as with purely organic molecules [110–113].

In this review, we focus on Cu(I) and Ag(I) complexes. For these materials, we have to address three crucial photophysical requirements:

1) The emitter compounds should exhibit high photoluminescence quantum yields ϕ_{PL}. After an electronic excitation, however, Cu(I) and Ag(I) complexes experience distinct flattening distortions with respect to the ground state geometries [114–123]. Usually, such geometry changes are related to an increase of nonradiative deactivation or even result in total quenching of the luminescence by vibrational relaxation. This is induced by a strong increase of the Franck–Condon factors of the low-lying vibrational modes of the excited electronic state and highly excited vibrational modes of the electronic ground state [124–126]. However, these shortcomings may be suppressed to a large extent by rigidifying the molecular structure either by sterically demanding ligands or by a rigid environment. This behavior has already been discussed frequently in the literature [9, 30, 31, 35, 92, 95, 96, 108, 109, 114, 119, 127]. We will address these properties in Sections 1.3 and 1.6. Interestingly, design of a material, a silver complex, with Φ_{PL} of 100% becomes possible

by following this strategy of rigidifying the molecular structure (see Refs [108, 109] and Section 1.6).

2) Well-designed TADF materials should exhibit relatively small energy separations $\Delta E(S_1-T_1)$. For organo-transition metal compounds, this is related to the occurrence of metal-to-ligand charge-transfer (MLCT) (and ligand-to-ligand charge-transfer (LL'CT)) states having frontier orbitals, highest occupied molecular orbital (HOMO) and lowest unoccupied molecular orbital (LUMO), which are spatially largely separated. This leads to a small exchange interaction [128–130] between the involved electrons and, hence, to the required small splitting between the singlet S_1 and triplet T_1 state. In particular, a relatively small $\Delta E(S_1-T_1)$ value is a necessary condition to obtain a short radiative TADF decay time, which is important to maximize the photoluminescence quantum yield Φ_{PL}. Moreover, for use in OLEDs, short decay times are important to minimize roll-off effects (for example, induced by saturation or triplet-polaron quenching) and to reduce device stability problems as well as undesired energy transfer processes from the emitter dopant to the host. In Sections 1.3 and 1.4, discussing case studies, we will present Cu(I) compounds and the dependence of photophysical properties on the $\Delta E(S_1-T_1)$ gap.

3) The TADF properties crucially depend on the allowedness of the $S_1 \rightarrow S_0$ fluorescence that is thermally activated from the lower-lying triplet state. The reason is that the corresponding radiative rate $k_r(S_1 \rightarrow S_0)$ also governs the TADF decay time. $k_r(S_1 \rightarrow S_0)$ should be as large as possible to obtain a short TADF decay time. However, basic quantum chemical considerations show that these rates $k_r(S_1 \rightarrow S_0)$ and $\Delta E(S_1-T_1)$ correlate. We address this behavior in Section 1.5.

Thus, designing of TADF compounds with short TADF decay time and high emission quantum yield is a challenge. For example, radiative TADF decay times of Cu(I) complexes of less than 3–5 μs have not been reported so far [30, 31, 96]. We will discuss this challenge in several sections of this investigation. In particular, in Section 1.6, where we focus on designing new Ag(I) complexes, we will show how to develop a breakthrough TADF material [108, 109] with a radiative TADF decay time of only 1.4 μs (at $\phi_{PL} = 100\%$), which is significantly shorter than so far reported.

This chapter is organized as follows: In Section 1.2, we introduce different parameters that have to be addressed for designing efficient TADF materials based on Cu(I) and Ag(I) complexes, and we present the materials studied in this chapter together with selected photophysical data. Sections 1.3 and 1.4 display case studies of Cu(I) complexes with large and small $\Delta E(S_1-T_1)$ energy separation, respectively. Furthermore, we discuss effects of SOC with respect to properties of the lowest triplet state, such as phosphorescence allowedness and zero-field splitting (ZFS). In Section 1.5, we show on a very simple quantum mechanical basis that the size of the energy gap $\Delta E(S_1-T_1)$ and the allowedness of the singlet $S_1 \rightarrow$ singlet S_0 transition are correlated. This result is clearly supported by experimental data. In Section 1.6, we present photophysical properties of a new TADF class of Ag(I) complexes, and we show how an extraordinarily efficient TADF material can be designed. Finally, in a conclusion, we will give a short summary and point to future perspectives.

1.2 TADF, Molecular Parameters, and Diversity of Materials

The molecular TADF effect was already reported more than five decades ago [131, 132]. It can be described by the use of Figure 1.1. By an optical excitation, one usually excites a singlet state, for example, higher-lying vibrational levels of the S_1 state. Subsequently, fast vibrational relaxation of the order of 10^{-12} s [124] proceeds to lower lying vibrational levels. Then, depending on the class of molecules, prompt fluorescence and/or down-ISC (as well as nonradiative relaxation to the electronic ground state) can occur. For Cu(I) and Ag(I) complexes, being in the focus of this contribution, ISC from S_1 to the T_1 state is very effective, since fast ISC occurring in the time range of 3–30 ps has been observed [118, 121, 122, 133, 134]. The individual value does not only depend on molecular properties, for example, on the extent of SOC of higher-lying singlets to the lowest triplet state, but also on the local environment, such as a fluid or a rigid matrix [134]. In general, a significant prompt fluorescence ($S_1 \rightarrow S_0$) is not detected, instead a very bright long-lived phosphorescence ($T_1 \rightarrow S_0$) is frequently observed at low temperature [30, 31, 33–35, 70, 92, 95–97, 135, 136]. At a higher temperature and in a situation of a fast thermal equilibration, population of the higher-lying singlet state is governed by the Boltzmann distribution. In this context, "fast" means that the down- and up-ISC processes are much faster than the $T_1 \rightarrow S_0$ and $S_1 \rightarrow S_0$ emission decay processes. As a consequence, the emission decays with a single decay time being a weighted average (Eq. (1.1)) of the $T_1 \rightarrow S_0$ and $S_1 \rightarrow S_0$ decay processes. Usually, this is fulfilled for Cu(I) and Ag(I) complexes with low-lying MLCT states. At lower temperatures population of the S_1 state due to the Boltzmann thermal distribution is frozen out. For completeness, it is remarked that below $T \approx 15$ K, effects of spin–lattice relaxation (SLR) between the triplet substates might strongly slow down relaxation processes [31, 33, 34, 88, 95, 137–139]. Consequently, frequently a multiexponential decay (from the different T_1 substates to the ground state S_0) is observed at very low temperature.

At higher temperatures, up-ISC processes efficiently depopulate the triplet T_1 state, i.e. the three triplet substates, and populate the singlet S_1 state. In particular, if the emission decay time of the T_1 state is long, for example, longer than several 100 μs, the phosphorescence is largely "quenched" and almost only the $S_1 \rightarrow S_0$ fluorescence is observed. Because the population of the S_1 state is fed from the long-living triplet state (triplet reservoir), this type of fluorescence is also long living compared with the prompt fluorescence that does not involve a triplet state. Hence, the emission is denoted as TADF.

For completeness, it is mentioned that for several Cu(I) complexes, SOC with respect to the T_1 state is significant. In this situation, the TADF process will not fully deplete the T_1 state during its much shorter population time, and a combined TADF/phosphorescence is observed. This property is not in the focus of the present contribution, but is discussed in detail in the literature [31, 33, 34, 97].

In the emission layer of an OLED device, the electron-hole recombination produces excitons of different spin multiplicity, that is, 75% are of triplet and 25% of singlet character [67, 68, 78, 79] (Figure 1.1). These excitons can be trapped in the emitter molecule. Such mechanisms are discussed in Ref. [67]. Subsequently, fast internal conversions, i.e. one singlet path and three triplet paths, populate the S_1 and the T_1 state, respectively [67]. Then similarly to the behavior after optical excitation, thermal activation takes place. In particular, in a situation of a forbidden $T_1 \rightarrow S_0$ transition and a relatively small $\Delta E(S_1-T_1)$ value, finally almost all excitations are transferred to the lowest excited singlet state, which then exhibits delayed $S_1 \rightarrow S_0$ fluorescence. According to this process, the molecular TADF effect as exploited in an OLED device has been denoted as *singlet harvesting mechanism* [9, 30, 31, 70, 94–96].

As already stressed, the TADF decay time $\tau(\text{TADF})$ should be as short as possible (obviously, at a high emission quantum yield), if the emitter is applied in an OLED. To achieve this goal, valuable guidelines can be deduced. Especially, a discussion of the temperature dependence of the emission decay time is helpful. This is easily explained by the use of a model discussion of a simplified molecule with two excited states, a singlet S_1 and a triplet T_1, being in fast thermal equilibrium, and the electronic ground state S_0 (compare Figure 1.1). In this situation, the

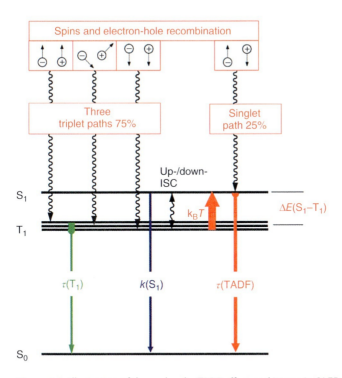

Figure 1.1 Illustration of the molecular TADF effect and its use in OLEDs (singlet harvesting). $\tau(T_1)$ and $\tau(\text{TADF})$ are the phosphorescence decay time and the TADF decay time, respectively. $k(S_1) = k_r(S_1 \rightarrow S_0)$ is the radiative rate of the $S_1 \rightarrow S_0$ transition (prompt fluorescence). Up-ISC is also often denoted as reverse intersystem crossing (rISC).

decay time $\tau(T)$ of the luminescent molecule is described by a Boltzmann-type relation [30] (compare also [140–143]):

$$\tau(T) = \frac{3 + \exp\left(-\frac{\Delta E(S_1 - T_1)}{k_B T}\right)}{3k(T_1) + k(S_1)\exp\left(-\frac{\Delta E(S_1 - T_1)}{k_B T}\right)} \quad (1.1)$$

$k(T_1) = 1/\tau(T_1)$ and $k(S_1) = 1/\tau(S_1)$ are the decay rates with the decay times $\tau(T_1)$ and $\tau(S_1)$ of the triplet and singlet excited state, respectively, and $\Delta E(S_1-T_1)$ is the energy separation between the S_1 and T_1 state. $\tau(T)$ represents the experimentally accessible emission decay time at a given temperature. For the subsequent discussion, it is assumed that the molecular parameters $\tau(T_1)$, $\tau(S_1)$, and $\Delta E(S_1-T_1)$ are temperature independent and that the splitting of the triplet state into three substates (ZFS) is small, i.e. much smaller than $k_B T$.

Thus, application of Eq. (1.1) to the measured decay times at different temperatures opens access to the molecular parameters given above. In particular, it becomes possible to determine very small energy separations being much below the attainable spectral resolution. For example, using a slightly modified Eq. (1.1) energy separations of only a few cm^{-1} can be resolved despite the fact that the MLCT emission bands are frequently as broad as several thousand cm^{-1} (compare [9, 31, 95, 97, 140]).

At very low temperature, the exponential terms in Eq. (1.1) are negligible, and the measured decay time $\tau(T)$ displays the phosphorescence decay time $\tau(T_1)$, while at high temperature (and long $\tau(T_1)$), the term containing $\tau(T_1)$ can be neglected, and one obtains essentially the decay time $\tau(TADF)$. In Sections 1.3, 1.4, and 1.6, several case studies, discussing the $\tau(T)$ temperature behavior, are presented.

Eq. (1.1) shows that three parameters crucially determine the emission decay time. These parameters can be deduced from fitting Eq. (1.1) to the measured emission decay times for a suitable temperature range. If this is carried out for a number of Cu(I) and/or Ag(I) complexes, valuable guidelines for molecular design rules can be extracted. Therefore, a more detailed discussion of the three parameters is illustrative.

Figure 1.2 schematically visualizes the photophysical background of the parameters that govern Eq. (1.1). For all three cases, subsequently discussed in detail, it is assumed that the respective molecule shows TADF.

1.2.1 TADF and Phosphorescence

Figure 1.2a displays an energy level diagram of a compound that exhibits significant SOC of the triplet state T_1 to a singlet state. Quantum mechanical considerations show that SOC between the triplet state T_1 and the singlet state S_1 both stemming from the same orbital configuration vanishes [9, 30, 31, 69–77]. However, mixing in of different, higher-lying singlets S_n (with $n > 1$) can be significant. If so, the triplet state of Cu(I) complexes, for example, exhibits a distinct ZFS of several cm^{-1} (a few 0.1 meV), and the $T_1 \to S_0$ transition rate for the phosphorescence can become as large as $5 \times 10^4\ s^{-1}$ (20 μs) [30, 34, 97, 135, 136].

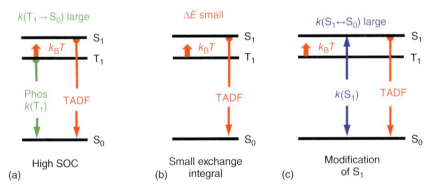

Figure 1.2 Different strategies for minimizing the emission decay time of TADF compounds for OLED applications. The zero-field splitting of the T_1 state into three substates is not shown in this diagram.

Accordingly, a second, effective radiative decay channel is opened in addition to the TADF decay path. These combined radiative decay paths can distinctly shorten the ambient temperature emission decay time. In this contribution, we do not focus further on this effect, but compare the literature reports [30, 31, 34, 97, 144].

1.2.2 Minimizing $\Delta E(S_1-T_1)$

As already addressed, the energy separation $\Delta E(S_1-T_1)$ between the lowest singlet S_1 and triplet state T_1 should be relatively small (Figure 1.2b). This can be well achieved with Cu(I) and Ag(I) complexes, if the lowest lying excited states are largely of ^{1,3}MLCT character. In this situation, a distinct charge separation between the unpaired electrons can occur. As a consequence, the quantum mechanical exchange interaction [128–130] and, hence, also the singlet–triplet splitting becomes small (Section 1.5). Because $\Delta E(S_1-T_1)$ enters in an exponential term in Eq. (1.1), $\Delta E(S_1-T_1)$ reduction has a dominating effect on the TADF decay time, τ(TADF). However, this seems to be limited, when $\Delta E(S_1-T_1)$ becomes lower than 200–300 cm^{-1} (25 to $\approx$40 meV), since at smaller splitting, decrease of the S_1-S_0 transition rate $k_r(S_1 \rightarrow S_0)$ might induce an opposite trend and lead to an increase of τ(TADF). (In Section 1.5 it will be shown that $\Delta E(S_1-T_1)$ and $k_r(S_1 \rightarrow S_0)$ are related to each other.) The variation of $\Delta E(S_1-T_1)$ can be achieved by a suitable molecular design as discussed below in this section and in the case studies presented in the next two sections.

1.2.3 Importance of $k_r(S_1-S_0)$

The allowedness of the $S_1 \leftrightarrow S_0$ transition can be expressed, for example, by the radiative fluorescence rate $k_r(S_1 \rightarrow S_0)$ (Figure 1.2c). According to Eq. (1.1), this rate also plays an important role at determining the TADF decay time. Frequently, this property is not adequately addressed. The rate should be as high as possible to obtain short TADF decay time. For this requirement also $\Delta E(S_1-T_1)$ should be as small as possible. However, the two photophysical parameters are correlated. Small splitting $\Delta E(S_1-T_1)$ requires a small exchange interaction between the unpaired electrons, at least when both T_1 and S_1 states are well described by a HOMO–LUMO excitation. For a small exchange interaction, small overlap of HOMO and LUMO is advantageous. At the same time, small HOMO–LUMO overlap leads to a small oscillator strength (small allowedness) of the $S_1 \to S_0$ transition and, thus, to a small $k_r(S_1 \to S_0)$ value and, hence, to a long fluorescence decay time. Indeed, experimental studies on Cu(I) complexes exhibiting TADF reveal that such a correlation exists for a large number of compounds (compare also Section 1.5) [31].

However, it is indicated that a close correlation between these two photophysical properties might not always be so strict. In particular, the S_1 state might be modified by a suitable molecular design. For example, quantum mechanical configuration interaction (CI) can be helpful in this respect. This means that a different, higher-lying singlet state, which is energetically proximate and carries high allowedness (high oscillator strength) with respect to the transition to the electronic ground state, can mix with the S_1 state and induce a higher $S_1 \to S_0$ allowedness. Presumably, this mechanism is important for the Ag(I) complexes discussed in Section 1.6.

Experimental access to the photophysical parameters as discussed above and in Figure 1.2 becomes possible by the use of a fitting procedure of Eq. (1.1) to the measured values of $\tau(T)$ over a large temperature range. The required range depends on the size of $\Delta E(S_1-T_1)$. For example, if $\Delta E(S_1-T_1)$ values are larger than about 700 cm^{-1} (87 meV), a temperature range of 77 K $\leq T \leq$ 300 K might usually be sufficient. However, for smaller splittings, the range has to be extended, for example, to $T = 30$ K to be able to characterize a compound with a splitting of only $\Delta E(S_1-T_1) = 370$ cm^{-1} (compound **2**, see Section 1.4) [31, 96]. Moreover, if the splitting of the T_1 state into substates of a few cm^{-1} (a few 0.1 meV) and the corresponding photophysical properties should be addressed, extension of the temperature range to about $T = 1.3$ K (and application of a slightly modified Eq. (1.1)) is required [58, 70, 88, 89, 97, 138, 145].

In Sections 1.3, 1.4, and 1.6, we will present case studies. In these we will show how to develop a deeper understanding of representative compounds. In Table 1.1, for a broader overview, we summarize a selection of photophysical data for a large number of complexes that are addressed in this contribution. The compounds studied were investigated as powder materials. In this respect, it should be remarked that very frequently, the decay behavior measured of powder materials is modified, for example, by processes of energy transfer or triplet–triplet annihilation. However, if the low-lying charge-transfer (CT) states of the complexes exhibit geometry distortions also in the relatively rigid crystalline environment, localization (self-trapping) can occur and prevent energy transfer effects. Thus, the emission of the powder material can display molecular properties. Accordingly, concentration quenching does not occur, and the decay behavior does not show any distinctive feature (such as shortening of decay times) with concentration increase [30, 31, 58, 95].

1.2 TADF, Molecular Parameters, and Diversity of Materials

Table 1.1 Photophysical data based on luminescence measurements of powder materials arranged according to increasing $\Delta E(S_1-T_1)$.

Compound	Photophysical data	References
Cu$_2$I$_2$(MePyrPHOS)(Pph$_2$)$_2$ **1**	$\Delta E(S_1-T_1) = 270$ cm^{-1} $\lambda_{max}(300\text{ K}) = 511$ nm $\Phi_{PL}(300\text{ K}) = 97\%$ $\tau(300\text{ K}) = 5$ µs $\tau(T_1 \rightarrow S_0) = 23$ µs $\tau(S_1 \rightarrow S_0) = 570$ ns	[31, 38]
Cu(dppb)(pz$_2$Bph$_2$) **2**	$\Delta E(S_1-T_1) = 370$ cm^{-1} $\lambda_{max}(300\text{ K}) = 535$ nm $\Phi_{PL}(300\text{ K}) = 70\%$ $\tau(300\text{ K}) = 3.3$ µs $\tau(T_1 \rightarrow S_0) = 1200$ µs $\tau(S_1 \rightarrow S_0) = 180$ ns	[43, 96, 146]
[Cu(µ-Cl)(PNMe$_2$)]$_2$ **3**	$\Delta E(S_1-T_1) = 460$ cm^{-1} $\lambda_{max}(300\text{ K}) = 506$ nm $\Phi_{PL}(300\text{ K}) = 45\%$ $\tau(300\text{ K}) = 6.6$ µs $\tau(T_1 \rightarrow S_0) = 250$ µs $\tau(S_1 \rightarrow S_0) = 210$ ns	[33]
[Cu(µ-Br)(PNMe$_2$)]$_2$ **4**	$\Delta E(S_1-T_1) = 510$ cm^{-1} $\lambda_{max}(300\text{ K}) = 490$ nm $\Phi_{PL}(300\text{ K}) = 65\%$ $\tau(300\text{ K}) = 4.1$ µs $\tau(T_1 \rightarrow S_0) = 1200$ µs $\tau(S_1 \rightarrow S_0) = 110$ ns	[33]

(Continued)

Table 1.1 (Continued)

Compound	Photophysical data	References
[Cu(μ-I)(PNMe$_2$)]$_2$ **5**	$\Delta E(S_1-T_1) = 570$ cm^{-1} $\lambda_{max}(300\ K) = 464$ nm $\Phi_{PL}(300\ K) = 65\%$ $\tau(300\ K) = 4.6$ μs $\tau(T_1 \rightarrow S_0) = 290$ μs $\tau(S_1 \rightarrow S_0) = 90$ ns	[33]
Cu$_2$Cl$_2$(dppb)$_2$ **6**	$\Delta E(S_1-T_1) = 600$ cm^{-1} $\lambda_{max}(300\ K) = 545$ nm $\Phi_{PL}(300\ K) = 35\%$ $\tau(300\ K) = 3$ μs $\tau(T_1 \rightarrow S_0) = 2200$ μs $\tau(S_1 \rightarrow S_0) = 70$ ns	[100, 107]
[Cu(μ-I)(PNpy)]$_2$ **7**	$\Delta E(S_1-T_1) = 630$ cm^{-1} $\lambda_{max}(300\ K) = 465$ nm $\Phi_{PL}(300\ K) = 65\%$ $\tau(300\ K) = 5.6$ μs $\tau(T_1 \rightarrow S_0) = 250$ μs $\tau(S_1 \rightarrow S_0) = 100$ ns	[33]
Cu(pop)(pz$_2$Bph$_2$) **8**	$\Delta E(S_1-T_1) = 650$ cm^{-1} $\lambda_{max}(300\ K) = 464$ nm $\Phi_{PL}(300\ K) = 90\%$ $\tau(300\ K) = 13$ μs $\tau(T_1 \rightarrow S_0) = 500$ μs $\tau(S_1 \rightarrow S_0) = 170$ ns	[9, 95, 147, 148]

Table 1.1 (Continued)

Compound	Photophysical data	References
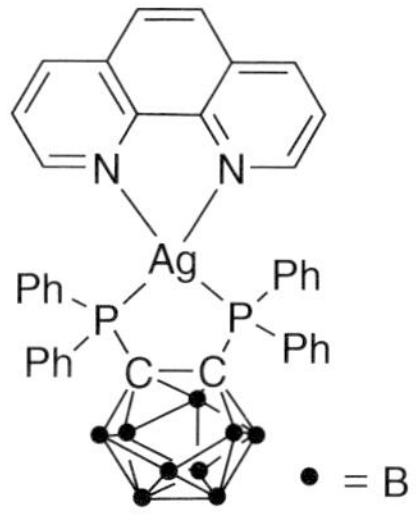 Ag(phen)(P$_2$-nCB) **9**	$\Delta E(S_1-T_1)$ –[a] $\lambda_{max}(300\,K) = 575$ nm $\Phi_{PL}(300\,K) = 36\%$ $\tau(300\,K) = 2$ μs $\tau(77\,K) = 270$ μs $\tau(S_1 \to S_0)$ –	[109] Table 1.7
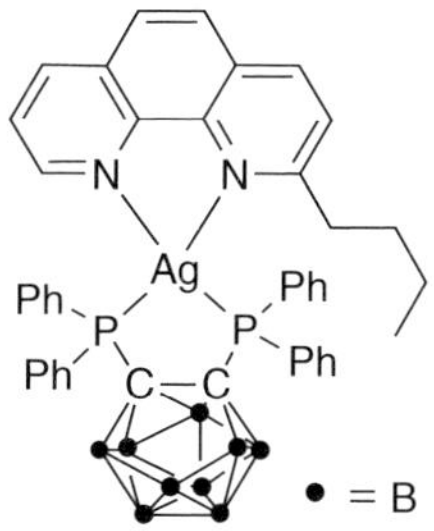 Ag(mbp)(P$_2$-nCB) **10**	$\Delta E(S_1-T_1) = 640$ cm^{-1} $\lambda_{max}(300\,K) = 535$ nm $\Phi_{PL}(300\,K) = 70\%$ $\tau(300\,K) = 2$ μs $\tau(T_1 \to S_0) = 1600$ μs $\tau(S_1 \to S_0) = 32$ ns	[109] Table 1.7
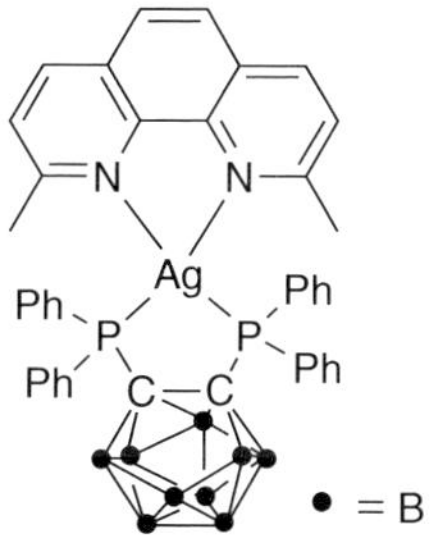 Ag(dmp)(P$_2$-nCB) **11**	$\Delta E(S_1-T_1) = 650$ cm^{-1} $\lambda_{max}(300\,K) = 537$ nm $\Phi_{PL}(300\,K) = 78\%$ $\tau(300\,K) = 2.8$ μs $\tau(T_1 \to S_0) = 890$ μs $\tau(S_1 \to S_0) = 36$ ns	[109] Table 1.6
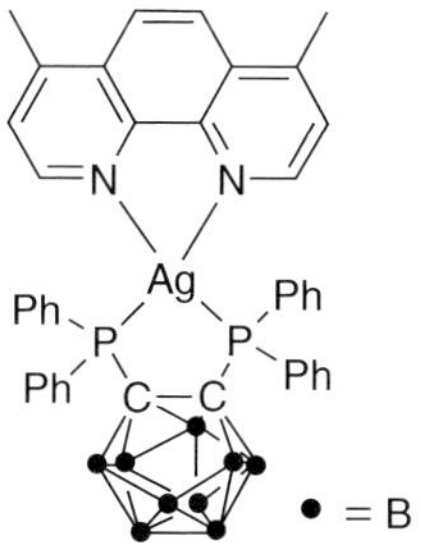 Ag(idmp)(P$_2$-nCB) **12**	$\Delta E(S_1-T_1)$[a] $\lambda_{max}(300\,K) = 562$ nm $\Phi_{PL}(300\,K) = 45\%$ $\tau(300\,K) = 1.7$ μs $\tau(T_1 \to S_0) = 475$ μs	[109]

(Continued)

Table 1.1 (Continued)

Compound	Photophysical data	References
Ag(dbp)(P$_2$-nCB) **13**	$\Delta E(S_1-T_1) = 650\ cm^{-1}$ $\lambda_{max}(300\ K) = 526\ nm$ $\Phi_{PL}(300\ K) = 100\%$ $\tau(300\ K) = 1.4\ \mu s$ $\tau(T_1 \to S_0) = 1570\ \mu s$ $\tau(S_1 \to S_0) = 18\ ns$	[108, 109]
Cu(tmbpy)(pop)$^+$ **14**[b]	$\Delta E(S_1-T_1) = 720\ cm^{-1}$ $\lambda_{max}(300\ K) = 555\ nm$ $\Phi_{PL}(300\ K) = 55\%$ $\tau(300\ K) = 11\ \mu s$ $\tau(T_1 \to S_0) = 84\ \mu s$ $\tau(S_1 \to S_0) = 160\ ns$	[35]
(IPr)Cu(py$_2$-BMe$_2$) **15**	$\Delta E(S_1-T_1) = 740\ cm^{-1}$ $\lambda_{max}(300\ K) = 475\ nm$ $\Phi_{PL}(300\ K) = 76\%$ $\tau(300\ K) = 11\ \mu s$ $\tau(T_1 \to S_0) = 34\ \mu s$ $\tau(S_1 \to S_0) = 160\ ns$	[34, 149]
[Cu(PNP^{t}Bu)]$_2$ **16**	$\Delta E(S_1-T_1) = 786\ cm^{-1}$ $\lambda_{max}(300\ K) = 512\ nm$ $\Phi_{PL}(300\ K) = 57\%$ $\tau(300\ K) = 11\ \mu s$ $\tau(T_1 \to S_0) = 343\ \mu s$ $\tau(S_1 \to S_0) = 79\ ns$	[46]

Table 1.1 (Continued)

Compound	Photophysical data	References
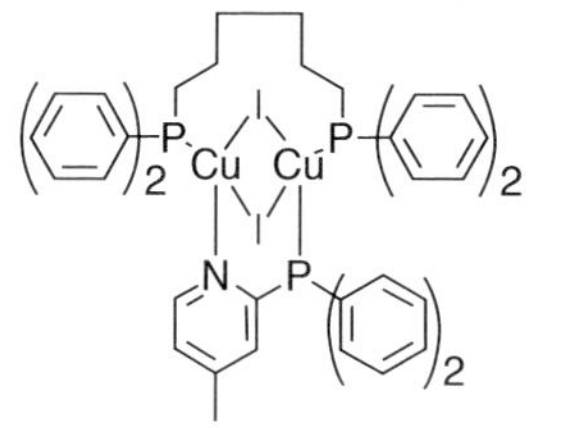 Cu$_2$I$_2$(MePyrPHOS)(dpph) **17**	$\Delta E(S_1-T_1) = 830$ cm^{-1} $\lambda_{max}(300\,K) = 519$ nm $\Phi_{PL}(300\,K) = 88\%$ $\tau(300\,K) = 24$ μs $\tau(T_1 \to S_0) = 110$ μs $\tau(S_1 \to S_0) = 190$ ns	[30, 41]
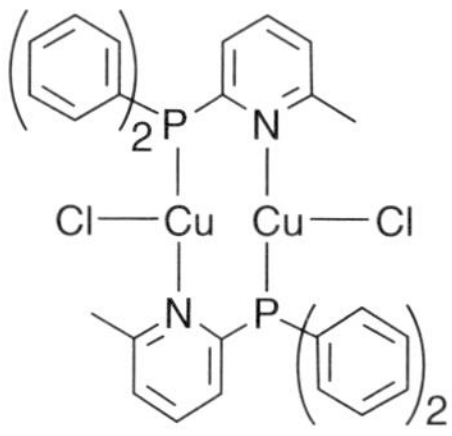 Cu$_2$Cl$_2$(N^P)$_2$ **18**	$\Delta E(S_1-T_1) = 930$ cm^{-1} $\lambda_{max}(300\,K) = 485$ nm $\Phi_{PL}(300\,K) = 92\%$ $\tau(300\,K) = 8.3$ μs $\tau(T_1 \to S_0) = 42$ μs $\tau(S_1 \to S_0) = 40$ ns	[96, 150]
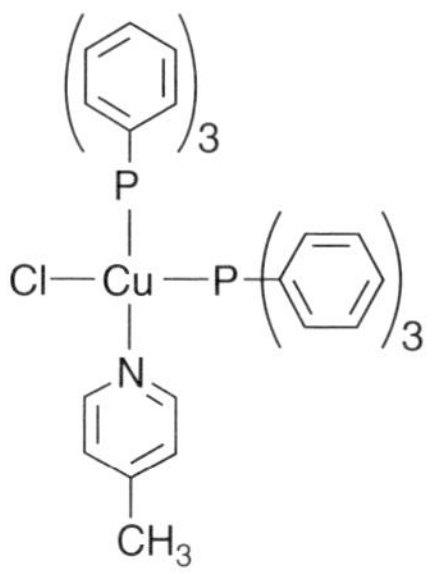 CuCl(Pph$_3$)$_2$(4-Mepy) **19**	$\Delta E(S_1-T_1) = 940$ cm^{-1} $\lambda_{max}(300\,K) = 468$ nm $\Phi_{PL}(300\,K) = 99\%$ $\tau(300\,K) = 9.4$ μs $\tau(T_1 \to S_0) = 34$ μs $\tau(S_1 \to S_0) = 47$ ns	[98]
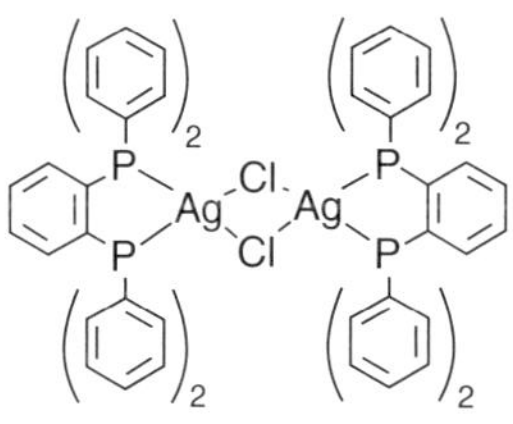 Ag$_2$Cl$_2$(dppb)$_2$ **20**	$\Delta E(S_1-T_1) = 980$ cm^{-1} $\lambda_{max}(300\,K) = 480$ nm $\Phi_{PL}(300\,K) = 93\%$ $\tau(300\,K) = 15$ μs $\tau(T_1 \to S_0) = 1100$ μs $\tau(S_1 \to S_0) = 45$ ns	[107]

(Continued)

Table 1.1 (Continued)

Compound	Photophysical data	References
Cu(dmp)(phanephos)$^+$ **21**^{c}	$\Delta E(S_1-T_1) = 1000$ cm^{-1} $\lambda_{max}(300\,K) = 530$ nm $\Phi_{PL}(300\,K) = 80\%$ $\tau(300\,K) = 14$ µs $\tau(T_1 \to S_0) = 240$ µs $\tau(S_1 \to S_0) = 40$ ns	[92, 151]
Cu(pop)(pz$_4$B) **22**	$\Delta E(S_1-T_1) = 1000$ cm^{-1} $\lambda_{max}(300\,K) = 447$ nm $\Phi_{PL}(300\,K) = 90\%$ $\tau(300\,K) = 22$ µs $\tau(T_1 \to S_0) = 450$ µs $\tau(S_1 \to S_0) = 80$ ns	[95, 148]
CuBr(Pph$_3$)$_2$(4-Mepy) **23**	$\Delta E(S_1-T_1) = 1070$ cm^{-1} $\lambda_{max}(300\,K) = 467$ nm $\Phi_{PL}(300\,K) = 95\%$ $\tau(300\,K) = 15$ µs $\tau(T_1 \to S_0) = 50$ µs $\tau(S_1 \to S_0) = 41$ ns	[98]
CuI(Pph$_3$)$_2$(4-Mepy) **24**	$\Delta E(S_1-T_1) = 1170$ cm^{-1} $\lambda_{max}(300\,K) = 455$ nm $\Phi_{PL}(300\,K) = 66\%$ $\tau(300\,K) = 9.5$ µs $\tau(T_1 \to S_0) = 49$ µs $\tau(S_1 \to S_0) = 14$ ns	[98]

Table 1.1 (Continued)

Compound	Photophysical data	References
Cu(pop)(pz$_2$BH$_2$) **25**	$\Delta E(S_1-T_1) = 1300\ \text{cm}^{-1}$ $\lambda_{max}(300\ \text{K}) = 436\ \text{nm}$ $\Phi_{PL}(300\ \text{K}) = 45\%$ $\tau(300\ \text{K}) = 20\ \mu\text{s}$ $\tau(T_1 \to S_0) = 610\ \mu\text{s}$ $\tau(S_1 \to S_0) = 10\ \text{ns}$	[70, 95, 148]

$\Delta E(S_1-T_1)$ and $\tau(S_1 \to S_0)$ values result from fitting procedures. A prompt fluorescence of the S_1 state has not been observed directly.
a) Similar $\Delta E(S_1-T_1)$ values are expected to occur for all Ag(phen-substituted)(P$_2$-nCB) complexes displayed in this table.
b) Investigated as [Cu(tmbpy)(pop)](BF$_4$) powder.
c) Investigated as [Cu(dmp)(phanephos)](PF$_6$) powder.
Source: Ref. [31]. Reproduced with permission of Elsevier.

1.3 Case Study: TADF of a Cu(I) Complex with Large $\Delta E(S_1-T_1)$

The luminescence properties of TADF compounds depend strongly on the energy separation between the lowest excited singlet and triplet state. In this case study, we discuss properties of Cu(dmp)(phanephos)$^+$ **21** (with dmp = 2,9-dimethyl-1,10-phenanthroline and phanephos = 4,12-bis(diphenylphosphino)-[2.2]-paracyclophane) (Figure 1.3).

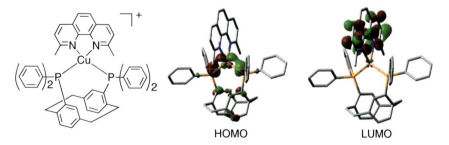

Figure 1.3 Chemical formula and frontier orbitals of Cu(dmp)(phanephos)$^+$ **21** obtained for the DFT-optimized triplet state (T$_1$) geometry. The calculations were performed at the B3LYP/def2-SVP level of theory [152, 153]. Hydrogen atoms are omitted for clarity. HOMO and LUMO exhibit distinctly different spatial distributions. The HOMO is mainly composed of the copper 3d and phosphorus sp^3 atomic orbitals, while the LUMO represents essentially a π^* orbital of the dmp ligand.

1.3.1 DFT and TD-DFT Calculations

Quantum mechanical density functional theory (DFT) computations reveal that for this compound, the lowest excited states result dominantly from HOMO to LUMO transitions from metal and phosphorus orbitals to dmp ligand orbitals [92] (Figure 1.3). Accordingly, the excited states have significant admixtures of ^{1,3}MLCT character. This assignment is also supported by the photophysical investigations as discussed below. Time-dependent density functional theory (TD-DFT) calculations, carried out in the T_1 state geometry, allow us to estimate an energy separation between the singlet and triplet MLCT states of $\Delta E(S_1-T_1) \approx 0.22$ eV. Since TD-DFT computations give energies of vertical transitions between Franck–Condon states, the calculated $\Delta E(S_1-T_1)$ value is overestimated as demonstrated experimentally (see below) but allows us to expect an occurrence of TADF at ambient temperature.

1.3.2 Flattening Distortions and Nonradiative Decay

An MLCT transition in Cu(I) complexes often leads to flattening distortions of the molecule in the excited state relative to the ground state geometry [114, 116–122, 154, 155]. Such distortions are usually connected with an increase of non-radiative deactivations or even quenching of the emission due to a strong increase of the Franck–Condon factors of the low-lying vibrational modes of the excited state and the highly excited vibrational modes of the electronic ground state [124–126]. Thus, engineering of a highly emissive compound requires that such geometry changes are minimized. This can be achieved by using matrix materials characterized by cages of rigid microenvironments. In a different approach, the excited state distortions can also be reduced or even largely suppressed at the molecular level.

Cu(dmp)(phanephos)$^+$ **21** represents an example of a Cu(I) complex in which the excited state distortions are hindered owing to a rational molecular design. In particular, the diphosphine phanephos with a wide P–Cu–P bite angle of 116° [92] forms a rigid "semicage" for the metal ion coordinated by the second ligand (Figure 1.4). Methyl groups in the 2- and 9-positions of dmp exert steric demands that further hinder flattening distortions. Thus, mutual steric interactions of the chelating ligands strongly reduce flattening distortions and as a consequence radiationless relaxations.

Indeed, Cu(dmp)(phanephos)$^+$ **21** displays intense green-yellow luminescence at ambient temperature even in solution (Figure 1.5). For instance, in dichloromethane (DCM) the quantum yield Φ_{PL} is 40% (Table 1.2). With this Φ_{PL} value and the measured decay time of $\tau(CH_2Cl_2, 300\,K) = 10\,\mu s$, the nonradiative rate k_{nr} can be estimated, using the relation

$$k_{nr} = (1 - \Phi_{PL})/\tau \tag{1.2}$$

The resulting rate of $k_{nr} = 6.0 \times 10^4\,s^{-1}$ represents one of the smallest k_{nr} values found for Cu(I) complexes in liquid solution so far [31]. This proves the validity of the molecular design strategy applied to Cu(dmp)(phanephos)$^+$ **21**. The excited state distortions and, thus, the extent of nonradiative relaxation can further be

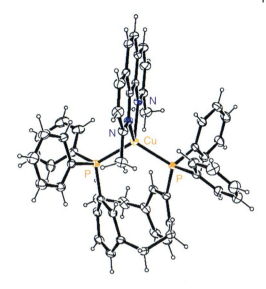

Figure 1.4 Perspective drawing of Cu(dmp)(phanephos)⁺ **21** (enantiomer R) resulting from X-ray crystallography studies. *Source*: Ref. [92]. Adapted with permission of the Royal Society of Chemistry.

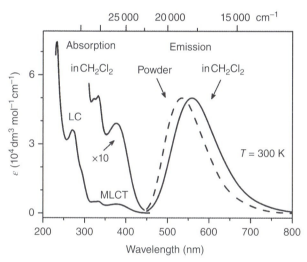

Figure 1.5 Ambient temperature absorption and emission spectra of Cu(dmp)(phanephos)⁺ **21** recorded in diluted ($c \approx 3 \times 10^{-5}$ M^{-1}) DCM solution (solid lines) and as [Cu(dmp)(phanephos)](PF$_6$) powder (dashed line). LC and MLCT denote ligand centered (π–π*) and metal-to-ligand charge-transfer (d–π*) transitions, respectively. *Source*: Ref. [92]. Adapted with permission of the Royal Society of Chemistry.

reduced by increasing the rigidity of the environment. Using the Φ_{PL} and τ data summarized in Table 1.2, the rates k_{nr} for a polymer matrix and a solid sample are found to be of $k_{nr} = 1.8 \times 10^4$ s^{-1} (poly(methyl methacrylate), PMMA) and $k_{nr} = 1.4 \times 10^4$ s^{-1} (powder), respectively. Interestingly, the changes of k_{nr} induced by the strongly different matrix rigidities are distinctly less than one order of magnitude. This is regarded as being relatively small, and it indicates that the excited state distortions are already partly suppressed at the level of molecular structure.

Table 1.2 Luminescence properties of [Cu(dmp)(phanephos)](PF$_6$) **21** in dichloromethane, PMMA (poly(methyl methacrylate)), and as powder.

	T = 300 K					T = 77 K				
	λ_{max} (nm)	τ (μs)	Φ_{PL} (%)	k_r (s^{-1})	k_{nr} (s^{-1})	λ_{max} (nm)	τ (μs)	Φ_{PL} (%)	k_r (s^{-1})	k_{nr} (s^{-1})
CH$_2$Cl$_2$	558	10	40	4.0×10^4	6.0×10^4	548	130	60	4.6×10^3	3.1×10^3
PMMA	535	20	65	3.3×10^4	1.8×10^4	567	170	60	3.5×10^3	2.4×10^3
Powder	530	14	80	5.7×10^4	1.4×10^4	562	240	70	2.9×10^3	1.3×10^3

Source: Ref. [92]. Reproduced with permission of Royal Society of Chemistry.

The discussed trend is also reflected by the small extent of spectral changes observed for the emission spectra. The emission maximum at $T = 300$ K of compound **21** is found at $\lambda_{max} = 558$ nm in DCM. For the compound doped in PMMA, it lies at $\lambda_{max} = 535$ nm and at 530 nm for a powder sample, respectively. Accordingly, the largest blue shift $\Delta\lambda_{max}$ (fluid solution → powder) amounts to $\Delta\lambda(max) = 28$ nm (≈ 950 cm^{-1}, 118 meV) only. For comparison, for the blue-green emitting Cu(pop)(pz$_2$BH$_2$) **25**, $\Delta\lambda(max)$ was found to be as large as 99 nm (≈ 4200 cm^{-1}) [95]. In the latter case, the large shift is related to the more flexible molecular structure that enables significant flattening distortions in the MLCT excited states. In compound **21**, such distortions are largely suppressed due to the large bite angle and bulky phanephos ligand.

1.3.3 TADF Properties

As expected for a TADF material, the emission decay time τ is strongly dependent on temperature (Table 1.2). Upon heating from $T = 77$ K to ambient temperature, the decay time becomes about one order of magnitude shorter with the quantum yield remaining approximately equally high. Thus, the change of τ is connected to a change of the radiative decay rate k_r. This rate is determined according to the following relation:

$$k_r = \Phi_{PL}/\tau \tag{1.3}$$

In Table 1.2, radiative decay rates are given for different environments and temperatures. In particular, for the powder material k_r amounts to 2.9×10^3 s^{-1} at $T = 77$ K and increases with temperature increase to 5.7×10^4 s^{-1} at ambient temperature, i.e. by a factor of almost 20. Thus, different emission mechanisms are active in the two temperature regimes.

The temperature dependence of the decay time is studied in more detail for a powder sample (Figure 1.6). Between 20 K and about 120 K, the decay time is almost constant and as long as $\tau \approx 240$ μs (plateau). The assignment of this emission to a $T_1 \rightarrow S_0$ phosphorescence is straightforward. Obviously, in this plateau range no other decay mechanism is activated (thermal energy at ≈ 120 K amounts to ≈ 83 cm^{-1} or ≈ 10 meV). However, with further temperature increase, a steep decrease of the decay time is observed, which is due to the

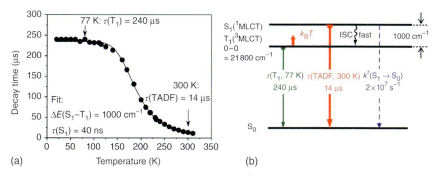

Figure 1.6 (a) Emission decay time of [Cu(dmp)(phanephos)](PF$_6$) **21** powder versus temperature. The emission was excited with a pulsed UV laser at $\lambda_{exc} = 355$ nm (pulse width 7 ns) and detected at $\lambda_{det} = 550$ nm. The solid line represents a fit of Eq. (1.1) to the experimental data fixing the phosphorescence decay time $\tau(T_1) = 240$ μs (plateau at 20 K < T < 120 K). The resulting fit parameters are $\Delta E(S_1-T_1) = 1000$ cm^{-1} and $\tau(S_1) = 40$ ns. $\tau(\text{TADF}) = 14$ μs is the decay time dominated by the delayed fluorescence at ambient temperature. (b) Energy level diagram for Cu(dmp)(phanephos)$^+$. The radiative rate for the $S_1 \to S_0$ transition $k_r(S_1 \to S_0)$ was determined according to Eq. (1.3) assuming $\tau(S_1) = 40$ ns (fit) and $\Phi_{PL} = 80\%$ (measured at ambient temperature). The energy of the $T_1 \to S_0$ 0–0 transition is estimated from the high-energy flank of the 77 K emission spectrum (not reproduced). *Source*: Ref. [92]. Reproduced with permission of Royal Society of Chemistry.

increase of the radiative decay rate k_r as discussed above. This change is related to a growing involvement of the higher-lying S_1 singlet state with its much higher decay rate of the transition to the electronic ground state S_0. The S_1 state is thermally activated from the lower-lying T_1 state. Hence, the ambient temperature emission represents (largely) a TADF (see also below). It exhibits a decay time of $\tau(\text{TADF}) = 14$ μs at 300 K.

The temperature dependence of $\tau = \tau(T)$, as displayed in Figure 1.6a, can be interpreted in terms of a three states kinetic model involving the electronic ground state S_0, the lowest triplet state T_1, and the lowest excited singlet state S_1, as expressed by Eq. (1.1). By fitting this equation to the measured decay times and inserting the measured decay time $\tau(T_1) = 240$ μs (plateau at 20 K < T < 120 K), values of $\tau(S_1) = 40$ ns and $\Delta E(S_1-T_1) = 1000$ cm^{-1} are obtained. The value of $\Delta E(S_1-T_1) = 1000$ cm^{-1} corresponds well to the spectral blue shift of 1070 cm^{-1} observed for the emission maximum with temperature increase from $T = 77$ K ($\lambda_{max} = 562$ nm) to 300 K ($\lambda_{max} = 530$ nm) (Table 1.2). This correspondence between the activation energy and the spectral shift upon temperature increase represents a further support for the assignment of the ambient temperature emission as TADF.

For completeness, it is remarked that the emission at ambient temperature frequently does not only represent TADF but also contains some $T_1 \to S_0$ (phosphorescence) contribution. (Prompt fluorescence is not observed due to fast ISC of the order of a few tens of picoseconds [118, 122, 133, 134, 154, 156–158].) According to Refs [33, 97], the intensity ratio $(S_1)/I(T_1)$ can be expressed by

$$\frac{I(S_1)}{I(T_1)} = \frac{k_r(S_1)}{k_r(\text{I}) + k_r(\text{II}) + k_r(\text{III})} \cdot e^{-\frac{\Delta E(S_1-T_1)}{k_B T}} \quad (1.4a)$$

with

$$k_r(T_1) = \frac{1}{3}(k_r(I) + k_r(II) + k_r(III)) \quad (1.4b)$$

one obtains

$$\frac{I(S_1)}{I(T_1)} = \frac{k_r(S_1)}{3 \cdot k_r(T_1)} \cdot e^{-\frac{\Delta E(S_1-T_1)}{k_B T}} \quad (1.4c)$$

$k_r(I)$, $k_r(II)$, and $k_r(III)$ are the rates of transitions from the triplet substates I, II, and III of the lowest triplet state (T_1) to the ground state (S_0), respectively. $k_r(T_1) = k_r(T_1 \to S_0)$ represents the average transition rate from the three triplet substates to the ground state. We will return to the average decay properties later in Section 1.4 (compare Eq. (1.9)). $k_r(S_1) = k_r(S_1 \to S_0)$ is the transition rate from the lowest excited singlet state (S_1) to the ground state, and $\Delta E(S_1 T_1)$ is the energy gap between states S_1 and T_1 (compare [31, 33]). For Cu(dmp)(phanephos)$^+$ **21** at $T = 300$ K with $\Delta E(S_1-T_1) = 1000$ cm^{-1} (fit, Figure 1.6b), $k_r(S_1) = 2 \times 10^7$ s^{-1} (fit), and $k_r(T_1) = 2.9 \times 10^3$ s^{-1} (measured at 77 K; Table 1.2), one obtains $I(S_1)/I(T_1) \approx 20$. Thus, the emission spectrum at ambient temperature is clearly dominated by TADF ($\approx$95% TADF, $\approx$5% phosphorescence from T_1).

1.3.4 Radiative $S_1 \to S_0$ Rate, Absorption, and Strickler–Berg Relation

The radiative rate $k_r(S_1)$ for the electronic transition between the excited singlet state S_1 and the ground state S_0 can be determined independently from an analysis of the absorption spectrum [92]. According to Figure 1.7, the absorption

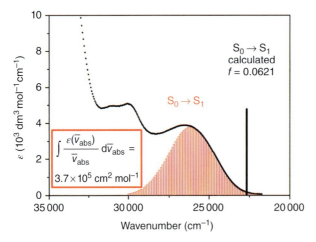

Figure 1.7 Absorption spectrum of Cu(dmp)(phanephos)$^+$ **21** in CH$_2$Cl$_2$ at ambient temperature. (Compare Figure 1.5.) The red-shaded area approximates the lowest absorption. The integrated intensity is determined to 3.7×10^5 cm^2 mol^{-1}. The vertical line at $\bar{\nu} = 22\,670$ cm^{-1} represents the calculated energy of the $S_0 \to S_1$ MLCT transition with a relatively small oscillator strength of $f = 0.0621$. The TD-DFT calculation was performed for the ground state optimized geometry at the B3LYP/def2-SVP level of theory. Source: Ref. [92]. Adapted with permission of Royal Society of Chemistry.

peak of lowest energy centered at ≈26 300 cm^{-1} (380 nm) and showing a slight spectral overlap with the emission is assigned to the $S_0 \rightarrow S_1$ transition. Thus, a (radiative) transition rate for the related emission $S_1 \rightarrow S_0$ can be estimated using the well-known Strickler–Berg relation between the radiative decay rate of the spontaneous emission and the strength of the corresponding absorption [159]. Accordingly, $k_r(S_1)$ can be expressed as

$$k_r(S_1 \rightarrow S_0) = 8 \cdot \ln 10 \cdot \pi \cdot c \cdot n^2 \cdot N_A^{-1} \cdot \langle \bar{\nu}_{fl}^{-3} \rangle_{av}^{-1} \cdot \int \frac{\varepsilon(\bar{\nu}_{abs})}{\bar{\nu}_{abs}} d\bar{\nu}_{abs} \quad (1.5)$$

where c is the speed of light in vacuum, N_A is the Avogadro number, and n is the refractive index of the medium. $\langle \bar{\nu}_{fl}^{-3} \rangle_{av}^{-1}$ displays the reciprocal of the mean value of the third power of the fluorescence energy $\bar{\nu}_{fl}$ (cm^{-1}) (weighted with the emission intensity at each $\bar{\nu}_{fl}$ value of the spectrum). The integral $\int \frac{\varepsilon(\bar{\nu}_{abs})}{\bar{\nu}_{abs}} d\bar{\nu}_{abs}$ represents the absorption strength of the $S_0 \rightarrow S_1$ band. $\varepsilon(\bar{\nu}_{abs})$ is the molar absorption (extinction) coefficient at a given energy $\bar{\nu}_{abs}$. If $\langle \bar{\nu}_{fl}^{-3} \rangle_{av}^{-1}$ is approximated by the third power of the emission maximum $\bar{\nu}_{max}^3$, Eq. (1.5) can be rewritten as

$$k_r(S_1 \rightarrow S_0) = \text{const} \cdot n^2 \cdot \bar{\nu}_{max}^3 \cdot \int \frac{\varepsilon(\bar{\nu}_{abs})}{\bar{\nu}_{abs}} d\bar{\nu}_{abs} \quad (1.6)$$

with const ≈ 2.88 × 10^{-12} s^{-1} mol cm.

From an integration of the lowest energy absorption band estimated by the shaded area as shown in Figure 1.7, a value of $\int \frac{\varepsilon(\bar{\nu}_{abs})}{\bar{\nu}_{abs}} d\bar{\nu}_{abs} = 3.7 \times 10^5$ cm^2 mol^{-1} is obtained. Thus, for $\bar{\nu}_{max} = 17\,500$ cm^{-1} (emission maximum in CH$_2$Cl$_2$ at ambient temperature and with $n = 1.42$), a spontaneous fluorescence rate of $k_r(S_1 \rightarrow S_0) \approx 1.2 \times 10^7$ s^{-1} is obtained. With respect to the different approximations made, this value corresponds reasonably well to the fluorescence decay rate of $k_r(S_1 \rightarrow S_0) \approx 2 \times 10^7$ s^{-1} as determined from the temperature dependence of the emission decay time as discussed above (Figure 1.6).

The value of k_r lying in the range of $1-2 \times 10^7$ s^{-1} corresponds to a moderately allowed transition, as it is expected for a $S_0 \rightarrow {}^1$MLCT transition. The moderate allowedness of this transition is also reflected in the value of the small oscillator strength resulting from TD-DFT calculations (Figure 1.7). For the optimized singlet ground state geometry at the B3LYP/def2-SVP level of theory, the oscillator strength for the $S_0 \rightarrow S_1$ transition of $f = 0.0621$ is obtained [92]. With a simple relation, the radiative rate can be estimated according to [124]:

$$k_r \cong \bar{\nu}^2 f \quad (1.7)$$

where $\bar{\nu}$ is the energy (in wave numbers) corresponding to the maximum wavelength of absorption. With the calculated value of $f = 0.0621$ and the $S_0 \rightarrow S_1$ transition energy of 22 670 cm^{-1}, a radiative rate of 3×10^7 s^{-1} is estimated, being in agreement with the value of $k_r(S_1 \rightarrow S_0) = 2 \times 10^7$ s^{-1} as determined experimentally.

With respect to the approximations made applying the different and independent methods, the $k_r(S_1 \rightarrow S_0)$ values (2×10^7 s^{-1} from the decay time analysis, 1.2×10^7 s^{-1} from the absorption strength analysis, and 3×10^7 s^{-1} from a

TD-DFT approach) are in good agreement. This is an important result, since it strongly supports the TADF assignment with respect to the involvement of the singlet S_1 state in the emission process at ambient temperature. Moreover, it is concluded that the geometry changes that still take place upon excitation do not significantly alter the $S_1 \rightarrow S_0$ transition rate determined for the relaxed geometry (from emission properties) as compared with the rate determined for the unrelaxed molecular geometry (from absorption).

1.4 Case Study: TADF of a Cu(I) Complex with Small $\Delta E(S_1-T_1)$

As already addressed, Cu(I) complexes display a large variety of TADF properties. In particular, this is related to a large range of energy separations $\Delta E(S_1-T_1)$ between the lowest singlet and triplet state being larger than 10^3 cm^{-1} (120 meV) or as small as a few hundred cm^{-1}. In Section 1.3, a complex with a relatively large $\Delta E(S_1-T_1)$ was presented. In the present case study, we will focus on a complex characterized by a small $\Delta E(S_1-T_1)$ value and discuss related TADF properties.

1.4.1 DFT and TD-DFT Calculations

In Figure 1.8, frontier orbital plots of Cu(dppb)(pz$_2$Bph$_2$) **2** (with dppb = 1,2-bis(diphenylphosphino)benzene and pz$_2$Bph$_2$ = diphenylbis(pyrazolylborate)) are reproduced together with the complex' chemical structure. The HOMO is derived from a metal 3d atomic orbital with significant contributions from the coordinating phosphorus and nitrogen atoms, whereas the LUMO is localized on the *o*-phenylene ring of the dppb ligand. Thus, similarly to Cu(dmp)(phanephos)$^+$ **21**, the orbitals display distinctly different spatial distributions, and the related HOMO $\rightarrow$ LUMO excitations are of CT character. It can be further shown by

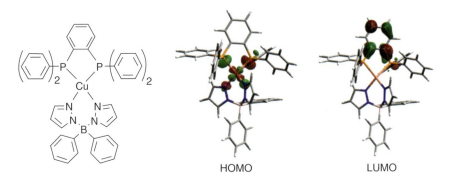

Figure 1.8 Chemical formula and contour plots of the HOMO and LUMO of Cu(dppb)(pz$_2$Bph$_2$) **2** resulting from DFT calculations for the triplet state geometry at the B3LYP/def2-SVP level of theory. The frontier orbitals exhibit distinctly different spatial distributions. The HOMO is mainly composed of the copper 3d and phosphorus sp^3 atomic orbitals, while the LUMO represents a π^* orbital of the dppb ligand. Compare [96].

TD-DFT calculations that the lowest excited singlet state S_1 and the triplet state T_1 are of more than 90% of HOMO–LUMO character [96]. Due to the distinct involvement of the metal, these states are assigned to largely represent ^{1}MLCT and ^{3}MLCT states, respectively. The significant spatial separation of HOMO and LUMO allows us to predict a relatively small exchange interaction and, thus, a small singlet–triplet energy separation $\Delta E(S_1–T_1)$. Indeed, TD-DFT calculations in the triplet state optimized geometry give a small value of $\Delta E(S_1–T_1) = 72$ meV (≈ 580 cm^{-1}), estimated as the energy difference between vertical $S_0 \rightarrow S_1$ and $S_0 \rightarrow T_1$ transitions [96] (Table 1.3), as being three times smaller than the value calculated for Cu(dmp)(phanephos)$^+$ **21**. Again, the $\Delta E(S_1–T_1)$ value calculated as the energy difference between the computed vertical excitations is overestimated. The experimentally determined activation energy amounts to $\Delta E(S_1–T_1) \sim 370$ cm^{-1} (46 meV).

According to the spatial separation of the molecular orbitals involved in the transition between the electronic ground state S_0 and the lowest excited singlet state S_1 (^{1}MLCT), it can be predicted that the transition dipole moment and, thus, the oscillator strength of the transition is relatively small. Indeed, TD-DFT calculations performed for the T_1 geometry at the B3LYP/def2-SVP level of theory result to $f = 0.0016$, being more than one order of magnitude smaller than the value 0.0201 calculated for the $S_0 \rightarrow S_1$ transition of complex [Cu(dmp)(phanephos)]$^+$ **21**. Further results of TD-DFT calculations are presented in Table 1.3 that are later used to explain effects of SOC in complex **2**.

1.4.2 Emission Spectra and Quantum Yields

Emission properties of Cu(dppb)(pz$_2$Bph$_2$) **2** were studied for powder samples over a wide temperature range from $T = 1.5$ to 300 K. Figure 1.9 displays representative emission spectra. Emission maxima, quantum yields, and decay times are collected in Table 1.4.

The compound shows intense green-yellow luminescence at all temperatures in the investigated range of 1.5–300 K, with quantum yields Φ_{PL} of 70% at ambient temperature and about 100% at $T = 77$ K. The spectra are broad and unstructured, which correlates with the predicted MLCT character of the corresponding electronic transitions. With temperature increase to $T \geq 30$ K, only a small blue shift is observed with the emission maximum λ(max) shifting from 548 nm at 1.5 K (not shown) and 30 K to 535 nm at 80 and 300 K. This is a consequence of the thermal activation of the TADF decay path via the higher-lying S_1 state above $T \approx 50$ K (see below).

1.4.3 The Triplet State T_1 and Spin–Orbit Coupling

The emission spectra do not display distinct changes with temperature change. However, the decay kinetics varies drastically even at very low temperature (below $T = 20$ K). This is related to the properties of the triplet state and its substates. Therefore, before discussing effects of thermal activation of the singlet state, the TADF effect, we want to focus on triplet state properties.

Table 1.3 Vertical transition energies, oscillator strengths, and main orbital contributions of selected electronic transitions of Cu(dppb)(pz$_2$Bph$_2$) **2** resulting from TD-DFT calculations for the optimized triplet state geometry (T$_1$ state) at the B3LYP/def2-SVP level of theory.

Transition	Energy (eV)	Oscillator strength	Main contribution
S$_0 \rightarrow$ T$_1$	1.303	0	HOMO → LUMO
S$_0 \rightarrow$ T$_2$	1.926	0	HOMO → LUMO + 1
S$_0 \rightarrow$ T$_3$	2.404	0	HOMO → LUMO + 2
S$_0 \rightarrow$ T$_4$	2.601	0	HOMO → LUMO + 3
S$_0 \rightarrow$ T$_5$	2.681	0	HOMO → LUMO + 4
S$_0 \rightarrow$ S$_1$	1.375	0.0016	HOMO → LUMO
S$_0 \rightarrow$ S$_2$	2.033	0.0038	HOMO → LUMO + 1
S$_0 \rightarrow$ S$_3$	2.623	0.0547	HOMO → LUMO + 2
S$_0 \rightarrow$ S$_4$	2.746	0.0253	HOMO → LUMO + 3

1.4 Case Study: TADF of a Cu(I) Complex with Small $\Delta E(S_1-T_1)$

Table 1.3 (Continued)

Transition	Energy (eV)	Oscillator strength	Main contribution
$S_0 \rightarrow S_5$	2.814	0.0727	HOMO → LUMO+4
$S_0 \rightarrow S_6$ SOC[a)]	2.848	0.0102	HOMO−1 → LUMO

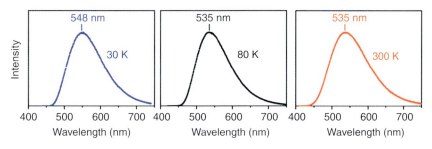

The lowest $S_0 \rightarrow S_n$ transition of HOMO−n → LUMO character that contains different Cu-3d character than the HOMO and that can exhibit SOC to the T_1 state is marked (see text).
a) SOC to T_1 possible.
Source: Refs [31, 96]. Reproduced with permission of Elsevier.

Figure 1.9 Luminescence spectra of Cu(dppb)(pz$_2$Bph$_2$) **2** powder recorded at different temperatures. *Source*: Ref. [96]. Reproduced with permission of American Chemical Society.

Table 1.4 Emission data for a powder sample of Cu(dppb)(pz$_2$Bph$_2$) **2**.

Temperature (K)	λ_{max} (nm)	Φ_{PL} (%)	τ (μs)	k_r (s^{-1})[a)]	k_{nr} (s^{-1})[a)]
300	535	70	3.3	210×10^3	9×10^4
80	535	≈100[b)]	300	3.3×10^3	—
30	548	≈100[b)]	1200	8.3×10^2	—

a) k_r and k_{nr} are determined by the use of Eqs. (1.2) and (1.3), respectively.
b) It is assumed that the quantum yield Φ_{PL} at $T = 30$ K amounts to 100%, as determined experimentally at $T = 77$ K.
Source: Ref. [96]. Reproduced with permission of American Chemical Society.

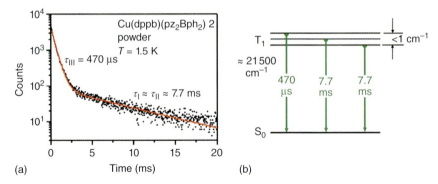

Figure 1.10 (a) Emission decay curve of Cu(dppb)(pz$_2$Bph$_2$) **2** at 1.5 K recorded for a powder sample after pulsed excitation at λ_{exc} = 372 nm and with signal detection at λ_{det} = 550 nm. Source: Ref. [96]. Reproduced with permission of American Chemical Society. (b) Energy level scheme showing the triplet substates and the related decay paths at T = 1.5 K at vanishing fast equilibration.

At T = 1.5 K, the decay curve is distinctly nonmonoexponential. The decay curve can be fitted with a bi-exponential function with the time constants of 7.7 ms and 470 μs (Figure 1.10). These different decay constants are assigned to emissions from the three individual triplet T$_1$ substates I, II, and III with $\tau_I \approx \tau_{II}$ = 7.7 ms and τ_{III} = 470 μs, respectively. It is remarked that for Cu(I) complexes, frequently a bi-exponential decay behavior is observed and not a tri-exponential one as expected for three triplet substates [57, 88, 160, 161]. This is due to the fact that the decay times of two substates are often very similar [33, 95–97]. Theoretical calculations based on SOC-TD-DFT (ADF2014) computations support this assignment [162].

At low temperature of T = 1.5 K, these T$_1$ substates are not thermally equilibrated due to slow SLR [57, 88, 137, 160, 161]. This behavior is related to small ZFSs of the T$_1$ state of less than 1 or 2 cm^{-1} (0.1 or 0.2 meV) [9, 57, 70, 88, 160, 161]. Such a small ZFS value is a consequence of weak SOC of the T$_1$ substates with higher-lying states. Furthermore, the emission decay time of $\tau(T_1)$ = 1200 μs (at 30 K) is extremely long if compared with Cu(dmp)(phanephos)$^+$ **21** (Section 1.3) and many other Cu(I) complexes [9, 31, 33, 35, 40, 42, 46, 50, 70, 92, 95–98, 100, 135, 163, 164]. Again, this is a consequence of the weak SOC with respect to the lowest triplet state.

Obviously, the (weak) allowedness of the T$_1 \to$ S$_0$ transition is not dominantly related to the SOC constant of copper, which is with ζ = 857 cm^{-1} [165] not very small. More important is the extent of mixing of energetically higher-lying singlet state(s). In a very simplified perturbational approach, the radiative rate can be described by [31, 69, 70, 166]:

$$k_r(T_1 \to S_0) \approx \text{const} \cdot \frac{|\langle S_m|H_{SO}|T_1\rangle|^2}{|E(T_1) - E(S_m)|^2} \cdot |\langle S_0|e\vec{r}|S_m\rangle|^2 \tag{1.8}$$

H_{SO} is the SO operator. $E(S_m)$ and $E(T_1)$ are the (unperturbed) energies of the (pure) singlet state S$_m$ and the (pure) triplet state T$_1$, respectively. In this simple model, it is assumed that one higher-lying singlet state S_m couples dominantly to the state T_1, i.e. to at least one T_1 triplet substate. $\langle S_0|e\vec{r}|S_m\rangle$ is the dipole matrix element with the dipole operator $e\vec{r}$.

A discussion of the energy denominator and its size is particularly helpful. Presumably, SOC with the energetically most proximate singlet state of adequate character represents a leading contribution to the radiative rate. Therefore, Eq. (1.8) shows only one mixing singlet state, being the state S_m, although several other singlet states may additionally contribute to the allowedness of the $T_1 \to S_0$ transition.

Quantum mechanical considerations show that SOC between a triplet state T_1 and a singlet state S_1 that both stem from the same orbital configuration, for example, from the HOMO $\to$ LUMO excitation, is negligible as for efficient SOC different d orbitals must be involved in the coupling states [9, 30, 31, 69–72, 75, 167].

The TD-DFT calculations presented above for compound **2** show that the energetically nearest singlet state that involves another d orbital than the T_1 state is the singlet state S_6. It originates from the HOMO $-1 \to$ LUMO electronic transition. According to Table 1.3, the energy separation that is responsible for dominant SOC amounts to $\Delta E(S_6-T_1) = 1.545$ eV ($\approx 12\,500$ cm^{-1}) (ΔE(HOMO $-$ (HOMO-1)) $= 1.4$ eV). As a consequence of this large energy denominator (Eq. 1.8), the triplet state does not experience effective SOC with state S_6. Hence, the phosphorescence decay time is expected to be very long. Indeed, this is found for compound **2** with $\tau(T_1 \to S_0) = 1200\,\mu$s. For comparison, the phosphorescence decay time of a compound that exhibits a much smaller energy denominator is given. For example, $Cu_2Cl_2(N^{\wedge}P)_2$ **18** with an energy separation between HOMO and HOMO-1 (that involve different d orbitals) amounts to only 0.378 eV (3000 cm^{-1}) [31, 97]. Since this HOMO$-1 \to$ LUMO excitation essentially defines the S_2 state, the energy separation to the T_1 state is much smaller than for compound **2**. Accordingly, the mixing of S_2 and T_1 becomes distinctly stronger (Eq. 1.8), and, hence, the triplet decay time amounts to only $\tau(T_1 \to S_0) = 43\,\mu$s [31, 97]. Thus, the $T_1 \to S_0$ allowedness of compound **18** is by a factor of almost 30 higher than found for compound **2**.

The size of the squared dipole matrix element $\langle S_0|e\vec{r}|S_m\rangle$ with the dipole operator $e\vec{r}$ is also of importance for the radiative rate $k_r(T_1 \to S_0)$ (or more exactly for the rate of the individual triplet substate that mixes with S_m). The corresponding value is proportional to the oscillator strength or the molar absorption coefficient of the singlet–singlet transition $S_0 \to S_m$, whereby S_m is the singlet state that can mix with the T_1 state via SOC [69, 72]. With respect to the corresponding allowedness, it is referred to Table 1.3.

For completeness, it is remarked that literature discussions frequently assume dominating SOC between T_1 and S_1. However, for compound **2**, the corresponding energy separation is very small amounting to only $\Delta E(S_1-T_1) = 370$ cm^{-1} (see below). As a consequence, relatively strong SOC would be expected to occur. Above, it has been demonstrated that this is not the case. Obviously, the simple literature approach is not suited.

1.4.4 Temperature Dependence of the Emission Decay Time and TADF

Let us focus on the temperature dependence of the emission decay time. With temperature increase, the SLR processes become faster (presumably according to a Raman process of SLR [137, 138, 160]), resulting in a fast thermalization of the three T_1 substates. At sufficiently high temperature, e.g. above $T = 10$ or 20 K, an average emission decay time τ_{av} is normally observed as expressed by Eq. (1.9) [9, 33, 57, 70, 88, 95, 96, 137, 139, 161]:

$$\tau_{av} = 3(\tau_I^{-1} + \tau_{II}^{-1} + \tau_{III}^{-1})^{-1} \tag{1.9}$$

Inserting the decay times of $\tau_I \approx \tau_{II} = 7.7$ ms and $\tau_{III} = 470$ μs, as determined at $T = 1.5$ K, one obtains $\tau_{av} = 1250$ μs. Almost the same value of $\tau(T_1) = 1200$ μs (monoexponential decay) is measured at $T = 30$ K (Figure 1.11). Thus, the assignments of the decay times we made above are validated.

Figure 1.11 displays emission decay curves and the temperature dependence of the decay time for the range of 20 K $\leq T \leq 300$ K. From $T = 20$ K to about 40 K, the decay time remains constant (plateau) with $\tau(T_1) = 1200$ μs. With further temperature increase, the decay time decreases drastically to $\tau(80\text{ K}) = 300$ μs and $\tau(300\text{ K}) = 3.3$ μs. The plot of the measured τ values versus temperature has a characteristic form of an s-shaped curve similar to the one obtained for Cu(dmp)(phanephos)$^+$ **21** (Figure 1.6a) but with the point of maximum slope shifted from 180 to 70 K for compound **2**. The radiative rate k_r, determined by the use of Eq. (1.3), rises from the low-temperature value of $k_r(30\text{ K}) = 8.3 \times 10^2$ s^{-1} to $k_r(80\text{ K}) = 3.3 \times 10^3$ s^{-1}, and to $k_r(300\text{ K}) = 2.1 \times 10^5$ s^{-1}, respectively (Table 1.4). The latter value represents a rate increase by a factor of 250 as compared with the $k_r(30\text{ K})$ value. This drastic increase of the radiative rate combined with the spectral blue shift of 13 nm with temperature increase (Figure 1.9) is explained by an involvement of a higher-lying state that carries a higher radiative rate, hence demonstrating the occurrence of TADF.

An analysis of the decay time data according to the Boltzmann-type Eq. (1.1) leads to an activation energy of $\Delta E(S_1 - T_1) = 370$ cm^{-1} and a prompt fluorescence decay rate of $k_r(S_1 \rightarrow S_0) = 3.9 \times 10^6$ s^{-1} (taking the emission quantum yield of $\Phi_{PL} = 70\%$ as measured at 300 K into account) (Figure 1.11).[1] Again, it is stressed that the prompt fluorescence could not be observed directly in our experiments conducted with nanosecond time resolution. This agrees with measured ISC time of the order of several picoseconds [118, 122, 134, 154, 156–158]. The resulting energy level diagram and the relevant rates are summarized in Figure 1.11c.

The value of $\Delta E(S_1 - T_1) = 370$ cm^{-1} represents one of the smallest splitting values found so far [31, 96]. Accordingly, Cu(dppb)(pz$_2$Bph$_2$) **2** shows a very short

---

[1] For completeness, it is remarked that the value of $k^r(S_1 \rightarrow S_0) = 3.9 \times 10^6$ s^{-1} represents a coarse estimate, since Eq. (1.1) is only valid, if the parameters in this equation, $k(T_1)$, $k(S_1)$, and $\Delta E(S_1 - T_1)$, are independent of temperature. This is not strictly the case, since the emission quantum yield decreases with temperature. However, an alternative fit can be applied when we restrict the temperature range for the fit procedure to $30 \leq T \leq 150$ K. Then it is reasonable to assume constant parameters, in particular, the emission quantum yield should be almost constant, i.e. $\approx 100\%$ as measured at $T = 80$ K [96]. For this restricted fit range, essentially the same fit parameters are obtained as discussed above. For this situation, we find $\tau(S_1) = 180$ ns at $\Phi_{PL} = 100\%$ and a rate of $k^r(S_1) = 5.6 \times 10^6$ s^{-1}.

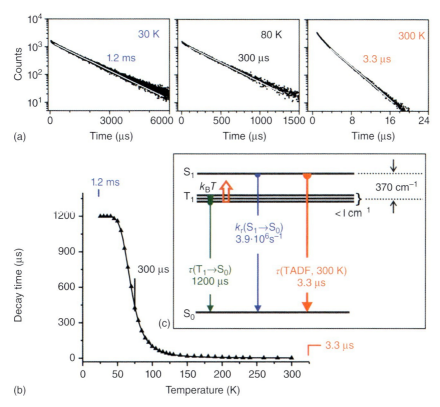

Figure 1.11 (a) Emission decay profiles of Cu(dppb)(pz$_2$Bph$_2$) **2** (powder) at 30, 80, and 300 K recorded after pulsed excitation at $\lambda_{exc} = 378$ nm and detected at $\lambda_{det} = 540$ nm. (b) Emission decay time τ versus temperature. The solid line represents a fit of Eq. (1.1) to the experimental $\tau(T)$ values fixing $\tau(T_1) = 1.2$ ms as measured at $T = 30$ K. The fit parameters are $\tau(S_1) = 180$ ns and $\Delta E(S_1-T_1) = 370$ cm^{-1}, respectively. (c) Energy level diagram of Cu(dppb)(pz$_2$Bph$_2$) **2**. Both competing emission processes are marked: Phosphorescence with a decay time of $\tau(T_1) = 1.2$ ms dominating the photophysical behavior at temperatures below about 50 K and TADF determining the emission properties at higher temperatures with a measured decay time at 300 K of 3.3 μs (with the emission quantum yield of $\Phi_{PL} = 70\%$). Source: Ref. [96]. Reproduced with permission of American Chemical Society.

TADF decay time of $\tau(\text{TADF}) = 3.3$ μs, being one of the shortest values reported so far for Cu(I) complexes (compare Section 1.6).

The experimental characterization of the luminescence behavior of Cu(dppb)(pz$_2$Bph$_2$) **2** supports the predictions based on model calculations, as developed above. According to the distinct spatial separation of the orbitals involved in the lowest excited states, the (formal) fluorescence decay time of $\tau(S_1) = 180$ ns (calculated from $k_r(S_1 \to S_0) = 3.9 \times 10^6$ s^{-1} and $\Phi_{PL} = 70\%$) is relatively long for a spin-allowed transition. For instance, it is about four times longer than the decay time of the S$_1$ state as determined for Cu(dmp)(phanephos)$^+$ **21** (Section 1.3). A large difference of the $\tau(S_1)$ lifetimes could be predicted by the TD-DFT calculations, as the oscillator strength of the corresponding $S_0 \to S_1$ transition (calculated for the T$_1$ state geometry), being 0.0016 for

Cu(dppb)(pz$_2$Bph$_2$) **2** (Table 1.3), is more than an order of magnitude smaller than calculated for Cu(dmp)(phanephos)$^+$ **21** with $f = 0.0201$ (Section 1.3).

Interestingly, the case studies presented in Sections 1.3 and 1.4 focusing on two complexes with very different allowedness of the S$_1 \leftrightarrow$ S$_0$ transitions is displayed inversely in the size of the energy splitting $\Delta E(\mathrm{S}_1-\mathrm{T}_1)$. For Cu(dmp)(phanephos)$^+$ **21**, it amounts to 1000 cm^{-1} (120 meV), while for Cu(dppb)(pz$_2$Bph$_2$) **2** a value of 370 cm^{-1} (46 meV) is found. This important relation will be addressed in Section 1.5.

Moreover, the photophysical studies presented in Sections 1.3 and 1.4 reveal an important practical conclusion concerning the assignments of emission processes. Both compounds show a phosphorescence plateau at low temperatures, at $T < 120$ K in the case of Cu(dmp)(phanephos)$^+$ **21** with $\Delta E(\mathrm{S}_1-\mathrm{T}_1) = 1000$ cm^{-1} and at $T < 50$ K in the case of Cu(dppb)(pz$_2$Bph$_2$) **2** with $\Delta E(\mathrm{S}_1-\mathrm{T}_1) = 370$ cm^{-1}, respectively. At ambient temperature, the emission of both compounds represents TADF. Importantly, in many laboratories only two temperature regimes of $T = 300$ K (ambient temperature) and $T = 77$ K (boiling point of nitrogen) are easily accessible. Therefore, characterization of new compounds is usually performed at these two temperatures. Based on these results, preliminary conclusions concerning the character of the emissive states and the emission mechanism(s) are drawn. For Cu(dmp)(phanephos)$^+$ **21**, with the measured decay time of $\tau(77\,\mathrm{K}) = 240$ µs (plateau range between 20 and 120 K) and $\tau(300\,\mathrm{K}) = 14$ µs, respectively, the emissions at low temperature and at ambient temperature would be assigned correctly as phosphorescence from the T$_1$ state and as TADF, respectively. For Cu(dppb)(pz$_2$Bph$_2$) **2** with $\tau(77\,\mathrm{K}) = 300$ µs and $\tau(300\,\mathrm{K}) = 3.3$ µs, a similar assignment would not be correct. As shown in Figure 1.11, the phosphorescence decay time $\tau(\mathrm{T}_1) = 1200$ µs (plateau for $T \leq 40$ K) is four times longer than found at 77 K. Thus, the emission at 77 K cannot be assigned as phosphorescence. On the contrary, it represents mainly delayed fluorescence (TADF) even at $T = 77$ K. The ratio of TADF to phosphorescence is estimated (by the use of Eq. (1.4c)) to 75% : 25%. Correspondingly, the emission spectra recorded at $T = 77$ K and 300 K are not shifted with respect to each other as for the two temperatures, the spectral maxima of compound **2** are found at 535 nm. Thus, conclusions made on the basis of 77 K and ambient temperature measurements must be taken with care. The risk of possible misinterpretation is particularly large when the energy separation $\Delta E(\mathrm{S}_1-\mathrm{T}_1)$ is small.

1.5 Energy Separation $\Delta E(\mathrm{S}_1-\mathrm{T}_1)$ and S$_1 \to$ S$_0$ Fluorescence Rate

In Section 1.2, it was already shortly discussed that a reduction of the energy separation between the lowest singlet S$_1$ and triplet T$_1$ state is connected with a decrease of the radiative singlet–singlet rate $k_\mathrm{r}(\mathrm{S}_1 \to \mathrm{S}_0)$. This relation has also been addressed in the two previous sections based on case studies. Since both photophysical parameters crucially determine the TADF behavior, in particular, the TADF decay time, we want to focus in this section on a simple model that may explain this relation.

1.5.1 Experimental Correlation Between $\Delta E(S_1-T_1)$ and $k_r(S_1 \rightarrow S_0)$ for Cu(I) Compounds

During the last years, the $\Delta E(S_1-T_1)$ energy separations and the $S_1 \rightarrow S_0$ fluorescence decay times have been determined for a large number of Cu(I) compounds that show TADF. Both parameters result from fit procedures by the use of Eq. (1.1). For two compounds, this has been discussed in detail in Sections 1.3 and 1.4. It is stressed again that the prompt fluorescence was not observed directly, but the corresponding $k_r(S_1 \rightarrow S_0)$ rate could be determined. In Table 1.5, adapted from Ref. [31], we summarize the corresponding data for Cu(I) compounds. The fitting procedure leads to the formal (prompt) decay time $\tau(S_1)$ or to the rate $k(S_1 \rightarrow S_0)$. From this information and by the use of the measured photoluminescence quantum yield (Φ_{PL}), one can easily determine the radiative rate $k_r(S_1 \rightarrow S_0)$ applying Eq. (1.3).

In Figure 1.12, the $k_r(S_1 \rightarrow S_0)$ data are plotted versus $\Delta E(S_1-T_1)$. It is obvious that with decreasing energy splitting, the allowedness of the $S_1 \rightarrow S_0$ transition decreases drastically. For example, if compound **25** is compared with compound **1**, $\Delta E(S_1-T_1)$ decreases from 1300 to 270 cm^{-1}, while the allowedness of the

Table 1.5 Energy separation $\Delta E(S_1-T_1)$ and radiative rate $k_r(S_1 \rightarrow S_0)$ determined by fitting procedures applying Eq. (1.1) to experimental decay time data. $\tau(S_1)$ is the (formal) prompt fluorescence decay time and Φ_{PL} the emission quantum yield.

	Compound	$\Delta E(S_1-T_1)$ (cm^{-1})	$\tau(S_1)$ (ns)	Φ_{PL} (300 K)	$k_r(S_1 \rightarrow S_0)$ ($10^6 \times$ s^{-1})	References
1	Cu$_2$I$_2$[MePyrPHOS)(Pph$_3$)$_2$	270	570	0.97	1.7	[30]
2	Cu(dppb)(pz$_2$Bph$_2$)	370	180	0.70	3.9	[96]
3	[Cu(μ-Cl)(PNMe$_2$)]$_2$	460	210	0.45	2.1	[33]
4	[Cu(μ-Br)(PNMe$_2$)]$_2$	510	110	0.65	5.9	[33]
5	[Cu(μ-I)(PNMe$_2$)]$_2$	570	90	0.65	7.2	[33]
6	Cu$_2$Cl$_2$(dppb)$_2$	600	70	0.35	5.0	[107]
7	[Cu(μ-I)(PNpy)]$_2$	630	100	0.65	6.5	[33]
8	Cu(pop)(pz$_2$BPh$_2$)	650	170	0.9	5.3	[9, 95]
14	Cu(tmbpy)(pop)$^+$	720	160	0.55	3.4	[35]
15	(IPr)Cu(py$_2$-BMe$_2$)	740	160	0.76	4.8	[34]
16	[Cu(PNP^{t}Bu)]$_2$	786	79	0.57	7.2	[46]
17	Cu$_2$I$_2$(MePyrPHOS)(dpph)	830	190	0.88	4.6	[30]
18	Cu$_2$Cl$_2$(N^P)$_2$	930	40	0.92	23	[97]
19	CuCl(Pph$_3$)$_2$(4-Mepy)	940	47	0.99	21	[98]
21	Cu(dmp)(phanephos)$^+$	1000	40	0.80	20	[92]
22	Cu(pop)(pz$_4$B)	1000	80	0.9	11	[95]
23	CuBr(Pph$_3$)$_2$(4-Mepy)	1070	41	0.95	23	[98]
24	CuI(Pph$_3$)$_2$(4-Mepy)	1170	14	0.66	47	[98]
25	Cu(pop)(pz$_2$BH$_2$)	1300	10	0.45	45	[70, 95]

Source: Ref. [31]. Adapted with permission of Elsevier.

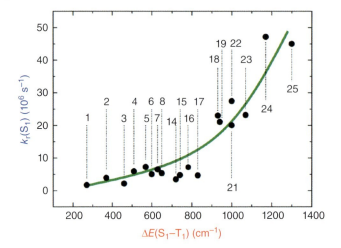

Figure 1.12 Radiative decay rate $k_r(S_1-S_0)$ plotted versus $\Delta E(S_1-T_1)$ for different Cu(I) complexes that show TADF at ambient temperature (data from Table 1.5). The fit curve represents an exponential function as guide for the eye. *Source*: Ref. [31]. Reproduced with permission of Elsevier.

$S_1 \rightarrow S_0$ transition decreases by a factor of about 26. In a simple consideration, using Eq. (1.1), it can be seen that the TADF decay time will not become shorter, when $\Delta E(S_1-T_1)$ reaches 300 to 200 cm^{-1} ($\approx$40 or 25 meV). Such a minimum decay time lies in the range of several microsecond [168], at least for the type of Cu(I) compounds discussed here.

1.5.2 Quantum Mechanical Considerations

In this section, we want to illustrate on a simple quantum mechanical basis, following [31], why a small energy separation $\Delta E(S_1-T_1)$ between the lowest singlet S_1 and triplet T_1 state is related to a fast radiative rate of the $S_1 \rightarrow S_0$ transition, i.e. a small $k_r(S_1 \rightarrow S_0)$ value. Let us assume that S_1 and T_1 can be described by a one-electron transition from HOMO φ_H to LUMO φ_L. In this situation, simple expressions can be given for the radiative rate $k_r(S_1-S_0)$ ($= k_r(S_1 \rightarrow S_0)$) and the energy splitting $\Delta E(S_1-T_1)$.

The radiative rate may be obtained from the transition dipole moment $\vec{\mu}(S_1 - S_0)$, which is approximately given by

$$\vec{\mu}_{H,L} = e \int \varphi_H(\vec{r}) \vec{r} \varphi_L(\vec{r}) \mathrm{d}^3 r = e \int \varphi_H(\vec{r}) \varphi_L(\vec{r}) \vec{r} \ \mathrm{d}^3 r \tag{1.10}$$

$\vec{r}$ is the dipole vector and e is the electron charge.

Thus, the radiative rate can be expressed by (see Ref. [169], p. 159), [31]:

$$k_r(S_1 - S_0) = 2C\nu^3 n^3 |\vec{\mu}_{HL}|^2 \tag{1.11}$$

$$k_r(S_1 - S_0) = 2e^2 C\nu^3 n^3 \left| \int \varphi_H(\vec{r}) \varphi_L(\vec{r}) \vec{r} \ \mathrm{d}^3 r \right|^2 \tag{1.12}$$

with the numerical constant $C = 16\pi^3/(3\varepsilon_0 hc^3)$, wherein ε_0 is the vacuum permittivity, h is Planck's constant, and c is the velocity of light. $\nu = \Delta E(S_1-S_0)/h$ is the transition frequency and n the refractive index.

The energy separation $\Delta E(S_1-T_1)$ can be expressed in this approximation by twice the exchange integral K_{HL} for HOMO and LUMO (see Ref. [[130], p. 86]) giving

$$\Delta E(S_1 - T_1) \approx 2K_{HL} \tag{1.13}$$

with

$$K_{HL} = \frac{e^2}{4\pi\varepsilon_0} \int \varphi_H(\vec{r}_1)\varphi_L(\vec{r}_2)\frac{1}{|\vec{r}_2 - \vec{r}_1|}\varphi_H(\vec{r}_2)\varphi_L(\vec{r}_1) d^3r_1 \, d^3r_2 \tag{1.14}$$

or

$$K_{HL} = \frac{e^2}{4\pi\varepsilon_0} \int \varphi_H(\vec{r}_1)\varphi_L(\vec{r}_1)\frac{1}{|\vec{r}_2 - \vec{r}_1|}\varphi_H(\vec{r}_2)\varphi_L(\vec{r}_2) d^3r_1 \, d^3r_2 \tag{1.15}$$

It is an important result that both the exchange interaction and hence the energy splitting $\Delta E(S_1-T_1)$ (Eq. (1.13)) as well as the radiative rate (Eq. (1.12)) depend quadratically on the product of $\varphi_H(r)\varphi_L(r)$ or the squared overlap of HOMO and LUMO.[2] Accordingly, it becomes obvious that $\Delta E(S_1-T_1)$ and $k_r(S_1 \to S_0)$ correlate. A small HOMO–LUMO overlap implies small $\Delta E(S_1-T_1)$ and small $k_r(S_1 \to S_0)$. This relation is schematically illustrated in Figure 1.13.

The simple qualitative model presented above allows us to understand the experimental results as displayed in Figure 1.12. The model's basic assumption is that the low-lying singlet S_1 and triplet T_1 states, originating from the HOMO → LUMO excitation, are energetically well separated from higher-lying energy states. Accordingly, in the scope of this material class (or this model), we cannot reduce the TADF decay time distinctly below a few microsecond. However, the simple model, being valid for the compounds shown in Table 1.5, does not contain mixing of the singlet state S_1 with a higher-lying singlet state that carries high oscillator strength with respect to the transition to the electronic ground state. In principle, such mixing, induced by CI, might

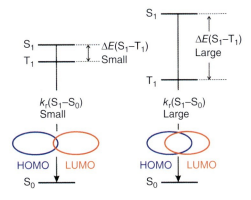

Figure 1.13 Schematic illustration of the relation between energy splitting $\Delta E(S_1-T_1)$ and the radiative decay rate $k_r(S_1-S_0)$ on the spatial overlap of HOMO and LUMO. Source: Ref. [31]. Reproduced with permission of Elsevier.

---

2 It is noted that the modified overlap $\int |\varphi_H||\varphi_L| dV$ is a quantitative measure for the spatial overlap [170].

significantly increase the radiative rate of the lowest singlet–singlet transition without strongly altering the energy splitting $\Delta E(S_1-T_1)$. Presumably, such Cu(I) complexes may be developed in the future. In Section 1.6, we will address this challenge shortly again.

1.6 Design Strategies for Highly Efficient Ag(I)-Based TADF Compounds

In contrast to Cu(I) complexes, TADF materials based on Ag(I) are rarely reported [104, 106–109, 171]. This is related to the higher oxidation potential of Ag^+ compared with Cu^+ [172]. Accordingly, the 4d orbitals of Ag(I) complexes lie mostly energetically below ligand-centered (LC) orbitals. As a consequence, low-lying states of ^{3}LC character determine the emission properties [106, 173–176]. Thus, Ag(I) complexes often do not exhibit TADF, but long-lived phosphorescence and sometimes even slow ISC [175]. Therefore, designing Ag(I) complexes that show TADF represents an optimization task. In this respect, it is required to destabilize the energetically lower-lying 4d orbitals by an organic ligand with good electron-donating ability. This may be attainable with electron-donating bidentate phosphine ligands. Indeed, it has already been demonstrated that this strategy is successful. For example, $Ag_2Cl_2(dppb)_2$ **20** represents a blue light-emitting material that shows efficient TADF with $\Phi_{PL} = 97\%$, though with a relatively long decay time of τ(TADF, 300 K) = 15 μs (Table 1.1) [107]. Another attractive ligand is *nido*-carborane-bis-(diphenylphosphine) (P_2-nCB) [177]. It coordinates via phosphine groups and, thus, induces substantial electron-donating character. Additionally, electron donation is strongly enhanced by the negative charge of the *nido*-carborane moiety. Suitable complexes can be built using the (P_2-nCB) ligand in combination with 1,10-phenanthroline (phen) or substituted phen ligands. Accordingly, a series of neutral Ag(I) complexes, referring to the numbers from **9** to **13**, is obtained (Table 1.1) [108, 109]. In this section, we want to focus on these Ag(phen)(P_2-nCB)-type compounds and to demonstrate key steps for designing a material that shows TADF behavior. Interestingly, by this strategy an efficiency breakthrough is reached [108, 109].

1.6.1 Ag(phen)(P_2-nCB): A First Step to Achieve TADF

DFT and TD-DFT calculations give an insight into electronic properties of Ag(phen)(P_2-nCB) **9**. The chemical structure as well as HOMO and LUMO plots are displayed in Figure 1.14. The calculations were performed for the optimized triplet T_1 state geometry. From the TD-DFT approach, it is indicated that the states S_1 and T_1 are dominated by HOMO → LUMO electronic transitions by 96% and 94%, respectively. Analysis of the frontier orbitals shows that the HOMO is mainly localized on the phosphines and the silver ion, whereas the LUMO is localized on the phen ligand. Accordingly, both states may be assigned to be of (metal + ligand L) to ligand L′ charge-transfer (^{1,3}MLL′CT) character. The small overlap of HOMO and LUMO suggests the occurrence

1.6 Design Strategies for Highly Efficient Ag(I)-Based TADF Compounds | 35

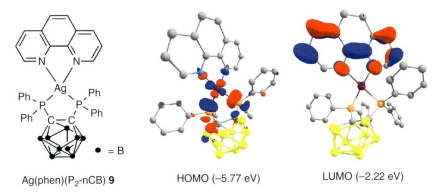

Ag(phen)(P$_2$-nCB) **9**   HOMO (−5.77 eV)   LUMO (−2.22 eV)

Figure 1.14 Chemical structure formula and calculated (M062X/def-2SVP) frontier orbital isosurface contour plots (iso-value = 0.05) for Ag(phen)(P$_2$-nCB) **9**. The calculations were carried out for the gas phase optimized (M06/def-2SVP) T$_1$ state geometry. Source: Ref. [109]. Reproduced with permission of Elsevier.

of a small exchange interaction between the unpaired electrons and, thus, a small $\Delta E(S_1-T_1)$ splitting. From TD-DFT calculations the energy gap of $\Delta E(S_1-T_1) = 1070$ cm^{-1} (133 meV) can be estimated from the energy difference between vertical electron transitions $S_0 \to S_1$ and $S_0 \to T_1$. This agrees approximately with the value of 725 cm^{-1} (90 meV) roughly estimated from the shift of the emission peak maxima upon cooling (see below). Hence, this Ag(I) complex represents an interesting TADF candidate.

Figure 1.15 displays emission and absorption spectra of Ag(phen)(P$_2$-nCB) **9**. The absorption peak of low molar extinction (3020 M^{-1} cm^{-1}) near 400 nm is

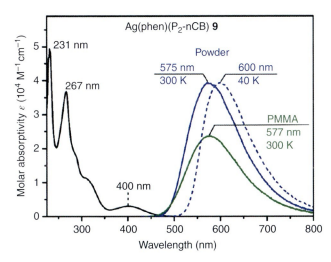

Figure 1.15 Emission and absorption spectra of Ag(phen)(P$_2$-nCB) **9** at different temperatures. The absorption spectrum is measured with a dichloromethane (DCM) solution of ≈10^{-5} M concentration at 300 K (black line). The emission spectra are shown in colored lines ($\lambda_{exc} = 410$ nm). The PMMA film was doped with ≈1 wt% of compound **9**. Source: Ref. [109]. Reproduced with permission of Elsevier.

assigned to the $S_0 \rightarrow S_1$ (^{1}MLL′CT) transition, while the structures of higher allowedness and higher energy are ascribed to LC transitions. (Compare also the TD-DFT results shown in the SI of Ref. [109].) The emission spectral bands are broad and unstructured as usually found for CT transitions. With temperature reduction from $T = 300$ to 40 K, a red shift of the peak maximum of the powder material from 575 to 600 nm (25 nm corresponding to 725 cm^{-1} or 90 meV) is observed. Such a behavior agrees well with the occurrence of TADF at 300 K. At ambient temperature, the emission stems dominantly from the singlet state S_1, while at $T = 40$ K, the TADF process is frozen out and only the lower-energy phosphorescence from the T_1 state occurs.

For completeness, it is noted that the slight red shift of the ambient temperature emission of the PMMA-doped emitter **9**, as compared with the powder material, is related to the lower rigidity of the PMMA film given by the emitter's environment. A corresponding behavior has already been discussed for Cu-based TADF compounds in Sections 1.3 and 1.4 as well as in the literature [31, 95].

1.6.2 Emission Quenching in Ag(phen)(P$_2$-nCB)

It is of particular importance for application of emitters in OLEDs that the photoluminescence quantum yield of the emitter should be as high as possible. However, for Ag(phen)(P$_2$-nCB), the values amount only to Φ_{PL}(powder, 300 K) = 36% and Φ_{PL}(PMMA doped, 300 K) = 26%, respectively (Table 1.6).[3] Therefore, it is of interest to understand why emission quenching is distinctly effective for this compound.

For Cu(I) complexes, it is well known that flattening distortions occur upon excitation of MLCT states (Section 1.3 and see Refs [31, 114, 117, 118, 121, 122, 178]). A similar behavior is also expected to be relevant for Ag(I) complexes. DFT geometry optimizations (M06/def2-SVP) show that the molecule in the lowest excited (relaxed) triplet state T_1 of Ag(phen)(P$_2$-nCB) **9** is distinctly twisted toward planarization as compared with the ground state geometry. This distortion cannot be characterized by one simple parameter, but in a rough description, one may take the change of the angle $\Delta\varphi$ between two planes placed into the molecular core of the complex (Figure 1.16). Related to such a distortion, the potential energy surfaces of the involved energy states are shifted significantly with respect to each other. This leads to an increase of the Franck–Condon factors between higher-lying vibrational states of the electronic ground state S_0 and the lower-lying vibrational states of the excited state. The larger the Franck–Condon factors, the more efficient are non-radiative relaxation processes [124–126]. Indeed, the emission quantum yield of compound **9** with a large value of $\Delta\varphi \approx 35°$ amounts only to $\Phi_{PL} = 36\%$, although the compound sits in a rigid crystalline (powder) environment.

These considerations lead to the suggestion to rigidify the emitter's structure by introducing intramolecular steric hindrances. Such an approach has already been successfully applied [9, 30, 31, 35, 92, 108, 109, 114, 119, 127]. (Compare

---

3 The emission quantum yield of Ag(phen)(P2-nCB) in degassed DCM solution is $\Phi_{PL} \ll 1\%$.

Table 1.6 Photophysical data that govern TADF properties of a series of Ag(I) complexes.

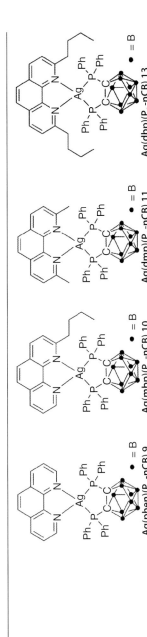

	Ag(phen)(P$_2$-nCB) 9	Ag(mbp)(P$_2$-nCB) 10	Ag(dmp)(P$_2$-nCB) 11	Ag(dbp)(P$_2$-nCB) 13
Φ_{PL} (300 K)	36%	70%	78%	100%
$f(S_1 \rightarrow S_0)$[a]	0.0258	0.0478	0.0423	0.0536
$k_r(S_1 \rightarrow S_0)$[b]	—	2.2×10^7 s^{-1}	2.2×10^7 s^{-1}	5.6×10^7 s^{-1}
τ^r(TADF, 300 K)[c]	5.3 μs	2.9 μs	3.2 μs	1.4 μs
$\Delta E(S_1 - T_1)$[d]	—	640 cm^{-1}	650 cm^{-1}	650 cm^{-1}
λ_{max} (300 K)	575 nm	535 nm	537 nm	526 nm

a) TD-DFT calculated (M062X/def2-SVP) oscillator strength based on gas phase optimized (M06/def2-SVP) T$_1$ state geometries.
b) Radiative decay rate of the prompt fluorescence determined from the fit of Eq. (1.1) to experimental decay times at various temperatures.
c) Radiative decay time measured at 300 K, essentially representing TADF.
d) Energy gap between the lowest excited singlet and triplet states as determined from the fit of Eq. (1.1) to experimental decay times over a temperature range of 30 ≤ T ≤ 300 K.

Source: Refs [108, 109]. Reproduced with permission of American Chemical Society.

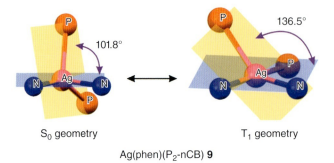

Ag(phen)(P$_2$-nCB) **9**

Figure 1.16 Schematic visualization of the geometry change between the electronic ground state S$_0$ and the triplet state T$_1$. Only the coordination core around the Ag ion is shown. The angles refer to the inclination between the P–Ag–P (orange) and the N–Ag–N (blue) planes. The charge transfer excitation induces a flattening distortion characterized by a model parameter of $\Delta\varphi \approx 35°$. The calculations were carried out on the M06/def2-SVP level of theory for the gas phase conditions. *Source:* Ref. [109]. Reproduced with permission of Elsevier.

also Section 1.3.) We will discuss this procedure in the next section, applying it to Ag(R-phen)(P$_2$-nCB), with R representing a substitution.

1.6.3 Sterical Hindrance. Tuning of the Emission Quantum Yield up to 100%

Several Ag(I) complexes comprising the P$_2$-nCB ligand in combination with a phenanthroline ligand, in each case differently substituted at the positions 2 and 9, are displayed in Table 1.6. The photophysical investigations show that with more bulky substituents on the phenanthroline ligand, the nonradiative decay rate decreases, and, hence, the photoluminescence quantum yield increases. For example, an increase from $\Phi_{PL} = 36\%$ found for Ag(phen)(P$_2$-nCB) **9** with an unsubstituted phen ligand to $\Phi_{PL} = 100\%$ for Ag(dbp)(P$_2$-nCB) **13** with the 2,9-di-*n*-butyl phenanthroline ligand is observed [108, 109]. The calculated model parameter $\Delta\varphi$, describing coarsely the flattening angle, correlates with this photophysical behavior, as was expected. Without sterical hindrance, the $\Delta\varphi$ parameter amounts to ≈35°, while for the compounds with sterical hindrance, the angle change is about three times smaller. For Ag(dbp)(P$_2$-nCB) **13**, for example, the $\Delta\varphi$ parameter amounts only to ≈12°. (Compare Figures 1.16 and 1.17.) According to the significantly lower geometry change upon CT excitation, the emission quantum yield increases drastically (Table 1.6).

The TD-DFT calculations (M062X/def2-SVP) carried out for the optimized T$_1$ state geometry reveal another important variation in the series of complexes shown in Table 1.6. The oscillator strength $f(S_0 \rightarrow S_1)$ of the $S_1 \rightarrow S_0$ electronic transition increases from $f = 0.0258$ for Ag(phen)(P$_2$-nCB) **9** to $f = 0.0536$ for Ag(dbp)(P$_2$-nCB) **13**, i.e. by a factor of more than two [109]. At the first sight, the reason for this increase is not obvious. However, in a computational model, as presented in Ref. [109], it can be shown that TD-DFT calculations carried out for Ag(phen)(P$_2$-nCB) **9**, fixed to the T$_1$ state geometry of Ag(dbp)(P$_2$-nCB)

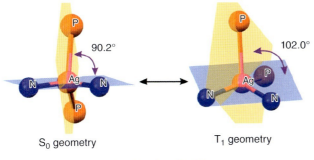

Ag(dbp)(P$_2$-nCB) **13**

Figure 1.17 Schematic visualization of the geometry change between the electronic ground state S$_0$ and the triplet state T$_1$. Only the coordination core around Ag is shown. The angles refer to the inclination between the P–Ag–P (orange) and the N–Ag–N (blue) planes. The charge-transfer excitation induces a flattening distortion characterized by a model parameter $\Delta\varphi \approx 12°$. The geometry optimizations were carried out for gas phase conditions at the M06/def2-SVP level of theory. (Compare Figure 1.16.) *Source*: Ref. [109]. Adapted with permission of Elsevier.

13, gives the S$_0 \to$ S$_1$ oscillator strength of $f = 0.0687$. This value is even higher than the value, calculated for compound Ag(dbp)(P$_2$-nCB) **13** itself. Obviously, the effect of the complex geometry on the S$_1 \to$ S$_0$ oscillator strength is superior to the electronic influence of the substituents on positions 2 and 9 of the 1,10-phenanthroline ligand [109].

The increase of the calculated oscillator strength of the series of Ag(I) complexes should also be displayed in the experimentally determined radiative rate (that is, proportional to the oscillator strength). Indeed, the radiative rates $k_r(S_1 \to S_0)$ (Table 1.6) show a similar increase as is seen, for example, when Ag(mbp)(P$_2$-nCB) with $k_r(S_1$–$S_0) = 2.2 \times 10^7$ s^{-1} is compared with Ag(dbp)(P$_2$-nCB) with $k_r(S_1$–$S_0) = 5.6 \times 10^7$ s^{-1}.

In summary, the drastic increase of the emission quantum yield in the series of Ag(I) complexes arranged in Table 1.6 is induced by two different effects: (i) Increasing sterical hindrance strongly reduces nonradiative relaxation. (ii) In parallel, the 2,9-substitutions stabilize a complex geometry that leads to a high radiative rate. Both effects are responsible in a very favorite way for attaining the high emission quantum yield of $\Phi_{PL} = 100\%$ for Ag(dpb)(P$_2$-nCB).

For completeness, we also calculated the S$_0 \to$ S$_1$ oscillator strengths for Cu(I) complexes with the same ligands. For example, TD-DFT calculation (M062X/def2-SVP) carried out for compound Cu(dbp)(P$_2$-nCB) in gas phase for the T$_1$ state geometry (M06/def2-SVP) gives an oscillator strength value of $f = 0.0660$, while the value for the corresponding Ag(dbp)(P$_2$-nCB) complex amounts to $f = 0.0536$. Obviously, the change of allowedness affected by replacement of Ag(I) through Cu(I) is not very distinct.[4]

Guided by this result, it seems to be justified to relate properties based on oscillator strengths of the S$_1 \leftrightarrow$ S$_0$ transitions of Ag(I) complexes to trends that

---

4 The emission quantum yield of Cu(dbp)(P$_2$-nCB) powder is relatively low ($\Phi_{PL} = 16\%$). Therefore, this material is not suitable for OLED applications.

are observed for Cu(I) complexes. In particular, comparison to the relation between the energy splitting $\Delta E(S_1-T_1)$ and the radiative rate $k_r(S_1-S_0)$, as displayed in Figure 1.12 (Section 1.5) for Cu(I) complexes, elucidates an interesting result. Inserting the data found for Ag(dbp)(P$_2$-nCB) with $\Delta E(S_1-T_1) = 650$ cm^{-1} and $k_r(S_1-S_0) = 5.6 \times 10^7$ s^{-1} (Table 1.6), it becomes obvious that these data do not fit. The rate of the Ag(I) complex is about one order of magnitude higher than expected from the relation shown in Figure 1.12. This result is highly interesting, since it indicates how to develop new materials that break the restrictions imposed by the simple quantum mechanical model presented in Section 1.5. Apparently, the singlet state S_1 wave function of the discussed Ag(I) complexes are not simply given by the HOMO–LUMO excitation, but are distinctly modified by configurational interaction. This means that higher-lying singlet states, resulting from other configurations, mix and thus induce significantly larger $S_1 \rightarrow S_0$ transition rates. Further quantum mechanical investigations have to be carried out in this respect. Nevertheless, these results represent a guideline for the development of materials with shorter TADF decay time, as it has already been demonstrated for Ag(dbp)(P$_2$-nCB) **13** [108].

1.6.4 Detailed Characterization of Ag(dbp)(P$_2$-nCB)

The TADF properties of Ag(dbp)(P$_2$-nCB) **13** are highly attractive due to its photoluminescence quantum yield of $\Phi_{PL} = 100\%$ and the very short decay time of $\tau(\text{TADF}) = 1.4$ μs. Therefore, in this section, we will focus on a detailed characterization of the compound's emission properties.

The lowest excited singlet S_1 and triplet T_1 states predominantly originate from the HOMO → LUMO transition (92%), slightly less than found for Ag(phen)(P$_2$-nCB) **9**. The HOMO is mainly composed of silver (13%) and phosphorus (47%) orbitals, while the LUMO represents a π^* orbital of the dbp ligand (Figure 1.18). Thus, we can assign the two lowest excited states as ^{1,3}(MLL'CT) states with L and L' representing P$_2$-nCB and dbp, respectively. According to the TD-DFT calculations carried out for the optimized T_1 state geometry of Ag(dbp)(P$_2$-nCB) **13**, the energy separation $\Delta E(S_1-T_1)$ can be estimated to 0.15 eV (≈ 1200 cm^{-1}). This represents an upper bound that is obtained from vertical excitations of $S_0 \rightarrow S_1$ and $S_0 \rightarrow T_1$ transitions, respectively. The calculated value is largely in agreement with the experimentally determined activation energy of $\Delta E(S_1-T_1) = 650$ cm^{-1} (see below).

The *n*-butyl substitutions at the 2,9-positions of the phen ligand lead to a distinct sterical hindrance with respect to a geometry change upon the CT excitation. DFT computations show that this flattening distortion is much less expressed than found for Ag(phen)(P$_2$-nCB). (Compare Figures 1.16 and 1.17.) As a consequence, it is expected that the emission quantum yield of the more rigid complex Ag(dbp)(P$_2$-nCB) **13** is higher than that of Ag(phen)(P$_2$-nCB), as already discussed above and as experimentally found.[5]

---

[5] The molecular rigidity of Ag(dbp)(P$_2$-nCB) is largely maintained, even if the complex is doped in PMMA, in contrast to most other TADF compounds (compare Section 1.3 and Ref. [95]) For Ag(dbp)(P$_2$-nCB) doped in PMMA, Φ_{PL} drops from 100% (powder) only to 85% (Table 1.7).

1.6 Design Strategies for Highly Efficient Ag(I)-Based TADF Compounds

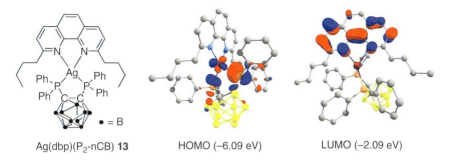

Ag(dbp)(P$_2$-nCB) **13**   HOMO (−6.09 eV)   LUMO (−2.09 eV)

Figure 1.18 Chemical structure formula and calculated (M062X/def2-SVP) isosurface contour plots (iso-value = 0.05) of the frontier orbitals of Ag(dbp)(P$_2$-nCB) **13**. The calculations were carried out for the gas phase optimized (M06/def2-SVP) T$_1$ state geometry. *Source*: Ref. [108]. Reproduced with permission of American Chemical Society.

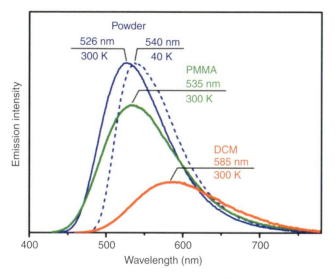

Figure 1.19 Emission spectra of Ag(dbp)(P$_2$-nCB) **13**. The measurements were carried out under different conditions, as marked in the diagram. Concentrations: PMMA: $c \approx 1$ wt%, dissolved in dichloromethane (DCM): $c \approx 10^{-5}$ M. $\lambda_{exc} = 410$ nm. *Source*: Ref. [108]. Reproduced with permission of American Chemical Society.

Figure 1.19 displays the emission spectra of Ag(dbp)(P$_2$-nCB) **13**. The spectra are broad and unstructured as expected for CT transitions. Even cooling to $T = 1.5$ K does not lead to any better resolution (not shown). However, application of methods based on the temperature dependence of the emission decay time reveals additional information, as will be demonstrated below. For the powder material, a blue shift is observed with temperature increase from $T = 40$ K (T$_1$ emission) to $T = 300$ K (S$_1$ emission) of $\Delta\lambda = 14$ nm corresponding to 490 cm^{-1} (60 meV). This value fits approximately to the activation energy of $\Delta E(S_1-T_1) = 650$ cm^{-1} (80 meV) as determined below. However, it is noted that the emission spectra represent transitions between Franck–Condon states, and

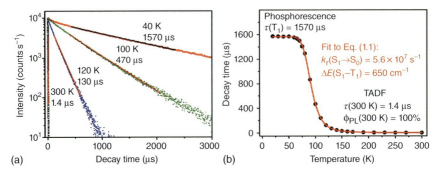

Figure 1.20 (a) Luminescence decay curves of Ag(dbp)(P$_2$-nCB) **13** powder measured at different temperatures. (b) Luminescence decay time (τ) plotted versus temperature. The values of $k(S_1 \rightarrow S_0) = 5.6 \times 10^7$ s^{-1} (18 ns) and $\Delta E(S_1-T_1) = 650$ cm^{-1} result from a fit of Eq. (1.1) to the experimental $\tau(T)$ data, with $\tau(T_1)$ fixed to 1570 μs as determined directly for $T < 60$ K (plateau). $\lambda_{exc} = 378$ nm, diode laser PB-375 L, and pulse width = 100 ps. Source: Adapted with permission from Ref. [108]. Copyright © 2017, American Chemical Society.

therefore the shift of emission spectra is not very reliable to assess the $\Delta E(S_1-T_1)$ gap, especially when the excited states are of CT character.

Detailed information on the electronic structure and the relevant decay processes can be obtained from time-resolving measurements, as already shown in previous sections. Figure 1.20a shows almost monoexponential emission decay curves measured at different temperatures. In Figure 1.20b, the decay times are plotted versus temperature. In the temperature range of $20 \text{ K} \leq T \leq 60 \text{ K}$, a constant value of 1570 μs is observed (plateau). With this value and the low-temperature emission quantum yield of $\Phi_{PL}(77 \text{ K}) = 87\%$ (Table 1.7), one can determine the radiative rate to $k_r(40 \text{ K}) = 5.5 \times 10^2$ s^{-1} (applying Eq. (1.3), whereby it is assumed that the quantum yield at $T = 40$ K is the same as the measured one at $T = 77$ K). With temperature increase to $T = 300$ K, the decay time decreases drastically to $\tau(300 \text{ K}) = 1.4$ μs (at $\Phi_{PL} = 100\%$). Accordingly, the radiative rate increases by a factor of about 1300 to $k_r(300 \text{ K}) = 7.1 \times 10^5$ s^{-1}. Obviously, such a change has to be related to the involvement of different electronic transitions at low and high temperature, respectively. At low temperature, the emission is a phosphorescence from the T_1 state, and at ambient temperature it represents TADF from the S_1 state.

The monoexponentiality of the decay curves indicates fast thermalization between the involved energy states due to fast up- and down-ISC and small inhomogeneities (small variations of $\Delta E(S_1-T_1)$) of the compounds in the powder material. In this situation, the emission decay time $\tau(T)$ of a molecular system of two excited energy states, T_1 and S_1, can be expressed by Eq. (1.1). The fit of this equation to the experimental data (Figure 1.20b) gives the activation energy of $\Delta E(S_1-T_1) = 650$ cm^{-1} (80 meV) and the radiative rate of the prompt fluorescence of $k_r(S_1 \rightarrow S_0) = 5.6 \times 10^7$ s^{-1}. Formally, this value corresponds to a fluorescence decay time of $\tau(S_1) = 18$ ns. However, the related prompt fluorescence is not directly observed, since the processes of ISC from S_1 to T_1 are about three orders of magnitude faster (compare Ref. [118, 121, 133]) [134].

Table 1.7 Emission data for Ag(I) complexes as powder materials and doped in PMMA measured at different temperatures.

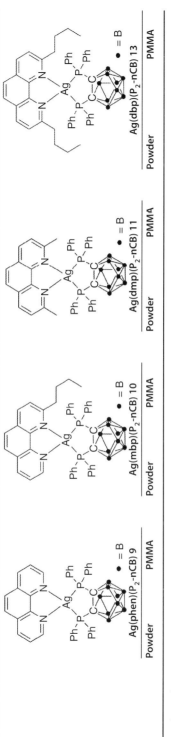

	Ag(phen)(P$_2$-nCB) 9		Ag(mbp)(P$_2$-nCB) 10		Ag(dmp)(P$_2$-nCB) 11		Ag(dbpp)(P$_2$-nCB) 13	
	Powder	PMMA	Powder	PMMA	Powder	PMMA	Powder	PMMA
λ_{max} (300 K)	575 nm	577 nm	535 nm	555 nm	537 nm	540 nm	526 nm	535 nm
Φ_{PL} (300 K)	36%	26%	70%	58%	78%	75%	100%	85%
τ (300 K)	2.0 µs		2.0 µs		2.8 µs		1.4 µs	
k_r (300 K)	1.8×10^5 s^{-1}		3.5×10^5 s^{-1}		2.8×10^5 s^{-1}		7.1×10^5 s^{-1}	
k_{nr} (300 K)	3.2×10^5 s^{-1}		1.5×10^5 s^{-1}		0.79×10^5 s^{-1}		$<0.21 \times 10^5$ s$^{-1\,b)}$	
Φ_{PL} (77 K)	15%		70%		68%		87%	
τ (77 K)	270 µs		1390 µs		804 µs		1300 µs	
k_r (77 K)	5.6×10^2 s^{-1}		5×10^2 s^{-1}		8.5×10^2 s^{-1}		6.7×10^2 s^{-1}	
k_{nr} (77 K)	3.1×10^3 s^{-1}		2.2×10^2 s^{-1}		3.9×10^2 s^{-1}		1×10^2 s^{-1}	
τ (T$_1$, 40 K)			1600 µs		885 µs		1570 µs	
k_r (S$_1 \to$ S$_0$)[a)]			2.2×10^7 s^{-1}		2.8×10^7 s^{-1}		5.6×10^7 s^{-1}	
ΔE(S$_1$−T$_1$)[a)]			640 cm^{-1}		650 cm^{-1}		650 cm^{-1}	

a) Determined from the fit of experimental luminescence decay times according to Eq. (1.1), measured for a powder sample at different temperatures.
b) Determined assuming 3% error for the measured Φ_{PL} value, which would allow Φ_{PL} = 97%.

The experimental value of $k(S_1-S_0) = 5.6 \times 10^7$ s^{-1} found for the prompt fluorescence rate is remarkably large. Cu(I) complexes investigated so far (and that have comparable $\Delta E(S_1-T_1)$ splittings) exhibit only rates that are one order of magnitude smaller [31]. (See Section 1.5, Figure 1.12.) Thus, the high (prompt) fluorescence rate can be identified as a key feature that leads to the exceptionally fast TADF decay time of Ag(dbp)(P$_2$-nCB). This behavior fits perfectly to the large oscillator strength that is calculated for the $S_1 \to S_0$ transition. (Compare the discussion presented above in this section.)

Photophysical data for Ag(dbp)(P$_2$-nCB) **13** are summarized in Table 1.7 and in an energy level diagram shown in Figure 1.21. Herein, the material's properties are highlighted: At low temperature ($T < 60$ K), one observes only long-living phosphorescence as the $T_1 \to S_0$ transition decays with $\tau = 1570$ μs. Such a long phosphorescence decay time is not unusual for Ag(I) or Cu(I) compounds [31, 33, 96, 107, 179]. It displays the spin forbiddenness of this transition. The reason is that SOC to singlet states is weak. According to the discussion presented in Refs [108, 109], the next higher-lying singlet state (S_4) that exhibits a different 4d orbital character than the T_1 state and that can induce SOC (following general quantum mechanical rules) [70–77] is energetically far ($\Delta E(S_4-T_1) = 1.57$ eV). Therefore, the singlet character mixed into the T_1 state is very small. As already mentioned in Section 1.4, SOC via the S_1 state can be neglected.

With temperature increase, the S_1 state is populated according to the thermal energy $k_B T$. As a consequence, the decay time decreases and a spectral blue shift is observed. This represents the TADF effect. The corresponding activation energy, as determined from the decay time plot (Figure 1.20b), amounts to $\Delta E(S_1-T_1) = 650$ cm^{-1}. This energy separation is not very small (compare compound **2**, Section 1.4), but the $S_1 \to S_0$ transition rate is much higher than for any other organometallic TADF material (with comparable $\Delta E(S_1-T_1)$ values). Thus, the TADF decay time drops to the record value of τ(TADF, 300 K) = 1.4 μs. Moreover, the Ag(I) complex represents the first TADF material with a radiative decay time comparable to those of Ir(III) complexes [1, 69, 70, 86, 89] that have become famous for OLED applications.

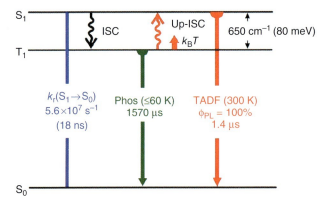

Figure 1.21 Energy level diagram and decay times/rates for Ag(dbp)(P$_2$-nCB) **13** powder. Frequently, up-ISC is also denoted as reverse ISC (rISC). *Source*: Ref. [108]. Adapted with permission of American Chemical Society.

This short TADF decay time or the related very large radiative decay rate of k_r(TADF, 300 K) $= 7.1 \times 10^5$ s^{-1} is responsible for the high quantum yield of $\Phi_{PL} = 100\%$, measured at $T = 300$ K. For completeness, it is mentioned that the quantum yield at $T = 77$ K amounts to only Φ_{PL}(77 K) $= 87\%$, since at that temperature the decay time is with τ(77 K) $= 1300$ μs relatively long, leading to a radiative rate of only k_r(77 K) $= 6.7 \times 10^2$ s^{-1}. Using Eq. (1.2), the nonradiative rate for this temperature is determined to $k_{nr} = 1 \times 10^2$ s^{-1}. Thus at 77 K, the non-radiative process can moderately compete with the radiative process. But at ambient temperature, the TADF rate predominates by about three orders of magnitude. Hence, nonradiative processes are no longer relevant.

1.7 Conclusion and Future Perspectives

In this chapter, we study TADF material design based on photophysical properties investigated for a large number of compounds, in particular, with respect to OLED applications. Especially, we focus on photoluminescence properties and on the crucial requirement of designing materials that exhibit short emission decay times (high radiative rates), obviously at high emission quantum yields. The decay time should be as short as possible in order to minimize nonradiative quenching, saturation effects, and, in particular, chemical reactions that might occur in the excited state. Thus, short TADF decay time will help to increase the OLED device lifetime. Here, we discuss important molecular or photophysical parameters and analyze their impact on the TADF decay time. For example, it is well known that the energy separation $\Delta E(S_1-T_1)$ between the lowest excited singlet state S_1 and triplet state T_1 should be as small as possible. Accordingly, we present detailed photophysical properties of two case studies referring to materials that exhibit a large $\Delta E(S_1-T_1)$ value of 1000 cm^{-1} (120 meV) and a small one of 370 cm^{-1} (46 meV), respectively. From these studies – extended by investigations of photophysical properties of many other Cu(I) TADF compounds – we can show, however, that small $\Delta E(S_1-T_1)$ is not a sufficient requirement for short TADF decay time. High allowedness of the transition between the emitting S_1 state and the electronic ground state S_0, expressed by the radiative rate $k_r(S_1 \rightarrow S_0)$, is also very important. This has often been disregarded. However, mostly small $\Delta E(S_1-T_1)$ is related to a small $k_r(S_1 \rightarrow S_0)$. As a consequence, a reduction of τ(TADF) to below a few microsecond might be problematic. This relation results from an investigation of a large number of Cu(I) complexes and basic quantum mechanical considerations. However, these studies are based on a situation, in fact a very frequent one, in which the involved states, S_1 and T_1, stem from the same HOMO–LUMO excitation. Other higher-lying singlet states from which the $S_1 \rightarrow S_0$ transition might borrow allowedness are energetically too far from the S_1 state. However, new materials can be designed for which this disadvantage is not prevailing. Very probably, the new TADF compound, Ag(dbp)(P$_2$-nCB), represents such an example. Indeed, we obtained a TADF record material with τ(TADF) $= 1.4$ μs at 100% emission quantum yield, as discussed in Section 1.6.

As a consequence, it is an important issue for future developments of TADF materials with even shorter decay times to focus on the different effects of (i) reducing the overlap of HOMO and LUMO, which leads to a smaller exchange interaction (small $\Delta E(S_1-T_1)$) and (ii) to provide other energetically low-lying singlet states from which the $S_1 \rightarrow S_0$ transition can borrow oscillator strength. For completeness, it is remarked that a different strategy to reduce the emission decay time can also be successful. It has been shown that an increase of SOC with respect to the T_1 state, leading to an increase of the $T_1 \rightarrow S_0$ phosphorescence rate, will open another radiative decay path. In this situation, the phosphorescence decay path is added to the TADF path [144]. Accordingly, the overall emission decay time is also significantly reduced [31, 97].

For completeness, it is mentioned that similar design rules (with the exception of increasing the phosphorescence decay rate) are also valid for purely organic TADF materials. Thus, recently it was possible to develop compounds by minimizing even a residual donor-acceptor hyper-conjugation. These compounds show extremely small $\Delta E(S_1-T_1)$ values of ≤ 10 cm^{-1} (≈ 1 meV) ($\ll k_B T$ at ambient temperature) and that carry sufficient $S_1 \rightarrow S_0$ allowedness to result in a decay time regime of only a few 100 ns. Interestingly, at this small energy separation, thermal activation is not a key property. All excitons that populate the triplet state (75%) are transferred directly by ISC to the singlet state S_1 (that is populated independently by 25%). This new mechanism of *Direct Singlet Harvesting* will lead us to beyond TADF and might be successful for next generation OLED applications [180–183]. An equivalent design strategy as developed for the organic materials could also be successful for organometallic TADF materials in future.

Acknowledgments

The authors thank the German Ministry of Education and Research for financial support in the scope of the cyCESH project (FKN 13N12668). R.C. thanks the European Research Council (ERC) for support in the framework of the MSCA RISE Project no. 645628. M.Z.S. thanks Professor Duncan Bruce (York) and the University of York for their help with computational facilities. Moreover, we thank our cooperation partners for fruitful collaborations. This includes the research groups of Prof. Dr. Chensheng Ma (Shenzhen University, China), Prof. Dr. Wai-Ming Kwok (Hong Kong Polytechnic University, Hong Kong), and Prof. Dr. Thomas A. Niehaus (Claude Bernard University, Lyon, France). In addition, we acknowledge the German Academic Exchange Service (DAAD) and the Bavaria California Technology Center (BaCaTec) for giving us the opportunity to establish and maintain our collaborations.

References

1 Yersin, H. (2008). *Highly Efficient OLEDs with Phosphorescent Materials*. Weinheim: Wiley-VCH.
2 Brütting, W. and Adachi, C. (2012). *Physics of Organic Semiconductors*. Weinheim: Wiley-VCH.

3 Kim, Y., Park, S., Lee, Y.H., Jung, J., Yoo, S., and Lee, M.H. (2016). Homoleptic Tris-cyclometalated iridium complexes with substituted o-carboranes: green phosphorescent emitters for highly efficient solution-processed organic light-emitting diodes. *Inorg. Chem.* 55: 909–917.

4 Liu, Z., Qiu, J., Wei, F., Wang, J., Liu, X., Helander, M.G., Rodney, S., Wang, Z., Bian, Z., Lu, Z., Thompson, M.E., and Huang, C. (2014). Simple and high efficiency phosphorescence organic light-emitting diodes with codeposited copper(I) emitter. *Chem. Mater.* 26: 2368–2373.

5 Zhang, X.Q., Xie, Y.M., Zheng, Y., Liang, F., Wang, B., Fan, J., and Liao, L.S. (2016). Highly phosphorescent platinum(II) complexes based on rigid unsymmetric tetradentate ligands. *Org. Electron. Phys. Mater. Appl.* 32: 120–125.

6 Kim, J., Lee, K.H., Lee, S.J., Lee, H.W., Kim, Y.K., Kim, Y.S., and Yoon, S.S. (2016). Red phosphorescent bis-cyclometalated iridium complexes with fluorine-, phenyl-, and fluorophenyl-substituted 2-arylquinoline ligands. *Chem. Eur. J.* 22: 4036–4045.

7 Wong, M.Y., Hedley, G.J., Xie, G., Kölln, L.S., Samuel, I.D.W., Pertegás, A., Bolink, H.J., and Zysman-Colman, E. (2015). Light-emitting electrochemical cells and solution-processed organic light-emitting diodes using small molecule organic thermally activated delayed fluorescence emitters. *Chem. Mater.* 27: 6535–6542.

8 Minaev, B., Baryshnikov, G., and Agren, H. (2014). Principles of phosphorescent organic light emitting devices. *Phys. Chem. Chem. Phys.* 16: 1719–1758.

9 Yersin, H., Rausch, A.F., and Czerwieniec, R. (2012). Organometallic emitters for OLEDs. triplet harvesting, singlet harvesting, case structures, and trends. In: *Physics of Organic Semiconductors*, 2e (ed. W. Brütting and C. Adachi), 371–425. Weinheim: Wiley-VCH.

10 Sun, Y., Giebink, N.C., Kanno, H., Ma, B., Thompson, M.E., and Forrest, S.R. (2006). Management of singlet and triplet excitons for efficient white organic light-emitting devices. *Nature* 440: 908–912.

11 Adachi, C., Baldo, M.A., Thompson, M.E., and Forrest, S.R. (2001). Nearly 100% internal phosphorescence efficiency in an organic light emitting device. *J. Appl. Phys.* 90: 5048–5051.

12 Lamansky, S., Djurovich, P., Murphy, D., Abdel-Razzaq, F., Lee, H.E., Adachi, C., Burrows, P.E., Forrest, S.R., and Thompson, M.E. (2001). Highly phosphorescent bis-cyclometalated iridium complexes: synthesis, photophysical characterization, and use in organic light emitting diodes. *J. Am. Chem. Soc.* 123: 4304–4312.

13 Kim, S.Y., Jeong, W.I., Mayr, C., Park, Y.S., Kim, K.H., Lee, J.H., Moon, C.K., Brütting, W., and Kim, J.J. (2013). Organic light-emitting diodes with 30% external quantum efficiency based on a horizontally oriented emitter. *Adv. Funct. Mater.* 23: 3896–3900.

14 Schmidt, T.D., Reichardt, L.J., Rausch, A.F., Wehrmeister, S., Scholz, B.J., Mayr, C., Wehlus, T., Ciarnáin, R.M., Danz, N., Reusch, T.C.G., and Brütting, W. (2014). Extracting the emitter orientation in organic light-emitting diodes from external quantum efficiency measurements. *Appl. Phys. Lett.* 105: doi: 10.1063/1.4891680.

15 Cherpak, V., Stakhira, P., Minaev, B., Baryshnikov, G., Stromylo, E., Helzhynskyy, I., Chapran, M., Volyniuk, D., Tomkuté-Lukšiené, D., Malinauskas, T., Getautis, V., Tomkeviciene, A., Simokaitiene, J., and Grazulevicius, J.V. (2014). Efficient "warm-white" OLEDs based on the phosphorescent bis-cyclometalated iridium(III) complex. *J. Phys. Chem. C* 118: 11271–11278.

16 Tuong Ly, K., Chen-Cheng, R.-W., Lin, H.-W., Shiau, Y.-J., Liu, S.-H., Chou, P.-T., Tsao, C.-S., Huang, Y.-C., and Chi, Y. (2016). Near-infrared organic light-emitting diodes with very high external quantum efficiency and radiance. *Nat. Photonics* 11: 63–69.

17 Ma, D., Zhang, C., Qiu, Y., and Duan, L. (2017). Sustainable phosphorescence based on solution-processable and vacuum-sublimable cationic ruthenium(II) complexes achieved by counter-ion control. *Org. Electron.* 42: 194–202.

18 Zhang, J., Zhu, X., Zhong, A., Jia, W., Wu, F., Li, D., Tong, H., Wu, C., Tang, W., Zhang, P., Wang, L., and Han, D. (2017). New platinum(II) one-armed Schiff base complexes for blue and orange PHOLEDs applications. *Org. Electron.* 42: 153–162.

19 Salehi, A., Ho, S., Chen, Y., Peng, C., Yersin, H., and So, F. (2017). Highly efficient organic light-emitting diode using a low refractive index electron transport layer. *Adv. Opt. Mater.* 5: 1700197.

20 Yang, C.H., Beltran, J., Lemaur, V., Cornil, J., Hartmann, D., Sarfert, W., Fröhlich, R., Bizzarri, C., and De Cola, L. (2010). Iridium metal complexes containing N-heterocyclic carbene ligands for blue-light-emitting electrochemical cells. *Inorg. Chem.* 49: 9891–9901.

21 Henwood, A.F. and Zysman-Colman, E. (2016). Luminescent iridium complexes used in light-emitting electrochemical cells (LEECs). *Top. Curr. Chem.* 374: 36.

22 Pal, A.K., Cordes, D.B., Slawin, A.M.Z., Momblona, C., Ortí, E., Samuel, I.D.W., Bolink, H.J., and Zysman-Colman, E. (2016). Synthesis, properties, and light-emitting electrochemical cell (LEEC) device fabrication of cationic Ir(III) complexes bearing electron-withdrawing groups on the cyclometallating ligands. *Inorg. Chem.* 55: 10361–10376.

23 Ertl, C.D., Bolink, H.J., Housecroft, C.E., Constable, E.C., Ortí, E., Junquera-Hernández, J.M., Neuburger, M., Shavaleev, N.M., Nazeeruddin, M.K., and Vonlanthen, D. (2016). Bis-sulfone- and bis-sulfoxide-spirobifluorenes: polar acceptor hosts with tunable solubilities for blue-phosphorescent light-emitting devices. *Eur. J. Org. Chem.* 2016: 2037–2047.

24 Henwood, A.F. and Zysman-Colman, E. (2017). Lessons learned in tuning the optoelectronic properties of phosphorescent iridium(III) complexes. *Chem. Commun.* 53: 807–826.

25 Ertl, C.D., Momblona, C., Pertegás, A., Junquera-Hernández, J.M., La-Placa, M.G., Prescimone, A., Ortí, E., Housecroft, C.E., Constable, E.C., and Bolink, H.J. (2017). Highly stable red-light-emitting electrochemical cells. *J. Am. Chem. Soc.* 139: 3237–3248.

26 Martir, D.R., Momblona, C., Pertegás, A., Cordes, D.B., Slawin, A.M.Z., Bolink, H.J., and Zysman-Colman, E. (2016). Chiral iridium(III) complexes in light-emitting electrochemical cells: exploring the impact of stereochemistry on the photophysical properties and device performances. *ACS Appl. Mater. Interfaces* 8: 33907–33915.

27 Armaroli, N., Accorsi, G., Holler, M., Moudam, O., Nierengarten, J.F., Zhou, Z., Wegh, R.T., and Welter, R. (2006). Highly luminescent Cu^I complexes for light-emitting electrochemical cells. *Adv. Mater.* 18: 1313–1316.

28 Tang, S. and Edman, L. (2016). Light-emitting electrochemical cells: a review on recent progress. *Top. Curr. Chem.* 374: 40. doi: 10.1007/s41061-016-0040-4.

29 Che, C.M., Kwok, C.C., Lai, S.W., Rausch, A.F., Finkenzeller, W.J., Zhu, N., and Yersin, H. (2010). Photophysical properties and OLED applications of phosphorescent platinum(II) schiff base complexes. *Chem. Eur. J.* 16: 233–247.

30 Leitl, M.J., Zink, D.M., Schinabeck, A., Baumann, T., Volz, D., and Yersin, H. (2016). Copper(I) complexes for thermally activated delayed fluorescence: from photophysical to device properties. *Top. Curr. Chem.* 374: 25.

31 Czerwieniec, R., Leitl, M.J., Homeier, H.H.H., and Yersin, H. (2016). Cu(I) complexes – thermally activated delayed fluorescence. Photophysical approach and material design. *Coord. Chem. Rev.* 325: 2–28.

32 Hong, X., Wang, B., Liu, L., Zhong, X.X., Li, F.B., Wang, L., Wong, W.Y., Qin, H.M., and Lo, Y.H. (2016). Highly efficient blue–green neutral dinuclear copper(I) halide complexes containing bidentate phosphine ligands. *J. Lumin.* 180: 64–72.

33 Leitl, M.J., Küchle, F.-R., Mayer, H.A., Wesemann, L., and Yersin, H. (2013). Brightly blue and green emitting Cu(I) dimers for singlet harvesting in OLEDs. *J. Phys. Chem. A* 117: 11823–11836.

34 Leitl, M.J., Krylova, V.A., Djurovich, P.I., Thompson, M.E., and Yersin, H. (2014). Phosphorescence versus thermally activated delayed fluorescence. Controlling singlet-triplet splitting in brightly emitting and sublimable Cu(I) compounds. *J. Am. Chem. Soc.* 136: 16032–16038.

35 Linfoot, C.L., Leitl, M.J., Richardson, P., Rausch, A.F., Chepelin, O., White, F.J., Yersin, H., and Robertson, N. (2014). Thermally activated delayed fluorescence (TADF) and enhancing photoluminescence quantum yields of [Cu^I(diimine)(diphosphine)]$^+$ complexes – photophysical, structural, and computational studies. *Inorg. Chem.* 53: 10854–10861.

36 Zhang, Q., Zhou, Q., Cheng, Y., Wang, L., Ma, D., Jing, X., and Wang, F. (2004). Highly efficient green phosphorescent organic light-emitting diodes based on CuI complexes. *Adv. Mater.* 16: 432–436.

37 Zink, D.M., Volz, D., Baumann, T., Mydlak, M., Flügge, H., Friedrichs, J., Nieger, M., and Bräse, S. (2013). Heteroleptic, dinuclear copper(I) complexes for application in organic light-emitting diodes. *Chem. Mater.* 25: 4471–4486.

38 Volz, D., Zink, D.M., Bocksrocker, T., Friedrichs, J., Nieger, M., Baumann, T., Lemmer, U., and Bräse, S. (2013). Molecular construction kit for tuning solubility, stability and luminescence properties: heteroleptic MePyrPHOS-copper iodide-complexes and their application in organic light-emitting diodes. *Chem. Mater.* 25: 3414–3426.

39 Volz, D., Baumann, T., Flügge, H., Mydlak, M., Grab, T., Bachle, M., Barner-Kowollik, C., and Bräse, S. (2012). Auto-catalysed crosslinking for next-generation OLED-design. *J. Mater. Chem.* 22: 20786–20790.

40 Wallesch, M., Volz, D., Zink, D.M., Schepers, U., Nieger, M., Baumann, T., and Bräse, S. (2014). Bright coppertunities: multinuclear Cu[I] complexes with N–P ligands and their applications. *Chem. Eur. J.* 20: 6578–6590.

41 Volz, D., Chen, Y., Wallesch, M., Liu, R., Fléchon, C., Zink, D.M., Friedrichs, J., Flügge, H., Steininger, R., Göttlicher, J., Heske, C., Weinhardt, L., Bräse, S., So, F., and Baumann, T. (2015). Bridging the efficiency gap: fully bridged dinuclear Cu(I)-complexes for singlet harvesting in high-efficiency OLEDs. *Adv. Mater.* 27: 2538–2543.

42 Osawa, M., Hoshino, M., Hashimoto, M., Kawata, I., Igawa, S., and Yashima, M. (2015). Application of three-coordinate copper(I) complexes with halide ligands in organic light-emitting diodes that exhibit delayed fluorescence. *Dalton Trans.* 44: 8369–8378.

43 Igawa, S., Hashimoto, M., Kawata, I., Yashima, M., Hoshino, M., and Osawa, M. (2013). Highly efficient green organic light-emitting diodes containing luminescent tetrahedral copper(I) complexes. *J. Mater. Chem. C* 1: 542–551.

44 Hashimoto, M., Igawa, S., Yashima, M., Kawata, I., Hoshino, M., and Osawa, M. (2011). Highly efficient green organic light-emitting diodes containing luminescent three-coordinate copper(I) complexes. *J. Am. Chem. Soc.* 133: 10348–10351.

45 Wei, F., Qiu, J., Liu, X., Wang, J., Wei, H., Wang, Z., Liu, Z., Bian, Z., Lu, Z., Zhao, Y., and Huang, C. (2014). Efficient orange–red phosphorescent organic light-emitting diodes using an in situ synthesized copper(I) complex as the emitter. *J. Mater. Chem. C* 2: 6333–6341.

46 Deaton, J.C., Switalski, S.C., Kondakov, D.Y., Young, R.H., Pawlik, T.D., Giesen, D.J., Harkins, S.B., Miller, A.J.M., Mickenberg, S.F., and Peters, J.C. (2010). E-type delayed fluorescence of a phosphine-supported $Cu_2(\mu\text{-}NAr_2)_2$ diamond core: harvesting singlet and triplet excitons in OLEDs. *J. Am. Chem. Soc.* 132: 9499–9508.

47 Zhang, Q., Komino, T., Huang, S., Matsunami, S., Goushi, K., and Adachi, C. (2012). Triplet exciton confinement in green organic light-emitting diodes containing luminescent charge-transfer Cu(I) complexes. *Adv. Funct. Mater.* 22: 2327–2336.

48 Cheng, G., So, G.K.-M., To, W.-P., Chen, Y., Kwok, C.-C., Ma, C., Guan, X., Chang, X., Kwok, W.-M., and Che, C.-M. (2015). Luminescent zinc(II) and copper(I) complexes for high-performance solution-processed monochromic and white organic light-emitting devices. *Chem. Sci.* 6: 4623–4635.

49 Zhang, Q., Chen, X.-L., Chen, J., Wu, X.-Y., Yu, R., and Lu, C.-Z. (2015). Four highly efficient cuprous complexes and their applications in solution-processed organic light-emitting diodes. *RSC Adv.* 5: 34424–34431.

50 Chen, X.-L., Yu, R., Zhang, Q.-K., Zhou, L.-J., Wu, X.-Y., Zhang, Q., and Lu, C.-Z. (2013). Rational design of strongly blue-emitting cuprous

complexes with thermally activated delayed fluorescence and application in solution-processed OLEDs. *Chem. Mater.* 25: 3910–3920.

51 Yersin, H., Monkowius, U., and Czerwieniec, R. (2010). Singulett-Harvesting mit löslichen Kupfer(I)-Komplexen für opto-elektronische Vorrichtungen. Patent DE 102010031831 A1.

52 Yersin, H., Monkowius, U., and Czerwieniec, R. (2011). Copper(I) complexes for opto-electronic devices. Patent WO 2012010650 A1.

53 Medina-Rodriguez, S., Orriach-Fernandez, F.J., Poole, C., Kumar, P., de la Torre-Vega, A., Fernandez-Sanchez, J.F., Baranoff, E., and Fernandez-Gutierrez, A. (2015). Copper(I) complexes as alternatives to iridium(III) complexes for highly efficient oxygen sensing. *Chem. Commun.* 51: 11401–11404.

54 Smith, C.S., Branham, C.W., Marquardt, B.J., and Mann, K.R. (2010). Oxygen gas sensing by luminescence quenching in crystals of Cu(xantphos)(phen)$^+$ complexes. *J. Am. Chem. Soc.* 132: 14079–14085.

55 Smith, C.S. and Mann, K.R. (2012). Exceptionally long-lived luminescence from [Cu(I)(isocyanide)$_2$(phen)]$^+$ complexes in nanoporous crystals enables remarkable oxygen gas sensing. *J. Am. Chem. Soc.* 134: 8786–8789.

56 Czerwieniec, R., Leitl, M., and Yersin, H. (2013). Optical oxygen sensors with copper(I) complexes. Patent DE 102012101067 A1, Patent WO 2013117460 A2.

57 Prokhorov, A.M., Hofbeck, T., Czerwieniec, R., Suleymanova, A.F., Kozhevnikov, D.N., and Yersin, H. (2014). Brightly luminescent Pt(II) pincer complexes with a sterically demanding carboranyl-phenylpyridine ligand: A new material class for diverse optoelectronic applications. *J. Am. Chem. Soc.* 136: 9637–9642.

58 Hofbeck, T., Lam, Y.C., Kalbáč, M., Záliš, S., Vlček, A., and Yersin, H. (2016). Thermally tunable dual emission of the d^8–d^8 dimer [Pt$_2$(μ-P$_2$O$_5$(BF$_2$)$_2$)$_4$]$^{4-}$. *Inorg. Chem.* 55: 2441–2449.

59 Mak, C.S.K., Pentlehner, D., Stich, M., Wolfbeis, O.S., Chan, W.K., and Yersin, H. (2009). Exceptional oxygen sensing capabilities and triplet state properties of Ir(ppy-NPh$_2$)$_3$. *Chem. Mater.* 21: 2173–2175.

60 Knorn, M., Rawner, T., Czerwieniec, R., and Reiser, O. (2015). [Copper(phenanthroline)(bisisonitrile)]$^+$-complexes for the visible-light-mediated atom transfer radical addition and allylation reactions. *ACS Catal.* 5: 5186–5193.

61 Tang, X.-J. and Dolbier, W.R. (2015). Efficient Cu-catalyzed atom transfer radical addition reactions of fluoroalkylsulfonyl chlorides with electron-deficient alkenes induced by visible light. *Angew. Chem. Int. Ed.* 54: 4246–4249.

62 Bagal, D.B., Kachkovskyi, G., Knorn, M., Rawner, T., Bhanage, B.M., and Reiser, O. (2015). Trifluoromethylchlorosulfonylation of alkenes: evidence for an inner-sphere mechanism by a copper phenanthroline photoredox catalyst. *Angew. Chem. Int. Ed.* 54: 6999–7002.

63 Wang, B., Shelar, D.P., Han, X.-Z., Li, T.-T., Guan, X., Lu, W., Liu, K., Chen, Y., Fu, W.-F., and Che, C.-M. (2015). Long-lived excited states of zwitterionic

copper(I) complexes for photoinduced cross-dehydrogenative coupling reactions. *Chem. Eur. J.* 21: 1184–1190.

64 Yang, Q., Dumur, F., Morlet-Savary, F., Poly, J., and Lalevée, J. (2015). Photocatalyzed Cu-based ATRP involving an oxidative quenching mechanism under visible light. *Macromolecules* 48: 1972–1980.

65 Baralle, A., Fensterbank, L., Goddard, J.-P., and Ollivier, C. (2013). Aryl radical formation by copper(I) photocatalyzed reduction of diaryliodonium salts: NMR evidence for a Cu^{II}/Cu^{I} mechanism. *Chem. Eur. J.* 19: 10809–10813.

66 Luo, S.-P., Mejía, E., Friedrich, A., Pazidis, A., Junge, H., Surkus, A.-E., Jackstell, R., Denurra, S., Gladiali, S., Lochbrunner, S., and Beller, M. (2013). Photocatalytic water reduction with copper-based photosensitizers: a noble-metal-free system. *Angew. Chem. Int. Ed.* 52: 419–423.

67 Yersin, H. (2004). Triplet emitters for OLED applications. Mechanisms of exciton trapping and control of emission properties. *Top. Curr. Chem.* 241: 1–26.

68 Helfrich, W. and Schneider, W.G. (1966). Transients of volume-controlled current and of recombination radiation in anthracene. *J. Chem. Phys.* 44: 2902–2909.

69 Yersin, H. and Finkenzeller, W.J. (2008). Triplet emitters for organic light emitting diodes: basic properties. In: *Highly Efficient OLEDs with Phosphorescent Materials* (ed. H. Yersin), 1–97. Weinheim: Wiley-VCH.

70 Yersin, H., Rausch, A.F., Czerwieniec, R., Hofbeck, T., and Fischer, T. (2011). The triplet state of organo-transition metal compounds. Triplet harvesting and singlet harvesting for efficient OLEDs. *Coord. Chem. Rev.* 255: 2622–2652.

71 Rausch, A.F., Homeier, H.H.H., and Yersin, H. (2010). Organometallic Pt(II) and Ir(III) triplet emitters for OLED applications and the role of spin–orbit coupling: a study based on high-resolution optical spectroscopy. *Top. Organomet. Chem.* 29: 193–235.

72 Rausch, A.F., Homeier, H.H.H., Djurovich, P.I., Thompson, M.E., and Yersin, H. (2007). Spin–orbit coupling routes and OLED performance – studies of blue-light emitting Ir(III) and Pt(II) complexes. *Proc. SPIE* 6655: doi: 10.1117/12.731225.

73 Azumi, T. and Miki, H. (1997). Spectroscopy of the spin sublevels of transition metal complexes. In: *Electronic and Vibronic Spectra of Transition Metal Complexes II* (ed. H. Yersin), 1–40. Berlin, Heidelberg: Springer.

74 Miki, H., Shimada, M., Azumi, T., Brozik, J.A., and Crosby, G.A. (1993). Effect of the ligand-field strength on the radiative properties of the ligand-localized $^3\pi\pi^*$ state of rhodium complexes with 1,10-phenanthroline. Proposed role of dd states. *J. Phys. Chem.* 97: 11175–11179.

75 Kimachi, S., Satomi, R., Miki, H., Maeda, K., Azumi, T., and Onishi, M. (1997). Excited-state properties of the ligand-localized $^3\pi\pi^*$ state of cyclometalated ruthenium(II) complexes. *J. Phys. Chem. A* 101: 345–349.

76 Obara, S., Itabashi, M., Okuda, F., Tamaki, S., Tanabe, Y., Ishii, Y., Nozaki, K., and Haga, M.-A. (2006). Highly phosphorescent iridium complexes containing both tridentate bis(benzimidazolyl)-benzene or -pyridine and bidentate

phenylpyridine: synthesis, photophysical properties, and theoretical study of Ir-Bis(benzimidazolyl)benzene complex. *Inorg. Chem.* 45: 8907–8921.

77 Abedin-Siddique, Z., Ohno, T., Nozaki, K., and Tsubomura, T. (2004). Intense fluorescence of metal-to-ligand charge transfer in [Pt(0)(binap)$_2$] [binap = 2,2′-bis(diphenylphosphino)-1,1′-binaphthyl]. *Inorg. Chem.* 43: 663–673.

78 Baldo, M.A., O'Brien, D.F., You, Y., Shoustikov, A., Sibley, S., Thompson, M.E., and Forrest, S.R. (1998). Highly efficient phosphorescent emission from organic electroluminescent devices. *Nature* 395: 151–154.

79 Baldo, M.A., Lamansky, S., Burrows, P.E., Thompson, M.E., and Forrest, S.R. (1999). Very high-efficiency green organic light-emitting devices based on electrophosphorescence. *Appl. Phys. Lett.* 75: 4–6.

80 Kawamura, Y., Goushi, K., Brooks, J., Brown, J.J., Sasabe, H., and Adachi, C. (2005). 100% phosphorescence quantum efficiency of Ir(III) complexes in organic semiconductor films. *Appl. Phys. Lett.* 86: 071104.

81 Sasabe, H., Takamatsu, J.-I., Motoyama, T., Watanabe, S., Wagenblast, G., Langer, N., Molt, O., Fuchs, E., Lennartz, C., and Kido, J. (2010). High-efficiency blue and white organic light-emitting devices incorporating a blue iridium carbene complex. *Adv. Mater.* 22: 5003–5007.

82 Li, K., Ming Tong, G.S., Wan, Q., Cheng, G., Tong, W.Y., Ang, W.H., Kwong, W.L., and Che, C.M. (2016). Highly phosphorescent platinum(II) emitters: photophysics, materials and biological applications. *Chem. Sci.* 7: 1653–1673.

83 Cheng, G., Kui, S.C.F., Ang, W.-H., Ko, M.-Y., Chow, P.-K., Kwong, C.-L., Kwok, C.-C., Ma, C., Guan, X., Low, K.-H., Su, S.-J., and Che, C.-M. (2014). Structurally robust phosphorescent [Pt(O^N^C^N)] emitters for high performance organic light-emitting devices with power efficiency up to 126 lm W^{-1} and external quantum efficiency over 20%. *Chem. Sci.* 5: 4819–4830.

84 Gildea, L.F. and Williams, J.A.G. (2013). Iridium and platinum complexes for OLEDs. In: *Organic Light-Emitting Diodes: Materials, Devices and Applications* (ed. A. Buckley), 77–113. Cambridge: Woodhead Publishing.

85 Zysman-Colman, E. (2017). *Iridium (III) in Optoelectronic and Photonics Applications*. Weinheim: Wiley-VCH.

86 Deaton, J.C. and Castellano, F.N. (2017). Archetypal iridium(III) compounds for optoelectronic and photonic applications. In: *Iridium (III) in Optoelectronic and Photonics Applications* (ed. E. Zysman-Colman), 1–69. Weinheim: Wiley-VCH.

87 Yersin, H. and Kratzer, C. (2002). Energy transfer and harvesting in [Ru$_{1-x}$Os$_x$(bpy)$_3$](PF$_6$)$_2$ and {Λ-[Ru(bpy)$_3$]Δ-[Os(bpy)$_3$]}(PF$_6$)$_4$. *Coord. Chem. Rev.* 229: 75–93.

88 Yersin, H. and Donges, D. (2001). Low-lying electronic states and photophysical properties of organometallic Pd(II) and Pt(II) compounds. Modern research trends presented in detailed case studies. In: *Transition Metal and Rare Earth Compounds: Excited States, Transitions, Interactions II* (ed. H. Yersin), 81–186. Berlin, Heidelberg: Springer.

89 Hofbeck, T. and Yersin, H. (2010). The triplet state of *fac*-Ir(ppy)$_3$. *Inorg. Chem.* 49: 9290–9299.

90 Hedley, G.J., Ruseckas, A., and Samuel, I.D.W. (2008). Ultrafast luminescence in Ir(ppy)$_3$. *Chem. Phys. Lett.* 450: 292–296.

91 Daniels, R.E., Culham, S., Hunter, M., Durrant, M.C., Probert, M.R., Clegg, W., Williams, J.A., and Kozhevnikov, V.N. (2016). When two are better than one: bright phosphorescence from non-stereogenic dinuclear iridium(III) complexes. *Dalton Trans.* 45: 6949–6962.

92 Czerwieniec, R., Kowalski, K., and Yersin, H. (2013). Highly efficient thermally activated fluorescence of a new rigid Cu(I) complex [Cu(dmp)(phanephos)]$^+$. *Dalton Trans.* 42: 9826–9830.

93 Volz, D., Wallesch, M., Flechon, C., Danz, M., Verma, A., Navarro, J.M., Zink, D.M., Bräse, S., and Baumann, T. (2015). From iridium and platinum to copper and carbon: new avenues for more sustainability in organic light-emitting diodes. *Green Chem.* 17: 1988–2011.

94 Yersin, H. and Monkowius, U. (2008). Komplexe mit kleinen Singulett-Triplett-Energie-Abständen zur Verwendung in opto-elektronischen Bauteilen (Singulett-Harvesting-Effekt). Internal patent filing, University of Regensburg 2006. German Patent DE 10 2008 033563.

95 Czerwieniec, R., Yu, J., and Yersin, H. (2011). Blue-light emission of Cu(I) complexes and singlet harvesting. *Inorg. Chem.* 50: 8293–8301.

96 Czerwieniec, R. and Yersin, H. (2015). Diversity of copper(I) complexes showing thermally activated delayed fluorescence: basic photophysical analysis. *Inorg. Chem.* 54: 4322–4327.

97 Hofbeck, T., Monkowius, U., and Yersin, H. (2015). Highly efficient luminescence of Cu(I) compounds: thermally activated delayed fluorescence combined with short-lived phosphorescence. *J. Am. Chem. Soc.* 137: 399–404.

98 Ohara, H., Kobayashi, A., and Kato, M. (2014). Simple and extremely efficient blue emitters based on mononuclear Cu(I)-halide complexes with delayed fluorescence. *Dalton Trans.* 43: 17317–17323.

99 Bergmann, L., Friedrichs, J., Mydlak, M., Baumann, T., Nieger, M., and Bräse, S. (2013). Outstanding luminescence from neutral copper(I) complexes with pyridyl-tetrazolate and phosphine ligands. *Chem. Commun.* 49: 6501–6503.

100 Tsuboyama, A., Kuge, K., Furugori, M., Okada, S., Hoshino, M., and Ueno, K. (2007). Photophysical properties of highly luminescent copper(I) halide complexes chelated with 1,2-bis(diphenylphosphino)benzene. *Inorg. Chem.* 46: 1992–2001.

101 Cuttell, D.G., Kuang, S.M., Fanwick, P.E., McMillin, D.R., and Walton, R.A. (2002). Simple Cu(I) complexes with unprecedented excited-state lifetimes. *J. Am. Chem. Soc.* 124: 6–7.

102 Kang, L., Chen, J., Teng, T., Chen, X.L., Yu, R., and Lu, C.Z. (2015). Experimental and theoretical studies of highly emissive dinuclear Cu(I) halide complexes with delayed fluorescence. *Dalton Trans.* 44: 11649–11659.

103 Zink, D.M., Bächle, M., Baumann, T., Nieger, M., Kühn, M., Wang, C., Klopper, W., Monkowius, U., Hofbeck, T., Yersin, H., and Bräse, S. (2013). Synthesis, structure, and characterization of dinuclear copper(I) halide complexes with P^N ligands featuring exciting photoluminescence properties. *Inorg. Chem.* 52: 2292–2305.

104 Chen, J., Teng, T., Kang, L., Chen, X.-L., Wu, X.-Y., Yu, R., and Lu, C.-Z. (2016). Highly efficient thermally activated delayed fluorescence in dinuclear Ag(I) complexes with a bis-bidentate tetraphosphane bridging ligand. *Inorg. Chem.* 55: 9528–9536.

105 Chen, X.L., Yu, R., Wu, X.Y., Liang, D., Jia, J.H., and Lu, C.Z. (2016). A strongly greenish-blue-emitting Cu_4Cl_4 cluster with an efficient spin–orbit coupling (SOC): fast phosphorescence: versus thermally activated delayed fluorescence. *Chem. Commun.* 52: 6288–6291.

106 Osawa, M., Kawata, I., Ishii, R., Igawa, S., Hashimoto, M., and Hoshino, M. (2013). Application of neutral d^{10} coinage metal complexes with an anionic bidentate ligand in delayed fluorescence-type organic light-emitting diodes. *J. Mater. Chem. C* 1: 4375–4383.

107 Yersin, H., Leitl, M.J., and Czerwieniec, R. (2014). TADF for singlet harvesting – next generation OLED materials based on brightly green and blue emitting Cu(I) and Ag(I) compounds. *Proc. SPIE* 9183: doi: 10.1117/12.2061010.

108 Shafikov, M.Z., Suleymanova, A.F., Czerwieniec, R., and Yersin, H. (2017). Design strategy for Ag(I)-based thermally activated delayed fluorescence reaching an efficiency breakthrough. *Chem. Mater.* 29: 1708–1715.

109 Shafikov, M.Z., Suleymanova, A.F., Czerwieniec, R., and Yersin, H. (2017). Thermally activated delayed fluorescence from Ag(I) complexes: a route to 100% quantum yield at unprecedently short decay time. *Inorg. Chem.* 56: 13274–13285.

110 Uoyama, H., Goushi, K., Shizu, K., Nomura, H., and Adachi, C. (2012). Highly efficient organic light-emitting diodes from delayed fluorescence. *Nature* 492: 234–238.

111 Zhang, Q., Li, B., Huang, S., Nomura, H., Tanaka, H., and Adachi, C. (2014). Efficient blue organic light-emitting diodes employing thermally activated delayed fluorescence. *Nat. Photonics* 8: 326–332.

112 Kaji, H., Suzuki, H., Fukushima, T., Shizu, K., Suzuki, K., Kubo, S., Komino, T., Oiwa, H., Suzuki, F., Wakamiya, A., Murata, Y., and Adachi, C. (2015). Purely organic electroluminescent material realizing 100% conversion from electricity to light. *Nat. Commun.* 6: 8476.

113 Cui, L.-S., Nomura, H., Geng, Y., Kim, J.U., Nakanotani, H., and Adachi, C. (2017). Controlling singlet–triplet energy splitting for deep-blue thermally activated delayed fluorescence emitters. *Angew. Chem. Int. Ed.* 56: 1571–1575.

114 McMillin, D.R. and McNett, K.M. (1998). Photoprocesses of copper complexes that bind to DNA. *Chem. Rev.* 98: 1201–1220.

115 Mara, M.W., Fransted, K.A., and Chen, L.X. (2015). Interplays of excited state structures and dynamics in copper(I) diimine complexes: implications and perspectives. *Coord. Chem. Rev.* 282–283: 2–18.

116 Chen, L.X., Jennings, G., Liu, T., Gosztola, D.J., Hessler, J.P., Scaltrito, D.V., and Meyer, G.J. (2002). Rapid excited-state structural reorganization captured by pulsed X-rays. *J. Am. Chem. Soc.* 124: 10861–10867.

117 Chen, L.X., Shaw, G.B., Novozhilova, I., Liu, T., Jennings, G., Attenkofer, K., Meyer, G.J., and Coppens, P. (2003). MLCT state structure and dynamics of a copper(I) diimine complex characterized by pump–probe X-ray and laser spectroscopies and DFT calculations. *J. Am. Chem. Soc.* 125: 7022–7034.

118 Iwamura, M., Watanabe, H., Ishii, K., Takeuchi, S., and Tahara, T. (2011). Coherent nuclear dynamics in ultrafast photoinduced structural change of bis(diimine)copper(I) complex. *J. Am. Chem. Soc.* 133: 7728–7736.

119 Lavie-Cambot, A., Cantuel, M., Leydet, Y., Jonusauskas, G., Bassani, D.M., and McClenaghan, N.D. (2008). Improving the photophysical properties of copper(I) bis(phenanthroline) complexes. *Coord. Chem. Rev.* 252: 2572–2584.

120 Armaroli, N., Accorsi, G., Cardinali, F., and Listorti, A. (2007). Photochemistry and photophysics of coordination compounds: copper. In: *Photochemistry and Photophysics of Coordination Compounds I* (ed. V. Balzani and S. Campagna), 69–115. Berlin, Heidelberg: Springer.

121 Iwamura, M., Takeuchi, S., and Tahara, T. (2015). Ultrafast excited-state dynamics of copper(I) complexes. *Acc. Chem. Res.* 48: 782–791.

122 Garakyaraghi, S., Danilov, E.O., McCusker, C.E., and Castellano, F.N. (2015). Transient absorption dynamics of sterically congested Cu(I) MLCT excited states. *J. Phys. Chem. A* 119: 3181–3193.

123 Siddique, Z.A., Yamamoto, Y., Ohno, T., and Nozaki, K. (2003). Structure-dependent photophysical properties of singlet and triplet metal-to-ligand charge transfer states in copper(I) bis(diimine) compounds. *Inorg. Chem.* 42: 6366–6378.

124 Turro, N.J., Ramamurthy, V., and Scaiano, J.C. (2010). *Modern Molecular Photochemistry of Organic Molecules*. University Science Books.

125 Siebrand, W. (1967). Radiationless transitions in polyatomic molecules. I. Calculation of Franck–Condon factors. *J. Chem. Phys.* 46: 440–447.

126 Robinson, G.W. and Frosch, R.P. (1963). Electronic excitation transfer and relaxation. *J. Chem. Phys.* 38: 1187–1203.

127 McCusker, C.E. and Castellano, F.N. (2013). Design of a long-lifetime, earth-abundant, aqueous compatible Cu(I) photosensitizer using cooperative steric effects. *Inorg. Chem.* 52: 8114–8120.

128 Atkins, P.W. (1991). *Quanta: A Handbook of Concepts*. Oxford University Press.

129 Barltrop, J.A. and Coyle, J.D. (1975). *Excited States in Organic Chemistry*. Wiley.

130 Szabo, A. and Ostlund, N.S. (1989). *Modern Quantum Chemistry: Introduction to Advanced Electronic Structure Theory*. Dover Publications.

131 Parker, C.A. and Hatchard, C.G. (1961). Triplet-singlet emission in fluid solutions. Phosphorescence of eosin. *Trans. Faraday Soc.* 57: 1894–1904.

132 Parker, C.A. and Hatchard, C.G. (1962). Triplet-singlet emission in fluid solution. *J. Phys. Chem.* 66: 2506–2511.

133 Bergmann, L., Hedley, G.J., Baumann, T., Bräse, S., and Samuel, I.D. (2016). Direct observation of intersystem crossing in a thermally activated delayed fluorescence copper complex in the solid state. *Sci. Adv.* 2: e1500889.

134 Ma, C., Kwok, W.-M., Czerwieniec, R., and Yersin, H. Manuscript in Preparation.

135 Gneuss, T., Leitl, M.J., Finger, L.H., Yersin, H., and Sundermeyer, J. (2015). A new class of deep-blue emitting Cu(I) compounds – effects of counter ions on the emission behavior. *Dalton Trans.* 44: 20045–20055.

136 Gneuss, T., Leitl, M.J., Finger, L.H., Rau, N., Yersin, H., and Sundermeyer, J. (2015). A new class of luminescent Cu(I) complexes with tripodal ligands – TADF emitters for the yellow to red color range. *Dalton Trans.* 44: 8506–8520.

137 Yersin, H. and Strasser, J. (2000). Triplets in metal–organic compounds. Chemical tunability of relaxation dynamics. *Coord. Chem. Rev.* 208: 331–364.

138 Yersin, H., Humbs, W., and Strasser, J. (1997). Characterization of excited electronic and vibronic states of platinum metal compounds with chelate ligands by highly frequency-resolved and time-resolved spectra. *Top. Curr. Chem.* 191: 153–249.

139 Tinti, D.S. and El-Sayed, M.A. (1971). New techniques in triplet state phosphorescence spectroscopy: application to the emission of 2,3-dichloroquinoxaline. *J. Chem. Phys.* 54: 2529–2549.

140 Harrigan, R.W. and Crosby, G.A. (1973). Symmetry assignments of the lowest CT excited states of ruthenium(II) complexes via a proposed electronic coupling model. *J. Chem. Phys.* 59: 3468–3476.

141 Hager, G.D. and Crosby, G.A. (1975). Charge-transfer exited states of ruthenium(II) complexes. I. Quantum yield and decay measurements. *J. Am. Chem. Soc.* 97: 7031–7037.

142 Azumi, T., O'Donnell, C.M., and McGlynn, S.P. (1966). On the multiplicity of the phosphorescent state of organic molecules. *J. Chem. Phys.* 45: 2735–2742.

143 Finkenzeller, W.J. and Yersin, H. (2003). Emission of Ir(ppy)$_3$. Temperature dependence, decay dynamics, and magnetic field properties. *Chem. Phys. Lett.* 377: 299–305.

144 Yersin, H., Leitl, M.J., Hofbeck, T., Czerwieniec, R., and Monkowius, U. (2013). Extended singlet harvesting for OLEDs and other electronic devices. Patent DE 102013106426 A1, Patent WO 2014202675 A1.

145 Yersin, H., Humbs, W., and Strasser, J. (1997). Low-lying electronic states of [Rh(bpy)$_3$]$^{3+}$, [Pt(bpy)$_2$]$^{2+}$, and [Ru(bpy)$_3$]$^{2+}$. A comparative study based on highly resolved and time-resolved spectra. *Coord. Chem. Rev.* 159: 325–358.

146 Czerwieniec, R., El-Naggar, A.M., Albassam, A.A., Kityk, I.V., Graf, M., and Yersin, H. (2015). Electric-field induced nonlinear optical materials based on a bipolar copper (I) complex embedded in polymer matrices. *J. Mater. Sci. Mater. Electron.* 26: 8394–8397.

147 Yersin, H., Czerwieniec, R., and Hupfer, A. (2012). Singlet harvesting with brightly emitting Cu(I) and metal-free organic compounds. *Proc. SPIE* 8435: doi: 10.1117/12.921372.

148 Yersin, H., Monkowius, U., Czerwieniec, R., and Yu, J. (2009). Patent DE 102008048336 A1 (2008) Patent WO 2010031485 A1 (2009).

149 Krylova, V.A., Djurovich, P.I., Conley, B.L., Haiges, R., Whited, M.T., Williams, T.J., and Thompson, M.E. (2014). Control of emission colour with N-heterocyclic carbene (NHC) ligands in phosphorescent three-coordinate Cu(I) complexes. *Chem. Commun.* 50: 7176–7179.

150 Yersin, H., Monkowius, U., and Hofbeck, T. (2013). Patent DE 102011080240 (2011), Patent WO 2013017675 (2013).

151 Czerwieniec, R. and Yersin, H. (2014). Patent DE 102013100181 (2013), Patent WO 2014108430 (2014).

152 Becke, A.D. (1993). A new mixing of Hartree–Fock and local density-functional theories. *J. Chem. Phys.* 98: 1372–1377.

153 Weigend, F. and Ahlrichs, R. (2005). Balanced basis sets of split valence, triple zeta valence and quadruple zeta valence quality for H to Rn: design and assessment of accuracy. *Phys. Chem. Chem. Phys.* 7: 3297–3305.

154 Shaw, G.B., Grant, C.D., Shirota, H., Castner, E.W., Meyer, G.J., and Chen, L.X. (2007). Ultrafast structural rearrangements in the MLCT excited state for copper(I) bis-phenanthrolines in solution. *J. Am. Chem. Soc.* 129: 2147–2160.

155 Vorontsov, I.I., Graber, T., Kovalevsky, A.Y., Novozhilova, I.V., Gembicky, M., Chen, Y.-S., and Coppens, P. (2009). Capturing and analyzing the excited-state structure of a Cu(I) phenanthroline complex by time-resolved diffraction and theoretical calculations. *J. Am. Chem. Soc.* 131: 6566–6573.

156 Hua, L., Iwamura, M., Takeuchi, S., and Tahara, T. (2015). The substituent effect on the MLCT excited state dynamics of Cu(I) complexes studied by femtosecond time-resolved absorption and observation of coherent nuclear wavepacket motion. *Phys. Chem. Chem. Phys.* 17: 2067–2077.

157 Papanikolaou, P.A. and Tkachenko, N.V. (2013). Probing the excited state dynamics of a new family of Cu(I)-complexes with an enhanced light absorption capacity: excitation-wavelength dependent population of states through branching. *Phys. Chem. Chem. Phys.* 15: 13128–13136.

158 Tschierlei, S., Karnahl, M., Rockstroh, N., Junge, H., Beller, M., and Lochbrunner, S. (2014). Substitution-controlled excited state processes in heteroleptic copper(I) photosensitizers used in hydrogen evolving systems. *ChemPhysChem* 15: 3709–3713.

159 Strickler, S.J. and Berg, R.A. (1962). Relationship between absorption intensity and fluorescence lifetime of molecules. *J. Chem. Phys.* 37: 814–822.

160 Schmidt, J., Wiedenhofer, H., von Zelewsky, A., and Yersin, H. (1995). Time-resolved vibrational structures of the triplet sublevel emission of Pd(2-thpy)$_2$. *J. Phys. Chem.* 99: 226–229.

161 Czerwieniec, R., Finkenzeller, W.J., Hofbeck, T., Starukhin, A., Wedel, A., and Yersin, H. (2009). Photophysical properties of Re(pbt)(CO)$_4$ studied by high resolution spectroscopy. *Chem. Phys. Lett.* 468: 205–210.

162 Niehaus, T. A. and Yersin, H. Unpublished results.

163 Wallesch, M., Volz, D., Fléchon, C., Zink, D.M., Bräse, S., and Baumann, T. (2014). Bright copportunities: efficient OLED devices with copper(I) iodide-NHetPHOS-emitters. *Proc. SPIE* 9183: doi: 10.1117/12.2060499.

164 Bizzarri, C., Strabler, C., Prock, J., Trettenbrein, B., Ruggenthaler, M., Yang, C.-H., Polo, F., Iordache, A., Brüggeller, P., and Cola, L.D. (2014). Luminescent dinuclear Cu(I) complexes containing rigid tetraphosphine ligands. *Inorg. Chem.* 53: 10944–10951.

165 Murov, S. L., Carmichael, I., and Hug, G.L. (1993). Handbook of Photochemistry, 2. Taylor & Francis, p 340.

166 McGlynn, S.P., Azumi, T., and Kinoshita, M. (1969). *Molecular Spectroscopy of the Triplet State*. Prentice-Hall.

167 El-Sayed, M.A. (1963). Spin–orbit coupling and the radiationless processes in nitrogen heterocyclics. *J. Chem. Phys.* 38: 2834–2838.

168 Leitl, M.J. (2015). *Photophysical Characterization of OLED relevant Cu(I) Complexes Exhibiting Thermally Activated Delayed Fluorescence (TADF)*. Universität Regensburg.

169 McHale, J. L. *Molecular Spectroscopy*. Prentice Hall: 1999.

170 Penfold, T.J. (2015). On predicting the excited-state properties of thermally activated delayed fluorescence emitters. *J. Phys. Chem. C* 119: 13535–13544.

171 Shafikov, M.Z., Suleymanova, A.F., Schinabeck, A., and Yersin, H. (2018). Dinuclear Ag(I) complex designed for highly efficient thermally activated delayed fluorescence. *J. Phys. Chem. Lett.* 9: 702–709.

172 Kaeser, A., Moudam, O., Accorsi, G., Séguy, I., Navarro, J., Belbakra, A., Duhayon, C., Armaroli, N., Delavaux-Nicot, B., and Nierengarten, J.F. (2014). Homoleptic copper(I), silver(I), and gold(I) bisphosphine complexes. *Eur. J. Inorg. Chem.* 1345–1355.

173 Hsu, C.-W., Lin, C.-C., Chung, M.-W., Chi, Y., Lee, G.-H., Chou, P.-T., Chang, C.-H., and Chen, P.-Y. (2011). Systematic investigation of the metal-structure–photophysics relationship of emissive d^{10}-complexes of group 11 elements: the prospect of application in organic light emitting devices. *J. Am. Chem. Soc.* 133: 12085–12099.

174 Igawa, S., Hashimoto, M., Kawata, I., Hoshino, M., and Osawa, M. (2012). Photoluminescence properties, molecular structures, and theoretical study of heteroleptic silver(I) complexes containing diphosphine ligands. *Inorg. Chem.* 51: 5805–5813.

175 Hsu, C.-C., Lin, C.-C., Chou, P.-T., Lai, C.-H., Hsu, C.-W., Lin, C.-H., and Chi, Y. (2012). Harvesting highly electronically excited energy to triplet manifolds: state-dependent intersystem crossing rate in Os(II) and Ag(I) complexes. *J. Am. Chem. Soc.* 134: 7715–7724.

176 Kunkely, H. and Vogler, A. (2006). Optical properties of Ag(tripod)X with tripod = 1,1,1-tris(diphenyl-phosphinomethyl)ethane and $X^- = Cl^-$ and I^-: Intraligand and ligand-to-ligand charge transfer. *Inorg. Chim. Acta* 359: 388–390.

177 Crespo, O., Gimeno, M.C., Jones, P.G., and Laguna, A. (1996). Silver complexes with the nido-diphosphine [7,8-$(PPh_2)_2$-7,8-$C_2B_9H_{10}$]. *Dalton Trans.* 4583–4588.

178 Capano, G., Rothlisberger, U., Tavernelli, I., and Penfold, T.J. (2015). Theoretical rationalization of the emission properties of prototypical Cu(I)–phenanthroline complexes. *J. Phys. Chem. A* 119: 7026–7037.

179 Matsumoto, K., Shindo, T., Mukasa, N., Tsukuda, T., and Tsubomura, T. (2010). Luminescent mononuclear Ag(I)–bis(diphosphine) complexes: correlation between the photophysics and the structures of mononuclear Ag(I)–bis(diphosphine) complexes. *Inorg. Chem.* 49: 805–814.

180 Yersin, H., Mataranga-Popa, L., and Czerwieniec, R. (2017). *Design of organic TADF molecules. The role of $\Delta E(S_1-T_1)$: from fluorescence to TADF and beyond – towards the fourth generation OLED mechanism*. Krutyn: 22nd International Krutyn Summer School http://www.excilight.com/node/203.

181 Yersin, H., Mataranga-Popa, L., and Czerwieniec, R. (2017). Organische Moleküle für Direktes Singulett-Harvesting mit kurzer Emissionsabklingzeit zur Verwendung in opto-elektronischen Vorrichtungen. Patent EP 17170682.3.

182 Yersin, H., Mataranga-Popa, L., and Czerwieniec, R. (2017). Patent DE 102017101432.2.

183 Yersin, H., Mataranga-Popa, L., Li, S.-W., and Czerwieniec, R. (2018). Design strategies for materials showing thermally activated delayed fluorescence and beyond: Towards the fourth-generation OLED mechanism. *J. Soc. Info. Display.* 26: 194–199. doi: 10.1002/jsid.654.

2

Highly Emissive d¹⁰ Metal Complexes as TADF Emitters with Versatile Structures and Photophysical Properties

Koichi Nozaki and Munetaka Iwamura

University of Toyama, Graduate School of Science and Engineering, 3190 Gofuku, Toyama 930-8555, Japan

2.1 Introduction

In the past few decades, luminescent metal complexes have been intensively studied because of their unique photophysical properties such as long-lived phosphorescent states with high emission quantum yields [1–4]. The search for luminescent metal complexes containing heavy metals such as Ir(III) and Pt(II) has recently intensified. This was a consequence of the demonstration that phosphorescent emitters possess advantages over conventional fluorescent emitters, when developing organic light-emitting diodes (OLEDs) [5–10]. Phosphorescent emitters can harvest singlet excitons (comprising 25% of the total), as well as triplet excitons (comprising 75%) generated during charge recombination in emitting layers. This results in near 100% internal quantum efficiency. Pioneering studies by Thompson and coworkers employed luminescent cyclometalated Ir(III) complexes, such as Ir(ppy)$_3$ [7, 11]. Much subsequent effort has focused on exploring highly emissive d⁶ or d⁸ metal complexes suitable for phosphorescent emitters in OLEDs [12–16].

In recent years, there has been increasing attention given to luminescent complexes of d¹⁰ metal ions such as Cu(I), Ag(I), and Au(I) [17–23]. The main advantages of these complexes over d⁶ or d⁸ metal complexes are (i) the absence of low-lying ³dd excited states that often quench the luminescent triplet states of d⁶ or d⁸ metal complexes and (ii) a large variety of coordination structures such as tetrahedral, trigonal planar, cubane, halogen-bridged diamond core, etc. In particular, Cu(I) complexes have attracted considerable interest. Copper is abundant and therefore cost effective, and Cu(I) complexes have potential in optoelectronics [20, 24–31], sensors [32, 33], photosensitizers [34–38], photocatalysts [39–45], and upconversion sensitizer [46]. Cu(I) complexes with chelating bisdiimine ligands (N^N), [Cu(NN)$_2$], have been widely investigated to understand the photophysics of pseudotetrahedral Cu(I) complexes. Promising OLED emitters include a family of [Cu(P^P)$_2$], heteroleptic [Cu(N^N)(P^P)] complexes with the chelating bisphosphine ligand (P^P), and a halogen-bridged dinuclear Cu(I) complex with a Cu$_2$(μ-X)$_2$ core. Three-coordinate Cu(I) complexes with trigonal planar structures have also attracted attention, because they can potentially avoid

Highly Efficient OLEDs: Materials Based on Thermally Activated Delayed Fluorescence,
First Edition. Edited by Hartmut Yersin.
© 2019 Wiley-VCH Verlag GmbH & Co. KGaA. Published 2019 by Wiley-VCH Verlag GmbH & Co. KGaA.

Jahn–Teller structural distortion. Such structural versatility of Cu(I) complexes results in their emission spanning a wide range of the visible spectrum.

Cu(I) complexes often exhibit thermally activated delayed fluorescence (TADF) (E-type delayed fluorescence). This is fluorescence from singlet states generated by thermally induced reverse intersystem crossing (rISC) from triplet states [28, 47–49]. The photophysical properties of TADF of Cu(I) complexes are comparable (and sometimes superior) to those of highly luminescent cyclometalated Ir(III) complexes. Cationic Cu(I) complexes require counter anions to balance their charge and maintain electrical neutrality. However, neutral Cu(I) complexes have been synthesized using monodentate or bidentate anionic ligands. Some of the cation-free Cu(I) complexes can be sublimed under vacuum, so are promising emitters for vacuum-deposited OLEDs.

Recent reviews have highlighted the photoluminescence and electroluminescence of Cu(I) complexes [20, 22, 28, 29]. The current review focuses on the theoretical background and photophysical properties of TADF-type d^{10} metal complexes as well as fast dynamics of flattening structural distortion upon photoexcitation of d^{10} metal complexes with a pseudotetrahedral structure.

2.2 Phosphorescence and TADF Mechanisms [50, 51]

Most transition metal complexes exhibit phosphorescence from their lowest triplet states [50, 51]. The intensity of this phosphorescence originates from intensity borrowing of the oscillator strength of fluorescence, through mixing of the T_1 state and higher-lying singlet states [10b, 50–54]. This arises from the strong spin–orbit coupling of the d electrons. The radiative rate constants of phosphorescence of third-row transition metal complexes have reportedly reached 10^6 s^{-1}, which is 10^4–10^7 times higher than those of organic compounds (typically in the order of 10^2–10^{-1} s^{-1} [55a]). The phosphorescent mechanism involves spin–orbit coupling of d electrons and dipole-allowed fluorescent transition. Thus, larger radiative rate constants are often observed in complexes where the T_1 and low-lying singlet excited states have large contributions from the metal-to-ligand charge-transfer (MLCT) electronic configuration. The radiative rate constants of phosphorescence ($k_{r,T}$) can be calculated from the transition dipole moments of the singlet state components involved in the lowest triplet states using the expression

$$k_{r,P} = \frac{16\pi^3 \times 10^6 \times \tilde{\nu}^3}{3h\varepsilon_0}|M_{T\to GS}|^2 \quad (2.1)$$

where $\tilde{\nu}$ is the emission energy of the T_1 state in wavenumber and h and ε_o are Planck's constant and the vacuum permittivity, respectively. The transition dipole moment ($M_{T\to GS}$) originates from intensity borrowing of fluorescence through mixing of the T_1 and higher-lying singlet states (S_n), due to the strong spin–orbit coupling of d electrons:

$$M_{T\to GS} = \sum_n M_{S_n \to GS} \times \frac{\langle \Psi_{S_n}|H_{SO}|\Psi_{T_1}\rangle}{E_{T_1} - E_{S_n}} \quad (2.2)$$

where $M_{S_n} \to GS$ denotes the transition dipole moment from S_n state to the ground state and the Ψ's and E's are the eigenfunctions and eigenvalues of the Hamiltonian without spin–orbit coupling, respectively.

The spin angular momentum must be exchanged with orbital angular momentum in the phosphorescent mechanism. This means that the MLCT electronic configurations in Ψ_{S_n} and Ψ_{T_1} must involve different types of d orbitals. Thus, when S_1 and T_1 have the same electronic configuration, the conservation rule of angular momentum diminishes the direct spin–orbit coupling between the two states. In other words, intensity borrowing and intersystem crossing (ISC) between the two states would be inhibited and weakly arrowed via vibronic spin–orbit coupling [56].

In the TADF mechanism, the T_1 and S_1 states are in thermal equilibrium, and triplet excitons are converted to emissive singlet excitons via rISC. Strong spin–orbit coupling is not required for the slow rISC process, which occurs in the microsecond or submicrosecond time scale.

When S_1 and T_1 are in equilibrium, the ratio of the populations of the S_1 and T_1 states can be approximated using a pseudo-equilibrium constant (K), which can be expressed using their degeneracy (g_1 and g_3) and the energy difference (ΔE_{ST}) between S_1 and T_1:

$$K = \frac{g_1}{g_3} \exp\left(-\frac{\Delta E_{ST}}{RT}\right) \tag{2.3}$$

By neglecting the prompt fluorescence component, the radiative rate constant ($k_{r,DF}$) for TADF is given by Eq. (2.4) using the lifetime (τ) and emission quantum yield (Φ) of the TADF:

$$k_{r,DF} = \frac{\Phi_{em}}{\tau_{DF}} = \frac{k_{r,S_1} K + k_{r,T_1}}{1 + K} \tag{2.4}$$

where k_{r,S_1} and k_{r,T_1} are the radiative rate constants of the S_1 and T_1 states, respectively. Eq. (2.4) shows that the radiative rate constant of TADF ($k_{r,DF}$) is temperature dependent, which is one of the important characteristics of TADF. Because $k_{r,DF}$ is approximately proportional to $k_{r,S1}$ and K, $k_{r,DF}$ increases with increasing $k_{r,S1}$ and sharply increases with decreasing ΔE_{ST}. Ignoring the differing geometries of the S_1 and T_1 states, ΔE_{ST} is equal to twice the exchange integral (J). When the electronic configurations of both S_1 and T_1 are formed from electronic transition between the highest occupied orbital (ϕ_H) and lowest unoccupied orbital (ϕ_L), the exchange integral J is defined as

$$J = \left\langle \phi_H(1)\phi_L(2) \left| \frac{e^2}{r_{12}} \right| \phi_L(1)\phi_H(2) \right\rangle \tag{2.5}$$

where e is the charge of an electron and r_{12} is the distance between two electrons.

Accordingly, the J integral is the coulombic repulsive energy between two electrons in the overlapped region between the highest occupied molecular orbital (HOMO) and lowest unoccupied molecular orbital (LUMO). Thus, the J integral increases with increasing spatial overlap of the HOMO and LUMO. The general strategy for reducing ΔE_{ST} is to employ a spatially separated HOMO and LUMO, such as in twisted intramolecular charge-transfer excited states. For

Cu(I) complexes, the electronic configuration of the lowest excited state would be charge transfer between orthogonal orbitals, i.e. dσ antibonding and π orbitals of aromatic ligands, resulting in a very small ΔE_{ST}.

The radiative rate constant $k_{r,S}$ is proportional to the square of the transition electric dipole moment (M), which is given as a matrix element of the dipole moment operator ($\hat{\mu}$) between S$_1$ and GS. The moment M for S$_1 \rightarrow$ GS is often approximated by the integral between ϕ_H and ϕ_L:

$$M = -\langle \Psi_{S1}|\hat{\mu}|\Psi_{GS}\rangle \approx -\langle \phi_H|\hat{\mu}|\phi_L\rangle \tag{2.6}$$

Accordingly, a small spatial overlap of the HOMO and LUMO decreases the integral M and therefore decreases $k_{r,S}$. However, the decrease in $k_{r,S}$ can be partially moderated by the larger mean distance (r) between the two electrons distributed on the spatially separated HOMO and LUMO. This theoretical argument has been experimentally validated, with various reported metal complexes and donor–acceptor systems involving charge transfer between the spatially separated HOMO and LUMO exhibiting efficient TADF.

The transition moment M for S$_1 \rightarrow$ GS is the origin of the lowest energy absorption. Thus, the radiative rate constant $k_{r,S}$ involved in TADF can be predicted from the intensity of the lowest energy absorption band using the Strickler–Berg equation: [55]

$$k_{r,S} = 2.880 \times 10^{-9} n^2 \langle \tilde{\nu}_{max}^{-3}\rangle_{av}^{-1} \int \varepsilon(\tilde{\nu}) d\ln\tilde{\nu} \tag{2.7}$$

where $\langle \tilde{\nu}_{max}^{-3}\rangle_{av} = \int I(\tilde{\nu})d\tilde{\nu} / \int \tilde{\nu}^3 I(\tilde{\nu})d\tilde{\nu}$, n is the refractive index, $\tilde{\nu}$ is the excitation energy in wavenumber, ε is the absorption coefficient, and $I(\tilde{\nu})$ is the emission spectrum measured in terms of relative numbers of quanta at each wavenumber.

TADF is characterized by a (i) strong temperature dependence of radiative rate constant as described by Eq. (2.4), (ii) small energy separation between S$_1$ and T$_1$, and (iii) large radiative rate constant ($k_{r,S} > 1 \times 10^6$ s^{-1}) of the upper emissive state.

The energy separation between S$_1$ and T$_1$ can be determined through analyzing the temperature-dependent radiative constants using Eq. (2.4). Alternatively, it can be estimated from the spectral shift of luminescence at ambient temperature, compared with that of the lower emitting state. It can also be estimated using theoretical calculations such as time-dependent density functional theory (TD-DFT) at the geometry optimized for T$_1$ or S$_1$. A large difference in the radiative rate constants at room temperature and 77 K is often employed as a convenient assignment of TADF. But this does not exclude the possibility of luminescence from thermally equilibrated triplet states. The radiative rate of the upper emissive state is the crucial factor for differentiating TADF from other mechanisms.

2.3 Structure-Dependent Photophysical Properties of Four-Coordinate [Cu(N^N)$_2$] Complexes

The ground state of the d^{10} complex [Cu(N^N)$_2$] prefers a pseudotetrahedral (D_{2d}) molecular geometry, with the two diimine ligands orientated perpendicular

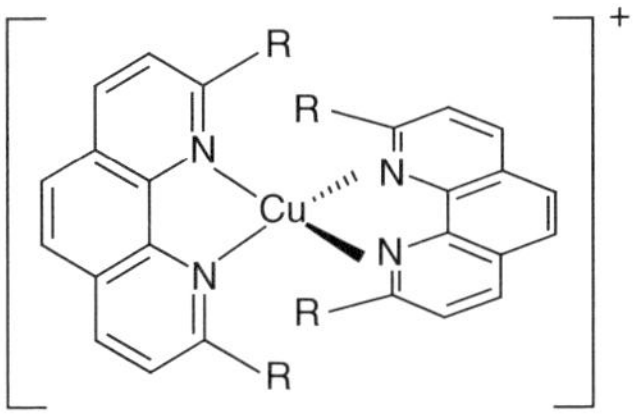

1: R = CH$_3$ (dmp)
2: R = n-C$_4$H$_9$
3: R = t-C$_4$H$_9$
4: R = sec-C$_4$H$_9$
5: R = n-C$_6$H$_{13}$
6: R – 4-n-butylphenyl

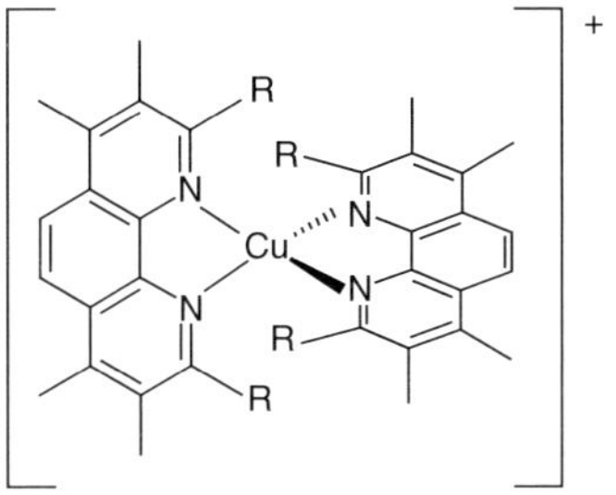

7: R = sec-C$_4$H$_9$

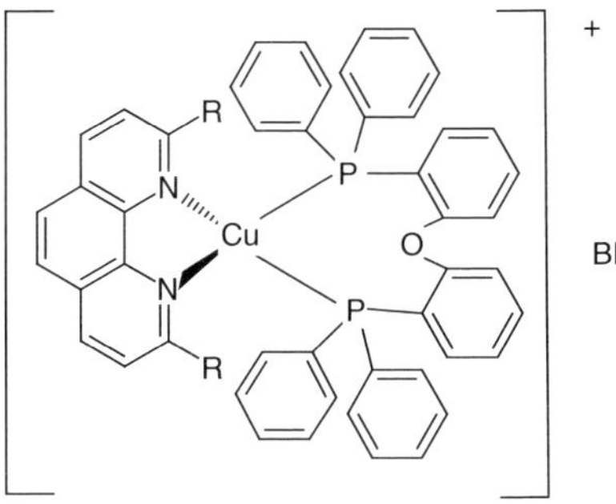

8 : R = H
9 : R = CH$_3$
10: R = n-C$_4$H$_9$

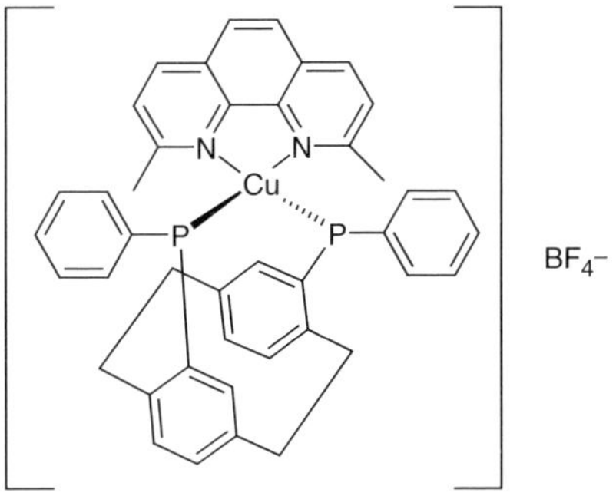

11

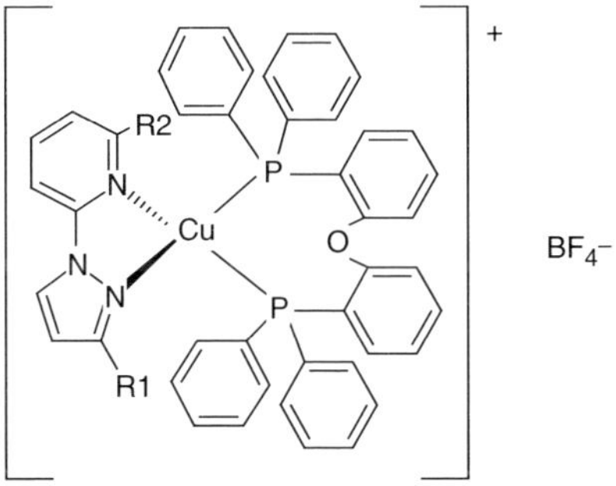

12 : R1 = H, R2 = H
13 : R1 = CH$_3$, R2 = H
14: R1 = CF$_3$, R2 = H
15: R1 = H, R2 = 9H-carbazolyl

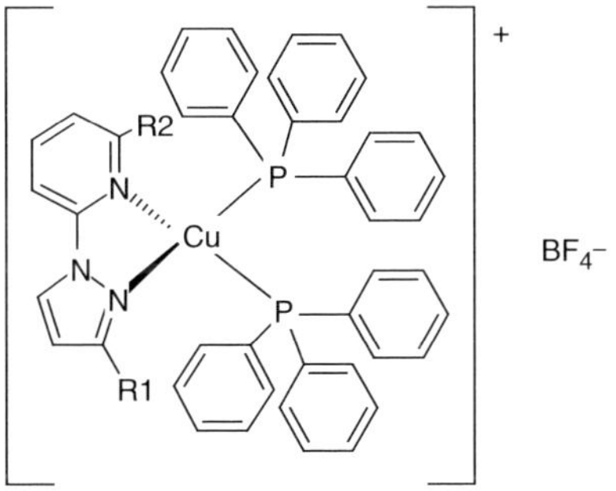

16: R1 = H, R2 = 9H-carbazolyl

Scheme 2.1 Molecular structures of Cu(I) complexes.

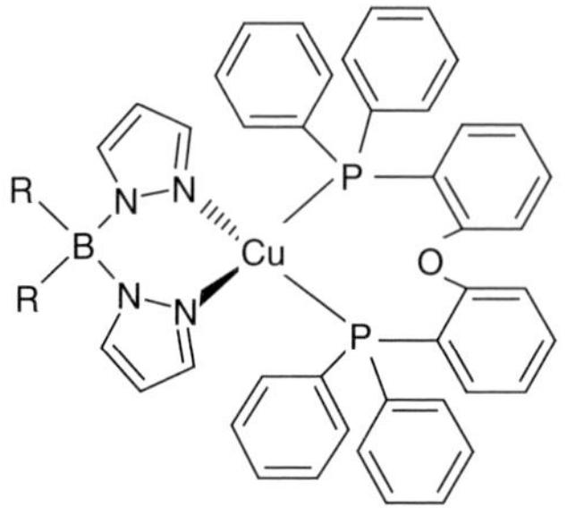

17 : R = H
18 : R = pyrazol-1-yl
19 : R = phenyl

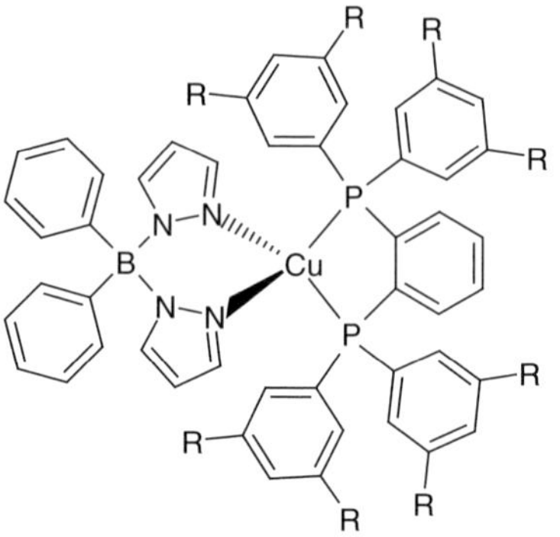

20 : R = H
21 : R = F
22 : R = CF$_3$

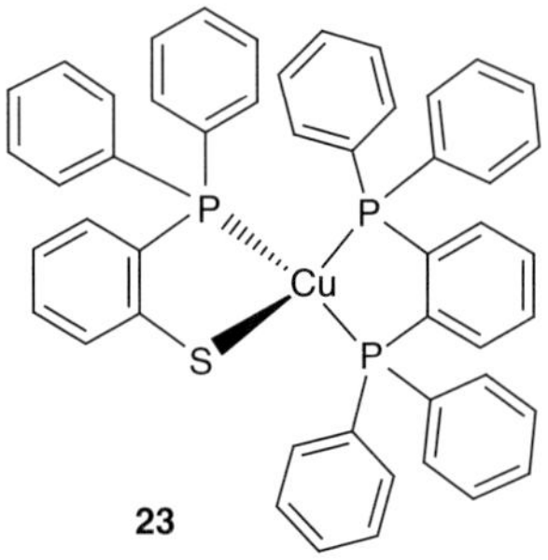

23

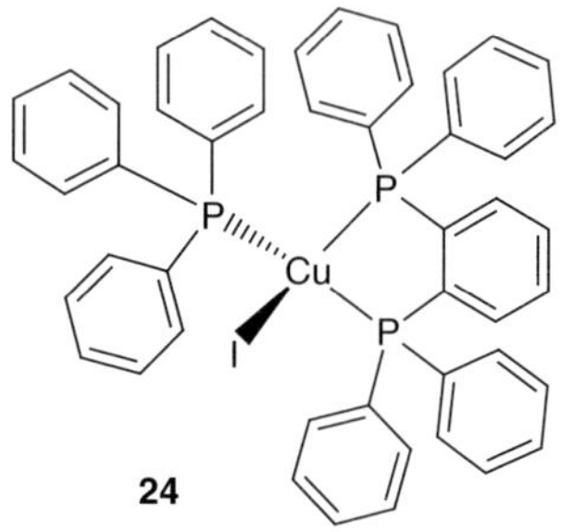

24

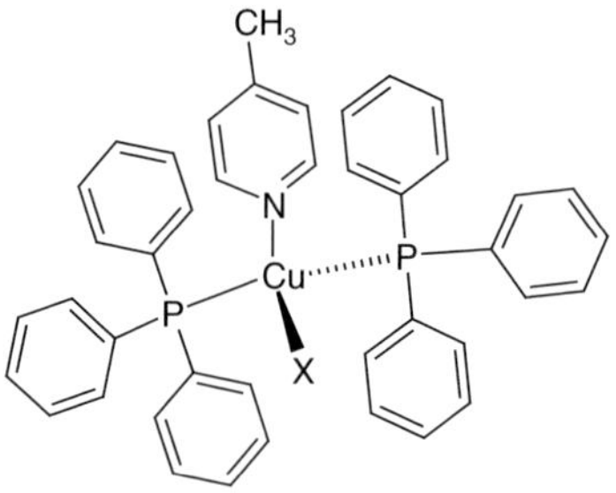

25 : X = Cl
26 : X = Br
27 : X = I

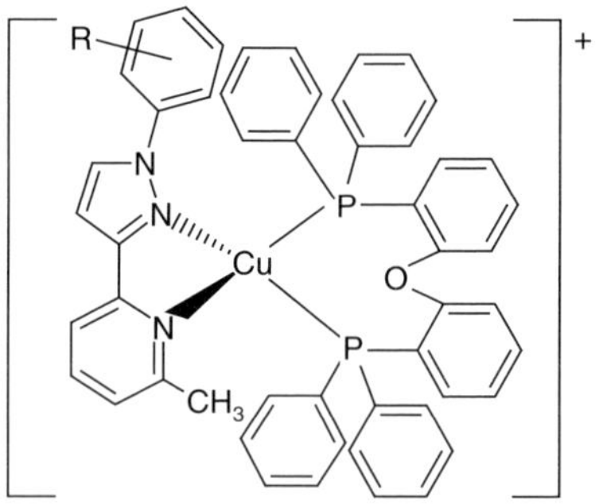

28 : R = H
29 : R = 2-CH$_3$
30 : R = 4-CH$_3$
31 : R = 2-CF$_3$
32 : R = 4-CF$_3$

Scheme 2.1 (Continued)

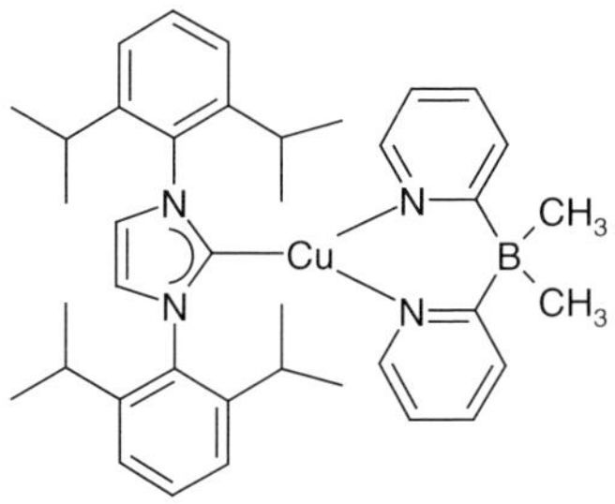

2.3 Structure-Dependent Photophysical Properties of Four-Coordinate [Cu(N^N)$_2$] Complexes | 67

33

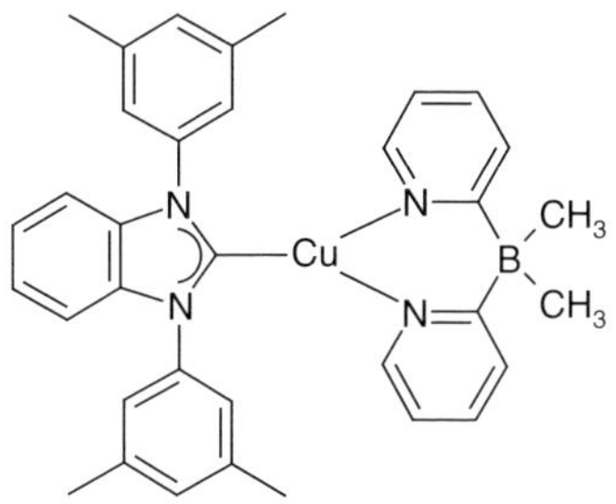

34

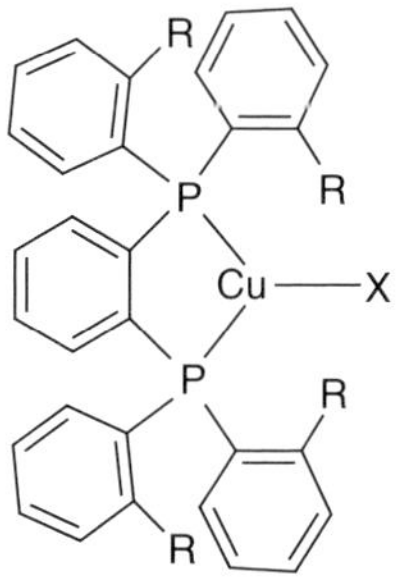

35 : X = Cl, R = CH$_3$
36 : X = Br, R = CH$_3$
37 : X = I, R = CH$_3$
38 : X = Br, R = Et
39 : X = Br, R = iPr
40 : X = SPh, R = CH$_3$
41 : X = SPh, R = iPr

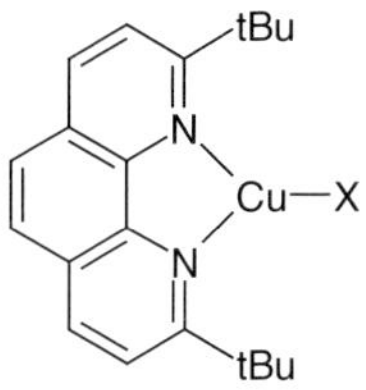

42 : X = Cl
43 : X = Br
44 : X = I

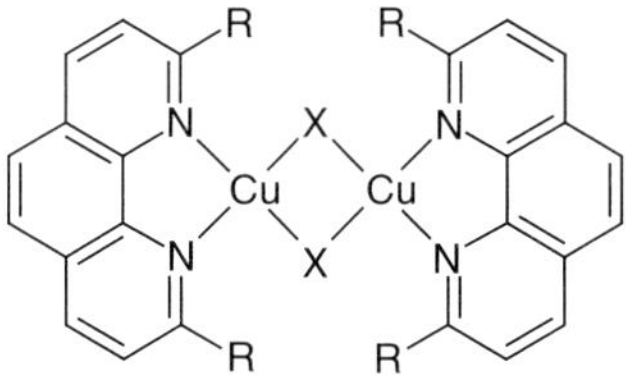

45 : X = I, R = CH$_3$

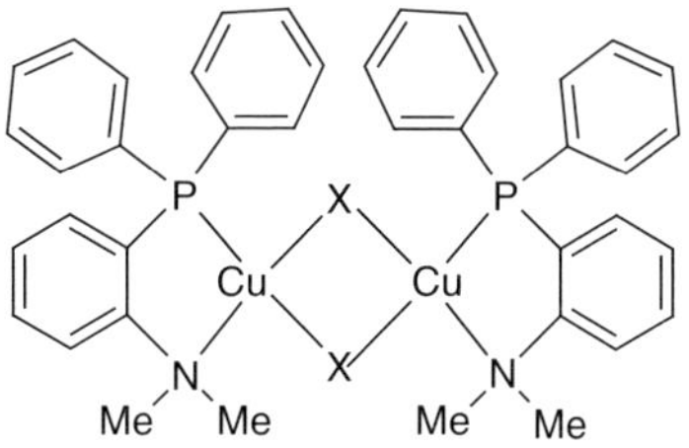

52 : X = Cl
53 : X = Br
54 : X = I

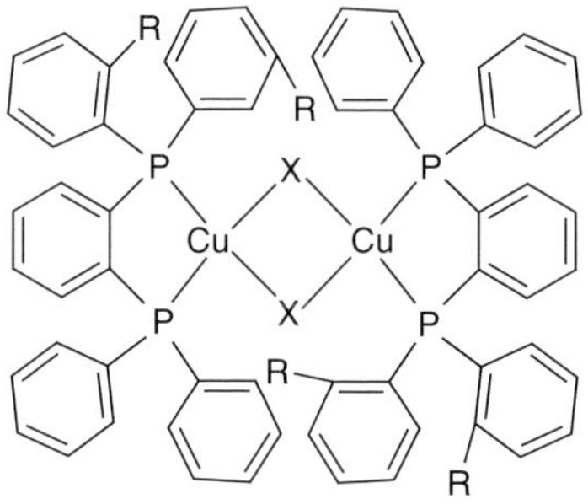

46 : X = Cl, R = H
47 : X = Br, R = H
48 : X = I, R = H

49 : X = Cl, R = CH$_3$
50 : X = Br, R = CH$_3$
51 : X = I, R = CH$_3$

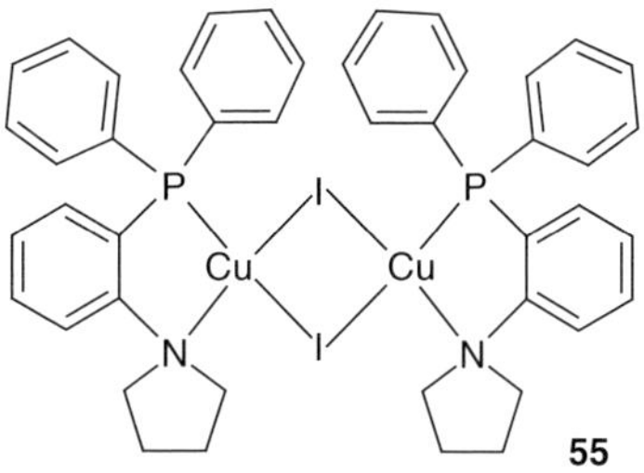

55

Scheme 2.1 (Continued)

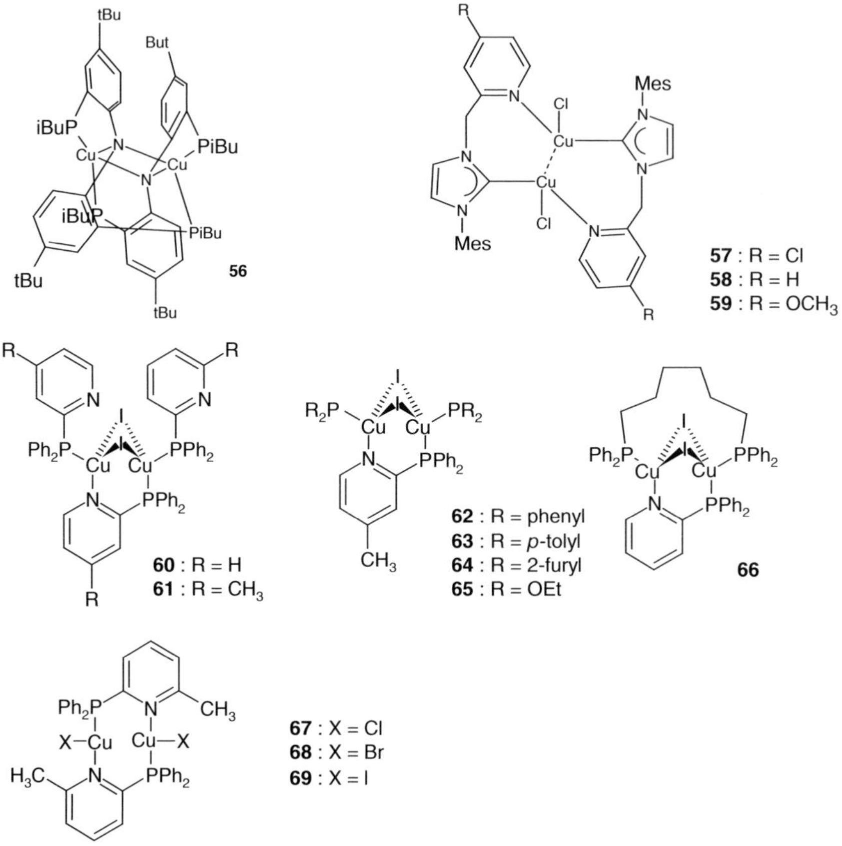

Scheme 2.1 (*Continued*)

to each other. Upon excitation to MLCT excited states, Cu(I) is oxidized to Cu(II) with a d^9 configuration, affording a large structural change from a tetrahedral-like structure to a flattened square-planar structure, because of the pseudo-Jahn–Teller (PJT) effect. Much of the pioneering work on copper(I) complexes was accomplished by McMillin and coworkers [57–61]. For example, [Cu(dmp)$_2$]$^+$ (dmp = 2,9-dimethyl-1,10-phenanthroline) (**1**, Scheme 2.1) exhibits weak broad luminescence in a deep-red region in noncoordinating solvents ($\Phi \sim 2 \times 10^{-4}$) (Figure 2.1). Complex **1** has been extensively investigated using spectroscopic techniques, such as ultrafast transient absorption [62, 63, 64b], time-correlated single-photon counting [50], fluorescence upconversion [64–67], and light-initiated time-resolved X-ray absorption spectroscopy [62a, 66] as well as theoretical methods [68–71]. [Cu(phen)$_2$]$^+$ (phen = 1,10-phenanthroline) does not exhibit emission under the same conditions as for complex **1**. The bulky substituents at the 2,9-positions of phen suppress the excited-state flattening distortion and nonemissive deactivation, via a short-lived exciplex formed with Lewis basic solvents.

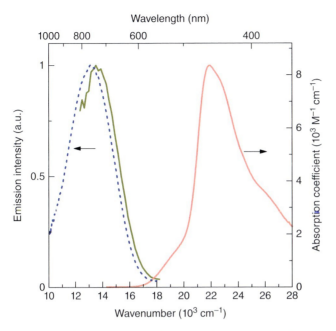

Figure 2.1 Absorption (red solid line), steady-state emission (blue dotted line), and short-lived component in the time-resolved emission (amber solid line) spectra in CH$_2$Cl$_2$ at 298 K for **1** time gated from −20 to 150 ps. *Source*: Ref. [50]. Reproduced with permission of American Chemical Society.

In 1979, McMillin tentatively assigned a short-lived (<1 ns) emission from [Cu(phen)(PPh$_3$)$_2$] (PPh$_3$ = triphenylphosphine) in the solid state to prompt fluorescence from a singlet MLCT state [57a]. From the temperature dependence of the emission lifetime, they proposed that the origin of the emission was thermally equilibrated ^{1,3}MLCT. Unfortunately, a more detailed analysis was complicated by difficulties in speciation by ligand disarrangement [72, 73]. In 1983, McMillin and coworkers attributed the emission of [Cu(dmp)$_2$]$^+$ in solution at ambient temperature to two thermally equilibrated excited states [58a]. The lower long lifetime state was assigned to ^{3}MLCT. Two assignments were proposed for the upper emissive state. McMillin and coworkers assigned the higher state to ^{1}MLCT, because of its high radiative rate constant of >1 × 10^7 s^{-1} [58a]. However, an alternative assignment to a higher-lying state spilt from a ^{3}E state in D_{2d} symmetry due to spin–orbit coupling was proposed, based on the emission lifetime and Stokes shift (i.e. the energy difference between the lowest absorption or excitation peak and the emission peak) [74].

Nozaki and coworkers directly determined the luminescence lifetimes and radiative rate constants of the upper state (13–16 ps) in **1** and its analogous complexes (**2**) using a time-correlated single-photon counting technique [50]. They assigned the upper state to the lowest ^{1}MLCT state, based on the large radiative rate constant (>1 × 10^6 s^{-1}) of the prompt fluorescence and detailed theoretical considerations. The observed lifetime of 10 ps corresponded to ISC from ^{1}MLCT to ^{3}MLCT and was significantly slower than those for d^6 metal complexes such as [Ru(bpy)$_3$]$^{2+}$ (<40 fs) (bpy = 2,2′-bipyridine) [75, 76]. The spin–orbit coupling of the 3d electron in copper(I) is as strong as that of the 4d electron in ruthenium(II). The slow ISC in the copper(I) complexes was attributed to structural distortion, because of the PJT effect in the MLCT state. The distortion from the tetrahedral structure caused a large energy splitting (>6.8 × 10^3 cm^{-1}) between the HOMO(21b$_2$) and HOMO-1(21b$_3$), with large orbital coefficients (Figure 2.2). Therefore, ultrafast ISC induced by a strong spin–orbit interaction (∼300 cm^{-1}) involving the HOMO(21b$_2$) and HOMO-1(21b$_3$), e.g. ISC from 1^{1}B$_1$(21b$_2$→22b$_3$) to 2^{3}A(21b$_3$→22b$_3$), was energetically unfavorable, and the ISC would occur from 1^{1}B$_1$(S$_1$) to 1^{3}A(T$_1$: 21b$_2$ → 22b$_2$) induced by relatively small spin–orbit coupling (<30 cm^{-1}) due to small d orbital coefficients in ligand-centered orbitals (22b$_2$ and 22b$_3$) (Figure 2.3a). The large energy splitting between HOMO and HOMO-1 also reduced the mixing due to spin–orbit coupling between the lowest ^{3}MLCT (1^{3}A) and ^{1}MLCT with large oscillator strength (2^{1}B$_1$, f = 0.1) (Figure 2.3b). This yielded a low radiative phosphorescence rate of ∼1 × 10^3 s^{-1}. Similar photophysical features were observed for Pt(0)(binap)$_2$ (**75**) that has a pseudotetrahedral structure [51].

A large energy separation between the HOMO and HOMO-1 with different d orbital characteristics also reduces the zero-field splitting (zfs) of the T$_1$ state. Owing to spin–orbit coupling of the 3d Cu electron, the T$_1$ state with MLCT character mixes with singlet and triplet MLCT excited states, which involve different d characteristics to that of the T$_1$ state. This is the origin of the radiative rate of phosphorescence. The spin–orbit coupling also produces the splitting of the T$_1$ sublevels (i.e. zfs). Yersin et al. reported that four-coordinate Cu(I) complexes had very small zfs values (<2 cm^{-1}) and low phosphorescence radiative rates (∼1 × 10^3 s^{-1}), c.f. complex **17** [10b]. This was attributed to the large energy separation of the HOMO and HOMO-1 (>1 eV). In contrast, complex **67** had a relatively large zfs value (15 cm^{-1}) and large radiative rate (∼1 × 10^4 s^{-1}), because the energy separation between the HOMO and HOMO-1 was only ∼0.3 eV [77b].

Fluorescence from ^{1}MLCT in complex **1** results in a large Stokes shift of ∼5.4 × 10^3 cm^{-1}. DFT calculations by Nozaki revealed that the greater part of this Stokes shift originated from the large flattening distortion, due to the PJT effect. Increasing the bulkiness of the 2,9-substituent phen ligands decreased the Stokes shift: 5400 cm^{-1} (**1**) > 5000 cm^{-1} (**2, 5**) > 4700 cm^{-1} (**4**) > 4200 cm^{-1} (**3**) > 3800 cm^{-1} (**7**) (Table 2.1). The lowest Stokes shift occurred when using the bulkiest *sec*-butyl and *tert*-butyl substituents [79].

The magnitude of the structural distortion significantly influences the radiative rate constant of phosphorescence, as well as the ISC rate between ^{1}MLCT and ^{3}MLCT. Calculation including spin–orbit coupling indicated that the radiative rate constant of phosphorescence in complex **1** decreased with decreasing dihedral angle between the two dmp planes [50]. This trend was consistent with the observation that copper(I) complexes with long alkyl chains at the 2,9-positions of phen (**5,6**) exhibited dramatically enhanced luminescence intensities at temperatures of ≤120 K [78]. This was interpreted to be a consequence of blocking the geometry of the ground state arrangement in the rigid matrix. Castellano and coworkers recently reported that the time constants of ISC increased in the

2.3 Structure-Dependent Photophysical Properties of Four-Coordinate [Cu(N^N)$_2$] Complexes

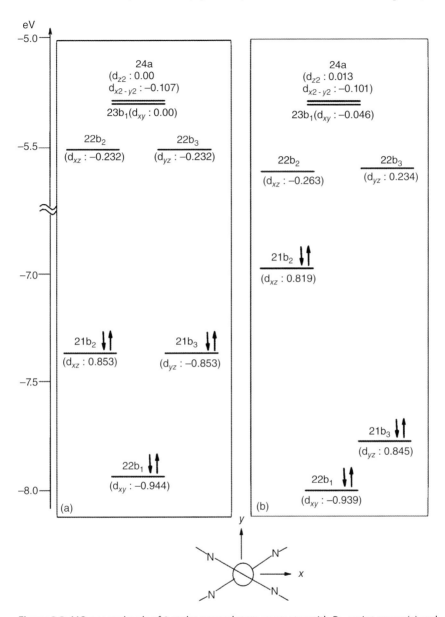

Figure 2.2 MO energy levels of **1** at the ground state geometry with D_{2d} point group (a) and those at the ^{1}MLCT geometry (b). The numbers in parentheses are coefficients of d orbitals in each MO. The orientation of the molecule in the coordinate system used for calculation is shown at the bottom. *Source*: Ref. [50]. Reproduced with permission of American Chemical Society.

Table 2.1 Photophysical properties of TADF of d^{10} metal complexes.

	In solution at r.t.[a]				In solid state or rigid matrix at r.t.				At 77 K			
Complex	λ_{em}, nm (SS, 10^3 cm^{-1})[c]	Φ	τ (μs)	$k_{r,DF}$ (10^4 s^{-1})	λ_{em}, nm (SS, 10^3 cm^{-1})[c]	Φ	τ (μs)	$k_{r,DF}$ (10^4 s^{-1})	τ (μs)	$k_{r,T}$ (10^4 s^{-1})	ΔE_{ST}[b] (cm^{-1})	References
[Cu(N^N)$_2$]$^+$												
1	748 (5400)	0.0004	0.09	0.44					0.82		1800	[58a, 78, 79b]
2	724 (5.0)	0.0010	0.132	0.76					2.40		800	[78]
3	615 (4.2)	0.06	3.2	1.9							790	[79a]
4	690 (4.7)	0.0045	0.38	1.13								[79b]
5	724 (5.0)	0.0007	0.98	0.76							800	[78]
6	718 (3.5)	0.0012	0.224	0.54							600	[78]
7	631 (3.8)	0.063	2.8	2.28					2.6		1150	[79b]
[Cu(N^N)(P^P)], [Cu(P^P)(P^X)], etc.												
8	700 (11.3)	0.0018	0.19	0.95								[73]
9	565 (8.4)	0.15	14.3	1.0							[d]	[73]
10	560 (8.6)	0.16	16.1	1.0							800	[25b, 73]
11	558 (8.8)	0.40	10	6.0	530	0.80	14	5.7	240	0.29	1000	[77a]
12	590	0.021	1.6	1.3	490	0.56	20.4	2.7			1450[e]	[80]
13	536	0.45	11.9	3.8	465	0.87	12.2	7.1			1370[e]	[80]
14	540 (9.9)	0.30	13.3	2.3	492	0.75	22.8	3.3	346		1450[e]	[80]
15					518 (7.7)	0.98	23	0.34	521	0.11	1450	[81]
16					495 (6.8)	0.45	134	4.3	671	0.092	1050	[81]
17	535	0.09	1.3	6.9	436	0.45	20	2.3	610		1300	[30]
18	500	0.02	0.5	4.0	447	0.90	22	4.1	450		1000	[30]

19	498	0.08	1.8	4.4	464	0.90	13	6.9	480	0.20	650	[30]
20	609 (9.9)	<0.001	0.08	1.3	535	0.70	3.3	21	1200	0.83	1294[e]	[82]
											370	[77b]
21	614 (10.0)	0.005	0.13	3.8					800	1.1	1623[e]	[82]
22	616 (8.8)	0.02	0.35	5.7					300	3.1	1511[e]	[82]
23	580 (9.1)	0.02			521	0.52	1.7, 0.33[f]		847	0.086	309[e], 467[g]	[83]
24	494 (6.0)	0.8	8.5	9.4	492 (6.0)	0.8	4.0, 8.5[f]		132	0.21	800	[84]
25					468	0.99	9.4[h]	11[h]	36[h]	2.3[h]	940	[85]
26					467	0.95	15[h]	6.3[h]	52[h]	1.8[h]	1070	[85]
27					455	0.66	9.5[h]	6.9[h]	52[h]	1.4[h]	1170	[85]
28	500 (8.6)				464	0.82	28	2.9	350		600	[86]
29	505 (9.2)				477	0.99	29	3.4	237		600	[86]
30	500 (8.2)				479	0.83	25	3.3	292		700	[86]
31	520 (9.6)				471	0.84	37	2.3	258		700	[86]
32	505 (8.0)				481	0.87	25	3.5	437		700	[86]
Three-coordinate [Cu(N^N)X], [Cu(P^P)X]												
33					475	0.76	11	6.9	34	2.7	740	[87]
34					575	0.73	18	4.1[i]	21	3.8	>3000[i]	[87]
35	534 (7.9)	0.43	4.9	8.8	517	0.38	4.6	8.3	2500	0.022	680	[88]
36	527 (7.7)	0.47	5.4	8.7	512	0.55	8.0	6.9	360	0.24	810	[88]
37	517 (7.3)	0.60	6.5	9.2	473	0.59	7.1	8.3	100	0.85	830	[88]
38					487	0.80	6.5	12	520		600	[89]
39	519 (7.4)	0.50	7.7	6.5	486	0.95	8.9	11	910		710	[89]
40	592	0.24	1.4	17.1	488 (9.6)	0.95	6.6	14	1100	0.96	690	[90]

(continued)

Table 2.1 (Continued)

	In solution at r.t.[a]				In solid state or rigid matrix at r.t.				At 77 K		ΔE$_{ST}$ [b]	References
Complex	λ$_{em}$, nm (SS, 10³ cm⁻¹)[c]	Φ	τ (μs)	k$_{r,DF}$ (10⁴ s⁻¹)	λ$_{em}$, nm (SS, 10³ cm⁻¹)[c]	Φ	τ (μs)	k$_{r,DF}$ (10⁴ s⁻¹)	τ (μs)	k$_{r,T}$ (10⁴ s⁻¹)	(cm⁻¹)	
41	546	0.15	1.0	15	500	0.95	5.0	19	1900	0.55	630	[90]
42					677 (4.5)	<0.01	0.9, 0.4[f]	<1.0				[91]
43					658 (4.4)	0.02	1.1, 0.6[f]	~2			1200[e]	[91]
44					624 (3.2)	0.15	3.7	~4				[91]
Binuclear {Cu$_2$(μ-X)$_2$}-core, etc.												
45					667 (3.5)	0.18	6.4	2.8	25[j]		[d]	[91]
46					533	0.6	1.5, 4.2[f]		2400			[84]
47					520	0.6	1.0, 4.3[f]		2300			[84]
48					502	0.8	2.5, 4.0[f]		50, 211[f]		664	[84]
49	559				490 (6.3)	0.42	6.3	6.7	405		450	[92]
50	551				482 (5.9)	0.50	6.4	7.8	193		440	[92]
51	541				488 (6.2)	0.95	4.9	19	115		450	[92]
52					506	0.45	6.6	6.8	220	0.45	460	[93]
53					490	0.65	4.1	16	930	0.11	510	[93]
54					464 (4.1)	0.65	4.6	14	270	0.37	570	[93]
55					465	0.65	5.6	12	250	0.4	630	[93]
56	524 (2.1)	0.57	11.5	4.9					336		786	[24]
57					550 (4.0)	0.49	11.0	4.5	90, 155[f]		740	[94]
58					520 (4.6)	0.59	11.0	5.4	97.6		710	[94]
59					538 (6.4)	0.67	15.3	4.4[i]	18.7		>1800[i]	[94]
60					537	0.81	6.5	12	32	2.8	400	[95]

#											Ref
61	581	0.02	0.2[h]	10[h]	515	0.89	2.3[h]	39[h]			[96]
62	572	0.24	2.4[h]	10[h]	510 (5.4)	0.99	1.92[h]	52[h]		480	[96]
63	589	0.14	1.7[h]	8.2[h]	542 (6.5)	0.74	3.73[h]	20[h]			[96]
64					500 (6.3)	0.84	1.76[h]	48[h]			[96]
65					507 (5.9)	0.80	2.34[h]	34[h]			[96]
66					538	0.77	3.6	26	115	725[g]	[97]
67					485	0.92	8.3	11	44	930	[31]
68					501	0.52	12.4	4.2	84	950[g]	[31]
69					484	0.76	7.3	10	51	1100[g]	[31]

Ag(I), Au(I), Pt(0), Pd(0)

#											Ref
70					505	0.32	2.2, 0.56[f]		574	199[e]	[83]
71					610	0.12	1.7, 0.47[f]		52	405[e]	[83]
72	760 (5.7)	0.12	1.25	9.9						1150	[51]
73	750 (5.8)	0.32	2.6	12						1300	[98]
74	690 (7.9)	0.27	2.4	11						1400	[98]
75	670 (8.2)	0.38	3.2	12						910	[98]

a) In deaerated solution.
b) Determined from temperature dependence of decay rate.
c) Values in parentheses are Stokes shift (SS) in 10^3 cm^{-1} estimated from reported spectra.
d) Tentatively assigned to TADF.
e) Calculated using TD-DFT.
f) Multiexponential decay.
g) Determined from spectral shift.
h) Averaged value of multicomponent.
i) Phosphorescence.
j) At 22 K.

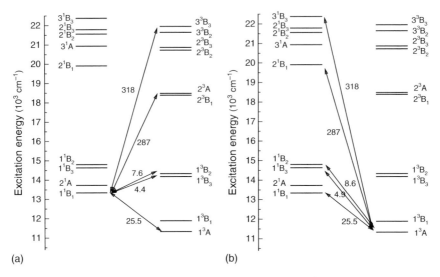

Figure 2.3 Spin–orbit integrals between the lowest singlet state (1^1B_1) and triplet states (a) and those between the lowest triplet state (1^3A) and singlet states at the flattened geometry of **1** (b). Values in figures are in cm^{-1}. *Source*: Ref. [50]. Reproduced with permission of American Chemical Society.

order [65]: 9–20 ps (**1**) > 5–14 ps (**4**) > 2–6 ps (**7**). This reflected the degree of PJT flattening distortion.

A large structural rearrangement due to the PJT effect is usually associated with an increase in nonradiative deactivation or even quenching of emission due to an increase in the Franck–Condon factors (i.e. the energy gap law [99]). Introducing sterically bulky groups at the 2,9-positions of phen suppresses flattening and also suppresses five-coordinate exciplex formation in homoleptic and heteroleptic diamine Cu(I) complexes [100].

2.4 Flattening Distortion Dynamics of the MLCT Excited State

Pioneering spectroscopic and computational studies of [Cu(N^N)$_2$] indicated that the overall dynamics of the MLCT state can be described by the following processes: (i) internal conversion occurring within 50 fs after photoexcitation; (ii) PJT distortion or flattening within 1 ps; (iii) ISC with a time constant of 2–20 ps, which is a function of the dihedral angle between the two diimine ligands; and (iv) ^{3}MLCT decay, which is occasionally accelerated by direct interaction of the Cu center with solvent molecules forming an exciplex [62c, 64–66].

The rapid PJT dynamics of complex **1** upon excitation to the S$_2$ state were investigated by Iwamura and coworkers using a femtosecond upconversion technique [64]. They assigned internal conversion from the Franck–Condon state of S$_2$ to that of S$_1$ (45 fs), PJT structural flattening (660 fs), and ISC to T$_1$ (7.4 ps). They found that the substituents at the 2,9-positions of phen affected the excited-state properties, including the lifetimes of the lowest

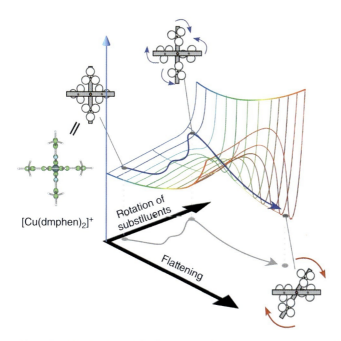

Figure 2.4 Mechanism of the flattening relaxation process of copper bis-phenanthroline complexes with substituents at 2,9-position. The case of **1** is taken as an example.
Source: Ref. [64c]. Reproduced with permission of Royal Society of Chemistry.

excited states and also earlier structural change dynamics. The time constant of the structural change of [Cu(phen)$_2$]$^+$ was 200 fs, while those of [Cu(dmp)$_2$]$^+$ and [Cu(2,9-dpphen)$_2$]$^+$ (dpphen = diphenyl-1,10-phenanthroline) were 660 and 920 fs, respectively. This indicated that the structural change processes slowed with increasing bulkiness of the 2,9-substituents of phen. The solvent dependence of flattening dynamics, and the absence of a substituent effect on flattening rate for [Cu(4,7-substituted phenanthroline)$_2$]$^+$, indicated that retardation due to the 2,9-substituents was due to steric effects (Figure 2.4), rather than to solvent friction [67].

2.5 Green and Blue Emitters: [Cu(N^N)(P^P)] and [Cu(N^N)(P^X)]

The [Cu(N^N)$_2$] complexes typically exhibit orange or deep-red emission. However, emission quantum yields reported to date are <0.1, and their radiative rates are relatively low (<2 × 10^4 s^{-1}) (Table 2.1). This is probably because the S$_1$ → GS radiative transition is symmetrically forbidden in D_{2d} symmetry or only weakly allowed in pseudo-D_{2d} symmetry [50]. A family of heteroleptic Cu(I) complexes with [Cu(N^N)(P^P)] or [Cu(N^N)(P^X)] structures was recently investigated. Interest in these complexes arose because of their higher emission energies, higher emission quantum yields (up to 0.9 in solid state), and higher $k_{r,DF}$ values compared with those for [Cu(N^N)$_2$] as shown in Table 2.1. Heteroleptic Cu(I) complexes are advantageous that they can be tuned from blue to green emitters in OLEDs.

For the heteroleptic [Cu(N^N)(P^P)] complexes, an equilibrium between the heteroleptic and homoleptic [Cu(N^N)$_2$] or [Cu(P^P)$_2$] species is sometimes observed in solution. This is a major limitation for preparing stable [Cu(N^N)(P^P)] derivatives. Detailed analyses revealed that the equilibrium was mainly influenced by the relative thermodynamic stabilities of these solution species [73]. McMillin and coworkers reported that the emission quantum yields of the [Cu(N^N)(P^P)] complexes (**8–10**) increased considerably, upon introducing sterically bulky substituents on the phosphine ligands [72b]. The emission quantum yields also increased when bridging by two phosphine ligands, e.g. 0.16 for [Cu(dbphen)(POP)] (POP = 2,2′-bis(diphenylphosphino)diphenyl ether) (**10**). An advantage of POP is that only a small excess of chelating phosphine was necessary to suppress the formation of [Cu(N^N)$_2$]. The [Cu(N^N)(POP)] complexes exhibited higher-energy CT states, and more positive Cu(II)/Cu(I) potentials, than the [Cu(N^N)$_2$] complexes. A larger P–Cu–P bite angle of ~116° (compared with 84° for N–Cu–N) and the more diffuse donor orbitals of the phosphine ligands probably stabilized Cu(I). Czerwieniec et al. synthesized complex **11**, which possessed a rigid semicage [77]. The combined steric effects of phanephos and dmp decreased the flattening distortion. This increased the emission quantum yield to 0.4 in solution and 0.8 in powder form.

In [Cu(N^N)(POP)]-type complexes, the HOMO is mainly dσ antibonding in character and is distributed over Cu(I) and the POP ligand. The LUMO is predominantly localized on the diamine ligands. The strongly green-blue emitting [Cu(pypz)(POP)]$^+$ (**12**), [Cu(pympz)(POP)]$^+$ (**13**), and [Cu(pytfmpz)(POP)]$^+$ (**14**) complexes were developed using an electron-rich five-membered diamine ligand [80]. This destabilized the LUMO and increased the HOMO–LUMO energy gap. Complexes **11**, **13**, and **14** exhibited TADF with high quantum yields of up to 0.45 in dichloromethane solution and up to 0.87 in the neat solid. Complex **13** exhibited a large bathochromic shift of 71 nm, but its emission quantum yield in solution was still as high as 0.45. This suggested that the electronic properties of the N^N ligand were important in determining the emission quantum yield.

Complexes **15** and **16** were synthesized by hybridizing a carbazole functional group into the N^N ligand (czpzpy), to improve the hole-transport properties of the subsequently prepared OLED. Complex **15** exhibited intense TADF, with a quantum yield of 0.9 in the solid state. An efficient solution-processed OLED was then fabricated by spin coating a mixture of czpzpy and [Cu(CH$_3$CN)$_2$(POP)]BF$_4$ [81].

Yersin and coworkers synthesized strongly luminescent neutral [Cu(N^N)POP] complexes (**17–19**) using anionic N^N ligands of tetrakis(pyrazol-1-yl)borate and related borate ligands [30, 77b]. These complexes exhibited strong blue-white luminescence, with a quantum yield of up to 0.9 in the solid state. Complex **18** exhibited a large bathochromic shift of 500 nm in solution, 407 nm in PMMA, and 447 nm in powder form. The respective quantum yields were 0.02, 0.3, and 0.9. These yields illustrated that distortion of the excited state was largely suppressed in the rigid matrix.

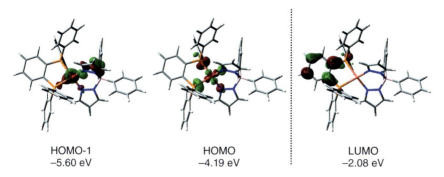

Figure 2.5 Frontier molecular orbitals of **20** resulting from DFT calculations for the triplet state geometry. *Source*: Ref. [77b]. Reproduced with permission of American Chemical Society.

Osawa and coworkers introduced F and CF_3 moieties at the meta position of the four peripheral phenyl groups of dppb. This increased the ease of sublimation of the resulting Cu(I) complexes [82]. Conventional bottom-emitting devices with a three-layer structure containing complex **22** exhibited bright green luminescence, with an external quantum efficiency of 17.7%.

Czerwieniec and Yersin reported the diverse emission behavior of the Cu(I) complexes **11**, **19**, and **20** [77b]. Complex **20** exhibited the smallest ΔE_{ST} of 370 cm^{-1} and the highest radiative rate of TADF ($k_{r,DF}$) in the Cu(I)(N^N)(P^P) family. Its radiative rate of phosphorescence was relatively low at 0.083×10^4 s^{-1}. As depicted in Figure 2.5, the HOMO and LUMO are distinctly separated in complex **20**, resulting in very small exchange integral and thus the small ΔE_{ST}.

Osawa et al. synthesized heteroleptic d^{10} metal complexes (**23**) containing P^P and P^S [83]. The origin of TADF in these complexes was attributed to ligand-to-ligand charge transfer. The thiolate ligand with strong electron-donating character (PS$^-$) reduced the contribution of the metal-centered orbital to the HOMO.

Kato and coworkers reported [CuX(PPh$_3$)$_2$(4-Mepy)] (X = Cl, Br, I) complexes (**25**–**27**) containing triphenylphosphine and halide ligands. These exhibited strong blue TADF emission, with a quantum yields approaching 100% in crystalline form. The HOMO was distributed over the Cu and halogen atom. The LUMO was localized over dppb or 4-Mepy. Thus, the lowest excited state was assigned to a mixed MLCT and XLCT [85].

Lu and coworkers synthesized five [Cu(N^N)(POP)] complexes (**28**–**32**) containing functional 3-C-linked 6-methylpyridine pyrazolate diamine ligands [86]. These complexes exhibited intense blue to blue-green TADF, with an emission quantum yield of ~100% and a small spectral shift between the solid, thin film, and solution phases.

2.6 Three-Coordinate Cu(I) Complexes

Tetrahedral four-coordinate Cu(I) complexes tend to suffer from large structure distortion, due to the PJT effect. Lotito and Peters reported emitting three-coordinate Cu(I) arylamidophosphine complexes [101a]. Thompson and coworkers reported phosphorescence from three-coordinate Cu(I) phenanthroline complexes containing monodentate N-heterocyclic carbine (NHC) ligands [101b]. The three-coordinate Cu(I) geometry may suppress PJT distortion in the MLCT states, although exciplex formation with the solvent may still occur.

TADF-type Cu(I) complexes with three-coordinate structures (**36**) were reported by Osawa and coworkers [88], and detailed parameters of their TADF were reported later. Introducing bulky substituents at the *ortho* position of the peripheral phenyl groups of dppb was found to stabilize the three-coordinate structure (**35–41**). In the absence of this, only halogen-bridged dinuclear copper complexes were formed. A vapor-deposited OLED doped with complex **36** exhibited a maximum external quantum efficiency of 21.3%. Jahn–Teller distortion was reported for the excited state of the trigonal planar Au(I)(PPh$_3$)$_3$ complex with pseudo-D_{3h} symmetry [101c]. When one electron was promoted from the degenerate HOMO and HOMO-1 (which have large contributions from the d$_{xy}$ and d$_{x2-y2}$ orbitals), the geometry of the structure changed from Y to T shaped. The three-coordinate Cu(I) complexes with one bidentate and one monodentate ligand had slightly distorted trigonal planar structures. They possessed C_{2v} symmetry, in which d$_{xy}$ and d$_{x2-y2}$ were no longer degenerate. DFT calculations for complex **36** indicated that structure change around Cu(I) between the ground and excited states was minimal, which greatly reduced the emission bandwidth.

Yersin and coworkers reported that the complex **33** exhibited TADF, whereas complex **34** exhibited pure phosphorescence, despite their similar chemical structures [87]. These three-coordinate Cu(I) complexes possessed bidentate and aromatic or halogen-monodentate ligands. Complexes **33** and **34** possessed three higher-lying occupied molecular orbitals containing d orbitals: dσ antibonding orbital with bidentate (dσ-B)*, dσ antibonding orbital with monodentate (dσ-M)*, and dπ antibonding orbital with monodentate (dπ-M)*. The (dσ-M)* orbital was stabilized in a distorted trigonal planar geometry. Thus, the HOMO was either (dσ-B)* or (dπ-M)* in character, depending on the electronic structure of the monodentate ligand. Variations in the photophysical properties and S$_1$–T$_1$ splitting with ligand orientation were also observed for complex **40** [90].

The three-coordinate geometry largely eliminates the possibility of flattening distortion. However, other distortions may still occur in the excited state. This is because the Stokes shifts of the three-coordinate complexes are not as small as those of other Cu(I) complexes. Calculations of the excited state of complex **36** predicted small changes from pyramidal to more planar geometries of the P(aryl)$_2$(phenylene) units bound to the Cu(I) center in the optimized S$_1$ and T$_1$ states.

2.7 Dinuclear Cu(I) Complexes

Copper(I) has a high affinity toward halogen-containing ligands and forms a large family of tetranuclear and dinuclear complexes of various nuclearity and structure [22c, 28]. Neutral Cu(I) complexes have been obtained from the reaction of cuprous halides with various P- and N-containing ligands. These have versatile structures and interesting photophysical properties. Ford and coworkers reported the luminescence properties of a {Cu$_4$(μ-I)$_4$} cubane unit, consisting of a copper tetrahedron with iodide ions capping the four faces [102]. This structure contained a pair of halogen-bridged {Cu$_2$(μ-I)$_2$} cores, with their Cu–Cu axes aligned perpendicular to each other. Various halogen-bridged Cu(I) complexes with a {Cu$_2$(μ-X)$_2$} core have been synthesized, and their emission properties have been investigated.

Tsuboyama and et al. reported the first TADF-type Cu(I) dinuclear complex. It contained a {Cu$_2$(μ-I)$_2$} core chelated by bulky diphosphine ligands (dppb) (**46–48**) [84]. The dinuclear complexes adopted two configurations in the excited states: a distorted tetrahedral (butterfly-type) geometry and a flattened geometry. The flattened geometry had a nonradiative rate of at least 2 orders of magnitude larger than that for the distorted tetrahedral geometry, leading to a much smaller TADF emission quantum yield (0.009) at room temperature. These complexes exhibited intense blue-green luminescence in the solid state (0.6–0.8). Emission arose from a (M + X)LCT state with a distorted tetrahedral geometry. Introducing methyl groups into the phenyl rings of one of the biphenyl phosphine groups of dppb caused an emission blue shift of 43 nm (**49–51**) [92]. Dinuclear Cu(I) complexes consisting of a butterfly-shaped {Cu$_2$(μ-X)$_2$} core and three P^N-type ligands are also reportedly highly luminescent(**52–55**) [93]. Their photoluminescent properties depended on the substituent on the pyridine, and the resulting emission could be tuned from blue to red. The emission of the complexes depended on the rigidity of the environment. A photophysical analysis of the Cu(I) complexes with {Cu$_2$(μ-X)$_2$} core chelated by two aminophosphane ligands indicated that the zfs values of the triplet (M + X)LCT state were smaller than 1 or 2 cm^{-1}. The triplet state properties were not obviously affected by the differing spin–orbit coupling of the two bridging halide atoms.

Deaton et al. reported that the Cu(I) complex with a {Cu$_2$(μ-N)$_2$} core (**56**) exhibited strong TADF emission, with a high quantum yield of 0.57 and relatively narrow emission bandwidth in solution as shown in Figure 2.6 [24]. The Stokes shift was only 2070 cm^{-1} even in solution, much smaller than those of other Cu(I) complexes (typically 5000–10 000 cm^{-1}). This is the smallest Stokes shift of any reported Cu(I) complex. As shown in Figure 2.7, the HOMO of complex **56** was delocalized over the four-membered {Cu$_2$(μ-N)$_2$} core, while the LUMO resided largely over the four ligand aryl groups. This delocalization of the HOMO and LUMO possibly suppressed structural distortion in the excited state. Vapor-deposited OLEDs doped with complex **56** in the emissive layer yielded a very high external quantum efficiency of 16.1%.

Steffen and coworkers synthesized emissive monomeric and dimeric Cu(I) halogen complexes (**42–44** and **45**, respectively) containing phen derivatives [91]. TDDFT calculations supported the TADF mechanism, in agreement with the low-temperature properties of complex **43** in a glassy matrix and of complex **45** in the solid state. However, the temperature dependence of lifetimes for complex **45** suggested that at least two triplet states were involved in the TADF.

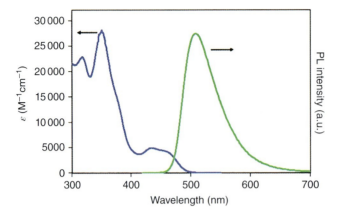

Figure 2.6 (a) Absorption (blue) and emission (green) of **56** in mTHF solution at 295 K. The emission bandwidth is relatively narrow, and the Stokes shift is only 2070 cm^{-1}, the smallest value of any reported Cu(I) complex. *Source*: Ref. [24]. Reproduced with permission of American Chemical Society.

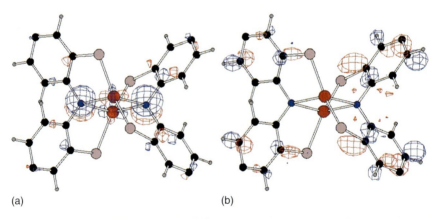

(a)　　　　　　　　　　　　(b)

Figure 2.7 Calculated HOMO (a) and LUMO (b) of **56** at the T_1 geometry. For clarity, the alkyl groups have been removed from the visualizations. *Source*: Ref. [24]. Reproduced with permission of American Chemical Society.

Cu(I) complexes with [Cu$_2$(μ-X)$_2$(N^P)$_2$] structures (**52–55**) exhibited bright blue-white luminescence [93]. The triplet state of these complexes had a high radiative rate, with a large zfs value of 10 cm^{-1}. As a result, ~80% of the emission intensity resulted from the singlet state via TADF, and the remaining 20% was contributed by the triplet state.

Reported dimeric Cu(I) chlorido NHC–picolyl complexes (**57–59**) possess cuprophilic interactions, with a Cu–Cu distance of 2.52–2.57 Å [94]. Complexes **57** and **58** exhibited TADF with a small ΔE_{ST} of 620–740 cm^{-1}. However, complex **59** containing a methoxy group exhibited an unusually short emission lifetime at 77 K and no TADF emission at room temperature, suggesting a ΔE_{ST} of >1800 cm^{-1}. DFT calculations indicated that the HOMO was distributed over two copper ions in complex **59** but localized over one copper ion in complexes **57** and **58**. The authors noted that relatively small energy splitting between HOMO and HOMO-1 with different d orbital character greatly enhanced spin–orbit coupling, resulting in a high phosphorescence radiative rate.

2-Di-phenylphosphinopyridine (PyrPHOS) is a common ligand used in luminescent Cu(I) complexes. It can be readily modified to change its luminescence color or solubility [96a]. Its straightforward chemical modification and intense solid state luminescence has led to a huge number of dinuclear Cu(I) halide complexes with PyrPHOS-type ligands being reported by Baumann and Brase [28, 95, 96]. These complexes consisted of a butterfly-shaped {Cu$_2$(μ-X)$_2$} core and two substituted PyrPHOS ligands. The small bite angle prohibited mononuclear complexes from forming with these ligands.

Dinuclear Cu(I) halide complexes containing N^P ligands (PyrPHOS type ligands) with a diphenylphosphino group consisting of N-heterocycles have been reported [95]. They exhibited emission from the blue to red regions, with high quantum yields of up to 0.96 in neat powders. The energy separation for complex **60** was experimentally determined to be 400 cm^{-1}. This fact combined with theoretical calculations indicated that the strong luminescence of these dinuclear Cu(I) complexes was due to TADF.

A family of heteroleptic PyrPHOS dinuclear complexes was developed by substituting monodentate ligands in homoleptic PyrPHOS complexes with various P-donor ligands (**60**–**66**) [95–97]. These complexes exhibited very high TADF, with quantum yields of up to 0.99 in neat powders and up to 0.85 in thin films, and were broadly soluble in many solvents, and most were thermally stable at above 250 °C. Heteroleptic Cu(I) complexes tend to undergo dissociation and rearrangement reactions. However, the homoleptic and heteroleptic PyrPHOS Cu(I) dinuclear complexes were sufficiently stable to be used in the solution processing of OLED devices.

In these PyrPHOS dinuclear complexes, the HOMO was largely localized on the metal halide center. The LUMO was mainly located on the bridging N^P ligand. Modifying the N-heterocycles mainly affected the LUMO energy of the complex. The HOMO energy was largely unchanged, which resulted in a blue shift in the luminescence. Adding different substituents to the PyrPHOS ligands enhanced the solubility in solvents such as toluene. Consequently, the solubility and luminescence color of the PyrPHOS dinuclear complexes could be independently tuned.

In most Cu(I) complexes, radiative processes from the T$_1$ states are not efficient ($k_{r,T} \sim 1 \times 10^3$ s^{-1}). This is because of the large energy separation of the HOMO and HOMO-1 containing d orbitals. Yersin and coworkers synthesized dinuclear complexes (**67**–**69**), in which two Cu(I) ions were bridged by two PyrPHOS ligands [31]. Complex **67** exhibited a large zfs value of 15 cm^{-1} and a large radiative rate of $\sim 1 \times 10^4$ s^{-1}. This was because of the small energy separation between the HOMO and HOMO-1 of $\sim$0.3 eV. Complex **67** exhibited $\sim$80% of its emission intensity from TADF. The remaining 20% was contributed from phosphorescence from T$_1$. This result hints at the possibility of new strategies for developing TADF-type Cu(I) emitters with additional radiative pathways.

2.8 Ag(I), Au(I), Pt(0), and Pd(0) Complexes

Other d^{10} metal complexes such as Ag(I) and Au(I) complexes also reportedly exhibit TADF (Scheme 2.2). Osawa et al. reported that heteroleptic coinage complexes of Ag(I) (**70**) and Au(I) (**71**) exhibited bluish green and orange TADF luminescence, with quantum yields of up to 0.32 in the solid state. However, the Au(I) complex was unstable in solution, because of rapid ligand exchange. The Ag(I) complex exhibited low solubility in organic solvents, making it difficult to employ in OLED fabrication [83].

Pt(0) and Pd(0) complexes (**72**–**75**) have been shown to luminesce in solution. Tsubomura et al. reported that the Pt(0) complex, $[Pt(binap)_2]$ (binap = 2,2′-bis(diphenylphosphino)-1,1′-binaphthyl), exhibited intense red luminescence, which originated from an MLCT state in benzene [98]. The bidentate binap has an extended π-electron system with a low-energy π^*-acceptor orbital. This makes it suitable for MLCT transitions in the visible region.

$[Pt(0)(binap)_2]$ (**74**) exhibited luminescence from an MLCT state. It had a quantum yield of 0.12 and lifetime of 1.2 μs in toluene at ambient temperature. Ultrafast spectroscopic measurements revealed that the intense luminescence was mainly due to TADF and that the lifetime of the prompt fluorescence was 3.2 ps [51]. The large energy difference between the HOMO and HOMO-2

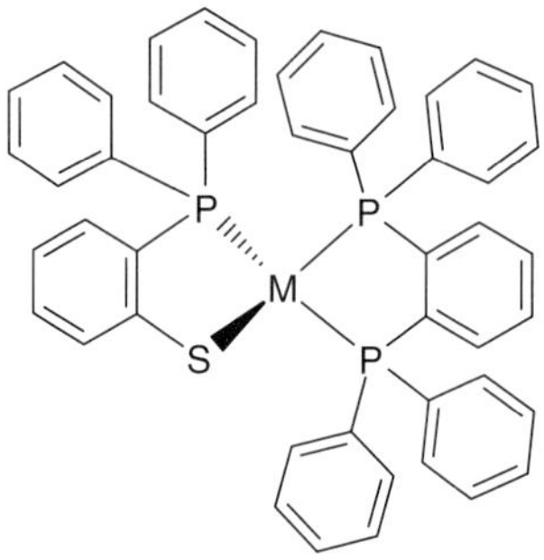

70 : M = Ag(I)
71 : M = Au(I)

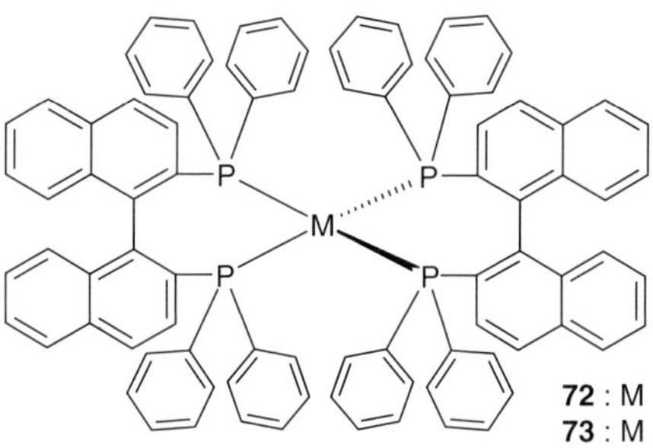

72 : M = Pt(0)
73 : M = Pd(0)

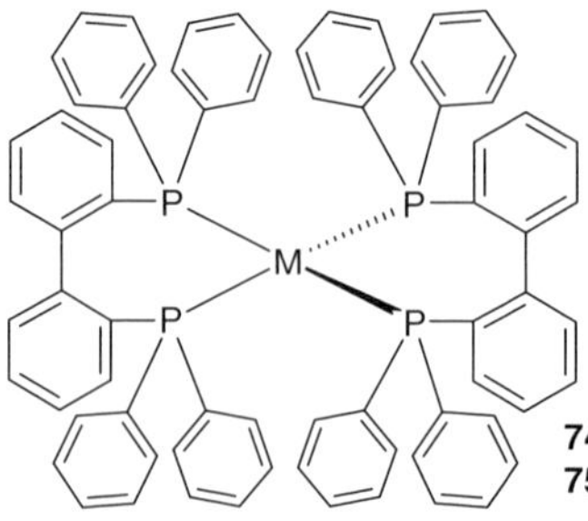

74 : M = Pt(0)
75 : M = Pd(0)

Scheme 2.2 Molecular structures of d^{10} metal complexes.

reduced the extent of mixing between T_1 and S_n, due to spin–orbit interactions. This decreased the radiative rate of phosphorescence.

2.9 Summary

This section highlights the TADF properties of d^{10} metal complexes with various nuclearity (e.g. mononuclear, dinuclear), coordination structures (e.g. three and four coordinate), and ligand type (e.g. X, N^N, P^P, N^P, P^X, etc.). The versatile structures of d^{10} metal complexes have led to the continuous development of Cu(I) complexes as effective TADF emitters over the last decade. These range from monomeric four-coordinate [Cu(N^N)$_2$] and [Cu(N^N)(P^P)] to dimeric four-coordinate {Cu$_2$(μ-X)$_2$} core, {Cu$_2$(μ-N)$_2$} core, and N^P bridged {Cu$_2$(μ X)$_2$} core, to three-coordinate [Cu(P^P)X] and [Cu(N^N)X] complexes.

The S_1 and T_1 states associated with charge transfer involving orthogonal orbitals (i.e. $d\sigma^* \to \pi^*$) are separated by a very small energy. This leads to some Cu(I) complexes exhibiting efficient TADF with relatively large radiative rate ($k_{r,DF} > 2 \times 10^5$ s^{-1}). The photophysical data compiled for TADF-type d^{10} metal complexes in Table 2.1 indicate that the coordination structure enabling smaller ΔE_{ST} is essential to achieve higher $k_{r,DF}$. Since the ΔE_{ST} value reduces with decreasing the special overlap between the HOMO and LUMO of the complexes (see Eq. (2.5)), the uses of appropriate ligands such as dppb, PyrPHOS, and/or dinuclearization are therefore effective strategy to improve the radiative properties of TADF-type Cu(I) complexes.

The TADF mechanism enables them to harvest both singlet and triplet excitons generated in OLEDs, affording internal quantum efficiencies of ~100%. The abundance, low cost, and low toxicity of copper makes Cu(I) complexes ideal as emitters in optoelectronics. Moderately strong spin–orbit coupling of the Cu 3d electron enhances the ISC rate from S_1 to T_1 by up to several picoseconds. This in turn increases the upper limit of the rISC rate and thus the upper limit for the radiative rate of TADF. The spin–orbit coupling also provides an additional radiative channel from the lowest triplet state ($k_{r,T} > 3 \times 10^4$ s^{-1}).

The major drawback of Cu(I) complexes is their large structural changes that occur in the excited state. This significantly lowers their emission yield and broadens their luminescence spectra because of the large Stokes shift. Some Cu(I) complexes presented herein exhibit more favorable luminescence properties, because this flattening distortion is suppressed by introducing sterically bulky groups on the ligands. Spectral broadening of Cu(I) complexes is detrimental to developing pure color emitters. However, delocalization of the excited state in complexes containing Cu$_2$N$_2$ cores can reduce the luminescence bandwidth and Stokes shift. This may provide an alternative design pathway for the structures of Cu(I) TADF emitters.

Most Cu(I) complexes exhibit low stability due to ligand dissociation or thermal decomposition. This is detrimental for fabricating OLEDs by solution processing or vacuum deposition. However, recently developed Cu(I) complexes are sufficiently stable in solution or at elevated temperature, so are potential emitting materials for low-cost OLED devices.

References

1 DeArmond, M.K. and Carlin, C.M. (1981). *Coord. Chem. Rev.* 36: 325.
2 Lees, A. (1987). *J. Chem. Rev.* 87: 711.
3 Juris, A., Balzani, V., Barigelletti, F., Campagna, S., Belser, P., and von Zelewsky, A. (1988). *Coord. Chem. Rev.* 84: 85.
4 Campagna, S., Puntoriero, F., Nastasi, F., Bergamini, G., and Balzani, V. (2007). *Top. Curr. Chem.* 280: 117.
5 (a) Baldo, M.A., Lamansky, S., Burrows, P.E., Thompson, M.E., and Forrest, S.R. (1999). *Appl. Phys. Lett.* 75: 4. (b) O'Brien, D.F., Baldo, M.A., Thompson, M.E., and Forrest, S.R. (1999). *Appl. Phys. Lett.* 74: 442.
6 Forrest, S.R. (2000). *J. Appl. Phys.* 87: 8049.
7 Adachi, C., Baldo, M.A., Thompson, M.E., and Forrest, S.R. (2001). *J. Appl. Phys.* 90: 5048.
8 Lamansky, S., Djurovich, P., Murphy, D., Abdel-Razzaq, F., Lee, H.-E., Adachi, C., Burrows, P.E., Forrest, S.R., and Thompson, M.E. (2001). *J. Am. Chem. Soc.* 123: 4304.
9 Nazeeruddin, M.K., Humphry-Baker, R., Berner, D., Rivier, S., Zuppiroli, L., and Grätzel, M. (2003). *J. Am. Chem. Soc.* 125: 8790.
10 (a) Yersin, H. (ed.) (2008). *Highly Efficient OLEDs with Phosphorescent Materials*. Weinheim: Wiley VCH. (b) Yersin, H., Rausch, A.F., Czerwieniec, R., Hofbeck, T., and Fischer, T. (2011). *Coord. Chem. Rev.* 255: 2622–2652.
11 (a) Baldo, M.A., O'Brien, D.F., You, Y., Shoustikov, A., Sibley, S., Thompson, M.E., and Forrest, S.R. (1998). *Nature* 395: 151. (b) Baldo, M.A., Thompson, M.E., and Forrest, S.R. (2000). *Nature* 403: 750.
12 Tamayo, A.B., Alleyne, B.D., Djurovich, P.I., Lamansky, S., Tsyba, I., Ho, N.N., Bau, R., and Thompson, M.E. (2003). *J. Am. Chem. Soc.* 125: 7377.
13 Namdas, E.B., Ruseckas, A., Samuel, I.D.W., Lo, S.-C., and Burn, P.L. (2004). *J. Phys. Chem. B* 108: 1570.
14 D'Andrade, B.W., Holmes, R.J., and Forrest, S.R. (2004). *Adv. Mater.* 16: 624.
15 Sun, Y.R., Giebink, N.C., Kanno, H., Ma, B.W., Thompson, M.E., and Forrest, S.R. (2006). *Nature* 440: 908.
16 Evans, R.C., Douglas, P., and Winscom, C.J. (2006). *Coord. Chem. Rev.* 250: 2093.
17 Armaroli, N., Accorsi, G., Cardinali, F., and Listorti, A. (2007). *Top. Curr. Chem.* 280: 69.
18 Wang, Y., Ding, B., Cheng, P., Liao, D.Z., and Yan, S.P. (2007). *Inorg. Chem.* 46: 2002.
19 Yam, V.W.-W. and Cheng, E.C.-C. (2007). *Top. Curr. Chem.* 281: 269.
20 Barbieri, A., Accorsi, G., and Armaroli, N. (2008). *Chem. Commun.* 19: 2185.
21 Zhang, X., Li, B., Chen, Z.H., and Chen, Z.N. (2012). *J. Mater. Chem.* 22: 11427–11441.

22 (a) Tsuge, K., Chishina, Y., Hashiguchi, H., Sasaki, Y., Kato, M., Ishizaka, S., and Kitamura, N. (2013). *Chem. Rev.* 113: 3686–3733. (b) Tsuge, K., Chishina, Y., Hashiguchi, H., Sasaki, Y., Kato, M., Ishizaka, S., and Kitamura, N. (2016). *Coord. Chem. Rev.* 306: 636–651. (c) Cariati, E., Lucenti, E., Botta, C., Giovanella, U., Marinotto, D., and Righetto, S. (2016). *Coord. Chem. Rev.* 306: 566–614.

23 (a) Hu, Z., Deibert, B.J., and Li, J. (2014). *Chem. Soc. Rev.* 43: 5815–5840. (b) Czerwieniec, R., Leitl, M.J., Homeier, H.H.H., and Yersin, H. (2016). *Coord. Chem. Rev.* 325: 2–28.

24 Deaton, J.C., Switalski, S.C., Kondakov, D.Y., Young, R.H., Pawlik, T.D., Giesen, D.J., Harkins, S.B., Miller, A.J.M., Mickenberg, S.F., and Peters, J.C. (2010). *J. Am. Chem. Soc.* 132: 9499–9508.

25 (a) Liu, Z., Qayyum, M.F., Wu, C., Whited, M.T., Djurovich, P.I., Hodgson, K.O., Hedman, B., Solomon, E.I., and Thompson, M.E. (2011). *J. Am. Chem. Soc.* 133: 3700–3703. (b) Zhang, Q., Komino, T., Huang, S., Matsunami, S., Goushi, K., and Adachi, C. (2012). *Adv. Funct. Mater.* 22: 2327–2336.

26 Zink, D.M., Volz, D., Baumann, T., Mydlak, M., Flügge, H., Friedrichs, J., Nieger, M., and Bräse, S. (2013). *Chem. Mater.* 25: 4471–4486.

27 Tao, Y., Yuan, K., Chen, T., Xu, P., Li, H., Chen, R., Zheng, C., Zhang, L., and Huang, W. (2014). *Adv. Mater.* 26: 7931–7958.

28 Wallesch, M., Volz, D., Zink, D.M., Schepers, U., Nieger, M., Baumann, T., and Bräse, S. (2014). *Chem. Eur. J.* 20: 6578–6590.

29 Xu, H., Chen, R., Sun, Q., Lai, W., Su, Q., Huang, W., and Liu, X. (2014). *Chem. Soc. Rev.* 43: 3259–3302.

30 Czerwieniec, R., Yu, J.-B., and Yersin, H. (2011). *Inorg. Chem.* 50: 8293–8301.

31 Hofbeck, T., Monkowius, U., and Yersin, H. (2015). *J. Am. Chem. Soc.* 137: 399–404.

32 (a) Smith, C.S., Branham, C.W., Marquardt, B.J., and Mann, K.R. (2010). *J. Am. Chem. Soc.* 132: 14079–14085. (b) Smith, C.S. and Mann, K.R. (2012). *J. Am. Chem. Soc.* 134: 8786–8789.

33 Liu, X., Sun, W., Zou, L., Xie, Z., Li, X., Lu, C., Wang, L., and Cheng, Y. (2012). *Dalton Trans.* 41: 1312–1319.

34 Sakaki, S., Kuroki, T., and Hamada, T. (2002). *J. Chem. Soc., Dalton Trans.* 840–842.

35 Lavie-Cambot, A., Cantuel, M., Leydet, Y., Jonusauskas, G., Bassani, D.M., and McClenaghan, N.D. (2008). *Coord. Chem. Rev.* 252: 2572–2584.

36 Bessho, T., Constable, E.C., Graetzel, M., Hernandez Redondo, A., Housecroft, C.E., Kylberg, W., Nazeeruddin, M.K., Neuburger, M., and Schaffner, S. (2008). *Chem. Commun.* 3717–3719.

37 Huang, J., Buyukcakir, O., Mara, M.W., Coskun, A., Dimitrijevic, N.M., Barin, G., Kokhan, O., Stickrath, A.B., Ruppert, R., Tiede, D.M., Stoddart, J.F., Sauvage, J.P., and Chen, L.X. (2012). *Angew. Chem., Int. Ed.* 51: 12711–12715.

38 Sandroni, M., Favereau, L., Planchat, A., Akdas-Kilig, H., Szuwarski, N., Pellegrin, Y., Blart, E., Le Bozec, H., Boujtita, M., and Odobel, F. (2014). *J. Mater. Chem. A* 2: 9944–9947.

39 Mao, Z., Chao, H.Y., Hui, Z., Che, C.M., Fu, W.F., Cheung, K.K., and Zhu, N. (2003). *Chem. Eur. J.* 9: 2885–2894.

40 Balzani, V. and Campagna, C. (ed.) (2007). *Coordination Compounds I*. Berlin: Springer.

41 Liu, B., Yu, Z.T., Yang, J., Hua, W., Liu, Y.Y., and Ma, J.F. (2011). *Inorg. Chem.* 50: 8967–8972.

42 Yuan, Y.J., Yu, Z.T., Zhang, J.Y., and Zou, Z.G. (2012). *Dalton Trans.* 41: 9594–9597.

43 Wen, T., Zhang, D.X., Liu, J., Lin, R., and Zhang, J. (2013). *Chem. Commun.* 49: 5660–5662.

44 Baralle, A., Fensterbank, L., Goddard, J.P., and Ollivier, C. (2013). *Chem. Eur. J.* 19: 10809–10813.

45 Kjnayzer, R.S., McCusker, C.E., Olaiya, B.S., and Castellano, F.N. (2013). *J. Am. Chem. Soc.* 135: 14068–14070.

46 McCusker, C.E. and Castellano, F.N. (2015). *Inorg. Chem.* 54: 6035–6042.

47 Boudin, S. (1930). *J. Chim. Phys.* 27: 285.

48 (a) Parker, C.A. and Hatchard, C.G. (1961). *Trans. Faraday Soc.* 57: 1894. (b) C. A. Parker, *Advances in Photochemistry*, Vol. 2, Ed. by W. A. Noyes, Jr., G. S. Hammond, J. N. Pitts, Jr., John Wiley and Sons, New York, 1964, 305.

49 (a) Uoyama, H., Goushi, K., Shizu, K., Nomura, H., and Adachi, C. (2012). *Nature* 492: 234. (b) Nakagawa, T., Ku, S.-Y., Wong, K.-T., and Adachi, C. (2012). *Chem. Commun.* 48: 9580.

50 Siddique, Z.A., Yamamoto, Y., Ohno, T., and Nozaki, K. (2003). *Inorg. Chem.* 42: 6366–6378.

51 Siddique, Z.A., Ohno, T., Nozaki, K., and Tsubomura, T. (2004). *Inorg. Chem.* 43: 663–673.

52 (a) Ceulemans, A. and Vanquickenborne, L.G. (1981). *J. Am. Chem. Soc.* 103: 2238. (b) Kober, E.M. and Meyer, T.J. (1984). *Inorg. Chem.* 23: 3877. (c) Miki, H., Shimada, M., Azumi, T., Brozik, J.A., and Crosby, G.A. (1993). *J. Phys. Chem.* 97: 11175.

53 Nozaki, K. (2006). *J. Chin. Chem. Soc.* 53: 101–113.

54 Obara, S., Itabashi, M., Okuda, F., Tamaki, S., Tanabe, Y., Ishii, Y., Nozaki, K., and Haga, M. (2006). *Inorg. Chem.* 45: 8907–8921.

55 (a) Turro, N.J., Ramamurthy, V., and Scaiano, J.C. (2009). *Principles of Molecular Photochemistry: An Introduction"*, Chapter 4. Sausalito, California: University Science Books. (b) Strickler, S.J. and Berg, R.A. (1962). *J. Chem. Phys.* 37: 814.

56 (a) Siebrand, W. (1970). *Chem. Phys. Lett.* 6: 192. (b) Lawetz, V., Orlandi, G., and Siebrand, W.J. (1972). *Chem. Phys.* 56: 4058.

57 (a) Buckner, M.T., Matthews, T.G., Lytle, F.E., and McMillin, D.R. (1979). *J. Am. Chem. Soc.* 101: 5846. (b) Blasse, G. and McMillin, D.R. (1980). *Chem. Phys. Lett.* 70: 1. (c) Blaskie, M.W. and McMillin, D.R. (1980). *Inorg. Chem.* 19: 3519. (d) Burke, P.J., McMillin, D.R., and Robinson, W.R. (1980). *Inorg. Chem.* 19: 1211.

58 (a) Kirchhoff, J.R., Gamache, R.E., Blaskie, M.W., Paggio, A.D., Lengel, R.K., and McMillin, D.R. (1983). *Inorg. Chem.* 22: 2380. (b) McMillin, D.R., Kirchhoff, J.R., and Goodwin, K.V. (1985). *Coord. Chem. Rev.* 64: 83.

References

59 Everly, R.M. and McMillin, D.R. (1991). *J. Phys. Chem.* 95: 9071.
60 McMillin, D.R. and McNett, K.M. (1998). *Chem. Rev.* 98: 1201.
61 (a) Cunningham, C.T., Cunningham, K.L.H., Michalec, J.F., and McMillin, D.R. (1999). *Inorg. Chem.* 38: 4388–4392. (b) Cunningham, C.T., Moore, J.J., Cunningham, K.L.H., Fanwick, P.E., and McMillin, D.R. (2000). *Inorg. Chem.* 39: 3638–3644.
62 (a) Chen, L.X., Jennings, G., Liu, T., Gosztola, D.J., Hessler, J.P., Scaltrito, D.V., and Meyer, G.J. (2002). *J. Am. Chem. Soc.* 124: 10861. (b) Chen, L.X., Shaw, G.B., Novozhilova, I., Liu, T., Jennings, G., Attenkofer, K., Meyer, G.J., and Coppens, P. (2003). *J. Am. Chem. Soc.* 125: 7022. (c) Shaw, G.B., Grant, C.D., Shirota, H., Castner, E.W., Meyer, G.J., and Chen, L.X. (2007). *J. Am. Chem. Soc.* 129: 2147. (d) Lockard, J.V., Kabehie, S., Zink, J.I., Smolentsev, G., Soldatov, A., and Chen, L.X. (2010). *J. Phys. Chem. B* 114: 14521–14527.
63 Gunaratne, T., Rodgers, M.A.J., Felder, D., Nierengarten, J.F., Accorsi, G., and Armaroli, N. (2003). *Chem. Commun.* 3010.
64 (a) Iwamura, M., Takeuchi, S., and Tahara, T. (2007). *J. Am. Chem. Soc.* 129: 5248–5256. (b) Iwamura, M., Watanabe, H., Ishii, K., Takeuchi, S., and Tahara, T. (2011). *J. Am. Chem. Soc.* 133: 7728–7736. (c) Iwamura, M., Takeuchi, S., and Tahara, T. (2014). *Phys. Chem. Chem. Phys.* 16: 4143–4154. (d) Hua, L., Iwamura, M., Takeuchi, S., and Tahara, T. (2015). *Phys. Chem. Chem. Phys.* 17: 2067–2077. (e) Iwamura, M., Takeuchi, S., and Tahara, T. (2015). *Acc. Chem. Res.* 48: 782–791.
65 Garakyaraghi, S., Danilov, E.O., McCusker, C.E., and Castellano, F.N. (2015). *J. Phys. Chem. A* 119: 3181–3193.
66 Mara, M.W., Fransted, K.A., and Chen, L.X. (2015). *Coord. Chem. Rev.* 282–283: 2–18.
67 Iwamura, M., Kobayashi, F., and Nozaki, K. (2016). *Chem. Lett.* 45: 167–169.
68 Sakaki, S., Mizutani, H., and Kase, Y. (1992). *Inorg. Chem.* 31: 4375.
69 Zgierski, M.Z. (2003). *J. Chem. Phys.* 118: 4045.
70 Robertazzi, A., Magistrato, A., de Hoog, P., Carloni, P., and Reedijk, J. (2007). *Inorg. Chem.* 46: 5873.
71 Capano, G., Chergui, M., Rothlisberger, U., Tavernelli, I., and Penfold, T.J. (2014). *J. Phys. Chem. A* 118: 9861–9869.
72 (a) Palmer, C.E.A. and McMillin, D.R. (1987). *Inorg. Chem.* 26: 3837–3840. (b) Kuang, S., Cuttell, D.G., McMillin, D.R., Fanwick, P.E., and Walton, R.A. (2002). *Inorg. Chem.* 41: 3313.
73 Kaeser, A., Mohankumar, M., Mohanraj, J., Monti, F., Holler, M., Cid, J.-J., Moudam, O., Nierengarten, I., Karmazin-Brelot, L., Duhayon, C., Delavaux-Nicot, B., Armaroli, N., and Nierengarten, J.-F. (2013). *Inorg. Chem.* 52: 12140–12151.
74 Parker, W.L. and Crosby, G.A. (1989). *J. Phys. Chem.* 93: 5692.
75 Bhasikuttan, A.C., Suzuki, M., Nakashima, S., and Okada, T. (2002). *J. Am. Chem. Soc.* 124: 8398–8405.
76 Cannizzo, A., van Mourik, F., Gawelda, W., Zgrablic, G., Bressler, C., and Chergui, M. (2006). *Angew. Chem.* 118: 3246–3248.

77 (a) Czerwieniec, R., Kowalski, K., and Yersin, H. (2013). *Dalton Trans.* 42: 9826–9830. (b) Czerwieniec, R. and Yersin, H. (2015). *Inorg. Chem.* 54: 4322–4327.

78 Felder, D., Nierengarten, J.F., Barigelletti, F., Ventura, B., and Armaroli, N. (2001). *J. Am. Chem. Soc.* 123: 6291.

79 (a) Asano, M.S., Tomiduka, K., Sekizawa, K., Yamashita, K., and Sugiura, K. (2010). *Chem. Lett.* 39: 376–378. (b) McCusker, C.E. and Castellano, F.N. (2013). *Inorg. Chem.* 52: 8114–8120.

80 Chen, X.-L., Yu, R., Zhang, Q.-K., Zhou, L.-J., Wu, X.-Y., Zhang, Q., and Lu, C.-Z. (2013). *Chem. Mater.* 25: 3910–3920.

81 Chen, X.-L., Lin, C.-S., Wu, X.-Y., Yu, R., Teng, T., Zhang, Q.-K., Zhang, Q., Yang, W.-B., and Lu, C.-Z. (2015). *J. Mater. Chem. C* 3: 1187–1195.

82 Igawa, S., Hashimoto, M., Kawata, I., Yashima, M., Hoshino, M., and Osawa, M. (2013). *J. Mater. Chem. C* 1: 542–551.

83 Osawa, M., Kawata, I., Ishii, R., Igawa, S., Hashimoto, M., and Hoshino, M. (2013). *J. Mater. Chem. C* 1: 4375–4383.

84 Tsuboyama, A., Kuge, K., Furugori, M., Okada, S., Hoshino, M., and Ueno, K. (2007). *Inorg. Chem.* 46: 1992–2001.

85 Ohara, H., Kobayashi, A., and Kato, M. (2014). *Dalton Trans.* 43: 17317–17323.

86 Zhang, Q., Chen, J., Wu, X.-Y., Chen, X.-L., Yu, R., and Lu, C.-Z. (2015). *Dalton Trans.* 44: 6706–6710.

87 Leitl, M.J., Krylova, V.A., Djurovich, P.I., Thompson, M.E., and Yersin, H. (2014). *J. Am. Chem. Soc.* 136: 16032–16038.

88 Hashimoto, M., Igawa, S., Yashima, M., Kawata, I., Hoshino, M., and Osawa, M. (2011). *J. Am. Chem. Soc.* 133: 10348–10351.

89 Osawa, M., Hoshino, M., Hashimoto, M., Kawata, I., Igawa, S., and Yashima, M. (2015). *Dalton Trans.* 44: 8369–8378.

90 Osawa, M. (2014). *Chem. Commun.* 50: 1801–1803.

91 Nitsch, J., Kleeberg, C., Froehlich, R., and Steffen, A. (2015). *Dalton Trans.* 44: 6944–6960.

92 Kang, L., Chen, J., Teng, T., Chen, X.-L., Yu, R., and Lu, C.-Z. (2015). *Dalton Trans.* 44: 11649–11659.

93 Leitl, M.J., Kuchle, F.-R., Mayer, H.A., Wesemann, L., and Yersin, H. (2013). *J. Phys. Chem. A* 117: 11823–11836.

94 Nitsch, J., Lacemon, F., Lorbach, A., Eichhorn, A., Cisnetti, F., and Steffen, A. (2016). *Chem. Commun.* 52: 2932–2935.

95 Zink, D.M., Bachle, M., Baumann, T., Nieger, M., Kuhn, M., Wang, C., Klopper, W., Monkowius, U., Hofbeck, T., Yersin, H., and Brase, S. (2013). *Inorg. Chem.* 52: 2292–2305.

96 (a) Volz, D., Zink, D.M., Bocksrocker, T., Friedrichs, J., Nieger, M., Baumann, T., Lemmer, U., and Bräse, S. (2013). *Chem. Mater.* 25: 3414–3426. (b) Volz, D., Wallesch, M., Grage, S.L., Göttlicher, J., Steininger, R., Batchelor, D., Vitova, T., Ulrich, A.S., Heske, C., Weinhardt, L., Baumann, T., and Bräse, S. (2014). *Inorg. Chem.* 53: 7837–7847.

97 Volz, D., Chen, Y., Wallesch, M., Liu, R., Flechon, C., Zink, D.M., Friedrichs, J., Flugge, H., Steininger, R., Gottlicher, J., Heske, C., Weinhardt, L., Brase, S., So, F., and Baumann, T. (2015). *Adv. Mater.* 27: 2538–2543.
98 Tsubomura, T., Ito, Y., Inoue, S., Tanaka, Y., Matsumoto, K., and Tsukuda, T. (2008). *Inorg. Chem.* 47: 481–486.
99 Kober, E.M., Caspar, J.V., Lumpkin, R.S., and Meyer, T.J. (1986). *J. Phys. Chem.* 90: 3722–3744.
100 Gothard, N.A., Mara, M.W., Huang, J., Szarko, J.M., Rolczynski, B., Lockard, J.V., and Chen, K.X. (2012). *J. Phys. Chem. A* 116: 1984–1992.
101 (a) Lotito, K.J. and Peters, J.C. (2010). *Chem. Commun.* 46: 3690–3692. (b) Krylova, V.A., Djurovich, P.I., Whited, M.T., and Thompson, M.E. (2010). *Chem. Commun.* 46: 6696–6698. (c) Barakat, K.A., Cundari, T.R., and Omary, M.A. (2003). *J. Am. Chem. Soc.* 125: 14228.
102 (a) Kyle, K.R., Palke, W.E., and Ford, P.C. (1990). *Coord. Chem. Rev.* 97: 35. (b) Ford, P.C., Cariati, E., and Bourassa, J. (1999). *Chem. Rev.* 99: 3625.

3

Luminescent Dinuclear Copper(I) Complexes with Short Intramolecular Cu–Cu Distances

Akira Tsuboyama

Materials R&D Center, R&D Headquarters, Canon, Inc., Tokyo, 146-8501, Japan

3.1 Introduction

In the past two decades, transition metal complexes have attracted much attention as luminescent dopants of OLED devices. In particular, synthesis of various cyclometalated iridium(III) complexes such as *fac*-Ir(ppy)$_3$ derivatives have been a subject of extensive studies to search phosphorescence emitters for fabrication of the efficient organic light-emitting diode (OLED) [1]. The phosphorescence emitters are able to capture both the singlet and triplet excitons generated electrochemically in the devices, leading to internal quantum efficiency of 100%. This value is much higher than the theoretical limit, 25%, of the fluorescence emitters in the devices. Actually, a number of OLED devices using the iridium complexes successfully have afforded the high internal quantum efficiency close to 100% [1c].

Organic compounds and metal complexes, which show thermally activated delayed fluorescence (TADF), have been recognized as promising alternatives of iridium complexes for fabrication of highly efficient OLED devices [2, 3a, g]. TADF materials with a small energy gap, ΔE_{S-T}, between the lowest excited triplet state, T_1, and singlet state, S_1, are known to give fluorescence emitted from S_1, which is thermally populated by a reverse intersystem crossing from T_1. Among the TADF materials, copper(I) complexes are the luminescent metal complexes investigated as TADF dopants in OLED devices.

Luminescence properties of copper(I) complexes have been systematically studied from 1980s. For example, studies on a series of the [Cu(N∧N)$_2$]$^+$ (N∧N = 1,10-phenanthroline derivatives) complexes carried out by McMillin's group have revealed that the luminescence originates from MLCT (charge transfer (CT) from Cu to the ligand) excited states [4]. They found that the weak emission intensity observed from the metal-to-ligand charge transfer (MLCT) is caused by the structural distortion from the tetrahedral to the flattening geometry in the excited states. The distortion accelerates the nonradiative processes occurring in the MLCT excited states, and, thus, the emission yield is inevitably low. With regard to the copper(I) complexes of 1,10-phenanthroline derivatives, the bulky substituents at the 2,9-positions are found to suppress strongly the flattening motion in the MLCT excited states, leading to high emission yields.

Highly Efficient OLEDs: Materials Based on Thermally Activated Delayed Fluorescence,
First Edition. Edited by Hartmut Yersin.
© 2019 Wiley-VCH Verlag GmbH & Co. KGaA. Published 2019 by Wiley-VCH Verlag GmbH & Co. KGaA.

The intensities and the peak energies of the emission spectra observed for [Cu(N∧N)₂]⁺ complexes are found to decrease by lowering the temperature. This observation is explained by assuming that the emissive ^{1}MLCT and ^{3}MLCT excited states have an energy separation, ΔE_{S-T}, small enough to achieve the thermal equilibrium between them [4b].

In the same period of time, luminescent cubane-type tetranuclear copper(I) halide complexes [Cu₄I₄py₄] (py = pyridine derivatives) have been reported by Ford's group [5a]. The complexes in the solid state exhibit visible photoluminescence emitting from two different excited states, the cluster-centered (^{3}CC) state and the halogen-to-ligand charge-transfer (^{3}XLCT) state. In the XLCT state, CT occurs from the {Cu₄X₄} core to the π^* orbital of the pyridine derivatives. The emission from the ^{3}CC state is observed when the Cu···Cu distance is shorter than the sum of van der Waals radius of the two copper(I) atoms (2.8 Å). These pioneering works mentioned above are suggestive for elucidation of the electronic nature of the copper(I) complexes in the emissive excited states.

Copper(I) ions react with a variety of chelating, bridging, and ancillary ligands to form luminescent copper(I) complexes: monomeric [3, 4], dimeric [6–12], trimeric [13], tetrameric [5, 14], and polymeric copper(I) complexes [5, 6a]. Among them, luminescent dinuclear copper(I) complexes have been extensively investigated as the TADF materials in OLED. Further, the emission studies of the dinuclear copper(I) complexes are very important to explain the intramolecular d¹⁰(Cu)–d¹⁰(Cu) interaction participating in the emissive excited states.

3.2 Overview of Luminescent Dinuclear Copper(I) Complexes

3.2.1 Structure

The luminescent dinuclear copper(I) complexes studied recently have a common dicopper(I) cyclic unit, {Cu₂(μ-L)₂} (L = bridging ligand). The rigid cyclic units of the dinuclear complexes are expected to give robustness necessary for manufacturing of OLEDs by the thermal vacuum deposition method.

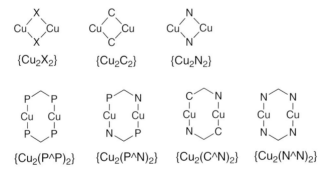

Figure 3.1 Sketch of cyclic unit geometries of dinuclear copper(I) complexes.

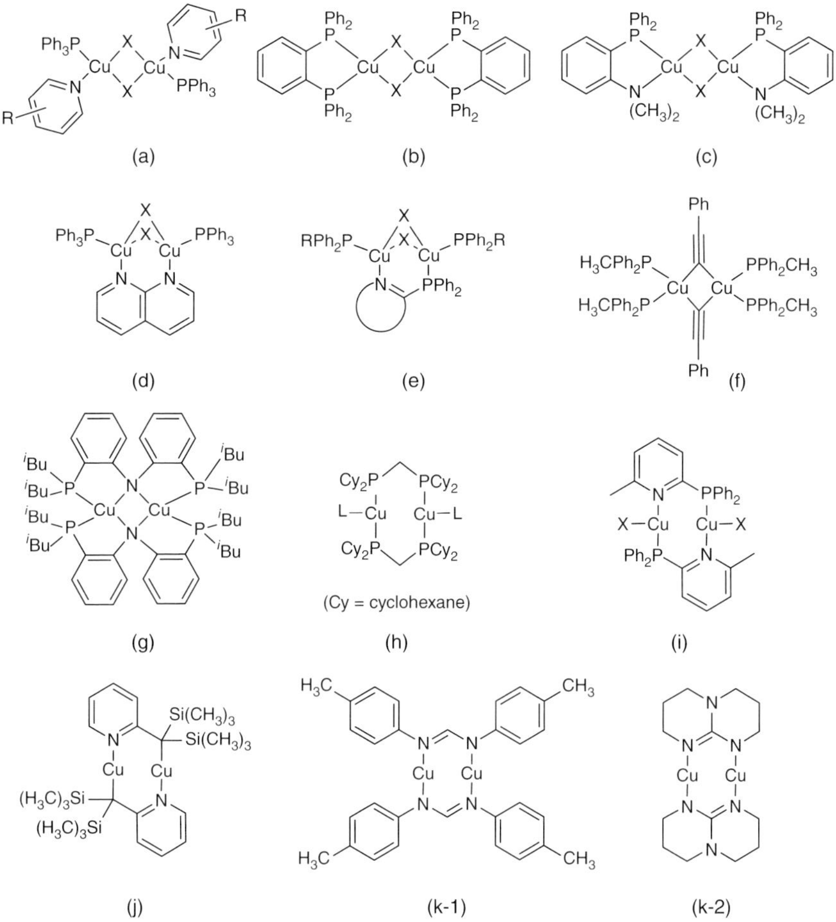

Figure 3.2 Dinuclear copper(I) complexes listed in Table 3.1.

Figure 3.1 illustrates the sketch of the cyclic units, and Figure 3.2 shows chemical formulas of the dinuclear copper complexes. In Table 3.1 are listed the coordination numbers, the cyclic unit types, and the Cu—Cu distances. Though complexes (k-1) and (k-2) [15] are not emissive, they are listed as references.

The complexes in Table 3.1 are classified into three groups according to the geometry: the four-coordinate tetrahedral geometry (a–g), the three-coordinate trigonal geometry (h, i), and the two-coordinate linear geometry (j, k).

The dinuclear copper complexes, (a–g), have the diamond-shaped four-membered cyclic units of $\{Cu_2X_2\}$ (X = halogen) [6, 7], $\{Cu_2C_2\}$ [8], and $\{Cu_2N_2\}$ [9]. The halogen-bridged dinuclear rings, $\{Cu_2X_2\}$, take both planar and butterfly geometries depending on the ancillary ligands. The butterfly geometry implies that the four-membered $\{Cu_2X_2\}$ ring is folded along the X···X axis. Since complexes (a–f) are composed of the charge-compensated units ($\{Cu_2X_2\}$ and $\{Cu_2C_2\}$) and neutral ancillary ligands (PPh$_3$, dppb, pyridine, and phosphine derivatives), these complexes are charge neutral. Practically, design of charge-neutral complexes is very important for us to fabricate OLED devices with the use of the thermal vacuum deposition method.

As listed in Table 3.1, dinuclear complexes (a) [6a, b], (b) [6c], and (c) [6d] have the {Cu$_2$X$_2$} units with the ancillary ligands of pyridine derivatives (dmap, 3-bzpy, and 4-bzpy), P∧P (o-bis(diphenylphosphino)benzene), and P∧N (Ph$_2$P-o-(C$_6$H$_4$)-N(CH$_3$)$_2$ and the derivatives), respectively. Complex (b) is thermally stable and sublimable, and, therefore, an OLED device using (b) as an emitter is fabricated by the thermal vacuum deposition method. The external quantum efficiency of the device amounts to 4.8%. With regard to the complexes (c) [6d], the geometries of the {Cu$_2$X$_2$} units vary depending on the halogen

Table 3.1 Structure and luminescence parameters of the dinuclear copper(I) complexes.

	Complex	Coord. no.	Core type	L or X	d(Cu-Cu)[a] (Å)
a	[Cu(μ-X)(PPh$_3$)(L)]$_2$	4	Planar{Cu$_2$X$_2$}	dmap 3-bzpy 4-bzpy	3.226 2.872 2.999
b	[Cu(μ-X)dppb]$_2$	4	Butterfly {Cu$_2$X$_2$}	Cl Br I	2.866 2.873 2.898
c	[Cu(μ-X)P∧N]$_2$	4	Planer/butterfly {Cu$_2$X$_2$}	Cl (pl.) Br (but.) I (but.)	2.983 2.559 2.574
d	[Cu$_2$(μ-X)$_2$-(μ-N∧N)(PPh$_3$)$_2$]	4	Butterfly {Cu$_2$X$_2$}	I Br	2.612 2.627
e-1	[Cu$_2$(μ-X)$_2$-(μ-P∧N)(P∧N)$_2$]	4	Butterfly {Cu$_2$X$_2$}	Cl, Br, I	2.720–2.829
e-2	[Cu$_2$(μ-X)$_2$-(μ-P∧N)(PPh$_3$)$_2$]	4	Butterfly {Cu$_2$X$_2$}	Cl, Br, I	2.759–2.886
f	[Cu(PPh$_2$Me)$_2$-(μ,η1-C≡CPh)]$_2$	4	Planer {Cu$_2$C$_2$}	—	2.454
g	[Cu(μ-NAr$_2$)]$_2$	4	Planer {Cu$_2$N$_2$}	—	2.728
h	[CuL(μ-P∧P)]$_2$	3	Distorted{Cu$_2$(P∧P)$_2$}	ClO$_4$ PF$_6$ I BF$_4$	2.685[e] 2.790 2.905[e] 2.691
i	[CuX(μ-P∧N)]$_2$	3	Distorted{Cu$_2$(P∧N)$_2$}	Cl Br I	3.078 2.666 2.666
j	[Cu(μ-C∧N)]$_2$	2	Planer{Cu$_2$(C∧N)$_2$}	—	2.412[h]
k-1 k-2	[Cu(μ-N∧N)]$_2$	2	Planer{Cu$_2$(N∧N)$_2$}	Form hpp	2.497 2.453

Table 3.1 (Continued)

	Luminescence					
	λ_{peak} [nm][b] RT/(77 K)	τ [μs][c] RT/(77 K)	Φ[d] RT/(77 K)	Emissive excited state	Phospho TADF	References
a	450/468 579/599 689/730	7.5/430[g] 0.75/0.90 0.16/4.0[g]	—	(M + X)-LCT	—	[6a, b]
b	533/537 520/524 502/505	4.2/2400 4.3/2300 4.0/211	0.6/— 0.6/— 0.8/0.44	(M + X)-LCT	TADF	[6c]
c	506/513 490/498 464/471	6.6/220 4.1/930 4.6/270	0.45/1.0 0.65/1.0 0.65/1.0	(M + X)-LCT	TADF	[6d]
d	670 720	0.83 0.22	0.2 <0.05	(M + X)-LCT	—	[6e]
e-1	481–713	—	0.03–0.96	—	TADF	[7b]
e-2	451–579	1.09–4.22[f]	0.36–0.99	(M + X)-LCT	—	[7e]
f	467 509/ 464 511	87/—	—	LMCT	—	[8]
g	560	10.9	0.68	CT	TADF	[9]
h	475/476 380,475/ 379,485 460/463 377,474/ 390,475	44/120 63/91[g] 8.3/11 58/89[g]	— — — 0.42	MC	Phosphorescence	[10b, c]
i	485/510 501/526 484/511	8.3/44 12.4/84 7.3/51	0.92/0.97 0.52/0.97 0.76/0.90	MLCT	TADF	[11]
j	520[i]	12[i]	0.75[i]	—	—	[8, 12]
k-1 k-2	—	—	—	—	—	[15]

a) Intramolecular Cu—Cu distance.
b) Emission peak wavelength.
c) Emission lifetime.
d) Emission quantum yield.
e) Average distance.
f) Weighted average lifetime.
g) Longer component of lifetime.
h) Reference [12a].
i) Reference [8].
j) Measured in this work.

species: planar (Cl) and butterfly geometries (Br and I). The butterfly-shaped complexes (Br and I) tend to have the Cu—Cu distance shorter than that of the planar one (Cl).

The halogen-bridged complexes (d, e1, and e2) with the additional bridging ligands (L), N∧N (1,8-naphthyridine) [6e], and P∧N (ex. 2-diphenylphosphinopyridine derivatives) [7] form the butterfly-shaped rings {Cu$_2$(μ-X$_2$)(μ-L)}. The additional bridging ligand 1,8-naphthyridine, with short bite distance [6e], brings about close contact between the two copper(I) atoms accompanying the subsequent distortion of the {Cu$_2$X$_2$} units, resulting in the short Cu—Cu distances: 2.612 Å for X = I and 2.627 Å for X = Br. A series of {Cu$_2$(μ-X$_2$)(μ-P∧N)}-type complexes has been investigated as the TADF materials, which consist of a variety of the P∧N bridging ligands and the phosphine-based ancillary ligands [7]. Since the Cu—Cu distances of these complexes are as long as 2.720–2.886 Å, the interaction between the two copper(I) atoms is suggested to be minimum. The solution-processed OLED devices using the complexes as an emitter are found to give the excellent external quantum efficiency, 23% [7a].

The acetylide–copper complex (f) bearing the {Cu$_2$C$_2$} ring [8] possesses the shortest Cu—Cu distance (2.454 Å) among the complexes with the diamond-shaped units. Complex (g) [9] synthesized with the use of the bulky monoanionic tridentate ligand (P∧N∧P$^-$) is rigid, thermally stable, charge neutral, and sublimable. The OLED device containing complex (g) as a TADF emitter displays high performance: The external quantum efficiency is as high as 16%.

As shown in Figure 3.2, several dinuclear copper(I) complexes with the eight-membered cyclic units (h–k) [10–12, 15] have been reported. The two copper(I) atoms in the complexes are bilaterally bridged with the two bridging ligands such as P∧P, P∧N, C∧N, and N∧N to form the eight-membered cyclic units. Among them, the three-coordinate complexes (h [10] and i [11]) with the monoanionic ligands, X$^-$ (halogen) or L (ClO$_4$$^-$, PF$_6$$^-$), have the nonplanar {Cu$_2$L$_2$} units because the 3-coordinate copper(I) atom prefers nonlinear geometry. In contrast, the 2-coordinate complexes (j [12] and k [15]) with two linear coordinate Cu(I) form the planar eight-membered cyclic units with the H-shaped coordination mode. These complexes consist of a pair of copper(I) atoms and monoanionic bridging ligands, C∧N$^-$ and N∧N$^-$, to afford charge-neutral species without counteranions.

The dinuclear complexes listed in Table 3.1 have the Cu—Cu distances in the range of 2.41–3.23 Å. The copper complexes (f) (2.454 Å), (j) (2.412 Å), and (k) (2.497 Å (form) and 2.453 Å (hpp)) possess the Cu—Cu distances shorter than others, suggesting the presence of the d^{10}(Cu)–d^{10}(Cu) interaction between the two copper(I) atoms. These results indicate that the dinuclear 2-coordinate copper complexes with the eight-membered planar unit structure tend to afford short Cu—Cu distances.

MO calculations carried out for the dinuclear Cu(I) complexes in the ground state indicate that the bond order of the Cu(I)–Cu(I) is very close to zero and, thus, the d^{10}–d^{10} interaction hardly stabilizes the Cu(I)—Cu(I) bond [15a]. However, the Cu—Cu bond is considered to be energetically stabilized in the MLCT excited states of the dinuclear Cu(I) complexes by the charge resonance interaction, Cu(I)–Cu(II) ↔ Cu(II)–Cu(I). Thus, the Cu—Cu distance is suggested to give significant effects on the emission properties of the dinuclear Cu(I) complexes.

3.2.2 Luminescence Properties

In Table 3.1 are listed the peak wavelengths (λ_{peak}), lifetimes (τ), and quantum yields (Φ) of emission measured for the dinuclear copper(I) complexes at room temperature and/or 77 K in the solid state. The abundant and reliable data of luminescence for the complexes in the solid state are very useful for an estimation of the device performance prior to manufacturing the OLED.

At room temperature, the emission peak wavelengths, λ_{peak}, from the complexes cover the wide range of visible light, 450–720 nm; the lifetimes, τ, are ranged from 0.16 to 87 μs; and the maximum Φ values are as high as 0.99. On going from room temperature to 77 K, emission peak wavelengths, λ_{peak}, are red-shifted by 1–33 nm, and lifetimes, τ, become longer by one to three orders of magnitude. This strong temperature dependence observed for λ_{peak} and τ suggests that the emission at ambient temperature is ascribed to TADF. Probably, the higher and lower emissive states, S_1 and T_1, are in the thermal equilibrium.

Most of the emissive excited states of the complexes in Table 3.1 have the CT character, such as (M + X)LCT, LMCT, and MLCT. Usually, the emission spectra from the CT states are broad and featureless in shape. MO calculations have been used for assignment of the emissive excited states of the complexes.

The emissions of the {Cu_2X_2} complexes, (a)–(e), originate from the (M + X)LCT excited state, which are formed via electronic transition from the {Cu_2X_2}-based HOMO to the π^*-natured LUMO located at the ancillary ligands with aromatic systems [6, 7]. On the other hand, the emissive excited state of complex (f) has been assigned to the LMCT state ($Cu_2 \leftarrow Ph-C\equiv C$) modified by the metal–metal interactions [8].

The photophysical studies on a series of the {Cu_2X_2} complexes (a [6a, b], e-1 [7b], and e-2 [7e]) reveal that the emission color of the complexes is systematically changed by changing the reduction potential of the ancillary ligand.

Complex (b), [Cu(μ-X)dppb]$_2$, indicates that the λ_{peak} values slightly vary with halogen species (X = Cl, Br and I), probably suggesting that the ligand field strength of the halogen species gives effects on the emission peak wavelengths.

The CT character in the excited states gives a small value of the exchange integral, leading to a small energy gap between S_1 and T_1, which is essential for TADF emitters. From measurements of the lifetimes, τ, at various temperatures, the emissions from complexes (b, c, e, g, and i) at ambient temperature have been attributed to the TADF. In contrast, because of lack of a low-lying unoccupied π^*-orbital as an electron acceptor, the emission from complexes (h) is considered to arise from the metal-centered (MC) excited state, $^3[4p\sigma, 3d\sigma^*]$ [10].

The complex (j) has the shortest Cu–Cu distance (2.412 Å) in the ground state among the complexes listed in Table 3.1. On the other hand, complexes (d) [6e] possess the moderately short Cu–Cu distances, 2.612 and 2.627 Å. Because of the fact that the complex (j) gives the emission intensity stronger than (d), it is suggested that the emission properties of these complexes are related to the

Cu—Cu distances. However, there is an argument over whether the Cu—Cu distance affects the emission from the dinuclear complexes [6e].

Since we have been interested in the effects of the short Cu—Cu distance on luminescence from the dicopper(I) complexes, we have studied the complexes, [Cu(μ-C∧N)]$_2$, as described in the next section.

3.3 Structural and Photophysical Studies of the Dinuclear Copper(I) Complexes: [Cu(μ-C∧N)]$_2$ (C∧N = 2-(bis(trimethylsilyl)methyl)pyridine Derivatives)

3.3.1 Outline

This section describes our recent studies of structures and photophysical properties for a series of luminescent dinuclear copper(I) complexes with short Cu—Cu distances. Figure 3.3 illustrates the molecular frameworks of the copper(I) complexes studied here. The complexes have a general structure, [Cu(μ-C∧N)]$_2$ [**1–4**], in which C∧N is a monoanionic bridging ligand: 2-(bis(trimethylsilyl)methyl)pyridine (1 L), 2-(bis(trimethylsilyl)methyl)-6-methylpyridine (2 L), 2-(bis(trimethylsilyl)methyl)-6-fluoropyridine (3 L), and 2-(bis(trimethylsilyl)methyl)quinoline (4 L). The complex, [Cu(μ-1 L)]$_2$ **1**, was first synthesized by Papasergio et al. in 1983 [12a]. Crystallographic analysis revealed that the intramolecular Cu—Cu distances (2.412–2.440 Å) observed in these complexes are shorter than 2.8 Å, indicating a substantial interaction between the two copper atoms. These complexes in the solid state exhibit intense green and orange photoluminescence at room temperature: $\lambda_{peak} = 500$–576 nm, $\Phi = 0.32$–0.72, and $\tau = 3.4$–$13.2\,\mu s$. From theoretical studies, the lowest excited state formed via the electronic transition, LUMO ($\pi^*_{ligand} + 4p\sigma_{Cu-Cu}$) ← HOMO ($3d\sigma^*_{Cu-Cu}$), is assigned to the metal–metal-to-ligand charge-transfer (MMLCT) state mixed with the metal–metal-centered (MMC) state. Kinetic studies revealed that the emission of these complexes at ambient temperature originates from TADF:

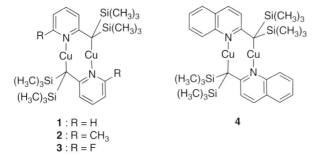

1 : R = H
2 : R = CH$_3$
3 : R = F

4

Figure 3.3 Copper(I) complexes studied in this work.

The emissive excited states, ^{1}[MMLCT+MMC] and ^{3}[MMLCT+MMC], are in equilibrium with the energy difference of 450 cm^{-1}. A multilayered OLED device that uses [Cu(μ-1 L)]$_2$ **1** as a green TADF dopant affords external quantum efficiency = 7.2% and luminous efficiency = 26.3 cd A^{-1} at 90 cd cm^{-2}.

3.3.2 X-ray Crystallographic Study

Major structural parameters of complexes **1–4** are summarized in Table 3.2. ORTEP plots of **3** and **4** are shown in Figure 3.4. These complexes exist as dimers with two linear coordinate copper sites bridged with the two C∧N bidentate ligands to form elongated hexagonal eight-membered {Cu$_2$(C∧N)$_2$} cyclic units. The {Cu$_2$(C∧N)$_2$} units are essentially planar. For example, the torsion angles of Cu N C C$_{Cu}$ are 2.0° and 2.2° for **3** and **4**, respectively. The copper(I) atoms adopt almost linear geometry for **1–4** with the C−Cu−N angles of 177.9–178.6°.

It should be noted that the intramolecular Cu−Cu distances (2.412–2.448 Å) observed in these complexes are significantly shorter than 2.8 Å, suggesting the presence of a substantial interaction between the two copper atoms. On the other hand, the minimum intermolecular Cu⋯Cu distances in the crystals are as long as 7.0 and 8.8 Å for **3** and **4**, respectively, probably indicating no interaction between the Cu(I) atoms of the neighboring molecules.

Table 3.2 Crystallographic data of **1–4**.

Parameter	**1**[a]	**2**[b]	**3**	**4**
Space group	—	—	P2$_1$/n (#14)	P2$_1$/c (#14)
Z value	—	—	2	2
T (K)	—	—	296 ± 1	296 ± 1
a (Å)	—	—	8.63 (2)	8.81 (2)
b (Å)	—	—	15.94 (2)	17.73 (2)
c (Å)	—	—	12.15 (2)	11.74 (2)
b (°)	—	—	108.58 (6)	101.40 (6)
V (Å^3)	—	—	1585 (4)	1797 (5)
R1, wR2 (I > 2.0 s(I))	—	—	0.0307	0.0692
R1, wR2 (all data)	—	—	0.089	0.112
Goodness of fit indicator	—	—	0.947	0.919
Cu−C (Å)	1.950 (4)	1.965 (3)	1.956 (3)	1.939 (5)
Cu−N (Å)	1.910 (3)	1.936 (3)	1.928 (3)	1.912 (4)
Cu⋯Cu (Å)	2.412 (1)	2.436 (1)	2.448 (3)	2.440 (2)
C−Cu−N (°)	178.0 (5)	177.9 (2)	178.64 (7)	178.21 (18)
Cu−N−C−C$_{Cu}$ (°)	—	—	−2.0 (3)	−2.2 (7)

a) Reference [12b].
b) Reference [12c].

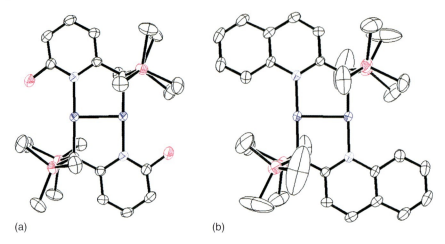

Figure 3.4 ORTEP diagram of the complex **3** (a) and **4** (b). Hydrogens are omitted for clarification. Ellipsoids denote 30% probability.

3.3.3 Photophysical Properties

3.3.3.1 Absorption Spectrum

Figure 3.5 shows UV–Vis absorption spectra of **1**–**4**. These complexes have intense absorption bands with the molar absorption coefficient $\varepsilon > 10^5$ M^{-1} cm^{-1} at 300–320 nm for **1**–**3** and 300–350 nm for **4**. These absorption bands are assigned to the spin-allowed π–π^* transition of the pyridine and quinoline rings in the ligands. Since the conjugated system of quinoline is larger than that of pyridine, the absorption band of **4** shows the bathochromic shift by 30 nm in comparison with those of **1**–**3**.

The weak absorption bands with $\varepsilon < 7000$ M^{-1} cm^{-1} at 350–470 nm for **1**–**3** and at 400–550 nm for **4** can be attributed to the electronic transition affected by the

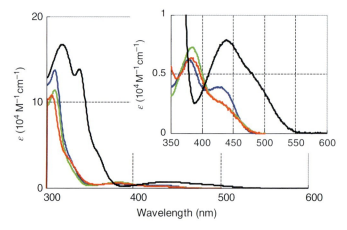

Figure 3.5 Absorption spectra of the copper(I) complexes in toluene: **1** (red), **2** (blue), **3** (green), and **4** (black). The inset shows a magnified view of the absorption edge.

two copper atoms. In order to investigate the lowest energy transition state, DFT calculation of **1** is carried out in the next section.

3.3.3.2 DFT Calculation

As shown in Figure 3.6, the DFT calculation [16] for complex **1** reveals that the electron density in the HOMO is mainly located at the two copper atoms, while that in the LUMO is distributed over both the copper atoms and the pyridine rings. Therefore, the calculation predicts that the lowest excited state formed via the LUMO ← HOMO transition is a mixed state of (i) the MMLCT state (CT state from the Cu—Cu center to the pyridine ring) and (ii) the MMC state (Cu—Cu-centered state).

Figure 3.7 displays a schematic molecular orbital diagram illustrating interaction between two copper(I) atoms. Because the Cu—Cu distances are definitely shorter than 2.8 Å, the occupied $3d_{z^2}$ orbitals of the two Cu(I) are able to interact each other, leading to the formation of the bonding and antibonding orbitals, $3d\sigma$ and $3d\sigma^*$, respectively. As a result, the HOMO corresponds to the Cu—Cu-centered antibonding $3d\sigma^*$ orbital. Similarly, with regard to the vacant orbitals, the interaction between the 4p orbitals of the two copper atoms leads to the formation of the bonding ($4p\sigma$) and antibonding orbitals ($4p\sigma^*$) [10]. Since the π^* orbital (pyridine) is close in energy to the 4p orbital (Cu(I)), the vacant LUMO is considered as the mixed molecular orbital of $4p\sigma$ and π^* orbitals.

On the basis of these consideration mentioned above, the HOMO–LUMO transition is described as $[4p\sigma + \pi^*] \leftarrow [3d\sigma^*]$, and, thus, the lowest excited state is ascribed to the metal–metal-to-ligand charge-transfer (MMLCT: $\pi^* \leftarrow 3d\sigma^*$) state mixed with the metal–metal-centered (MMC: $4p\sigma \leftarrow 3d\sigma^*$) state. From the

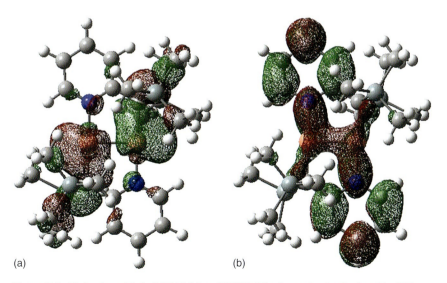

(a)   (b)

Figure 3.6 Molecular orbitals (HOMO (a) and LUMO (b)) of complex **1** calculated by DFT method using package program GAUSSIAN09. Source: Ref. [16]. Reproduced with permission of Royal Society of Chemistry.

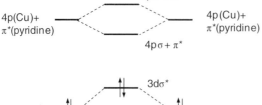
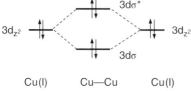

Figure 3.7 Molecular orbital diagram illustrating interaction between two copper(I) atoms.

DFT calculation, the contribution ratios of the MMLCT and the MMC states to the lowest excited state are evaluated as 42% and 5%, respectively.

3.3.3.3 Emission Properties

Emission spectra, quantum yields, and lifetimes of **1**–**4** were measured using solid crystalline samples. From the X-ray crystallographic studies, the intermolecular distances between the eight-membered units, {$Cu_2(C\wedge N)_2$}, are found to be longer than 7 Å, suggesting that the intermolecular interaction between the units is negligibly small in the solid crystalline. In fact, neither the excimer formation nor the triplet–triplet annihilation is observed by measurements of emission spectra and the lifetimes.

Figure 3.8 shows the emission spectra of **1**–**4** in the solid state at 298 K. The spectra are broad without showing any vibrational progressions, suggesting that, in agreement with the DFT calculations, the emissive excited states have CT character. The emission peak wavelengths are found to shift by changing the substituents of the ligands.

Although the substituent of the organic ligand is assumed to give the minimum effect on the metal orbitals, the π^* state of the ligand is affected by the electronic nature of the substituent, resulting in the peak shift of the emission

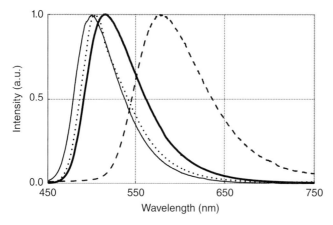

Figure 3.8 Photoluminescence spectra of the copper complexes in the solid state at 298 K: **1** (bold solid), **2** (solid), **3** (dotted), and **4** (broken).

Table 3.3 Photophysical parameters of **1–4** in the solid state.

Complex	λ_{peak} (nm)[a]		τ (µs)[b]		Φ[c]
	298 K	77 K	298 K	77 K	298 K
1	516	540	13.2	647	0.75
2	500	526	8.1	695	0.58
3	502	535	8.0	611	0.52
4	576	581	3.4	100	0.32

a) Emission peak wavelength.
b) Emission lifetime (experimental errors are ±5%).
c) Emission quantum yield (experimental errors are ±5%).

from the MLCT excited state. As shown in Figure 3.8, the emission peak wavelengths of both **2** having CH_3 and **3** having F as substituents are found to be slightly blueshifted by ca. 10 nm in comparison with that of the nonsubstituted complex **1**. The blueshift of the emission peak observed for **2** is ascribed to the electron-donating effect of the CH_3 group on the π^* state of the ligand. However, the blueshift of the emission peak observed for **3** is unable to be explained only by the electron-drawing effect of the F atom on the π^* state. Presumably, the emissive excited state, MMLCT+MMC, was influenced by both the geometry around the Cu−Cu center and the electronic nature of the substituents. Complex **4** has the emission peak wavelength (576 nm) significantly longer than that of complex **1** (516 nm) because of lowering of LUMO in energy by the effects of the strong electron-acceptor quinoline.

Emission peak wavelengths (λ_{peak}), emission lifetimes (τ), and emission quantum yields (Φ) of the complexes in the solid state at 298 and 77 K are summarized in Table 3.3. At 298 K, the complexes exhibit green to orange luminescence (λ_{peak} = 500–576 nm) with microsecond lifetimes (τ = 3.4–13.2 µs) and high emission quantum yields (Φ = 0.32–0.75). The emission lifetimes, τ, at 298 K are significantly shorter than those at 77 K, and the emission peak wavelengths are redshifted by 5–24 nm with a decrease in temperature from 298 to 77 K.

3.3.3.4 Emission Decay Kinetic Analysis

Figure 3.9 shows emission decay curves of the copper(I) complexes **1–4** in the solid state at 298 and 77 K, obtained with 355-nm laser pulses. The time-resolved emission spectroscopy reveals that the spectra detected immediately after the laser pulse uniformly decay in the whole wavelength region observed. The emission intensity, I, is found to decay according to first-order kinetics in the whole temperature range, 77–298 K, studied:

$$I = A \exp(-t/\tau) \tag{3.1}$$

where τ and A are the lifetime and preexponential factor, respectively.

As shown in Figure 3.10, the emission quantum yield, Φ, and the emission lifetime, τ, are measured as a function of temperature, T, obtained for complex **1**. While the yield, Φ, gradually decreases on going from 298 to 77 K, the lifetime, τ, is found to increase dramatically.

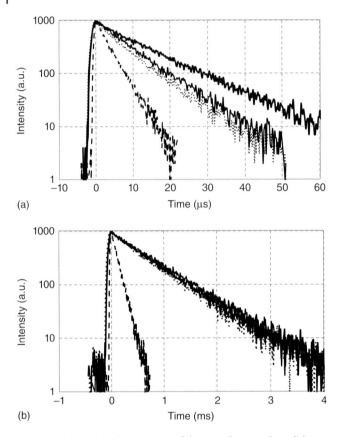

Figure 3.9 Emission decay curves of the complexes in the solid state at 298 K (a) and 77 K (b) in the solid state: **1** (bold solid), **2** (solid), **3** (dotted), and **4** (broken).

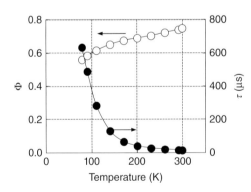

Figure 3.10 Temperature dependence of emission quantum yield (Φ) and lifetime (τ) of **1** in the solid state.

The usual expression of the emission lifetime, τ, and the emission yield, Φ, are given by

$$\tau^{-1} = k_r + k_{nr} \tag{3.2}$$

$$\Phi = k_r/(k_r + k_{nr}) = k_r \tau \tag{3.3}$$

Here, k_r and k_{nr} denote the radiative and nonradiative rate constant, respectively. Both k_r and k_{nr} are calculated from Eqs. (3.2) and (3.3). When the emissive excited state is identical in the temperature range studied, the k_r value is inevitably independent of temperature. However, the k_r value obtained at 298 K is ca. 50 times larger than that at 77 K. We, thus, concluded that the emissive excited state at 298 K differs from that at 77 K. These results suggest that the emission occurs from the two emissive excited states.

As described in earlier papers [6c], the two emissive states are assumed to be in the thermal equilibrium: a high-energy state with the decay rate constant k_H and a low-energy one with k_L. With the use of the equilibrium constant K between the two states, the lifetime, τ, is expressed as

$$\tau^{-1} = (k_L + k_H K)/(1 + K) \tag{3.4}$$

$$K = A \exp(-\Delta H/RT) \tag{3.5}$$

Here, ΔH is the enthalpy change between the low- and high-energy states, and A is the preexponential factor depending on both the spin multiplicity and the conformations of the two states (discussed later in this section).

According to Eq. (3.4), the radiative rate constant of complex **1**, $k_r (= \tau_r^{-1})$, is represented as

$$\tau_r^{-1} = k_r = (k_{rL} + k_{rH} K)/(1 + K) \tag{3.6}$$

where k_{rL} and k_{rH} are the radiative constants of the low- and high-energy states, respectively. The emission yield, Φ, is defined as

$$\Phi = k_r \tau \tag{3.7}$$

Figure 3.11a (filled circles) displays the plot of k_r versus T obtained for complex **1**. With the use of a weighted linear least-square fitting method, the parameters k_{rL}, k_{rH}, ΔH, and factor A in Eq. (3.6) are determined from the plot. The solid line in Figure 3.11a with the filled circles is the calculated one with $k_{rL} = 8.4 \times 10^2$ s^{-1}, $k_{rH} = 1.0 \times 10^7$ s^{-1}, $\Delta H = 450$ cm^{-1}, and $A = 0.048$ (see Table 3.4).

The temperature dependence of the emission yield is explained with the use of radiative and nonradiative rate constants, which are expressed as a function of temperature. The nonradiative rate constant, $k_{nr} = \tau_{nr}^{-1}$, is formulated as

$$\tau_{nr}^{-1} = k_{nr} = (k_{nrL} + k_{nrH} K)/(1 + K) \tag{3.8}$$

where k_{nrL} and k_{nrH} are the nonradiative rate constants of the low- and high-energy states, respectively. The k_{nr} values are readily calculated by subtracting k_r from τ^{-1}. Figure 3.11a (unfilled circles) shows the plot of k_{nr} versus T obtained with complex **1**. The nonradiative rate constant k_{nr} is generally composed of the temperature-dependent and temperature-independent terms. Thus, the k_{nrL} and k_{nrH} are written as

$$k_{nrL} = k_{nrL}(0) + k'_{nrL} \exp(-\Delta E_L/RT) \tag{3.9}$$

$$k_{nrH} = k_{nrH}(0) + k'_{nrH} \exp(-\Delta E_H/RT) \tag{3.10}$$

We have carried out the curve fitting of the plot k_{nr} vs. T with the use of Eqs. (3.8)–(3.10) and the linear least-squares method. Table 3.4 lists the parameters

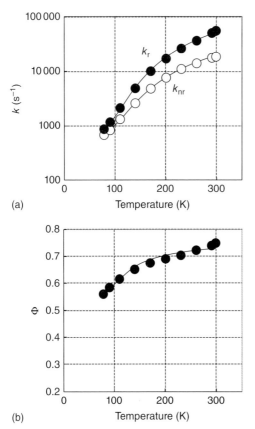

Figure 3.11 Radiative (k_r) and nonradiative rate constants (k_{nr}) (a) and emission quantum yield (Φ) (b) of **1** in the solid state, represented as a function of T: The open and closed circles are the data observed at the temperature, T, and the solid lines are the calculated ones with the use of Eqs. (3.4)–(3.10) and parameters listed in Table 3.4.

Table 3.4 Kinetic parameters for complex **1** in the solid state.

ΔH (cm^{-1})	450
k_{rL} (s^{-1})	8.4×10^2
k_{rH} (s^{-1})	1.0×10^7
A	0.048
ΔE_L (cm^{-1})	0
ΔE_H (cm^{-1})	420
k'_{nrL} (s^{-1})	0
$k_{nrL}(0)$ (s^{-1})	7.4×10^3
k'_{nrH} (s^{-1})	1.4×10^5
$k_{nrH}(0)$ (s^{-1})	6.6×10^2

$k_{nrL}(0)$, k'_{nrL}, ΔE_L, $k_{nrH}(0)$, k'_{nrH}, and ΔE_H obtained for complex **1**. These values are found to reproduce well the plot of k_{nr} vs. T (see Figure 3.11a).

As mentioned above, all the kinetic parameters necessary to elucidate the temperature dependence of Φ are determined. As shown in b of Figure 3.11, the plots of Φ vs. T, obtained experimentally, fit well with the calculated curve with the use of Eqs. (3.4)–(3.10) and the parameters listed in Table 3.4.

Temperature-dependent Φ and τ are explained by assuming that the two electronic states, high- and low-energy states, are in the thermal equilibrium. The radiative rate constant, k_{rH}, for the high-energy state is determined as $k_{rH} = 1.0 \times 10^7$ s^{-1}, which is four orders of magnitude larger than k_{rL} for the low-energy state ($k_{rL} = 8.4 \times 10^2$ s^{-1}). From this result, the high- and low-energy states are assigned to the spin-allowed ^{1}[MMLCT+MMC] state and the spin-forbidden ^{3}[MMLCT+MMC] state, respectively.

The energy separation, ΔH, obtained from K is the energy difference between the two states. We found that ΔH (880 cm^{-1}) evaluated from difference in the peak energies of the emission spectra observed at 77 and 298 K is in moderate agreement with the ΔH value (450 cm^{-1}) obtained with kinetic analysis.

As described above, the high- and low-energy states have been ascribed to S_1 and T_1, respectively [3b, 6d]. When the population ratio between S_1 and T_1 simply follows the Boltzmann distribution, the preexponential factor A in Eq. (3.5) should be 1/3 on the assumption that the molecular conformations of the two states are identical in the solid state. However, we obtained $A = 0.048$. This result suggests that the factor A is governed by not only the spin multiplicity but also the conformations of the two states. A is usually expressed as

$$A = 1/3 \, \exp(\Delta S/R). \tag{3.11}$$

Here ΔS is the difference in the entropy term between the singlet and the triplet state. From Eq. (3.11) and $A = 0.048$, we obtain $\Delta S/R = -1.94$, suggesting that the triplet state has freedom little more than the singlet state even in the solid state.

Figure 3.12 illustrates the schematic energy-state diagram of complex **1**. In summary, luminescence of complex **1** at ambient temperature is concluded to be TADF.

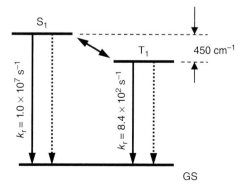

Figure 3.12 Schematic energy-state diagram of complex **1**.

3.3.4 OLED Device

Complex **1**, which is thermally stable and sublimable, is suitable for fabrication of OLED device with a thermal vacuum deposition method. Figure 3.13a shows the multilayered structure of the OLED device using complex **1** as an emissive dopant. The layers in the device are composed of a hole-transporting layer (HTL), an emissive layer (EML), an electron-transporting layer (ETL), and an electron-injection layer (EIL). The materials used for the device are PF01 (4,4′-bis[phenyl(9,9′-dimethylfluorenyl)amino]biphenyl) for HTL (40 nm), **1** (11 wt%) in CBP (4,4′-N,N'-dicarbazolebiphenyl) for EML (20 nm), Bphen (4,7-biphenyl-1,10-phenanthroline) for ETL (50 nm), and KF for EIL (1 nm). The concentration of **1** in the EML is 11 wt%, which affords the optimum external efficiency η_{ex}. At concentrations of higher than 11 wt%, η_{ex} tends to decrease.

Figure 3.13b displays the plots of luminance and luminous efficiency vs. current density. The maximum luminance is obtained as 8190 cd m^{-2} at 159 mA cm^{-2}. Luminous efficiency, power efficiency, and external quantum

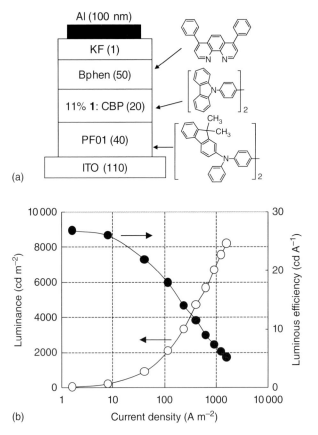

Figure 3.13 The OLED device construction and molecular formulas of the compounds used in each layer (a) and plots of luminance and luminous efficiency *versus* current density of the device (b).

efficiency are 25.9 cd A^{-1}, 18.7 lm W^{-1}, and 7.7%, respectively, at 210 cd m^{-2}. The electroluminescence (EL) spectrum has a maximum intensity at 529 nm. In comparison with the photoluminescence spectrum, which has the peak intensity at 516 nm in the solid state, the EL spectrum is redshifted, probably because of facts that (i) the EL spectrum suffers microcavity effects depending on the layer thickness and/or (ii) the emission spectra in soft matrices usually show redshift in comparison with those in solid powder materials. The OLED device gives the external quantum efficiency, 7.2%. This value is much higher than the theoretical limitation of the fluorescence device, 5%. It is clear that the dinuclear complexes studied here are useful as TADF emitters in OLED devices.

3.3.5 Experimental

3.3.5.1 Synthesis

2-fluoro-6-methylpyridine, quinaldine, trimethylsilyl chloride, copper(I) chloride, n-butyllithium (1.6 M hexane solution), and all solvents were commercially available (Aldrich) and were used as received. The ligands 3 L and 4 L were prepared according to the conventional manner from the previous paper [12]. The dinuclear complexes **3** and **4** were synthesized in the following general procedure [12]. All procedures were carried out under nitrogen flow. The ligand (4 mmol) was dissolved in THF (20 ml). To the solution, n-butyllithium 1.6 M hexane solution (2.5 ml, 4 mmol) was added dropwise at −30 °C, and the mixture was stirred for 30 min at room temperature. CuCl (4 mmol, 396 mg) was added to the solution at −30 °C and stirred at room temperature for 2 h. After THF was removed in vacuo, hexane (30 ml) was added to the reaction mixture. The reaction mixture was filtered, and the solvent was removed in vacuo to give pale color powder. The powder was purified by train sublimation with argon stream under the pressure of 10^{-3} Torr at 160–180 °C to obtain yellow and orange block crystals of **3** and **4**, respectively.

$Cu_2(3L)_2$ (**3**): yellow crystals (yield after train sublimation: 611 mg, 24%): ^{1}H NMR (CDCl$_3$, 298 K) δ (ppm) 7.57 (dd, 2H, $J = 8.4$ Hz), 7.12 (d, 2H, $J = 6.0$ Hz), 6.60 (d, 2H, $J = 8.4$ Hz), 0.16 (s, 36H). Anal. Calcd for $C_{24}H_{42}Cu_2F_2N_2Si_4$: C, 45.32; H, 6.66; N, 4.40. Found: C, 45.12; H, 6.81; N, 4.42.

$Cu_2(4L)_2$ (**4**): orange crystals (yield after train sublimation: 420 mg, 15%): ^{1}H NMR(CDCl$_3$, 298 K) δ (ppm) 8.62 (d, 2H, $J = 7.8$ Hz), 7.91 (d, 2H, $J = 8.7$ Hz), 7.75 (d, 2H, $J = 7.8$ Hz), 7.65 (t, 2H, $J = 7.6$ Hz), 7.50 (d, 2H, $J = 8.7$ Hz), 7.45 (t, 2H, $J = 7.6$ Hz), 0.22(s, 36H). Anal. Calcd for $C_{32}H_{48}Cu_2N_2Si_4$: C, 54.89; H, 6.91; N, 4.00. Found: C, 54.98; H, 6.68; N, 4.12.

3.3.5.2 Measurement, Calculation, and Device

Measurement: Elemental analyses were carried out with an elemental analyzer Vario EL CHNOS from Elementar Co. Photoluminescence spectra were recorded on a Hitachi F4500 fluorescence spectrometer. Spectral data were corrected with the use of a commercial standard light provided by Hitachi Co. Solid-state emission quantum yields (Φ) were determined with a Hamamatsu Photonics C9920-02 absolute PL quantum yield measurement system equipped with an integrated sphere. Emission lifetimes were measured by a Hamamatsu Photonics

C4334 streakscope with excitation light ($\lambda = 354.7$ nm) from an Nd:YAG laser (Surelite-II from Continuum Co.). UV–visible absorption spectra were recorded on a Shimadzu UV3100S spectrophotometer. ^{1}H-NMR (500 MHz) spectra were recorded on a Bruker Avance 500 NMR spectrometer.

3.3.5.3 X-ray Structure Analysis

Cubic crystals of **3** and **4** suitable for X-ray analysis were obtained with the train sublimation method in argon stream. Diffraction data were collected at room temperature on a Rigaku RAXIS-RAPID imaging plate diffractometer equipped with graphite monochromated Mo-Kα radiation ($\lambda = 0.71069$ Å). The crystal-to-detector distance was 127.40 mm. Readout was performed in the 0.100 mm pixel mode. The data were collected to a maximum 2θ value of 55.0°. A total of 44 oscillation images were collected. The exposure rates were 60.0 sec for **3** and 120.0 sec for **4** per degree. The crystal structures were solved by using direct methods (SIR92) for **3** and Patterson methods (DIRDIF99 PATTY) for **4**. The crystal structures were refined by the full-matrix least-squares method on F^2. All nonhydrogen atoms were refined anisotropically, while hydrogen atoms were included but not refined. All analyses were performed by the crystallographic software package CrystalStructure 4.0.

3.3.5.4 DFT Calculation

Density functional calculations using the Gaussian09 quantum chemical program [16] have been carried out with the B3LYP/6-31G*.

3.3.5.5 OLED Device

OLED devices were fabricated by the conventional vacuum deposition method. The devices were made on an indium tin oxide (ITO) film (15 Ω cm^{-2}, thickness 120 nm, from Nippon Sheet Glass Co.) with a 3.14 mm^2 round-patterned area. The organic materials for the EL devices were vacuum deposited in turn on the ITO film at chamber pressures of less than 10^{-4} Pa, and aluminum was deposited over a KF layer as a cathode. The emissive layer was formed by codeposition of the emissive dopant, **1**, and the host molecule, 4,4'-N,N'-dicarbazolebiphenyl (CBP).

3.4 Conclusion

The copper(I) complexes with the dinuclear cyclic units are the promising materials as the robust and highly efficient TADF emitters in OLEDs. The intense TADFs observed at ambient temperature originate from emissive CT excited states, which are formed via electronic transition from the d orbital localized on the dicopper(I) units to the low-lying π^* orbital of the ligands. The CT character in the emissive states leads to the small exchange integral of electrons, resulting in the small energy gap ΔE_{S-T} necessary for the complexes to emit TADF.

Dinuclear copper complexes with a planar eight-membered cyclic ring can be synthesized with the use of the bidentate bridging ligands. Some of the complexes have the Cu–Cu distance significantly shorter than the sum of the van der Waals radius of two Cu atoms. The short Cu–Cu distance results in the overlapping of the 3d–3d and 4p–4p orbitals; the former gives rise to the formation of 3dσ and 3dσ^* orbitals, and the latter, 4pσ and 4pσ^*, respectively. Since the electronic transition responsible for the emission is (4p$\sigma + \pi^*$) ← 3dσ^*, the emissive excited states of the dicopper complexes are concluded to possess the MC, 4pσ ← 3dσ^*, mixed with the MLCT character, π^* ← 3dσ^*. When the distance between the two Cu(I) atoms is reduced, the orbital energy of the 4pσ decreases, and, thus, the contribution of the 4pσ orbital to the LUMO (4p$\sigma + \pi^*$) becomes large. Accordingly, the shorter distance between the two copper atoms results in the larger contribution of the MC excited state to the emissive excited state. The d^{10}–d^{10} interaction presumably plays an important role in the luminescence properties especially in dicopper complexes with the short Cu–Cu distance. Overviewing the structures of the complexes presented in this section, the planar eight-membered cyclic unit seems to be beneficial for the dicopper complexes to form the short Cu–Cu distances.

We have studied structures and photophysics of a series of the luminescent dinuclear copper(I) complexes [Cu(μ-C∧N)]$_2$ (**1–4**) with short Cu–Cu distances to elucidate the origin of emission and also fabricated an OLED device using complex **1** as a TADF emitter:

1) From the decay analysis, emission from the complexes at ambient temperature is ascribed to TADF, in which the emissive excited states, S_1 and T_1, are in the thermal equilibrium with the energy difference, 450 cm^{-1}.
2) The complexes investigated in the present work have the Cu–Cu distance (2.41–2.44 Å) shorter than the sum of van der Waals radius of the copper atoms. Theoretical studies reveal that the "cuprophilic" interaction in the excited states is very important to elucidate the luminescence properties of the dinuclear Cu(I) complexes. TADF from the complexes are concluded to result from the ^{1}[MMLCT+MMC] and ^{3}[MMLCT+MMC] states in the thermal equilibrium. The MMLCT transition involves CT from an occupied Cu–Cu antibonding orbital (3dσ^*_{Cu-Cu}) to an unfilled ligand-based π^* orbital.
3) The complexes possess distinctive characteristics as TADF emitters: (i) bright luminescence in the solid state at ambient temperature, (ii) a variety of emission colors, (iii) the thermal stability and sublimability endurable to the thermal vacuum deposition method for preparing OLED devices, and (iv) moderate EL efficiency of the OLED device. We consider that the complexes studied here are promising candidates for TADF emitters in OLED devices.

Acknowledgment

This research was carried out in the material group of Corporate R&D Headquarters in Canon Inc. The author is very grateful to all colleagues for intensive collaboration. In particular, the author thanks Prof. Mikio Hoshino (RIKEN, The Institute of Physical & Chemical Research) for fruitful discussion on photophysics. I also like to thank Dr. Kazunori Ueno (CSIRO) for supportive advice, Dr. Isao Kawata (Canon) for theoretical study and DFT calculation, Mr. Jun Kamatani for technical support of synthesis, and Mr. Taihei Mukaide and Dr. Katsuaki Kuge (Canon) for X-ray crystallographic analysis.

References

1 (a) Yersin, H. (ed.) (2008). *Highly Efficient OLEDs with Phosphorescent Materials*. Weinheim, Germany: Wiley-VCH Ltd. (b) Baldo, M.A., Lamansky, S., Burrows, P.E., Thompson, M.E., and Forrest, S.R. (1999). Very high-efficiency green organic light-emitting devices based on electrophosphorescence. *Appl. Phys. Lett.* 75 (1): 4–6. (c) Adachi, C., Baldo, M.A., Thompson, M.E., and Forrest, S.R. (2001). Nearly 100% internal phosphorescence efficiency in an organic light emitting device. *J. Appl. Phys.* 90 (10): 5048–5051. (d) Lamansky, S., Djurovich, P., Murphy, D., Abdel-Razzaq, F., Lee, H.-E., Adachi, C., Burrows, P.E., Forrest, S.R., and Thompson, M.E. (2001). Highly phosphorescent bis-cyclometalated iridium complexes: synthesis, photophysical characterization, and use in organic light emitting diodes. *J. Am. Chem. Soc.* 123: 4304–4312. (e) Lamansky, S., Djurovich, P., Murphy, D., Abdel-Razzaq, F., Kwong, R., Tsyba, I., Bortz, M., Mui, B., Bau, R., and Thompson, M.E. (2001). Synthesis and characterization of phosphorescent cyclometalated iridium complexes. *Inorg. Chem.* 40: 1704–1711. (f) Tsuboyama, A., Iwawaki, H., Furugori, M., Mukaide, T., Kamatani, J., Igawa, S., Moriyama, T., Miura, S., Takiguchi, T., Okada, S., Hoshino, M., and Ueno, K. (2003). Homoleptic cyclometalated iridium complexes with highly efficient red phosphorescence and application to organic light-emitting diode. *J. Am. Chem. Soc.* 125: 12971–12979. (g) Okada, S., Okinaka, K., Iwawaki, H., Furugori, M., Hashimoto, M., Mukaide, T., Kamatani, J., Igawa, S., Tsuboyama, A., Takiguchi, T., and Ueno, K. (2005). Substituent effects of iridium complexes for highly efficient red OLEDs. *Dalton Trans.* 1583–1590.

2 (a) Uoyama, H., Goushi, K., Shizu, K., Nomura, H., and Adachi, C. (2012). Highly efficient organic light-emitting diodes from delayed fluorescence. *Nature* 492: 234–238. (b) Mehes, G., Nomura, H., Zhang, Q., Nakagawa, T., and Adachi, C. (2012). Enhanced electroluminescence efficiency in a spiro-acridine derivative through thermally activated delayed fluorescence. *Angew. Chem. Int. Ed.* 51: 11311–11315. (c) Zhang, Q., Li, J., Shizu, K., Huang, S., Hirata, S., Miyazaki, H., and Adachi, C. (2012). Design of efficient thermally activated delayed fluorescence materials for pure blue organic light emitting diodes. *J. Am. Chem. Soc.* 134: 14706–14709. (d) Yersin, H., Rausch, A.F., Czerwieniec, R., Hofbeck, T., and Fischer, T. (2011). The triplet state of organo-transition metal compounds. Triplet harvesting and singlet harvesting for efficient OLEDs. *Coord. Chem. Rev.* 255: 2622–2652. (e) Czerwieniec, R., Leitl, M.J., Homeier, H.H.H., and Yersin, H. (2016). Cu(I) complexes – thermally activated delayed fluorescence: photophysical approach and material design. *Coord. Chem. Rev.* 325: 2–28.

3 (a) Czerwieniec, R., Yu, J., and Yersin, H. (2011). Blue-light emission of Cu(I) complexes and singlet harvesting. *Inorg. Chem.* 50: 8293–8301. (b) Czerwieniec, R. and Yersin, H. (2015). Diversity of copper(I) complexes

showing thermally activated delayed fluorescence: basic photophysical analysis. *Inorg. Chem.* 54: 4322–4327. (c) Leitl, M.J., Krylova, V.A., Djurovich, P.I., Thompson, M.E., and Yersin, H. (2014). Phosphorescence versus thermally activated delayed fluorescence. controlling singlet-triplet splitting in brightly emitting and sublimable Cu(I) compounds. *J. Am. Chem. Soc.* 136: 16032–16038. (d) Hashimoto, M., Igawa, S., Yashima, M., Kawata, I., Hoshino, M., and Osawa, M. (2011). Highly efficient green organic light-emitting diodes containing luminescent three-coordinate copper(I) complexes. *J. Am. Chem. Soc.* 133: 10348–10351. (e) Osawa, M., Hoshino, M., Hashimoto, M., Kawata, I., Igawa, S., and Yashima, M. (2015). Application of three-coordinate copper(I) complexes with halide ligands in organic light-emitting diodes that exhibit delayed fluorescence. *Dalton Trans.* 44: 8369–8378. (f) Barbieri, A., Accorsi, G., and Armaroli, N. (2008). Luminescent complexes beyond the platinum group: d^{10} avenue. *Chem. Commun.* 2185–2193. (g) H. Yersin, German Patent DE10 2008, 033563.

4 (a) McMillin, D.R. and McNett, K.M. (1998). Photoprocesses of copper complexes that bind to DNA. *Chem. Rev.* 98: 1201–1219. (b) Kirchhoff, J.R., Gamache, R.E., Blaskie, M.W., Del Paggio, A.A., Lengel, R.K., and McMillin, D.R. (1983). Temperature dependence of luminescence from Cu(NN)$^{2+}$ systems in fluid solution. Evidence for the participation of two excited states. *Inorg. Chem.* 22: 2380–2384. (c) Everly, R.M. and McMillin, D.R. (1995). Reinvestigation of the absorbing and emitting charge-transfer excited states of [Cu(NN)$_2$]$^+$ systems. *J. Phys. Chem.* 95: 9071–9075. (d) Everly, R.M., Ziessel, R., Suffert, J., and McMillin, D.R. (1991). Steric influences on the photoluminescence from copper(I) phenanthrolines in rigid media. *Inorg. Chem.* 30: 559–561. (e) Eggleston, M.K. and McMillin, D.R. (1997). Steric effects in the ground and excited states of Cu(NN)$_2^+$ systems. *Inorg. Chem.* 36: 172–176. (f) Cunningham, C.T., Moore, J.J., Cunningham, K.L.H., Fanwick, P.E., and McMillin, D.R. (2000). Structural and photophysical studies of Cu(NN)$_2^+$ systems in the solid state. Emission at last from complexes with simple 1,10-phenanthroline ligands. *Inorg. Chem.* 39: 3638–3644.

5 (a) Ford, P.C., Cariati, E., and Bourassa, J. (1999). Photoluminescence properties of multinuclear copper(I) compounds. *Chem. Rev.* 99: 3625–3647. (b) Yam, V.W.-W. and Lo, K.K.-W. (1999). Luminescent polynuclear d^{10} metal complexes. *Chem. Soc. Rev.* 28: 323–334. (c) Peng, R., Li, M., and Li, D. (2010). Copper(I) halides: a versatile family in coordination chemistry and crystal engineering. *Coord. Chem. Rev.* 254: 1–18. (d) Lane, A.C., Vollmer, M.V., Laber, C.H., Melgarejo, D.Y., Chiarella, G.M., Fackler, J.P. Jr., Yang, X., Baker, G.A., and Walensky, J.R. (2014). Multinuclear copper(I) and silver(I) amidinate complexes: synthesis, luminescence, and CS$_2$ insertion reactivity. *Inorg. Chem.* 53: 11357–11366. (e) Knorr, M., Guyon, F., Khatyr, A., Strohmann, C., Allain, M., Aly, S.M., Lapprand, A., Fortin, D., and Harvey, P.D. (2012). Construction of (CuX)$_{2n}$ cluster-containing (X = Br, I; n = 1, 2) coordination polymers assembled by dithioethers ArS(CH$_2$)$_m$SAr (Ar = Ph, p-Tol; m = 3, 5): effect of the spacer length, aryl group, and metal-to-ligand ratio on the dimensionality, cluster nuclearity, and the luminescence properties of the metal-organic frameworks. *Inorg. Chem.* 51: 9917–9934.

(f) Rajput, G., Yadav, M.K., Drew, M.G.B., and Singh, N. (2015). Impact of ligand framework on the crystal structures and luminescent properties of Cu(I) and Ag(I) clusters and a coordination polymer derived from thiolate/iodide/dppm ligands. *Inorg. Chem.* 54: 2572–2579. (g) Brown, E.C., Bar-Nahum, I., York, J.T., Aboelella, N.W., and Tolman, W.B. (2007). Ligand structural effects on Cu_2S_2 bonding and reactivity in side-on disulfido-bridged dicopper complexes. *Inorg. Chem.* 46: 486–496. (h) Carvajal, M.A., Alvarez, S., and Novoa, J.J. (2004). The nature of intermolecular $Cu^I \cdots Cu^I$ interactions: a combined theoretical and structural database analysis. *Chem. Eur. J.* 10: 2117–2132. (i) Maderlehner, S., Leitl, M.J., Yersin, H., and Pfitzner, A. (2015). Halocuprate(I) zigzag chain structures with *N*-methylated DABCO cations – bright metal-centered luminescence and thermally activated color shifts. *Dalton Trans.* 44: 19305–19313.

6 (a) Araki, H., Tsuge, K., Sasaki, Y., Ishizuka, S., and Kitamra, N. (2005). Luminescence ranging from red to blue: a series of copper(I)-halide complexes having rhombic {$Cu_2(\mu$-X$)_2$} (X = Br and I) units with N-heteroaromatic ligands. *Inorg. Chem.* 44: 9667–9675. (b) Tsuge, K. (2013). Luminescent complexes containing halogeno-bridged dicopper(I) unit {$Cu_2(\mu$-X$)_2$}(X = Cl, Br, and I). *Chem. Lett.* 42: 204–208. (c) Tsuboyama, A., Kuge, K., Furugori, M., Okada, S., Hoshino, M., and Ueno, K. (2007). Photophysical properties of highly luminescent copper(I) halide complexes chelated with 1,2-bis(diphenylphosphino)benzene. *Inorg. Chem.* 46: 1992–2001. (d) Leitl, M.J., Küchle, F.-R., Mayer, H.A., Wesemann, L., and Yersin, H. (2013). Brightly blue and green emitting Cu(I) dimers for singlet harvesting in OLEDs. *J. Phys. Chem. A* 117: 11823–11836. (e) Araki, H., Tsuge, K., Sasaki, Y., Ishizuka, S., and Kitamra, N. (2007). Synthesis, structure, and emissive properties of copper(I) complexes[$Cu^I_2(\mu$-X$)_2(\mu$-1,8-naphthyridine)(PPh$_3)_2$] (X = I, Br) with a butterfly-shaped dinuclear core having a short Cu–Cu distance. *Inorg. Chem.* 46: 10032–10034.

7 (a) Volz, D., Chen, Y., Wallesch, M., Liu, R., Flechon, C., Zink, D.M., Friedrichs, J., Flügge, H., Steininger, R., Gottlicher, J., Heske, C., Weinhardt, L., Bräse, S., So, F., and Baumann, T. (2015). Bridging the efficiency gap: fully bridged dinuclear Cu(I) complexes for singlet harvesting in high-efficiency OLEDs. *Adv. Mater.* 27 (15): 2538–2543. (b) Zink, D.M., Bächle, M., Baumann, T., Nieger, M., Kühn, M., Wang, C., Klopper, W., Monkowius, U., Hofbeck, T., Yersin, H., and Bräse, S. (2013). Synthesis, structure, and characterization of dinuclear copper(I) halide complexes with P∧N ligands featuring exciting photoluminescence properties. *Inorg. Chem.* 52: 2292–2305. (c) Volz, D., Nieger, M., Friedrichs, J., Baumann, T., and Bräse, S. (2013). How the quantum efficiency of a highly emissive binuclear copper complex is enhanced by changing the processing solvent. *Langmuir* 29: 3034–3044. (d) Volz, D., Zink, D.M., Bocksrocker, T., Friedrichs, J., Nieger, M., Baumann, T., Lemmer, U., and Bräse, S. (2013). Molecular construction kit for tuning solubility, stability and luminescence properties: heteroleptic MePyrPHOS-copper iodide-complexes and their application in organic light-emitting diodes. *Chem. Mater.* 25: 3414–3426. (e) Zink, D.M., Volz, D., Baumann, T., Mydlak, M., Flügge, H., Friedrichs, J., Nieger, M., and Bräse, S. (2013). Heteroleptic, dinuclear copper(I) complexes for application in organic

light-emitting diodes. *Chem. Mater.* 25: 4471–4486. (f) Zink, D.M., Baumann, T., Friedrichs, J., Nieger, M., and Bräse, S. (2013). Copper(I) complexes based on five-membered P∧N heterocycles: structural diversity linked to exciting luminescence properties. *Inorg. Chem.* 52: 13509–13520. (g) Volz, D., Wallesch, M., Grage, S.L., Göttlicher, J., Steininger, R., Batchelor, D., Vitova, T., Ulrich, A.S., Heske, C., Weinhardt, L., Baumann, T., and Bräse, S. (2014). Labile or stable: can homoleptic and heteroleptic PyrPHOS–copper complexes be processed from solution? *Inorg. Chem.* 53: 7837–7847. (h) H. Yersin, U. Monkowius, T. Fischer, T. Hofbeck, German Patent DE10 2009, 030475A1.

8 Yam, V.W.-W., Lee, W.-K., Cheung, K.K., Lee, H.-K., and Leung, W.-P. (1996). Photophysics and photochemical reactivities of organocopper(I) complexes. Crystal structure of [$Cu_2(PPh_2Me)_4(\mu,\eta^1\text{-}C{\equiv}CPh)_2$]. *J. Chem. Soc. Dalton Trans.* 2889–2891.

9 (a) Harkins, S.B. and Peters, J.C. (2005). A highly emissive Cu_2N_2 diamond core complex supported by [PNP]$^-$ ligand. *J. Am. Chem. Soc.* 127: 2030–2031. (b) Deaton, J.C., Switalski, S.C., Kondakov, D.Y., Young, R.H., Pawlik, T.D., Giesen, D.J., Harkins, S.B., Miller, A.J.M., Mickenberg, S.F., and Peters, J.C. (2010). E-Type delayed fluorescence of a phosphine-supported $Cu_2(\mu\text{-}NAr_2)_2$ diamond core: harvesting singlet and triplet excitons in OLEDs. *J. Am. Chem. Soc.* 132: 9499–9508. (c) Harkins, S.B., Mankad, N.P., Miller, A.J.M., Szilagyi, R.K., and Peters, J.C. (2008). Probing the electronic structures of [$Cu_2(\mu\text{-}XR_2)$]$^{n+}$ diamond cores as a function of the bridging X atom (X = N or P) and charge (n = 0, 1, 2). *J. Am. Chem. Soc.* 130: 3478–3485.

10 (a) Che, C.-M., Mao, Z., Miskovski, V.M., Tse, M.-C., Chan, C.-K., Cheung, K.-K., Phillips, D.L., and Leung, K.-H. (2000). Cuprophilicity: spectroscopic and structural evidence for Cu–Cu bonding interactions in luminescent dinuclear copper(I) complexes with bridging diphosphane ligands. *Angew. Chem. Int. Ed.* 39: 4084–4088. (b) Mao, Z., Chao, H.-Y., Hui, Z., Che, C.-M., Fu, W.-F., Cheung, K.-K., and Zhu, N. (2003). $^3[(d_{x2-y2}, d_{xy})(p_z)]$ Excited states of binuclear copper(I) phosphine complexes: effect of copper–ligand and copper–copper interactions on excited state properties and photocatalytic reductions of the 4,4'-dimethyl-2,2'-bipyridinium ion in alcohols. *Chem. Eur. J.* 9: 2885–2897. (c) Fu, W.-F., Gan, X., Che, C.-M., Cao, Q.-Y., Zhou, Z.-Y., and Zhu, N.N.-Y. (2004). Cuprophilic interactions in luminescent copper(I) clusters with bridging bis(dicyclohexylphosphino)methane and iodide ligands: spectroscopic and structural investigations. *Chem. Eur. J.* 10: 2228–2236. (d) Che, C.-M. and Lai, S.-W. (2005). Structural and spectroscopic evidence for weak metal–metal interactions and metal–substrate exciplex formations in d^{10} metal complexes. *Coord. Chem. Rev.* 249: 1296–1309.

11 Hofbeck, T., Monkowius, U., and Yersin, H. (2015). Highly efficient luminescence of Cu(I) compounds: thermally activated delayed fluorescence combined with short-lived phosphorescence. *J. Am. Chem. Soc.* 137: 399–404.

12 (a) Papasergio, R.I., Raston, C.L., and White, A.H. (1983). Synthesis of pyridine functionalised, sterically hindered lithium and copper(I) alkyls; crystal structures of [{2-($Me_3Si)_2C(M)C_5H_4N$}$_2$] (M = Li or Cu), dimeric compounds free of multicentre bonding. *J. Chem. Soc. Chem. Commun.* 1419–1420. (b) Papasergio, R.I., Raston, C.L., and White, A.H. (1987). Syntheses and crystal structures of complexes [M_2R_2] [M = Cu, Ag, or Au; R = 2-C($SiMe_3)_2C_5H_4N$]

and [Cu$_4$R′$_4$] [R′ = 2-CH(SiMe$_3$)C$_5$H$_4$N]; electrochemical generation of [Cu$_2$R$_2$]$^{2+}$. *J. Chem. Soc. Dalton Trans.* 3085–3091. (c) van den Ancker, T.R., Bhargava, S.K., Mohr, F., Papadopoulos, S., Raston, C.L., Skelton, B.W., and White, A.H. (2001). Syntheses and crystal structures of binuclear gold(I), silver(I) and copper(I) complexes containing bulky pyridyl functionalised alkyl ligands. *J. Chem. Soc. Dalton Trans.* 3069–3072.

13 (a) Rasika Dias, H.V., Diyabalanage, H.V.K., Rawashdeh-Omary, M.A., Franzman, M.A., and Omary, M.A. (2003). Bright phosphorescence of a trinuclear copper(I) complex: luminescence thermochromism, solvatochromism, and "concentration luminochromism". *J. Am. Chem. Soc.* 125: 12072–12073. (b) Omary, M.A., Rawashdeh-Omary, M.A., Diyabalanage, H.V.K., and Rasika Dias, H.V. (2003). Blue phosphors of dinuclear and mononuclear copper(I) and silver(I) complexes of 3,5-bis(trifluoromethyl)pyrazolate and the related bis(pyrazolyl)borate. *Inorg. Chem.* 42: 8612–8614. (c) Rasika Dias, H.V., Diyabalanage, H.V.K., Eldabaja, M.G., Elbjeirami, O., Rawashdeh-Omary, M.A., and Omary, M.A. (2005). Brightly phosphorescent trinuclear copper(I) complexes of pyrazolates: substituent effects on the supramolecular structure and photophysics. *J. Am. Chem. Soc.* 127: 7489–7501.

14 (a) James, A.M., Laxman, R.K., Fronczek, F.R., and Maverick, A.W. (1998). Phosphorescence and structure of a tetrameric copper(I)-amide cluster. *Inorg. Chem.* 37: 3785–3791. (b) Noto, M., Goto, Y., and Era, M. (2003). Electrophosphorescence from tetrameric copper(I)-amide cluster. *Chem. Lett.* 32 (1): 32–33. (c) Manbeck, G.F., Brennessel, W.W., Evans, C.M., and Eisenberg, R. (2010). Tetranuclear copper(I) iodide complexes of bis(1-benzyl-1*H*-1,2,3-triazole) ligands: structural characterization and solid state photoluminescence. *Inorg. Chem.* 49: 2834–2843.

15 (a) Cotton, F.A., Feng, X., Matusz, M., and Poli, R. (1988). Experimental and theoretical studies of the copper(I) and silver(I) dinuclear N,N′-Di-p-tolylformamidinato complexes. *J. Am. Chem. Soc.* 110: 7077–7083. (b) Cotton, F.A., Feng, X., and Timmons, D.J. (1998). Further study of very close nonbonded Cu^I-Cu^I contacts. Molecular structure of a new compound and density functional theory calculations. *Inorg. Chem.* 37: 4066–4069.

16 Frisch, M.J., Trucks, G.W., Schlegel, H.B., Scuseria, G.E., Robb, M.A., Cheeseman, J.R., Scalmani, G., Barone, V., Mennucci, B., Petersson, G.A., Nakatsuji, H., Caricato, M., Li, X., Hratchian, H.P., Izmaylov, A.F., Bloino, J., Zheng, G., Sonnenberg, J.L., Hada, M., Ehara, M., Toyota, K., Fukuda, R., Hasegawa, J., Ishida, M., Nakajima, T., Honda, Y., Kitao, O., Nakai, H., Vreven, T., Montgomery, J.A. Jr., Peralta, J.E., Ogliaro, F., Bearpark, M., Heyd, J.J., Brothers, E., Kudin, K.N., Staroverov, V.N., Kobayashi, R., Normand, J., Raghavachari, K., Rendell, A., Burant, J.C., Iyengar, S.S., Tomasi, J., Cossi, M., Rega, N., Millam, J.M., Klene, M., Knox, J.E., Cross, J.B., Bakken, V., Adamo, C., Jaramillo, J., Gomperts, R., Stratmann, R.E., Yazyev, O., Austin, A.J., Cammi, R., Pomelli, C., Ochterski, J.W., Martin, R.L., Morokuma, K., Zakrzewski, V.G., Voth, G.A., Salvador, P., Dannenberg, J.J., Dapprich, S., Daniels, A.D., Farkas, Ö., Foresman, J.B., Ortiz, J.V., Cioslowski, J., and Fox, D.J. (2009). *Gaussian 09, Revision D.01*. Wallingford, CT: Gaussian, Inc.

4

Molecular Design and Synthesis of Metal Complexes as Emitters for TADF-Type OLEDs

Masahisa Osawa and Mikio Hoshino

RIKEN, 2-1 Hirosawa, Wako, 351-0198 Saitama, Japan

4.1 Introduction

Nowadays advances in the organic light-emitting diode (OLED) technology have promoted the commercialization of new light sources for mercury-free lighting, flat-panel displays, and smartphones. Practical applications of OLEDs have been realized not only because of the successful synthesis of new emissive substances and peripheral materials but also because of the significant progress in processing technology. Furthermore, extensive research in device chemistry and physics has accelerated the commercialization of OLEDs.

In 1987, Tang and Van Slyke reported the fabrication of the first efficient OLED [1], which consists of the hole-transport and emissive layers sandwiched between oppositely charged electrodes. The system, which uses an organic molecule as an emitting dopant, was a fluorescent OLED, and thus, only singlet excitons were responsible for the emission.

It has been well established that both the singlet and triplet excited states are generated by charge recombination in OLEDs. The relative yield of the triplet state is three times higher than that of the singlet state [2, 3]. The triplet state of organic molecules scarcely emits at room temperature because of the very small radiative rate constants. Therefore, the electrochemically formed triplet excited states (75%) in OLEDs, which contain fluorescent organic molecules as emitting dopants, hardly participate in emission at room temperature. Hence, in the case of the fluorescent OLEDs, 75% of the excited states generated by charge recombination are wasted as heat. By taking the light-extraction efficiency into account, the theoretical upper limit of external quantum efficiency (EQE) (photons per electron) for fluorescent OLEDs is evaluated to be as small as 5%.

In order to achieve large EQE values for OLEDs, it is necessary for the triplet state to participate in the emission processes. Fortunately, there are many metal complexes that show strong phosphorescence emission from the triplet state at room temperature. Heavy metal complexes, such as iridium and platinum complexes, doped in the emissive layer have been found to emit strong phosphorescence, leading to remarkable improvement of the EQE (>20%) [4, 5]. Heavy metals in the complexes are known to increase the rate constants for the

Highly Efficient OLEDs: Materials Based on Thermally Activated Delayed Fluorescence,
First Edition. Edited by Hartmut Yersin.
© 2019 Wiley-VCH Verlag GmbH & Co. KGaA. Published 2019 by Wiley-VCH Verlag GmbH & Co. KGaA.

intersystem crossing (ISC) process, $S_1 \to T_1$, and the phosphorescence process, $T_1 \to S_0$, due to a large spin–orbit interaction induced by the heavy atoms. In some cases, the internal quantum efficiencies of the devices reach 100% owing to the unique ability of the metal complexes to harvest both singlet and triplet excitons. Reverse intersystem crossing (rISC) process from T_1 to S_1 is frequently observed because of the thermal activation of T_1, and thus, relatively large S_1–T_1 gaps (greater than several thousand cm^{-1}) are essential for strongly phosphorescent materials to avoid rISC.

Up to now, a number of tris- or bis(cyclometalated) iridium(III) complexes with octahedral geometries have been successfully synthesized as efficient emitters in phosphorescence-type organic light-emitting diodes (PHOLEDs) [5–9]. The high photoluminescence quantum yields (PLQYs > 80%) and relatively short lifetimes (a few microsecond) are very important to avoid the roll-off effect, which is frequently observed in the plot of illuminance versus current density. The roll-off effect sometimes results from triplet–triplet annihilation (TTA). The unique photophysical properties of the metal complexes mentioned above can be elucidated from the nature of the three triplet sublevels with characteristic zero-field splitting (ZFS) parameters, which are similar to those of [Ru(bpy)$_3$]$^{2+}$ (bpy = 2,2′-bipyridine) [10–12]. In fact, the highest sublevel of the metal-to-ligand charge-transfer (^{3}MLCT) triplet state of the iridium complex has a very large radiative rate constant (k_r) because of the facile mixing with ^{1}MLCT via effective spin–orbit coupling (SOC). According to the studies on comprehensive emission mechanisms of these Ir-based phosphorescent materials [13–16], the strong MLCT character of the Ir complexes in the emitting excited states is found to contribute to the large emission yield and the short lifetime of phosphorescence. Further, an investigation on the symmetry effects in the molecule reveals that the complexes, which have degenerate d orbitals, increase the MLCT character of the emissive excited state. After appropriate optimization of the device that uses the Ir complex as a dopant, high EQEs over 30% in PHOLEDs have been successfully obtained [17, 18].

The most efficient phosphorescent emitters are the complexes containing iridium or platinum, which are very expensive and highly localized on the earth. In addition, Ir(III) or Pt(II) with d^6 or d^8 configurations have low-lying d–d* states. These facts are an inevitable impediment for Ir(III) or Pt(II) complexes to achieve low-priced pure blue phosphorescent OLEDs.

Recently, new materials, which exhibit thermally activated delayed fluorescence (TADF), have been investigated for application as emitting dopants in the OLEDs. Delayed fluorescence (DF) has been originally observed for organic molecules [19]. Although the spectra of DF are the same as those of normal fluorescence (NF), the lifetimes of DF and NF are very different. DF has been divided into two groups depending on the origin of emission: P- and E-types. A representative molecule for the P-type is pyrene, which exhibits DF from the excited singlet state produced by TTA. An example for an E-type molecule is eosin. It exhibits DF from the excited singlet state, which is thermally populated from the triplet state of the molecule. Thus, lifetimes of DF for both types are markedly affected by the triplet lifetimes, generally resulting in much longer emission lifetimes than those of NF.

With regard to the E-type molecules, the S–T energy gaps are small, and consequently, the excited singlet state is thermally accessible from the triplet state by an rISC process. The E-type DF is synonymous with the TADF.

When (i) the radiative rate constant, k_r, of S_1 is much larger than the nonradiative rate constant (k_{nr}) and (ii) the rate to achieve the equilibration between S_1 and T_1 is sufficiently faster than the triplet decay rate, fluorescence yields via the TADF process are necessarily very large. Needless to say, the rate for the achievement of the equilibration is governed by the S_1–T_1 energy gap: The smaller the gap, the faster the rate.

TADF-type OLEDs with the use of organic emitters were reported by Adachi and coworkers [20]. The organic emitters studied were composed of an electron donor and an acceptor moiety in a molecule that exhibits strong intramolecular charge-transfer (CT) emission. These emitters inevitably have small S–T energy gaps that are necessary to emit efficient TADF. The S–T energy gaps are governed by the electron exchange terms in the excited state. Since the strong CT interaction in the excited state results in the small electron exchange term, J, the organic emitters reported give intensive CT fluorescence via the TADF process.

In general, luminescent materials that emit phosphorescence at room temperature contain heavy metals to gain the large radiative rate constant at T_1 by the effective mixing of T_1 with higher S_n ($n \geq 2$) states through SOC.

In the past two decades, some complexes containing light metals have been found to exhibit TADF [21–25]. These complexes have an S–T energy gap small enough to attain thermal equilibrium between the singlet and the triplet state at room temperature. The use of these metal complexes as guest molecules enables us to construct effective OLEDs that show emission by harvesting both the singlet and triplet excitons of host molecules generated by charge recombination inside the devices. The singlet and the triplet excitons of host molecules undergo energy transfer to the guest molecules, the metal complexes, leading to the formation of both the S_1 and T_1 states of the guest molecules. The S_1 state of the metal complexes is readily converted to the T_1 state by ISC, and the T_1 state generates the thermally activated S_1 state responsible for TADF. Accordingly, the excited states harvested by the metal complexes in the devices are mostly transformed to the triplet state, which emits highly efficient TADF.

The TADF from tetrahedral copper(I) complexes $[Cu(NN)_2]^+$ with diimine ligands (NN) in solution was first observed by McMillin and coworkers [26]. The emission occurs from the MLCT excited states. The plots of the emission lifetimes versus temperature, T, measured for $[Cu(NN)_2]^+$ have been elucidated on the basis of the Boltzmann distribution between the S_1 and T_1 states, and thus, the so-called two-emitting-state model of this complex was proposed. The energy gap between the two states was estimated as c. 1800 cm^{-1}. The radiative rate constant, k_r, from the upper level, S_1, was estimated as $\sim 10^7$ s^{-1}, and that from the lower T_1 level, $\sim 10^3$ s^{-1}. As predicted by the two-state model, the S–T energy gap is found to be close to the difference in energy between the emission peaks at room temperature and 77 K. The two-state model proposed for $[Cu(NN)_2]^+$ is consistent with recent results obtained by femtosecond spectroscopic methods and theoretical calculations [27–29]. The lifetimes of greenish TADF from $[Cu(NN)P_2]^+$ with aryl phosphine ligands (P) in the solid state were measured as a function of temperature to determine the S–T energy gap. It is found that the S–T energy gap (1000 cm^{-1}) [30] between two states is much smaller than that of [Cu(NN)$_2$]$^+$.

The first application of tetrahedral Cu(I) complexes as an emitter in OLEDs was reported by Wang and coworkers [31]. At that time, efficient emission from the copper(I) complexes was not recognized as TADF. However, recent studies on emission from the Cu(I) complexes revealed that many complexes with tetrahedral structures would be TADF emitters. Nowadays, many TADF-type emitters of Cu(I) complexes doped in devices are found to afford high EQEs in comparison with those of the well-established Ir(III)-based materials.

In this chapter, we introduce metal complex emitters exhibiting TADF for OLEDs, particularly d^{10} metal complexes. In Section 4.2, we discuss the emissive electronic states of tetrahedral copper(I) complexes from the perspective of their molecular orbital (MO) configuration. The correlation between the MO character and the coordinating atoms like P (phosphorus ligands) and N (nitrogen ligands) is briefly described. In Section 4.3, we mention important properties of the mononuclear TADF-type Cu(I) emitters and summarize the guidelines for the fabrication of efficient OLEDs based on these materials. The conventional OLEDs containing these emitters in the emitting layer exhibit a high efficiency comparable with that of cyclometalated iridium(III)-based devices, which are the current standards used as a measure of efficiency. In Section 4.4, representative examples of the dinuclear TADF-type Cu(I) emitters applied in OLEDs are briefly expounded. Other d^{10} metal complexes (Ag(I) and Au(I)) exhibiting TADF have been presented in Section 4.5. The last section is a short conclusion.

4.2 Cu(I) Complexes for OLEDs

Luminescent Cu(I) complexes have been extensively studied since the 1980s. A vast number of investigations have been carried out for understanding the structural changes occurring in the excited states of [Cu(NN)$_2$]$^+$ due to the pseudo-Jahn–Teller effect since it has been suggested that the tetrahedral structure of [Cu(NN)$_2$]$^+$ in the excited state transforms into a "flattened" square-planar-like form [32–37]. The conformational changes occurring in the excited state are an interesting target for fundamental research. However, such conformational changes create problems for Cu(I) complexes with two diimine ligands to achieve high luminescence quantum efficiency. This is because the structural changes occurring in the excited states are commonly accompanied by acceleration in the rate of nonradiative decay processes, and thus, most of the tetrahedral Cu(I) complexes tend to exhibit weak emission. A major challenge in the design and synthesis of efficient emitters based on tetrahedral Cu(I) complexes in the 1990s was to establish a strategy to prevent structural changes in the excited states of the complexes [38–42].

4.2.1 Energy Levels of Molecular Orbitals in Tetrahedral Geometries

Electronic transitions in tetrahedral Cu(I) metal complexes have been theoretically elucidated using DFT calculations. The simple orbital diagrams of

Figure 4.1 Geometries' orbital diagram of tetrahedral Cu(I) complexes.

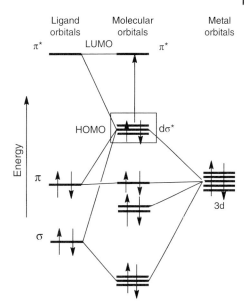

these complexes are shown in Figure 4.1. The highest occupied molecular orbital (HOMO: dσ*) is explained as a combination of the three degenerate valence orbitals ($3d_{xy}$, $3d_{yz}$, $3d_{xz}$) of a central metal with the four orbitals of the coordinating atoms of the ligands located at the vertices of a tetrahedron. The electronic transitions responsible for emission of tetrahedral Cu(I) complexes are principally governed by the nature of the HOMOs and lowest unoccupied molecular orbitals (LUMOs). The former is antibonding in nature because the filled 3d orbitals of Cu(I) interact with the lone pair electrons of the coordinating atoms of the ligand while the latter is the antibonding orbital (π^*) of the aromatic group of the ligand. Thus, the first electronic transition of the tetrahedral Cu(I) complexes is roughly assumed to be due to a d → π^* transition. This HOMO–LUMO transition is suggested to be MLCT [43]. The changes of constituent orbitals of HOMO depending on the kind of ligands are discussed later.

4.2.2 Ligand Variation

The character of the HOMO (dσ*) in a tetrahedral geometry depends on the nature of the ligands. Here we present two typical examples, [Cu(dmp)$_2$]$^+$ (dmp = 2,9-dimethyl-1,10-phenanthroline) complex **1** and [Cu(dppbz)$_2$]$^+$ (dppbz = 1,2-bis(diphenylphosphino)benzene complex **2** [44, 45]. Both the ligands, dmp and dppbz, have been frequently used for synthesizing emissive Cu(I) complexes. As shown in Figure 4.2 (A), nonbonding orbitals of the coordinating P atoms in the diphosphine ligand have energies higher than those of the N atoms in the diimine ligand because of the fact that the valence orbitals of phosphorus and nitrogen atoms are 3p and 2p orbitals, respectively.

The LUMOs of the complexes are the π^* orbitals of the ligands, dmp and dppbz, while the HOMOs are principally composed of the fully occupied 3d orbitals

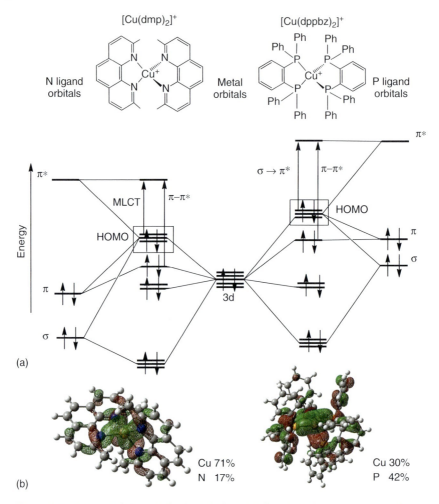

Figure 4.2 [Cu(dmp)$_2$]$^+$ (**1**) and [Cu(dppbz)$_2$]$^+$ (**2**). (a) Chemical structures and orbital diagrams. (b) NTO pairs for the lowest triplet excited state at the S$_0$ optimized geometry.

of Cu(I) and the nonbonding valence orbitals of the coordinating atoms of the ligands. Since the 3d orbitals of Cu(I) in **1** contribute to the HOMO to a greater extent than in **2**, the HOMO–LUMO transition has a large MLCT character in comparison with the latter, and thus, the transition is considered to be (d–π*). Unlike the case of **1**, the HOMO of **2** is mainly composed of 3p orbitals of the lone pair electrons on the P atoms, and therefore, the HOMO–LUMO transition is regarded as (σ–π*).

With regard to the HOMOs, the natural transition orbital (NTO) maps obtained from excited-state calculations clearly demonstrate the difference in orbital compositions between **1** and **2** as shown in Figure 4.2 (B). This figure illustrates the hole (approximately HOMO) distribution for the T$_1$ states of **1** and **2** with geometries optimized at S$_0$. The contributions of Cu(I) and the

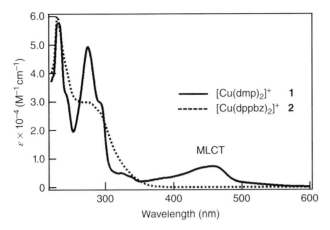

Figure 4.3 Absorption spectra of [Cu(dmp)$_2$]$^+$ (**1**) and [Cu(dppbz)$_2$]$^+$ (**2**) in 2-MeTHF. *Source*: M. Osawa, unpublished data.

nitrogen atom in **1** to the hole distribution are 71% and 17%, respectively. In the hole distribution of **2**, the contribution of Cu(I) is as small as 30%, while that of the P atoms of the ligands is as large as 42%. As shown in Figure 4.3, these results explain well the absorption spectra of **1** and **2**. The absorption spectrum of **1** shows intense ligand-centered bands responsible for the π–π^* transitions of the dmp ligands in the higher-energy region and the relatively weak MLCT absorption bands with a λ_{max} of 460 nm in the visible region. In contrast, the electronic absorption spectrum of **2** shows only a broad band with $\lambda_{max} = 280$ nm in the UV region, which resembles the characteristic spectrum of the dppbz ligand. This is probably because the contribution from MLCT to the electronic transition is decreased and that from intraligand charge transfer (ILCT) is increased by changing from the NN to PP ligands. In agreement with the HOMO–LUMO transition mentioned above, the contribution of MLCT to the excited states is very different between complexes **1** and **2**. Despite this difference, both complexes exhibit the pseudo-Jahn–Teller effect in their excited states. Complex **2** exhibits a weak red emission with $\lambda_{max} = 680$ nm in 2-methyltetrahydrofuran (2-MeTHF) (M. Osawa, unpublished data). Upon excitation, complex **1** forms the MLCT excited state in which an electron is transferred from Cu(I) to the π^* orbital of the ligands. The central Cu(I) atom is formally oxidized to Cu(II) (d^9), leading to a change in structure from tetrahedral to a "flattened" square-planar-like structure. In fact, the structural change in the excited states in solutions causes a marked redshift of the emission peak and a decrease in the emission yield. Obviously, the HOMO energy levels of the Cu(I) complexes are sensitive to the deformation of the tetrahedral structure. The reduction in symmetry from tetrahedral to square-planar-like structure by flattening of the ligands results in the destabilization of the HOMO, and thus, we observe the redshift of the emission maximum.

In summary, the character of the HOMO in Cu(I) complexes changes according to the nature of the lone pair of electrons located on the N or P atoms of the ligands, and the energy level is sensitive to the symmetry of the structure.

4.3 Mononuclear Cu(I) Complexes for OLEDs

Figure 4.4 exhibits (I) the diagram for the emission processes of phosphorescent emitters and (II) that of TADF materials in devices. The large SOC interaction gives a key effect for phosphorescent materials to achieve high yields of T_1 and a large radiative rate constant (k_r) of T_1 resulting from the mixing with higher S_n ($n \geq 2$) excited states. However, DF materials give a high emission yield without any large SOC interaction caused by heavy atoms. When k_r in the S_1 state of the TADF materials is much larger than k_{nr}, photoluminescence (PL) quantum efficiencies are as high as 100%.

According to a simple excited-state model, one electron is located in the HOMO, while another one in the LUMO. Using the distance (r_{12}) between the two electrons, the electron exchange term J is expressed as

$$J = \langle \Phi_{LUMO}(1)\Phi_{HOMO}(2) \mid e^2/(r_{12}) \mid \Phi_{LUMO}(2)\Phi_{HOMO}(1) \rangle \tag{4.1}$$

Here, Φ_{LUMO} and Φ_{HOMO} are the wave functions of the LUMO and HOMO, respectively.

The energy gap, ΔE_{ST}, between S_1 and T_1 is written as

$$\Delta E_{ST} = 2J \tag{4.2}$$

The J value decreases when the distance between the two electrons (r_{12}) becomes larger. Equations (4.1) and (4.2) indicate that the TADF occurs more efficiently as the J value becomes smaller. This situation is occasionally satisfied in the case of molecules that possess an electron donor (D) and an acceptor (A) unit. By absorbing light, the excited singlet state of D, D*, interacts with the ground state of A, which leads to the formation of an intramolecular exciplex:

$$D^* - A \rightarrow (D^+ - A^-)^* \tag{4.3}$$

When D and A are almost separated electronically, the HOMO in the molecule, D–A, is confined to D, and the LUMO to A. In this case, the J value decreases with an increase in the distance, r_{AD}, between A and D. Thus, the energy gap between S_1 and T_1 in the exciplex tends to decrease with an increase in r_{AD}. A strong emission of TADF from D-A is presumably observable when (i) the radiative rate constant of the exciplex at S_1 is much larger than the nonradiative rate constant,

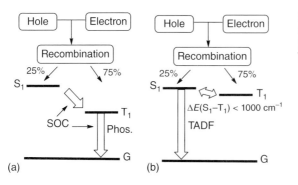

Figure 4.4 Electroluminescence processes of (a) phosphorescent and (b) TADF emitters in devices.

(ii) the energy of the triplet exciplex is lower than those of the triplet states of D and A, and (iii) the rate for attainment of the equilibrium between S_1 and T_1 is faster than that for the decay of the triplet exciplex.

The organic TADF emitter hardly shows any phosphorescence because of the small radiative rate constant of the excited triplet state. In contrast to organic emitters, the Cu(I) complexes afford both phosphorescence and TADF. Tetrahedral Cu(I) complexes with an MLCT character in its excited states are known to have small ΔE_{ST} [26, 30], and thus, the S_1 and T_1 are in thermal equilibrium at a given temperature. The population ratio, $[S_1]/[T_1]$, is given by the Boltzmann equation:

$$[S_1]/[T_1] = (g_{S_1}/g_{T_1}) \exp(-\Delta E_{ST}/RT) \tag{4.4}$$

Here g_{S_1} and g_{T_1} are the degeneracies of the S_1 and the T_1 states, respectively. The spin multiplicity of S_1 and T_1 gives $g_{S_1}/g_{T_1} = 1/3$. Because of the thermal equilibrium between S_1 and T_1, the fluorescence lifetime from S_1 is the same as that of the phosphorescence from T_1. It is frequently observed that the emission peak of the Cu(I) complexes is redshifted ongoing from high to low temperatures. The origin of the redshift of the emission peak observed by lowering the temperature is interpreted in terms of the increase in the population of the triplet state, T_1: The emission changes from DF to phosphorescence with a decrease in temperature.

The emission mechanism of Cu(I) complexes is usually expressed as follows:

$$T_1 \xrightleftharpoons{K} S_1 \tag{4.5}$$

$$S_1 \xrightarrow{k_{rS}} S_0 + h\nu_F \tag{4.6}$$

$$S_1 \xrightarrow{k_{nS}} S_0 \tag{4.7}$$

$$T_1 \xrightarrow{k_{rT}} S_0 + h\nu_P \tag{4.8}$$

$$T_1 \xrightarrow{k_{nT}} S_0 \tag{4.9}$$

Here, K is the equilibrium constant between the S_1 and T_1 states. S_1 gives fluorescence with the radiative constant, k_{rS}, and is thermally deactivated with the nonradiative rate constant, k_{nrS}. The triplet state, T_1, emits phosphorescence with a rate constant, k_{rT}, and is deactivated with a nonradiative rate constant, k_{nrT}. With the use of the Eqs. (4.5)–(4.9), the rate constant, k, for the decay rate of emission is formulated as

$$k = (k_{S_1} + k_{T_1} K)/(1 + K) \tag{4.10}$$

Here, k_{S_1} and k_{T_1} are the rate constants for the decay of S_1 and T_1, respectively. The equilibrium constant, K, is given by

$$K = [S_1]/[T_1] = A \exp(-\Delta H/RT) = (1/3) \exp(-\Delta G/RT) \tag{4.11}$$

$$\Delta G = \Delta H - T\Delta S \tag{4.12}$$

where ΔG, ΔS, and ΔH are the free energy change, the entropy change, and the enthalpy change between the S_1 and T_1 states, respectively. When the conformational change between S_1 and T_1 is assumed to be negligibly small, Eq. (4.11) is equivalent to Eq. (4.4), i.e. $\Delta S = 0$ and $\Delta H = \Delta E_{ST}$. This assumption is found to be satisfied in many cases of TADF observed for the crystals of Cu(I) complexes [16, 46–48].

The rate constants, k_{S_1} and k_{T_1}, are rewritten as

$$k_{S_1} = k_r(S_1) + k_{nr}(S_1) \qquad (4.13)$$

$$k_{T_1} = k_r(T_1) + k_{nr}(T_1) \qquad (4.14)$$

Thus, Eq. (4.10) is formulated as

$$k = \{k_r(S_1) + k_{nr}(S_1) + (k_r(T_1) + k_{nr}(T_1))K\}/(1+K)$$
$$= \{k_r(S_1) + k_r(T_1)K\}/(1+K) + \{k_{nr}(S_1) + k_{nr}(T_1)K\}/(1+K) \qquad (4.15)$$

Equation (4.15) implies that the radiative and nonradiative rate constants, k_r and k_{nr}, respectively, of the Cu(I) complexes are expressed as

$$k_r = \{k_r(S_1) + k_r(T_1)K\}/(1+K) \qquad (4.16)$$

$$k_{nr} = \{k_{nr}(S_1) + k_{nr}(T_1)K\}/(1+K) \qquad (4.17)$$

On the assumption that the formation yield of the triplet state is close to 1, the emission yield, Φ, is written as

$$\Phi = k_r/k = k_r(1+K)/(k_{S_1} + k_{T_1}K) \qquad (4.18)$$

In cases where k_{S_1} and k_{T_1} are independent of temperature, the A and ΔH values in Eq. (4.11) are readily obtained by measuring the temperature dependence of k, which is given by Eq. (4.10). The k_{S_1} and k_{T_1} values in Eq. (4.10) must be the same as those obtained from the Φ values given by Eq. (4.18).

In general, the rate constants, $k_r(S_1)$ and $k_r(T_1)$, are independent of temperature, while the nonradiative rate constants, $k_{nr}(S_1)$ and $k_{nr}(T_1)$, are represented as a function of temperature:

$$k_{nr}(S_1) = k_{nr}^0(S_1) + k_{nr}^S(S_1)\exp(-\Delta E_S/RT) \qquad (4.19)$$

$$k_{nr}(T_1) = k_{nr}^0(T_1) + k_{nr}^T(T_1)\exp(-\Delta E_T/RT) \qquad (4.20)$$

Here, $k_{nr}^0(S_1)$ and $k_{nr}^0(T_1)$ are the temperature-independent terms, and $k_{nr}^S(S_1)$ and $k_{nr}^T(T_1)$ denote the preexponential factors of the temperature-dependent terms, respectively. As shown in Eqs. (4.19) and (4.20), S_1 and T_1 have the temperature-dependent nonradiative processes with activation energies represented by ΔE_S and ΔE_T, respectively. With the use of Eqs. (4.10)–(4.20), the rate constant, k, for the decay of emission is strictly expressed as a function of temperature.

It is noted that the electronic property and spin multiplicity of the excited state are reflected in the magnitude of the radiative rate constant. Thus, the determination of the k_r values for the emission process is very important for the elucidation of the electronic nature of the emissive excited state.

Photophysical properties of crystalline Cu(I) complexes, which emit TADF, are very sensitive to the molecular structure. As noted previously, the HOMO energy level is markedly affected by the symmetry of the molecular structure in crystals. The LUMO (mainly π^*) level is also changed by intermolecular and/or intramolecular interactions (π–π, CH–π, and hydrogen bond interactions) [49, 50]. The tetrahedral TADF-type Cu(I) complexes studied hitherto are roughly classified into two groups: (i) the complexes that have emissive NN ligands, which indicates that the LUMO is located on these NN ligands, and (ii) the complexes that have the LUMO on PP ligands as shown in Figure 4.5 and Figure 4.6. The chemical structures of the studied ligands are summarized in Figure 4.7.

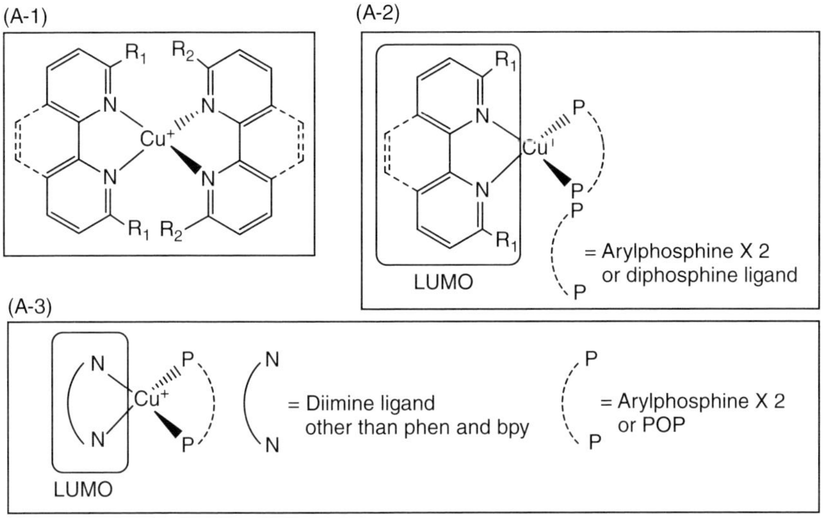

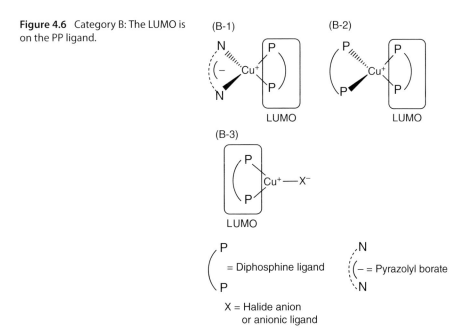

Figure 4.5 Category A: The LUMO is on the NN ligand.

Figure 4.6 Category B: The LUMO is on the PP ligand.

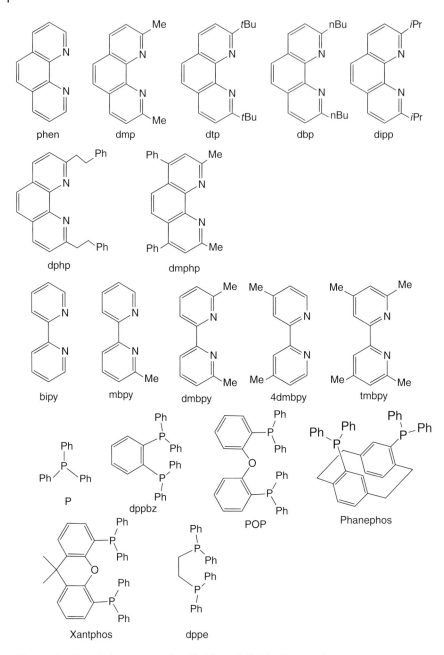

Figure 4.7 Chemical structures of studied ligands (A-2 in Figure 4.5).

The Cu(I) complexes afford little detectable NF because of the very fast ISC process ($\sim 10^{-11}$ s) occurring from S_1 to T_1 [27, 28]. Thus, the triplet yield, Φ_{ST}, is estimated to be ~ 1. In spite of the fact that the SOC interaction of Cu(I) is weak, the Φ_{ST} value of the Cu(I) complexes is outstandingly large.

4.3.1 Bis(diimine) Type

TADF from bisdiimine-type copper(I) complexes ($[Cu(NN)_2]^+$) in solution was first observed in the early 1980s [26]. This type of complexes generally exhibits TADF from the MLCT excited states. Recent experimental results and MO calculations clearly prove the small energy gap between S_1 and T_1 [42, 51, 52]. Based on extensive research concerning the highly emissive complexes, it has been established that the incorporation of bulky substituents into the diimine ligands on the side of the metal center (A-1; Figure 4.5) is very effective in preventing the structural changes of the excited states in both the solutions and solid states. For instance, sterically congested alkyl substituents are generally introduced at the 2- and 9-positions of phenanthroline (or the 6,6′-positions of 2,2′-bipyridine), resulting in a high Φ_{PL} value. Among these complexes, the homoleptic copper(I) bisdiimine complex $[Cu(dtp)_2]BF_4$ **3** (dtp = 2,9-di-*tert*-butyl-1,10-phenanthroline) displays the maximum TADF performance: $\Phi_{PL} = 0.056$ and $\tau = 3.26\,\mu s$ in CH_2Cl_2 under a N_2 atmosphere [53]. Evidently, introduction of the congested alkyl groups into the ligands improves the photophysical properties of the Cu(I) complexes.

4.3.2 $[Cu(NN)(PP)]^+$ Complexes with phen or bipy Derivatives as Ligands

Heteroleptic complexes composed of the NN (phen- or bipy-based ligand) and PP ligands generally exhibit better TADF performance than bis(diimine) complexes (category A-2 in Figure 4.5) [54]. These ionic complexes possess counter anions. Exchange of one diimine ligand in $[Cu(NN)_2]^+$ for the PP ligand generates σ bonds between the Cu(I) and P atoms. Strong back-donation of the coordinated P atoms results in electron withdrawing from Cu(I) in the complexes. Hence, the MLCT interaction between Cu(I) and the diimine ligand in $[Cu(NN)(PP)]^+$ becomes weak, resulting in the shift of the emission peak toward blue in comparison with the emission peak of $[Cu(NN)_2]^+$. The emission yield of $[Cu(NN)(PP)]^+$ is much larger than that of $[Cu(NN)_2]^+$, probably due to the energy gap law [54]. However, the emission yields of $[Cu(NN)(PP)]^+$ in solutions are usually not so high ($\Phi < 0.01$), and the pseudo-Jahn–Teller effect in the excited states is suggested to be responsible for the low emission yields. In 2002, $[Cu(dmp)(POP)]BF_4$ (**8**) containing the *bis*[2-(diphenylphosphino)phenyl]ether (POP) ligand was first reported to display efficient emission with $\Phi = 0.16$ and $\tau = 16.1\,\mu s$ in solutions by McMillin [55]. Since this report, Cu(I) complexes with various combinations of diimine and diphosphine ligands have been synthesized to provide strong TADF materials. In contrast to the discussion mentioned above, recent theoretical studies on $[Cu(NN)(PP)]^+$ complexes suggest that the emissive transition from $[Cu(NN)(PP)]^+$ is dominated by MLCT [56, 57].

Table 4.1 Emission properties of [Cu(NN)(P$_2$ or PP)]$^+$ (A-2 in Figure 4.5).[a]

Compound	λ_{max} (nm)	Φ_{PL} (%)	τ (μs)	Condition	References
[Cu(phen)P$_2$]BF$_4$ **4**	543	14	8.1		
[Cu(dmp)P$_2$]BF$_4$ **5**	509	32	18.1		
[Cu(dbp)P$_2$]BF$_4$ **6**	504	57	32.9	PMMA (20 wt%)	[31]
[Cu(phen)(POP)]BF$_4$ **7**	555	16	4.6		
[Cu(dmp)(POP)]BF$_4$ **8**	527	49	13.2		
[Cu(dbp)(POP)]BF$_4$ **9**	519	69	20.3		
	523	71	2.5, 12.2	Neat film	[58]
[Cu(bpy)P$_2$]BF$_4$ **10**	550	0.4	10.0		
[Cu(4dmbpy)P$_2$]BF$_4$ **11**	537	0.2	10.7		
[Cu(bpy)(POP)]BF$_4$ **12**	559	3.3	9.0		
[Cu(4dmbpy)(POP)]BF$_4$ **13**	544	0.3	11.2	PMMA (5 wt%)	[59]
[Cu(dmbpy)(POP)]BF$_4$ **14**	528	14.5	15.7		
[Cu(bpy)(xantphos)]BF$_4$ **15**	559	3.3	9.0		
[Cu(tmbpy)(xantphos)]BF$_4$ **16**	528	14.5	15.7		
[Cu(tmbpy)(POP)]BF$_4$ **17**	555 (555)	55 (74)	13 (11)	Powder (after grinding)	[60]
[Cu(mbpy)(POP)]PF$_6$ **18**	550 (567)	10.7 (9.5)	6.0 (2.6)	Powder (PMMA)	[61]
[Cu(dmbpy)(POP)]PF$_6$ **19**	529 (535)	38.4 (43.2)	10.9 (10.5)		
[Cu(dmp)(POP)]tfpb **20**	517[b]	88[b]	26[b]		
[Cu(dmp)(xantphos)]tfpb **21**	540[b]	66[b]	30.2[b]	Crystalline film	[62]
[Cu(dipp)(xantphos)]tfpb **22**	513[a]	95[b]	38.5[b]		
[Cu(phen)(dppbz)]ClO$_4$ **23**	553	18.33	2.15, 7.42	Powder	[63]
[Cu(bpy)(dppe)]PF$_6$ **24**	601[b]	<0.5[b]	0.28, 0.53[b]	Powder	[64]
[Cu(4dmbpy)(dppe)]PF$_6$ **25**	583[b]	<0.5[b]	0.19[b]	Powder	
[Cu(dmp)(phanephos)]PF$_6$ **26**	530	80	14	Powder	[65]

a) λ_{max} is the peak wavelength of emission spectrum, Φ_{PL} is the photoluminescence quantum yield, and τ is the emission lifetime.
b) Under Ar.

Photophysical properties are markedly affected by the changes in the environment surrounding the emissive metal complexes. Actually, Cu(I) complexes in the solid state or dissolved in films afford large emission yields owing to the suppression of the structural distortions in the excited state. The emission data obtained with [Cu(NN)(P$_2$ or PP)]$^+$ complexes in the solids and the films are summarized in Table 4.1 (see the chemical structures of the ligands and the complexes in Figure 4.7 and the Appendix 4.A.1) [31, 58–65]. Even in rigid environments, substitution of the diimine ligand with bulky alkyl or phenyl groups at positions close to the central Cu(I) is found to exhibit remarkable increase in emission efficiencies of [Cu(NN)(P$_2$ or PP)]$^+$ (**4–26**). The substituents of the N–N ligands are responsible for suppression of the structural changes in the MLCT excited states. In fact, as shown in Table 4.1, the emission yields tend to increase when the substituent becomes bulky. Further, the TADF from [Cu(NN)(P$_2$ or PP)]$^+$ with large substituent groups in the N–N ligands are found to be blueshifted more than that with small groups. This finding indicates that the large substituents clearly suppress the structural changes occurring from a tetrahedral to a square-planar-like geometry. These facts are explained by assuming that a small geometry distortion of the Cu(I) complexes occurs even in films, leading to the significant effects on TADF.

[Cu(dbp)P$_2$]BF$_4$ **5** with *n*-butyl groups at the 2- and 9-positions of the phenanthroline ligand gives high quantum yield of 57% in PMMA films. Interestingly, [Cu(dbp)(POP)]BF$_4$ (**9**) shows a Φ value of 69%. Complexes **7** and **9** were the first examples of tetrahedral TADF-type emitters used for device fabrication [31]. The best efficiency of 11.0 cd A^{-1} at 1.0 mA cm^{-2} was obtained with the use of an emitting layer of poly(vinylcarbazole) (PVK) containing complexes **7** or **9** (23 wt%) in 2004. The configuration of the OLED devices is as follows: ITO/PEDOT/PVK: **7** or **9** (23 wt%)/BCP/Alq$_3$/LiF/Al. In 2006, light-emitting electrochemical cells (LECs: ITO/**9**/Al) were reported [58]. A current efficiency of 56 cd A^{-1} at 4 V and an EQE of 16% were achieved by the device with the use of complex **9**. In 2012, further optimization, specifically by using the high triplet energy charge transport material as a host and an exciton-blocking layer, successfully enhanced the performance of the devices based on complex **9**. The best efficiency of 49.5 cd A^{-1} and an EQE of 15% were achieved [66]. This high EQE value suggests that complex **9** is a TADF-type emitter. The device structure is as follows: ITO/PEDOT: PSS/PYD2: **9** (10 wt%)/DPEPO/LiF/Al.

Luminescent characteristics of structurally similar [Cu(tmbpy)(POP)]BF$_4$ (**17**) have been investigated in detail [60]. Complex **17** shows efficient yellow luminescence with a λ_{max} of 555 nm and $\Phi_{PL} = 0.55$ in the solid state at 300 K. By cooling down the sample from 300 to 77 K, the emission peak shifts to a longer wavelength by 20 nm (630 cm^{-1}). The S$_1$–T$_1$ gap of less than 720 cm^{-1} for complex **17** was determined by a curve fitting of the emission lifetimes measured at various temperatures. The small S$_1$–T$_1$ gap and the temperature-dependent behavior of the emission maximum indicate that complex **17** in the solid state at room temperature gives rise to TADF. The computational results reveal that the origin of TADF is mainly the MLCT excited state: The LUMOs are localized on the π system of the diimine ligand, while the HOMOs are on Cu(I).

Numerous studies on the related heteroleptic Cu(I) complexes have been carried out to elucidate the relationship between the emission properties and molecular structures in both crystals and solutions. It has been reported that complex **4** crystallized from solutions gives a single crystal in which two different conformers are confined. The fact that the single crystal shows no TADF at room temperature is explained by the π-stacking effects and the intermolecular energy transfer from one conformer to the other [67]. In films, complex **4** is free from the stacking effects and intermolecular energy transfer, resulting in the emission by TADF [31].

As mentioned above, the emission from crystals is quenched by the π-stacking effects, which effectively take place the nonradiative process in the excited states. Thus, the emission yield in crystals is expected to be increased by removal of the π-stacking effect. Actually, complex **7** in crystals is free from the π-stacking effects, and thus, the emission yield is reported to be high [68].

[Cu(NN)(PP)]$^+$, which has the substituent groups on the metal side of the diimine ligands, is unstable in solution and readily yields homoleptic [Cu(NN)$_2$]$^+$

and [Cu(PP)$_2$]$^+$ complexes by a disproportionation reaction [69]. Although isolated heteroleptic complexes [Cu(dmp)(PP)]$^+$ containing two methyl groups in the diimine ligands are very stable in the solid state, the formation of homoleptic complexes via ligand exchange reactions is observed in solutions. The homoleptic/heteroleptic ratio is basically dependent on their relative stabilities. Thus, selection of diphosphine ligands that affect the stability of [Cu(PP)$_2$]$^+$ is very important for the synthesis of heteroleptic Cu(I) complexes with high yields.

Among the PP ligands, the POP ligands are the most popular diphosphine ligands to prepare heteroleptic Cu(I) complexes, because the homoleptic complex [Cu(POP)$_2$]$^+$ is very unstable in solution due to the large bite angle (P–Cu–P = ~115°) [45]. In fact, the isolated homoleptic complex possesses a trigonal geometry with an uncoordinated phosphorus atom more preferentially than a tetrahedral one. As shown in Table 4.1, the emission yields of the [Cu(NN)(POP)]$^+$ complexes are generally larger than those of the heteroleptic complexes with xantphos, dppbz, and dppe ligands.

Recently, [Cu(dmp)(phanephos)]PF$_6$ **26** has been found to exhibit efficient TADF [65]. This complex has the rigid diphosphine ligand, phanephos, which has a large bite angle. The P–Cu–P angle of 116° in complex **26** is very similar to that of [Cu(dmp)(POP)]$^+$. Complex **26** displays a strong green emission with a peak maximum at 530 nm. The value of PLQY obtained with powder is as high as 0.80 at 300 K. The energy gap, $\Delta E(S_1-T_1)$, between S_1 and T_1 was obtained as 1000 cm^{-1} by measuring the rate constants for the decay of emission at various temperatures, indicating that complex **26** is probably a TADF-type emission material. These studies indicate that the highly luminescent Cu(I) complexes have the rigid structures to reduce the rates for the nonradiative processes in the excited states.

4.3.3 [Cu(NN)(PP)]$^+$ Complexes with NN Ligands Other Than phen or bipy Derivatives

The energy levels of LUMOs of the heteroleptic ionic Cu(I) complexes with a phen- or bipy-based ligand are very similar to each other because the Cu(I) complex in this category has a LUMO localized on the π^* system of the diimine ligands. Replacement of a phen- or bipy-based ligand in [Cu(NN)(PP)]$^+$ complexes with other diimine ligands is a common approach to search for highly luminescent Cu(I) emitters. In this context, such heteroleptic copper complexes, which are used in OLEDs as emitters, have been extensively studied and compiled. The chemical structures of the complexes are shown as category A-3 in Figure 4.5, and the chemical structures of the NN ligands investigated are represented in Figures 4.8 and 4.9 (see the chemical structures of the complexes in Appendix 4.A.1). The emission and device data of the complexes described in the literature are summarized in Table 4.2. The PP ligands used for the Cu(I) complexes are mostly limited to POP and two P ligands.

By extending the π system of diimine ligands, the LUMO levels in NN1–NN3 (Figure 4.8) become lower in energy, and the distortions in the excited states are reduced [70]. Complexes **27c** and **28c** with the most bulky and rigid NN3 ligands are found to exhibit high PLQYs in films, and their emission maxima are located

Table 4.2 Emission properties and device structure from [Cu(NN)(P$_2$ or POP)]$^+$ (A-3 in Figure 4.5).[a)]

Compound	λ_{PL} (nm)	Φ_{PL} (%)	Device structure	λ_{EL} (nm)	η_{ext} (%)	References
[Cu(NN1)P$_2$]BF$_4$ **27a**	618	8.0	ITO/PEDOT:PSS/PVK: **27a** or **27b** (10 wt%)/BCP/Alq$_3$/LiF/Al	617	0.08[b)]	[70]
[Cu(NN2)P$_2$]BF$_4$ **27b**	623	10		626	0.08[b)]	
[Cu(NN3)P$_2$]BF$_4$ **27c**	606	56	ITO/PEDOT:PSS/TCCz: **27c** (10 wt%)/TPBI/LiF/Al	606	1.7[b)]	
[Cu(NN1)(POP)]BF$_4$ **28a**	623	6.0	ITO/PEDOT:PSS/PVK: **28a** or **28b** (10 wt%)/BCP/Alq$_3$/LiF/Al	625	0.3[b)]	
[Cu(NN2)(POP)]BF$_4$ **28b**	628	10		529	0.6[b)]	
[Cu(NN3)(POP)]BF$_4$ **28c**	617	43	ITO/PEDOT:PSS/TCCz: **28c** (15 wt%)/TPBI/LiF/Al	518	4.5[b)]	
[Cu(NN4)(POP)]BF$_4$ **29**	585	—	ITO/2-TNATA/NPB/CBP: **29** (6 wt%)/TPBI/LiF/Al	572	—	[71]
[Cu(NN5)P$_2$]BF$_4$ **30**	500[c)]	0.1[c)]	ITO/PEDOT:PSS/PVK: **30** (3.4 wt%)/BCP/Alq$_3$/LiF/Al	497	—	[72]
[Cu$_2$(NN6)P$_4$](BF$_4$)$_2$ **31**	550	15	ITO/PVK: **31** (20 wt%)/F-TBB/Alq$_3$/LiF/Al	589	—	[73]
[Cu(NN7)(POP)]BF$_4$ **32a**	552	—	ITO/m-MTDATA/NPB/CBP: **32a** (7 wt%)/Bphen/Alq$_3$/LiF/Al	572	—	[74]
[Cu(NN8)(POP)]BF$_4$ **32b**	521	—	ITO/m-MTDATA/NPB/CBP: **32b** (15 wt%)/Bphen/Alq$_3$/LiF/Al	528	—	
[Cu(NN9)P$_2$]BF$_4$ **33a**	568	—	ITO/MoO$_3$/NPB/CBP: **33a** (15 wt%)/BCP/LiF/Al	570	—	[75]
[Cu(NN9)(POP)]BF$_4$ **33b**	568	—	ITO/MoO$_3$/NPB/CBP: **33b** (8 wt%)/BCP/LiF/Al	573	—	
[Cu(NN10)(POP)]BF$_4$ **34a**	579	12	ITO/PEDOT:PSS/TCCz: **34** (10 wt%)/BCP/Alq$_3$/LiF/Al	547	—	[76]
Cu(NN11)(POP) **34b**	564	16		555	—	
Cu(NN14)(POP) **35**	481	35	ITO/TAPC/mCP: **35** (8 wt%)/3TPyMB/TmPyPB/LiF/Al	~530	6.6	[77]

(continued)

Table 4.2 (Continued)

Compound	λ_{PL} (nm)	Φ_{PL} (%)	Device structure	λ_{EL} (nm)	η_{ext} (%)	References
[Cu(NN15)(POP)]BF$_4$ **36a**	530	25		—	2.0	[78]
[Cu(NN16)(POP)]BF$_4$ **36b**	549	27	ITO/PEDOT:PSS/PYD2: **36** (5 wt%)/DPEPO/LiF/Al	—	6.1	
[Cu(NN17)(POP)]BF$_4$ **36c**	544	36		—	7.4	
[Cu(NN17)(POP)] BF$_4$ **37a**	470	8	ITO/m-MTDATA/NPB/CBP: **37a** (23 wt%) or **37b** (18 wt%)/Bphen/Alq$_3$/LiF/Al	480	—	[79]
[Cu(NN18)(POP)] BF$_4$ **37b**	525	34		532	—	
[Cu(NN19)(POP)]BF$_4$ **38**	525[c]	0.25[c]	ITO/m-MTDATA/NPB/CBP: **38** (9 wt%)/Bphen/Alq$_3$/LiF/Al	525	—	[80]
[Cu(NN20)(POP)]BF$_4$ **39a**	490[d]	56[d]		516	3.18	[81]
[Cu(NN21)(POP)]BF$_4$ **39b**	465[d]	87[d]	ITO/PEDOT:PSS/26mCPy: **39** (20 wt%)/DPEPO/LiF/Al	504	1.59	
[Cu(NN22)(POP)]BF$_4$ **39c**	492[d]	75[d]		508	8.47	
[Cu(NN23)(POP)]BF$_4$ **40**	518[d]	98[d]	ITO/PEDOT:PSS/czpzpy: **40** (20 wt%)/DPEPO/TPBI/LiF/Al	516	6.36	[82]
[Cu(NN24)(POP)]BF$_4$ **41a**	532	16		537	3.5	[83]
[Cu(NN25)(POP)]BF$_4$ **41b**	537	14	ITO/PEDOT:PSS/PYD2: **41** (5 wt%)/DPEPO/LiF/Al	546	4.6	
[Cu(NN26)(POP)]BF$_4$ **41c**	516	48		526	6.7	
[Cu(NN27)(POP)]BF$_4$ **41d**	517	37		516	8.7	

a) See the Appendix 4.A.1 and 4.A.2 for abbreviations and molecular structures of materials for OLEDs; λ_{PL} is the peak wavelength of photoluminescence spectra of films, Φ_{PL} is the photoluminescence quantum yield in films, and η_{ext} is the external quantum efficiency. Measured at 10.0 mA cm^{-2}.
b) In CH$_2$Cl$_2$ solutions.
c) In the solid state.

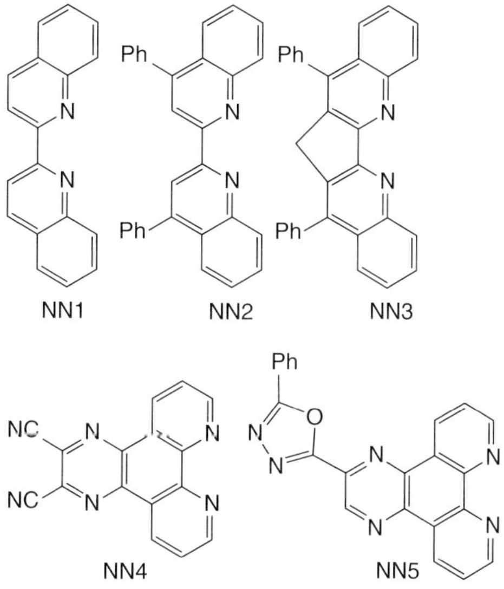

Figure 4.8 Chemical structures of NN ligands 1 (A-3 in Figure 4.5).

at long wavelengths around 620 nm. The device performance of **28c** possessing a POP ligand is much better than that of **27c** with two P ligands. After optimization of the device structures and the dopant concentration, the device with the structure ITO/PEDOT/TCCz: **28c**/TPBI/LiF/Al gives a current efficiency up to 6.4 cd A^{-1} and an EQE of 4.5%.

NN4 in the complex, **29**, and NN5 in the complex, **30**, are both dppz derivatives. The device containing **29** fabricated by the vacuum vapor deposition technique gives a turn-on voltage as low as 4 V, a maximum current efficiency of 11.3 cd A^{-1}, and a peak brightness of 2322 cd m^{-2} [71]. The oxadiazole unit attached to the NN5 ligand is expected to act not only as an electron-transporting role but also as a fence for prevention of π stacking between the molecules in the deposited films. The devices made of **30** as an emitter afford the brightness of 47 cd m^{-2} at 50 A cm^{-2} [72].

Metal complexes having the heterocyclic aromatic ligand, dipyrido[3,2-a:2',3'-c]phenazine (dppz), have been known to show interesting emission properties [84, 85]. Re(I) complexes with dppz were first used as emitters in OLEDs [86].

Diimines consisting of a pyridine and a five-membered heterocyclic compound, e.g. imidazole, benzimidazole, pyrazole, triazole, and tetrazole (NN6–NN27 in Figure 4.9), are commonly used as ligands for the preparation of luminescent metal complexes. The nitrogen atom of these heterocyclic rings is a stronger electron donor than that of pyridine, resulting in shorter Cu—N bond lengths in the Cu(I) complexes. The stability of the complexes increases by using this type of ligand.

In 2005, the dinuclear Cu(I) complex [Cu$_2$(NN6)$_2$P$_4$](BF$_4$)$_2$ **31**, possessing two 2-(2'-pyridyl)benzimidazole ligands, was reported for the first time [73]. With an increase in temperature from 77 K to ambient temperature, complex **31** shows a blueshift of the emission maximum from 563 to 550 nm in a PMMA film. This result suggests that the emission is ascribed to TADF. The MO of the related

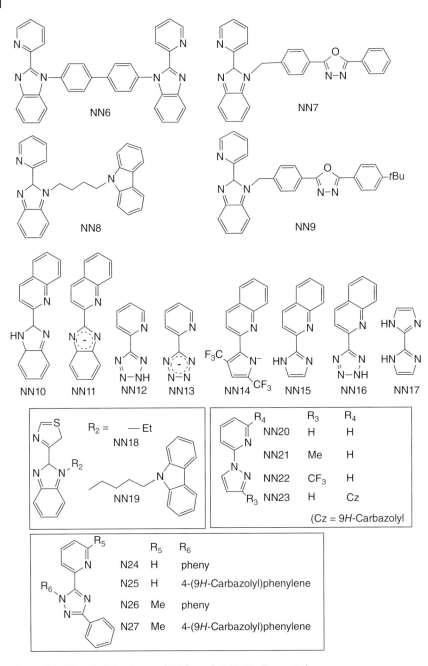

Figure 4.9 Chemical structures of NN ligands 2 (A-3 in Figure 4.5).

mononuclear complex indicate that the electronic nature of the HOMO is principally dominated by 3d orbitals of the Cu(I) ion. However, the contribution from the POP ligand cannot be ignored. The LUMO is mainly distributed on the 2-(2′-pyridyl)benzimidazole unit. Accordingly, the lowest electronic transition is considered to be an MLCT mixed with ligand-to-ligand' charge transfer (LL'CT) (CT from P atoms of the POP ligand to π^* orbitals of the diimine ligand) [87]. The luminescence efficiency of the prototype device [ITO/PVK: **31**(20 wt%)/F-TBB/Alq$_3$/LiF/Al] is low, and yellow-orange electroluminescence (EL) is observed. Most complexes in this category have counterions because the two bidentate ligands are neutral. These complexes are unstable toward sublimation and poorly soluble and/or unstable in nonpolar solvents, and hence, these are not amenable to vacuum deposition or solution processing methods for the preparation of OLEDs. However, the 2-(2′-pyridyl)benzimidazole ligand is attractive because of thermal stability. Thermal analyses demonstrate that the 2-(2′-pyridyl)benzimidazole ligand and NN8 are decomposed at 671 and 588 K, respectively [74]. On the other hand, there seems no obvious sign of decomposition with regard to NN7 at these temperatures. No thermal decomposition of both NN7 and NN8 takes place below 873 K. The Cu(I) complexes, **32a** and **32b**, containing the NN7 and the NN8 ligands are found to be thermally stable below 627 and 595 K, respectively. Therefore, the two complexes are stable enough for us to construct OLED devices by the vacuum deposition method at 573 K. Furthermore, the oxadiazolyl and carbazolyl arms incorporated into NN7 and NN8, respectively, are useful for the electron- and hole-transport processes in the devices.

The OLEDs utilizing **32a** and **32b** as dopants in the CBP emissive layer were fabricated by the vacuum deposition method. The 7 wt% **32a**- and 15 wt% **32b**-doped devices afford the peak EL efficiency of 2.8 cd A^{-1} at 1.1 mA cm^{-2} and 2.2 cd A^{-1} at 1.4 mA cm^{-2}, respectively. The maximum brightness of the device with **32a** was 8669 cd m^{-2} at 14 V. This value is much higher than previously reported values of the devices made from other Cu(I) complexes.

Recently, the relationship between the structural rigidity of the Cu(I) complexes and the performance of devices has been examined with the use of similar complexes, **33a** and **33b** [75]. Both complexes having the NN9 ligand are stable enough to be sublimated during the course of the EL device fabrication. The device fabricated using **33b**, which contains a rigid POP ligand, showed a better performance in comparison with that using **33a**, which is composed of two P ligands having a rigid structure less than POP. The rigidity of the copper emitters seems to be one of the important factors that dominate the device performance. In fact, the device doped with 8 wt% **33b** shows a strong yellow EL with a maximum brightness of 4758 cd m^{-2} at 12.3 V.

Luminescence properties and the device performance are expected to differ between the neutral and ionic Cu(I) complexes. 2-(2′-Pyridyl)benzimidazole ligands are readily transformed from a charge neutral to an anionic ligand by losing a proton in the presence of bases. Thus, it is possible to make both the neutral and ionic Cu(I) complexes with the use of these ligands. Neutral mononuclear Cu(I) complexes (**34b**) and ionic complexes with counterions (**34a**) were synthesized using the NN10 and NN11 ligands in order to compare their luminescence properties and device performance [76]. The neutral complex, **34b**, shows a blueshifted emission with longer lifetimes in comparison with the ionic complex. This result is explained by assuming that although both **34b** and **34a** have radiative transition based on MLCT plus ligand-centered $\pi-\pi^*$ transition (LC), the former possesses LC character much larger than the latter. By doping **34** in TCCz, OLEDs are fabricated with the device structure of ITO/PEDOT: PSS/TCCz: **34**(10 wt%)/BCP/Alq$_3$/LiF/Al. The device with the charge-neutral ligand **34b** exhibits a higher current efficiency than that with **34a**.

A series of neutral Cu(I) complexes with 5-(2-pyridyl)tetrazolate (NN13) and various phosphine ligands is found to show better luminescence properties than the ionic Cu(I) complexes containing the neutral NN12 ligand [88]. Although PLQYs of the ionic complexes are as small as 4–46%, those of the neutral complexes are 89% in the solid state. MO calculations indicate that the neutral complexes emit from the excited state, (ML + IL)CT. With an increase in the temperature from 77 K to ambient temperature, the neutral complex shows a blueshift of the emission maximum in the solid state, which implies that this emission would be TADF.

The charge-neutralized Cu(I) complex affords a high PLQY and undergoes facile sublimation under vacuum. The charge-neutral complex **35** having the anionic ligand NN14 is readily doped in OLEDs by the vacuum deposition method [77], and the peak EL efficiency of the device is obtained as 6.6%. It is suggested that **35** emits phosphorescence because of its strong MLCT character in the excited state, a substantially high triplet yield, Φ_{ISC}, and large radiative rate constant, k_r [88].

An approach to enhance the emission efficiencies of cationic complexes is to suppress the C–H vibrations [78]. According to the energy gap law, high-frequency vibrational modes such as C–H effectively induce nonradiative processes [89, 90]. By replacing these modes with those of lower frequency, the PLQY is expected to increase. The effects of suppression of the excited-state distortion and the C–H vibrational quenching on the PL quantum efficiency have been studied with the use of the complexes, **36a–c**. The ligands of these complexes are systematically changed from N15 to N17 to examine the effects of the C–H vibrational modes on the PLQY. The films doped with the complexes give the PLQY as 0.25 for **36a**, 0.27 for **36b**, and 0.36 for **36c**, respectively. From the Stokes shifts observed for these complexes, it was assumed that the large PLQY obtained with **36c** could be due to the suppression of C–H vibrational quenching. The increase in PLQYs improves the performance of OLEDs: The maximum EQE is 2.0% for **36a**, 6.1% for **36b**, and 7.4% for **36c** [78].

Blue emitters made of Cu(I) complexes are an attractive target to produce full-color displays. A good strategy to create a blue emitter is to synthesize the Cu(I) complex with a large energy separation between the HOMO and LUMO. It is known that (i) the HOMOs of the Cu(I) complexes are mainly confined to the d orbitals of the Cu(I) ion and (ii) the LUMOs are generally distributed over the π^* orbitals of the NN ligand. Thus, a high energy of the LUMOs of the N–N ligand is necessarily required to synthesize the blue emitter. In particular, electron donors of N and/or S introduced into the NN ligands at the α- and α'-positions of a C=N bond raise the energy level of LUMOs as is seen in NN17 and NN18. Actually, emission maxima are observed at 470 nm for **37a** having NN17 and at 525 nm for **37b** having NN18 [79]. These peaks are certainly shifted to high energy in comparison with those of complexes **33–35**, which have low-energy LUMOs. The OLEDs utilizing **37a** and **37b** as dopants were fabricated with a general structure of ITO/*m*-MTDATA/NPB/CBP:**37a** (23 wt%) or **37b** (18 wt%)/Bphen/Alq$_3$/LiF/Al. The EL emission peaks are located at 480 nm for the **37a**-based device and at 532 nm for the **37b**-based device. Although the **37a**-based device affords blue emission color sufficient for practical use, the peak EL efficiency is as low as 1.47 cd A^{-1}. In order to improve the EL efficiency, a structurally similar ligand, NN19, which has the carbazolylbutyl unit as an additional electron-donating group, was prepared. However, contrary to expectations, complex **38** with NN19 gave green emission: The peak wavelength of the emission spectrum is located at 525 nm. The OLEDs utilizing **38** are fabricated by the vacuum deposition method. The 9 wt% **38**-doped device gives a peak EL efficiency of 1.71 cd A^{-1} and a maximum brightness of 1500 cd m^{-2} [80].

As mentioned above, Cu(I) complexes, which emit blue light, have sufficient energy separation between the HOMO and LUMO. Electron-rich pyridylpyrazole ligands NN20–NN22, which are considered to have high-energy LUMOs, were prepared for the synthesis of Cu(I) complexes as blue emitters [81]. Powdered complexes **39a–c** with NN20–NN22 afford emission maxima at 490, 465, and 492 nm with high PLQYs of 56%, 87%, and 45%, respectively. Detailed investigation of the emission lifetimes at various temperatures reveals that these complexes are typical TADF materials with small energy gaps between S$_1$ and T$_1$ (~1400 cm^{-1}). The MO calculations carried out for these complexes support an observation that the energy gaps between S$_1$ and T$_1$ are small enough to show TADF [91]. The solution-processed OLEDs using **39c** exhibit an emission maximum at 508 nm, an EQE of 8.47%, a peak current efficiency of 23.68 cd A^{-1}, and a maximum brightness of 2033 cd m^{-2}.

A new solution-processed method to make OLEDs containing complex **40** is reported [82]. The emissive layers of **40** are readily prepared by spin-coating of the CH$_2$Cl$_2$ solution-dissolved [Cu(CH$_3$CN)$_4$(POP)]BF$_4$ and excess NN23. In this process, NN23 having the carbazole moiety acts as both a ligand and the host material, and thus, no purification of **40** is necessary. The OLED fabricated by this process affords an emission maximum at 514 nm, an EQE of 6.36%, a peak current efficiency of 17.53 cd A^{-1}, and a maximum brightness of 3251 cd m^{-2} at 14.3 V. This emission efficiency is comparable with that of the device made by conventional solution processes using the isolated complex **40**. From the examination of the emission lifetimes at various temperatures, complex **40** was assumed to emit TADF: The energy gap between S$_1$ and T$_1$ was estimated as 1049 cm^{-1}.

Complexes **41a–d** with triazolylpyridine ligands N24–N27 have been used as emitters in OLEDs [83]. The methyl group introduced into the pyridine moiety at a position in the metal side improves the photophysical properties of the Cu(I) complexes: The structural distortion of the complexes is suppressed by the methyl group in the excited state as seen in **41a–d**. However, the carbazole unit introduced into the triazole ring exhibits no effect on the photophysical properties. The OLED using **41d** gives the highest current efficiency of 26.2 cd A^{-1} and EQE of 8.7%.

In summary, owing to the rapid development of TADF-type heteroleptic copper(I) emitters, [Cu(NN)(PP)]$^+$, prototype OLEDs have provided new light sources for practical applications. Interestingly, the HOMO and LUMO of [Cu(NN)(PP)]$^+$ are localized separately: The LUMO distribution is largely confined to the π^* of the NN ligand, whereas the HOMO distribution is essentially confined to the Cu(I) atom and partly to the two P atoms of PP ligand. Thus, localization of the HOMO and LUMO in the Cu(I) complexes is one of the useful guiding principles for the synthesis of the Cu(I) complexes that show TADF.

4.3.4 Tetrahedral Cu(I) Complexes with the LUMO on the PP Ligand

As described in the previous sections, the key units in the [Cu(NN)(PP)]$^+$ complexes responsible for TADF are the aromatic groups of the NN ligand. The LUMO is principally localized on the NN ligand and the HOMO on the Cu(I) atom. The emission occurs from the MLCT (d,π^*) excited states. This section describes mononuclear Cu(I) complexes with the LUMO localized on diphosphine ligands as given in categories B-1 and B-2 in Figure 4.6. The structures of the complexes are shown in Figure 4.10, and the PL and the corresponding device data are summarized in Table 4.3.

Charge-neutral Cu(I) complexes are very attractive as promising emitters for vacuum-deposited OLEDs because they can be readily sublimed under vacuum. From this viewpoint, Cu(I) complexes **42–44** have been prepared [92]. These tetrahedral Cu(I) complexes were composed of dppbz derivatives and the anionic bidentate ligand diphenyl-bis(pyrazol-1-yl)borate (pz$_2$Bph$_2$$^-$) as shown in Figure 4.10a.

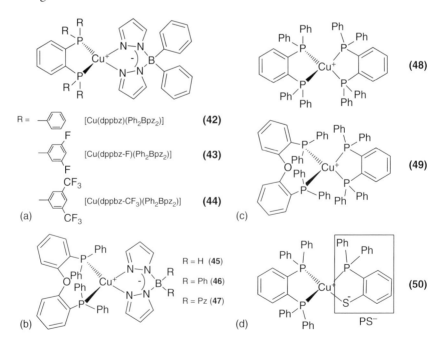

Figure 4.10 Chemical structures of studied Cu(I) complexes (B-1 and B-2 in Figure 4.6).

Table 4.3 Emission properties and device structures from [Cu(NN)$_n$(PP)$_{2-n}$]$^{+ \text{ or } 0}$ (B-1 in Figure 4.6).[a]

Compound	λ_{PL} (nm)	Φ_{PL} (%)	Device structure	λ_{EL} (nm)	η_{ext} (%)	References
[Cu(Ph$_2$Bpz$_2$)(dppbz)] **42**	545	50		552	11.9	
[Cu(Ph$_2$Bpz$_2$)(dppbz-F)] **43**	534	63	ITO/TAPC/mCP: **42–44** (10 wt%)/ 3TPYMB/LiF/Al	545	16.0	[92]
[Cu(Ph$_2$Bpz$_2$)(dppbz-CF$_3$)] **44**	523	68		528	17.7	
Cu(H$_2$Bpz$_2$)(POP) **45**	436[b]	45	—	—	—	[25]
Cu(Bpz$_4$)(POP) **46**	447[b]	90	—	—	—	
Cu(Ph$_2$Bpz$_2$)(POP) **47**	464[b]	90	—	—	—	
[Cu(dppbz)$_2$]BF$_4$ **18**	497[b]	56	ITO/PVK: **48** or **49** (12.5 wt%)/Al	~590	—	[44] (M. Osawa, unpublished data)
[Cu(dppbz)(POP)]BF$_4$ **49**	494[c]	2		~620	—	
Cu(dppbz)(PS) **50**	545	23	ITO/PEDOT:PSS/ PVK/mCP: **50** (10 wt%), TAPC(30 wt%)/ 3TPYMB/LiF/Al	550	7.8	[93]

a) See the Appendix 4.B for abbreviations and molecular structures of materials for OLEDs; λ_{PL} is the peak wavelength of photoluminescence spectra of films, Φ_{PL} is the photoluminescence quantum yield in films, and η_{ext} is the external quantum efficiency.
b) In the solid state.
c) In CH$_2$Cl$_2$ solutions.

Dppbz and POP are common diphosphine ligands used to prepare transition metal complexes. It contains two types of aromatic groups: a bridging o-phenylene group and auxiliary phenyl groups. The pz$_2$Bph$_2^-$ ligand of complexes **42–44** possesses a relatively high-energy π^* orbital in comparison with that of dppbz derivatives, and thus, the LUMOs of these complexes are the π^* orbitals of dppbz derivatives. As indicated in Figure 4.2, there is a strong relationship between the photophysical properties of dppbz and that of the Cu(I) complex because the transition responsible for TADF is regarded as a $\sigma \rightarrow \pi^*$ [MLCT + ILCT].

Thermogravimetric analyses of complexes **42–44** under vacuum indicated that, with an increase in the fluorine contents of the ligands, the Cu(I) complexes tend to become more sublimable. Complexes **42–44** in vacuum-deposited amorphous films have been found to show a strong green emission. Figure 4.11 displays the emission spectra of **42–44** in films at 293 and 77 K. Most probably, the excited-state structures of **42–44** are immobilized in the amorphous films due to the rigidity of the 1,3-bis(carbazol-9-yl)benzene (mCP) host. The bright green emission is considered to arise from an excited state with a tetrahedral structure. Although the half-widths and λ_{max} of emission spectra at 77 K are very similar to those at 293 K, the emission edges of **42–44** on the short wavelength side have been redshifted. Furthermore, the lifetimes of emission at 77 K are

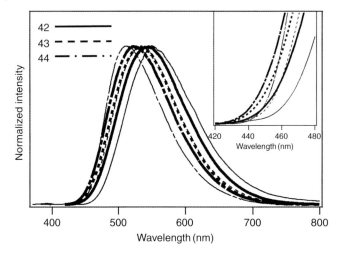

Figure 4.11 Corrected emission spectra of complexes **42–44** in films at 300 (thick line) and 77 K (thin line).

one or two orders of magnitude longer than those at 293 K. These observations indicate that the emission from **42** to **44** at room temperature would be a TADF. An NTO analysis demonstrated that the origin of green luminescence from **42** to **44** is mainly due to a $\sigma \rightarrow \pi^*$ transition (MLCT + **ILCT**).

Bottom-emitting devices with a conventional three-layer structure of ITO (110 nm)/TAPC (30 nm)/mCP + 10% **42–44** (25 nm)/3TPYMB (50 nm)/LiF (0.5 nm)/Al (100 nm) were fabricated. Properties of the devices are shown in Figure 4.12. All the three devices emit bright green light with the emission peak wavelength at 552 nm for **42**, 545 nm for **43**, and 528 nm for **44**. The maximum current efficiencies are determined as 34.6 for **42**, 46.7 for **43**, and 54.1 cd A^{-1} for **44**. These values are obtained with the current density of 0.02 mA cm^{-2} for **42**, 0.20 mA cm^{-2} for **43**, and 0.02 mA cm^{-2} for **44**. The maximum EQEs are evaluated as 11.9%, 16.0%, and 17.7% for devices containing **42**, **43**, and **44**, respectively. These high EQEs are close to those of the PHOLEDs based on rare metal complexes.

There are some fabricated examples of TADF-type OLEDs containing Cu(I) complexes as emitters [21, 22]. However, each of the devices including ours is still prototype. As shown in Figure 4.12, roll-off is serious for our devices. We have not measured the lifetimes of the devices. Such efforts are certainly future works for us after improvement of the devices' performance, e.g. the optimization of peripheral materials.

The complexes in the excited state show an MLCT character that causes flattening motions of the complex in the excited state, resulting in an increase in the rate of nonradiative decay. Thus, the solutions of **42–44** exhibited weak luminescence with $\Phi < 0.02$ at 293 K. However, as mentioned above, complexes **42–44** in rigid films afford strong emissions at room temperature. The EL performance of the devices is strongly related to the PL properties of Cu(I) complexes in the amorphous films. Thus, the detailed studies on the PL properties of the complexes

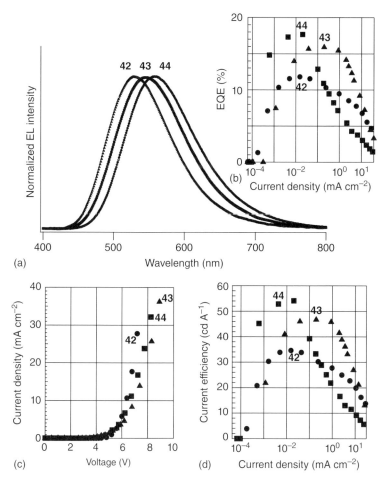

Figure 4.12 Properties of OLEDs containing complexes **42–44**. (a) Electroluminescence spectra. (b) The dependence of the EQE on the current density. (c) I–V characteristics. (d) The dependence of the current efficiency on the current density.

in amorphous films are inevitable prior to manufacturing the devices. The copper(I) complexes presented here possess simple structures and are attractive as emitters for OLEDs because of the fact that these complexes are readily sublimed *in vacuo* for facile doping in the emissive layer of the devices.

Strong blue TADF from structural analogues of the Cu(I) complexes **45–47** was reported [25]. The chemical structures of **45–47** are shown in Figure 4.10b. It is expected that these Cu(I) complexes would be applicable to OLEDs. Heteroleptic or heteroleptic bis(diphosphine) Cu(I) complexes, [Cu(dppbz)$_2$]BF$_4$ (**48**) and [Cu(dppbz)(POP)]BF$_4$ (**49**), represented in Figure 4.10c were applied to OLEDs as emitters [44]. Although complex **48** showed a strong emission with the PLQY of 56% in the solid state, the corresponding device performances are moderate at 490 cd m^{-2} for **48** and 330 cd m^{-2} for **49**. In both cases, the EL spectra are redshifted and significantly broader in comparison with the PL spectra observed in the solid states. This result suggests that the EL performance of the devices is not directly related to the PL properties in the solid states.

Figure 4.10d shows a charge-neutral tetrahedral Cu(I) complex, **50**, containing dppbz and a bidentate anionic ligand, 2-diphenylphosphinobenzenethiolate (PS$^-$) [93]. The luminescence from an amorphous film of mCP doped with 10% complex **50** was investigated. The emission peak wavelength observed at 293 K ($\lambda_{max} = 545$ nm) is redshifted to $\lambda_{max} = 561$ nm at 77 K, and the lifetime at 293 K ($\tau = 2.1\,\mu$s) becomes c. 350 times shorter than that at 77 K ($\tau = 780\,\mu$s). These observations suggest that luminescence from **50** in the amorphous film at 293 K is ascribed to TADF. In fact, the energy gap between S$_1$ and T$_1$ in **50** was estimated to be around 600 cm^{-1} from the temperature dependence of the decay rate constants of emission. NTO analyses of the MO along with the hole and electron maps reveal that the major transitions responsible for TADF are concerned with two types of LL'CT: One is the CT transition of an electron from sulfur to an empty antibonding π orbital on the phenylene and phenyl rings in the dppbz ligand (S → π^*), and another is that from the π orbital on the phenylene ring in the PS ligand to an empty antibonding π orbital on the phenylene and phenyl rings in the PP ligand ($\pi \to \pi^*$). The thiolate ligand (PS$^-$) with electron-donating character reduces the contribution of the metal orbitals to the HOMOs of the complexes, resulting in the decrease in the MLCT character of the excited states. It is interesting that complex **50** exhibits TADF from an LL'CT transition, but not from the MLCT or $\sigma \to \pi^*$ transitions. The quantum efficiency of complex **50** in films is 23% at most.

Complex **50** is thermally stable with a decomposition temperature higher than 300 °C. However, vacuum deposition of **50** was unsuccessful because of the extremely low vapor pressure of the complex. Thus, a prototype OLED containing **50** was prepared via a wet process. This device containing **50** exhibited green luminescence with a current efficiency of 21.3 cd A^{-1} and a maximum EQE of 7.8%. Since Φ_{PL} of complex **50** in films is 23%, the EQE value is reasonable as shown in Table 4.3.

4.3.5 Charge-Neutral Three-Coordinate Cu(I) Complexes

In this section, we describe the preparation and the photophysical properties of charge-neutral three-coordinate Cu(I) complexes containing phosphine ligands. The structure of the complex is shown as category B-3 in Figure 4.6.

Monodentate halide anions such as Cl$^-$, Br$^-$, and I$^-$ are very popular ligands to prepare luminescent multinuclear Cu(I) complexes, particularly dinuclear and tetranuclear complexes possessing diamond $\{Cu_2X_2\}_n$ units (X = Cl, Br and I, $n = 1$ or 2). These units are naturally occurring core structures. Hitherto, over a thousand complexes composed of $\{Cu_2X_2\}_n$ cores and various organic ligands have been reported [50, 94, 95]. In 2007, dinuclear Cu(I) complexes, [(dppbz)CuX]$_2$, were found to exhibit efficient green emission with $\Phi_{PL} = 0.6$–0.9 in the solid state (see next Section 4.4 in detail) [96]. This finding encouraged us to prepare a mononuclear three-coordinate Cu(I) complex with the use of dppbz derivatives having congested structures to avoid the formation of dinuclear Cu(I) complexes.

The three-coordinate complexes having molecular weights lower than those of the dinuclear complexes are expected to sublime readily in vacuum and are suitable for the preparation of OLEDs by the vacuum deposition method. We initially prepared simple three-coordinate Cu(I) complexes [(L$_{Me}$)CuX] [X = Cl (**51**), Br (**52**), or I(**53**)] by using a chelating diphosphine ligand [L$_{Me}$ = 1,2-bis(o-ditolylphosphino)benzene] [97]. Further, diphosphine ligands L$_{Et}$ and L$_{iPr}$, possessing the ethyl and isopropyl substituents, respectively, were synthesized [98]. The luminescence properties of diphosphine ligands, L$_{Me}$, L$_{Et}$, and L$_{iPr}$, were found to be almost identical. The structures of Cu(I) complexes [(L$_{Et}$)CuBr] **54** and [(L$_{iPr}$)CuBr] **55** are also shown in Figure 4.13. Listed in Table 4.4 are the emission peak wavelengths, λ_{max}, the quantum yields, Φ, and the lifetime, τ, observed for the Cu(I) complexes in films at 293 K.

Single-crystal X-ray diffraction studies reveal that complexes **51–53** and **55** possess monomeric three-coordinate structures. The molecular structure of **52** is shown in Figure 4.3b as an example. The coordination geometries of the copper centers in **51–55** are trigonal planar. The sums of the angles around the Cu(I) center are 359.66° for **51**, 359.37° for **52**, 359.43° for **53**, and 360.00° for **55**. The o-methyl groups of L$_{Me}$ are essential for the formation of the three-coordinate complexes. In fact, dppbz, which lacks methyl groups, only forms the halogen-bridged binuclear copper complex [(dppbz)Cu(μ-X)]$_2$. This is probably because the Cu$_2$X$_2$ diamond core in [(L$_{Me}$)Cu(μ-X)]$_2$ would be highly unstable due to the steric hindrance of the o-methyl groups located on the side

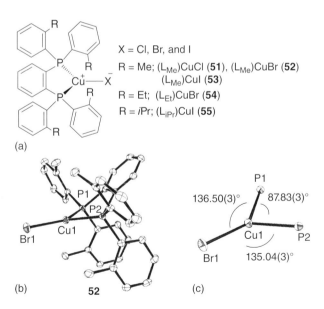

Figure 4.13 (a) Molecular structures of **51–55**. (b) ORTEP view of **52**. Thermal ellipsoids are drawn at the 50% probability level, and H atoms have been omitted for clarity. (c) Core structure of **52**.

Table 4.4 Emission properties of charge-neutral three-coordinate Cu(I) complexes (B-2 in Figure 4.6).[a]

Compound	T = 293 K			T = 77 K			References
	λ_{max} (nm)	Φ (%)	τ (μs)	λ_{max} (nm)	Φ (%)	τ (ms)	
(L$_{Me}$)CuCl **51**	517	67	6.1, 4.6	518	84	2.8, 0.67	
(L$_{Me}$)CuBr **52**	513	71	5.5, 3.9	506	95	1.3, 0.16	
(L$_{Me}$)CuI **53**	504	57	3.2, 1.4	487	83	0.51, 0.15	[97, 98]
(L$_{Et}$)CuBr **54**	510	66	8.8, 3.4	—	—	—	
(L$_{iPr}$)CuBr **55**	508	64	8.4, 2.6	—	—	—	
(L$_{Me}$)CuSPh **56**	488[b]	95[b]	6.6[b]	481	95	1.1	
(L$_{iPr}$)CuSPh **57**	500[b]	95[b]	5.0[b]	504	95	1.9	[99]
P$_2$Cucbz **58**	461[c]	24[c]	11.7[c]	—	—	—	
P$_2$CuNPh$_2$ **59**	521[c]	23[c]	3.17[c]	—	—	—	[100]
(POP)CuNPh$_2$ **60**	563[c]	18[c]	1.70[c]	—	—	—	
[(NHC1)Cu(phen)]OTf **61**	—	1.5	0.23, 1.1	—	—	—	
(NHC1)Cu(NN12) **62**	630	35	24.7	—	—	—	
(NHC1)Cu(NN28) **63**	560[d]	17[d]	10[d]	570[c]	58[c]	—	
(NHC1)Cu(NN29) **64**	594[d]	3.2[d]	1.3[d]	574[c]	68[c]	—	[101–104]
(NHC1)Cu(NN30) **65**	592[d]	1.4[d]	0.5[d]	560[c]	77[c]	—	
(NHC1)Cu(NN31) **66**	590[d]	2.4[d]	1.1[c]	555[c]	61[c]	—	
(NHC1)Cu(NN32) **67**	475[b]	76[b]	11[b]	490[b]	91[b]	34 μs[b]	
(NHC2)Cu(NN32) **68**	575[b]	73[b]	34[b]	585[b]	80[b]	21 μs[b]	

a) λ_{PL} is the peak wavelength of photoluminescence spectra of films, Φ_{PL} is the photoluminescence quantum yield in films, and τ is the decay time of emission.
b) In the solid state.
c) In methylcyclohexane solutions.
d) In cyclohexane solutions.

of the metal centers, and thus, unusual monomeric three-coordinate structures are produced. The introduction of a bulky substituent on the ligand appears necessary to produce the three-coordinate copper complexes.

The PL properties of **51–55** are presented in Table 4.4. Complexes **51–55** emit intense green phosphorescence in degassed CH$_2$Cl$_2$ solutions at 293 K. The luminescence quantum yields (Φ) of 0.43, 0.47, 0.60, 0.43, and 0.50 were obtained for **51**, **52**, **53**, **54**, and **55**, respectively. The phosphorescent lifetimes (τ) of the three complexes were measured by laser excitation at 355 nm: τ = 4.9, 5.4, 6.5, 3.8, and 8.3 μs for **51**, **52**, **53**, **54**, and **55**, respectively. Emission maxima (λ_{max}) of **51–53** are in the order **51** < **52** < **53**, suggesting that λ_{max} is affected by the ligand field strength (I$^-$ < Br$^-$ < Cl$^-$). Presumably the electronic nature of the triplet excited state of **51–53** is influenced to some extent by X$^-$ → π*(L$_{Me}$) CT transitions.

NTO analysis reveals that these transitions can be described as single hole–electron pairs and reproduces over 95% of the change in electron density upon excitation. The hole (approximately HOMO) and electron (approximately LUMO) distributions of **52** in the excited triplet state with the T$_1$ optimized

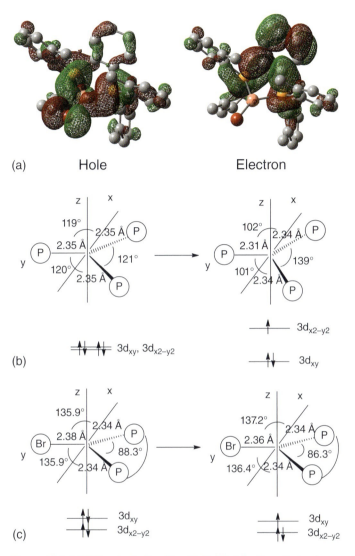

Figure 4.14 (a) NTO pairs for T_1 of $(L_{Me})CuBr$ **52** at the T_1 optimized geometry. (b,c) The optimized core structure of models in S_0 (left) and T_1 (right) and molecular orbital diagrams (below). (b) $[Cu(PMe_3)_3]^+$ and (c) $(L_{Me})CuBr$ **52**.

geometry are shown in Figure 4.14 (a). The hole distribution of **52** is essentially confined to the halogen atom but also extended slightly to the σ orbitals between the Cu(I) and P atoms. It is known that Cu d orbitals mix well with p orbitals of the Br atoms as in the case of the $\{Cu_2X_2\}$ core [105, 106]. Thus, the hole distribution spreads over the Br, Cu, and P atoms. On the other hand, the electron distribution is largely confined to the o-phenylene group in the L_{Me} ligand. Excited-state calculations carried out for **51–53** indicate that the emission results from the transition $(\sigma + X) \leftarrow \pi^*$.

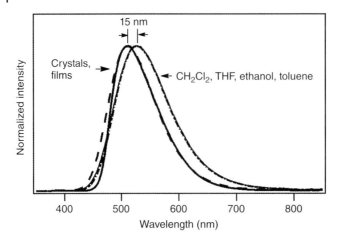

Figure 4.15 Corrected emission spectra of (L$_{Me}$)CuBr **52** in various matrices.

Omary observed a Jahn–Teller distortion from Y- to T-shaped geometry in the triplet excited state of trigonal planar Au(I) complexes containing three monophosphine ligands ([Au(PR$_3$)$_3$]$^+$) [107, 108]. Our calculations suggest that the Cu(I) complex, [Cu(PMe$_3$)$_3$]$^+$, with a structure analogous to that of [Au(PMe$_3$)$_3$]$^+$, undergoes similar distortion in the excited state. The optimized core structures of [Cu(PMe$_3$)$_3$]$^+$ and complex **52** in the ground state and triplet excited state are shown in Figure 4.14b,c. This type of distortion seems to be inevitable for [Cu(PMe$_3$)$_3$]$^+$ with an equilateral triangle structure because the two orbitals, 3d$_{xy}$ and 3d$_{x2-y2}$, are degenerate in the ground state. Since, upon excitation, an electron is transferred from the HOMO to the LUMO (4p$_z$ plus 4 s), the degeneracy of the two orbitals is resolved in the excited state, leading to the T shape of the complex. On the other hand, complex **52** has an isosceles triangle structure, suggesting that the two 3d orbitals are not fully degenerate in the ground state. Besides, the P–Cu–P angle is tightly fixed to c. 90° by the rigid L$_{Me}$ ligand. The structural features of **51–53** largely prevent the distortion. The small distortion of the excited states of **51–53** is assumed to reduce the rate of nonradiative decay, leading to a high Φ.

Complex **52** emits intense green phosphorescence in various matrices, dichloromethane, tetrahydrofuran, ethanol, toluene, films, and crystals at 293 K as shown in Figure 4.15. Spectral shifts hardly depend on the nature of matrices, suggesting that there is little structural change between the ground and excited states. This observation is consistent with the NTO analysis.

Figure 4.16 shows the emission lifetime, τ, measured for complex **52** in crystals between 77 and 300 K. The plot of τ versus T is well explained by the model of two excited states, S$_1$ and T$_1$. The S$_1$–T$_1$ energy gap, $\Delta E(S_1-T_1)$, of complex **52** is obtained as 810 cm^{-1} by the curve fitting of the plot of τ versus T with the use of Eq. (4.21) [98]:

$$\tau = \frac{3 + \exp(-\Delta E(S_1 - T_1)/k_B T)}{3/\tau_T + 1/\tau_S \exp(-\Delta E(S_1 - T_1)/k_B T)} \tag{4.21}$$

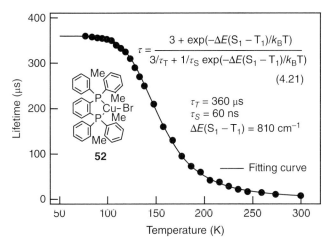

Figure 4.16 Emission decay time of (L$_{Me}$)CuBr **52** in crystals versus temperature. The parameters described in the inset were determined from a fit of Eq. (4.21) for (L$_{Me}$)CuBr **52**.

Complexes **51**–**55** afford small S$_1$–T$_1$ energy gaps, 600–830 cm^{-1}, in the solid state, indicating that the emission at room temperature is probably due to TADF.

The spectrum of phosphorescence from T$_1$ is commonly located at wavelengths longer than that of fluorescence from S$_1$. The TADF mechanism claims that the emission spectrum shows a redshift upon cooling [46]. Since the thermal population of the lower excited state, T$_1$, becomes dominant at low temperatures, phosphorescence from the T$_1$ state precedes TADF from the S$_1$ state, resulting in the redshift of emission from TADF materials. However, complexes **52**–**53** show a small blueshift in the emission spectra when the temperature is decreased from 293 to 77 K, as shown in Table 4.4. Such a phenomenon has been occasionally found for the materials with TADF [97–99]. The blueshift of the emission observed by lowering the temperature is presumably related to the energy relaxation processes in the excited states. However, further studies are necessary for full understanding of the blueshift of emission observed for TADF systems at low temperatures.

Conventional bottom-emitting devices containing **51**–**55** were fabricated using the vacuum deposition method. The devices had a three-layer structure of indium tin oxide ITO/TAPC/mCP: 10 wt% complex **51**–**55**/3TPYMB/LiF/Al. All of the devices containing complexes **51**–**55** emit bright green light with a maximum emission peak at 513–529 nm. The emission spectra of the OLEDs are in good agreement with the PL spectra of the complexes in the films. The maximum current efficiencies of OLEDs using complexes **51**–**55** were 55.6–69.5 cd A^{-1} at a current density of 0.01 mA cm^{-2}. At the maximum current efficiencies, the maximum EQEs were obtained as 18.6–2.5%. The excellent performance of the devices can be ascribed to the TADF-type emission in these systems. The Cu(I) complexes **51**–**55** emit from a thermally populated excited singlet state at an ambient temperature because of a small singlet–triplet (S$_1$–T$_1$) energy gap. Figure 4.17 shows (i) the EL spectra of the devices, (ii) I–V

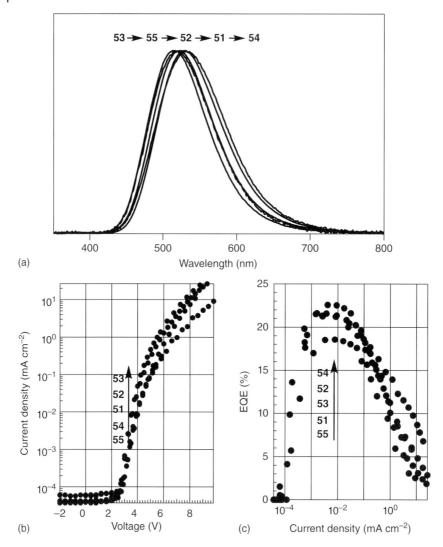

Figure 4.17 Properties of OLEDs containing **51–55**. (a) EL spectra. (b) I–V characteristics. (c) The dependence of the EQE on the current density.

characteristics, and (iii) the dependence of the EQE on the current density. Table 4.5 lists V_{on}, λ_{max}, EQE_{max}, $\eta_{c,max}$, and CIE.

There are some fabricated examples of TADF-type OLEDs containing Cu(I) complexes as emitters [21, 22]. Each of the devices including ours, however, is still prototype. We believe that further studies are necessary to manufacture practical devices with high luminance and long operating life after improvement of the devices' performance, e.g. the optimization of peripheral materials.

The three-coordinate Cu(I) complexes, Cu(L$_{Me}$)(SPh) **56** and Cu(L$_{iPr}$)(SPh) **57**, which afford TADF from the LL'CT excited state, are prepared by replacing halides with arylthiolate anions [99]. The chemical structures of **56** and **57**

Table 4.5 OLEDs' performance characteristics.

Complex	V_{on} (V)[a]	EL emission, λ_{max} (nm)	EQE_{max} (%)[b]	$\eta_{c,max}$ (cd A^{-1})[c]	CIE coord. x/y
51	3.3	527	21.1	67.7	0.30/0.55
52	3.0	517	21.3	65.3	0.29/0.54
53	3.1	513	21.2	62.4	0.26/0.51
54	3.3	529	22.5	69.4	0.32/0.54
55	3.3	515	18.6	55.6	0.26/0.51

a) The voltage required to reach a brightness of 1 cd m^{-2}.
b) Maximum external quantum efficiency.
c) Maximum current efficiency

are shown in Figure 4.18a. The Cu(I) complexes produce intense blue-green emission with the quantum yields as high as approximately 1.0 at both 293 and 77 K in the solid state. Small S_1-T_1 gaps ($\Delta E(S_1-T_1) = 690$ cm^{-1} for **56** and 630 cm^{-1} for **57**) indicate that the emission from **56** and **57** in the solid state at room temperature is presumably ascribed to TADF.

MO calculations reveal that the major transitions (~95%) that contribute to the emission from **56** and **57** are attributed to the two types of LL'CT. One is the CT from the sulfur atom to an empty antibonding π orbital on the phenylene and tolyl rings of the diphosphine ligand (S → π^*), and another is from the π orbital on the aryl ring of the thiolate ligand to an empty antibonding π orbital on the phenylene and tolyl rings of the diphosphine ligand ($\pi \to \pi^*$). Conversely, MLCT contributions are very small: ~2–3% for **56** and **57**. Although the transitions responsible for the emission in complexes **56** and **57** are almost the same as that of complex **50**, the PL efficiencies are very different: PLQY of complex **50** is as low as 0.23. Further studies are necessary to elucidate this difference. Complexes **56** and **57** were thermally unstable, and thus, we could not make OLEDs with them using the vacuum deposition method.

LL'CT emission from other three-coordinate Cu(I) complexes, **58–60**, composed of P(P) and amide ligands has been reported [100]. The chemical structures of **58–60** are shown in Figure 4.18b, and their photophysical data are summarized in Table 4.4. The calculated HOMO is localized over the entire amide ligand, while the LUMO is localized on the π^* orbital of the phosphine ligands. The emission, therefore, is assumed to occur from the LL'CT excited states. On the other hand, the MLCT character is very small. It seems possible to tune the emission color by a careful choice of amide and phosphine ligands as is exemplified by complexes **58**, **59**, and **60**, which exhibit blue, green, and yellow emission, respectively. Relatively short lifetimes imply that the emission could be ascribed to TADF.

N-Heterocyclic carbene (NHC) Cu(I) complexes with three-coordinate structures (**61–68**) have been reported [101–104]. The chemical structures of **61–68** are shown in Figure 4.18c and the Appendix 4.A.1., and their photophysical data are summarized in Table 4.4. The carbene complexes possessing a LUMO confined in the NN ligands emit both phosphorescence and TADF. According to the MO calculations, the structural changes due to the rotational motion around

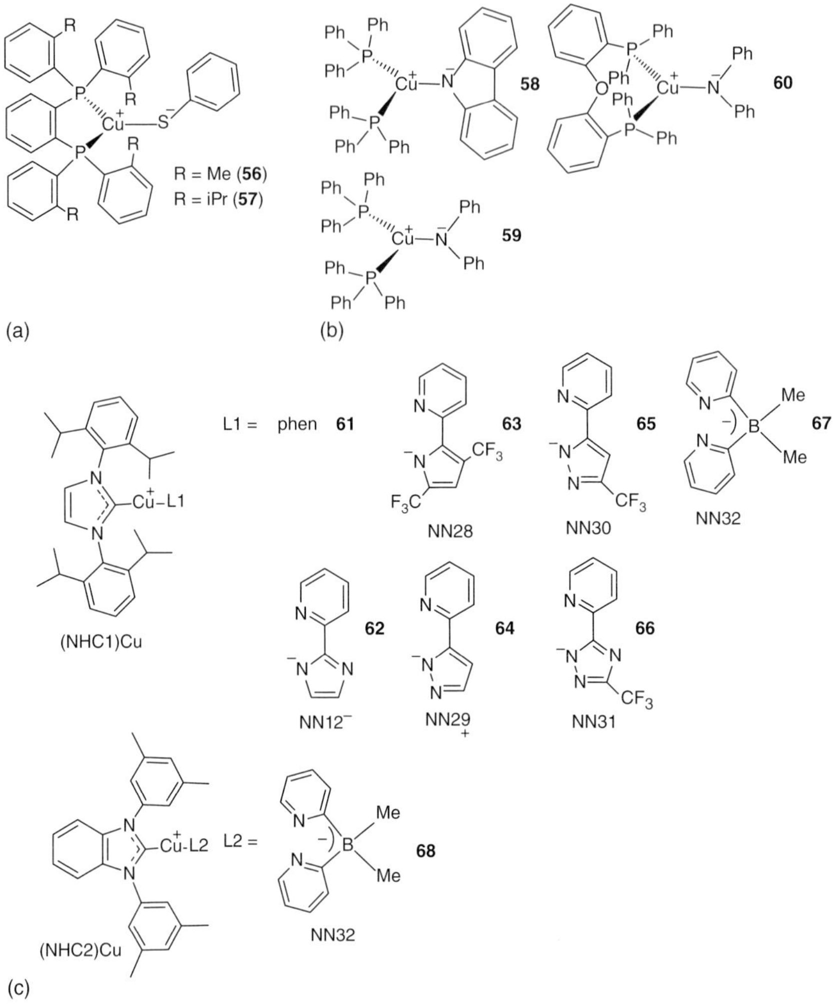

Figure 4.18 Chemical structures of three-coordinate Cu(I) complexes **56–68**.

the C_{NHC}—Cu bond are closely related to the luminescent characteristics. It was found that the change in the torsion angle between the two ligands induces the change in the energy gap between S_1 and T_1. A large torsion angle results in a large energy gap ($\Delta E(S_1-T_1)$), leading to an increase in the intensity of phosphorescence and a decrease in the intensity of TADF. Actually, complex **68** with a large torsion angle exhibits pure phosphorescence ($\tau = 18$ μs) at ambient temperature, whereas complex **67** with a small torsion angle displays emission ($\tau = 11$ μs; phosphorescence 38% and TADF 62%). The energy gaps of **67** and **68** have been estimated as 740 and >3000 cm^{-1}, respectively. This system gives a good guideline for designing TADF-type materials [104].

The synthesis and photophysical properties of emissive charge-neutral three-coordinate Cu(I) complexes were described in this section. These results suggest that three-coordinate copper(I) complexes are promising EL materials in terms of emission efficiency and thermal stability.

4.4 Dinuclear Cu(I) Complexes for OLEDs

Since the early 1970s, multinuclear, in particular di-, tri-, and tetranuclear, Cu(I) complexes displaying intriguing emission characteristics have been extensively studied [50, 94, 96, 109, 110]. The $\{Cu_2(\mu\text{-}X)_2\}_n$ complexes (where X = Cl, Br, and I, n = 1 or 2) have a common core structure, $Cu_2(\mu\text{-}X)_2$, frequently found in such multinuclear Cu(I) complexes, which exhibit luminescence thermochromism and luminescence rigidchromism caused by the structural changes in the excited state. Such sensitivity to external stimuli is an unfavorable character for optoelectronic materials. Thus, prevention of structural changes in the excited state might be necessary to synthesize strongly luminescent multinuclear Cu(I) complexes. In this section, dinuclear Cu(I) complexes that are used in OLEDs have been briefly surveyed. The chemical structures of the complexes examined are illustrated in Figure 4.19.

4.4.1 Dinuclear Cu(I) Complexes Possessing $\{Cu_2(\mu\text{-}X)_2\}$ Cores

In 2007, the preparation of dinuclear copper(I) complexes [(dppbz)Cu(μ-X)]$_2$ (X=Cl (**69**), Br (**70**), and I (**71**)) and their application to the vapor-deposited OLEDs were reported for the first time [96]. Complexes exhibited an efficient green emission with Φ_{PL} = 0.6–0.9 in the solid state, and the origin of emission was regarded as (M + X)LCT from DFT calculations. By measuring the temperature-dependent emission spectra and lifetimes at various temperatures, the energy gap between the ^{1}(M + L)CT and ^{3}(M + L)CT states in complex **71** was found to be as small as ~2.0 kcal mol^{-1}. Thus, the efficient green emission is probably ascribed to TADF. Since the EL peak (565 nm) is redshifted in comparison with the PL peak (502 nm) observed in the solid state, the EL is assumed to occur from the "flattened" square-planar-like structure in the excited state. Probably, the viscosity of the amorphous host is lower than that of the solid state, allowing the structural change from tetrahedral to a flattened one at room temperature. The configuration of OLED devices was as follows: ITO/PF01/CBP: **71** (10 wt%)/Bphen/KF/Al. The current efficiency of 10.4 cd A^{-1}, power efficiency with 4.2 lm W^{-1} at 93 cd m^{-2}, and maximum EQE with 4.8% were obtained with the device. In 2012, further optimization was carried out by using the high triplet energy charge transport material as a host with an exciton-blocking layer. The best efficiency of 30.6 cd A^{-1} and an EQE of 9.0% were achieved [66]. These results suggest that the dinuclear Cu(I) complexes are promising emitters of OLED.

In 2011, a simple method to fabricate devices via coevaporation of (CuI)$_2$ and pyridine-based ligands was reported and is shown in Figure 4.19 [111]. The distinctive feature of this process is that the excess organic ligand for Cu(I) also serves as a host matrix. The chemical structure of the Cu(I) complex

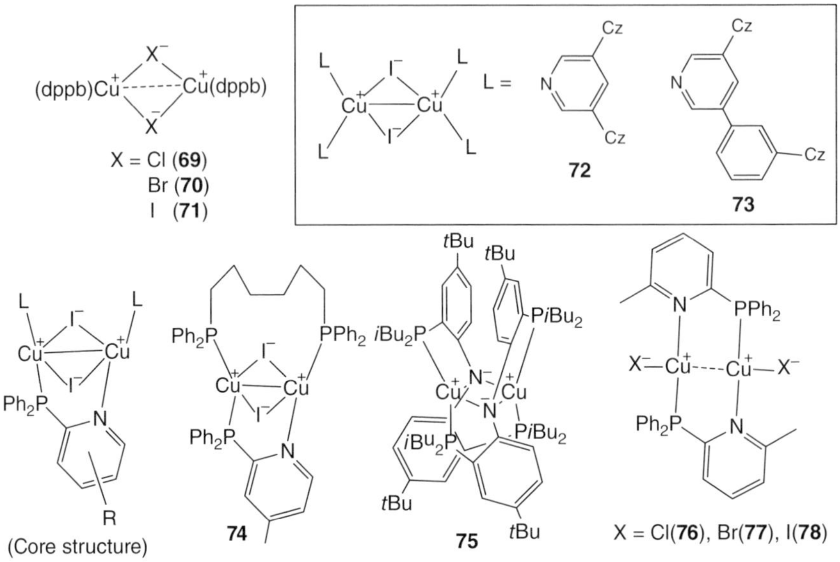

Figure 4.19 Chemical structures of dinuclear Cu(I) complexes **69–78**.

in a codeposited CuI ligand film was characterized by X-ray absorption fine structure (XAFS) including X-ray absorption near edge structure (XANES) and extended EXAFS. By using the codeposited film as an emissive layer, the best OLED (ITO/MoO$_3$/CBP/CuI: CPPyC **73** (4 wt%)/TPBi/LiF/Al) gives a maximum EQE as high as ∼15.7% at a luminance of 100 cd m^{-2} [112].

The application of halogen-bridged dinuclear Cu(I) complexes possessing an additional bridging ligand, diphenylphosphinopyridine, in OLEDs was reported in 2012 [113]. The core structure of the complex is shown in Figure 4.19. The geometry of the {Cu$_2$(μ-X)$_2$} unit in this core is not planar but butterfly-shaped due to the bridging ligand [114]. The combination of the {Cu$_2$(μ-X)$_2$} core with various mono- and bidentate N-heteroaromatic ligands and an ancillary ligand, PPh$_3$, is a well-known method to synthesize efficiently luminescent Cu(I) complexes [95, 110]. This synthetic strategy was also applied to prepare luminescent mononuclear Cu(I) complexes [115]. Nonetheless, the core structure is well designed to have a separated HOMO and LUMO. The LUMO is localized on the pyridine moiety, whereas the HOMO is distributed on the {Cu$_2$(μ-X)$_2$} unit. The separation of the HOMO and LUMO is an important factor for the complexes to afford highly efficient TADF (see Section 4.3) and enables us to change the emission color by the introduction of substituents into the pyridine ring (R group in the core structure in Figure 4.19). A series of Cu(I) complexes having this core structure with various substituents on the pyridine ring and other series of complexes possessing the bridging ligand bearing five-membered heterocyclic moieties (see Section 4.3.3) have been reported [116–121]. This strategy mentioned above has been successfully applied to the synthesis of Cu(I) complexes exhibiting emission in a wide range of color from 458 to 713 nm.

A double-bridged dinuclear Cu(I) complex, **74**, was recently demonstrated to give an internal quantum efficiency close to unity in a solution-processed film. The $\Delta E(S_1-T_1)$ of **74** is estimated as 726 cm^{-1} by measuring the emission peaks at room temperature and at 77 K. The chemical structure of **74** was identified by performing X-ray absorption spectroscopy. Furthermore, it is reported that an optimized device (ITO/PEDOT:PSS/PLEXCORE UT-314/PYD2:**74**(30 wt%)/3TPYMB/LiF/Al) gives a maximum EQE of 23% (73 cd A^{-1}) [122].

4.4.2 Other Dinuclear Cu(I) Complexes

TADF-type OLEDs containing the neutral dinuclear Cu(I) complex **75** having the two amido-bridged core were reported in 2010 and are shown in Figure 4.19 [123]. The rigid structure of complex **75** with two bis(diisobutylphenylphosphino) amido (PNP) ligands certainly enhances the PLQY up to 68% in solutions. The small S$_1$–T$_1$ energy gap of 740 cm^{-1} determined by fitting of the plot of τ (emission lifetimes) versus T (80–295 K) indicates that the efficient emission from **75** is TADF. A calculation of the HOMO and LUMO of **75** reveals that the HOMO is mainly distributed on the N atoms of the amido groups and slightly on the Cu atoms. The LUMO distribution is largely confined to the phenylene groups on the PNP ligands. Thus, the HOMO–LUMO transition is almost pure CT (MLCT + ILCT) in nature, leading to the small energy gap between S$_1$ and T$_1$ of complex **75**. Vapor-deposited OLEDs (ITO/CFx/TAPC/CBP: TAPC(25 wt%)/CBP: TAPC(20 wt%): **75**(0.2 wt%)/CBP/BAlq-13/LiF/Al) doped with the complex in the emissive layer gave a maximum EQE of 16.1%. The roll-off observed for the EL devices with increasing current density may be explained by the quenching of the excited state by holes and/or ionization of the excited state under the influence of the electric field on the basis of the examination of PL intensity from the hole-only device (ITO/CFx/TAPC/TAPC: **75**(8 wt%)/TAPC/Al).

Dinuclear Cu(I) complexes (**76–78**) exhibiting relatively short-lived emissions from both S$_1$ and T$_1$ in the thermal equilibrium at an ambient temperature have been reported [124]. These complexes are similar to complex **67** (see Section 4.3.4) displaying mixed emission (τ = 11 μs; phosphorescence 38% and TADF 62%). This emission mechanism can be regarded as a four-excited-state model in which S$_1$ and three triplet substates are in the thermal equilibrium. Complex **76** in powder form shows bright blue emission at 300 K with an emission peak wavelength at 485 nm, a decay time of 8.3 μs, and a quantum yield of 0.92. The emission consists of 20% phosphorescence and 80% TADF. Obviously a large SOC is necessary for the triplet state to emit phosphorescence. The large splittings of triplet substates of **76** are determined as 15 and 7 cm^{-1} for ΔE(III–I) and ΔE(II–I), respectively. On the other hand, the energy gap between S$_1$ and T$_1$ is 930 cm^{-1}. A balance of ΔE(ZFS) and $\Delta E(S_1-T_1)$ is very important to design this type of complexes.

4.5 Another Group of Metal Complexes Exhibiting TADF

In Sections 4.3.4 and 4.3.5, Cu(I) complexes exhibiting efficient TADF with dppbz derivative ligands have been described. The complexes are divided into

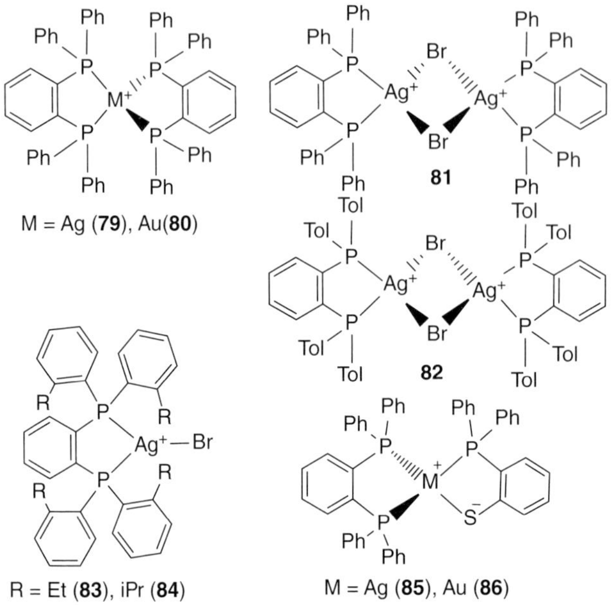

Figure 4.20 Chemical structures of silver(I) and gold(I) complexes **79–86**.

two groups: One is the complexes emitting from an ILCT excited state, and another is the complexes from an LL′CT excited state. Both transitions do not require the d orbitals of the copper(I) atom with regard to emissive transitions, indicating that the central metal need not be limited to the copper(I) atom.

Figure 4.20 shows the molecular structures of silver(I) or gold(I) complexes (**79–86**) with dppbz derivative ligands [93, 125–129] (M. Osawa, unpublished data). The photophysical properties are summarized in Table 4.6. All these complexes exhibit blue-orange luminescence with λ_{max} of 447–610 nm and $\Phi_{PL} = 0.12$–0.98 in the solid state at 300 K. At 77 K, the emission peaks of these complexes shift to the longer wavelength by 6–25 nm compared with those at 300 K. These results suggest that the emission from **79** to **85** in the solid state at room temperature is ascribed to TADF. According to the DFT calculations, the emissive excited states of complexes **79**, **80**, and **82–86** have a poor MLCT character (<15%). With regard to complexes **79** and **80**, the P atoms are found to contribute largely to the HOMO. The HOMO of three-coordinate Ag(I) complexes **83** and **84** essentially distributes on both the halogen and P atoms. Complexes **85** and **86** have the HOMOs mainly distributed on the S atom and the phenylene moiety. Thus, TADF is originating from the ILCT excited state for **79** and **80**, the (X + I)LCT excited state for **82–84**, and the LL′CT excited state for **85** and **86**, respectively. It is likely that the TADF of **79–84** is principally dependent on the electronic nature of the dppbz derivatives coordinated to Cu(I).

The temperature dependence of the emission decay time of the gold(I) complex **80** is shown in Figure 4.21. The emission is concluded to be TADF on the basis of the small energy gap $\Delta E(S_1 - T_1) = 630$ cm^{-1} and natural decay times of S_1 (70 ns) and T_1 (19 μs). It is well known that, because of the heavy atom

4.5 Another Group of Metal Complexes Exhibiting TADF

Table 4.6 Emission properties of Ag(I) and Au(I) complexes **79–85**.

Compound	T = 300 K			T = 77 K			$\Delta E(S_1-T_1)$ (cm^{-1})	References
	λ_{max} (nm)	Φ (%)	τ (μs)	λ_{max} (nm)	Φ (%)	τ (μs)		
[Ag(dppbz)$_2$]BF$_4$ **79**	447	45	1.0, 4.9	453	85	1200	—	[126, 127]
[Au(dppbz)$_2$]NO$_3$ **80**	486	95	3.7	492	95	18.6	630	[125, 128]
[(dppbz)CuCl]$_2$ **81**	480	93	15		96	1100	980	[129]
[(L$_{Me}$)AgBr]$_2$ **82**	487	56	2.6, 9.2	500	90	1200	—	(M. Osawa, unpublished data)
(L$_{Et}$)AgBr **83**	463	70	2.8, 12	482	98	2, 1000	—	
(L$_{iPr}$)AgBr **84**	463	98	5.7, 19	479	98	300, 1300	—	
[Ag(dppbz)(PS)] **85**	505	32	0.6, 2.2	530	83	570	—	[93]
[Au(dppbz)(PS)] **86**	610	12	8.4, 2.6	630	18	52	—	

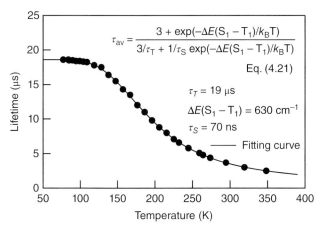

Figure 4.21 Emission decay time of **80** in crystals versus temperature. The parameters described in the inset were determined from a fit of Eq. (4.21) for **80**.

effects, the Au(I) complexes often emit phosphorescence from the aromatic ligands at room temperature. Au(I) brings forth an increase in the rate constant of spin-forbidden processes, S → T and T → S, due to SOC induced by the heavy gold atoms [130–132]. The TADF emission also suffers similar heavy atom effects. Both the ISC and rISC rates would be increased by the effects of Au(I), and thus, **80** exhibits intense blue TADF with a λ_{max} of 486 nm and $\Phi_{PL} = 0.95$ in the solid state at 300 K.

4.6 Conclusion

Over the past three decades, the tetrahedral d^{10} copper(I) complexes having diimine and/or diphosphine (or two monophosphine) ligands have been found to exhibit TADF. These Cu(I) complexes have been extensively studied as luminescent materials for OLEDs over the last 10 years due to the low cost and stable supply of copper metal.

In this chapter, studies on TADF-type Cu(I) complexes have been summarized and described. TADF is principally observed from the Cu(I) complexes having a small energy gap between T_1 and S_1. The nature of the excited states responsible for TADF is classified into three groups: (i) the MLCT and (M + X)LCT excited states, (ii) the ILCT excited state, and (iii) the LL'CT excited state.

A drawback of the TADF-type Cu(I) complexes is the structural changes occurring in the excited states due to the pseudo-Jahn–Teller effect. These structural changes bring forth low emission quantum yields for these complexes. This problem is resolved by constructing a "rigid structure" for both mono- and dinuclear Cu(I) complexes, which suppresses the structural changes in the excited states. With regard to the $[Cu(NN)(PP)]^+$ complexes, an effective method is to introduce bulky alkyl or phenyl groups into the (NN) ligands at positions facing toward the central metal. Similarly, the bridging ligands of the dinuclear copper(I) complexes with $\{Cu_2(\mu\text{-}X)_2\}$ units (X = Cl, Br, and I) are found to increase the quantum efficiencies of TADF in amorphous films.

There are still problems for the application of the TADF-type Cu(I) complexes to an OLED display. For example, an improvement in the thermal and redox stability of the Cu(I) emitter is a key requirement to ensure a long lifetime for the OLEDs fabricated by the vacuum deposition method. For thermal stability, the phosphine ligands seem to be more suitable than imine ligands. However, the metal-centered oxidation potential of phosphine-based copper(I) complexes is often close to that of the phosphine ligands, and thus, the complexes are readily decomposed by oxidation. At present, emissive Cu(I) complexes with both high redox stability and thermal stability have not yet been thoroughly researched, but this is an unavoidable challenge for the manufacturing of OLEDs for practical use.

Acknowledgments

The Integrated Collaborative Research Program between RIKEN and Canon Inc. is acknowledged for the funding of our research. M.O. is grateful to the Ministry of Education, Culture, Sports, Science and Technology of the Japanese Government (Grant-in-Aid for Scientific Research, No.24410080) for financial support of this research. We are grateful to Dr. S. Okada, Dr. A. Tsuboyama, Dr. K. Suzuki, Mr. M. Yashima, Dr. I. Kawata, Mr. S. Igawa, Dr. M. Hashimoto, and Mr. R. Ishii (Canon Inc.) for the fruitful collaboration.

Appendix

4.A.1 Schematic Structures of 1–86

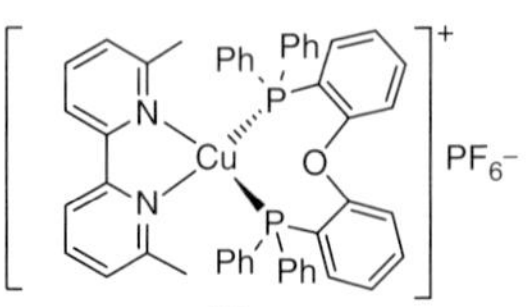

19

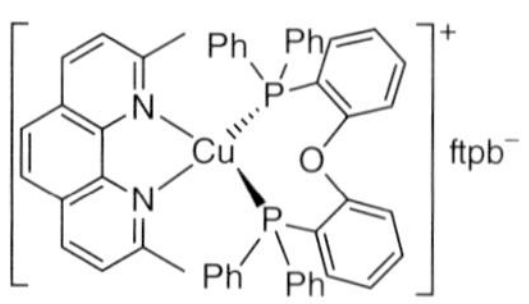

20

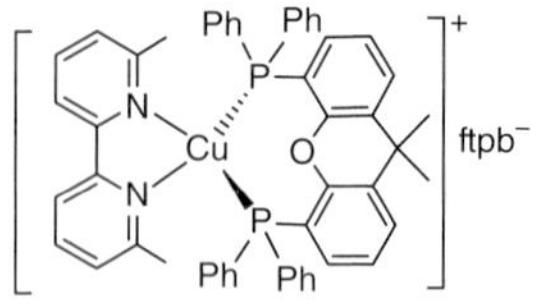

21

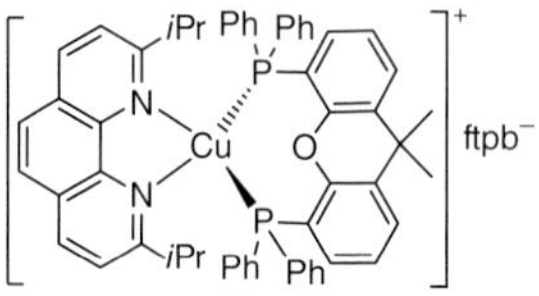

22

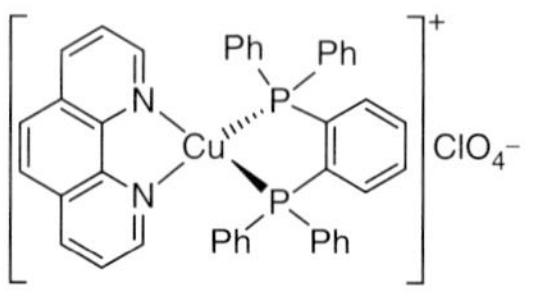

23

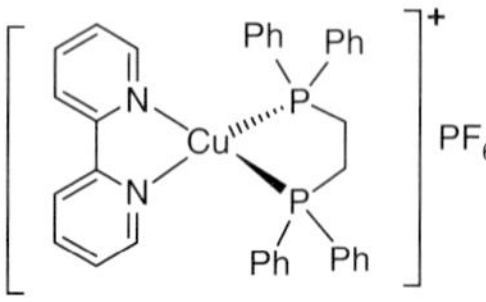

24

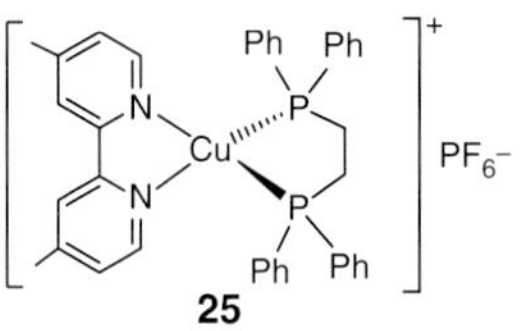

25

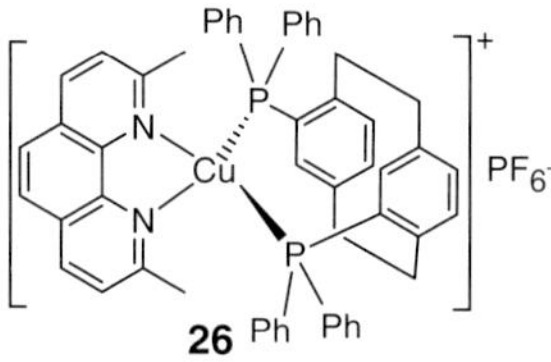

26

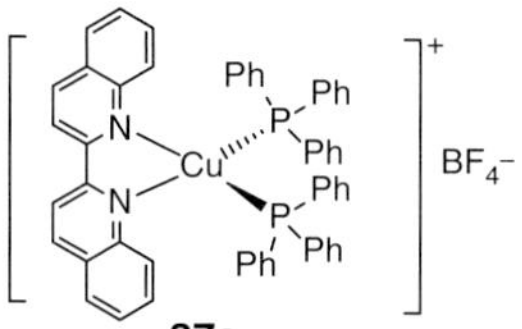

27a

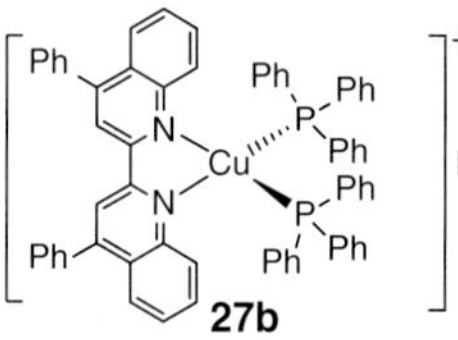

27b

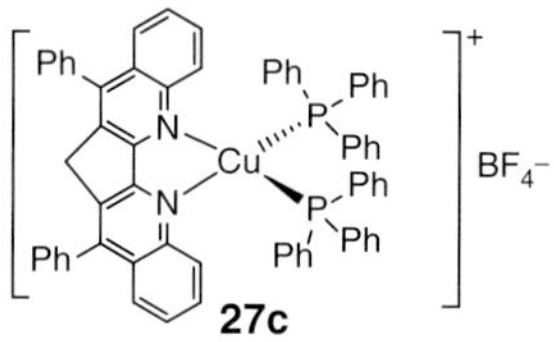

27c

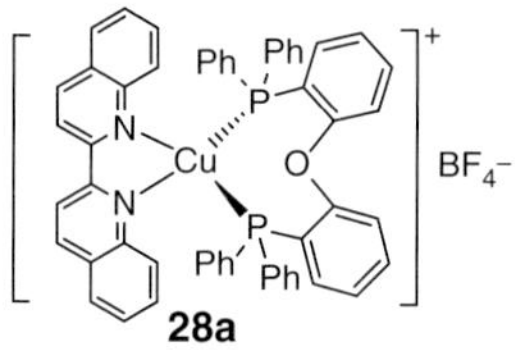

28a

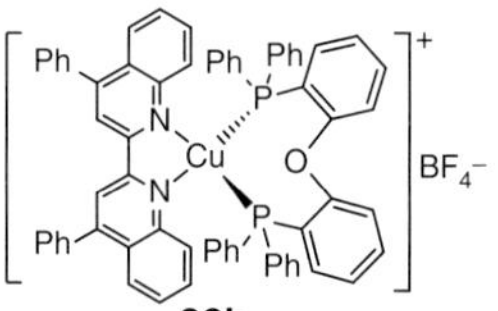

28b

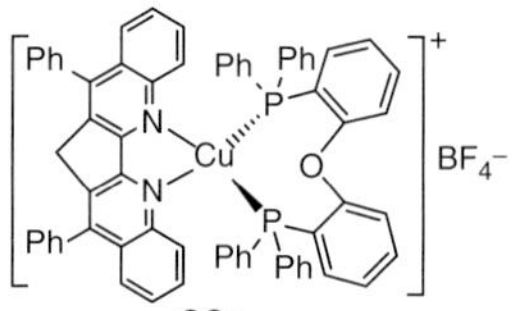

28c

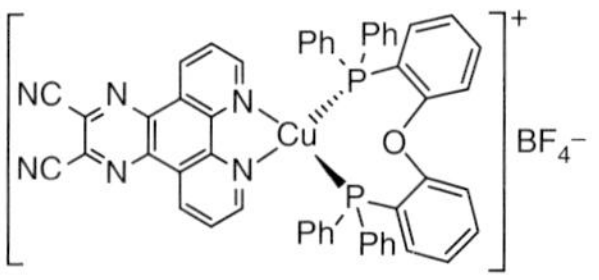

29

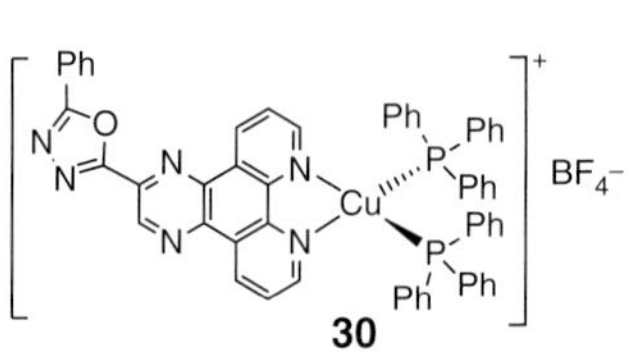

30

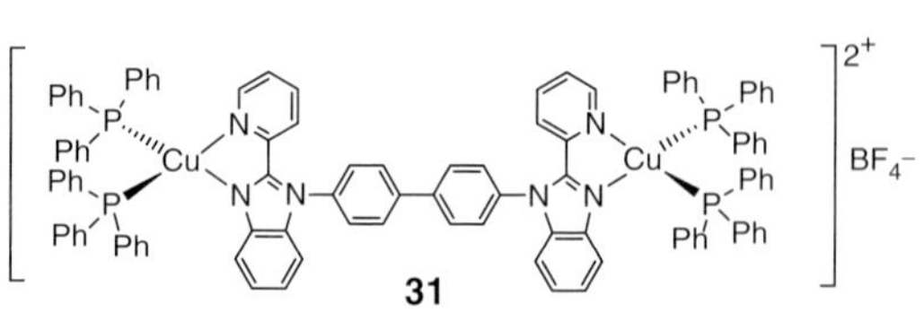

31

4.A.1 Schematic Structures of 1–86

164 | *4 Molecular Design and Synthesis of Metal Complexes as Emitters for TADF-Type OLEDs*

4.A.1 Schematic Structures of 1–86

4.A.1 Schematic Structures of 1–86

4.A.2 Abbreviations and Molecular Structures of Materials for OLEDs

ITO: indium tin oxide

PEDOT:PSS: poly(3,4-ethylene-dioxythiophene)/poly(styrene sulfonate)

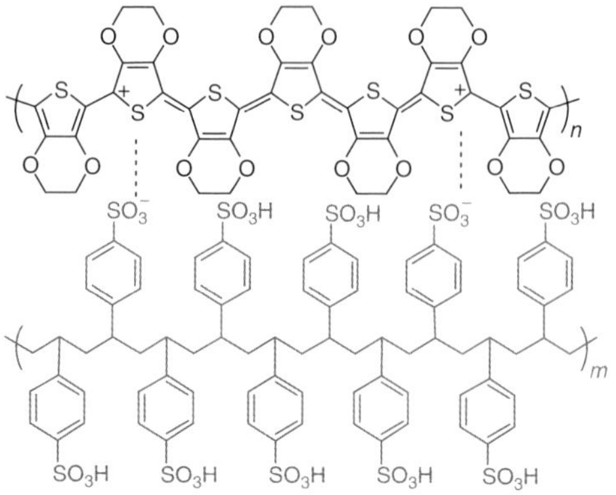

PVK: poly(vinylcarbazole)

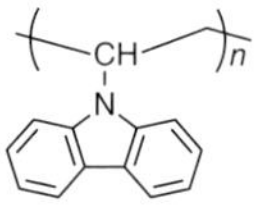

Alq$_3$ or AlQ: aluminum 8-hydroxyquinolinate

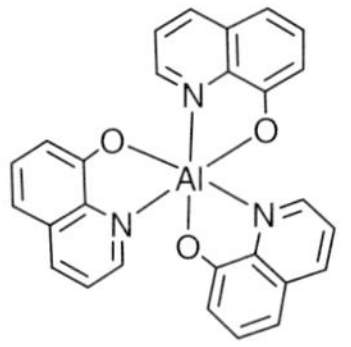

BCP: 2,9-dimethyl-4,7-diphenyl-1,10-phenanthroline

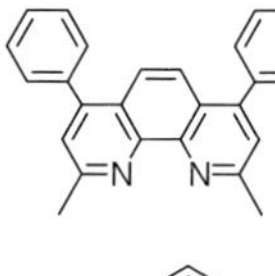

PYD2 or 26mCPy: 2,6-bis(N-carbazolyl) pyridine

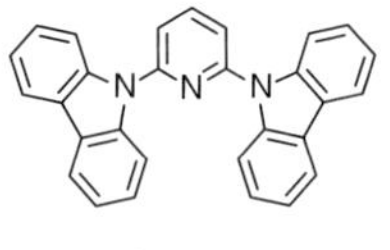

DPEPO: bis[2-(diphenylphosphino) phenyl] ether oxide

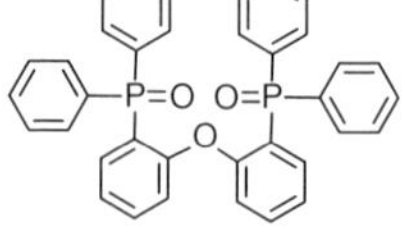

TCCz: N-(4-(carbazol-9-yl)phenyl)-3,6-bis(carbazol-9-yl) carbazole

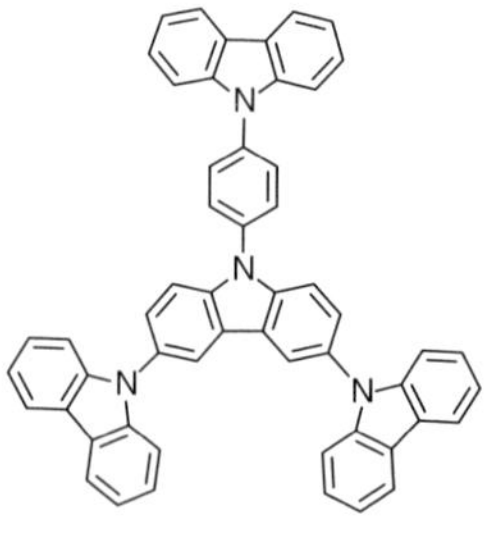

TPBI or TBi: 1,3,5-tris(N-phenyl benzimidazol-2-yl)benzene

2-TNATA: 4,4′,4″-tris[(2-naphthyl)phenylamino]triphenylamine

CBP: 4,4′-N,N′-dicarbazole-biphenyl

F-TBB: 1,3,5-tris(4′-fluorobiphenyl-4-yl)benzene

NPB: N,N′-di(1-naphthyl)-N,N′-diphenyl-(1,1′-biphenyl)-4,4′-diamine

m-MTDATA: 4,4′,4″-tris[(3-methylphenyl)phenylamino]triphenylamine

Bphen: 4,7-biphenyl-1,10-phenanthroline

MoO_3: molybdenum trioxide

TAPC: di-[4-(N,N-ditolyl-amino)-phenyl]cyclohexane

mCP: 1,3-bis(carbazol-9-yl)benzene

3TPYMB: tris(2,4,6-trimethyl-3-(pyridine-3-yl)phenyl)borane

TmPyPB: 1,3,5-tri[(3-pyridyl)phen-3-yl]benzene

Czpzpy: 2-(9H-carbazolyl)-6-(1H-pyrazolyl)pyridine

CDBP: 4,4′-bis(9-carbazolyl)-2,2′-dimethyl-biphenyl

PF01: 4,4′-bis[phenyl(9,9′-dimethylfluorenyl)amino]biphenyl

CPPyC: 3-(carbazol-9-yl)-5-((3-carbazol-9-yl)phenyl)pyridine

CFx: plasma polymerization of CHF_3 at low frequencies

BAlq-13: bis(2-methyl-quinolin-8-olato)(2,6-diphenylphenolato)aluminum(III)

References

1 Tang, C.W. and VanSlyke, S.A. (1987). *Appl. Phys. Lett.* 51: 913–915.
2 Baldo, M.A., Lamansky, S., Burrows, P.E., Thompson, M.E., and Forrest, S.R. (1999). *Appl. Phys. Lett.* 75: 4–6.
3 Baldo, M.A., Thompson, M.E., and Forrest, S.R. (2000). *Nature* 403: 750–753.
4 Baldo, M.A., O'Brien, D.F., You, Y., Shoustikov, A., Sibley, S., Thompson, M.E., and Forrest, S.R. (1998). *Nature* 395: 151–154.
5 Adachi, C., Baldo, M.A., Thompson, M.E., and Forrest, S.R. (2001). *J. Appl. Phys.* 90: 5048–5051.
6 Tanaka, D., Sasabe, H., Li, Y.-J., Su, S.-J., Takeda, T., and Kido, J. (2007). *Jpn. J. Appl. Phys., Part 2* 46: L10–L12.
7 Tanaka, D., Agata, Y., Takeda, T., Watanabe, S., and Kido, J. (2007). *Jpn. J. Appl. Phys., Part 2* 46: L117–L119.
8 Kondakova, M.E., Pawlik, T.D., Young, R.H., Giesen, D.J., Kondakov, D.Y., Brown, C.T., Deaton, J.C., Lenhard, J.R., and Klubek, K.P. (2008). *J. Appl. Phys.* 104: 094501.
9 Chopra, N., Lee, J., Zheng, Y., Eom, S.-H., Xue, J., and So, F. (2008). *Appl. Phys. Lett.* 93: 143307.
10 Harrigan, R.W., Hager, G.D., and Crosby, G.A. (1973). *Chem. Phys. Lett.* 21: 487–490.
11 Nozaki, K., Takamori, K., Nakatsugawa, Y., and Ohno, T. (2006). *Inorg. Chem.* 45: 6161–6178.
12 Yersin, H., Humbs, W., and Strasser, J. (1997). *Coord. Chem. Rev.* 159: 325–358.
13 Lacky, D.E., Pankuch, B.J., and Crosby, G.A. (1980). *J. Phys. Chem.* 84: 2068–2074.
14 Marchetti, A.P., Deaton, J.C., and Young, R.H. (2006). *J. Phys. Chem.* 110: 9828–9838.
15 Hofbeck, T. and Yersin, H. (2010). *Inorg. Chem.* 49: 9290–9299.
16 Yersin, H., Rausch, A.F., and Czerwieniec, R. (2012). *Organometallic Emitters for OLEDs: Triplet Harvesting, Singlet Harvesting, Case Structures, and Trendsin Physics of Organic Semiconductors*, 2e (ed. W. Brutting and C. Adachi), 371–424. Wiley-VCH.
17 Kim, K.-H., Moon, C.-K., Lee, J.-H., Kim, S.-Y., and Kim, J.-J. (2014). *Adv. Mater.* 26: 3844–3847.
18 Lee, C.W. and Lee, J.Y. (2013). *Adv. Mater.* 25: 5450–5454.
19 Parker, C.A. (1964). *Advanced in Photochemistry*, vol. 2 (ed. W.A. Noyes Jr., G.S. Hammond and J.N. Pitts Jr.), 305–383. New York: Wiley.
20 Endo, A., Ogasawara, M., Takahashi, A., Yokoyama, D., Kato, Y., and Adachi, C. (2009). *Adv. Mater.* 21: 4802–4806.
21 Tao, Y., Yuan, K., Chen, T., Xu, P., Li, H., Chen, R., Zheng, C., Zhang, L., and Huang, W. (2014). *Adv. Mater.* 26: 7931–7958.
22 Dumur, F. (2015). *Org. Electron.* 21: 27–39.

23 Czerwieniec, R., Leitl, M.J., Homeier, H.H.H., and Yersin, H. (2016). *Coord. Chem. Rev.* 325: 2–28.
24 Leitl, M.J., Schinabeck, A., Yersin, H., Zink, D.M., Baumann, T., and Volz, D. (2016). *Top. Curr. Chem.* 374: 25.
25 Czerwieniec, R., Yu, J.-B., and Yersin, H. (2011). *Inorg. Chem.* 50: 8293–8301.
26 Kirchhoff, J.R., Gamache, R.E. Jr., Blaskie, M.W., Del Paggio, A.A., Lengel, R.K., and McMillin, D.R. (1983). *Inorg. Chem.* 22: 2380–2384.
27 Iwamura, M., Takeuchi, S., and Tahara, T. (2015). *Acc. Chem. Res.* 48: 782–791.
28 Iwamura, M., Takeuchi, S., and Tahara, T. (2007). *J. Am. Chem. Soc.* 129: 5248–5256.
29 Iwamura, M., Watanabe, H., Ishii, K., Takeuchi, S., and Tahara, T. (2011). *J. Am. Chem. Soc.* 133: 7728–7736.
30 Blasse, G. and McMillin, D.R. (1980). *Chem. Phys. Lett.* 70: 1–3.
31 Zhang, Q., Zhou, Q., Cheng, Y., Wang, L., Ma, D., Jing, X., and Wang, F. (2004). *Adv. Mater.* 16: 432–436.
32 McMilllin, D.R. and McNett, K.M. (1998). *Chem. Rev.* 98: 1201–1219.
33 Everly, R.M., Ziessel, R., Suffert, J., and McMillin, D.R. (1991). *Inorg. Chem.* 30: 559–561.
34 Sakaki, S., Mizutani, H., and Kase, Y. (1992). *Inorg. Chem.* 31: 4575–4581.
35 Shinozaki, K. and Kaizu, Y. (1994). *Bull. Chem. Soc. Jpn.* 67: 2435–2439.
36 Eggleston, M.K., Fanwick, P.E., Pallenberg, A.J., and McMillin, D.R. (1997). *Inorg. Chem.* 36: 4007–4010.
37 Eggleston, M.K., McMillin, D.R., Koenig, K.S., and Pallenberg, A.J. (1997). *Inorg. Chem.* 36: 172–176.
38 Miller, M.T., Gantzel, P.K., and Karpishin, T.B. (1998). *Inorg. Chem.* 37: 2285–2290.
39 Miller, M.T., Gantzel, P.K., and Karpishin, T.B. (1999). *J. Am. Chem. Soc.* 121: 4292–4293.
40 Miller, M.T., Gantzel, P.K., and Karpishin, T.B. (1999). *Inorg. Chem.* 38: 3414–3422.
41 Miller, M.T. and Karpishin, T.B. (1999). *Inorg. Chem.* 38: 5246–5249.
42 Felder, D., Nierengarten, J.F., Barigelletti, F., Ventura, B., and Armaroli, N. (2001). *J. Am. Chem. Soc.* 123: 6291–6299.
43 Kutal, C. (1990). *Coord. Chem. Rev.* 99: 213–252.
44 Moudam, O., Kaeser, A., Delavaux-Nicot, B., Duhayon, C., Holler, M., Accorsi, G., Armaroli, N., Seguy, I., Navarro, J., Destruel, P., and Nierengarten, J.-F. (2007). *Chem. Commun.* 3077–3079.
45 Kaeser, A., Moudam, O., Accorsi, G., Seguy, I., Navarro, J., Belbakra, A., Duhayon, C., Armaroli, N., Delavaux-Nicot, B., and Nierengarten, J.-F. (2014). *Eur. J. Inorg. Chem.* 2014: 1345–1355.
46 Yersin, H., Rausch, A.F., Czerwieniec, R., Hofbeck, T., and Fischer, T. (2011). *Coord. Chem. Rev.* 255: 2622–2652.
47 Leitl, M.J., Kuechle, F.-R., Mayer, H.A., Wesemann, L., and Yersin, H. (2013). *J. Phys. Chem. A* 117: 11823–11836.
48 Czerwieniec, R. and Yersin, H. (2015). *Inorg. Chem.* 54: 4322–4327.

49 Zhang, X., Chi, Z., Zhang, Y., Liu, S., and Xu, J. (2013). *J. Mater. Chem. C* 1: 3376–3390.
50 Cariati, E., Lucenti, E., Botta, C., Giovanella, U., Marinotto, D., and Righetto, S. (2016). *Coord. Chem. Rev.* 306: 566–614.
51 Siddique, Z.A., Yamamoto, Y., Ohno, T., and Nozaki, K. (2003). *Inorg. Chem.* 42: 6366–6378.
52 Asano, M.S., Tomiduka, K., Sekizawa, K., Yamashita, K.-I., and Sugiura, K.-I. (2010). *Chem. Lett.* 39: 376–378.
53 Green, O., Gandhi, B.A., and Burstyn, J.N. (2009). *Inorg. Chem.* 48: 5704–5714.
54 Armaroli, N., Accorsi, G., Cardinali, F., and Listorti, A. (2007). *Top. Curr. Chem.* 280: 69–115.
55 Cuttell, D.G., Kuang, S.-M., Fanwick, P.E., McMillin, D.R., and Walton, R.A. (2002). *J. Am. Chem. Soc.* 124: 6–7.
56 Yang, L., Feng, J.-K., Ren, A.-M., Zhang, M., Ma, Y.-G., and Liu, X.-D. (2005). *Eur. J. Inorg. Chem.* 2005 (10): 1867–1879.
57 Zou, L.-Y., Cheng, Y.-X., Li, Y., Li, H., Zhang, H.-X., and Ren, A.-M. (2014). *Dalton Trans.* 43: 11252–11259.
58 Zhang, Q., Zhou, Q., Cheng, Y., Wang, L., Ma, D., Jing, X., and Wang, F. (2006). *Adv. Funct. Mater.* 16: 1203–1208.
59 Andres-Tome, I., Fyson, J., Baiao Dias, F., Monkman, A.P., Iacobellis, G., and Coppo, P. (2012). *Dalton Trans.* 41: 8669–8674.
60 Linfoot, C.L., Leitl, M.J., Richardson, P., Rausch, A.F., Chepelin, O., White, F.J., Yersin, H., and Robertson, N. (2014). *Inorg. Chem.* 53: 10854–10861.
61 Keller, S., Constable, E.C., Housecroft, C.E., Neuburger, M., Prescimone, A., Longo, G., Pertegas, A., Sessolo, M., and Bolink, H.J. (2014). *Dalton Trans.* 43: 16593–16596.
62 Smith, C.S., Branham, C.W., Marquardt, B.J., and Mann, K.R. (2010). *J. Am. Chem. Soc.* 132: 14079–14085.
63 Li, X.-L., Ai, Y.-B., Yang, B., Chen, J., Tan, M., Xin, X.-L., and Shi, Y.-H. (2012). *Polyhedron* 35: 47–54.
64 Nishikawa, M., Wakita, Y., Nishi, T., Miura, T., and Tsubomura, T. (2015). *Dalton Trans.* 44: 9170–9181.
65 Czerwieniec, R., Kowalski, K., and Yersin, H. (2013). *Dalton Trans.* 42: 9826–9830.
66 Zhang, Q., Komino, T., Huang, S., Matsunami, S., Goushi, K., and Adachi, C. (2012). *Adv. Funct. Mater.* 22: 2327–2336.
67 Coppens, P., Sokolow, J., Trzop, E., Makal, A., and Chen, Y. (2013). *J. Phys. Chem. Lett.* 4: 579–582.
68 Zhang, L., Li, B., and Su, Z. (2009). *Langmuir* 25: 2068–2074.
69 Kaeser, A., Mohankumar, M., Mohanraj, J., Monti, F., Holler, M., Cid, J.-J., Moudam, O., Nierengarten, I., Karmazin-Brelot, L., Duhayon, C., Delavaux-Nicot, B., Armaroli, N., and Nierengarten, J.-F. (2013). *Inorg. Chem.* 52: 12140–12151.
70 Zhang, Q., Ding, J., Cheng, Y., Wang, L., Xie, Z., Jing, X., and Wang, F. (2007). *Adv. Funct. Mater.* 17: 2983–2990.

71 Che, G., Su, Z., Li, W., Chu, B., Li, M., Hu, Z., and Zhang, Z. (2006). *Appl. Phys. Lett.* 89: 103511–103513.

72 Walsh, P.J., Lundin, N.J., Gordon, K.C., Kim, J.-Y., and Lee, C.-H. (2009). *Opt. Mater.* 31: 1525–1531.

73 Jia, W.L., McCormick, T., Tao, Y., Lu, J.-P., and Wang, S. (2005). *Inorg. Chem.* 44: 5706–5712.

74 Si, Z., Li, J., Li, B., Liu, S., and Li, W. (2008). *J. Lumin.* 129: 181–186.

75 Yu, T., Liu, P., Chai, H., Kang, J., Zhao, Y., Zhang, H., and Fan, D. (2014). *J. Fluoresc.* 24: 933–943.

76 Min, J., Zhang, Q., Sun, W., Cheng, Y., and Wang, L. (2011). *Dalton Trans.* 40: 686–693.

77 Hsu, C.-W., Lin, C.-C., Chung, M.-W., Chi, Y., Lee, G.-H., Chou, P.-T., Chang, C.-H., and Chen, P.-Y. (2011). *J. Am. Chem. Soc.* 133: 12085–12099.

78 Wada, A., Zhang, Q., Yasuda, T., Takasu, I., Enomoto, S., and Adachi, C. (2012). *Chem. Commun.* 48: 5340–5342.

79 Zhang, L., Li, B., and Su, Z. (2009). *J. Phys. Chem. C* 113: 13968–13973.

80 Zhang, D. (2010). *J. Lumin.* 130: 1419–1424.

81 Chen, X.-L., Yu, R., Zhang, Q.-K., Zhou, L.-J., Wu, X.-Y., Zhang, Q., and Lu, C.-Z. (2013). *Chem. Mater.* 25: 3910–3920.

82 Chen, X.-L., Lin, C.-S., Wu, X.-Y., Yu, R., Teng, T., Zhang, Q.-K., Zhang, Q., Yang, W.-B., and Lu, C.-Z. (2015). *J. Mater. Chem. C* 3: 1187–1195.

83 Zhang, Q., Chen, X.-L., Chen, J., Wu, X.-Y., Yu, R., and Lu, C.-Z. (2015). *Dalton Trans.* 44: 10022–10029.

84 Lundin, N.J., Walsh, P.J., Howell, S.L., McGarvey, J.J., Blackman, A.G., and Gordon, K.C. (2005). *Inorg. Chem.* 44: 3551–3560.

85 Lundin, N.J., Walsh, P.J., Howell, S.L., Blackman, A.G., and Gordon, K.C. (2008). *Chem. Eur. J.* 14: 11573–11583.

86 David, G., Walsh, P.J., and Gordon, K.C. (2004). *Chem. Phys. Lett.* 383: 292–296.

87 McCormick, T., Jia, W.-L., and Wang, S. (2006). *Inorg. Chem.* 45: 147–155.

88 Li, E.Y.-T., Jiang, T.-Y., Chi, Y., and Chou, P.-T. (2014). *Phys. Chem. Chem. Phys.* 16: 26184–26192.

89 Stein, G. and Wurzberg, E. (1975). *J. Chem. Phys.* 62: 208–213.

90 Haas, Y., Arein, G., and Wurzberg, E. (1974). *J. Chem. Phys.* 60: 258–263.

91 Li, Q., Xu, S.-X., Wang, J.-L., Xia, H.-Y., Zhao, F., and Wang, Y.-B. (2014). *Int. J. Quantum Chem.* 114: 1685–1691.

92 Igawa, S., Hashimoto, M., Kawata, I., Yashima, M., Hoshino, M., and Osawa, M. (2013). *J. Mater. Chem. C* 1: 542–551.

93 Osawa, M., Kawata, I., Ishii, R., Igawa, S., Hashimoto, M., and Hoshino, M. (2013). *J. Mater. Chem. C* 1: 4375–4383.

94 Kyle, K.R., Palke, W.E., and Ford, P.C. (1990). *Coord. Chem. Rev.* 97: 35–46.

95 Tsuge, K., Chishina, Y., Hashiguchi, H., Sasaki, Y., Kato, M., Ishizaka, S., and Kitamura, N. (2016). *Coord. Chem. Rev.* 306: 566–614.

96 Tsuboyama, A., Kuge, K., Furugori, M., Okada, S., Hoshino, M., and Ueno, K. (2007). *Inorg. Chem.* 46: 1992–2001.

97 Hashimoto, M., Igawa, S., Yashima, M., Kawata, I., Hoshino, M., and Osawa, M. (2011). *J. Am. Chem. Soc.* 133: 10348–10351.

98 Osawa, M., Hoshino, M., Hashimoto, M., Kawata, I., Igawa, S., and Yashima, M. (2015). *Dalton Trans.* 44: 8369–8378.
99 Osawa, M. (2014). *Chem. Commun.* 50: 1801–1803.
100 Lotito, K.J. and Peters, J.C. (2010). *Chem. Commun.* 46: 3690–3692.
101 Krylova, V.A., Djurovich, P.I., Whited, M.T., and Thompson, M.E. (2010). *Chem. Commun.* 46: 6696–6698.
102 Krylova, V.A., Djurovich, P.I., Aronson, J.W., Haiges, R., Whited, M.T., and Thompson, M.E. (2012). *Organometallics* 31: 7983–7993.
103 Krylova, V.A., Djurovich, P.I., Conley, B.L., Haiges, R., Whited, M.T., Williams, T.J., and Thompson, M.E. (2014). *Chem. Commun.* 50: 7176–7179.
104 Leitl, M.J., Krylova, V.A., Djurovich, P.I., Thompson, M.E., and Yersin, H. (2014). *J. Am. Chem. Soc.* 136: 16032–16038.
105 Araki, H., Tsuge, K., Sasaki, Y., Ishizaka, S., and Kitamura, N. (2005). *Inorg. Chem.* 44: 9667–9675.
106 Araki, H., Tsuge, K., Sasaki, Y., Ishizaka, S., and Kitamura, N. (2007). *Inorg. Chem.* 46: 10032–10034.
107 Barakat, K.A., Cundari, T.R., and Omary, M.A. (2003). *J. Am. Chem. Soc.* 125: 14228–14229.
108 Sinha, P., Wilson, A.K., and Omary, M.A. (2005). *J. Am. Chem. Soc.* 127: 12488–12489.
109 Tsuge, K. (2013). *Chem. Lett.* 42: 204–208.
110 Liu, Z., Djurovich, P.I., Whited, M.T., and Thompson, M.E. (2012). *Inorg. Chem.* 51: 230–236.
111 Liu, Z., Qayyum, M.F., Wu, C., Whited, M.T., Djurovich, P.I., Hodgson, K.O., Hedman, B., Solomon, E.I., and Thompson, M.E. (2011). *J. Am. Chem. Soc.* 133: 3700–3703.
112 Liu, Z., Qiu, J., Wei, F., Wang, J., Liu, X., Helander, M.G., Rodney, S., Wang, Z., Bian, Z., Lu, Z., Thompson, M.E., and Huang, C. (2014). *Chem. Mater.* 26: 2368–2373.
113 Volz, D., Baumann, T., Fluegge, H., Mydlak, M., Grab, T., Baechle, M., Barner-Kowollik, C., and Brase, S. (2012). *J. Mater. Chem.* 22: 20786–20790.
114 Volz, D., Nieger, M., Friedrichs, J., Baumann, T., and Brase, S. (2013). *Langmuir* 29: 3034–3044.
115 Ohara, H., Kobayashi, A., and Kato, M. (2014). *Dalton Trans.* 43: 17317–17323.
116 Volz, D., Zink, D.M., Bocksrocker, T., Friedrichs, J., Nieger, M., Baumann, T., Lemmer, U., and Brase, S. (2013). *Chem. Mater.* 25: 3414–3426.
117 Zink, D.M., Volz, D., Baumann, T., Mydlak, M., Fluegge, H., Friedrichs, J., Nieger, M., and Brase, S. (2013). *Chem. Mater.* 25: 4471–4486.
118 Zink, D.M., Baechle, M., Baumann, T., Nieger, M., Kuehn, M., Wang, C., Klopper, W., Monkowius, U., Hofbeck, T., Yersin, H., and Braese, S. (2013). *Inorg. Chem.* 52: 2292–2305.
119 Zink, D.M., Baumann, T., Friedrichs, J., Nieger, M., and Braese, S. (2013). *Inorg. Chem.* 52: 13509–13520.
120 Volz, D., Wallesch, M., Grage, S.L., Goettlicher, J., Steininger, R., Batchelor, D., Vitova, T., Ulrich, A.S., Heske, C., Weinhardt, L., Baumann, T., and Braese, S. (2014). *Inorg. Chem.* 53: 7837–7847.

121 Wallesch, M., Volz, D., Zink, D.M., Schepers, U., Nieger, M., Baumann, T., and Braese, S. (2014). *Chem. Eur. J.* 20: 6578–6590.
122 Volz, D., Chen, Y., Wallesch, M., Liu, R., Flechon, C., Zink, D.M., Friedrichs, J., Fluegge, H., Steininger, R., Goettlicher, J., Heske, C., Weinhardt, L., Braese, S., So, F., and Baumann, T. (2015). *Adv. Mater.* 27: 2538–2543.
123 Deaton, J.C., Switalski, S.C., Kondakov, D.Y., Young, R.H., Pawlik, T.D., Giesen, D.J., Harkins, S.B., Miller, A.J.M., Mickenberg, S.F., and Peters, J.C. (2010). *J. Am. Chem. Soc.* 132: 9499–9508.
124 Hofbeck, T., Monkowius, U., and Yersin, H. (2015). *J. Am. Chem. Soc.* 137: 399–404.
125 Osawa, M., Kawata, I., Igawa, S., Hoshino, M., Fukunaga, T., and Hashizume, D. (2010). *Chem. Eur. J.* 16: 12114–12126.
126 Osawa, M. and Hoshino, M. (2008). *Chem. Commun.* 6384–6386.
127 Igawa, S., Hashimoto, M., Kawata, I., Hoshino, M., and Osawa, M. (2012). *Inorg. Chem.* 51: 5805–5813.
128 Osawa, M., Kawata, I., Igawa, S., Tsuboyama, A., Hashizume, D., and Hoshino, M. (2009). *Eur. J. Inorg. Chem.* 2009: 3708–3711.
129 Yersin, H., Leitl, J.M., and Czerwieniec, R. (2014). Proceedings Volume 9183, Organic Light Emitting Materials and Devices XVIII, 91830N/1-91830N/11
130 Osawa, M., Hoshino, M., Akita, M., and Wada, T. (2005). *Inorg. Chem.* 44: 1157–1159.
131 Osawa, M., Hoshino, M., and Hashizume, D. (2007). *Chem. Phys. Lett.* 436: 89–93.
132 Osawa, M., Hoshino, M., and Hashizume, D. (2008). *Dalton Trans.* 2248–2252.

5

Ionic [Cu(NN)(PP)]$^+$ TAD9727 F Complexes with Pyridine-based Diimine Chelating Ligands and Their Use in OLEDs

Rongmin Yu and Can-Zhong Lu

Fujian Institute of Research on the Structure of Matter, Chinese Academy of Sciences, Fuzhou, 350002, Fujian, PR China

5.1 Introduction

In 1998, Forrest and coworkers discovered that all the excitons (both singlet and triplet) can be harvested and transferred into light in an electroluminescent device with a Pt phosphorescent complex as the emitting dopant [1]. Since then, phosphorescent materials, especially those of the third-row transition metals with strong spin–orbit interaction, have been studied extensively, driven majorly by their wide range of applications in electroluminescent devices, such as organic light-emitting diodes (OLEDs) and light-emitting electrochemical cells (LEECs) [2–11]. The phosphorescent materials of Ir(III) and Pt(II) complexes are the most attractive because their high emission quantum yields of up to almost 100%, and short emission decay times of 2 μs can be achieved. For these materials, the central metals can induce strong spin–orbit coupling (SOC), resulting in the fast intersystem crossing (ISC) and the effective radiative transition from the first triplet excited state to the ground state. The emission mechanism of these phosphors allows the use of all excitons for the generation of light in OLEDs. However, there still exist materials limitations, for example, low efficient blue-light emission for d^6 and d^8 transition metal complexes and material costs of the precious metal sources. Thus, it is desirable to look for alternative materials to solve these problems. Cuprous emissive complexes have been paid great attentions for their cost-effectiveness, attractive emissions, and the d^{10} configuration of Cu(I) [12–14].

Cuprous emissive complexes have been investigated intensively since 1970 and become an important class of the luminescent materials [15–18]. Except a few examples of three-coordinate emissive Cu(I) complexes [19–25], the majority of the Cu(I) luminescent complexes have a distorted tetrahedral geometry around the Cu(I) atoms defined by two diimine ligands ([Cu(NN)$_2$]$^+$) or a diimine ligand and a diphosphine ligand ([Cu(NN)(PP)]$^+$) [25–36]. As compared with the homoleptic [Cu(NN)$_2$]$^+$ complexes, the heteroleptic [Cu(NN)(PP)]$^+$ complexes with phosphine ligands exhibit much higher photoluminescence quantum yields (PLQYs).

Highly Efficient OLEDs: Materials Based on Thermally Activated Delayed Fluorescence,
First Edition. Edited by Hartmut Yersin.
© 2019 Wiley-VCH Verlag GmbH & Co. KGaA. Published 2019 by Wiley-VCH Verlag GmbH & Co. KGaA.

In contrast to the heavy-metal elements, the SOC of the copper element is much less efficient. Consequently, the cuprous phosphorescent complexes usually possess long emission lifetimes of up to hundred microseconds. The long decay times of triplet states facilitate triplet–triplet annihilation (TTA) in electroluminescence devices. However, this detrimental effect can be avoided by the fact that many emissive cuprous complexes can emit efficient thermally activated delayed fluorescence (TADF) at room temperature because of the existence of a small energy gap between the S_1 and T_1 states [7, 13, 18]. Displaying TADF at ambient temperature, the cuprous complexes have an observed lifetime as short as comparable with those of the heavy-metal phosphorescent materials. Capable of harvesting both singlet and triplet excitons as well as having compatible lifetimes, the TADF cuprous complexes are believed to have realistic potential application in OLEDs that are currently dominant by the heavy-metal phosphorescent materials [12, 37].

5.2 The Influence of Molecular and Electronic Structure on Emissive Properties of Cu(I) Complexes

Copper has two common oxidation states: Cu(I) and Cu(II). The Cu(II) ion has a d^9 electronic configuration. Its compounds have relatively intense colors due to metal-centered (MC) absorption bands. The deactivation of the MC excited states is usually through nonradiative pathways, so the luminescence of Cu(II) compounds is very rare and unattractive. In contrast to the Cu(II) complexes, the Cu(I) complexes with a d^{10} electronic configuration have much more interesting photophysical properties that have caused considerable attention and have been investigated intensively for decades. Many emissive Cu(I) complexes with different types of ligands have been reported, among which, four-coordinate complexes with homoleptic $[Cu(NN)_2]^+$ or heteroleptic $[Cu(NN)(PP)]^+$ basic configuration, are the most extensively investigated. Much effort has been made to elucidate fundamentally the relationship between the emissive performance of the Cu(I) complexes and their structures. It has been found that the emission performance of the Cu(I) complexes is strongly affected by the steric and electronic structures of the ligands, as well as the overall structural rigidity and steric protection of the Cu(I) complex structures.

For the cationic $[Cu(NN)_2]^+$ and $[Cu(NN)(PP)]^+$ complexes, their highest occupied molecular orbital (HOMO) contains usually substantial contribution from d orbitals of Cu atom due to the low oxidation potential of Cu(I) ion. As a result, the lowest excited states have the character of MLCT, leading to the broad and unstructured emission spectra that are very common for the emissive Cu(I) complexes. Upon excitation, the lowest MLCT excited state is populated; thus the metal center changes its formal oxidation state from Cu(I) to Cu(II). Because the structural preference for Cu(I) and Cu(II) is different, structural rearrangement undergoes after excitation occurs. The emissive copper complex performs a flattening distortion from the Cu(I)-preferred tetrahedral structure toward the Cu(II)-favorite square-planar structure. The structural distortion leads to a better overlap of the ground state and the excited-state vibrational wave functions, which causes an increase of the radiationless deactivation and a smaller energy gap between the ground state and excited state. Meanwhile, the resulting smaller energy gap induces a further increase of the radiationless decay rate according to energy gap law. Furthermore, in the excited-state flattened tetrahedral structure, a fifth coordination site becomes available for the Cu(II) ion. The interaction of the available site with solvent molecules or counterions, forming exciplex, often causes severely nonradiative quenching [38]. Such structural rearrangement is especially severe in a nonrigid environment. Many Cu(I) complexes exhibit poor emissive properties because of the structural rearrangement. Thus, in the design of highly luminescent Cu(I) complexes, it is important to choose appropriate ligands that can construct a rigid and sterically protected environment around the Cu(I) ion that can keep the complex from structural distortion and exciplex quenching upon excitation.

The homoleptic [Cu(NN)$_2$]$^+$ complexes have been investigated continually for decades. The influence of the steric factors on the emissive properties of the Cu(I) complexes has been demonstrated in many studies. The homoleptic [Cu(phen)$_2$]$^+$ complex with pristine phenanthroline ligand does not emit luminescence in CH$_2$Cl$_2$. The substitutions on the 2 and 9 positions with a simple methyl group in [Cu(dmp)$_2$]$^+$ have been shown to improve the emissive properties of the [Cu(NN)$_2$]$^+$ complexes [39]. In the solution of [Cu(dmp)$_2$]$^+$ in degassed CH$_2$Cl$_2$, a luminescence with $\Phi \approx 0.1\%$ is observed. The introduction of the methyl group into the pristine phenanthroline ligands gives a more rigid [Cu(NN)$_2$]$^+$ structure that suppresses the degree of the structural distortion upon excitation, so the emissive performance is improved. A better result ($\Phi = 1\%$) has been found in the heteroleptic complex, [Cu(dmp)(dtbp)]$^+$ (dtbp = 2,9-di-*tert*-butyl-1,10-phenanthroline), in which the excited-state distortion is further limited by the bulky *tert*-butyl substituent [40]. Even though the PLQYs in these examples are very small, these examples show the importance of the steric factors in affecting the emissive properties of the Cu(I) complexes. Many phenanthroline-based ligands (as shown in Figure 5.1) with various substituents have been designed and used in the synthesis of the homoleptic [Cu(NN)$_2$]$^+$ complexes [15]. Some of them, such as the sterically encumbered phenanthroline ligands with bulky aryl substituents (3,5-di-tert-butyl-4-methoxyphenyl, 2,4,6-trimethylphenyl) in the 2,9-position, have been shown to be very effective in protecting the Cu(I) excited states from the exciplex quenching, and the homoleptic [Cu(NN)$_2$]$^+$ complexes with these robust ligands have emissive lifetimes that are almost unaffected by the presence of molecular oxygen and only slightly shortened in the nucleophilic solvents [41]. However, the emission performance of the [Cu(NN)$_2$]$^+$ complexes is generally poor. In CH$_2$Cl$_2$ at ambient temperature, the emission quantum yields of the complexes are typically below 0.001, which limited strongly their application in OLEDs.

In recent years increasing attention has been paid to the heteroleptic [Cu(NN)(PP)]$^+$ complexes because they exhibit greatly increased PLQYs as compared with the homoleptic [Cu(NN)$_2$]$^+$ complexes due to the electronic and steric effect of the P ligands [42–44]. The history of the heteroleptic [Cu(NN)(PP)]$^+$ complexes can be dated back to late 1970s when McMillin and coworkers

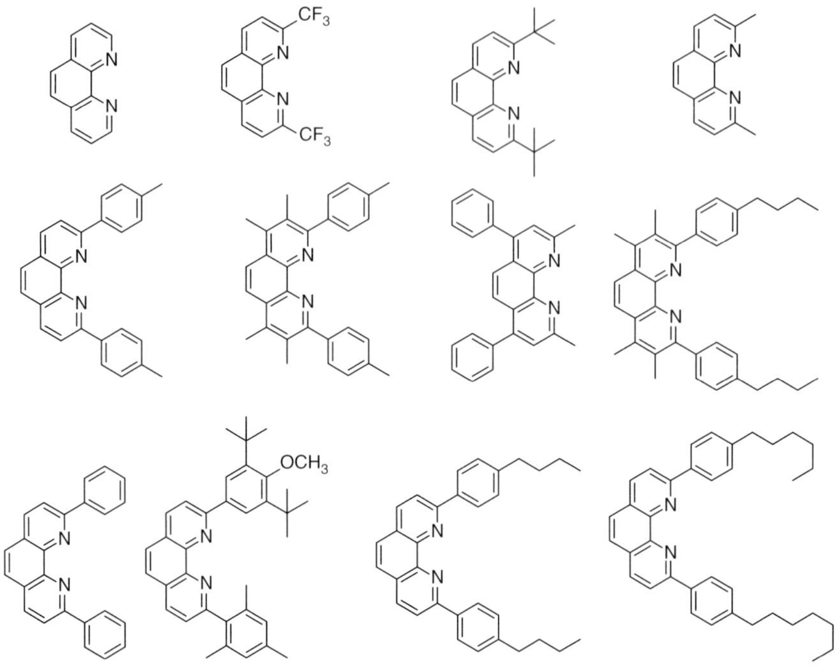

Figure 5.1 Phenanthroline-based ligands.

reported the spectral studies of a series of Cu(I) complexes with mixed ligands of diimine ligands and triphenylphosphine [42, 45]. However, these studies have not caused much attention until 2002 when McMillin and coworkers obtained the first example of the highly emissive heteroleptic [Cu(NN)(PP)]$^+$ complexes, [Cu(dmp)(POP)]$^+$ (dmp = 2,9-dimethyl-1,10-phenanthroline; POP = bis[2-(diphenylphosphino)phenyl]ether) [46, 47]. The emission quantum yield and lifetime of the complex in deaerated CH_2Cl_2 are reported to be 0.15 and 14.3 μs, respectively. In this compound, the combination of the diimine ligand with alkyl substituents at 2,9 positions and the chelating diphosphine POP ligand imparts the Cu(I) complexes with an exceptionally sterically protected and rigid structure, which is necessary for a highly emissive Cu(I) complexes. In addition, the electronic effect of the POP ligand stabilizes the HOMO and thus leads to the higher-energy MLCT excited states, compared with [Cu(NN)$_2$]$^+$ complexes, which disfavors nonradiative deactivations according to the energy law. The structural rigidity and effective steric protection of the complex obstruct efficiently the structural flattening distortion and the exciplex quenching at the excited states, along with the electronic effect of the POP ligand, which leads to the improvement of the emissive efficiencies for the heteroleptic Cu(I) complex. The uses of this complex and its analogs in electroluminescent devices display very promising results [48–51]. Ever since, the research on the emissive Cu(I) complexes has reclaimed tremendous interest, especially in the area of the electroluminescence devices.

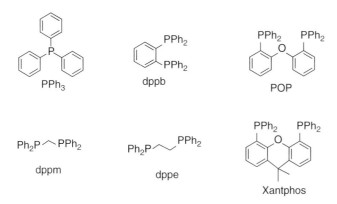

Figure 5.2 Examples of phosphine ligands used in the synthesis of heteroleptic [Cu(NN)(PP)]$^+$ complexes.

A variety of phosphine ligands has been used to prepare Cu(I) complexes together with phenanthroline-type units: bis[2-(diphenylphosphino)phenyl]ether (POP), triphenylphosphine (PPh$_3$), bis(diphenylphosphino)ethane (dppe), and bis(diphenylphosphino)methane (dppm) (Figure 5.2) [26, 46, 47, 50, 52–55]. Meanwhile, various diimine ligands other than phenanthroline have been designed and utilized [26, 28–30, 35, 56–59]. Currently, it is not uncommon that cuprous complexes exhibit high PLQYs over 80% in the solid state [30, 57, 60–62]. The application of these materials as the emitting dopants in the fabrication of OLEDs and LEECs has also been studied [35, 47, 51, 63–67]. The importance of the ligand choice for the emissive Cu(I) complexes has been shown in many cases. Properly tuning the steric and electronic properties of the ligands has been shown effective to improve emission efficiency somehow. However, it is not easy to draw a general rule to predict the photophysical behavior of the emissive cuprous complexes. Hard work on trial and errors in the optimization of the emissive properties of the Cu(I) complexes is still unavoidable.

Next we will introduce the recent development of the cationic emissive complexes Cu(NN)(PP)$^+$ with pyridyl derivative diimine ligands and their uses in the fabrication of OLEDs.

5.3 Heteroleptic Diimine/Diphosphine [Cu(NN)(PP)]$^+$ Complexes with Pyridine-Based Ligand

5.3.1 [Cu(NN)(PP)]$^+$ Complexes with 2,2′-bipyridyl-based Ligands

Bipyridyl derivative ligands have been used extensively in the preparation of emissive complexes of Ru(II) and Ir(III) [68, 69]. In contrast, the emissive Cu(I) complexes with bipyridyl derivative ligands have been seldom studied, possibly due to the typically lower quantum yields observed for Cu(I) complexes with these ligands in comparison with analogous phenanthroline Cu(I)

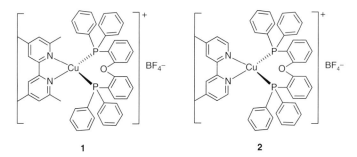

Figure 5.3 Chemical structures of heteroleptic Cu(I) complexes **1** and **2** with 2,2′-bipyridyl diimine ligands.

complexes [42, 45, 70–72]. Yersin and coworkers reported a strongly emissive Cu(I) complex with 4,4′,6,6′-tetramethyl-2,2′-bipyridyl ligand, complex **1**, as shown in Figure 5.3 [33]. This complex displays a bright green emission ($\lambda_{max} = 555$ nm, $\tau = 13$ μs) at room temperature as powder with high quantum yield ($\Phi = 74\%$). In comparison, complex **2** emits much weakly ($\lambda_{max} = 575$ nm, $\Phi = 9\%$) under the same conditions. The improvement of the emission for complex **1** is ascribed to the steric effect of 6,6′-methyl groups on the bipyridyl ligand. The 4,4′,6,6′-tetramethyl-2,2′-bipyridyl ligand joining with POP ligand imparts complex **1** a more rigid and sterically protected structure than the 4,4′-dimethyl-2,2′-bipyridyl ligand does in complex **2**. A detailed analysis of the temperature-dependent decay behavior of **1** reveals that this complex emits efficient TADF at room temperature. At $T = 77$ K, the emission decay time is 87 μs, and the decay time decreases drastically to 11 μs at ambient temperature. The energy separation between the first excited singlet and the first triplet state (ΔE_{ST}) estimated from the shift of the emission spectra from 77 to 300 K is 630 cm^{-1} that is consistent well to the fit value of 770 cm^{-1} from Eq. (5.1) [33]:

$$\tau_{obs} = \frac{3 + \exp\left(\frac{-\Delta E_{ST}}{k_B T}\right)}{\frac{3}{\tau_T} + \frac{1}{\tau_S} \exp\left(\frac{-\Delta E_{ST}}{k_B T}\right)} \tag{5.1}$$

5.3.1.1 [Cu(NN)(PP)]⁺ Complexes with 2-(2′-pyridyl)benzimidazole and 2-(2′-pyridyl)imidazole-based Ligands

In 2005, Wang and coworkers reported a series of binuclear or trinuclear Cu(I) emissive complexes with 2-(2′-pyridyl)imidazolyl (pbb) derivatives as the bridging ligand and triphenylphosphine as the terminal ligands, as shown in Figure 5.4 [56]. The emission of these complexes in THF or CH_2Cl_2 solution is very weak. The films of these complexes in PVK or PMMA display yellow-orange luminescence ($\lambda_{max} = 535$–550 nm) when irradiated by UV light at ambient temperature. The absolute emission quantum yield of complexes **3–6** (doped in PMMA, 20 : 80 wt%) was measured to be 11%, 17%, 12%, and 15%, respectively, which are comparable with that of [Cu(Phen)(PPh$_3$)$_2$][BF$_4$] (14%). Complex **6** was used as emitting dopant in the fabrication of OLED:ITO/PEDOT/PVK:20 wt% Cu(I)

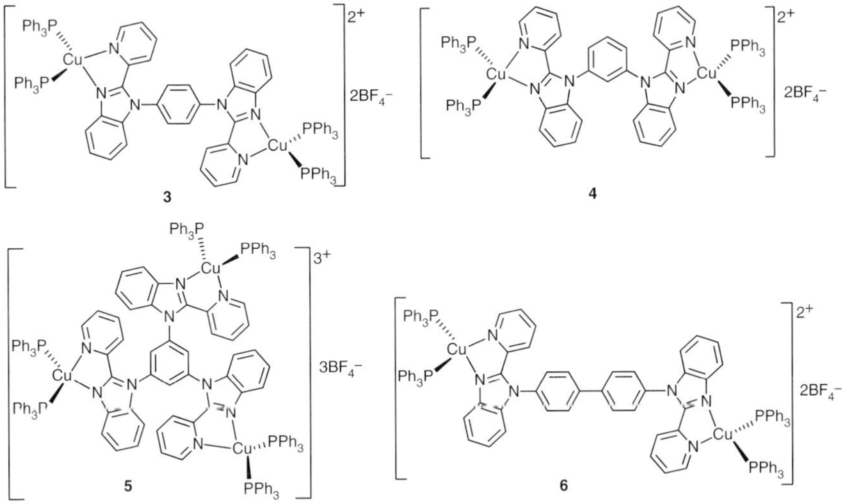

Figure 5.4 Chemical structures of heteroleptic Cu(I) complexes **3–6** with 2-(2′-pyridyl)benzimidazolyl diimine ligands.

(60 nm)/F-TBB (15 nm)/Alq$_3$ (15 nm)/LiF (1 nm)/Al. The turn-on voltage of the OLED is 9 V. The OLED emits weak yellow-orange light with the highest current efficiency of 0.2 cd A^{-1} and a maximum brightness less than 60 cd m^{-2}. The poor performance of the OLED is attributed to the poor quantum yield of complex **6** and its long decay time.

The impact of the phosphine ligands, PPh$_3$, dppe, POP, and DPPMB, on the electronic properties of the Cu(I) complexes (**7–10**) containing the 2-(2′-pyridyl)benzimidazolylbenzene ligand, has been studied (Figure 5.5) [56]. The optical bandgap between HOMO and lowest unoccupied molecular orbital (LUMO) using the edge of the lowest-energy absorption band was determined to be 2.80, 2.50, 2.60, and 2.10 eV for complexes **7–10**, respectively. Density function calculation for these complexes shows that the HOMO is dominated by the d orbital of the copper(I) ion, together with the significant contributions from the phosphine ligands and a small contribution from the pbb ligand. The LUMO is almost totally composed of the atomic orbitals of the pbb ligand. The molecular orbital (MO) calculation results are consistent that the lowest electronic transition in complexes **7–10** is MLCT/ILCT in nature. The complexes **7–10** are not emissive neither in solution nor as pure solids at ambient temperature, but emit when doped into PMMA polymer with the maximum values at 536~593 nm and very long decay time (>200 μs). It is surprising that the complex with PPh$_3$ (**7**) has the best emissive performance among these complexes. In contrast to the phenanthroline-based Cu(I) complexes, where the ligands increase substantially the emission quantum efficiency as compared with the PPh$_3$ ligand, the POP ligand decreases evidently the emission quantum efficiency in this series. The PLQY for **7** and **9** is 13% and 1.04%, respectively. This study shows again that the phosphine ligands have great influence on the emissive properties of Cu(I) complexes, but it is difficult to draw a general rule to

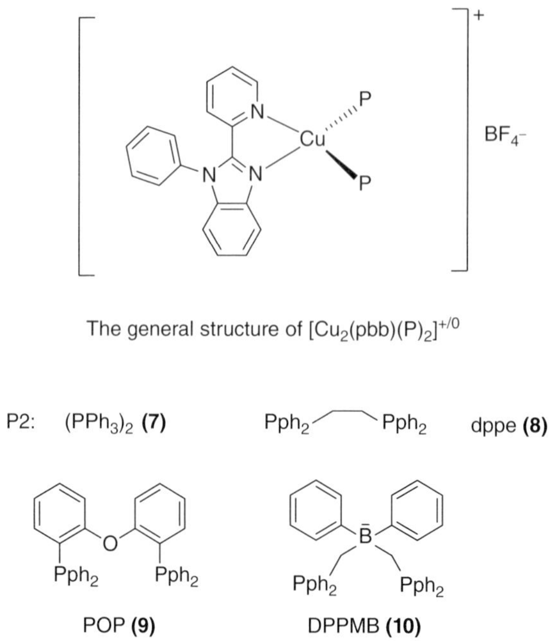

Figure 5.5 Chemical structures of heteroleptic Cu(I) complexes **7–10** with 2-(2′-pyridyl)benzimidazolyl-benzene.

The general structure of [Cu$_2$(pbb)(P)$_2$]$^{+/0}$

P2: (PPh$_3$)$_2$ **(7)**   Pph$_2$⌒Pph$_2$ dppe **(8)**

Pph$_2$–O–Pph$_2$ Pph$_2$–B̄–Pph$_2$
POP **(9)**   DPPMB **(10)**

predict the effect of a phosphine ligand beforehand not considering the diimine ligand. Systematic check and optimization of the PP ligand is necessary to find a suitable combination of the ligands for a specified NN ligand and vice versa.

Si et al. modified the 2-(2′-pyridyl)benzimidazolyl ligand with hole-transporting or electron-transporting group and obtained Cu(I) complexes (**11–13**) as shown in Figure 5.6 [73]. The complexes are emissive in a CH_2Cl_2 solution with the maximum value at 532 nm for **11**, 552 nm for **12**, and 521 nm for **13**. OLEDs with the structure of ITO/m- MTDATA(30 nm)/NPB(20 nm)/dopant:CBP(30 nm)/Bphen(20 nm)/Alq$_3$(20 nm)/LiF(0.8 nm)/Al(200 nm) were fabricated. The device efficiency turns out to be poor, but the brightness of the devices is impressive. The maximum brightness achieved from the devices based on 1 wt% of complex **12** and 7 wt% of complex **13** reaches 5543 cd m^{-2} at 16 V and 8669 cd m^{-2} at 14 V, respectively.

Adachi and coworkers reported a series of emissive Cu(I) complexes with pyridyl- or quinolyl-imidazole ligands as shown in Figure 5.7 [74]. DFT calculations for these complexes indicate that the electron density in the HOMOs is mainly localized at the Cu(I) ion and Cu–P σ-bonding orbital, while that in the LUMOs is localized over the diimine ligands. The calculations confirm that the lowest excited states of these complexes contain the contributions of MLCT. Films containing 10 wt% of these complexes in 2,6-dicarbazolo–1,5-pyridine (PYD2) emit green (λ_{max} = 530 nm for **14**) or yellow (λ_{max} = 549 nm for **15** and λ_{max} = 544 nm for **16**) luminescence when irradiated with light of 365 nm, and the PLQYs are 25% for **14**, 27% for **15**, and 36% for **16**, respectively. The decay lifetimes for **15** and **16** are very long, 174.7 μs for **15** and 243.0 μs for **16**, respectively. The Stokes-like shifts for **14**, **15**, and **16** are 178, 156, and 157 nm,

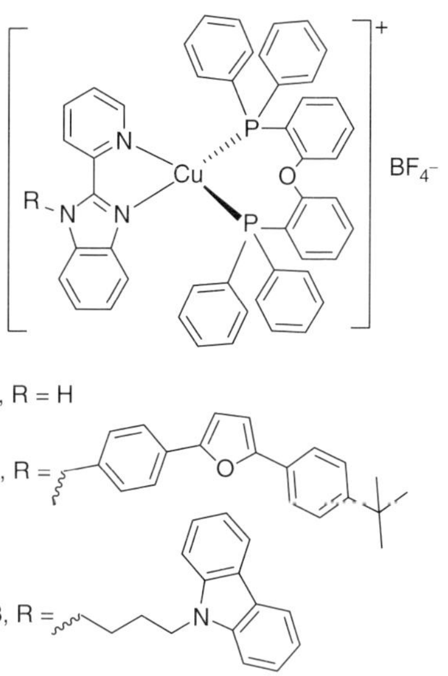

Figure 5.6 Chemical structures of heteroleptic Cu(I) complexes **11–13** with 2-(2′-pyridyl)benzimidazolyl diimine ligands.

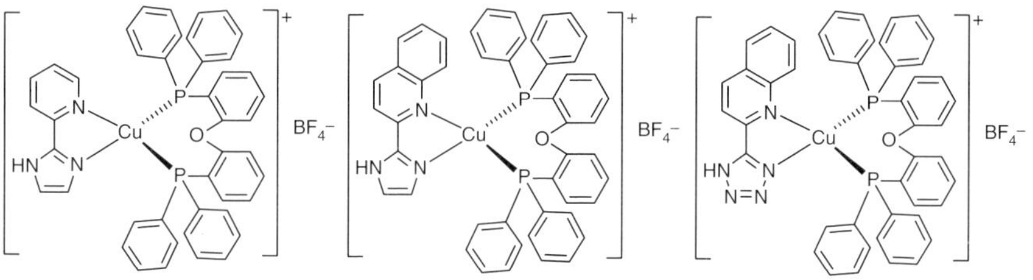

11, R = H

12, R =

13, R =

Figure 5.7 Chemical structures of heteroleptic Cu(I) complexes **14–16**.

respectively, which indicate that the excited-state distortion in **15** and **16** is suppressed more effectively. The electroluminescent performance of these complexes was tested in OLEDs with structure ITO/PEDOT:PSS (40 nm)/10 wt% of **Cu(I)**: PYD2 (30 nm)/DPEPO (50 nm)/LiF (0.7 nm)/Al (100 nm). The maximum EQE is 2.0% for **14**, 6.1% for **15**, and 7.4% for **16**, respectively. The enhancement of the PLQY and the electroluminescence efficiency for complex **16** is contributed to the suppression of not only the excited-state distortion but also C–H vibrational quenching.

5.3.2 [Cu(NN)(PP)]⁺ Complexes with 5-(2-pyridyl)tetrazole-based Ligands

The mononuclear neutral and ionic Cu(I) complexes with 5-(2-pyridyl)tetrazolate and various phosphine ligands were reported by Bräse and coworkers [60] (Figure 5.8). Their photophysical properties were studied and compared in the solid state (powder). The neutral complexes show strong green luminescence in the region from 502 to 545 nm with very high PLQYs (76–89%) and lifetimes of 17.8–26.6 μs. The cationic complexes exhibit redshifted emission bands from

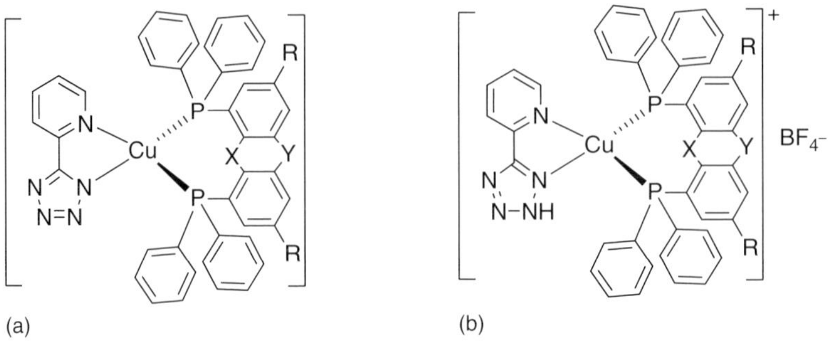

Figure 5.8 Chemical structures of heteroleptic Cu(I) complexes **17–20** with 5-(2-pyridyl)tetrazolate ligand.

(**17**) R = H, X = H_2, Y = H_2
(**18**) R = H, X = O, Y = H_2
(**19**) R = H, X = O, Y = CMe_2
(**20**) R = Me, X = O, Y = H_2

518 to 569 nm, significantly smaller PLQYs of 4–46%, and shortened lifetimes (5.2–15.3 μs). The less efficient luminescence of the ionic complexes is ascribed to the vibrational quenching of the N—H bond on the tetrazole ring, the negative influence of counterions, and the relatively weaker binding of the neutral diimine ligand. The influence of the PP ligands on the PLQYs is not substantial in the neutral species, but is evident in the ionic species. The PPh_3 ligand has the best emission performance when combined with neutral 5-(2-pyridyl)tetrazolate rather than the chelating diphosphine ligands. DFT calculations reveal that the localization of the frontier orbitals of the ionic complexes is very different from that of the neutral complexes. The HOMO of ionic complex **17b** is mainly located on the Cu(I) atoms and the phosphorous atoms, while the LUMO is located on the neutral diimine ligand. In contrast, in the neutral complex **17a**, the HOMO is mainly confined to the Cu(I) atom and the tetrazole moiety together with small contributions from the pyridine N atom, and the LUMO is located primarily on the pyridine ring and to a smaller extent on the tetrazole ring. The phosphines do not contribute to the frontier orbitals, i.e. they should not influence the emission wavelength directly. The calculation results are consistent to the experimental observation that the phosphine ligands have different effect on the photophysical properties of the ionic complexes and the neutral complexes.

Three mononuclear or binuclear Cu(I) heteroleptic complexes with 2-(2-tert-butyl-2H-tetrazol-5-yl)pyridine and various P^P ligands as shown in Figure 5.9 were reported [75]. These complexes are weakly luminescent in CH_2Cl_2 solution but emit bright green light in a rigid matrix. The emission maxima and PLQYs of the PMMA films containing 25% wt of the complexes are 495 nm and 35% for **20**, 515 nm and 7% for **21**, and 515 nm and 17% for **22**, respectively.

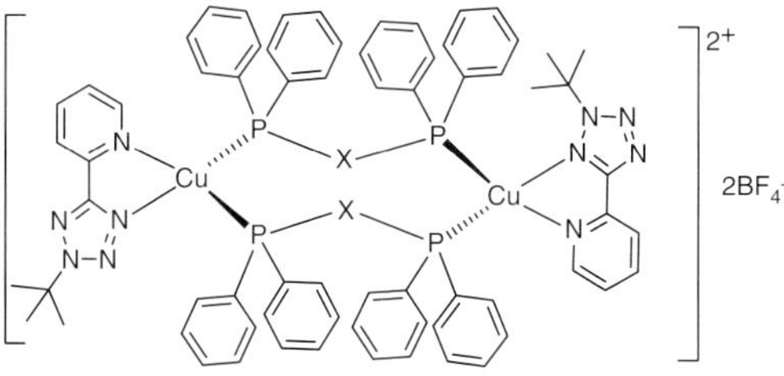

5.3 Heteroleptic Diimine/Diphosphine [Cu(NN)(PP)]+ Complexes with Pyridine-Based Ligand

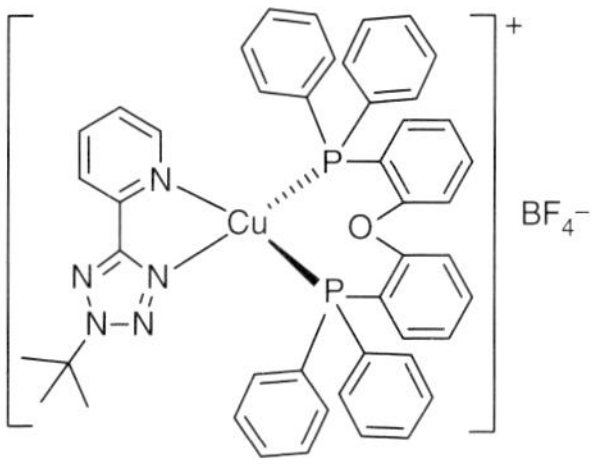

[Figure showing structures of complexes 21, 22, 23 with X = CH$_2$ (21), CH$_2$CH$_2$ (22)]

Figure 5.9 Chemical structures of heteroleptic Cu(I) complexes **21–23** with 5-(2-pyridyl)tetrazolate derivative ligand.

5.3.3 [Cu(NN)(PP)]+ Complexes with 3-(2'-pyridyl)-1,2,4-triazole-based Ligands

A series of heteroleptic Cu(I) complexes based on 3-(2'-pyridyl)-1,2,4-triazole diimine ligands are shown in Figure 5.10 [76]. Complexes **24–26** are ionic and contain 5-tert-butyl-3-(2-pyridyl)-1H-1,2,4-triazole as diimine ligand together with various phosphine ligands. In the solid state at room temperature, complex **25** emits a green luminescence ($\lambda_{max} = 525$ nm, $\tau = 15.9$ μs, $\Phi = 49.1\%$), and complexes **24** and **26** emit strongly a blue luminescence ($\lambda_{max} = 456$ nm, $\tau = 107$ μs, $\Phi = 41.8\%$ for **24** and $\lambda_{max} = 468$ nm, $\tau = 26.9$ μs, $\Phi = 40.4\%$ for **26**, respectively). In degassed CH$_2$Cl$_2$ solution at ambient temperature, complexes **24–26** show a broad single emission band centered at 504, 581, and 525 nm with quantum yields of 9.7%, 1.6,% and 14%, respectively. The quantum yield of **24** is comparable with that of the [Cu(dmp)(POP)]+ [47]. Neutral complex **27** and ionic complex **28** are

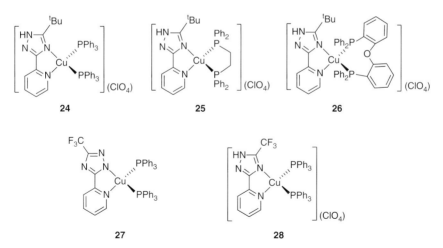

Figure 5.10 Chemical formula of heteroleptic Cu(I) complexes **24–28** with 3-(2'-pyridyl)-1,2,4-triazole diimine ligands.

a pair of Cu(I) complexes based on 5-trifluoromethyl-3-(2-pyridyl)-1,2,4-triazole ligand that can readily be interconverted through protonation or deprotonation. It is interesting that the coordinating pattern of the ligand changes upon protonation or deprotonation, as shown in Figure 5.9. Complex **27** exhibits a slightly higher emission quantum yields than complex **28** both in the solid state and in degassed CH_2Cl_2 solution. The redshifts of the emission, caused by a change of the environment from powder (crystalline environment) to solution, are 65 and 33 nm for **27** and **28**, respectively. This result shows that the coordinating pattern in **28** imparts the Cu(I) complex a more rigid molecular framework than that in **27**. Similar to many TADF complexes, a blueshift of the emission occurs as temperature increases for **27** and **28**. Surprisingly, a remarkable blueshift of 64 nm (from 536 to 472 nm) as temperature increases from 77 to 298 K was observed for **27** in the solid state, whereas a blueshift of only 4 nm (from 500 to 496 nm) was found for **28** under the same conditions, indicating that the ring inversion and N–H deprotonation of the 1,2,4-triazolyl have a significant impact on the emissive properties. DFT calculations reveal that the frontier orbitals of the neutral **27** and ionic **28** have very similar orbital composition. In **27** and **28**, the HOMO is primarily composed of the contribution from Cu and two PPh_3, and the LUMO is essentially localized on the diimine ligand. Unlike in similar neutral Cu(I) complex **17a**, the phosphines in **27** contribute largely to the frontier orbitals.

5.3.4 [Cu(NN)(PP)] Complexes with 2-(2-pyridyl)-pyrrolide-based Ligands

Two neutral [Cu(NN)(PP)] complexes based on 2-pyridyl pyrrolide-diimine ligands are shown in Figure 5.11 [76]. In the degassed CH_2Cl_2 solutions at RT, complexes **29** and **30** show a single emission band maximized at 559 and 551 nm with quantum yields of 34% and 28%, respectively. In solid state at room temperature, 78 and 60 nm blueshifts of the emission as compared with that in the CH_2Cl_2 solutions are observed for **29** and **30**, respectively. For most of the [Cu(NN)(PP)]$^+$ complexes, the PLQY is improved as the rigidity of the matrix increases; however, the PLQY of **30** in the solid state (5%) is much smaller than that in CH_2Cl_2 solution. More interestingly, the emission spectra of these complexes in the 77 K rigid medium have well-resolved vibronic progressions. This is very uncommon for the [Cu(NN)(PP)]$^+$ complexes in which the S_1 and T_1 states have MLCT character. Theoretical calculations reveal that the S_1

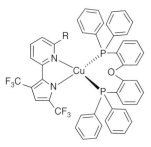

Figure 5.11 Chemical structures of heteroleptic Cu(I) complexes **29** and **30** with 2-pyridyl pyrrolide- diimine ligands.

R = H (**29**), ph (**30**)

state for these complexes has a substantial MLCT(>20%) contribution, while the T_1 state has $\pi\pi^*$ configuration in nature. The calculated $\pi\pi^*$ character of the T_1 state is consistent to the well-resolved vibronic progressions of the emission spectra of these complexes in the 77 K rigid medium. Complex **29** was utilized in the fabrication of OLEDs with structure as ITO/TAPC(40 nm)/mCP doped with 8 wt% of **29** (30 nm)/3TPyMB (3 nm)/TmPyPB(37 nm)/LiF (0.8 nm)/Al (150 nm), attaining peak EL efficiencies of 6.6%, 20.0 cd A^{-1}, and 14.9 lm W^{-1} for the forward directions.

5.3.5 [Cu(NN)(PP)]$^+$ Complexes with 1-(2-pyridyl)-pyrazole-based Ligands

Over the past decade, the uses of the pyrazole derivatives as diimine ligands instead of traditional 2,2-bipyridyl and 2,10-phenanthroline derivatives in the syntheses of emissive complexes have caused attention [16, 18, 20, 68, 69, 77–84]. In 2008, Qiu and coworkers reported two blue-emitting cationic iridium complexes with pypz ((pypz = 1-(2-pyridyl)pyrazole)) as ancillary diimine ligands and demonstrated that the pypz ligand destabilizes significantly the LUMO orbital of the complexes and, therefore, shows much blueshifted emission with respect to the complex with 2,2-bipyridyl ligand [85]. With the thought that the pypz ligand is more electron rich than the ligands of phenanthrolines, the pypz ligand and its derivatives have been used to the synthesis of cationic Cu(NN)(POP)$^+$ complexes **31**–**33**, as shown in Figure 5.12 [30]. The energy gaps between the HOMO and LUMO of these complexes **31**–**33** estimated from the onset wavelengths of the absorption spectra measured in CH_2Cl_2 are 3.18, 3.23, and 3.04 eV that are indeed higher than the value found for [Cu(dmp)(POP)]$^+$ (2.7 eV). These complexes have been found to be TADF materials and highly emissive with very high PLQYs both in the degassed CH_2Cl_2 solutions and in the solid states at ambient temperature. High performances of the complexes as emitting dopant have been realized in a series of cost-effective solution-processed OLEDs.

Complexes of **31**–**33** in argon-saturated CH_2Cl_2 display emissions with the maxima at 590, 536, and 540 nm with quantum yields of 2.1%, 45%, and 30% and decay times of 1.6, 11.9, and 13.3 μs, respectively. As compared with complex [Cu(phen)(POP)](BF$_4$), complex **31** exhibits improved PLQY due to the electron-donating effect of the pypz ligand [47]. For complex **32**, an

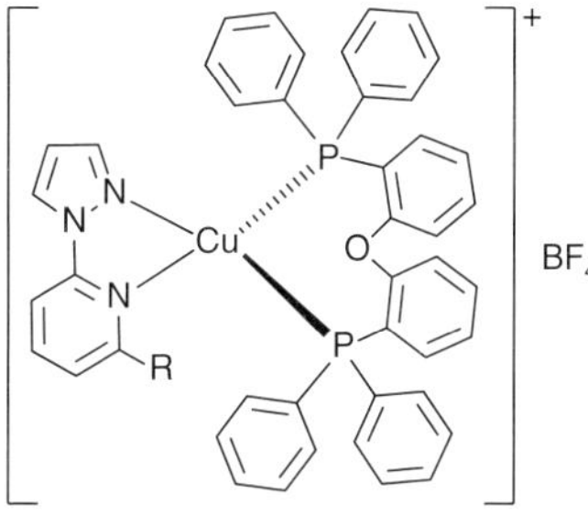

Figure 5.12 Chemical structures of heteroleptic Cu(I) complexes **31**–**33** with 1-(2-pyridyl)-pyrazole-diimine ligands.

R = H (**31**), CH$_3$ (**32**), CF$_3$ (**33**)

evident blueshift and a much higher PLQY are observed for its emission. This observation is due to the cooperative effect of the electron-donating character and the steric demanding effect of the methyl group in the ligand. Complex **33** represents a typical example that shows the importance of steric effect in the Cu(I) emissive complexes. As compared with the pristine pypz ligand in **31**, the electron-withdrawing group of CF_3 at the diimine ligand in **3** leads to a redshift effect and a lower efficiency of the emission, whereas the steric demand of the CF_3 group leads to a smaller Stokes shift and a higher efficiency of the emission. Obviously, the steric effect of CF_3 group dominates its electronic effect as indicated by the blueshift and higher PLQY of complex **33** as compared with complex **31**.

In the solid state, complexes **31**, **32**, and **33** display the emission maxima of 490, 465, and 492 nm with quantum yields of 56%, 87%, and 75% and decay times of 20.4, 12.2, and 22.8 μs, respectively. Improved PLQYs are observed for these complexes in the solid states as compared with those in the CH_2Cl_2 solutions. With an increase of the matrix rigidity, the freedom for changes of the molecular geometries upon MLCT excitation is decreased, and the geometric distortion from the tetrahedral-like ground state (d^{10}) to the flattened excited state (d^9) is constrained [22, 47, 86]. Consequently, shorter emission maxima, higher quantum yields, and longer emissive lifetime are observed. The largest redshift of the emission, caused by a change of the environment from solid state to the CH_2Cl_2 solution, is found for complex **31** that has the pristine pypz ligand, while for complexes **32** and **33** with relatively bulkier ligands, the shifts of the emission maxima are significantly smaller. This observation shows that the extent of the steric effect varies in different environments.

With temperature increasing from 77 to 298 K, a blueshift of the emission maximum by 18, 28, and 19 nm is observed for **31**–**33**, respectively. Calculated from the onset energy of the emission spectra at 77 and 298 K, the energy gaps between S_1 and T_1 (ΔE_{ST}), 0.18, 0.17, and .018 eV, are obtained, for **31**, **32**, and **33**, respectively. The observed emissive lifetimes of **33** reduce drastically by about 16 times from 358 μs at 77 K to 22.8 μs at 298 K. The emissive lifetimes at varied temperatures were measured and fit using Eq. (5.1). The fitted $\tau(T_1)$ value (346 μs) approximates to the value (358 μs) observed at 77 K, and the ΔE_{ST} value (0.17 eV) agrees well with the energy differences (0.18 eV) between the onsets of emission spectra at 77 and 298 K. These observations are consistent with the properties of TADF complexes that are often observed in Cu(I) and Ag(I) complexes with small ΔE_{ST} [7, 19, 31, 32, 87].

Solution-processed OLEDs containing these complexes were fabricated with the configuration ITO/PEDOT:PSS (40 nm)/20 wt% of Cu(I): 26mCPy (30 nm)/DPEPO (50 nm)/LiF (0.7 nm)/Al (100 nm). Using **33** as dopant, excellent efficiencies up to 23.68 cd A^{-1}, 8.47%, and a peak brightness of 2033 cd cm^{-2} were realized. OLEDs based on **31** and **32** exhibited relatively inferior performances although the PL quantum yields of these complexes are comparable. The efficiencies of the OLEDs using **31** as dopant (CE = 8.82 cd A^{-1}, EQE = 3.18%) and using **32** as dopant (CE = 4.07 cd A^{-1}, EQE = 1.59%) are much lower than those of **33**-containing OLED. Better performances in the devices based on these complexes are expected to be obtained via further device optimization.

5.3.6 [Cu(NN)(PP)]$^+$ Complexes with Carbazolyl-modified 1-(2-pyridyl)-pyrazole-based Ligands

The carbazole group is a well-known unit that has been used extensively as the host materials in the OLEDs, organometallic iridium(III) complexes, and pure organic TADF complexes for its high triplet level and good hole-transporting property [88, 89]. However, highly emissive cuprous complexes with carbazole-containing ligands have been far less studied. Liu and coworkers reported a codeposition approach in which CuI and a carbazole-based compound were deposited simultaneously under vacuum to produce an emissive cuprous complex in situ in the resulting emitting layer [90, 91]. Highly efficient vacuum-deposited OLEDs were fabricated, and the carbazole-based compound served as both ligand and host matrix in the devices. Recently, the photophysical properties of two strongly luminescent Cu(I) complexes containing carbazole modifying 1-(2-pyridyl)-pyrazole (czpypz)-based diimine ligand as shown in Figure 5.13 have been synthesized and used in OLED fabrication [29].

Complexes **34** and **35** are weakly luminescent in degassed CH$_2$Cl$_2$ at room temperature with $\lambda_{max} = 540$ nm for **34** and $\lambda_{max} = 569$ nm for **35**, respectively. However, the complexes emit efficiently in the solid state at room temperature with quantum yields of 45% and 98%, emission maxima at 495 and 518 nm, and decay times of 134 and 23 μs, for **34** and **35**, respectively. From the comparison between the emissive properties of complex **31** and **35**, a conclusion that can be drawn is that the introduction of the carbazolyl substituent into pypz ligand causes a redshift of the emission maximum and an enhancement of the emission efficiency in the solid state. Complexes **34** and **35** display TADF. With temperature increasing from 77 to 298 K, a blueshift of the emission maximum and a decreased lifetime are observed. The energy gaps between the S$_1$ and T$_1$ (ΔE_{ST}) estimated from the onsets of emission spectra at 77 and 298 K are 0.15 eV for **34** and 0.12 eV for **35**, respectively.

Interestingly, in addition to acting as chelating diimine ligand in the formation of the complexes **34** and **35**, the czpypz can also be used as an excellent host material in the fabrication of OLEDs for the Cu(I) complexes. High efficient OLED was fabricated conveniently by preparing the emitting layer from directly spin-coating a mixture of czpypz and the Cu(I) precursor, [Cu(CH$_3$CN)$_2$(POP)]BF$_4$, in which the emissive dopant forms in situ and the excess czpypz acts as the host material. The device configuration is ITO/PEDOT:PSS (40 nm)/2

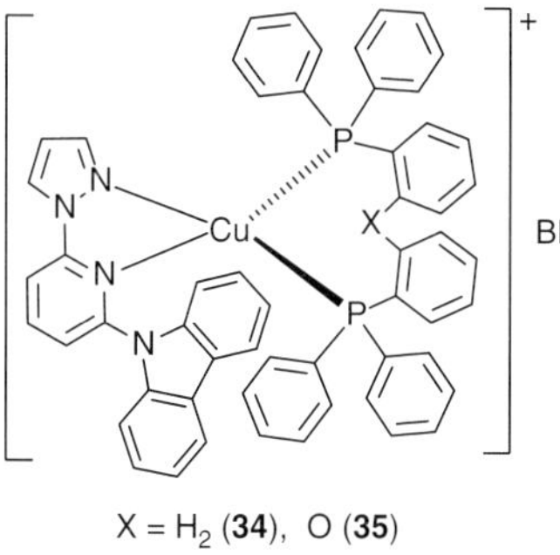

Figure 5.13 Chemical structures of heteroleptic Cu(I) complexes **34** and **35** with 1-(2-pyridyl)-pyrazole-based diimine ligands.

X = H$_2$ (**34**), O (**35**)

(20 wt%):czpzpy (30 nm)/DPEPO (10 nm)/TPBI (50 nm)/LiF (0.8 nm)/Al (100 nm). A comparison of the device performances with the emitting dopant prepared in advance and formed in situ shows essentially identical device properties. The maximum brightness, the peak efficiency, and the external quantum yield of the device with the czpypz ligand as the host and the Cu(I) complex as the dopant are 2939 cd m^{-2}, 7.34 cd A^{-1} (CE), and 6.34% (EQE), respectively. The device with the formation of the Cu(I) complex in situ displays a maximum brightness of 3251 cd m^{-2}, a peak efficiency of 17.53 cd A^{-1} (CE), and 6.36% (EQE).

5.3.7 [Cu(NN)(PP)]$^+$ Complexes with 1-phenyl-3-(2-pyridyl)pyrazole-based Ligands

Four cationic Cu(NN)(POP)]$^+$ complexes **36–39** with various 1-phenyl-3-(2-pyridyl)pyrazole-diimine ligands are shown in Figure 5.14 [26]. The energy gaps (ΔE_g) between HOMO and LUMO for **36** (3.04 eV), **37** (3.08 eV), **38** (3.04 eV), and **39** (3.00 eV) estimated from the onset wavelengths of the absorption spectra in CH$_2$Cl$_2$ are very close, which indicates that the substituent groups on the phenyl ring have little influence on the electronic properties of these complexes. These complexes are highly emissive in the degassed CH$_2$Cl$_2$ solution, in the films of PMMA, and in the solid states. Blueshifted emission maximum, increased lifetime, and improved PLQYs are observed for these complexes as the matrix rigidity of the environment increases, which is very common for Cu(I) complexes because of the structural rearrangement upon excitation. In the CH$_2$Cl$_2$ solutions, these complexes emit bright yellow-orange luminescence with maximum at 556–580 nm and lifetime in the range of 4.2–5.7 μs. In the PMMA films of 10% wt concentration, the complexes emit green light with maximum at 506–515 nm and PLQYs of 46–54%. In the solid states, the complexes are highly emissive with PLQYs up to 91% for complex **37**. The redshift of the emission for these complexes is similar (62–86 nm), caused by a change of the environment from solid state to CH$_2$Cl$_2$ solution. From the onsets of the emission spectra of these complexes in PMMA at 77 and 298 K, small energy gaps ΔE_{ST} (0.16–0.24 eV) between S$_1$ and T$_1$ were obtained. The small ΔE_{ST} together with the dramatic decrease of the decay lifetime observed for complex **36** as the temperature increases indicates that the complexes are TADF

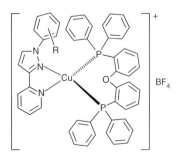

Figure 5.14 Chemical structures of heteroleptic Cu(I) complexes **36–39** with 1-phenyl-3-(2-pyridyl)pyrazole-diimine ligands.

R = H (**36**), 2-CH$_3$ (**37**), 2-CF$_3$ (**38**), 4-CF$_3$ (**39**)

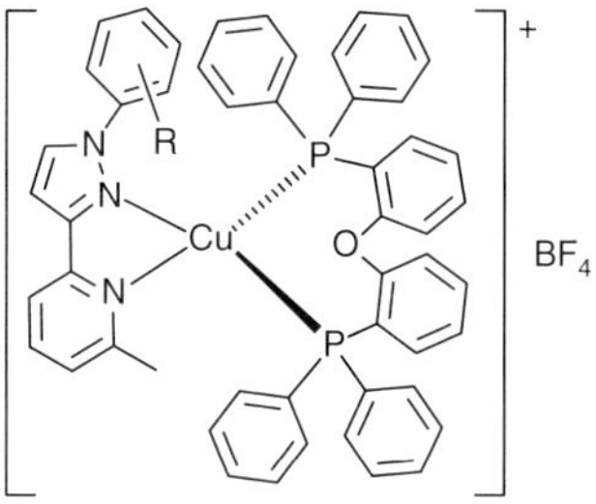

Figure 5.15 Chemical structures of heteroleptic Cu(I) complexes **40–44** with 1-phenyl-3-(6-methyl-2-pyridyl)pyrazole-diimine ligands.

R = H (**40**), 2-CH$_3$ (**41**), 4-CH$_3$ (**42**), 2-CF$_3$ (**43**), 4-CF$_3$ (**44**)

materials. OLEDs with configuration as ITO/PEDOT:PSS(40 nm)/10 wt% of Cu(I): host(30 nm)/DPEPO(50 nm)/LiF(0.7 nm)/Al(100 nm) were made. The device with **36** as emitting dopant gives the best device performance with the peak current efficiency of 17.8 cd A^{-1} and EQE of 6.4%.

Figure 5.15 shows the chemical configuration of a series of ionic [Cu(NN)(POP)]$^+$ complexes (**40–45**) with 1-phenyl-3-(6-methyl-2-pyridyl)pyrazole-diimine ligands [92]. These complexes are highly emissive both in the CH$_2$Cl$_2$ solution and in the rigid matrices with the emission maximum in the blue- to bluish-green region. It is remarkable that the PLQYs of these complexes are very high and up to 99% in the solid states. The PLQYs of the PMMA films of 10 wt% concentration of these complexes are also very impressive and up to 59% that is very high for the reported Cu(I) complexes. The blueshifted emission maximum and the improved PLQYs are also observed for these complexes as the matrix rigidity of the environment increases, but it is surprising that the lifetimes of these complexes in PMMA films are longer than those in the solid states. The redshifts of the emissions for these complexes caused by a change of the environment from solid powder to CH$_2$Cl$_2$ solutions are much smaller as compared with those for complexes **36–39**. This result indicates that the diimine ligands in **40–44** combined with POP can suppress better the structural rearrangement and exciplex quenching of these Cu(I) complexes upon excitation than the ligands in **36–39**. The smallest redshift is 16 nm observed for complex **42**. For comparison, the redshift observed in **36** is 83 nm. The studies of varied temperature emission spectra and decay behaviors of these complexes indicate that these complexes are TADF complexes with small ΔE_{ST} (0.08–0.20 eV).

5.3.8 [Cu(NN)(PP)]$^+$ Complexes with 3-phenyl-5-(2-pyridyl)-1H-1,2,4-triazole-based Ligands

Four [Cu(NN)(POP)]$^+$ Cu(I) complexes containing 3-phenyl-5-(2-pyridyl)-1H-1,2,4-triazole derivatives as diimine ligands, **45–48**, are shown in Figure 5.16 [27]. The energy gaps (ΔE_g) between HOMO and LUMO estimated from the onset wavelengths of the absorption spectral of these complexes are 2.7 eV for **45** and **46** and 2.85 eV for **47** and **48**. The redshifts of these caused by the change of matrix from PMMA films to degassed CH$_2$Cl$_2$ are 69, 64, 51, and 52 nm, for **45–48**, respectively. These results demonstrate that the methyl group at the pyridine ring increases the energy gap of the

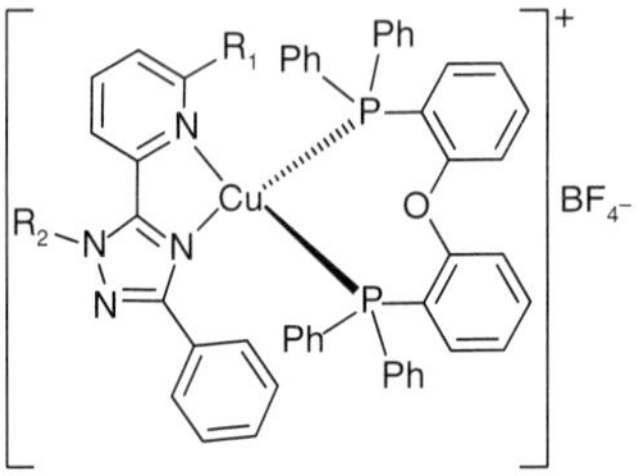

Figure 5.16 Chemical structures of heteroleptic Cu(I) complexes 45–48 with 3-phenyl-5-(2-pyridyl)-1H-1,2,4-triazole-based diimine ligands.

(45) R_1 = H, R_2 = phenyl
(46) R_1 = H, R_2 = 4-carbazolylphenyl
(47) R_1 = Me, R_2 = phenyl
(48) R_1 = Me, R_2 = 4-carbazolylphenyl

frontier orbitals and the structural rigidity of the resulting Cu(I) complexes and thus improves the photoluminescence of the complexes. As compared, the carbazolyl group does not have obvious influence on the energy gap and the structural rigidity of the complexes and thus displays very limited influence on the photoluminescence of the complexes. The OLEDs with the configuration ITO/PEDOT:PSS (40 nm)/20 wt% of **47** or **48**:PYD2 (2,6-bis(N-carbazolyl)pyridine, 30 nm)/DPEPO (bis[2-(di-(phenyl)phosphino)-phenyl] ether oxide, 50 nm)/LiF (0.7 nm)/Al (100 nm) (where ITO denotes indium tin oxide, PEDOT:PSS is poly (3, 4-ethylenedioxythiophene)–poly(styrenesulfonic acid) were fabricated. The device with **47** displays a maximum brightness of 2096 cd m^{-2}, a peak efficiency of 14.2 cd A^{-1} (CE), and 5.9% (EQE). The device with **48** displays a maximum brightness of 2447 cd m^{-2}, a peak efficiency of 16.7 cd A^{-1} (CE), and 6.7% (EQE). These results suggest that the dopant with carbazolyl group gives a better utilization of the excitons in the OLED.

5.4 Conclusion and Perspective

In this chapter, the recent progress of the emissive [Cu(NN)(PP)]$^+$ complexes with pyridine-based diimine ligands has been briefly summarized. The rationalization of the relationship between the structure and the photophysical property of the [Cu(NN)(PP)]$^+$ complex has made much progress over the last decade, and it has been shown that the electronic and steric effect of the ligands have substantial influence on the photophysical property of the [Cu(NN)(PP)]$^+$ complexes. The key principle in designing the Cu(I) complexes is to construct an overall structure with rigid and steric conformation that can prevent the structure from distortion rearrangement and exciplex quenching upon excitation through judicious choice and adjustment of the ligands, including the ligand topology, bulk, and rigidity. Under the guide of designing principle, many [Cu(NN)(PP)]$^+$ complexes based on pyridyl diimine ligands with interesting properties have been designed and optimized. Clearly, the present progress of the research on this field attests the ability of pyridyl diimine ligands to obtain highly emissive Cu(I) TADF materials, especially those highly blue-emitting complexes that are problematic for Ir, Pt, and Os complexes because of the existence of dd* orbitals. Designing new diimine and diphosphine ligands, development of new TADF Cu(I) materials with high PLQY and short decay lifetime, and full assessment of their device stability and light output in OLEDs will be still the main tasks in this area in the near future [93].

References

1 Baldo, M.A., O'Brien, D.F., You, Y., Shoustikov, A., Sibley, S., Thompson, M.E., and Forrest, S.R. (1998). *Nature* 395: 151–154.
2 Song, G.F.W.M.X., Wang, J., Wang, Y.H., Bai, F.Q., and Qin, Z.K. (2015). *Spectrochim. Acta. Part A* 134: 406–412.
3 Moral, L.M.M., Son, W.J., Olivier, Y., and Sancho-García, J.C. (2015). *J. Chem. Theory Comput.* 11: 168–177.
4 Kessler, F., Costa, R.D., Di Censo, D., Scopelliti, R., Orti, E., Bolink, H.J., Meier, S., Sarfert, W., Gratzel, M., Nazeeruddin, M.K., and Baranoff, E. (2012). *Dalton Trans.* 41: 180–191.
5 Du, B.-S., Liao, J.-L., Huang, M.-H., Lin, C.-H., Lin, H.-W., Chi, Y., Pan, H.-A., Fan, G.-L., Wong, K.-T., Lee, G.-H., and Chou, P.-T. (2012). *Adv. Funct. Mater.* 22: 3491–3499.
6 Lin, C.-H., Hsu, C.-W., Liao, J.-L., Cheng, Y.-M., Chi, Y., Lin, T.-Y., Chung, M.-W., Chou, P.-T., Lee, G.-H., Chang, C.-H., Shih, C.-Y., and Ho, C.-L. (2012). *J. Mater. Chem.* 22: 10684.
7 Yersin, H., Rausch, A.F., Czerwieniec, R., Hofbeck, T., and Fischer, T. (2011). *Coord. Chem. Rev.* 255: 2622–2652.
8 Baranoff, E., Fantacci, S., De Angelis, F., Zhang, X., Scopelliti, R., Gratzel, M., and Nazeeruddin, M.K. (2011). *Inorg. Chem.* 50: 451–462.
9 Murawski, C., Liehm, P., Leo, K., and Gather, M.C. (2014). *Adv. Funct. Mater.* 24: 1117–1124.
10 Tang, D.P.T.M., Wong, Y., Chan, M., Wong, K., and Yam, V.W. (2014). *J. Am. Chem. Soc.* 136 (51): 17861–17868.
11 Yersin, H. and Finkenzeller, W.J. (2008). *Highly Efficient OLEDs with Phosphorescent Materials*, 1–97. Wiley-VCH.
12 Volz, D., Wallesch, M., Flechon, C., Danz, M., Verma, A., Navarro, J.M., Zink, D.M., Brase, S., and Baumann, T. (2015). *Green Chem.* 17: 1988–2011.
13 Dumur, F. (2015). *Org. Electron.* 21: 27–39.
14 Czerwieniec, R., Leitl, M.J., Homeier, H.H.H., and Yersin, H. (2016). *Coord. Chem. Rev.* 325: 2–28.
15 Laviecambot, A., Cantuel, M., Leydet, Y., Jonusauskas, G., Bassani, D., and McClenaghan, N. (2008). *Coord. Chem. Rev.* 252: 2572–2584.
16 Xiang, H., Cheng, J., Ma, X., Zhou, X., and Chruma, J.J. (2013). *Chem. Soc. Rev.* 42: 6128–6185.
17 McMillin, D.R. and McNett, K.M. (1998). *Chem. Rev.* 98: 1201–1219.
18 Tao, Y., Yuan, K., Chen, T., Xu, P., Li, H., Chen, R., Zheng, C., Zhang, L., and Huang, W. (2014). *Adv. Mater.* 26: 7931–7958.
19 Osawa, M., Kawata, I., Ishii, R., Igawa, S., Hashimoto, M., and Hoshino, M. (2013). *J. Mater. Chem. C* 1: 4375.
20 Krylova, V.A., Djurovich, P.I., Whited, M.T., and Thompson, M.E. (2010). *Chem. Commun. (Camb.)* 46: 6696–6698.

21 Lotito, K.J. and Peters, J.C. (2010). *Chem. Commun. (Camb.)* 46: 3690–3692.
22 Hashimoto, M., Igawa, S., Yashima, M., Kawata, I., Hoshino, M., and Osawa, M. (2011). *J. Am. Chem. Soc.* 133: 10348–10351.
23 Krylova, V.A., Djurovich, P.I., Conley, B.L., Haiges, R., Whited, M.T., Williams, T.J., and Thompson, M.E. (2014). *Chem. Commun. (Camb.)* 50: 7176–7179.
24 Marion, R., Sguerra, F., Di Meo, F., Sauvageot, E., Lohier, J.F., Daniellou, R., Renaud, J.L., Linares, M., Hamel, M., and Gaillard, S. (2014). *Inorg. Chem.* 53: 9181–9191.
25 Leitl, M.J., Krylova, V.A., Djurovich, P.I., Thompson, M.E., and Yersin, H. (2014). *J. Am. Chem. Soc.* 136: 16032–16038.
26 Zhang, Q., Chen, X.-L., Chen, J., Wu, X.-Y., Yu, R., and Lu, C.-Z. (2015). *RSC Adv.* 5: 34424–34431.
27 Zhang, Q., Chen, X.L., Chen, J., Wu, X.Y., Yu, R., and Lu, C.Z. (2015). *Dalton Trans.* 44: 10022–10029.
28 Qin, L., Zhang, Q., Sun, W., Wang, J., Lu, C., Cheng, Y., and Wang, L. (2009). *Dalton Trans.* 9388–9391.
29 Chen, X.-L., Lin, C.-S., Wu, X.-Y., Yu, R., Teng, T., Zhang, Q.-K., Zhang, Q., Yang, W.-B., and Lu, C.-Z. (2015). *J. Mater. Chem. C* 3: 1187–1195.
30 Chen, X.L., Yu, R.M., Zhang, Q.K., Zhou, L.J., Wu, C.Y., Zhang, Q., and Lu, C.Z. (2013). *Chem. Mater.* 25: 3910–3920.
31 Hofbeck, T., Monkowius, U., and Yersin, H. (2015). *J. Am. Chem. Soc.* 137: 399–404.
32 Czerwieniec, R. and Yersin, H. (2015). *Inorg. Chem.* 54: 4322–4327.
33 Linfoot, C.L., Leitl, M.J., Richardson, P., Rausch, A.F., Chepelin, O., White, F.J., Yersin, H., and Robertson, N. (2014). *Inorg. Chem.* 53: 10854–10861.
34 Yersin, H. Czerwieniec, R. and Hupfer, A. (2012). Proceedings of SPIE, Volume 8435, SPIE Photonics Europe, 16–19 April 2012, Brussels, Belgium.
35 Czerwieniec, R., Yu, J.B., and Yersin, H. (2011). *Inorg. Chem.* 50: 8293–8301.
36 Kang, L., Chen, J., Teng, T., Chen, X.L., Yu, R., and Lu, C.Z. (2015). *Dalton Trans.* 44: 11649–11659.
37 Volz, D., Chen, Y., Wallesch, M., Liu, R., Fléchon, C., Zink, D.M., Friedrichs, J., Flügge, H., Steininger, R., Göttlicher, J., Heske, C., Weinhardt, L., Bräse, S., So, F., and Baumann, T. (2015). *Adv. Mater.* 27: 2538–2543.
38 Chen, L.X., Jennings, G., Liu, T., Gosztola, D.J., Hessler, J.P., Scaltrito, D.V., and Meyer, G.J. (2002). *J. Am. Chem. Soc.* 124: 10861–10867.
39 Fasina, T.M., Collings, J.C., Burke, J.M., Batsanov, A.S., Ward, R.M., Albesa-Jové, D., Porrès, L., Beeby, A., Howard, J.A.K., Scott, A.J., Clegg, W., Watt, S.W., Viney, C., and Marder, T.B. (2005). *J. Mater. Chem.* 15: 690–697.
40 Feng, D., Gu, Z.Y., Chen, Y.P., Park, J., Wei, Z., Sun, Y., Bosch, M., Yuan, S., and Zhou, H.C. (2014). *J. Am. Chem. Soc.* 136: 17714–17717.
41 Kalsani, V., Schmittel, M., Listorti, A., Accorsi, G., and Armaroli, N. (2006). *Inorg. Chem.* 45: 2061–2067.
42 Rader, R.A., McMillin, D.R., Buckner, M.T., Matthews, T.G., Casadonte, D.J., Lengel, R.K., Whittaker, S.B., Darmon, L.M., and Lytle, F.E. (1981). *J. Am. Chem. Soc.* 103: 5906–5912.
43 Breddels, P.A., Berdowski, P.A.M., Blasse, G., and McMillin, D.R. (1982). *J. Chem. Soc., Faraday Trans. 2* 78: 595–601.
44 Sheldrick, G.M. (2014). *Institute for Inorganic Chemistry*. Göttingen, Germany: University of Göttingen.

45 Dolbier, W.R., Xie, P., Zhang, L., Xu, W., Chang, Y., and Abboud, K.A. (2008). *J. Org. Chem.* 73: 2469–2472.
46 Kuang, S.M., Cuttell, D.G., McMillin, D.R., Fanwick, P.E., and Walton, R.A. (2002). *Inorg. Chem.* 41: 3313–3322.
47 Cuttell, D.G., Kuang, S.M., Fanwick, P.E., McMillin, D.R., and Walton, R.A. (2002). *J. Am. Chem. Soc.* 124: 6–7.
48 Zhang, Q., Zhou, Q., Cheng, Y., Wang, L., Ma, D., Jing, X., and Wang, F. (2006). *Adv. Funct. Mater.* 16: 1203–1208.
49 Zhang, Q., Li, J., Shizu, K., Huang, S., Hirata, S., Miyazaki, H., and Adachi, C. (2012). *J. Am. Chem. Soc.* 134: 14706–14709.
50 Armaroli, N., Accorsi, G., Holler, M., Moudam, O., Nierengarten, J.F., Zhou, Z., Wegh, R.T., and Welter, R. (2006). *Adv. Mater.* 18: 1313–1316.
51 Zhang, Q., Zhou, Q., Cheng, Y., Wang, L., Ma, D., Jing, X., and Wang, F. (2004). *Adv. Mater.* 16: 432–436.
52 Tsukuda, T., Nakamura, A., Arai, T., and Tsubomura, T. (2006). *B. Chem. Soc. Jpn.* 79: 288–290.
53 Saito, K., Arai, T., Takahashi, N., Tsukuda, T., and Tsubomura, T. (2006). *Dalton Trans.* 4444–4448.
54 Tsubomura, T., Takahashi, N., Saito, K., and Tsukuda, T. (2004). *Chem. Lett.* 33: 678–679.
55 Zhang, Q., Ding, J., Cheng, Y., Wang, L., Xie, Z., Jing, X., and Wang, F. (2007). *Adv. Funct. Mater.* 17: 2983–2990.
56 McCormick, T., Jia, W.L., and Wang, S.N. (2006). *Inorg. Chem.* 45: 147–155.
57 Chen, J.L., Cao, X.F., Wang, J.Y., He, L.H., Liu, Z.Y., Wen, H.R., and Chen, Z.N. (2013). *Inorg. Chem.* 52: 9727–9740.
58 Min, J., Zhang, Q., Sun, W., Cheng, Y., and Wang, L. (2011). *Dalton Trans.* 40: 686–693.
59 Sun, W., Zhang, Q., Qin, L., Cheng, Y., Xie, Z., Lu, C., and Wang, L. (2010). *Eur. J. Inorg. Chem.* 2010: 4009–4017.
60 Bergmann, L., Friedrichs, J., Mydlak, M., Baumann, T., Nieger, M., and Brase, S. (2013). *Chem. Commun. (Camb.)* 49: 6501–6503.
61 Volz, D., Zink, D.M., Bocksrocker, T., Friedrichs, J., Nieger, M., Baumann, T., Lemmer, U., and Bräse, S. (2013). *Chem. Mater.* 25: 3414–3426.
62 Czerwieniec, R., Kowalski, K., and Yersin, H. (2013). *Dalton Trans.* 42: 9826–9830.
63 Jia, W.L., McCormick, T., Tao, Y., Lu, J.-P., and Wang, S. (2005). *Inorg. Chem.* 44: 5706–5712.
64 Tsuboyama, A., Kuge, K., Furugori, M., Okada, S., Hoshino, M., and Ueno, K. (2007). *Inorg. Chem.* 46: 1992–2001.
65 Zink, D.M., Bächle, M., Baumann, T., Nieger, M., Kühn, M., Wang, C., Klopper, W., Monkowius, U., Hofbeck, T., Yersin, H., and Bräse, S. (2012). *Inorg. Chem.* 52: 2292–2305.
66 Harkins, S.B. and Peters, J.C. (2005). *J. Am. Chem. Soc.* 127: 2030–2031.
67 Xin, X.L., Chen, M., Ai, Y.B., Yang, F.L., Li, X.L., and Li, F. (2014). *Inorg. Chem.* 53: 2922–2931.
68 Costa, R.D., Orti, E., Bolink, H.J., Monti, F., Accorsi, G., and Armaroli, N. (2012). *Angew. Chem. Int. Ed.* 51: 8178–8211.
69 Zhang, Z., Zhao, W., Ma, B., and Ding, Y. (2010). *Catal. Commun.* 12: 318–322.

70 Linfoot, C.L., Richardson, P., Hewat, T.E., Moudam, O., Forde, M.M., Collins, A., White, F., and Robertson, N. (2010). *Dalton Trans.* 39: 8945–8956.
71 Costa, R.D., Tordera, D., Ortí, E., Bolink, H.J., Schönle, J., Graber, S., Housecroft, C.E., Constable, E.C., and Zampese, J.A. (2011). *J. Mater. Chem.* 21: 16108.
72 Li, X.-L., Ai, Y.-B., Yang, B., Chen, J., Tan, M., Xin, X.-L., and Shi, Y.-H. (2012). *Polyhedron* 35: 47–54.
73 Si, Z., Li, J., Li, B., Liu, S., and Li, W. (2009). *J. Lumin.* 129: 181–186.
74 Wada, A., Zhang, Q., Yasuda, T., Takasu, I., Enomoto, S., and Adachi, C. (2012). *Chem. Commun. (Camb.)* 48: 5340–5342.
75 Femoni, C., Muzzioli, S., Palazzi, A., Stagni, S., Zacchini, S., Monti, F., Accorsi, G., Bolognesi, M., Armaroli, N., Massi, M., Valenti, G., and Marcaccio, M. (2013). *Dalton Trans.* 42: 997–1010.
76 Hsu, C.W., Lin, C.C., Chung, M.W., Chi, Y., Lee, G.H., Chou, P.T., Chang, C.H., and Chen, P.Y. (2011). *J. Am. Chem. Soc.* 133: 12085–12099.
77 Tung, Y.L., Chen, L.S., Chi, Y., Chou, P.T., Cheng, Y.M., Li, E.Y., Lee, G.H., Shu, C.F., Wu, F.I., and Carty, A.J. (2006). *Adv. Funct. Mater.* 16: 1615–1626.
78 You, Y. and Park, S.Y. (2009). *Dalton Trans.* 1267–1282.
79 Yang, C.H., Cheng, Y.M., Chi, Y., Hsu, C.J., Fang, F.C., Wong, K.T., Chou, P.T., Chang, C.H., Tsai, M.H., and Wu, C.C. (2007). *Angew. Chemie Int. Ed.* 46: 2418–2421.
80 Kalinowski, J., Fattori, V., Cocchi, M., and Williams, J.A.G. (2011). *Coord. Chem. Rev.* 255: 2401–2425.
81 Shavaleev, N.M., Monti, F., Scopelliti, R., Baschieri, A., Sambri, L., Armaroli, N., Grätzel, M., and Nazeeruddin, M.K. (2013). *Organometallics* 32: 460–467.
82 Chi, Y. and Chou, P.T. (2010). *Chem. Soc. Rev.* 39: 638–655.
83 Fan, C. and Yang, C. (2014). *Chem. Soc. Rev.* 43: 6439–6469.
84 Xu, H., Chen, R., Sun, Q., Lai, W., Su, Q., Huang, W., and Liu, X. (2014). *Chem. Soc. Rev.* 43: 3259–3302.
85 He, L., Duan, L., Qiao, J., Wang, R., Wei, P., Wang, L., and Qiu, Y. (2008). *Adv. Funct. Mater.* 18: 2123–2131.
86 Armaroli, N. (2001). *Chem. Soc. Rev.* 30: 113–124.
87 Gneuss, T., Leitl, M.J., Finger, L.H., Rau, N., Yersin, H., and Sundermeyer, J. (2015). *Dalton Trans.* 44: 8506–8520.
88 Wong, W.Y. and Ho, C.L. (2009). *Coord. Chem. Rev.* 253: 1709–1758.
89 Yook, K.S. and Lee, J.Y. (2012). *Adv. Mater.* 24: 3169–3190.
90 Liu, Z.W., Qayyum, M.F., Wu, C., Whited, M.T., Djurovich, P.I., Hodgson, K.O., Hedman, B., Solomon, E.I., and Thompson, M.E. (2011). *J. Am. Chem. Soc.* 133: 3700–3703.
91 Liu, Z., Qiu, J., Wei, F., Wang, J., Liu, X., Helander, M.G., Rodney, S., Wang, Z., Bian, Z., Lu, Z., Thompson, M.E., and Huang, C. (2014). *Chem. Mater.* 26: 2368–2373.
92 Zhang, Q., Chen, J., Wu, X.-Y., Chen, X.-L., Yu, R., and Lu, C.-Z. (2015). *Dalton Trans.* 44: 6706–6710.
93 Leitl, M.J., Zink, D.M., Schinabeck, A., Baumann, T., Volz, D., and Yersin, H. (2016). *Top. Curr. Chem.* 374 (3): 1–34.

6

Efficiency Enhancement of Organic Light-Emitting Diodes Exhibiting Delayed Fluorescence and Nonisotropic Emitter Orientation

Tobias D. Schmidt and Wolfgang Brütting

University of Augsburg, Institute of Physics, Augsburg, 86135, Germany

6.1 Introduction

In 1963 the first observation of electroluminescence (EL) of a 5 mm thick anthracene crystal was presented by Pope et al. [1]. Thereafter, it took more than 20 years until Tang and van Slyke demonstrated the first efficient, low-voltage-driven, thin film organic light-emitting diode (OLED) in 1987 [2]. Three years later the first solution processed polymer-based OLED was developed by Burroughes et al. [3] and therewith the starting signal for a new, rapidly growing field of research had been given. These early OLEDs used fluorescent (so-called first-generation) emitter materials, which have the fundamental limitation that only singlet excitons, and, thus, only 25% of all electrically excited states are able to decay radiatively producing photons [4]. In 1998, Baldo et al. [5] presented the first OLED with nearly 100% internal quantum efficiency where they used phosphorescent emitters for the first time. In this second-generation emitter materials, the energy of all excited states is finally transferred to triplet states that show efficient radiative decay with lifetimes in the (sub)microsecond regime [6].

In the last few years, technologies based on OLEDs made tremendous progress [7, 8]. Nowadays, several applications using OLEDs are commercially available, e.g. in the general lighting sector as well as in the display market, especially, for smartphones and tablet computers. However, most of the commercial applications for white light emission are still using hybrid stack systems with a fluorescent blue and phosphorescent red and green emitting materials, because sufficiently stable and efficient phosphorescent molecules in the blue spectral range are still challenging [9, 10]. Thus, a promising new class of emitting compounds is coming more and more into the focus of the OLED community, namely, devices incorporating delayed fluorescence (so-called third-generation) emitter materials [11].

Delayed fluorescence can be enabled by two different mechanisms, viz triplet–triplet annihilation (TTA) [12–18] and thermally activated delayed fluorescence (TADF) [11, 19–26], originally termed p- and e-type delayed fluorescence [27, 28]. Although their working principles differ in detail, both phenomena increase

Highly Efficient OLEDs: Materials Based on Thermally Activated Delayed Fluorescence,
First Edition. Edited by Hartmut Yersin.
© 2019 Wiley-VCH Verlag GmbH & Co. KGaA. Published 2019 by Wiley-VCH Verlag GmbH & Co. KGaA.

device efficiency compared to standard fluorescent OLEDs and sometimes even reach the efficiency of phosphorescent applications by recycling triplet excitons and converting them into singlet excited states [29–32].

In many cases, however, we find that arguments related to spin multiplicity of the excited states alone (and their potential interconversion by these processes) are not sufficient to explain the observed efficiency boost enabled by these materials. Rather, some of the emitter molecules exhibit a strongly nonisotropic distribution of their optical transition dipole vectors (TDVs), which in the case of horizontal alignment is additionally increasing the light output from these OLEDs [33]. Therefore, this chapter aims at giving the required background for a thorough efficiency analysis of OLEDs disentangling both effects and presents exemplary case studies of OLEDs exhibiting these features. In the context of this article, we restrict ourselves to purely organic emitter materials; however, most of the statements are equally valid for metal–organic complexes containing e.g. Cu(I).

6.2 OLED Basics

6.2.1 Working Principle

For the sake of simplicity, we will elucidate the basic operation processes for the simplest OLED only consisting of one organic layer sandwiched between two electrodes (see Figure 6.1). In order to facilitate light emission to the outside world of the OLED, one of these two electrodes has to be (semi)transparent, e.g. a thin indium tin oxide (ITO) anode on a glass substrate. The most important energy values in such a simplified device are the work functions of the used electrodes and the highest occupied molecular orbital (HOMO) as well as the lowest unoccupied molecular orbital (LUMO) of the organic solid. The work functions of anode and cathode should be close to the positions of the HOMO and the LUMO levels, respectively, in order to enable charge carrier injection from the electrodes into the organic material.

In Figure 6.1 a schematic sketch of the important energy levels and the four different steps required for light emission are illustrated [7, 34, 35]. First, electrons are injected from the cathode side to the LUMO of the organic material, while simultaneously electrons are extracted at the anode side from the HOMO of the organic material (or in other words holes are injected into the HOMO) as soon as a sufficiently high external voltage is applied to the electrodes. Second, due to the external applied electric field, the charges are transported toward the electrode with the opposite polarity. Third, if electrons and holes come close enough together – less than the Coulomb radius – on their way through the OLED, they will recombine under the generation of an exciton, which describes a bound excited state on one organic molecule. Finally, the exciton can decay radiatively under creation of a photon, which can thereafter be extracted to the outside world of the OLED.

State-of-the-art OLEDs, of course, use more than one organic material to improve device performance [36]. There are several different layers, each

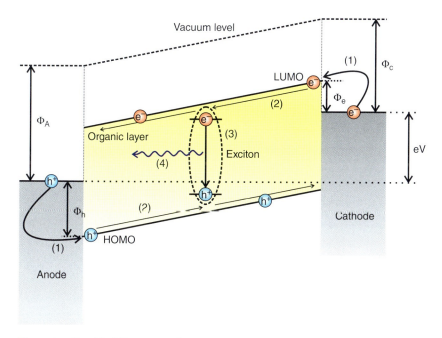

Figure 6.1 Simplified illustration of a single-layer OLED stack. In order to obtain charge carrier injection and subsequent light emission, an external voltage is applied to the electrodes. Light emission occurs following four subsequent steps: (1) charge carrier injection, (2) charge carrier transport, (3) charge carrier recombination, generating an exciton, and, finally, (4) radiative decay of the exciton creating a (visible) photon.

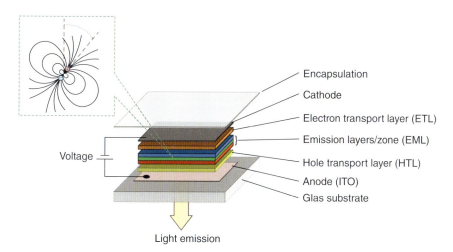

Figure 6.2 Simplified illustration of a three-layer OLED stack with an emission layer containing three subunits emitting light in different color. The light emission process from the excited state of a molecule can be described as electric dipole radiation with a characteristic emission pattern depending on the direction of the transition dipole vector (TDV).

with special properties. Thus, a typical OLED stack, as depicted in Figure 6.2, starts with an ITO anode on a glass substrate, subsequently followed by a (conductivity-doped) hole transport layer (HTL) and the emission layer (EML) consisting of two or sometimes even three different materials forming a so-called guest/host system. On top, the (conductivity-doped) electron transport layer (ETL) and a metallic cathode are completing the OLED structure electrically, and an encapsulation protects it from atmospheric moisture and oxygen. Often, the ETL and the HTL are simultaneously acting as blocking layer for the oppositely charged carrier species (HBL and EBL, respectively). In this case a small region next to the EML is not conductivity doped to avoid exciton quenching processes.

6.2.2 Electroluminescence Quantum Efficiency

The most important figure to benchmark OLEDs is the electroluminescence (or external) quantum efficiency (EQE), which is defined as the number of emitted photons divided by the number of injected charges and is given by four individual factors [4]:

$$\eta_{ext} = \gamma \cdot \eta_r \cdot q_{eff} \cdot \eta_{out} = \eta_{int} \cdot \eta_{out} \tag{6.1}$$

Therein, the first three factors yield the internal quantum efficiency η_{int}, giving the fraction of produced photons per injected carriers that are available for being extracted from the thin film structure, as quantified by the light-outcoupling factor η_{out}.

The factor γ represents the charge carrier balance of the device. Only if every electron finds a hole to recombine with (and vice versa), this value becomes unity. In state-of-the-art OLEDs, using appropriate transport and blocking layers γ can be assumed to be close to one [37, 38].

Next, η_r stands for the radiative exciton fraction and is determined by quantum-mechanical selection rules. In first approximation, only excitons with total spin equal to zero (singlet excitons) are allowed to decay radiatively. By contrast, excited states with total spin 1 (triplet excitons) can only decay nonradiatively. This selection rule can be overcome if spin–orbit coupling is strong for the excited molecule, such as for phosphorescent metal–organic complexes based on heavy metal central atoms, e.g., iridium or platinum [6]. However, for ordinary fluorescent molecules the spin–orbit coupling is too weak to change this selection rule significantly at room temperature.

The third factor q_{eff} is the effective radiative quantum efficiency (RQE) of the emitting system. It depends on the competition between radiative and nonradiative decay channels in the excited state, whereby the transition rate for the former is modified via the Purcell effect by the photonic density of states in an OLED microcavity (for details see Refs [39–43]). Additionally, in an electrically operating OLED, this factor decreases at high current densities due to nonradiative exciton quenching processes [44–47]. We want to emphasize here in the context of delayed fluorescence emitters that q_{eff} stands for (spontaneous) radiative decay processes from singlet excited states. Changes in their population by TADF or TTA would not change this value but enter only in the radiative exciton fraction η_r.

Finally, the remaining factor η_{out} is the light-outcoupling efficiency, which quantifies how many photons are actually able to leave the OLED through the transparent electrode without being trapped and dissipated in waveguided modes or surface plasmons or simply being reabsorbed. Depending, of course, on the details of the OLED layer stack, this fraction is typically around 20% for emissive dipoles with random orientation [42, 43]. Note that the light that is trapped in the glass substrate can easily be extracted by microlense arrays or scattering particles [48] and is, thus, not being considered as a loss channel.

As a result, the classical efficiency limits for OLEDs with first- and second-generation emitter materials are given as

$$\eta_{\text{EQE}} = 1 \times 0.25 \times 1 \times 0.2 = 5\% \quad \text{(fluorescent)} \tag{6.2}$$

$$\eta_{\text{EQE}} = 1 \times 1 \times 1 \times 0.2 = 20\% \quad \text{(phosphorescent)} \tag{6.3}$$

6.2.3 Delayed Fluorescence

In the simple spin statistical picture, η_r does not include effects like TTA and TADF, changing the fraction of singlet excited states after charge carrier recombination [5, 7, 11, 49–52]. Figure 6.3 illustrates the schematic mechanisms behind both processes resulting in delayed fluorescence and therewith an enhanced radiative exciton fraction.

In the case of TADF, the exchange energy between the first excited singlet (S_1) and triplet state (T_1) is in the range of the thermal energy (or a few times $k_B T$) resulting in a back transfer – i.e. reverse intersystem crossing (rISC) – of triplet into singlet excitons. As this process is slow compared with the prompt emission from singlet excitons generated by direct charge carrier recombination, a delayed fluorescence can be observed. For sufficiently low singlet–triplet splitting and high enough temperatures, this mechanism can theoretically result in a radiative exciton fraction of unity like for phosphorescent emitters [5, 7, 11, 49–52].

In contrast, TTA cannot reach this high value for η_r, because it is a bimolecular process where at least one triplet is lost instead of forming the desired singlet

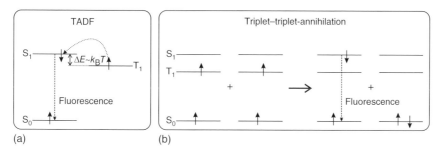

Figure 6.3 Simplified illustration of the underlying mechanisms of (a) thermally activated delayed fluorescence (TADF) and (b) triplet–triplet annihilation (TTA) enhancing the radiative exciton fraction in fluorescent materials after charge carrier recombination. For TADF materials, the exchange energy between the first excited singlet (S_1) and triplet state (T_1) is in the range of the thermal energy enabling the back transfer from the triplet to the singlet state resulting in delayed fluorescence. For the TTA process, two excited triplet states interact ideally forming one excited singlet and one ground state (S_0) leading to delayed fluorescence, too.

excited state [35, 53, 54]. TTA takes place if two excited triplet states encounter each other on neighboring molecules, where, in the ideal case, they form one singlet excited state (S_1) and one ground state (S_0) [53]:

$$T_1 + T_1 \rightarrow S_1 + S_0 \tag{6.4}$$

For this to happen, the excited T_1 state must at least have half of the excitation energy of the corresponding S_1 state ($E_{T_1} \geq \frac{1}{2} E_{S_1}$). The so generated excited singlet state can thereafter decay under light emission and can be detected as delayed fluorescence. Thus, ideally one-half of the (originally) nonemissive triplet excitons can be recycled as singlet excitons, leading to an upper limit for the radiative exciton fraction of $\eta_r = 0.25 + 0.5 \times 0.75 = 62.5\%$ [13]. However, there is another pathway for TTA that does not result in an excited singlet state and delayed fluorescence, but just in a deactivation of one of the two triplet excitons:

$$T_1 + T_1 \rightarrow T_1 + S_0 \tag{6.5}$$

Basically, the energy of the two triplets is used to excite one of both to a higher-lying triplet state (T_2, T_3, T_4, ...) relaxing by internal conversion again to the first excited triplet state without enhancing the radiative exciton fraction. This can be a loss channel in both fluorescent and phosphorescent OLEDs at high current density [45–47, 55, 56]. Furthermore, TTA is an unwanted decay channel for TADF emitting systems as it is reducing the amount of excited triplet states. We note that, in principle, two triplets with spin 1 each can form not only entities with total spin 0 ($S_1 + S_0$) and 1 ($T_1 + S_0$) but also a quintet state with total spin 2 ($Q_1 + S_0$), but due to its high energy, this process is usually not relevant [53].

6.2.4 Nonisotropic Emitter Orientation

Nonisotropic emitter orientation, i.e. horizontal alignment of the TDVs with respect to the substrate plane, is a powerful tool to increase OLED efficiency [33, 43, 57–59]. Treating the radiative transitions as classical oscillating dipoles (see Figure 6.2) makes it obvious that horizontal orientation is the preferred alignment for enhanced light outcoupling due to the fact that a dipole emits most of the energy perpendicular to its axis. Hence, the outcoupling factor can be significantly increased from ca. 20% (for the isotropic case) to over 30% (when perfectly horizontal alignment is achieved) in standard devices [43, 59] or from 25% to even 45% if very thin ITO anodes and birefringent transport layers are used [60, 61]. Noteworthy, such high values are attained without using (complex) outcoupling enhancements such as nanostructured electrodes [62] or scattering particles [48].

To have a measure quantifying the degree of anisotropy for the TDVs of an emissive guest/host system, the orientation parameter Θ is defined as the ratio of energy radiated by vertically aligned transition dipole moments to the total radiated power [43, 59, 63]. Based on this definition an isotropic distribution of the TDVs results in a Θ value of 0.33, a perfect horizontal alignment is identified by $\Theta = 0.0$, and a completely vertical orientation yields $\Theta = 1.0$.

In general the connection between Θ and an ensemble of molecules, each having different possible orientations of the TDVs, is

$$\Theta = \frac{\sum_i a_i \sum_j b_j p_{z,ij}^2}{\sum_i a_i \sum_j b_j |\vec{p}_{ij}|^2} \tag{6.6}$$

where a_i denotes the relative contribution of each dye molecule, b_j describes the relative contribution of the jth transition dipole moment $\vec{p}_{ij}$ on the ith molecule, and $p_{z,ij}$ denotes the vertical component of the corresponding TDV [64].

6.2.5 Optical Modeling

In this chapter we use optical modeling for a comprehensive efficiency analysis of OLEDs exhibiting delayed fluorescence and nonisotropic emitter orientation. Here we summarize only the most important concepts; for a detailed discussion we refer to the literature [41–43, 59].

Light emission in OLEDs is treated as dipole radiation of an ensemble of incoherent emitters embedded in a weak microcavity formed by the substrate and the multilayer stack [40]. Depending on the details of the used layers, such as their thicknesses, their optical constants, the emission spectrum, and the position as well as the orientation of the emissive dipoles, the EQE of an OLED is affected in two ways by the optical cavity via the product $q_{\text{eff}} \cdot \eta_{\text{out}}$.

First, the radiative decay rate (Γ_r) from the excited state is modified by the Purcell effect [39, 40, 65], i.e. the coupling to the photonic density of states of the cavity, while the nonradiative decay rate (Γ_{nr}) is not affected. Thus, the "free-space" value of the RQE q of the emitter in the given host is modified by introducing the Purcell factor F acting on the radiative rate to yield the *effective* RQE for a given OLED cavity [66]:

$$q_{\text{eff}} = \frac{F \cdot \Gamma_r}{F \cdot \Gamma_r + \Gamma_{nr}} \tag{6.7}$$

This, of course, also modifies the excited state lifetime τ with respect to its value in an infinite medium consisting of the dye and the matrix material τ_0 [63]:

$$\frac{\tau}{\tau_0} = \frac{\Gamma_r + \Gamma_{nr}}{F \cdot \Gamma_r + \Gamma_{nr}} \tag{6.8}$$

Second, depending on the cavity length and, in particular, on the distance (z) of the emitter to the highly reflecting metallic back electrode and on the emitter orientation Θ, the generated optical power is redistributed between different optical modes. In this context, the light-outcoupling factor η_{out} quantifies the fraction of the desired direct light emission that leaves the OLED through the transparent substrate ($\tilde{P}_{\text{air}}^{\text{out}}$) with respect to all generated optical power ($\tilde{P}_{\text{tot}}$) inside the OLED stack [43]:

$$q_{\text{eff}} \cdot \eta_{\text{out}} = \frac{\tilde{P}_{\text{air}}^{\text{out}}}{\tilde{P}_{\text{tot}}} \tag{6.9}$$

Therefore, the total power $\tilde{P}_{tot}$ is given by integrating the position- and orientation-dependent power dissipation spectrum $P(k_\parallel, \lambda, z, \Theta)$ of the dipoles inside the cavity over all in-plane wave vectors $k_\parallel$ and all wavelengths λ with the emission spectrum $S(\lambda)$ as weighting function:

$$\tilde{P}_{tot}(z, \Theta) = (1-q) + q \cdot \int_{\lambda_1}^{\lambda_2} S(\lambda) \int_0^\infty P(k_\parallel, \lambda, z, \Theta) dk_\parallel d\lambda \quad (6.10)$$

$$\equiv (1-q) + q \cdot F(z, \Theta) \quad (6.11)$$

where the double integral at the same time is the definition of the Purcell factor $F(z, \Theta)$, which – to emphasize this point again – depends particularly on the position z of the emitter in the cavity and its orientation Θ. The actually outcoupled fraction is obtained from a similar expression, where the integration over $k_\parallel$ is only performed over the accessible light-output cone, e.g. for direct and/or substrate emission.

For the following, it is important to recall that the RQE (q) is an intrinsic property of the emissive system, i.e. the dye and its dielectric environment (the matrix where the dye is embedded in with a certain concentration), which is *independent* of the cavity. The effective RQE (q_{eff}), however, is the particular value of the RQE for a given OLED stack, where the emitter has a certain position and orientation. Thus, when we systematically change the emitter position by a layer thickness variation, the whole series of OLEDs is characterized by a single value of q, while both q_{eff} and η_{out} are layer thickness and orientation dependent.

6.3 Comprehensive Efficiency Analysis of OLEDs

For a comprehensive efficiency analysis of OLEDs, one needs to determine the four individual factors of the external quantum efficiency separately from each other [44, 63, 67]. The first problem hereby is that at the moment there exists no method to determine the charge carrier balance γ of a device accurately. However, if appropriate blocking layers and conductivity-doped transport layers are used, this factor can be assumed to be very close to unity in state-of-the-art OLEDs [36]. In contrast to phosphorescent emitting systems, where the radiative exciton fraction can also be assumed as $\eta_r = 1$, for delayed fluorescence, one has to determine all three remaining factors contributing to the EL quantum efficiency, namely, η_r, q (or q_{eff} in the OLED cavity), and η_{out}. Thus, three independent experiments are performed on a set of OLEDs with the same emitting system but having different distances of their EML to the metallic mirror of the device (typically by variable ETL thicknesses) in order to get access to the unknown parameters (see Figure 6.4 for a graphical illustration):

1) Time-resolved EL and/or photoluminescence (PL) spectroscopy.
2) Polarized angular dependent emission spectroscopy under steady-state conditions.
3) External quantum efficiency measurements without and with macroextractor (at sufficiently low current densities to avoid current-induced quenching processes).

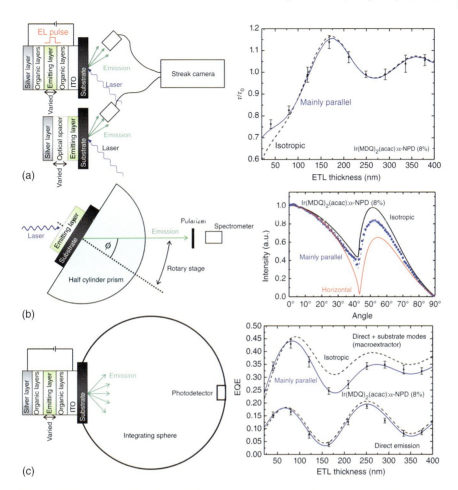

Figure 6.4 Illustration of the three different measurement techniques combined for a comprehensive efficiency analysis (the shown experimental and simulation data are taken from an investigation of a red phosphorescent emitting system). (a) Time-resolved electroluminescence (EL) and/or photoluminescence (PL) spectroscopy using a streak camera system to get access to the excited-state lifetimes of the emitting system for different interference conditions inside a microcavity. (b) Polarized angular dependent emission spectroscopy under steady-state conditions to determine the orientation parameter of the emitting system. (c) External quantum efficiency measurements without and with macroextractor for different emitter-to-cathode distances. In case of (a) and (c), the distance between the emitting and the metal layer is systematically varied. *Source*: Ref. [43]. Reproduced with permission of John Wiley & Sons.

Time-resolved EL and PL spectroscopy is a powerful tool to analyze excitonic processes in OLEDs. For phosphorescent emitting systems, EL and PL should basically yield the same information; however, differences can arise due to the fact that the spatial exciton generation profile for optical and electrical excitation can be different: PL usually probes a more or less uniformly excited EML, whereas EL originates from a spatially inhomogeneous recombination profile, and EL is affected by exciton quenching processes at high currents

[46, 68]. Thus, often electrical excitation is the better choice to determine the excited-state lifetime of the guest/host system, as it is closer to the real situation during electrical operation. The situation is somewhat different for fluorescent dyes, because the prompt decay time is typically much shorter than the RC time of the actual device and, hence, cannot be resolved with electrical excitation. In this case, the prompt decay can only be detected by optical excitation, if a suitable detection window (typically in the ns range) is chosen. On the other hand, mechanisms such as TTA or TADF will lead to delayed fluorescence signals (typically in the microsecond range), which can be observed by both PL and EL measurements. However, time-resolved EL has the advantage that a quantitative estimation of the relative contributions of both prompt and delayed fluorescence is possible, yielding direct information on the radiative exciton fraction η_r [13]. Furthermore, laser excitation does not directly populate the triplet states of the emitting molecules (except for special cases as Cu(I) complexes), in contrast to electrical excitation where three triplet excitons are created for each singlet exciton. Hence, delayed fluorescence signals are typically weaker in a PL experiment compared with an EL investigation.

As mentioned before, the Purcell effect strongly affects the effective radiative rate of the emitting system as a function of the cavity length, i.e. the distance of the emitter from the mirror-like metallic cathode. However, this is only true for the prompt decay but not the delayed component, because intersystem crossing (ISC) and rISC are not affected by the photon field in the OLED cavity and are acting on completely different time scales as the direct singlet emission. Thus, from a systematic variation of the OLED ETL thickness, it is possible to determine the RQE of the emitting system by comparing the oscillation of the excited-state lifetime of the prompt decay with numerical simulations (see Figure 6.4a).

As has been shown by simulation, the Purcell factor exhibits a certain dependence on the orientation of the transition dipole moments of the emitting species [59, 63, 69]. Thus, the knowledge of the emitter orientation is a crucial point to ensure a consistent determination of the emitter's effective RQE inside the OLED stack. Therefore, polarization- and angular-dependent PL emission pattern analysis is performed for the guest/host system under investigation and is then compared with optical simulations yielding the orientation parameter Θ (see Figure 6.4b). Here, two different experiments are possible. First, the analysis can be applied to a simplified stack only consisting of a thin layer ($\approx$10 nm) of the guest/host system on a glass substrate attached to a half-cylinder prism extracting the substrate modes, which is connected to a rotary stage. The film is then excited with an ultraviolet laser diode (375 nm), and the p-polarized emission containing contributions of both, vertical and horizontal TDVs, is detected as a function of the emission angle [70]. Alternatively, one can extract the orientation parameter from a complete OLED stack where the emitting system is positioned in the first node of the OLED cavity (at a distance $d \approx \lambda/4$ of the emission peak wavelength from the metallic cathode if the refractive index of the organic material is close to 2) with the same procedure as mentioned for the former method [71] by using electrical excitation. And, OLED stacks without metallic cathode can be investigated by PL spectroscopy, too. Note that it is mandatory to determine the exact layer thicknesses and optical constants for all used layers and to take into account possible anisotropies of the refractive indices (birefringence) to ensure an appropriate analysis.

However, nonisotropic emitter orientation is crucial not only for the changes of the excited state lifetime for varying emitter-to-cathode distances but also for the overall outcoupling efficiency of the device as a function of the ETL thickness. Thus, in order to complete the efficiency analysis, EL quantum efficiency measurements, again for a set of OLEDs with varying ETL thickness, in a calibrated integrating sphere (see Figure 6.4c) have to be performed for a sufficiently low current density to avoid current-induced quenching effects [44]. Note that measuring the zero degree emission intensity and weighting the obtained results with a lambertian emission profile is not an appropriate way to determine EQE values as, e.g. for nonisotropic emitter orientation the emission pattern can deviate from the lambertian case.

In principle, EQE measurements with systematically varied cavity length are sufficient to determine q and η_{out} in phosphorescent OLEDs [71], however, if no macroextractor is used, ambiguities concerning the orientation parameter Θ and thus η_{out} can arise [72]. Thus, it is mandatory to measure EQEs in both configurations (i.e. direct emission from the plane glass substrate and substrate mode emission using a macroextractor) [73]. This becomes even more important in the fluorescent case, as there are three individual factors determining the EQE. As will be seen in the following examples, information obtained from different experiments, as mentioned above, have to be combined to yield reliable results.

6.4 Case Studies

6.4.1 Treating the OLED as a Black Box

In this first example, we had a series of blue fluorescent OLEDs at hand [67], where only the stack layout as shown in Figure 6.5a including layer thicknesses, optical constants, and the emission spectrum, but no details of the used materials were known. In that sense this state-of-the-art blue fluorescent OLEDs are kind of a "black box," which will be used in the following to demonstrate how a comprehensive efficiency analysis has to be performed. The OLEDs exhibit an external quantum efficiency of about 5% for optimized layer thickness, which is the classical limit for a fluorescent device, if one assumes $\gamma = 1, q = 100\%, \eta_r = 25\%$, and $\eta_{out} = 20\%$.

Figure 6.5b illustrates the measured EQE values for varying ETL thickness for direct emission and with a macroextractor (half-ball lens) attached to the substrate giving access to the captured substrate modes. As mentioned before, for thickness optimized devices the EQE for direct emission reaches 5% for ETL thicknesses of 31 and 184 nm, respectively. Although it is possible to simulate the EQE for direct emission with isotropic emitter orientation (see lower curve in Figure 6.5b), an ideal charge carrier balance of one and the classical value of the radiative exciton fraction of 25%, this requires a value of 100% for the RQE of the emitting system – which is kind of very optimistic. More severe, however, is the fact that it is impossible to simultaneously describe the measured EQE

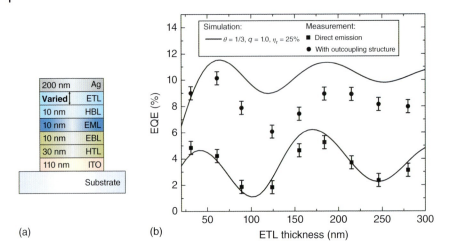

Figure 6.5 (a) Stack layout of a blue fluorescent OLED with varying ETL thickness. (b) External quantum efficiency measurements for direct emission (squares) and with a macroextractor attached to the substrate (dots) at a current density of 2 mA cm^{-2}. The solid lines represent optical simulations for isotropic emitter orientation and an assumed radiative exciton fraction of 25%, a charge carrier balance of unity, and a radiative quantum efficiency of 100%. *Source*: Ref. [67]. Reproduced with permission of AIP Publishing LLC.

obtained with the macroextractor by optical simulation based on the same set of parameters (see upper curve in Figure 6.5b). This clearly demonstrates that the efficiency analysis fails and simultaneously emphasizes the importance of taking the substrate mode extraction into account.

As discussed briefly in previous sections, the shape of the EQE oscillations with emitter/cathode distance, i.e. the relative heights of the first and the second maximum of the direct emission and the boost in efficiency, if the substrate modes are extracted, is strongly influenced by the RQE as well as the orientation of the transition dipole moments, whereas the charge carrier balance and the radiative exciton fraction are only linear scaling factors that do not change with ETL thickness. Hence, from Figure 6.5b it is obvious that one (or both) of the mentioned factors influencing the shape of the EQE oscillations was not considered properly. Thus, angular-dependent p-polarized PL emission pattern analysis was performed for an OLED without metallic cathode at an ETL thickness of 67 nm to get information about the emitter orientation. Figure 6.6 illustrates the measured angular-dependent intensity for an emission wavelength of 470 nm normalized to zero degree intensity and optical simulations for isotropic and horizontal alignment of the transition dipole moments. The best fit yields an orientation parameter of $\Theta = 0.09$, which is in fact nearly completely horizontal and far away from the initially assumed isotropic emitter orientation.

With this information it is now possible to repeat the analysis of the external quantum efficiency measurements for the OLEDs with varying ETL thickness taking the emitter orientation into account. Figure 6.7 illustrates the simulation of both the direct emission values and the substrate mode extraction, using the obtained orientation parameter and a reduced RQE of $q = 70\%$. However,

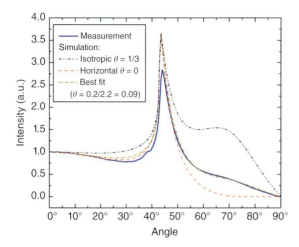

Figure 6.6 Determination of the orientation parameter of the blue fluorescent emitting system using an OLED stack with ETL thickness of 67 nm but without metallic cathode. The p-polarized angular-dependent emission pattern is compared to optical simulations yielding an orientation parameter $\Theta = 0.09$, i.e. nearly perfectly horizontal alignment of the transition dipole moments. *Source*: Ref. [67]. Reproduced with permission of AIP Publishing LLC.

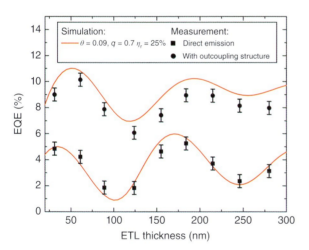

Figure 6.7 Efficiency analysis of the measured EQE values with and without macroextractor taking the strong horizontal orientation of the TDVs of the emitting species into account and assuming an unchanged radiative exciton fraction. The fit process (solid red line) yields a radiative quantum efficiency of 70%, much lower compared to the isotropic results. However, the fit for substrate mode emission still cannot describe the measured data in an appropriate way. *Source*: Ref. [67]. Reproduced with permission of AIP Publishing LLC.

there are still some deviations for the substrate mode emission indicating that something is missing. This could either be a charge balance factor γ smaller than one or a radiative exciton fraction η_r larger than 25%. While the former is very unlikely (and would actually require again a higher q value), an enhancement of η_r by delayed fluorescence is possible.

In order to prove this hypothesis, time-resolved PL and EL spectroscopy is performed in the microsecond range. The excitation was accomplished by a short laser pulse (337 nm, 0.7 ns) or a rectangular electrical pulse (variable height, 20 µs), respectively. The typical decay time for fluorescent emitting molecules is in the range of nanoseconds, hence, if one is detecting a long-lived emission response in the microsecond range, this gives strong evidence that a mechanism causing delayed fluorescence is present. Figure 6.8 illustrates the results of these measurements and clearly demonstrates the delayed fluorescence decay in both experiments. Note that the RC time of the used OLED is about 500 ns and thus cannot be responsible for the long-lived decay in the electrical excitation case. Interestingly, the delayed component is decreasing continuously and finally disappears in the photoluminescent experiment, if the sample is cooled to liquid helium temperature. This gives strong evidence that a thermally activated process is responsible for the delayed emission and therewith TTA can be ruled out.

To reduce the number of free parameters of the fit procedure for the external quantum efficiency, it is mandatory to obtain direct information about the RQE of the prompt fluorescence decay. Therefore, time-resolved PL spectroscopy in the nanosecond range was carried out. Figure 6.9 illustrates an exemplary measurement and the analysis of the prompt fluorescence lifetimes for different ETL thicknesses. Fitting theses values with optical simulations – taking the already determined dipole orientation into account – results in a RQE q of 45% only, which is clearly much less than both of the previously obtained values.

Turning back to the efficiency analysis of the measured EQE values, it is now possible to perform optical simulations with $q = 0.45$ as starting value of the fit process, $\Theta = 0.09$ as fixed input, and the radiative exciton fraction as free variable.

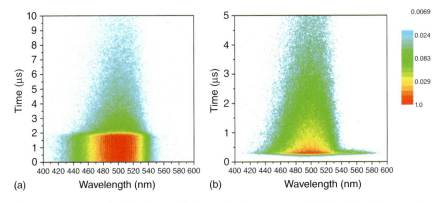

Figure 6.8 Time-resolved emission ((a) electroluminescence and (b) photoluminescence) in the microsecond range. Both measurements show a strong delayed fluorescence component. Note that the intensity scale is logarithmic. *Source*: Ref. [67]. Reproduced with permission of AIP Publishing LLC.

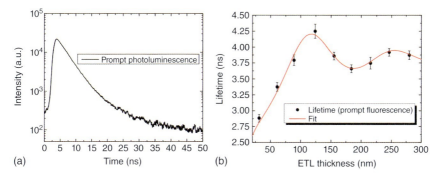

Figure 6.9 Time-resolved emission spectroscopy of the prompt fluorescence of the OLEDs. (a) Exemplary photoluminescence decay for an OLED with 184 nm ETL thickness resulting in an excited-state lifetime of 3.66 ns. (b) Determination of the radiative quantum efficiency using the prompt emission decay times yielding a radiative quantum efficiency $q = 45\%$ and an intrinsic fluorescence lifetime $\tau_0 = 3.75$ ns. Source: Ref. [67]. Reproduced with permission of AIP Publishing LLC.

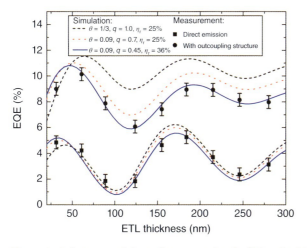

Figure 6.10 Summary of the earlier approaches for fitting the EQE values (black and red line) together with the fit using the radiative exciton fraction η_r as free fit parameter and a radiative quantum efficiency of 45%. The new simulation (blue line) can describe both measured data sets of the EQE with and without macroextractor in a reasonable way, resulting in an enhanced value of $\eta_r = 36\%$. Source: Ref. [67]. Reproduced with permission of AIP Publishing LLC.

Figure 6.10 shows that a consistent description of both contributions, direct and substrate emission, can be obtained with one parameter set, yielding $\eta_r = 0.36$, which is almost 50% higher than the classical value of 0.25. Note that this is only a lower limit for the radiative exciton fraction, as the charge carrier balance was assumed to be unity for the used device.

This example clearly demonstrates that even in cases where the EQE of an OLED does not exceed the classical limit of 5% for a fluorescent emitter, delayed fluorescence may contribute significantly to the radiative exciton fraction.

Moreover, to disentangle these effects and potential implications of nonisotropic emitter orientation, a comprehensive efficiency analysis using complementary experimental techniques combined with optical simulation is required.

6.4.2 Highly Efficient Thermally Activated Delayed Fluorescence Device

In this section we will investigate an OLED with a blue TADF emitter exceeding the classical efficiency limit of 5% by far. Again, by combining different techniques we will disentangle the contributions of the individual factors to the observed high EQE values [74].

The stack layout of the OLED under investigation together with the molecular structure of the used emitting guest/host system is given in Figure 6.11. From previous investigations, CC2TA (2,4-bis3-(9H-carbazol-9-yl)-9H-carbazol-9-yl-6-phenyl-1,3,5-triazine) was known to show strong TADF as it exhibits a very small singlet-to-triplet energy gap of only 60 meV [22]. The molecular structure follows the donor–acceptor–donor design principle to reduce the exchange splitting of purely organic fluorescent dyes. In this case, two carbazole containing donor groups and one triazine acceptor unit are used. The matrix material DPEPO (bis[2-(diphenylphosphino)phenyl]ether) is also used as hole blocker, while mCP (1,3-bis(N-carbazolyl)benzene) acts as electron blocker. Both of them have high triplet energies so that energy transfer from the triplet state of CC2TA to the surrounding matrix and/or blocker materials is strongly suppressed. For further details we refer to Ref. [22].

The OLED shows a typical diode behavior with sky blue emission as can be seen in Figure 6.12. The external quantum efficiency of the device reaches a maximum value of $11 \pm 1\%$ at 0.01 mA cm^{-2}, which clearly exceeds the classical limit for fluorescent emitting systems. For higher current densities a moderate roll-off is observed. In the following, we determine the different contributions to the efficiency including emitter orientation and TADF as independent contributions.

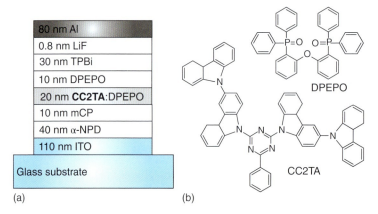

Figure 6.11 (a) OLED stack under investigation. (b) Molecular structure of the emitting system comprising DPEPO (matrix material) and CC2TA (fluorescent dye).

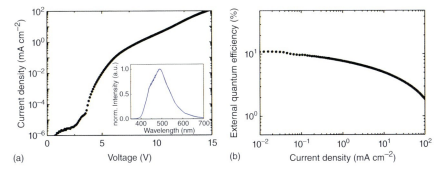

Figure 6.12 (a) Current density–voltage characteristics of the OLED depicted in Figure 6.11. The inset shows the sky blue emission spectrum for a current density of 1 mA cm^{-2}. (b) External quantum efficiency as a function of current density. The maximum EQE of 11 ± 1% is achieved for 0.01 mA cm^{-2}. *Source*: Ref. [74]. Reproduced with permission of John Wiley & Sons.

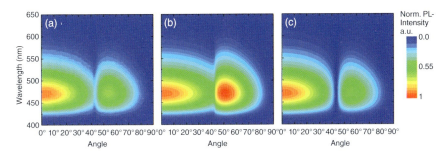

Figure 6.13 Angular-dependent PL emission pattern of a thin layer of CC2TA in DPEPO on a glass substrate connected to a half-cylinder prism. (a) Measured spectra, (b) simulation for isotropic emitter orientation, and (c) simulation for horizontal emitter orientation. All spectra are normalized to zero degree emission. *Source*: Ref. [74]. Reproduced with permission of John Wiley & Sons.

Figure 6.13 illustrates the measured angular-dependent PL emission pattern of a simplified stack, which only consists of a thin emitting layer on a glass substrate attached to a half-cylinder prism to extract the captured substrate modes, together with the results of optical simulations for isotropic and horizontal orientation of the TDVs. Comparing simulations with measurements clearly shows large deviations for the isotropic case for emission angles higher than 40°, while the horizontal simulation comes quite close to the measured data.

However, in order to determine the orientation parameter for CC2TA in DPEPO, it is useful to analyze the cross section of the angular emission pattern for the peak wavelength of 470 nm. Figure 6.14 shows the results of this investigation. The intensity of the optical simulations as well as of the PL emission is normalized to the zero degree values. The most interesting part of this analysis is the angular range between 40° and 90°, as the emission for extracted substrate modes is located in this region. The fit of the measured data results in an orientation parameter of $\Theta = 0.08$, which means, in a simplistic picture, that

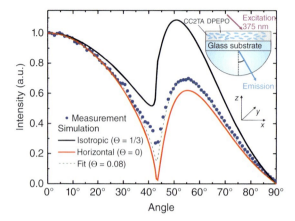

Figure 6.14 Cross section of the angular-dependent photoluminescence emission pattern for the peak wavelength of 470 nm normalized to zero degree intensity. The dots represent the measurement, while the solid lines show simulations for isotropic (black) and horizontal (red) emitter orientation. The fit (green dashed line) results in an orientation parameter Θ of 0.08, which means nearly perfect horizontal orientation of the TDVs of the emitting species CC2TA. Source: Ref. [74]. Reproduced with permission of John Wiley & Sons.

92% of the TDVs of the emitting species are aligned parallel to the substrate plane. This horizontal alignment enhances the outcoupling factor and, thus, contributes significantly to the high EQE of the OLEDs.

As next step it is necessary to determine the RQE of the emitting system. Here, we use a simplified stack and not a complete OLED structure with different ETL thicknesses as in the previous section. The layer stack only consists of a glass substrate, a thin layer of the emitting system, the optical spacer UGH-2 (1,4-bis(triphenylsilyl)benzene, and a thick silver layer acting as mirror, as depicted in the inset of Figure 6.15. The UGH-2 layer is implemented as wide bandgap material with high triplet energy to change the distance between the emitting system and the metallic mirror and, thus, to systematically vary the interference conditions for the radiative exciton decay. Figure 6.15 displays the changes of the excited-state lifetime of the prompt fluorescence emission in the nanosecond range for varying UGH-2 thickness together with optical simulations for isotropic as well as the present horizontal emitter orientation. Therefore, the behavior for the isotropic transition dipole distribution is only shown for completeness to emphasize the importance of taking emitter orientation into account even for the determination of the RQE of the emitting system. Apparently, the analysis with the correct horizontal orientation of the TDV of the emitting species results in a RQE q of 0.55 ± 0.04, which is consistent with the PL quantum efficiency of 0.62 measured in an integrating sphere, if the nonisotropic emitter orientation is taken into account. For more information about these calculations, we refer to [74].

Finally, it is possible to analyze the measured external quantum efficiency of the thickness optimized device taking all achieved information into account. Figure 6.16 contains three sets of optical simulations for $q = 55\%$ together with the measured EQE value. It is obvious that the experimental value can only

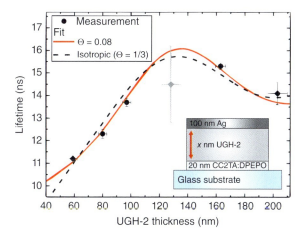

Figure 6.15 Determination of the radiative quantum efficiency of CC2TA in DPEPO via the prompt fluorescence lifetime for a simplified stack design shown as inset. The dots represent the measured decay times for different UGH-2 thicknesses, acting as optical spacer between the emitting system and a metallic mirror. The red line is the fit using the horizontal emitter orientation while the black dashed line is, for completeness, the corresponding simulation for isotropic distribution of the TDVs. Note that the gray data point was not considered for the fit due to very low intensity in the cavity minimum. *Source*: Ref. [74]. Reproduced with permission of John Wiley & Sons.

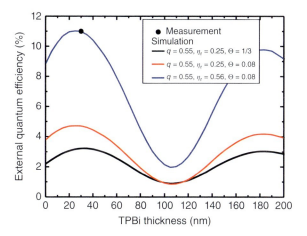

Figure 6.16 Measured EQE for a thickness optimized device together with optical simulations for a radiative quantum efficiency of 55% and variable ETL (TPBi) thickness. The black line represents a classical simulation with isotropic emitter orientation and a radiative exciton fraction of 25% (without TADF). The blue line corresponds to an enhanced outcoupling factor induced by the horizontal alignment of the TDVs. The red curve illustrates the optical simulation for horizontal emitter orientation and an enhanced radiative exciton fraction of 56% due to TADF, which is in good agreement with the measured data. *Source*: Ref. [74]. Reproduced with permission of John Wiley & Sons.

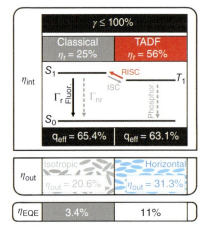

Figure 6.17 Summary of the efficiency analysis for the blue TADF emitter CC2TA in DPEPO. The left column represents a classical fluorescent model with isotropic emitter orientation, while the right column accounts for an enhanced radiative exciton fraction and outcoupling factor due to TADF and horizontal alignment of the TDVs of the emitting species, respectively. *Source*: Ref. [74]. Reproduced with permission of John Wiley & Sons.

be reproduced, if both the nonisotropic emitter orientation and an enhanced radiative exciton fraction η_r originating from TADF are considered. Moreover, with the assumption of a charge carrier balance of unity, one can state a lower limit for η_r of about 56%.

For better visualization, Figure 6.17 summarizes the contributions of the individual factors to the measured high EQE value of 11%. It is only possible to properly obtain the internal quantum efficiency, if the correct light-outcoupling factor (determined by the emitter orientation) is taken. For example, Lee et al. have previously assumed isotropic orientation and, thus, concluded on $\eta_{int} = \gamma \times \eta_r \times q_{eff} \approx 53.4\%$ (while the true value is only 35.1%). As a consequence, they largely overestimated the radiative exciton fraction as being more than 80% [22].

This impressively demonstrates the necessity of a comprehensive efficiency analysis in the case of TADF emitters, which are prone to exhibit nonisotropic orientation distribution of their TDVs in evaporated guest/host systems. For completeness, we note that the nature of the optical transition leading to the observed light emission was not subject of our investigations and also not the orientation of the involved TDV with respect to the molecular frame. However, we want to add that the calculations reported in Ref. [74] were probably oversimplified, as they did not take into account the charge-transfer character of the optical transitions.

6.4.3 Low Efficiency Roll-Off Triplet–Triplet Annihilation Device

Apart from TADF, there is a second possibility for an enhanced radiative exciton fraction of fluorescent emitters, namely, TTA. Hence, in this section we will discuss the advantages of this mechanism, in particular, in terms of reduced efficiency roll-off for high current densities.

We analyzed the well-known green fluorescent emitter 2,3,6,7-tetrahydro-1,1,7,7-tetramethyl-1H,5H,11H-10-(2-benzothiazolyl)quinolizino[9,9a,1gh]coumarin (C545T) doped at 1 wt% into the matrix material aluminum-tris(8-hydroxyquinoline) (Alq_3) [75]. For this emitting system external quantum efficiencies exceeding the classical limit of 5% have already been reported in literature [16, 76].

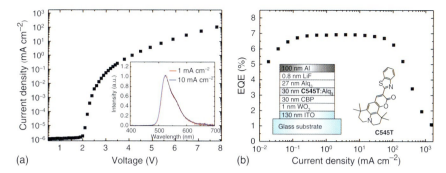

Figure 6.18 (a) Current density–voltage characteristics of a thickness optimized TTA OLED together with the electroluminescence emission spectrum for two different current densities. (b) External quantum efficiency values as a function of the current density. The stack design as well as the chemical structure of C545T is depicted as inset. The measured EQE stays constant over a wide current density region (three orders of magnitude). *Source*: Ref. [75]. Reproduced with permission of AIP Publishing LLC.

However, in the analysis of Pu et al. [16], there is a discrepancy of the determined radiative exciton fraction achieved by EQE measurements assuming a standard outcoupling factor of ≈ 20% and time-resolved EL investigations, resulting in η_r values of 54% and 31%, respectively. This gives strong evidence that the assumption of isotropic emitter orientation may not be valid in this case.

For this reason, we first fabricated thickness-optimized devices to reproduce the high external quantum efficiencies exceeding the classical limit of 5%. The current density–voltage characteristics as well as the stack layout, the EL emission spectrum, the chemical structure of C545T, and the EQE for a wide range of current densities are depicted in Figure 6.18. This OLED exhibits a very low onset voltage of only 2 V and shows strong green emission with a peak intensity at 523 nm. The maximum EQE (measured in a calibrated integrating sphere) of roughly 7%, exceeding the classical limit by almost 40%, is reached for a current density of 0.1 mA cm^{-2} and remains constant until 100 mA cm^{-2}. Thereafter, a fast decrease of the EQE to 1% for a current density of 1000 mA cm^{-2} is observed. The initial increase of the EQE is caused by unbalanced charge carrier flow for small current densities, while the efficiency roll-off for high current densities is suggested to be induced by mechanisms such as singlet- or triplet-polaron quenching [47]. Because of the formation of a negative interface polarization due to partial alignment of the permanent dipole moments of the Alq$_3$ molecules and therewith hole accumulation at the CBP/Alq$_3$ interface [77–79], a narrow emission profile located at this position is very likely. Hence, the emitter to cathode distance for this device is given by the sum of the thicknesses of the emitting system and the neat Alq$_3$ layer acting as ETL.

Next, we analyzed the orientation of the transition dipole moments of C545T doped into Alq$_3$. Therefore, we used again a simplified stack design with a thin layer (15 nm) of the emitting system on a glass substrate connected to a half-cylinder prism, and the p-polarized emission was detected as a function of the emission angle. Figure 6.19 shows the results of this investigation, which

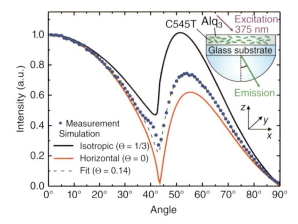

Figure 6.19 Angular-dependent photoluminescence cross section for the peak emission intensity at a wavelength of 523 nm for a 15 nm thick film of the emitting system C545T:Alq$_3$ attached to a half-cylinder prism. The dots are the measured data points, while the solid lines represent optical simulations for different emitter orientations: isotropic (black) and horizontal (red). The dashed line illustrates the fit resulting in an orientation parameter $\Theta = 0.14$. Source: Ref. [75]. Reproduced with permission of AIP Publishing LLC.

yields an orientation parameter Θ of 0.14 indicating a strong horizontal alignment of the TDVs of C545T. Thus, the classical assumption of $\eta_{out} \approx 20\%$ is not valid. Optical simulations reveal that the outcoupling factor of a thickness optimized device as used in this investigation is enhanced by a factor of 1.5 and reaches its maximum of $\approx 30\%$ for an emitter to cathode distance of 65 nm.

From previous publications it is known that OLEDs with C545T show delayed fluorescence from TTA [14]. Therefore, we performed time-resolved EL spectroscopy with voltage pulses of variable amplitudes and a length of 50 µs, required to achieve steady-state conditions, and analyzed the contribution of TTA to the overall EL. Figure 6.20 shows the normalized emission intensities as a function of time for different pulse amplitudes. Note that a reverse bias of −4 V was applied to the devices directly after shutting down the forward bias to prevent delayed charge carrier recombination from, e.g. trapped charge carriers, which would also lead to a delayed EL signal. As the device exhibits an RC time of less than 1 µs and the decay time of the delayed fluorescence is much larger than this value, it is possible to determine the contribution of TTA to the total EL by calculating the intersection of the electrical pulse with the extrapolation of the delayed fluorescence signal as depicted in Figure 6.20 by the two dashed lines. The quantitative analysis (for details see Ref. [75]) reveals an enhancement of η_r by a factor of 1.18 yielding an absolute value of 29.5% instead of the classical assumption of 25%. Remarkably, the relative ratios of steady state and delayed components are not affected by variable voltage pulse amplitudes, which means that the contribution of TTA is constant in this range, which in turn is in good agreement with the measured EQE values being more or less constant between 0.1 and 100 mA cm^{-2}.

In order to complete the efficiency analysis for the presented OLED type, we finally compare the measured EQE values for two OLEDs with different emitter to cathode distances (ETL thicknesses) with optical simulation for variable

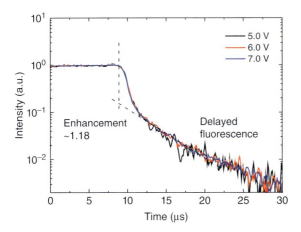

Figure 6.20 Time-resolved electroluminescence measurements showing a strong delayed fluorescence component after the electrical pulse. The relative fraction of delayed to prompt fluorescence does not change for different excitation voltages. From the indicated extrapolations (dashed lines), one can calculate the enhancement of the radiative exciton fraction η_r to be a factor of 1.18. Source: Ref. [75]. Reproduced with permission of AIP Publishing LLC.

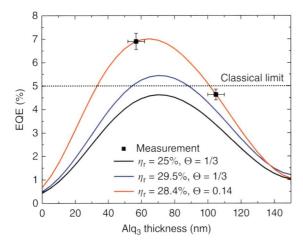

Figure 6.21 Measured EQE values for two OLEDs exhibiting different ETL thicknesses (dots) together with optical simulations for different assumptions for the radiative exciton fraction and emitter orientation. The comparison results in a radiative exciton fraction η_r of 28.4% that is in good agreement with the value obtained from time-resolved electroluminescence investigations. Source: Ref. [75]. Reproduced with permission of AIP Publishing LLC.

assumptions for the radiative exciton fraction and emitter orientation. We used a RQE for the emitting system derived from its PL quantum efficiency of $q = 0.8$ [80], and the charge carrier balance of the device was set to unity to obtain a lower limit for η_r. Figure 6.21 illustrates the measured EQE values for both OLEDs together with three different optical simulations. It can clearly be seen that the high external quantum efficiencies can only be reproduced by the optical simulation using the determined horizontal emitter orientation and an enhanced radiative exciton fraction η_r of 28.4%, which is in good agreement with the value obtained from the time-resolved EL measurements presented before. Thus, the importance of taking nonisotropic emitter orientation into account is again clearly demonstrated, because otherwise η_r would be strongly overestimated.

In total, the main boost in efficiency originates from the horizontal alignment of the transition dipole moments of the emissive species (40% relative enhancement), and the increase in the radiative exciton fraction induced by TTA only plays a minor role (14% relative enhancement).

6.5 Conclusion

More than 15 years after the first presentation of second-generation emitter materials based on phosphorescent metal–organic complexes, a third generation of emitter materials deploying triplet harvesting by delayed fluorescence is gaining more and more interest. OLEDs with extraordinary external quantum efficiency – comparable to their phosphorescent counterparts – clearly corroborate the huge potential of this family of materials.

In this chapter we presented an approach for a comprehensive efficiency analysis of OLEDs incorporating these third-generation emitting systems. In contrast to phosphorescent devices, for which two of the four factors determining the external quantum efficiency are known (the charge carrier balance and the radiative exciton fraction are unity for these OLEDs), one has to determine three factors independently from each other for delayed fluorescence OLEDs as the radiative exciton fraction becomes a free parameter. Furthermore, since the chemical structure of this new emitter class is based on the donor–acceptor principle, the geometrical shape of these molecules promotes nonisotropic orientation of their transition dipole moments when evaporated from the gas phase. Hence, this effect has to be taken carefully into account in a comprehensive efficiency analysis. If the anisotropy of the TDVs distribution is ignored, the radiative exciton fraction is clearly overestimated.

In the presented case studies, all investigated third-generation emitting materials exhibit strong horizontal emitter orientation enhancing the light-outcoupling factor by almost 50% resulting in a huge boost of the external quantum efficiency of devices using these emitting systems. This clearly demonstrates, on the one hand, the importance of taking nonisotropic emitter orientation properly into account for the efficiency analysis, i.e. for determining the increased radiative exciton fraction due to TADF or TTA, as well as, on the other hand, the room for further improvement of such devices as the radiative exciton fraction is still not at its limits in the considered devices. If efficient triplet harvesting and high radiative efficiency can be combined with horizontal orientation of the TDVs of these molecules, these third-generation emitting materials have the potential to match their phosphorescent counterparts as, for the moment, these molecules only exhibit moderate horizontal emitter orientation. Nevertheless, these third-generation emitter materials also need to be further improved to reach stability levels of phosphorescent emitters [81–83].

Acknowledgments

The authors thank the German Ministry of Education and Research (BMBF, contract No. 13N12240), Deutsche Forschungsgemeinschaft (DFG, contract No. BR 1728/13-1), and Bayerische Forschungsstiftung for funding. Furthermore, we acknowledge support by the Japanese Society for the Promotion of Science (JSPS) within their Summer Programs. We are grateful to OSRAM OLED GmbH (Regensburg, Germany) and the research group of Prof. Chihaya Adachi (Kyushu University, Fukuoka, Japan) for providing samples and fruitful discussions. Moreover, we acknowledge the contributions of Christian Mayr, Thomas Lampe, Bert J. Scholz and Jörg Frischeisen (University of Augsburg, Germany), Daniel S. Setz, Andreas F. Rausch, Thomas Wehlus and Thilo C. G. Reusch (Osram OLED GmbH, Germany), Michael Flämmich, Dirk Michaelis and Norbert Danz (Fraunhofer IOF, Germany), and Sae Youn Lee, Takuma Yasuda, and Chihaya Adachi (Kyushu University, Japan) to part of the results presented in this chapter.

References

1 Pope, M., Kallmann, H.P., and Magnante, P. (1963). Electroluminescence in organic crystals. *J. Chem. Phys.* 38 (8): 2042–2043.
2 Tang, C.W. and VanSlyke, S.A. (1987). Organic electroluminescent diodes. *Appl. Phys. Lett.* 51 (12): 913–915.
3 Burroughes, J.H., Bradley, D.D.C., Brown, A.R., Marks, R.N., Mackay, K., Friend, R.H., Burns, P.L., and Holmes, A.B. (1990). Light-emitting diodes based on conjugated polymers. *Nature* 347 (6293): 539–541.
4 Tsutsui, T., Aminaka, E., Lin, C.P., and Kim, D.U. (1997). Extended molecular design concept of molecular materials for electroluminescence: sublimed-dye films, molecularly doped polymers and polymers with chromophores. *Philos. Trans. R. Soc. London, Ser. A* 355 (1725): 801–814.
5 Baldo, M.A., O'Brien, D.F., You, Y., Shoustikov, A., Sibley, S., Thompson, M.E., and Forrest, S.R. (1998). Highly efficient phosphorescent emission from organic electroluminescent devices. *Nature* 395 (6698): 151–154.
6 Yersin, H. ed. (2007). *Highly Efficient OLEDs with Phosphorescent Materials*. Wiley-VCH.
7 Brutting, W. and Adachi, C. ed. (2012). *Physics of Organic Semiconductors*. Wiley-VCH.
8 Mullen, K. and Scherf, U. ed. (2005). *Organic Light Emitting Devices*. Wiley-VCH.
9 Yook, K.S. and Lee, J.Y. (2012). Organic materials for deep blue phosphorescent organic light-emitting diodes. *Adv. Mater.* 24 (24): 3169–3190.
10 Lee, J., Chen, H.-F., Batagoda, T., Coburn, C., Djurovich, P.I., Thompson, M.E., and Forrest, S.R. (2016). Deep blue phosphorescent organic light-emitting diodes with very high brightness and efficiency. *Nat. Mater.* 15 (1): 92–98.
11 Uoyama, H., Goushi, K., Shizu, K., Nomura, H., and Adachi, C. (2012). Highly efficient organic light-emitting diodes from delayed fluorescence. *Nature* 492: 234–238.

12 Kondakov, D.Y. (2007). Characterization of triplet–triplet annihilation in organic light-emitting diodes based on anthracene derivatives. *J. Appl. Phys.* 102 (11): 114504.

13 Kondakov, D.Y., Pawlik, T.D., Hatwar, T.K., and Spindler, J.P. (2009). Triplet annihilation exceeding spin statistical limit in highly efficient fluorescent organic light-emitting diodes. *J. Appl. Phys.* 106 (12): 124510.

14 Luo, Y. and Aziz, H. (2010). Correlation between triplet–triplet annihilation and electroluminescence efficiency in doped fluorescent organic light-emitting devices. *Adv. Funct. Mater.* 20 (8): 1285–1293.

15 Yokoyama, D., Park, Y., Kim, B., Kim, S., Pu, Y.-J., Kido, J., and Park, J. (2011). Dual efficiency enhancement by delayed fluorescence and dipole orientation in high-efficiency fluorescent organic light-emitting diodes. *Appl. Phys. Lett.* 99 (12): 123303.

16 Pu, Y.-J., Nakata, G., Satoh, F., Sasabe, H., Yokoyama, D., and Kido, J. (2012). Optimizing the charge balance of fluorescent organic light-emitting devices to achieve high external quantum efficiency beyond the conventional upper limit. *Adv. Mater.* 24 (13): 1765–1770.

17 Suzuki, T., Nonaka, Y., Watabe, T., Suzuki, T., Nonaka, Y., Watabe, T., Nakashima, H., Seo, S., Shitagaki, S., and Yamazaki, S. (2014). Highly efficient long-life blue fluorescent organic light-emitting diode exhibiting triplet-triplet annihilation effects enhanced by a novel hole-transporting material. *Jpn. J. Appl. Phys.* 53: 052102.

18 Chiang, C.-J., Kimyonok, A., Etherington, M.K., Griffiths, G.C., Jankus, V., Turksoy, F., and Monkman, A.P. (2013). Ultrahigh efficiency fluorescent single and bi-layer organic light emitting diodes: the key role of triplet fusion. *Adv. Funct. Mater.* 23 (6): 739–746.

19 Endo, A., Ogasawara, M., Takahashi, A., Yokoyama, D., Kato, Y., and Adachi, C. (2009). Thermally activated delayed fluorescence from sn^{4+}–porphyrin complexes and their application to organic light emitting diodes – a novel mechanism for electroluminescence. *Adv. Mater.* 21 (47): 4802–4806.

20 Endo, A., Sato, K., Yoshimura, K., Kai, T., Kawada, A., Miyazaki, H., and Adachi, C. (2011). Efficient up-conversion of triplet excitons into a singlet state and its application for organic light emitting diodes. *Appl. Phys. Lett.* 98 (8): 083302.

21 Czerwieniec, R., Yu, J., and Yersin, H. (2011). Blue-light emission of Cu(I) complexes and singlet harvesting. *Inorg. Chem.* 50 (17): 8293–8301.

22 Lee, S.Y., Yasuda, T., Nomura, H., and Adachi, C. (2012). High-efficiency organic light-emitting diodes utilizing thermally activated delayed fluorescence from triazine-based donor–acceptor hybrid molecules. *Appl. Phys. Lett.* 101 (9): 093306.

23 Zhang, Q., Li, B., Huang, S., Nomura, H., Tanaka, H., and Adachi, C. (2014). Efficient blue organic light-emitting diodes employing thermally activated delayed fluorescence. *Nature* 8: 326–332.

24 Dias, F.B., Bourdakos, K.N., Jankus, V., Moss, K.C., Kamtekar, K.T., Bhalla, V., Santos, J., Bryce, M.R., and Monkman, A.P. (2013). Triplet harvesting with 100% efficiency by way of thermally activated delayed fluorescence in charge transfer OLED emitters. *Adv. Mater.* 25 (27): 3707–3714.

25 Jankus, V., Data, P., Graves, D., McGuinness, C., Santos, J., Bryce, M.R., Dias, F.B., and Monkman, A.P. (2014). Highly efficient TADF OLEDs: how the emitter–host interaction controls both the excited state species and electrical properties of the devices to achieve near 100% triplet harvesting and high efficiency. *Adv. Funct. Mater.* 24 (39): 6178–6186.

26 Nobuyasu, R.S., Ren, Z., Griffiths, G.C., Batsanov, A.S., Data, P., Yan, S., Monkman, A.P., Bryce, M.R., and Dias, F.B. (2016). Rational design of TADF polymers using a donor–acceptor monomer with enhanced TADF efficiency induced by the energy alignment of charge transfer and local triplet excited states. *Adv. Opt. Mater.* 4 (5): 653.

27 Parker, C.A. (1963). Sensitized p-type delayed fluorescence. *Proc. R. Soc. London, Ser. A* 276 (1364): 125–135.

28 Jankus, V., Chiang, C. J., Dias, F., and Monkman, A.P. (2013). Deep blue exciplex organic light-emitting diodes with enhanced efficiency; P-type or E-type triplet conversion to singlet excitons? *Adv. Mater.* 25 (10): 1455–1459.

29 Cho, Y.J., Yook, K.S., and Lee, J.Y. (2014). A universal host material for high external quantum efficiency close to 25% and long lifetime in green fluorescent and phosphorescent OLEDs. *Adv. Mater.* 26 (24): 4050–4055.

30 Kim, B.S. and Lee, J.Y. (2014). Engineering of mixed host for high external quantum efficiency above 25% in green thermally activated delayed fluorescence device. *Adv. Funct. Mater.* 24 (25): 3970–3977.

31 Sun, J.W., Lee, J.-H., Moon, C.-K., Kim, K.-H., Shin, H., and Kim, J.-J. (2014). A fluorescent organic light-emitting diode with 30% external quantum efficiency. *Adv. Mater.* 26: 5684–5688.

32 Kaji, H., Suzuki, H., Fukushima, T., Shizu, K.I., Suzuki, K., Kubo, S., Komino, T., Oiwa, H., Suzuki, F., Wakamiya, A., Murata, Y., and Adachi, C. (2015). Purely organic electroluminescent material realizing 100% conversion from electricity to light. *Nat. Commun.* 6: 8476.

33 Frischeisen, J., Yokoyama, D., Endo, A., Adachi, C., and Brütting, W. (2011). Increased light outcoupling efficiency in dye-doped small molecule organic light-emitting diodes with horizontally oriented emitters. *Org. Electron.* 12 (5): 809–817.

34 Kohler, A. and Bassler, H. ed. (2015). *Electronic Processes in Organic Semiconductors*. Wiley-VCH.

35 Schwoerer, M. and Wolf, H.C. ed. (2006). *Organic Molecular Solids*. Wiley-VCH.

36 Lüssem, B., Riede, M., and Leo, K. (2013). Doping of organic semiconductors. *Phys. Status Solidi A* 210 (1): 9–43.

37 Pfeiffer, M., Leo, K., Zhou, X., Huang, J.S., Hofmann, M., Werner, A., and Blochwitz-Nimoth, J. (2003). Doped organic semiconductors: physics and application in light emitting diodes. *Org. Electron.* 4 (2): 89–103.

38 Adachi, C., Baldo, M.A., Thompson, M.E., and Forrest, S.R. (2001). Nearly 100% internal phosphorescence efficiency in an organic light-emitting device. *J. Appl. Phys.* 90 (10): 5048–5051.

39 Purcell, E.M. (1946). Spontaneous emission probabilities at radio frequencies. *Phys. Rev.* 69: 681.

40 Barnes, W.L. (1998). Fluorescence near interfaces: the role of photonic mode density. *J. Mod. Opt.* 45 (4): 661–699.

41 Neyts, K.A. (1998). Simulation of light emission from thin-film microcavities. *J. Opt. Soc. Am. A* 15 (4): 962–971.
42 Furno, M., Meerheim, R., Hofmann, S., Lüssem, B., and Leo, K. (2012). Efficiency and rate of spontaneous emission in organic electroluminescent devices. *Phys. Rev. B* 85: 115205.
43 Brütting, W., Frischeisen, J., Schmidt, T.D., Scholz, B.J., and Mayr, C. (2012). Device efficiency of organic light-emitting diodes: progress by improved light outcoupling. *Phys. Status Solidi A* 210: 44–65.
44 Setz, D.S., Schmidt, T.D., Flämmich, M., Nowy, S., Frischeisen, J., Krummacher, B.C., Dobbertin, T., Heuser, K., Michaelis, D., Danz, N., Brütting, W., and Winnacker, A. (2011). Comprehensive efficiency analysis of organic light-emitting devices. *J. Photonics Energy* 1 (1): 011006.
45 Murawski, C., Leo, K., and Gather, M.C. (2013). Efficiency roll-off in organic light-emitting diodes. *Adv. Mater.* 25 (47): 6801–6827.
46 Wehrmeister, S., Jäger, L., Wehlus, T., Rausch, A.F., Reusch, T.C.G., Schmidt, T.D., and Brütting, W. (2015). Combined electrical and optical analysis of the efficiency roll-off in phosphorescent organic light-emitting diodes. *Phys. Rev. Appl.* 3: 024008.
47 Zhang, Y. and Forrest, S.R. (2012). Triplets contribute to both an increase and loss in fluorescent yield in organic light emitting diodes. *Phys. Rev. Lett.* 108: 267404.
48 Gather, M.C. and Reineke, S. (2015). Recent advances in light outcoupling from white organic light-emitting diodes. *J. Photonics Energy* 5 (1): 057607.
49 Baldo, M.A., Lamansky, S., Burrows, P.E., Thompson, M.E., and Forrest, S.R. (1999). Very high-efficiency green organic light-emitting devices based on electrophosphorescence. *Appl. Phys. Lett.* 75 (1): 4–6.
50 Baldo, M.A., O'Brien, D.F., Thompson, M.E., and Forrest, S.R. (1999). Excitonic singlet-triplet ratio in a semiconducting organic thin film. *Phys. Rev. B* 60 (20): 14422–14428.
51 Wohlgenannt, M., Tandon, K., Mazumdar, S., Ramasesha, S., and Vardeny, Z.V. (2001). Formation cross-sections of singlet and triplet excitons in [pi]-conjugated polymers. *Nature* 409 (6819): 494–497.
52 Barford, W., Bursill, R.J., and Makhov, D.V. (2010). Spin-orbit interactions between interchain excitations in conjugated polymers. *Phys. Rev. B* 81: 035206.
53 Pope, M. and Swenberg, C.E. ed. (1999). *Electronic Processes in Organic Crystals and Polymers*. Oxford University Press.
54 Köhler, A. and Bässler, H. (2009). Triplet states in organic semiconductors. *Mater. Sci. Eng., R* 66 (4–6): 71–109.
55 van Eersel, H., Bobbert, P.A., Janssen, R.A.J., and Coehoorn, R. (2014). Monte Carlo study of efficiency roll-off of phosphorescent organic light-emitting diodes: evidence for dominant role of triplet-polaron quenching. *Appl. Phys. Lett.* 105 (14): 143303.
56 Coehoorn, R., van Eersel, H., Bobbert, P., and Janssen, R. (2015). Kinetic Monte Carlo study of the sensitivity of OLED efficiency and lifetime to materials parameters. *Adv. Funct. Mater.* 25: 2024–2037.
57 Yokoyama, D. (2011). Molecular orientation in small-molecule organic light-emitting diodes. *J. Mater. Chem.* 21: 19187–19202.

58 Flämmich, M., Frischeisen, J., Setz, D.S., Michaelis, D., Krummacher, B.C., Schmidt, T.D., Brütting, W., and Danz, N. (2011). Oriented phosphorescent emitters boost OLED efficiency. *Org. Electron.* 12 (10): 1663–1668.

59 Schmidt, T.D., Scholz, B.J., Mayr, C., and Brütting, W. (2013). Efficiency analysis of organic light-emitting diodes based on optical simulations. *IEEE J. Sel. Top. Quantum Electron.* 19 (5): 1–12.

60 Kim, S.-Y., Jeong, W.-I., Mayr, C., Park, Y.-S., Kim, K.-H., Lee, J.-H., Moon, C.-K., Brütting, W., and Kim, J.-J. (2013). Organic light-emitting diodes with 30% external quantum efficiency based on a horizontally oriented emitter. *Adv. Funct. Mater.* 23 (31): 3896–3900.

61 Callens, M.K., Yokoyama, D., and Neyts, K. (2015). Anisotropic materials in OLEDs for high outcoupling efficiency. *Opt. Express* 23 (16): 21128–21148.

62 Qu, Y., Slootsky, M., and Forrest, S.R. (2015). Enhanced light extraction from organic light-emitting devices using a sub-anode grid. *Nat. Photonics* 9: 758–763.

63 Schmidt, T.D., Setz, D.S., Flämmich, M., Frischeisen, J., Michaelis, D., Krummacher, B.C., Danz, N., and Brütting, W. (2011). Evidence for non-isotropic emitter orientation in a red phosphorescent organic light-emitting diode and its implications for determining the emitter's radiative quantum efficiency. *Appl. Phys. Lett.* 99 (16): 163302.

64 Jurow, M.J., Mayr, C., Schmidt, T.D., Lampe, T., Djurovich, P.I., Brütting, W., and Thompson, M.E. (2016). Understanding and predicting the orientation of heteroleptic phosphors in organic light-emitting materials. *Nat. Mater.* 15: 85–91.

65 Chen, X.-W., Choy, W.C.H., Liang, C.J., Wai, P.K.A., and He, S. (2007). Modifications of the exciton lifetime and internal quantum efficiency for organic light-emitting devices with a weak/strong microcavity. *Appl. Phys. Lett.* 91 (22): 221112.

66 Nowy, S., Krummacher, B.C., Frischeisen, J., Reinke, N.A., and Brütting, W. (2008). Light extraction and optical loss mechanisms in organic light-emitting diodes: influence of the emitter quantum efficiency. *J. Appl. Phys.* 104 (12): 123109.

67 Schmidt, T.D., Setz, D.S., Flämmich, M., Frischeisen, J., Michaelis, D., Mayr, C., Rausch, A.F, Wehlus, T., Scholz, B.J., Reusch, T.C.G., Danz, N., and Brütting, W. (2013). Comprehensive efficiency analysis of organic light-emitting diodes featuring emitter orientation and triplet-to-singlet up-conversion. *Appl. Phys. Lett.* 103 (9): 093303.

68 Giebink, N.C. and Forrest, S.R. (2008). Quantum efficiency roll-off at high brightness in fluorescent and phosphorescent organic light emitting diodes. *Phys. Rev. B* 77 (23): 235215.

69 Schmidt, T.D., Flämmich, M., Scholz, B.J., Michaelis, D., Mayr, C., Danz, N., and Brütting, W. (2012). Non-isotropic emitter orientation and its implications for efficiency analysis of organic light-emitting diodes. Proceedings of SPIE, Volume 8435, p. 843513.

70 Frischeisen, J., Yokoyama, D., Adachi, C., and Brütting, W. (2010). Determination of molecular dipole orientation in doped fluorescent organic thin films by photoluminescence measurements. *Appl. Phys. Lett.* 96 (7): 073302.

71 Flämmich, M., Gather, M.C., Danz, N., Michaelis, D., Bräuer, A.H., Meerholz, K., and Tünnermann, A. (2010). Orientation of emissive dipoles in OLEDs: quantitative in situ analysis. *Org. Electron.* 11 (6): 1039–1046.

72 Meerheim, R., Furno, M., Hofmann, S., Lüssem, B., and Leo, K. (2010). Quantification of energy loss mechanisms in organic light-emitting diodes. *Appl. Phys. Lett.* 97 (25): 253305.

73 Schmidt, T.D., Reichardt, L.J., Rausch, A.F., Wehrmeister, S., Scholz, B.J., Mayr, C., Wehlus, T., Ciarnain, R.M., Danz, N., Reusch, T.C.G., and Brütting, W. (2014). Extracting the emitter orientation in organic light-emitting diodes from external quantum efficiency measurements. *Appl. Phys. Lett.* 105 (4): 043302.

74 Mayr, C., Lee, S.Y., Schmidt, T.D., Yasuda, T., Adachi, C., and Brütting, W. (2014). Efficiency enhancement of organic light-emitting diodes incorporating a highly oriented thermally activated delayed fluorescence emitter. *Adv. Funct. Mater.* 24 (33): 5232–5239.

75 Mayr, C., Schmidt, T.D., and Brütting, W. (2014). High-efficiency fluorescent organic light-emitting diodes enabled by triplet–triplet annihilation and horizontal emitter orientation. *Appl. Phys. Lett.* 105 (18): 183304.

76 Okumoto, K., Kanno, H., Hamaa, Y., Takahashi, H., and Shibata, K. (2006). Green fluorescent organic light-emitting device with external quantum efficiency of nearly 10%. *Appl. Phys. Lett.* 89 (6): 063504.

77 Schmidt, T.D., Setz, D.S., Flämmich, M., Scholz, B.J., Jaeger, A., Diez, C., Michaelis, D., Danz, N., and Brütting, W. (2012). Degradation induced decrease of the radiative quantum efficiency in organic light-emitting diodes. *Appl. Phys. Lett.* 101 (10): 103301.

78 Noguchi, Y., Miyazaki, Y., Tanaka, Y., Sato, N., Nakayama, Y., Schmidt, T.D., Brütting, W., and Ishii, H. (2012). Charge accumulation at organic semiconductor interfaces due to a permanent dipole moment and its orientational order in bilayer devices. *J. Appl. Phys.* 111: 114508.

79 Nowy, S., Ren, W., Elschner, A., Lövenich, W., and Brütting, W. (2010). Impedance spectroscopy as a probe for the degradation of organic light-emitting diodes. *J. Appl. Phys.* 107 (5): 054501.

80 Tian, M., Luo, J., and Liu, X. (2009). Highly efficient organic light-emitting devices beyond theoretical prediction under high current density. *Opt. Express* 17 (24): 21370–21375.

81 Noguchi, Y., Kim, H.-J., Ishino, R., Goushi, K., Adachi, C., Nakayama, Y., and Ishii, H. (2015). Charge carrier dynamics and degradation phenomena in organic light-emitting diodes doped by a thermally activated delayed fluorescence emitter. *Org. Electron.* 17 (0): 184–191.

82 Sandanayaka, A.S.D., Matsushima, T., and Adachi, C. (2015). Degradation mechanisms of organic light-emitting diodes based on thermally activated delayed fluorescence molecules. *J. Phys. Chem. C* 119 (42): 23845–23851.

83 Nakanotani, H., Masui, K., Nishide, J., Shibata, T., and Adachi, C. (2013). Promising operational stability of high-efficiency organic light-emitting diodes based on thermally activated delayed fluorescence. *Sci. Rep.* 3: 2127.

7

TADF Kinetics and Data Analysis in Photoluminescence and in Electroluminescence

Tiago Palmeira and Mário N. Berberan-Santos

Universidade de Lisboa, CQFM-IN and IBB – Institute of Bioengineering and Biosciences, Instituto Superior Técnico, 1049-001 Lisboa, Portugal

7.1 TADF Kinetics

7.1.1 Introduction

The basic model for thermally activated delayed fluorescence (TADF) kinetics in the condensed phases is a three-state scheme, involving excited-state interconversion between S_1 and T_1, with the ground state S_0 as the initial and final state (Scheme 7.1):where k_F and k_P are the radiative rate constants for fluorescence and phosphorescence, respectively, k_G^S and k_G^T are the internal conversion rate constant for $S_1 \to S_0$ deactivation and the intersystem crossing (ISC) rate constant for $T_1 \to S_0$ deactivation, and k_{ISC}^S and k_{ISC}^T are the direct (ISC) and reverse intersystem crossing (rISC) rate constants for transitions between S_1 and T_1. k_{ISC}^S and k_{ISC}^T are also denoted in the literature as k_{ISC} and k_{rISC}, respectively. The rISC rate constant k_{ISC}^T is temperature dependent and is given by [1–3]

$$k_{ISC}^T(T) = \frac{\sum_v k_v \exp\left(-\frac{E_v}{k_B T}\right)}{\sum_v \exp\left(-\frac{E_v}{k_B T}\right)} \tag{7.1}$$

where k_v is the rISC rate constant of the vth vibrational level of T_1 (v representing the full set of vibrational quantum numbers) and E_v is the respective vibrational energy. Assuming that k_v is a step function, equal to a constant A for $E_v \geq \Delta E_{ST}$, where ΔE_{ST} is the S_1–T_1 energy gap, and zero otherwise, and further assuming that the energy difference between consecutive vibronic levels is much smaller than $k_B T$ and that the density of states is approximately constant, Eq. (7.1) becomes the simple Arrhenius equation [2, 3]:

$$k_{ISC}^T = A \exp\left(-\frac{\Delta E_{ST}}{k_B T}\right) \tag{7.2}$$

which, owing to the absence of detailed information on k_v and on the density of vibrational states, is the commonly used form, empirically validated, for the rISC rate constant. The approximate nature of Eq. (7.2) may explain, in part, why

Highly Efficient OLEDs: Materials Based on Thermally Activated Delayed Fluorescence,
First Edition. Edited by Hartmut Yersin.
© 2019 Wiley-VCH Verlag GmbH & Co. KGaA. Published 2019 by Wiley-VCH Verlag GmbH & Co. KGaA.

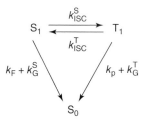

Scheme 7.1 Three-state kinetic scheme for TADF.

the recovered ΔE_{ST} does not always exactly match the spectroscopic value (when available). For this and other reasons, ΔE_{ST} should, in fact, be regarded as an activation energy for rISC that has a value close – but not necessarily identical – to the S_1–T_1 energy gap.

In the above analysis, it is assumed that a relatively fast equilibration exists among the triplet sublevels that can therefore be treated as a single entity whose intrinsic decay rate is the Boltzmann-weighted average of the sublevel decay rates [4]. This is valid for all temperatures of interest where TADF is operative, as the zero-field splitting in aromatic organic molecules (typically tenths of cm^{-1}) and in organometallic complexes (at most a few cm^{-1}) is much smaller than $k_B T$ when T exceeds a few kelvin, and the kinetics of triplet sublevel equilibration are usually fast when compared with radiative and nonradiative triplet relaxation processes [4a, 4b]. It is also assumed that the upper triplet states either do not contribute significantly to the TADF process or can be grouped together with T_1 for TADF analysis purposes, although in at least one case experimental results are compatible with a temperature-dependent T_2 contribution [4c].

At sufficiently high temperatures Eq. (7.2) gives $k_{ISC}^T = A$. Assuming that the molecule is stable under such conditions and that S_1 and T_1 are in fast equilibrium,

$$k_{ISC}^T(\infty)[T_1] = k_{ISC}^S[S_1] \qquad (7.3)$$

Furthermore, at these temperatures and in the absence of significant structural differences between S_1 and T_1 molecules, the relative populations follow the respective spin statistical weights [5, 6]:

$$\frac{[T_1]}{[S_1]} = 3 \qquad (7.4)$$

hence

$$k_{ISC}^T(\infty) = A = \frac{k_{ISC}^S}{3} = \frac{\Phi_T}{3\tau_F} \qquad (7.5)$$

where Φ_T is the quantum yield of triplet formation (also called Φ_{ISC}),

$$\Phi_T = \frac{k_{ISC}^S}{k_F + k_G^S + k_{ISC}^S} = k_{ISC}^S \cdot \tau_F \qquad (7.6)$$

and $\tau_F = 1/(k_F + k_G^S + k_{ISC}^S)$ is the singlet state lifetime in the absence of TADF (see Section 7.1.3.2).

If k_{ISC}^S is essentially temperature independent, then Eq. (7.5) holds for all temperatures; hence an approximate relation for the rISC rate constant is [5, 6]

$$\boxed{k_{ISC}^T = \frac{\Phi_T}{3\tau_F} \exp\left(-\frac{\Delta E_{ST}}{k_B T}\right)} \qquad (7.7)$$

Owing to the typical values of ΔE_{ST} for TADF molecules, the rISC rate constant is in most cases strongly temperature dependent.

7.1.2 Excitation Types

In the case of optical excitation (photoluminescence), the $T_1 \leftarrow S_0$ (radiative) transition is forbidden, and only S_1 is generated by photon absorption (Scheme 7.2):

The function I_{exc} may correspond to pulsed, modulated, or continuous excitation.

On the other hand, in the case of excitation by electrical current (electroluminescence, usually continuous excitation), both S_1 and T_1 are produced (Scheme 7.3), in a statistical ratio of 1 : 3, according to the respective spin multiplicities [7, 8], where the quantity I_{exc} is defined with respect to excitation by electron-hole recombination.

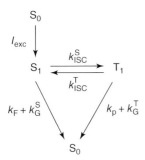

Scheme 7.2 TADF with optical excitation.

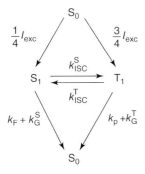

Scheme 7.3 TADF with electric excitation.

7.1.3 Photoexcitation

7.1.3.1 Rate Equations

The rate equations for weak photoexcitation (i.e. nonsaturating and avoiding triplet–triplet interaction) are

$$\frac{d[S_1]}{dt} = I_{exc}(t) - \frac{1}{\tau_F}[S_1] + k_{ISC}^T[T_1] \qquad (7.8)$$

$$\frac{d[T_1]}{dt} = k_{ISC}^S[S_1] - \frac{1}{\tau_P}[T_1] \qquad (7.9)$$

where $\tau_P = 1/(k_P + k_G^T + k_{ISC}^T)$ is called here the phosphorescence lifetime but, owing to TADF, does not correspond to a real decay time (see Section 7.1.3.2). This system of coupled equations can be solved exactly by a number of methods. The solution is well known (see e.g. [9]) as the set of differential equations is mathematically identical to that of monomer–excimer kinetics [10, 11].

7.1.3.2 Fluorescence and Phosphorescence Decays

In the case of delta-pulse excitation, $I_{exc}(t) = I_0 \delta(t)$, the singlet decay is given by a sum of two exponentials of time and the triplet decay as a difference of the same two exponentials [9–11]:

$$[S_1] = \frac{[S_1]_0}{\lambda_2 - \lambda_1}[(\lambda_2 - X)\exp(-\lambda_1 t) + (X - \lambda_1)\exp(-\lambda_2 t)] \qquad (7.10)$$

$$[T_1] = \frac{k_{ISC}^S[S_1]_0}{\lambda_2 - \lambda_1}[\exp(-\lambda_1 t) - \exp(-\lambda_2 t)] \qquad (7.11)$$

where

$$\lambda_{1,2} = \frac{1}{2}\left\{X + Y \mp \sqrt{(Y-X)^2 + 4k_{ISC}^S k_{ISC}^T}\right\} \qquad (7.12)$$

with

$$X = \frac{1}{\tau_F} \qquad (7.13)$$

$$Y = \frac{1}{\tau_P} = \frac{1}{\tau_P^0} + k_{ISC}^T \qquad (7.14)$$

where $\tau_P^0 = 1/(k_P + k_G^T)$ is formally identical to the low-temperature phosphorescence lifetime but refers to the temperature of the system.

In both TADF kinetics and monomer–excimer kinetics, the intensity of the higher-energy emitter (excited singlet state and monomer, respectively) increases with temperature, owing to an increase of the rate of the back step (reverse intersystem crossing, excimer dissociation). There is nevertheless one important difference between the two kinetics (apart from the molecularity of the direct step): Monomer and excimer intrinsic lifetimes are usually not very different, whereas singlet and triplet excited-state intrinsic lifetimes differ by several orders of magnitude. For this reason, the decay constants given by Eq. (7.12) can be simplified in the TADF case to [9]

$$\lambda_1 = \frac{\frac{1}{\tau_P^0} + (1-\Phi_T)k_{ISC}^T}{1+k_{ISC}^T \tau_F} \quad (7.15)$$

$$\lambda_2 = \frac{1}{\tau_F} + k_{ISC}^T \quad (7.16)$$

It is seen that the fluorescence decay has a short component with a lifetime $1/\lambda_2$, that is, smaller than the fluorescence lifetime τ_F and a long component (delayed fluorescence (DF) lifetime) with a lifetime $\tau_{DF} = 1/\lambda_1$, that is, smaller than the low-temperature phosphorescence lifetime. The higher the temperature, the shorter these two lifetimes are.

For $k_{ISC}^T \tau_F \ll 1$, as is usually the case, Eq. (7.15) further reduces to [5, 6, 9, 12–14]

$$\boxed{\lambda_1 = \frac{1}{\tau_{DF}} = \frac{1}{\tau_P^0} + (1-\Phi_T)k_{ISC}^T} \quad (7.17)$$

where τ_{DF} is the DF lifetime, associated with the slow component of both fluorescence and phosphorescence and which does not coincide with the phosphorescence lifetime τ_P defined above. For a simple derivation of Eq. (7.17), see ref. [14].

Equation (7.16), in turn, becomes

$$\lambda_2 = \frac{1}{\tau_F} \quad (7.18)$$

and defines the prompt fluorescence (PF) lifetime.

It also follows from Eq. (7.10) that the relative amplitude of the fluorescence slow component (DF) is [9]

$$\frac{\lambda_2 - X}{X - \lambda_1} = k_{ISC}^T \tau_F \quad (7.19)$$

and it is thus always very small as $k_{ISC}^T \tau_F \ll 1$.

7.1.3.3 Steady-state Fluorescence and Phosphorescence Intensities

In the case of continuous excitation (steady-state experiment, denoted as ss), writing $I_{exc} = I_0$, where I_0 is the number of moles of photons (einstein) absorbed per unit time and unit volume, and setting the time derivatives in Eqs. (7.8) and (7.9) equal to zero, gives

$$[S_1]_{ss} = \frac{\tau_F I_0}{1-\Phi_S \Phi_T} \quad (7.20)$$

$$[T_1]_{ss} = \frac{\Phi_T \tau_P I_0}{1-\Phi_S \Phi_T} \quad (7.21)$$

where

$$\Phi_S = \frac{k_{ISC}^T}{k_P + k_G^T + k_{ISC}^T} = k_{ISC}^T \tau_P \quad (7.22)$$

is defined as the quantum yield of singlet formation by rISC [13] (also called Φ_{rISC} in the more recent literature), compare Eq. (7.6). The corresponding fluorescence and phosphorescence intensities (wavelength integrated) are

$$I_F = k_F[S_1]_{ss} = \frac{\Phi_{PF} I_0}{1 - \Phi_S \Phi_T} \tag{7.23}$$

$$I_P = k_P[T_1]_{ss} = \frac{\Phi_T \theta_P I_0}{1 - \Phi_S \Phi_T} = \frac{\Phi_P I_0}{1 - \Phi_S \Phi_T} \tag{7.24}$$

where the PF quantum yield Φ_{PF} is the fluorescence yield in the absence of TADF, the phosphorescence quantum yield is $\Phi_P = \Phi_T \theta_P$, and θ_P is the phosphorescence quantum efficiency, given by $\theta_P = k_P/(k_P + k_G^T + k_{ISC}^T)$.

In the absence of reverse intersystem crossing ($\Phi_S = 0$, e.g. owing to triplet quenching by oxygen along with negligible singlet quenching), all the fluorescence is PF:

$$I_{PF} = \Phi_{PF} I_0 \tag{7.25}$$

hence

$$I_F = \frac{I_{PF}}{1 - \Phi_S \Phi_T} \tag{7.26}$$

$$I_{DF} = I_F - I_{PF} = \frac{\Phi_S \Phi_T}{1 - \Phi_S \Phi_T} I_{PF} \tag{7.27}$$

and therefore [5, 6, 9, 13, 15, 16]

$$\boxed{\frac{I_{DF}}{I_{PF}} = \frac{\Phi_{DF}}{\Phi_{PF}} = \frac{1}{\frac{1}{\Phi_S \Phi_T} - 1} = \frac{1}{\frac{1}{\Phi_T} - 1 + \frac{1}{\Phi_T \tau_P^0 k_{ISC}^T}} = \Phi_T k_{ISC}^T \tau_{DF}} \tag{7.28}$$

and [17–19]

$$\boxed{\frac{I_{DF}}{I_P} = \frac{\Phi_{DF}}{\Phi_P} = \frac{\Phi_{PF}}{k_P} k_{ISC}^T} \tag{7.29}$$

where Φ_{DF} in Eqs. (7.28) and (7.29) is the DF quantum yield. Equation (7.28) can be used to compute the rISC rate constant (or at least ΔE_{ST}) from experimental data [5, 13, 16, 20], e.g. in the form $k_{ISC}^T = (I_{DF}/I_{PF})/(\Phi_T \tau_{DF})$ (see also Section 7.2.4.2). Equation (7.29), derived and used by Parker [18, 19], was previously obtained by Rosenberg and Shombert [17].

The maximum possible fluorescence yield corresponds to $\Phi_S = 1$, and Eq. (7.26) gives

$$\boxed{\Phi_F^{max} = \frac{\Phi_{PF}}{1 - \Phi_T} = \frac{k_F}{k_F + k_G^S}} \tag{7.30}$$

In this way, strong TADF effectively eliminates the ISC nonradiative channel by always returning the excited molecule to S_1 (see in the next section the TADF cycle perspective).

7.1.3.4 Excited-state Cycles

In the customary description of the TADF mechanism, it is said that after photoexcitation to S_n ($n \geq 1$) and once attained S_1, ISC to the triplet manifold occurs, followed by rISC from T_1 back to S_1 and then by fluorescence emission. However, this description of TADF is incomplete. It was shown that the excited molecule may go through several S_1–T_1–S_1 cycles before fluorescence finally takes place [9, 21], as exemplified in Figure 7.1 for a single molecule undergoing three excited-state cycles.

In Figure 7.1 both ISC and rISC are for simplicity depicted as vertical lines. However, intrinsic intersystem crossing steps connect isoenergetic levels. In the $S_1 \rightarrow T_1$ case, direct ISC is quickly followed by vibrational relaxation, whereas in the $S_1 \leftarrow T_1$ case, thermal activation (according to the Boltzmann distribution) precedes the rISC step.

The existence of excited-state cycles is compatible with the kinetic results already derived. This can be explicitly shown using a convolution approach [22], where the evolution equations are directly written in integral form. The S_1 and T_1 populations are given by the following coupled equations [9]:

$$[S_1] = I_{exc}(t) \otimes \exp(-t/\tau_F) + k_{ISC}^T[T_1] \otimes \exp(-t/\tau_F) \tag{7.31}$$

$$[T_1] = k_{ISC}^S[S_1] \otimes \exp(-t/\tau_P) \tag{7.32}$$

where $\otimes$ stands for the convolution between two functions, $f \otimes g = \int_0^t f(u)g(t-u)du$, and τ_F and τ_P were previously defined (both lifetimes only have direct experimental meaning in the absence of reversibility).

The general solution can be obtained either by Laplace transforms or by insertion of Eq. (7.32) into Eq. (7.31) and then by repeated substitution of the left-hand

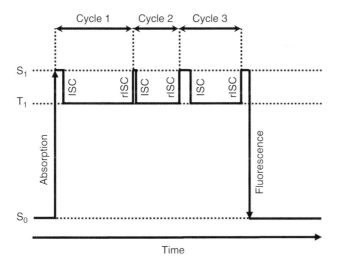

Figure 7.1 Example of several S_1–T_1–S_1 cycles for a single TADF molecule. Before decaying by fluorescence after three cycles (in this example), the excited molecule jumps at random times between the two states by intersystem crossing (ISC, rISC). The waiting times are, for each state, random variables whose averages correspond to the respective lifetimes τ_F and τ_P.

side on the right-hand side [9]:

$$[S_1] = I_{exc}(t) \otimes \exp(-t/\tau_F) + k_{ISC}^S k_{ISC}^T [S_1] \otimes \exp(-t/\tau_P) \otimes \exp(-t/\tau_F) \tag{7.33}$$

$$[S_1] = I_{exc}(t) \otimes \exp(-t/\tau_F)$$
$$+ k_{ISC}^S k_{ISC}^T I_{exc}(t) \otimes \exp(-t/\tau_F) \otimes \exp(-t/\tau_P) \otimes \exp(-t/\tau_F)$$
$$+ (k_{ISC}^S k_{ISC}^T)^2 I_{exc}(t) \otimes \exp(-t/\tau_F) \otimes \exp(-t/\tau_P) \otimes \exp(-t/\tau_F)$$
$$\otimes \exp(-t/\tau_P) \otimes \exp(-t/\tau_F)$$
$$+ \cdots \tag{7.34}$$

Hence the first term for the singlet decay can be associated with PF (zero $S_1 \to T_1 \to S_1$ cycles) and the remaining terms with DF, the nth term resulting from $n-1$ $S_1 \to T_1 \to S_1$ cycles. Summation of the terms of Eq. (7.34) leads to Eq. (7.10) (convolved with I_{exc}). Analogous results are obtained for the triplet decay.

We now turn to the steady-state situation. Again, for strong TADF to occur, the following inequalities must hold: $k_{ISC}^S \gg k_F + k_G^S$ and $k_{ISC}^T \gg k_P + k_G^T$. In most cases it is also observed that $k_{ISC}^S \gg k_{ISC}^T$ and $k_G^T \gg k_P$. Interconversion of the singlet and triplet emissive states then occurs several times before photon emission or nonradiative decay can take place. In this way, a pre-equilibrium between S_1 and T_1 exists, and the cycle $S_1 \to T_1 \to S_1$ repeats a number of times before fluorescence emission occurs. It is interesting to consider the following question: For a given set of rate constants, how many times is the cycle $S_1 \to T_1 \to S_1$ completed on the average, before return to the ground state occurs? Clearly, for a pre-equilibrium to exist, this cycling must occur many times. In order to quantitatively answer the above question, and related aspects, it is convenient to view TADF as the sequential process depicted in Scheme 7.4.

One then has

$$\Phi_F = \Phi_{PF}[1 + \Phi_T \Phi_S + (\Phi_T \Phi_S)^2 + \cdots] = \frac{\Phi_{PF}}{1 - \Phi_S \Phi_T} \tag{7.35}$$

compare Eq. (7.26). The first term corresponds to PF (0 cycles), and the remaining terms correspond to DF, the nth term resulting in general from $n-1$ $S_1 \to T_1 \to S_1$ cycles. Equation (7.35) can also be derived from Eq. (7.34).

The probability for fluorescence emission to occur after exactly n $S_1 \to T_1 \to S_1$ cycles obeys a geometric probability distribution [9, 21]:

$$p_n = (1 - \Phi_T \Phi_S)(\Phi_T \Phi_S)^n \tag{7.36}$$

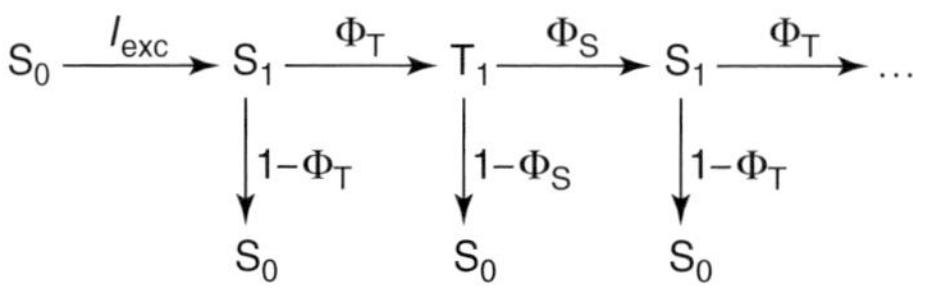

$$S_0 \xrightarrow{I_{exc}} S_1 \xrightarrow{\Phi_T} T_1 \xrightarrow{\Phi_S} S_1 \xrightarrow{\Phi_T} \cdots$$
$$\downarrow 1-\Phi_T \quad \downarrow 1-\Phi_S \quad \downarrow 1-\Phi_T$$
$$S_0 \quad\quad S_0 \quad\quad S_0$$

Scheme 7.4 TADF as a sequential process.

The average number of cycles $\bar{n}$ is thus given by [9, 21]

$$\bar{n} = \sum_{n=0}^{\infty} n p_n = \frac{\Phi_S \Phi_T}{1 - \Phi_S \Phi_T} = \frac{1}{\frac{1}{\Phi_S \Phi_T} - 1} \tag{7.37}$$

Comparison of Eq. (7.37) with Eq. (7.28) gives immediately

$$\bar{n} = \frac{\Phi_{DF}}{\Phi_{PF}} = \frac{I_{DF}}{I_{PF}} \tag{7.38}$$

and, using Eq. (7.27),

$$\bar{n} + 1 = \frac{\Phi_F}{\Phi_{PF}} = \frac{I_F}{I_{PF}} \tag{7.39}$$

Hence the increase in fluorescence intensity owing to TADF is a direct measure of the average number of $S_1 \to T_1 \to S_1$ cycles performed [9]. This result follows from the fact that each return to S_1 brings a new opportunity for fluorescence emission.

In the absence of reversibility, $\bar{n} = 0$. On the other hand, for the fastest possible excited-state equilibration ($k_{ISC}^T \to A, \Phi_S \simeq 1$), one has

$$\boxed{\bar{n}_{max} \simeq \frac{1}{\frac{1}{\Phi_T} - 1}} \tag{7.40}$$

and the maximum possible fluorescence intensification factor is $1/(1 - \Phi_T)$, as already discussed, cf. Eq. (7.30).

Using the photophysical parameters for fullerene $^{13}C_{70}$ in a polymer matrix (Zeonex) [4], $\Phi_T = 0.997$, $\tau_F = 700$ ps, $\tau_P^0 = 96$ ms, and $\Delta E_{ST} = 33$ kJ mol^{-1} (340 meV), the maximum average number of cycles is estimated to be 332 and the maximum fluorescence intensification factor to be 333. The average number of cycles as a function of temperature, computed using Eq. (7.27), is displayed in Figure 7.2 (for simplicity, the temperature dependence of τ_P^0 [4] is neglected).

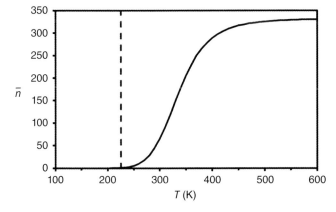

Figure 7.2 Computed average number of excited-state cycles versus temperature for $^{13}C_{70}$ in Zeonex. The TADF onset temperature (see Eq. (7.44)), shown by the dashed vertical line, is 225 K.

It is seen that TADF sets in at about 225 K and a large number of excited-state cycles are already effected at room temperature. The maximum value is expected to be attained at about 500 K; however this can happen only with a suitable matrix (not Zeonex) and in the absence of triplet quenching and thermal reactions.

7.1.3.5 TADF Onset Temperature

Efficient TADF can occur only in the absence of quenching by molecular oxygen (or other triplet state quencher); otherwise the excited-state loop will be broken by T_1 deactivation. Given that back intersystem crossing is always thermally activated, degassing or oxygen diffusion blocking is not enough to set in TADF: A minimum temperature is also required. Indeed, rISC ($S_1 \leftarrow T_1$) competes effectively with $T_1 \to S_0$ deactivation channels only above a certain temperature, characteristic of each molecule (in a given medium), T_0. This may be expressed quantitatively by imposing a certain I_{DF}/I_{PF} ratio, e.g. $I_{DF}/I_{PF} = 1$, meaning a doubling of the total fluorescence (that is given by $I_F = I_{DF} + I_{PF}$) owing to TADF. According to Eq. (7.38), this also means that the average number of excited-state cycles is equal to 1 at such temperature.

Using Eqs. (7.28) and (7.38),

$$\overline{n} = \frac{1}{\frac{1}{\Phi_T} - 1 + \frac{1}{\Phi_T \tau_P^0 k_{ISC}^T}} \quad (7.41)$$

Using Eq. (7.7) and solving for T, Eq. (7.41) becomes

$$T = \frac{T_g}{\ln\left\{\left(\frac{\Phi_T \tau_P^0}{3\tau_F}\right)\left[\left(1 + \frac{1}{\overline{n}}\right)\Phi_T - 1\right]\right\}} \quad (7.42)$$

where T_g is the singlet–triplet gap characteristic temperature:

$$T_g = \frac{\Delta E_{ST}}{R} \quad (7.43)$$

The TADF onset temperature T_0 (for which $\overline{n} = 1$) can thus be defined as

$$T_0 = \frac{T_g}{\ln\left[\frac{(2\Phi_T - 1)\Phi_T \tau_P^0}{3\tau_F}\right]} \quad (7.44)$$

It is seen that T_0 is controlled by three parameters: Φ_T, ΔE_{ST}, and the ratio τ_P^0/τ_F. For the example given in Figure 7.2, $T_0 = 225$ K ($-48\,°C$).

Equation (7.44) also shows that T_0 as defined does not exist for $\Phi_T \leq 0.5$. There cannot be efficient TADF for $\Phi_T \leq 0.5$, in the sense that fluorescence cannot be doubled, whatever the temperature (see Eq. (7.40)), given that only a minor fraction of S_1 goes to T_1. The same holds if $(2\Phi_T - 1)\left(\frac{\Phi_T \tau_P^0}{3\tau_F}\right) \leq 1$. Nevertheless, usually $\tau_P^0 \gg \tau_F$.

7.1.3.6 Conditions for Efficient TADF

As discussed in the previous sections, fluorescence enhancement by TADF (efficient TADF) means that the PF quantum yield is moderate in its absence, implying a high Φ_T, that is, $k_{ISC}^S \gg k_F + k_G^S$. The absence of TADF can also be due to low temperature or to triplet quenching. Suppression of the direct ISC by TADF raises the fluorescence quantum yield Φ_F up to $\Phi_F^{max} = k_F/(k_F + k_G^S)$; hence ideally $k_F \gg k_G^S$. In order to have fast reversibility, Φ_S must be close to 1; hence $k_{ISC}^T \gg 1/\tau_P^0$ or, equivalently, $\frac{\Phi_T}{3\tau_F} \exp\left(-\frac{\Delta E_{ST}}{RT}\right) \gg \frac{1}{\tau_P^0}$, as expressed by Eq. (7.41).

The discussion of TADF efficiency can also be based on Eq. (7.26), rewritten as

$$\boxed{\frac{\Phi_F}{\Phi_F^{max}} = \frac{1 - \Phi_T}{1 - \Phi_S \Phi_T}} \qquad (7.45)$$

This function is plotted in Figure 7.3. It is seen that Φ_F approaches the ceiling value Φ_F^{max} not only when Φ_T is small but also when Φ_T is significant, if Φ_S is also important (significant rISC).

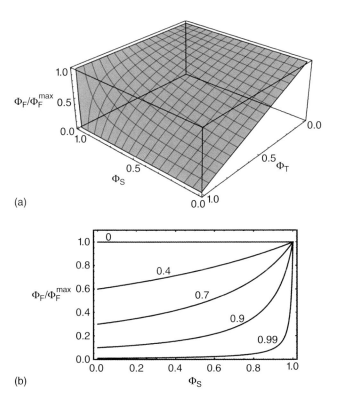

Figure 7.3 Fluorescence efficiency as a function of Φ_S and Φ_T, showing the effect of TADF: 3D plot (a); as a function of Φ_S, for several values of Φ_T, shown next to each curve (b).

For a given fluorescence yield ratio $R = \Phi_F/\Phi_F^{max}$, the value of Φ_S – hence the average number of cycles – is defined by the value of Φ_T:

$$\Phi_S = \frac{R - (1 - \Phi_T)}{\Phi_T R} \tag{7.46}$$

$$\bar{n} = \frac{R}{1 - \Phi_T} - 1 \tag{7.47}$$

For instance, if $\Phi_T = 0.90$ and the ratio target value is $R = 0.90$, then $\Phi_S = 0.988$, cf. Figure 7.3, for which $\bar{n} = 8$.

Equation (7.44) can be used to estimate a maximum permissible value for ΔE_{ST}, given a certain T_0:

$$\Delta E_{ST}^{max} = RT_0 \ln\left[\frac{(2\Phi_T - 1)\Phi_T \tau_P^0}{3\tau_F}\right] \tag{7.48}$$

Let us take $\Phi_T = 0.90$, $\tau_F = 5$ ns, and $\tau_P^0 = 50$ μs. Considering that, for practical purposes, TADF should be effective above 0 °C ($T_0 = 273$ K), one gets $\Delta E_{ST}^{max} = 18$ kJ mol^{-1} = 190 meV.

7.1.4 Electrical Excitation

7.1.4.1 Steady State

In the case of electroluminescence (Scheme 7.3), S_1 and T_1 are produced in a 1 : 3 ratio [7, 8].

Under steady-state conditions, the rate equations for weak excitation (i.e. non-saturating and avoiding triplet–triplet interaction) are

$$\frac{d[S_1]_{ss}}{dt} = \frac{1}{4}I_0 - \frac{1}{\tau_F}[S_1]_{ss} + k_{ISC}^T[T_1]_{ss} = 0 \tag{7.49}$$

$$\frac{d[T_1]_{ss}}{dt} = \frac{3}{4}I_0 + k_{ISC}^S[S_1]_{ss} - \frac{1}{\tau_P}[T_1]_{ss} = 0 \tag{7.50}$$

It follows that

$$[S_1]_{ss} = \tau_F \frac{1 + 3\Phi_S}{1 - \Phi_S \Phi_T} \frac{I_0}{4} \tag{7.51}$$

$$[T_1]_{ss} = \tau_P \frac{3 + \Phi_T}{1 - \Phi_S \Phi_T} \frac{I_0}{4} \tag{7.52}$$

$$I_F = \Phi_{PF} \frac{1 + 3\Phi_S}{1 - \Phi_S \Phi_T} \frac{I_0}{4} = \frac{1 + 3\Phi_S}{1 - \Phi_S \Phi_T} I_{PF} \tag{7.53}$$

$$I_P = \Phi_P \frac{3 + \Phi_T}{1 - \Phi_S \Phi_T} \frac{I_0}{4} \tag{7.54}$$

$$I_{DF} = \frac{\Phi_S(3 + \Phi_T)}{1 - \Phi_S \Phi_T} I_{PF} \tag{7.55}$$

$$\frac{I_{DF}}{I_{PF}} = \frac{1 + \frac{3}{\Phi_T}}{\frac{1}{\Phi_S \Phi_T} - 1} = (3 + \Phi_T) k_{ISC}^T \tau_{DF} \qquad (7.56)$$

$$\frac{I_{DF}}{I_P} = \frac{\Phi_{PF}}{k_P} k_{ISC}^T \qquad (7.57)$$

confer Eqs. (7.20)–(7.29). Equations (7.56) and (7.57) are especially noteworthy. Compared with their photostationary counterparts, Eqs. (7.28) and (7.29), it is seen that I_{DF}/I_{PF} is always higher in the case of electroluminescence, by a minimum factor of 4 (when Φ_T is close to 1), and is much higher when Φ_T is close to 0. This is understandable, as in electroluminescence S_1 is (formally) obtained from T_1 even when $\Phi_T = 0$, which does not happen under photoexcitation conditions. On the other hand, I_{Df}/I_P is identical for both excitation mechanisms.

7.1.4.2 Conditions for Efficient Electroluminescence

The intensification of fluorescence owing to rISC is obtained from Eq. (7.53) [23]:

$$\Phi_F = \frac{1 + 3\Phi_S}{1 - \Phi_S \Phi_T} \Phi_{PF} \qquad (7.58)$$

compare Eq. (7.26). Here, the maximum hypothetical intensification factor ($\Phi_S = 1$) is $4/(1 - \Phi_T)$, four times the photostationary one, owing to the contribution of directly excited triplets. The effective fluorescence quantum yield may attain 4 (if $k_F \gg k_G^S$), as three emitting singlets (out of four) originate in the triplet manifold and direct ISC is effectively suppressed. When considering the overall electroluminescence efficiency Φ_{EL} and assuming negligible phosphorescence, Eq. (7.58) must be divided by 4:

$$\boxed{\Phi_{EL} = \frac{1}{4} \left(\frac{1 + 3\Phi_S}{1 - \Phi_S \Phi_T} \right) \Phi_{PF}} \qquad (7.59)$$

Equation (7.59) can be rearranged to give

$$\Phi_{EL} = \frac{\Phi_{PF}}{4} + \frac{\Phi_S \Phi_T}{1 - \Phi_S \Phi_T} \Phi_{PF} + \frac{3}{4} \left(\frac{\Phi_S \Phi_T}{1 - \Phi_S \Phi_T} \right) \frac{\Phi_{PF}}{\Phi_T} \qquad (7.60)$$

and, using also the *photoluminescence* DF quantum yield, Eq. (7.27), Eq. (7.60) becomes

$$\boxed{\Phi_{EL} = \frac{1}{4}(\Phi_{PF} + \Phi_{DF}) + \frac{3}{4} \frac{\Phi_{DF}}{\Phi_T}} \qquad (7.61)$$

as given by Adachi [24, 25]. The first term corresponds to the contribution of directly excited singlets (containing prompt and delayed components), whereas the second is the (delayed) contribution from directly excited triplets, Φ_{DF}^T:

$$\Phi_{DF}^T = \Phi_S [1 + \Phi_S \Phi_T + (\Phi_S \Phi_T)^2 + \cdots] \Phi_{PF} = \frac{\Phi_{DF}}{\Phi_T} \qquad (7.62)$$

Note that Eq. (7.61) can be written as $\Phi_{EL} = \Phi_{PF}^{EL} + \Phi_{DF}^{EL}$, with $\Phi_{PF}^{EL} = \Phi_{PF}/4$ and $\Phi_{DF}^{EL} = (\Phi_{DF} + 3\Phi_{DF}^T)/4$.

Equation (7.61) allows concluding that there are two extreme cases for which the electroluminescence efficiency can be high. In both, Φ_F^{max} and Φ_S must be close to 1, implying that $k_F \gg k_G^S$ and that the rISC rate constant must dominate over the other triplet decay channels whose rate is $1/\tau_P^0$ and in particular $k_{ISC}^T \gg k_G^T$.

In the first case, Φ_{PF} is low owing to a high ISC rate constant, $k_{ISC}^S \gg k_F$; hence Φ_T is close to 1. However, efficient rISC (Φ_S close to 1 assumed) effectively eliminates the ISC channel, making Φ_{DF} approach 1; thus Φ_{EL} is also close to 1. In this case both ISC and rISC are fully operative, and the average number of $S_1 \rightarrow T_1 \rightarrow S_1$ cycles is high. This is the situation observed in efficient TADF under photoexcitation, and it can occur in electroluminescence as well (see Figure 7.4). Most TADF emitters intended for organic light-emitting diode (OLED) applications fall in this case [26].

However, the second term in the r.h.s. of Eq. (7.61), specific of electrical excitation, allows a second solution: If Φ_{PF} is close to 1, implying $k_F \gg k_{ISC}^S$, then Φ_T is low, and the average number of cycles is small to negligible. PF dominates the contribution from directly excited singlets and accounts for ¼ of the yield (see Figure 7.4). For very low Φ_T, Eq. (7.62) reduces to $\Phi_S\Phi_{PF}$, meaning that

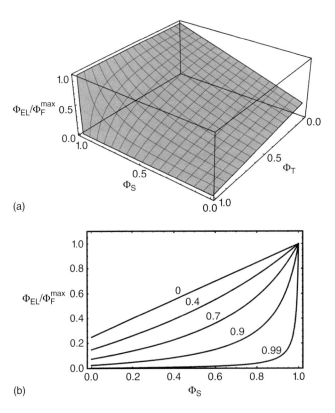

Figure 7.4 Electroluminescence efficiency as a function of Φ_S and Φ_T, displaying the effects of ISC and of rISC: 3D plot (a); as a function of Φ_S, for several values of Φ_T, shown next to each curve (b).

a single $T_1 \to S_1$ step occurs before emission. The dependence on Φ_S is linear (see Figure 7.4). For Φ_S close to 1, the triplet contribution approaches ¾ (see Figure 7.4).

Equation (7.59) can be rewritten as (compare Eq. (7.45))

$$\boxed{\frac{\Phi_{EL}}{\Phi_F^{max}} = \frac{(1 + 3\Phi_S)(1 - \Phi_T)}{4(1 - \Phi_S \Phi_T)}} \tag{7.63}$$

This function is plotted in Figure 7.4. For very low Φ_S, only the PF contribution exists, and the ratio is, at most, 1/4 (for $\Phi_T = 0$), as mentioned.

However, for increasing values of Φ_S, the relative luminescence yield increases, owing to DF coming from both singlet and triplet. The important point is that this increase occurs for all values of Φ_T and is fast for small values of Φ_T (see Figure 7.4).

Equation (7.63) gives the value of Φ_S for a given ratio $R_{EL} = \Phi_{EL}/\Phi_F^{max}$ and a given Φ_T:

$$\boxed{\Phi_S = \frac{\left(\frac{4R_{EL}}{1-\Phi_T}\right) - 1}{\left(\frac{4R_{EL}}{1-\Phi_T}\right)\Phi_T + 3}} \tag{7.64}$$

The required value of the rISC rate constant can next be computed:

$$\boxed{k_{ISC}^T = \frac{\Phi_S}{(1 - \Phi_S)\tau_P^0}} \tag{7.65}$$

Equation (7.7) finally relates this value with the remaining parameters (fluorescence lifetime, singlet–triplet gap, temperature). Several combinations of these three parameters correspond to a given value of the rISC rate constant.

It is thus mathematically viable to attain high values of Φ_{EL}/Φ_F^{max}, even in the near absence of cycles ($\Phi_T = 0$), that is, of significant TADF.

At this point, the question arises: Is it possible to reconcile a significant rISC (leading to a high Φ_S) with a low Φ_T, knowing that direct and reverse ISC rate constants are proportional?

The following example shows that this is indeed feasible: Let us assume a potentially highly fluorescent molecule, with $\Phi_F^{max} = 0.90$. Using a typical radiative lifetime for an allowed transition, 5 ns, one gets $k_F = 2.0 \times 10^8$ s^{-1} and $k_G^S = 2.2 \times 10^7$ s^{-1}. Let us also impose $\Phi_S = 0.90$. In order to proceed, we consider a phosphorescence lifetime $\tau_P^0 = 100$ μs. This value gives $k_{ISC}^T = 9.0 \times 10^4$ s^{-1}. Assuming further that $T = 300$ K and that $\Delta E_{ST} = 100$ meV, the pre-exponential factor A in Eq. (7.2) is 4.3×10^6 s^{-1}, and $k_{ISC}^S = 3A = 1.3 \times 10^7$ s^{-1}. The computed values for the direct ISC and rISC rate constants are not unrealistic and give $\Phi_{PF} = 0.85$, $\Phi_{DF} = 0.044$, and $\Phi_T = 0.055$. The quantum yield of triplet formation is thus quite small. From these parameters, the DF lifetime is $\tau_{DF} = 11$ μs and $\Phi_{EL} = 0.83$; hence $R_{EL} = \Phi_{EL}/\Phi_F^{max} = 0.92$. It is thus in principle possible, in electroluminescence, to have efficient triplet harvesting with a low Φ_T (implying at the same time that the molecule will display very weak TADF under photoexcitation in the example $\Phi_{DF}/\Phi_{PF} = 0.052$).

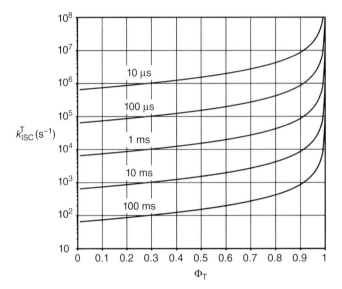

Figure 7.5 Computed rISC rate constant, according to Eqs. (7.64) and (7.65), as a function of Φ_T, for $R_{EL} = 0.90$ and the values of τ_P^0 shown next to each curve.

Efficient emitters with photophysical parameters matching this situation (low Φ_T) were indeed recently reported [27–29].

As a more systematic approach, let us now proceed according to Eqs. (7.64) and (7.65). Imposing $R_{EL} = 0.90$, a plot of k_{ISC}^T versus Φ_T is obtained (Figure 7.5). A similar plot can be drawn for any desired R_{EL} value.

Figure 7.5 clearly shows that the demands on rISC increase with the value of Φ_T and are also more stringent for short τ_P^0. Molecules with low Φ_T and long τ_P^0 require lower k_{ISC}^T for the same efficiency R_{EL}. In addition, Φ_F^{max} should be as high as possible, in order to have a high Φ_{EL}. These seem to be important guidelines in the design of third-generation OLED molecules.

7.1.5 More Complex Schemes

The three-state kinetic scheme implies that the S_1 Franck–Condon and emissive states are relatively similar. This is not always the case, especially with donor–acceptor molecules with strong charge transfer (CT) in the excited state [30–33]. In such situations, a distinction must be made between the photoexcited state and the relaxed CT state. The latter is the one participating in the photophysical mechanism represented by Scheme 7.1, although in many cases the full picture is still wanting.

If the photoexcited state to CT state conversion is fast enough, then the three-state kinetic scheme still holds, otherwise more complex kinetics ensues, with at least an additional rate constant corresponding to the conversion.

7.2 TADF Data Analysis

7.2.1 Introduction

Analysis of photophysical observables, by themselves or combined, like fluorescence and phosphorescence intensities (steady-state measurements) and fluorescence and phosphorescence decay times (time-resolved measurements) allows determining all kinetic parameters of Scheme 7.1. From the temperature dependence of some of the observables, the TADF activation energy ΔE_{ST} can also be estimated. There are several possible methods, depending on the quantities for which data was measured (owing not only to experimental techniques available but also to system's properties, e.g. phosphorescence can be essentially undetectable in some cases). These methods are described in the next three sections, with examples of application to two different systems, both degassed, eosin in glycerol and fullerene C_{70} in a cycloalkane polymer, Zeonex.

7.2.2 Steady-state Data

7.2.2.1 Delayed Fluorescence and Phosphorescence Intensities as a Function of Temperature: Rosenberg–Parker Method

This method, first used by Rosenberg and Shombert [17] and shortly afterward by Parker and Hatchard [18, 19], relies on Eqs. (7.29) and (7.2) combined, in a linearized form:

$$\ln\left(\frac{I_{DF}}{I_P}\right) = C - \frac{\Delta E_{ST}}{k_B T} \quad (7.66)$$

where C is a constant. As an example, a plot for eosin in glycerol is shown in Figure 7.6 and gives a very good straight line, in support of the form of Eq. (7.2). The recovered ΔE_{ST} is 40 kJ mol^{-1} (0.41 eV). The original measurements by Parker and Hatchard [18] for the same system gave 42 kJ mol^{-1} (0.43 eV). The spectroscopic value is 43 kJ mol^{-1} (0.45 eV). The method was applied to a number of molecules, including xanthene dyes [18, 34, 35], ketones [6, 36], thiones [37], polycyclic aromatic hydrocarbons [38], and fullerenes [13].

7.2.2.2 Prompt and Delayed Fluorescence Intensities as a Function of Temperature

In several cases, the phosphorescence is undetectable in the temperature range of interest, and the previous method cannot be used. Furthermore, in such cases the spectroscopic estimation of ΔE_{ST} is also not possible. To deal with this situation, a method using only prompt and DF steady-state intensities at several temperatures was devised [13]. This method is based on Eq. (7.28), rewritten as [13, 39]

$$\ln\left[\frac{I_{PF}}{I_{DF}} - \left(\frac{1}{\Phi_T} - 1\right)\right] = -\ln(A\tau_P^0 \Phi_T) + \frac{\Delta E_{ST}}{k_B T} \quad (7.67)$$

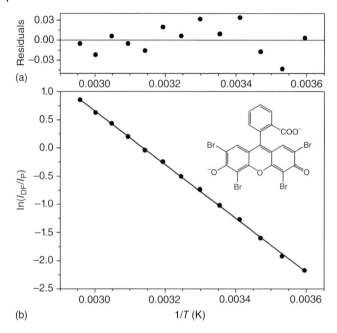

Figure 7.6 Rosenberg–Parker plot for 5.5 × 10⁻⁶ M eosin in glycerol, for temperatures between 5 °C and 65 °C. Emission wavelengths were 540 nm (delayed fluorescence) and 674 nm (phosphorescence).

It is therefore possible to obtain ΔE_{ST} from the temperature dependence of the ratio I_{PF}/I_{DF}. However, the correct value of Φ_T (assumed temperature independent) is required for a linear least-squares fit. The shape of the plot is a very sensitive function of Φ_T, not being, in general, a straight line. Variation of this parameter in the search for maximum linearity yields its best value and, simultaneously, ΔE_{ST}. Application of this method to fullerene C_{70} in Zeonex [4] (Figure 7.7) gives $A\tau_P^0 = 3.4 \times 10^7$, $\Phi_T = 0.995$, and $\Delta E_{ST} = 34$ kJ mol⁻¹ (0.35 eV). This type of plot has been frequently used in the OLED field [24].

To estimate a minimum value for Φ_T in a simple way, Eq. (7.28) is rewritten as [13]

$$\Phi_T = \frac{1}{1 + \frac{I_{PF}}{I_{DF}}} \frac{1}{\Phi_S} \tag{7.68}$$

and because Φ_S is smaller than or equal to unity, a lower bound for Φ_T is

$$\Phi_T^{min} = \frac{1}{1 + \frac{I_{PF}}{I_{DF}}} \tag{7.69}$$

Φ_T^{min} Z is closest to the fitted value of Φ_T, the closest Φ_S is to one, and the highest is the temperature. In the above example, taking the I_{DF}/I_{PF} ratio for the highest

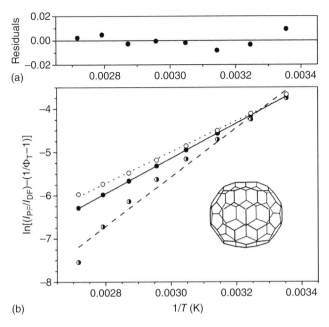

Figure 7.7 Linearized plot according to Eq. (7.67), for 7.5 × 10⁻³ M fullerene C_{70} in Zeonex [4], between 25 °C and 95 °C. Emission wavelength: 700 nm. The best straight line (filled circles) is obtained for $\Phi_T = 0.995(3)$ ($r^2 = 0.999$). The sensitivity with respect to Φ_T is demonstrated by the upper ($\Phi_T = 0.996$, $r^2 = 0.998$) and lower curves ($\Phi_T = 0.994$, $r^2 = 0.978$).

temperature measured (95 °C), $\Phi_T^{\min} = 0.994$ is obtained, already very close to the fitted value.

Alternatively (or subsequently), a nonlinear fitting can be performed [40]. Rewriting again Eq. (7.28):

$$\boxed{\frac{I_{DF}}{I_{PF}} = \left(a + be^{\frac{c}{T}}\right)^{-1}} \qquad (7.70)$$

with

$$a = \frac{1}{\Phi_T} - 1 \qquad (7.71)$$

$$b = \frac{1}{A\tau_P^0 \Phi_T} \qquad (7.72)$$

$$c = \frac{\Delta E_{ST}}{k_B} \qquad (7.73)$$

A fitting to the data shown in Figure 7.7 is displayed in Figure 7.8 and gives the same photophysical parameters.

248 | *7 TADF Kinetics and Data Analysis in Photoluminescence and in Electroluminescence*

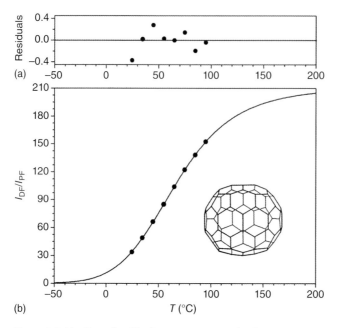

Figure 7.8 Nonlinear fit of I_{DF}/I_{PF} vs temperature for C_{70} in Zeonex, according to Eq. (7.70).

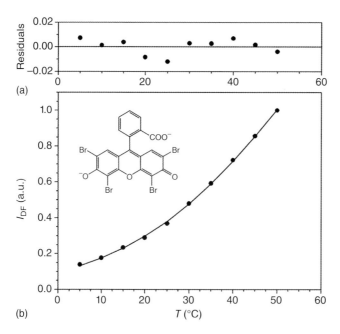

Figure 7.9 Nonlinear fit of I_{DF} vs temperature for eosin in glycerol, according to Eq. (7.74).

7.2.2.3 Delayed Fluorescence Intensity as a Function of Temperature

In case PF is too weak but assumed to be independent of temperature, Eq. (7.70) can be slightly modified to read

$$I_{DF} = \left(\alpha + \beta e^{\frac{c}{T}}\right)^{-1} \tag{7.74}$$

where $\beta/\alpha = b/a$ and ΔE_{ST} can still be obtained from parameter c. An example, eosin in glycerol, is shown in Figure 7.9. The fitting gives 40 kJ mol^{-1} (0.41 eV), in agreement with the value obtained using the Rosenberg–Parker method (see Section 7.2.2.1).

7.2.3 Decay Data

Measurement of the PF decay (fast decay) gives τ_F. Measurement of the DF decay (slow decay) provides the DF lifetime, τ_{DF} (Eq. (7.17)), rewritten as

$$\boxed{\frac{1}{\tau_{DF}} = \frac{1}{\tau_P^0} + B\exp\left(-\frac{\Delta E_{ST}}{k_B T}\right)} \tag{7.75}$$

where $B = (1 - \Phi_T)A$. In the case of TADF, this is also the phosphorescence decay time, as discussed in Section 7.1.3.2.

Determination of τ_{DF} versus temperature allows obtaining ΔE_{ST} and also to estimate τ_P^0 (if assumed to be temperature independent in a relatively narrow range). This was the first method used in TADF analysis, going back to Lewis, Lipkin, and Magel [41], in the approximate form

$$\frac{1}{\tau_{DF}} \simeq C \exp\left(-\frac{\Delta E_{ST}}{k_B T}\right) \tag{7.76}$$

where C is a constant.

Determination of all three parameters in Eq. (7.75) for a limited temperature range is difficult, owing to parameter correlation. It is preferable to set ΔE_{ST} at the steady-state value (Eq. (7.67) or (7.70)) and then carry out the fitting for the remaining two parameters. An example of this procedure is shown in Figure 7.10. It refers to C_{70} in Zeonex in the temperature range 30–95 °C. Using the steady-state values (see Section 7.2.2.2) $\Delta E_{ST} = 34$ kJ mol^{-1} (0.35 eV) and $\Phi_T = 0.995$, the following values are obtained: $A = 8 \times 10^8$ s^{-1} and $\tau_P^0 = 30$ ms. In studies covering more extended temperature ranges, parameter correlation is no longer a problem [4]. An approximate equation that has also been used to analyze TADF decay data [4c, 33] is based on the assumption of fast thermal equilibrium [14, 33] between S$_1$ and T$_1$ and can be written as

$$\tau_{DF} = \frac{3 + \exp\left(-\frac{\Delta E_{ST}}{k_B T}\right)}{\frac{3}{\tau_P^0} + \frac{1-\Phi_T}{\tau_F}\exp\left(-\frac{\Delta E_{ST}}{k_B T}\right)} \tag{7.77}$$

This equation reduces to Eq. (7.15) when Φ_T is close to 1 and to Eq. (7.17) when, additionally, $\Delta E_{ST} \gg k_B T$ (as is usually the case).

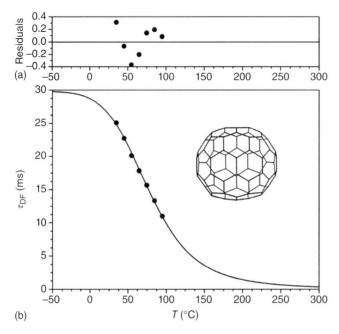

Figure 7.10 Delayed fluorescence lifetime vs temperature for C_{70} in Zeonex and respective fitting with Eq. (7.75).

7.2.4 Combined Steady-state and Decay Data

7.2.4.1 Linear Relation Between Delayed Fluorescence Lifetime and Intensity Ratio

Elimination of k_{ISC}^T from Eqs. (7.17) and (7.28) leads to [9]

$$\tau_{DF} = \tau_P^0 - \tau_P^0 \left(\frac{1}{\Phi_T} - 1 \right) \left(\frac{I_{DF}}{I_{PF}} \right) \tag{7.78}$$

This relation allows the determination of τ_P^0 and of Φ_T from a linear plot of τ_{DF} versus I_{DF}/I_{PF}, assuming that τ_P^0 is constant in the temperature range in question. An example is shown in Figure 7.11, C_{70} in Zeonex. The obtained photophysical parameters are $\tau_P^0 = 32$ ms and $\Phi_T = 0.995$.

7.2.4.2 Linearized Relation for the Determination of ΔE_{ST}

Equation (7.28), written as

$$\frac{I_{DF}}{I_{PF}} = \frac{\Phi_{DF}}{\Phi_{PF}} = \Phi_T k_{ISC}^T \tau_{DF} \tag{7.79}$$

was probably first obtained by Callis et al. [5]. It has been frequently used (including in the OLED field, e.g. [20]) to obtain k_{ISC}^T from the remaining parameters:

$$k_{ISC}^T = \frac{(\Phi_{DF}/\Phi_{PF})}{\Phi_T \tau_{DF}} \tag{7.80}$$

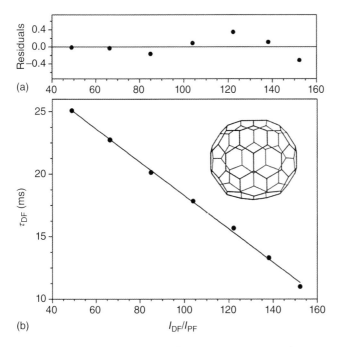

Figure 7.11 Delayed fluorescence lifetime vs the I_{DF}/I_{PF} ratio for C_{70} in Zeonex, in the temperature range 25–95 °C, and respective fitting with Eq. (7.78).

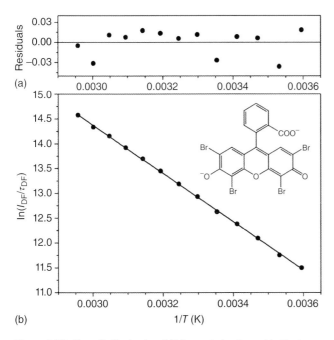

Figure 7.12 Plot of $\ln(I_{DF}/\tau_{DF})$ vs $1/T$ for eosin in glycerol in the temperature range 5–60 °C and respective fitting with Eq. (7.82).

With data obtained at several temperatures, Eq. (7.79) allows the determination of ΔE_{ST} [16]. Using Eq. (7.2), Eq. (7.80) becomes

$$\boxed{\ln\left(\frac{I_{DF}/I_{PF}}{\tau_{DF}}\right) = \ln(A\Phi_T) - \frac{\Delta E_{ST}}{k_B T}} \quad (7.81)$$

Again, if I_{PF} is not available but can be assumed to be constant, Eq. (7.80) gives

$$\ln\left(\frac{I_{DF}}{\tau_{DF}}\right) = C - \frac{\Delta E_{ST}}{k_B T} \quad (7.82)$$

where C is a constant. An example of this plot, for eosin in glycerol, is shown in Figure 7.12, for which $\Delta E_{ST} = 39$ kJ mol^{-1} (0.40 eV) is obtained.

7.3 Conclusion

The study of TADF started almost a century ago, when the fluorescence and phosphorescence mechanisms were still unclear [42]. After significant fundamental work, establishing the nature and relevance of TADF (see refs. [19], [41], and references therein), applications in the field of temperature [40] and trace oxygen [43, 44] sensing appeared. More recently, third-generation OLEDs relying on TADF were proposed [15, 45] and are under active development, with an already vast and rapidly growing literature.

In this chapter, a general view of the kinetics of TADF was presented, stressing the difference between photoluminescence and electroluminescence and discussing optimal conditions for both situations, from the point of view of photophysics. The methods of TADF analysis used for the determination of several photophysical parameters were also described, with examples given for each case.

Acknowledgment

This project was carried out within projects RECI/CTM-POL/0342/2012 (FCT, Portugal) and FAPESP 2017/2014 (FCT, Portugal).

References

1 Kono, H., Lin, S.H., and Schlag, E.W. (1988). On the role of low-frequency modes in the energy or temperature dependence of intersystem crossing. *Chem. Phys. Lett.* 145: 280–285.

2 Fukumura, H., Kikuchi, K., Koike, K., and Kokubun, H. (1988). Temperature effect on inverse intersystem crossing of anthracenes. *J. Photochem. Photobiol. A* 42: 283–291.

3 Tanaka, F., Okamoto, M., and Hirayama, S. (1995). Pressure and temperature dependences of the rate constant for S_1-T_2 intersystem crossing of anthracene compounds in solution. *J. Phys. Chem.* 99: 525–530.

4 (a) McGlynn, S.P., Azumi, T., and Kinoshita, M. (1969). *Molecular Spectroscopy of the Triplet State*. Englewood Cliffs: Prentice-Hall. (b) Strasser, J., Homeier, H.H.H., and Yersin, H. (2000). Role of zero-field-splitting for transition metal compounds without magnetic fields. *Chem. Phys.* 255: 301–316. (c) Palmeira, T., Fedorov, A., and Berberan-Santos, M.N. (2014). Temperature dependence of the phosphorescence and of the thermally activated delayed fluorescence of $^{12}C_{70}$ and $^{13}C_{70}$ in amorphous polymer matrices. Is a second triplet involved? *Methods Appl. Fluoresc.* 2: 035002.

5 Callis, J.B., Gouterman, M., Jones, Y.M., and Henderson, B.H. (1971). Porphyrins XXII: fast fluorescence, delayed fluorescence, and quasiline structure in palladium and platinum complexes. *J. Mol. Spectrosc.* 39: 410–420.

6 Jones, P.F. and Calloway, A.R. (1971). Temperature effects on the intramolecular decay of the lowest triplet state of benzophenone. *Chem. Phys. Lett.* 10: 438–443.

7 (a) Yersin, H., Rausch, A.F., Czerwieniec, R., Hofbeck, T., and Fischer, T. (2011). The triplet state of organo-transition metal compounds. Triplet harvesting and singlet harvesting for efficient OLEDs. *Coord. Chem. Rev.* 255: 2622–2652. (b) Czerwieniec, R., Leitl, M.J., Homeier, H.H.H., and Yersin, H. (2016). Cu(I) complexes – thermally activated delayed fluorescence. Photophysical approach and material design. *Coord. Chem. Rev.* 325: 2–28.

8 Uoyama, H., Goushi, K., Shizu, K., Nomura, H., and Adachi, C. (2012). Highly efficient organic light-emitting diodes from delayed fluorescence. *Nature* 492: 234–238.

9 Baleizão, C. and Berberan-Santos, M.N. (2007). Thermally activated delayed fluorescence as a cycling process between excited singlet and triplet states. Application to the fullerenes. *J. Chem. Phys.* 126: 204510.

10 Birks, J.B. (1970). *Photophysics of Aromatic Molecules*. London: Wiley.

11 Martelo, L., Fedorov, A., and Berberan-Santos, M.N. (2015). Phasor representation of monomer-excimer kinetics: general results and application to pyrene. *J. Phys. Chem. B* 119: 15023–15029.

12 Jovin, T.M., Bartholdi, M., Vaz, W.L.C., and Austin, R.H. (1981). Rotational diffusion of biological molecules by time-resolved delayed luminescence (phosphorescence, fluorescence) anisotropy. *Ann. N.Y. Acad. Sci.* 366: 176–196.

13 Berberan-Santos, M.N. and Garcia, J.M.M. (1996). Unusually strong delayed fluorescence of C_{70}. *J. Am. Chem. Soc.* 118: 9391–9394.

14 Rae, M. and Berberan-Santos, M.N. (2002). Pre-equilibrium approximation in chemical and photophysical kinetics. *Chem. Phys.* 280: 283–293.

15 Endo, A., Ogasawara, M., Takahashi, A., Yokoyama, D., Kato, Y., and Adachi, C. (2009). Thermally activated delayed fluorescence from Sn^{4+} – porphyrin complexes and their application to organic light-emitting diodes – A novel mechanism for electroluminescence. *Adv. Mater.* 21: 4802–4806.

16 Baleizão, C. and Berberan-Santos, M.N. (2011). The brightest fullerene. A new isotope effect in molecular fluorescence and phosphorescence. *ChemPhysChem* 12: 1247–1250.

17 Rosenberg, J.L. and Shombert, D.J. (1960). The phosphorescence of adsorbed acriflavine. *J. Phys. Chem.* 82: 3252–3257.

18 Parker, C.A. and Hatchard, C.G. (1961). Triplet-singlet emission in fluid solutions. Phosphorescence of eosin. *Trans. Faraday Soc.* 57: 1894–1904.
19 Parker, C.A. (1968). *Photoluminescence of Solutions*. Amsterdam: Elsevier.
20 Goushi, K., Yoshida, K., Sato, K., and Adachi, C. (2012). Organic light-emitting diodes employing efficient reverse intersystem crossing for triplet-to-singlet state conversion. *Nature Photonics* 6: 253–258.
21 Baleizão, C. and Berberan-Santos, M.N. (2009). How fast is a fast equilibrium? A new view of reversible reactions. *ChemPhysChem* 10: 199–205.
22 Berberan-Santos, M.N. (2010). Extending the convolution method: a general integral formalism for chemical kinetics. Application to enzymatic reactions. *MATCH: Commun. Math. Comput. Chem.* 63: 603–622.
23 Palmeira, T. and Berberan-Santos, M.N. (2017). Kinetic criteria for optimal thermally activated delayed fluorescence in photoluminescence and in electroluminescence. *J. Phys. Chem. C* 121: 701–708.
24 Adachi, C. (2014). Third-generation organic electroluminescence materials. *Jpn. J. Appl. Phys.* 53: 060101.
25 Shizu, K., Lee, J., Tanaka, H., Nomura, H., Yasuda, T., Kaji, H., and Adachi, C. (2015). Highly efficient electroluminescence from purely organic donor–acceptor systems. *Pure Appl. Chem.* 87: 627–638.
26 Cai, X., Li, X., Xie, G., He, Z., Gao, K., Liu, K., Chen, D., Cao, Y., and Su, S. (2016). "Rate-limited effect" of reverse intersystem crossing process: the key for tuning thermally activated delayed fluorescence lifetime and efficiency roll-off of organic light emitting diodes. *Chem. Sci.* 7: 4264–4275.
27 Hirata, S., Sakai, Y., Masui, K., Tanaka, H., Lee, S.Y., Nomura, H., Nakamura, N., Yasumatsu, M., Nakanotani, H., Zhang, Q., Shizu, K., Miyazaki, H., and Adachi, C. (2015). Highly efficient blue electroluminescence based on thermally activated delayed fluorescence. *Nature Materials* 14: 330–336.
28 Taneda, M., Shizu, K., Tanaka, H., and Adachi, C. (2015). High efficiency thermally activated delayed fluorescence based on 1,3,5-tris(4-(diphenylamino)phenyl)-2,4,6-tricyanobenzene. *Chem. Comm.* 51: 5028–5031.
29 Shizu, K., Noda, H., Tanaka, H., Taneda, M., Uejima, M., Sato, T., Tanaka, K., Kaji, H., and Adachi, C. (2015). Highly efficient blue electroluminescence using delayed-fluorescence emitters with large overlap density between luminescent and ground states. *J. Phys. Chem. C* 119: 26283–26289.
30 Dias, F.B., Bourdakos, K.N., Jankus, V., Moss, K.C., Kamtekar, K.T., Bhalla, V., Santos, J., Bryce, M.R., and Monkman, A.P. (2013). Triplet harvesting with 100% efficiency by way of thermally activated delayed fluorescence in charge transfer OLED emitters. *Adv. Mater.* 25: 3707–3714.
31 Tao, Y., Yuan, K., Chen, T., Xu, P., Li, H., Chen, R., Zheng, C., Zhang, L., and Huang, W. (2014). Thermally activated delayed fluorescence materials towards the breakthrough of organoelectronics. *Adv. Mater.* 26: 7931–7958.
32 Dias, F.B., Santos, J., Graves, D.R., Data, P., Nobuyasu, R.S., Fox, M.A., Batsanov, A.S., Palmeira, T., Berberan-Santos, M.N., Bryce, M.R., and Monkman, A.P. (2016). The role of local triplet excited states and D-A relative orientation in thermally activated delayed fluorescence: photophysics and devices. *Adv. Sci.* 1600080.

33 Bergmann, L., Zink, D.M., Bräse, S., Baumann, T., and Volz, D. (2016). Metal–organic and organic TADF-materials: status, challenges and characterization. *Top. Curr. Chem (Z)* 374 (22).
34 Levy, D. and Avnir, D. (1991). Room temperature phosphorescence and delayed fluorescence of organic molecules trapped in silica sol–gel glasses. *J. Photochem. Photobiol. A* 57: 41–63.
35 Duchowicz, R., Ferrer, M.L., and Acuña, A.U. (1998). Kinetic spectroscopy of erythrosine phosphorescence and delayed fluorescence in aqueous solution at room temperature. *Photochem. Photobiol.* 68: 494–501.
36 Carlson, S.A. and Hercules, D.M. (1971). Delayed thermal fluorescence of anthraquinone in solutions. *J. Am. Chem. Soc.* 93: 5611–5616.
37 Maciejewski, A., Szymanski, M., and Steer, R.P. (1986). Thermally activated delayed S_1 fluorescence of aromatic thiones. *J. Phys. Chem.* 90: 6314–6318.
38 Kropp, J.L. and Dawson, W.R. (1967). Radiationless deactivation of triplet coronene in plastics. *J. Phys. Chem.* 71: 4499–4506.
39 Salazar, F.A., Fedorov, A., and Berberan-Santos, M.N. (1997). A study of thermally activated delayed fluorescence in C_{60}. *Chem. Phys. Lett.* 271: 361–366.
40 Baleizão, C., Nagl, S., Borisov, S.M., Schäferling, M., Wolfbeis, O.S., and Berberan-Santos, M.N. (2007). An optical thermometer based on the delayed fluorescence of C_{70}. *Chem. Eur. J.* 13: 3643–3651.
41 Lewis, G.N., Lipkin, D., and Magel, T.T. (1941). Reversible photochemical processes in rigid media. A study of the phosphorescent state. *J. Am. Chem. Soc.* 63: 3005–3018.
42 Valeur, B. and Berberan-Santos, M.N. (2012). *Molecular Fluorescence. Principles and Applications*, 2e. Weinheim: Wiley-VCH.
43 Nagl, S., Baleizão, C., Borisov, S.M., Schäferling, M., Berberan-Santos, M.N., and Wolfbeis, O.S. (2007). Optical sensing and imaging of trace oxygen with record response. *Angew. Chem. Int. Ed.* 46: 2317–2319.
44 Kochmann, S., Baleizão, C., Berberan-Santos, M.N., and Wolfbeis, O.S. (2013). Sensing and imaging of oxygen with ppb limits of detection and based on the quenching of the delayed fluorescence of $^{13}C_{70}$ fullerene in polymer hosts. *Anal. Chem.* 85: 1300–1304.
45 H. Yersin, U. Monkowius, DE 10 2008 033 563 A1.

8

Intersystem Crossing Processes in TADF Emitters

Christel M. Marian[1], *Jelena Föller*[1], *Martin Kleinschmidt*[1], *and Mihajlo Etinski*[2]

[1] *Heinrich-Heine-University Düsseldorf, Institute of Theoretical and Computational Chemistry, Universitätsstr. 1, 40225 Düsseldorf, Germany*
[2] *University of Belgrade, Faculty of Physical Chemistry, Studentski Trg 12-16, 11000 Belgrade, Serbia*

8.1 Introduction

This chapter gives a brief overview over quantum chemical methods for computing rate constants of radiative and nonradiative molecular excited-state processes and summarizes our recent theoretical research on the photophysics of thermally activated delayed fluorescence (TADF) emitters.

8.1.1 Electroluminescent Emitters

In the first organic light-emitting diodes (OLEDs), the electroluminescence of fluorescent dyes such as 8-hydroxyquinoline aluminum (Alq_3) was exploited [1]. Dyes of this first generation are highly fluorescent but possess slow intersystem crossing (ISC) and negligible phosphorescence rates. Therefore only the singlet excitons, that means, only about 25% of the generated excitons, can be harvested [2]. The limited internal quantum efficiency appears to be their greatest disadvantage. Advantages are clear colors due to narrow emission bands and good operational stability due to fast radiative decay of the dyes (nanosecond regime) [3, 4].

The second generation of small-molecule OLEDs employs phosphorescent dopants instead of fluorescent ones. Typically, the emitters are organometallic complexes with Ir or Pt cores [5–9]. Their excited singlet states undergo fast ISC to the lowest triplet state. Thus, in addition to the triplet excitons, the singlet excitons can be harvested in phosphorescent organic light-emitting diodes (PHOLEDs), leading to an internal quantum yield of up to 100% [2, 10]. The main limitation of the phosphorescent dyes is their comparatively long radiative lifetime (microsecond regime) that leads to undesirable side effects, namely, quenching processes and bleaching reactions. The latter are particularly pronounced for blue PHOLED emitters. To avoid nonradiative decay via low-lying nonemissive metal-centered states, complexes with strongly σ donating and π accepting ligands were devised [11–13], but there is still room for improvement. Because of the very limited operational stability of blue PHOLEDs, hybrid fluorescent–PHOLEDs were employed for generating white light incorporating fluorescent blue and phosphorescent green to red emitters in one device [14].

The third generation of OLED emitters comprises organic donor–acceptor systems [4] as well as transition metal (TM) complexes [15–17] with small singlet–triplet energy gap ΔE_{ST} that lies within the range of thermal energy. Because of the small ΔE_{ST}, reverse intersystem crossing (rISC) from the lowest triplet to the lowest singlet is reasonably fast, and therefore TADF, also called E-type delayed fluorescence (DF) in the older literature [18], is possible in addition to direct fluorescence. Like PHOLEDs, TADF-based OLEDs show an internal quantum efficiency of up to 100% [19]. Furthermore, cheaper first-row TMs such as Cu instead of Ir or Pt can be used. A disadvantage of many presently available TADF OLEDs that they share with the PHOLEDs is the rather low intrinsic radiative transition rate ($k_r = 10^6$–10^7 s^{-1}) of the emitters that makes them sensitive to nonradiative decay processes such as triplet–triplet annihilation. Also, the emission from states with charge-transfer (CT) character typically is rather broad, which is not favorable for application in displays [4].

The latest class of OLEDs aims to combine high internal quantum efficiency and long operational stability by using assistant dopants for the harvesting of triplet- and singlet excited states in addition to fluorescence emitters. The assistant dopant transfers its excitation energy nonradiatively to the fluorescent acceptor by Förster resonant energy transfer (FRET). If the donor is sufficiently phosphorescent, it is even possible to induce FRET from triplet to singlet states or vice versa [20, 21]. This mechanism was exploited by Baldo et al. [22] for improving the efficiency of red fluorescence in OLEDs by using the green phosphor Ir(ppy)$_3$ as sensitizer. Fukagawa et al. [23] utilized singlet-to-triplet FRET from TADF assistant dopants to phosphorescent Ir and Pt complexes. In this way, the amount of phosphorescent emitter could be greatly reduced. Adachi and coworkers combined purely organic, sublimable TADF assistant dopants and fluorescence emitters in one layer, thus uniting the advantages of both [24].

8.1.2 Thermally Activated Delayed Fluorescence

TADF is looked upon as a significant emerging technology for generating highly performant electroluminescent devices for displays and lighting systems [25]. Despite the fact that TADF has been shown to give highly efficient OLEDs, the underlying mechanisms are still not clearly understood. Ideally, thermally stable dyes with small singlet–triplet energy gap, substantial S_1–T_1 (r)ISC, high fluorescence but minimal nonradiative decay to the electronic ground state are required. However, these conditions for efficient TADF emission are not easily met simultaneously. In addition to the intrinsic emitter properties, emitter–host interactions play an essential role for the luminescence properties of a device [23, 26].

The energy difference between a singlet- and triplet-coupled open-shell configuration depends on the exchange interaction of the unpaired electrons. This interaction is small when the density distributions of the orbitals involved in the excitation do not overlap substantially. Typically, this requirement is fulfilled by CT states where the unpaired electrons are far apart. Very small singlet–triplet splittings can also be achieved in molecular systems where the electron clouds in the half-occupied orbitals are not strongly displaced with respect to each other, but where their electron density distributions peak at different atoms and hence are disjunct [27, 28]. Such a situation occurs, for example, in nonalternant hydrocarbons with azulene as a well-known representative. Unfortunately, the overlap of orbital densities between the initial and final states plays also a decisive role for the magnitude of the electronic spin–orbit coupling (SOC) and for the fluorescence rate. Electronic SOC – a further prerequisite for efficient (r)ISC – is a fairly short-ranged interaction. Furthermore, SOC between singlet and triplet configurations with equal occupation of the spatial orbitals vanishes for symmetry reasons. As a consequence, SOC is in general very weak between singlet and triplet CT states. Owing to the near-degeneracy of d orbitals with different magnetic moments, the situation might be more favorable in TM complexes with metal-to-ligand charge-transfer (MLCT) excited states. The interplay of all the factors influencing the probability of TADF is not yet fully understood and needs further investigation. It seems to be clear, however, that a small singlet–triplet energy gap alone is not sufficient for enabling efficient TADF.

8.2 Intersystem Crossing Rate Constants

ISC is a nonradiative transition between states of different electronic spin multiplicity. Hence, a spin-dependent interaction operator is required to mediate the transition. In most cases, electronic SOC will dominate the interaction, but in cases in which this interaction is very weak, electronic spin–spin coupling (SSC) might come into play.

Assuming the coupling of the initial and final states to be small compared with their energy difference (which will be the case in typical TADF emitters), the ISC rate can be evaluated in the framework of perturbation theory (*Fermi's golden rule*). The rate constant for an ISC from a manifold of thermally populated initial vibronic states $|\Psi_a, \{v_{aj}\}\rangle$ to a quasi-continuum of final vibronic states $\langle\Psi_b, \{v_{bk}\}|$, caused by spin–orbit interaction, is then given by

$$k_{\text{ISC}} = \frac{2\pi}{\hbar Z} \sum_{j,k} e^{-\beta E_j} |\langle\Psi_b, \{v_{bk}\} | \hat{H}_{\text{SO}} | \Psi_a, \{v_{aj}\}\rangle|^2 \, \delta(E_{aj} - E_{bk}) \qquad (8.1)$$

where $Z = \sum_j e^{-\beta E_j}$ is a canonical partition function for vibrational motion in the initial electronic state, β is the inverse temperature, and E_j is the energy of the vibrational level in the initial electronic state.

In many articles relating to TADF emitters, the ISC from a singlet to a triplet state colloquially is called a downhill process, whereas the reverse transition, rISC, is called an uphill process. As may be seen from the delta distribution in Eq. (8.1), the energy is strictly conserved during the nonradiative transition, i.e. the initial and final states are isoenergetic. What people have in mind when speaking of downhill and uphill processes is the difference between the adiabatic energies of the initial and final electronic states, possibly including zero-point vibrational energy corrections. If that energy difference is positive, the transition is dubbed a downhill process and may occur at any temperature. If that energy difference is negative, thermal energy is required in addition to bridge the gap.

The efficiency of ISC and rISC is controlled by several factors. Intrinsically molecular factors are the magnitude of the spin–orbit coupling matrix element (SOCME), the adiabatic energy difference, and the coordinate displacement of the singlet and triplet potential energy surfaces as well as further factors such as the Duschinsky rotation of the respective vibrational modes. The most important external factor – aside from environment effects – is the temperature.

8.2.1 Condon Approximation

In the Condon approximation, where it is assumed that the electronic and vibrational degrees of freedom can be separated, the ISC rate is given by a product of the electronic and vibrational parts (direct SOC):

$$k_{ISC}^{dir} = \frac{2\pi}{\hbar Z} |\langle \Psi_b | \hat{H}_{SO} | \Psi_a \rangle|^2_{\mathbf{q}_0} \sum_{j,k} e^{-\beta E_j} |\langle \{v_{bk}\} | \{v_{aj}\} \rangle|^2 \, \delta(E_{aj} - E_{bk}) \quad (8.2)$$

In principle, the origin of the Taylor expansion, $\mathbf{q}_0$, can be chosen at will. It is common practice, however, to choose the minimum geometry of the initial state to determine the electronic SOCME for ISC.

In first-order perturbation theory, each Cartesian component of the spin–orbit Hamiltonian couples the singlet state to one and only one Cartesian triplet sublevel [29]. For this reason, phase factors do not play any role in the calculation of the total ISC rate from a given singlet state to all triplet sublevels in Condon approximation. Furthermore, a common set of vibrational wave functions is chosen for all triplet fine-structure levels. Hence, the squared contributions from all three components can just be summed up yielding

$$k_{ISC}^{dir}(S \rightarrow T) = \frac{2\pi}{\hbar Z} \left(\sum_{\kappa} |\langle T_b^{\kappa} | \hat{H}_{SO} | S_a \rangle|^2_{\mathbf{q}_0} \right)$$
$$\times \sum_{j,k} e^{-\beta E_j} |\langle \{v_{bk}\} | \{v_{aj}\} \rangle|^2 \, \delta(E_{aj} - E_{bk}) \quad (8.3)$$

The situation is slightly more complicated for the reverse transition from a triplet to a singlet state. In general, the fine-structure levels of a triplet state are separated by a zero-field splitting (ZFS). If the ZFS is large in relation to the temperature, individual rISC rate constants would have to be determined for every fine-structure level. Fortunately, ZFSs of TADF emitters are typically very small ($\ll 10$ cm^{-1}) compared with thermal energies (298 K $\simeq$ 207cm^{-1}) so that the rISC rate constants can be averaged. Hence, in first-order perturbation theory, the total rate constant of rISC for a molecule in the triplet state is given by

$$k_{\text{rISC}}^{\text{dir}}(\text{T} \to \text{S}) = \frac{2\pi}{3\hbar Z} \left(\sum_\kappa |\langle S_b | \hat{\mathcal{H}}_{\text{SO}} | T_a^\kappa \rangle|^2_{q_0} \right)$$
$$\times \sum_{j,k} e^{-\beta E_j} |\langle \{v_{bk}\} | \{v_{aj}\} \rangle|^2 \, \delta(E_{aj} - E_{bk}) \tag{8.4}$$

where the factor of 3 in the denominator of Eq. (8.4) takes care of the degeneracy of the triplet sublevels.

8.2.1.1 Electronic Spin–Orbit Coupling Matrix Elements

Microscopic spin–orbit Hamiltonians contain vector products between the electronic momentum and the derivatives of the one- and two-electron Coulomb potentials [30, 31]. Because these derivatives drop off like $1/r^3$, SOC is a fairly short-ranged interaction. Denoting the operator for the angular momentum of electron i with respect to nucleus K by $\vec{\ell}_{iK}$ and the corresponding operator for the angular momentum of electron i with respect to electron j by $\vec{\ell}_{ij}$, the Breit–Pauli spin–orbit Hamiltonian is given by

$$\hat{\mathcal{H}}_{\text{SO}}^{\text{BP}} = \frac{1}{4}\alpha^2 g_e \sum_i \left\{ \sum_K \frac{Z_K}{\hat{r}_{iK}^3} \vec{\ell}_{iK} \cdot \hat{\vec{s}}_i - \sum_{j \neq i} \frac{1}{\hat{r}_{ij}^3} \vec{\ell}_{ij} \cdot (\hat{\vec{s}}_i + 2\hat{\vec{s}}_j) \right\} \tag{8.5}$$

Herein, g_e is the gyromagnetic factor of the electron and α is the fine-structure constant. The two-electron terms of the spin–orbit Hamiltonian contribute roughly 50% to the SOCME in molecules composed of light elements and can therefore not be neglected. They can, however, be combined in good approximation with the true one-electron terms to form an effective one-electron mean-field operator [32]. Whether the mean-field approximation is sufficiently accurate to compute the typically very small SOCMEs of purely organic TADF emitters is not clear at present.

The El-Sayed rules state that ISC in organic molecules is fast between excited states of different orbital types such as $^1(n\pi^*) \rightsquigarrow {}^3(\pi\pi^*)$, whereas ISC between states of the same orbital type such as $^1(\pi'\pi^*) \rightsquigarrow {}^3(\pi\pi^*)$ is slow [33]. These qualitative rules are easily understood. To this end, we consider the one-electron spin–orbit Hamiltonian as a compound tensor operator of rank 0:

$$a_{\text{SO}} \vec{\ell} \cdot \hat{\vec{s}} = a_{\text{SO}}(\hat{\ell}_0 \hat{s}_0 - \hat{\ell}_{+1} \hat{s}_{-1} - \hat{\ell}_{-1} \hat{s}_{+1}) \tag{8.6}$$

where a_{SO} is a system-specific parameter and the subindices $0, \pm 1$ denote the tensor components of the spatial and spin angular momentum operators, respectively. Like the more familiar ladder operators, these tensor operators can shift the magnetic quantum numbers of electrons. (See Ref. [30] for more details.) Consider $\langle {}^1(n\pi^*)|\hat{\mathcal{H}}_{\text{SO}}|^3(\pi\pi^*)\rangle$, for example. The two states are related by a single excitation from π to n. While $\hat{\ell}_{-1}$ can be used to transform an out-of-plane p_π orbital to an in-plane n orbital, $\hat{s}_{+1}$ shifts the spin state of the electron from β to α. The spin–orbit operator in Eq. (8.6) does not contain any combination that changes the spin magnetic quantum number of the electron, but leaves its spatial angular momentum quantum number untouched. For that reason, $\langle{}^1(\pi'\pi^*)|\hat{H}_{SO}|^3(\pi\pi^*)\rangle$ will be small. Moreover, in spatially nondegenerate states (as is the case in organic compounds), angular momentum operators do not have diagonal matrix elements because they are purely imaginary. Therefore, the electronic SOCMEs between a singlet and a triplet state with equal spatial orbital occupation vanish. This means in particular that the electronic SOCMEs between two CT states of similar wave function characteristics, $\langle{}^1CT|\hat{H}_{SO}|^3CT\rangle$, which play a prominent role in TADF emitters, will be very small. However, ISC between states of equal orbital type is not always slow. In order to have appreciable SOC between states of the same orbital type, it is necessary to go beyond the Condon approximation (see also Section 8.2.2).

A further obstacle for efficient SOC in TADF emitters is the shortrangedness of the spin–orbit interaction. Because of its r^{-3} dependence, the largest contribution to the SOCME comes from one-center terms. Combining this criterion with the El-Sayed rules, one finds that in Condon approximation appreciable spin–orbit integrals may arise only if the involved molecular orbitals (MOs) exhibit electron densities at the same center and if the atomic orbitals have different magnetic quantum numbers. This excludes the typical pair of ^{1}CT and ^{3}CT states where the unpaired electrons typically are far apart, i.e. their spin densities have little overlap. Consequently, the two-electron exchange integral that largely determines the singlet–triplet energy gap is very small while at the same time also their mutual spin–orbit interaction is tiny. Hypothetically, substantial SOC can be imagined even for CT states, however, namely, if more than two electronic states are involved. Consider, for example, purely organic donor–acceptor systems in which the $(\pi_D\pi_A^*)$ and $(n_D\pi_A^*)$ excitations are energetically near degenerate. Herein, π_D and n_D represent occupied π MO and lone-pair orbitals of the donor, respectively, and π_A^* an unoccupied π MO of the acceptor. Comparing configurations, it is seen that $(\pi_D\pi_A^*)$ and $(n_D\pi_A^*)$ differ from each other by a local $(\pi_D \to n_D)$ replacement at the donor that might in turn yield large SOC. Likewise, in MLCT excited states of Cu(I) complexes, a $(d_\pi\pi^*)$ state might be located energetically close to a $(d_\sigma\pi^*)$ and could make use of the large SOC in the 3d shell.

8.2.1.2 Overlap of Vibrational Wave Functions

When deriving qualitative rules for probabilities of radiationless transitions in large molecules, Jortner and coworkers [34, 35] differentiated between two major cases: the weak and the strong coupling cases (Figure 8.1).

In the weak coupling case (Figure 8.1a), the coordinate displacement for each normal mode is assumed to be relatively small. In this case, the transition probability depends exponentially on the adiabatic energy difference ΔE, i.e. the smaller the energy gap, the larger the transition probability [34]. This relation is commonly called the *energy gap law*. People tend to forget, however, that this qualitative rule applies only for a pair of nested states.

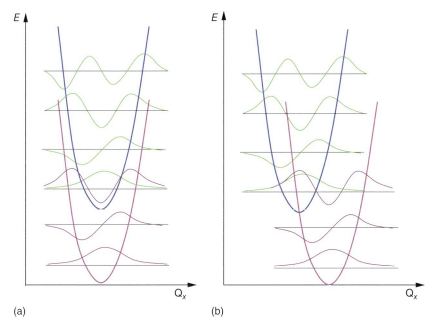

Figure 8.1 Schematic representation of the vibrational overlaps in the (a) weak and (b) strong coupling cases of nonradiative transitions. (a) Nested harmonic oscillators. (b) Displaced harmonic oscillators.

The strong coupling case (Figure 8.1b) is characterized by large relative displacements in some coordinates so that an intersection of the potential energy surfaces can be expected. The probability of the radiationless transition then exhibits a Gaussian dependence on the energy parameter $\Delta E - E_M$ where E_M is the molecular rearrangement energy that corresponds to half the Stokes shift for the two electronic states under consideration [34]. Taking, additionally, temperature effects into account by assuming a Boltzmann distribution resulted in a generalized activated rate equation similar to Marcus theory [35, 36]. In agreement with this model case, occasionally an inverse relationship between the transition probability and ΔE is observed, i.e. there exist cases where the transition probability increases with increasing energy gap [37].

8.2.2 Beyond the Condon Approximation

In order to have appreciable SOC between states of the same orbital type, it is necessary to go beyond the Condon approximation. Henry and Siebrand [38] were the first who discussed various contributions of different couplings to the ISC rate. In addition to the so-called direct spin–orbit interaction, they considered spin–orbit interaction induced by Herzberg–Teller vibronic coupling and spin–orbit interaction induced by Born–Oppenheimer vibronic coupling. In practice, the latter two types are difficult to tell apart. Similar to the Herzberg–Teller expansion of the vibronic interaction, the SOC can be expanded in a Taylor series with respect to the nuclear coordinates about an appropriately chosen reference point $\mathbf{q}_0$, for example, the equilibrium geometry of the initial state. It is plausible to use normal modes for nuclear coordinates. Then, up to the linear term in normal mode coordinates, the SOCME is [29]

$$\langle T_b^{\kappa}, \{v_{bk}\} | \hat{H}_{SO} | S_a, \{v_{aj}\}\rangle = \langle T_b^{\kappa} | \hat{H}_{SO} | S_a \rangle|_{\mathbf{q}_0} \langle \{v_{bk}\} | \{v_{aj}\}\rangle$$
$$+ \langle \{v_{bk}\} | \mathbf{b}^{\dagger} \mathbf{Q}_S | \{v_{aj}\}\rangle \qquad (8.7)$$
$$+ \cdots$$

where the vector $\mathbf{Q}_S$ contains normal mode coordinates of the singlet electronic state and $\mathbf{b}^{\dagger}$ is the adjoint of the vector comprising the first-order derivative couplings:

$$b_m = \left. \frac{\partial \langle T_b^{\kappa} | \hat{H}_{SO} | S_a \rangle}{\partial (\mathbf{Q}_S)_m} \right|_{\mathbf{q}_0} \qquad (8.8)$$

If the Taylor expansion is truncated after the linear term, the ISC rate for a singlet–triplet transition is a sum of three contributions due to (i) a direct term k_{ISC}^{dir}, (ii) a mixed direct-vibronic $k_{ISC}^{dir/vib}$, and (iii) a vibronic coupling term k_{ISC}^{vib}, where the direct term is identical to the expression in Eq. (8.3) and the latter two are given by

$$k_{ISC}^{dir/vib,\alpha} = \frac{4\pi}{\hbar Z} \langle T_b^{\kappa} | \hat{H}_{SO} | S_a \rangle|_{\mathbf{q}_0} \sum_{j,k} e^{-\beta E_j} \langle \{v_{bk}\} | \{v_{aj}\}\rangle$$
$$\times \langle \{v_{bk}\} | \mathbf{b}^{\dagger} \mathbf{Q}_S | \{v_{aj}\}\rangle \, \delta(E_{aj} - E_{bk}) \qquad (8.9)$$

$$k_{ISC}^{vib,\alpha} = \frac{2\pi}{\hbar Z} \sum_{j,k} e^{-\beta E_j} |\langle \{v_{bk}\} | \mathbf{b}^{\dagger} \mathbf{Q}_S | \{v_{aj}\}\rangle|^2 \, \delta(E_{aj} - E_{bk}) \qquad (8.10)$$

Note that all SOCMEs between the singlet state and the Cartesian components of the triplet state are purely imaginary. This is of particular importance when computing the mixed direct-vibronic contributions. Again, due to spin symmetry, there are no cross terms between different Cartesian components in first order. Save for a factor of 1/3 that takes account of the degeneracy of the three triplet fine-structure levels, a similar expression is obtained for the rISC starting from the triplet state.

Numerous examples have been found in heteroaromatic molecules where El-Sayed forbidden $^1(\pi\pi^*) \rightsquigarrow {}^3(\pi\pi^*)$ ISC processes have rate constants that are nearly as large as those of El-Sayed allowed transitions [39–45]. In most cases, vibronic interaction with an energetically close-lying $(n\pi^*)$ excited state through out-of-plane molecular vibrations enhances the transition probability. Even if no n-type orbitals are available, as, e.g. in pure hydrocarbons, pyramidalization of unsaturated carbon centers in the excited state can lead to a substantial increase of electronic SOCMEs [46–48].

8.2.3 Computation of ISC and rISC Rate Constants

ISC and rISC rate constants are highly sensitive with respect to the relative location of the singlet and triplet states. It is, therefore, of utter importance to employ reliable electronic structure methods for the computation of the excited-state potentials (see Section 8.3).

8.2.3.1 Classical Approach

In the high-temperature limit, the expression for the singlet–triplet ISC rate constant in Eq. (8.2) reduces to [49]

$$k_{ISC}^{class.} = \frac{\sqrt{\pi}}{\hbar\sqrt{\lambda RT}} |\langle \Psi_b | \hat{H}_{SO} | \Psi_a \rangle|_{\mathbf{q}_0}^2 \exp\left[-\frac{(\Delta E_{ab} - \lambda)^2}{4\lambda RT}\right] \quad (8.11)$$

Herein, ΔE_{ab} is the adiabatic energy difference between the singlet and the triplet state and λ denotes the Marcus reorganization energy that can – to first approximation – be equated with the energy variation in the initial singlet excited state when switching from the singlet equilibrium geometry to the triplet equilibrium geometry [49, 50]. In TADF compounds, the reorganization energy λ can adopt minuscule values because ^{1}CT and ^{3}CT states often exhibit similar equilibrium geometries.

8.2.3.2 Statical Approaches

The method for computing ISC rate constants, applied in our laboratories, is based on the generating function formalism and the multimode harmonic oscillator approximation [51]. Herein, the triplet state mass-weighted normal modes $\mathbf{Q}_T$ with frequencies ω_{T_i} are related to their singlet counterparts by a Duschinsky transformation [52]. The Duschinsky transformation $\mathbf{Q}_T = \mathbf{J}\mathbf{Q}_S + \mathbf{D}$, where $\mathbf{J}$ is the Duschinsky rotation matrix and $\mathbf{D}$ the displacement vector, is particularly important for pairs of states with strongly displaced minimum geometries.

Rates are obtained by numerical integration of the autocorrelation function in the time domain. This approach can even be applied to molecules with a large number of normal modes or to pairs of states that exhibit a large adiabatic energy gap. In these cases, the density of states becomes enormous, and a direct summation over all final vibrational states – even in a small energy interval around the initial state – is prohibitive. The generating function formalism is also applicable to finite-temperature conditions that are essential for uphill processes such as rISC. Herein, a Boltzmann population of vibrational levels in the initial state is assumed [53]. The derivation of the formulas for the direct, mixed direct-vibronic, and vibronic ISC rates in the finite-temperature case can be found in Ref. [29] A similar correlation function approach for computing rates constants of direct and vibronic ISC has been pursued by Shuai and coworkers [43].

8.2.3.3 Dynamical Approaches

Alternatively, nonadiabatic nuclear dynamics methods have been employed for determining the kinetic constants of ISC and rISC processes [54–56]. Herein, a diabatization scheme has been used to avoid the explicit calculation of nonadiabatic coupling matrix elements [57], and wavepacket dynamics simulations have been carried out within the framework of the multiconfiguration time-dependent Hartree (MCTDH) method [58]. We refrain from going into details here because these methods are reviewed in Chapter 9.

8.3 Excitation Energies and Radiative Rate Constants

8.3.1 Time-Dependent Density Functional Theory

It is well known that time-dependent density functional theory (TDDFT) yields substantial errors for the excitation energies of CT states, when approximate standard exchange-correlation functionals are used [59]. The balanced description of CT and locally excited (LE) states remains to be a challenge for TDDFT methods, even if used in conjunction with modern hybrid and range-separated density functionals. Huang et al. [60] systematically correlated calculated and experimental singlet- and triplet-transition energies of 17 CT compounds with the aim to find a recipe for the computational prediction of these quantities. They employed TDDFT using density functionals with varying amount of Hartree–Fock (HF) exchange ranging from 0% (BLYP) to 100% (M06-HF). They propose to determine the optimal percentage of HF exchange semiempirically from a comparison of the calculated vertical S_1 energy with the measured absorption maximum and to use this value for scaling the HF contribution to the exchange-correlation functional. Within the chosen set of molecules, optimal HF contributions between approximately 5% and 40% were found. This semiempirical procedure seems to work well, but it becomes very involved if also LE states play a role. In that case, the authors recommend employing distinct HF contributions for the different types of states.

Moral et al. [61] advocate the use of TDDFT in Tamm–Dancoff approximation (TDA) instead of full linear response TDDFT. They tested the performance of these approaches on a small set of organic molecules with experimentally known singlet–triplet splitting. Among them were three typical host materials with moderately high ΔE_{ST} values (0.5–0.7 eV) as well as three TADF emitters with low ΔE_{ST} values (0.1–0.3 eV). In this series, TDDFT–TDA yields a smaller root-mean-square deviation (RMSD) than TDDFT, leading the authors to conclude that TDA is better suited for computing singlet–triplet splittings. It appears questionable, however, whether the unweighted RMSD really represents a good measure for assessing the performance of different methods on this property. Owing to their significantly larger ΔE_{ST} values, the host materials dominate the error analysis. Looking at the raw data of these authors, a different picture emerges. The PBE0 functional is the only one for which both types of calculations have been carried out. Indeed, TDDFT–TDA reproduces the singlet–triplet splittings of the three host materials to a better extent than TDDFT, whereas TDDFT performs better for the three TADF emitters. If the focus is laid on vertical singlet excitation energies, the performance of TDA is very unsatisfactory. In conjunction with the B2-PLYP functional [62], the excitation energies of TADF materials are underestimated by up to 0.7 eV, whereas those of the host materials are somewhat overestimated. TDDFT–TDA performs better for TADF molecules if the B2GP-PLYP functional [63] is employed instead, but then the excitation energies of the host materials are largely overestimated (by up to 0.5 eV). Further investigations are necessary for a sound judgment because the set is too small for being really representative.

8.3.2 DFT-Based Multireference Configuration Interaction

The combined density functional theory and multireference configuration interaction (DFT/MRCI) method of Grimme and Waletzke is a well-established semiempirical quantum chemical method for efficiently computing excited-state properties [64]. The MRCI expansion is based on MOs from a closed-shell Kohn–Sham DFT calculation employing the BHLYP hybrid functional [65]. In the Hamiltonian, BHLYP orbital energies are utilized to incorporate parts of the dynamical electron correlation. Parameters that scale the Coulomb and exchange integrals and damp off-diagonal matrix elements have been introduced in the Hamiltonian to avoid double counting of the electron correlation. These parameters were fitted to experimental data. Independent benchmark studies on a representative set of organic molecules confirmed that the mean absolute error for DFT/MRCI electronic excitation energies lies below 0.2 eV [66]. A distributed memory parallel code facilitates the calculation of electronic spectra of larger molecules [67]. The average deviation is somewhat larger for first- and second-row TM complexes, but Escudero and Thiel found the DFT/MRCI method to be superior to the tested TDDFT approaches and thus recommended it for exploring the excited-state properties of TM complexes [68]. This even holds true for third-row TM complexes, if spin–orbit interaction is included that cannot be neglected in heavy-element compounds [69, 70]. Furthermore, DFT/MRCI is one of the few electronic structure methods applicable to large systems that gives the correct order of excited states in extended polyenes and polyacenes where doubly excited configurations play an essential role [67, 71].

While the method performs very well in general, it may be problematic when treating the donor–acceptor systems with small orbital density overlap that are key components of metal-free TADF OLEDs. Caution is advised if double excitations with four open shells contribute to the DFT/MRCI wave function with substantial weight or if singlet-coupled CT excitations exhibit lower energies than their triplet counterparts. Recently, an alternative form of correcting the matrix elements of an MRCI Hamiltonian that is built from a Kohn–Sham set of orbitals was devised in our laboratory [72]. The new parameterization is spin invariant and incorporates less empirism compared with the original formulation while preserving its high computational efficiency. The robustness of the original and redesigned Hamiltonians has been tested on experimentally known vertical excitation energies of organic molecules yielding similar statistics for the two parameterizations [72, 73]. Besides that, the new formulation is free from artifacts related to doubly excited states with four open shells, producing qualitatively correct and consistent results for excimers and covalently linked multichromophoric systems.

Long-range interactions are not well represented by either of the two parameterized Hamiltonians. Asymptotically, DFT/MRCI performs like the underlying BHLYP functional. For charge-separated systems this means that the energy increases with $1/2R$ instead of $1/R$ where R is the distance between the two charged subsystems. Furthermore, dispersion interactions are not properly taken care of by DFT/MRCI. The latter problem may be easily remedied, for example, by adding the semiempirical Grimme D3 dispersion correction [74].

8.3.3 Fluorescence and Phosphorescence Rates

Similar to ISC and rISC rate constants, also rate constants for radiative transitions can be derived in the framework of time-dependent perturbation theory. Herein, the vector potential for the motion of the electrons in the external electromagnetic radiation field is used as perturbation operator instead of $\hat{H}_{SO}$. Typically, one proceeds by a multipole expansion of the interaction, with the electric dipole operator as the leading term. The rate constant for spontaneous emission from a manifold of thermally populated initial vibronic states $|\Psi_a, \{v_{aj}\}\rangle$ to a quasi-continuum of final vibronic states $\langle \Psi_b, \{v_{bk}\}|$ due to electric dipole interaction is then given by

$$k_{rad} = \frac{4\alpha\omega^3}{3c^2 Z} \sum_{j,k} e^{-\beta E_j} |\langle \Psi_b, \{v_{bk}\} | \hat{\mu} | \Psi_a, \{v_{aj}\}\rangle|^2 \, \delta(E_{aj} - E_{bk} - \hbar\omega) \quad (8.12)$$

where ω is the radiation frequency, α is the fine-structure constant, c is the speed of light, and the other symbols have the same meaning as in Eq. (8.1). The procedure for simplifying the expression in Eq. (8.12) further by a Taylor expansion with respect to mass-weighted normal coordinates is similar to that described in Section 8.2. Formulas for radiative transition rates in Franck–Condon (FC) or Herzberg–Teller approximation, respectively, can readily be derived.

Once the electronic wave functions have been obtained, it is straightforward to compute the electric dipole coupling matrix elements for fluorescence emission. If wave functions are not available – which is the case in TDDFT and coupled-cluster approaches – the electronic transition rates can be computed by means of linear response theory [75]. Typical fluorescence rate constants for emission from LE $^1(\pi\pi^*)$ states are of the order of 10^8–10^9 s^{-1}, whereas they are several orders of magnitude smaller for $^1(n\pi^*)$ states. Because of competing nonradiative processes, the latter states are optically dark in most cases. The fluorescence rates of ^{1}CT states depend critically on the overlap of the electron density distributions of the orbitals involved in the transition. If that overlap is small, fluorescence rates of 10^5–10^6 s^{-1} are expected at most.

In the context of spin-forbidden transitions, the coupling matrix element in Eq. (8.12) requires a bit of attention. In this case, multiplicity-mixed electronic wave functions need to be employed. They are complex-valued in general. The fundamentals regarding selection rules and intensity borrowing from spin-allowed transitions have been worked out in detail in Ref. [30] and need not be repeated here. Nevertheless, it is instructive to compare phosphorescence from a $^3(n\pi^*)$, a $^3(\pi\pi^*)$, and a ^{3}CT state to the electronic ground state S$_0$. According to the El-Sayed rules (see Section 8.2.1.1), the $^3(n\pi^*)$ state exhibits sizeable SOCMEs with the electronic ground state and electronically excited $^1(\pi\pi^*)$ states. Large contributions to the electric transition dipole matrix element can originate from two terms: (i) the mutual spin–orbit interaction of $^3(n\pi^*)$ and S$_0$ multiplied by the difference of the static dipole moments of these states and (ii) the spin–orbit interaction of $^3(n\pi^*)$ with optically bright $^1(\pi\pi^*)$ states from which intensity can be borrowed. In heteroaromatic compounds, thus phosphorescence rates of 10^3 s^{-1} can be achieved. For a $^3(\pi\pi^*)$ state, the direct spin–orbit interaction with S$_0$ is very small so that term (i) can be neglected. With regard to term (ii), it is seen that $^3(\pi\pi^*)$ exhibits sizeable SOCMEs with $^1(n\pi^*)$ states, but the latter are optically dark. Therefore, the probability of a spin-forbidden radiative decay is much smaller for a $^3(\pi\pi^*)$ state, with rates typically below 1 s^{-1}. The same applies to ^{3}CT states of purely organic donor–acceptor systems. The shortrangedness of the spin–orbit interaction makes the contributions of type (i) vanish, despite the pronounced static dipole moment difference of the ^{3}CT and S$_0$ states. In contrast, phosphorescence from ^{3}MLCT states of TM complexes can acquire substantial probability through configuration interaction. The transition dipole moment may adopt sizable values originating from the combination of $\langle S_0|\hat{\mu}|^1(d'\pi^*)\rangle$ and $\langle ^1(d'\pi^*)|\hat{H}_{SO}|^3(d\pi^*)\rangle$ in MLCT transitions. Also the direct, first (i)-type interaction of the ^{3}MLCT state with the electronic ground state can play a role, because the SOCMEs $\langle S_0|\hat{H}_{SO}|^3(d\pi^*)\rangle$ may not be negligible. Depending on the amount of configuration mixing and the size of the SOCMEs, phosphorescence rates of 10^4–10^6 s^{-1} can be reached in these complexes.

Although the concept of intensity borrowing from spin-allowed transitions is very transparent if Rayleigh–Schrödinger perturbation theory is applied to expand the multiplicity-mixed wave functions in terms of pure $\hat{S}^2$ eigenfunctions, this is not the procedure followed in practice. The reason is that the (in principle) infinite perturbation sums converge very slowly with respect to the number of electronic states. In an actual calculation, it is more advantageous to use methods that avoid an explicit summation over states such as multireference spin–orbit configuration interaction (MRSOCI) [76] or quadratic response theory [77, 78].

8.4 Case Studies

In the following, a brief literature survey on the results of quantum chemical studies on TADF emitters will be given before turning to a review of our own recent (and still ongoing) work in more detail. For a few prototypical cases, spin-dependent multiconfigurational electronic structure methods have been employed to describe the electronically excited-state potentials and their couplings. These data are used as input for Fourier transform methods that enable us to determine rate constants of radiative and nonradiative transitions and thus may further the understanding of the photophysical processes in TADF emitters.

8.4.1 Copper(I) Complexes

Various organometallic complexes based upon d^{10} metal ions were found to show TADF, the most abundant ones being Cu(I) complexes [15–17]. Among them are four-coordinated but also three-coordinated bis-phosphine complexes [79–87] and bridged bimetallic Cu(I) complexes of type Cu$_2$X$_2$(N^P)$_2$ (X = Hal) [88, 89]. On the basis of combined DFT and TDDFT calculations, it was concluded that restricted flexibility of four-coordinated Cu(I)–bis-imine–bis-phosphine complexes leads to a reduction of nonradiative deactivation and thus an increase of emission quantum yield [83]. In threefold-coordinated Cu(I) complexes with a sterically demanding monodentate N-heterocyclic carbene (NHC) ligand and a heterocyclic bidentate ligand, the relative orientation of the ligands seems to decide whether TADF or phosphorescence is observed [90–93]. In the phosphorescent complexes, the conformational analysis indicates a nearly free rotation about the C_{NHC}—Cu bond in solution [91]. Conformational flexibility appears to be also the key to understanding the photophysical properties of three-coordinate thiolate Cu(I) complexes that give bright blue emission at 77 K and orange emission at ambient temperature [94]. Very recently, even highly efficient blue luminescence of two-coordinated Cu(I) complexes has been reported [95].

Most experimental work on these complexes is accompanied by some Kohn–Sham DFT calculations that focus on the nature of the highest occupied molecular orbital (HOMO) and the lowest unoccupied molecular orbital (LUMO). This bears some danger because numerous examples are known in the literature showing that the lowest electronically excited state does not necessarily originate from a HOMO–LUMO transition [44, 71]. Proceeding with due caution, it is preferable to use TDDFT or – even better – approximate coupled-cluster methods or DFT/MRCI to characterize the excited states. In several cases, TDDFT has been employed to determine singlet–triplet energy gaps ΔE_{ST} of Cu(I) complexes [83, 84, 88, 93, 96, 97]. Gneuß et al. found a good correlation between computed triplet emission wavelength, as obtained from TDDFT in conjunction the B3LYP functional, and measured peak maxima in a new class of luminescent mononuclear copper(I) halide complexes with tripodal ligands [96]. Quantum chemical studies of Cu(I) complexes that explicitly take account of spin–orbit interaction are very rare, however. To our knowledge, there is only a series of papers studying the emission properties of four-coordinated Cu(I)–bis-phenanthroline complexes [54, 55, 98, 99] and our own work on luminescent Cu(I)–NHC complexes [100]. In the following, these cases will be analyzed in more detail.

8.4.1.1 Three-Coordinated Cu(I)–NHC–Phenanthroline Complex

Using the methods described in Sections 8.2 and 8.3, recently Föller et al. [100] conducted a thorough quantum chemical study on the photophysical behavior of a luminescent Cu(I) complex comprising an NHC and a phenanthroline ligand (Figure 8.2). This complex had been investigated experimentally by Krylova et al. [90] who also performed DFT calculations and assigned the luminescence to originate from an MLCT state. The bulky isopropylphenyl substituents on the imidazol-2-ylidene ligand are essential for the relative orientation of the NHC and phenanthroline ligands. Substitution by methyl or even phenyl substituents in 1,3-position of the NHC leads to a barrierless torsional relaxation of the excited triplet and singlet states yielding a perpendicular conformation of the two ligands. Dispersion, included in the calculations by means of the semiempirical Grimme D3 correction [74], is seen to have a small but differential effect on the torsion potentials in the ground and excited states. It preferentially lowers the coplanar arrangement of the ligands and increases the barrier between the two minima on the excited-state potential energy surface. Qualitatively, this trend is easily understood. In the coplanar nuclear arrangement, the hydrogen atoms

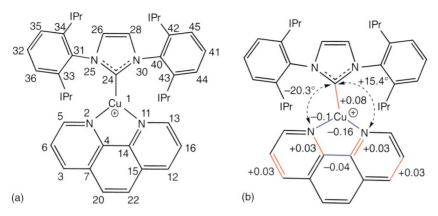

Figure 8.2 (a) Chemical structure of the Cu(I)–NHC–phenanthroline complex and (b) important differences in the coplanar S_0 and T_1 minimum nuclear arrangements according to Ref. [100]. Source: Ref. [100]. Reproduced with permission of American Chemical Society.

in positions 5 and 13 of the phenanthroline ligand (Figure 8.2a) directly point toward the aromatic π system of the isopropylphenyl substituents of the NHC ligand, whereas these are far apart when the NHC and phenanthroline ligands are oriented in a perpendicular fashion.

The spin-free vertical excitation spectra were calculated by means of the original DFT/MRCI method [64, 67]. SOC was included in the calculation of absorption spectra at the level of quasi-degenerate perturbation theory (QDPT) using the in-house SPOCK [101, 102]. Fluorescence and phosphorescence rates were obtained at the MRSOCI level [76]. FC emission profiles and temperature-dependent ISC and rISC rates were determined in harmonic approximation by means of a Fourier transform approach [51, 103].

The experimental absorption spectrum [90] and calculated spectra at scalar relativistic level and including SOC effects are displayed in Figure 8.3. They are seen to match perfectly, showing that the applied quantum chemical methods are very well suited for studying these complexes. Very weak bands between 400 nm and 500 nm were assigned to triplet MLCT states. In the calculated FC spectrum, the electronic excitations to the S_1, T_1, and T_2 states lie between 400 nm and 420 nm but have small oscillator strengths so that they are barely visible in Figure 8.3. At the ground-state equilibrium geometry, S_1 and T_2 may be characterized as d/σ → π*(phen) single excitations where d/σ is a linear combination of a d_{xy}-like orbital of copper with in-plane p orbitals of the phenanthroline (phen) nitrogens. At this point, T_1 results from a d/π → π*(phen) excitation where d/π is a d_{xz}-like orbital with π contributions located at the phenanthroline and NHC ligands. Following the line of arguments in Section 8.2.1.1, substantial spin–orbit interaction is expected between the T_1 and S_1 states because their coupling involves a change of orbital angular momentum. This is the case, indeed, at the FC point. However, a slight geometry distortion is sufficient to reverse the order of the two triplet states. Henceforth, we renumerate the states according to the order of their adiabatic minima. The main configurations of the relaxed S_1 and T_1 states

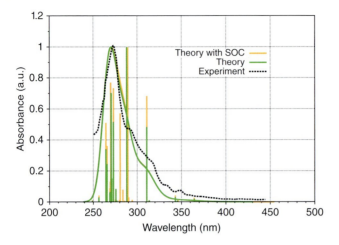

Figure 8.3 Absorption spectrum of the Cu(I)–NHC–phenanthroline complex shown in Figure 1.2a. The experimental data points were read from Ref. [90]. Note that the theoretical spectra have not been shifted but are displayed as calculated.

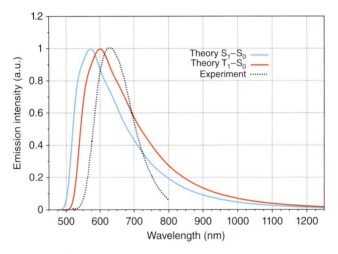

Figure 8.4 Emission spectra of the Cu(I)–NHC–phenanthroline complex shown in Figure 1.2b. The experimental data points were read from Ref. [90].

are d/σ → π*(phen) excitations. The excited-state minima with torsion angle of 0° show a T-shaped distortion of the three-coordinated Cu(I), as suggested by Krylova et al. [90].

The perpendicular arrangement constitutes a saddle-point on the electronic ground-state potential energy surface located approximately 0.35 eV above the minimum. In the excited states, this arrangement of the ligands yields a local minimum. It is separated from the global minimum with coplanar arrangement of the ligands only by a shallow barrier (c. 0.12 eV) that is easily overcome by thermal activation. However, comparison of the theoretical and experimental emission spectra (Figure 8.4) clearly shows that the complex emits preferentially in a coplanar arrangement of the NHC and phen ligands.

The singlet–triplet splitting changes only slightly along the path, namely, from 650 cm^{-1} at 0° to 830 cm^{-1} at 90°, in contrast to the observations of Leitl et al. [93] in a related Cu(I)–NHC–dipyridyldimethylborate complex. These authors report an increase from 540 cm^{-1} at 0° to 3700 cm^{-1} at 70° torsion angle. The different behavior of the two complexes can be explained by the different electron density distributions of the frontier orbitals. In the Cu(I)–NHC–phen complex, the density of the HOMO is mainly located at copper and the LUMO at the phen ligand, which does not change so much during the torsion (Figure 8.5). Correspondingly, only a small increase of the energy gap is found. In contrast, the orbital overlaps of the Cu(I)–NHC–dipyridyldimethylborate increases considerably upon torsion [93]. In that case, the LUMO is located at the NHC ligand and the HOMO of the 90° geometry has substantial additional density at the NHC ligand. Adiabatically, the singlet–triplet splitting exhibits a value of merely 0.08 eV in the Cu(I)–NHC–phen complex. As the energy gap between the S_1 and T_1 states is so small, TADF might be possible, and therefore phosphorescence and fluorescence as well as ISC and rISC rates were investigated for both the coplanar (torsion angle 0°) and perpendicular (torsion angle 90°) arrangement of the ligands by Föller et al. [100].

Figure 8.6 provides an overview over the computed rate constants in the coplanar arrangement of the ligands at room temperature. Kirchhoff et al. [104] carefully analyzed the kinetics of a three-level system relating to TADF. They concluded that the steady-state emission properties of the three-level system depend upon the relative values of the various rate constants as well as the relative energies of the levels and considered two limiting cases, the kinetic limit and the equilibrium limit. The basic assumption in the kinetic limit case is that the S_1 state achieves a steady-state concentration that is negligibly affected by the rISC process. Most of the photons would then appear from prompt fluorescence or phosphorescence. The decay of the Cu(I)–NHC–phen complex reaches

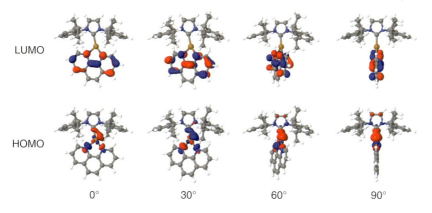

Figure 8.5 Frontier orbital densities of the Cu(I)–NHC–phenanthroline complex for different torsion angles according to Ref. [100]. *Source*: Ref. [100]. Reproduced with permission of American Chemical Society.

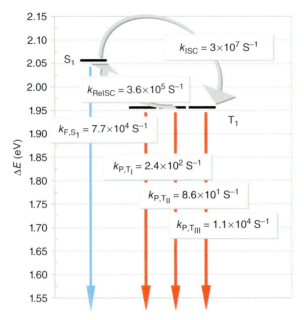

Figure 8.6 Computed rate constants (298 K) of the Cu(I)–NHC–phenanthroline complex in the coplanar T_1 minimum nuclear arrangement according to Ref. [100]. Source: Ref. [100]. Reproduced with permission of American Chemical Society.

nearly the kinetic limit at 77 K, as k_{rISC} is several orders of magnitude smaller than k_F and k_{ISC}. In contrast, the equilibrium limit case where the steady-state populations of the S_1 and T_1 states are determined by Boltzmann statistics appears to be more appropriate at room temperature. For this limit case to be adequate, $k_{\text{ISC}} \gg k_\text{F} + k_{\text{IC}_0}$ and $k_{\text{rISC}} \gg k_\text{P} + k_{\text{ISC}_0}$. We have not calculated the rate constants for the internal conversion (IC) of S_1 to S_0 (k_{IC_0}) and for the ISC from T_1 to S_0 (k_{ISC_0}). Because of the substantial energy gap between the S_1 and T_1 states on the one hand and the electronic ground state on the other hand, these processes are assumed to be much slower than the radiative decay rates and will be neglected in the following. The first condition for the equilibrium limit, namely, $k_{\text{ISC}} \gg k_\text{F}$ is certainly fulfilled at all temperatures, but the second condition $k_{\text{rISC}} \gg k_\text{P}$ is not (Figure 8.6). In such case, formula

$$\tau(T) = \frac{3 + \exp[-\Delta E(S_1 - T_1)/k_\text{B}T]}{\frac{3}{\tau(T_1)} + \frac{1}{\tau(S_1)} \exp[-\Delta E(S_1 - T_1)/k_\text{B}T]} \tag{8.13}$$

used by many experimentalists to fit the energy gap between the S_1 and T_1 states $\Delta E(S_1-T_1)$ and the decay times of the individual states from the temperature dependence of the total emission decay time [84] is not valid at low temperatures.

The respective quantum yields were calculated instead following the kinetic analysis of Hirata et al. [105]. In deriving their expression for the TADF quantum yield Φ_{TADF} in relation to the triplet quantum yield Φ_T, the authors assumed that the IC from the S_1 state to the electronic ground-state S_0 can be neglected and that the ISC from S_1 to T_1 is much faster than the reverse process. Neglecting the nonradiative deactivation of the T_1 sublevels, but taking account of their phosphorescence decay through the averaged high-temperature limit of the individual rate constants, one arrives at

$$\frac{\Phi_{TADF}}{\Phi_P} = \frac{1}{\frac{k_{P,av}}{k_{rISC}\Phi_{PF}} + 1} \quad (8.14)$$

For 298 K our calculated quantum yield of prompt fluorescence Φ_{PF} is quite small, only 0.3%. The quantum yield for the deactivation via phosphorescence is 77.7%, and the remaining 22% are the quantum yield for TADF. This may explain why Krylova et al. [90] classify this complex as phosphorescent emitter.

The emission quantum yield Φ_{em} is related to the radiative and nonradiative decay rate constants by

$$\Phi_{em} = \frac{k_r}{k_{nr} + k_r} = \frac{\tau_r}{\tau} \quad (8.15)$$

Using this relation, intrinsic radiative lifetimes τ_r can be deduced from the knowledge of the experimental emission decay time constant τ and the quantum yield Φ_{em}. The τ_r can directly be compared with theoretical values. In the present example, Krylova et al. [90] had determined values of $\tau = 0.08$ μs with $\Phi_{em} < 0.001$ in CH$_2$Cl$_2$ and of $\tau = 1.2$ μs with $\Phi_{em} = 0.026$ in crystals at room temperature. Using Eq. (8.15), intrinsic radiative lifetimes of >80 μs in CH$_2$Cl$_2$ at 300 K and of ≈46 μs in the crystalline state can be derived. These values compare better with the calculated fluorescence lifetime of $\tau_F = 11$ μs than with the calculated phosphorescence lifetime of $\tau_P = 267$ μs [100].

Summarizing, it is found that the excited singlet and triplet populations of the three-coordinated Cu(I)–NHC–phen complex have time to equilibrate before they decay radiatively to the electronic ground state. TADF is possible, but it competes with phosphorescence that is the dominating radiative decay channel. As discussed above, torsion of the ligands has only small impact on the singlet–triplet gap. However, the electronic coupling between the S_1 and T_1 states – and hence the probability for (reverse)ISC – is seen to increase substantially when moving from a coplanar to a perpendicular arrangement of the ligands. The quantum chemical analysis by Föller et al. [100] suggests that a perpendicular arrangement of the ligands in a three-coordinate NHC–Cu(I)–(N^N) complex is not a hindrance *per se* for observing TADF (in addition to phosphorescence), provided that the electron is transferred to the (N^N) ligand in the MLCT transition and not to the NHC ligand.

8.4.1.2 Four-coordinated Cu(I)–bis-Phenanthroline Complexes

In the 1980s, McMillin and coworkers investigated the luminescence of a series of cationic Cu(I)–bis-phen complexes in solution [104]. The emission intensity was found to decrease as the temperature of the solution is lowered, accompanied by a slight redshift in the positions of the emission maxima. The authors interpreted the strong temperature dependence of the solution emission of the cationic bis(2,9-dimethyl-l,10-phenanthroline)copper(I) complex (Cu(dmp)$_2^+$, Figure 8.7a) in terms of emission from two thermally equilibrated excited states separated by ca. 1800 cm^{-1} whereof the lower one was ascribed triplet MLCT

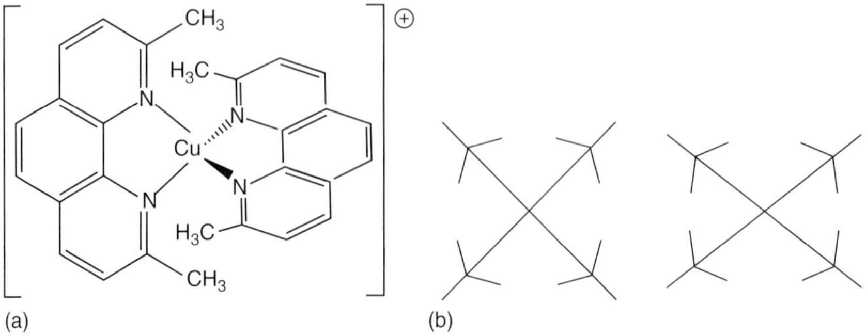

Figure 8.7 (a) Chemical structure of the cationic bis(2,9-dimethyl-l,10-phenanthroline) copper(I) complex and (b) schematic view of the pseudo-Jahn-Teller (PJT) distortion in the MLCT states according to Ref. [98]. *Source*: Ref. [98]. Reproduced with permission of American Chemical Society.

and the upper one singlet MLCT character. It was suggested that DF dominates the emission at room temperature.

Owing to a pseudo-Jahn–Teller (PJT) distortion of the copper d^9 electronic configuration, these complexes undergo a fast flattening structural change in the MLCT excited state upon photo excitation (Figure 8.7b). TDDFT calculations revealed that there are four closely spaced triplet states in energetic proximity to the lowest singlet state [106]. Siddique et al. [98] carried out a combined experimental and theoretical study of the Cu(dmp)$_2^+$ complex including a rough estimate of the spin-forbidden transition probabilities (semiempirical spin–orbit Hamiltonian on Cu only, single-configuration approximation of the wave functions). The dihedral angle of the two ligand planes was found to change from 90° in the electronic ground state to about 75° in the MLCT states. Siddique et al. noticed that the photophysical properties depend strongly on that dihedral angle, the flattening distortion reducing not only the magnitude of the mutual SOCMEs of the lower states but also the transition dipole moment of spin-allowed transition from which the phosphorescence borrows intensity.

In very elaborate theoretical investigations of the photophysics of cationic Cu(I)–bis-phen complexes, Capano et al. [54, 55, 99] used methods for determining the kinetic constants of the excited-state processes differing from those described above. They applied the vibronic coupling Hamiltonian [57] and the quantum dynamics within the framework of the MCTDH method [58] to study the primary excited-state nonadiabatic dynamics following the photoexcitation. To this end, they identified eight important vibrational modes and determined the nonadiabatic coupling coefficients for IC in a linear coupling model [54, 55] The results of the wavepacket dynamics show that the IC from the initially photoexcited S_3 state to the S_1 state takes place with a time constant of about 100 fs and that ultrafast ISC occurs and competes with structural dynamics associated with the PJT distortion [55]. Approximately 80% of the wave packets are found to cross into the triplet state within 1 ps. Further analysis of the MCTDH dynamics results emphasizes the importance of vibronic interactions for the ISC. It suggests that the ISC occurs close to the FC point – despite the moderate size of the SOCMEs – from S_1 to T_2 and T_3 promoted by the motion along a vibrational mode that drives the S_1 state into curve crossings with T_2 and T_3.

In their latest work, Capano et al. [99] used DFT, TDDFT, as well as classical and quantum mechanics/molecular mechanics (QM/MM) molecular dynamics (MD) simulations to investigate the influence of the geometric and electronic structure, SOC, singlet–triplet gap, and the solvent environment on the emission properties of a series of four-coordinated cationic bis-phen complexes. They systematically varied the parent structure by adding electron-withdrawing or donating substituents. Furthermore, they investigated the influence of long alkyl chains at 2,9-positions of phenanthrolines because they were said [104] to retard the excited-state PJT distortion and to increase the luminescence. In agreement with earlier work by Siddique et al. [98], Capano et al. conclude that the magnitude of the SOC matrix elements depends critically on the dihedral angle between the ligands, thus explaining the bi-phasic ISC observed experimentally to originate from the initially excited and relaxed S_1 structures, respectively. Furthermore, they show that the singlet–triplet gap of the MLCT states is governed by inductive effects of the substituents that also control the oscillator strength of the fluorescence.

It might be interesting to carry out similar theoretical studies for related (P^P) complexes in the future. Experimentally, it was shown that the quantum efficiencies of cationic mononuclear copper(I) complexes based on phenanthroline ligands could be increased from about 1% to around 60% when exchanging one of the phenanthrolines by a diphosphineether [108].

8.4.2 Metal-Free TADF Emitters

Electroluminescence from purely organic molecules was detected more than 60 years ago using acridine orange as the emitter [109]. Despite the substantial singlet–triplet splitting of ≈ 2900 cm^{-1}, acridine orange shows relatively strong DF [110]. Most of present-day's metal-free TADF emitters are bi- or multi-chromophoric systems consisting of electronically weakly coupled donor (D) and acceptor (A) subunits that undergo intramolecular charge-transfer (ICT) processes upon electronic excitation. The donor and acceptor subunits are covalently linked in such a way that the respective π systems are (nearly) orthogonal, either by twisted single bonds [19, 105, 111–120] or by a spiro-junction [121–123].

Many TADF molecules exhibit highly twisted structures, because the energetic splitting of a singlet–triplet pair depends on the degree to which the electron density distributions of the involved half-occupied orbitals overlap. However, in an ICT compound also the transition density, and hence the oscillator strength of the emission depends on this quantity. Recognizing that small, but nonnegligible overlap densities between the electronic wave functions of the S_0 and S_1 states may lead to a substantial increase of the fluorescence rate without significantly increasing ΔE_{ST}, Shizu et al. rationally designed highly efficient green and blue TADF emitters [117, 124].

A very instructive experimental investigation of donor–acceptor–donor (D–A–D) and donor–donor–donor (D–D–D) type emitters was carried out by Dias et al [111]. With one exception, all compounds had a disubstituted heterocyclic core, with the substituents attached either in an angular or linear fashion. In the D–D–D structures, the lowest-lying triplet state is an LE state with $^3(\pi\pi^*)$ characteristics. Nevertheless, all emitters show DF in ethanol solution at room temperature, though with different kinetics. Some exhibit a linear relation of the DF intensity with exciton dose, indicating thermal activation of the DF, whereas others show a quadratic dependence, implicating DF caused by triplet fusion. Particularly striking is the observation of TADF for a compound with a singlet–triplet splitting of 0.84 eV. To explain their observations, Dias et al. [111] postulate an intermediate $^3(n\pi^*)$ state that bridges the gap between the lowest triplet and singlet states in the heteronuclear compounds. Gibson et al. [56] predicted second-order SOC of the singlet and triplet CT states to an intermediate LE $^3(\pi\pi^*)$ state to enhance the ISC and rISC rate constants of a D–A molecule, composed of a phenothiazine (PTZ) donor and a dibenzothiophene-S-S-dioxide DBTO2 acceptor, by several orders of magnitude. The crucial role of vibronic SOC to an energetically close-lying ^{3}LE state for the TADF efficiency of the corresponding D–A–D compound was impressively demonstrated by Etherington et al. [120]. These authors employed host material with temperature-sensitive polarity to tune the LE and CT states of the PTZ-DBTO2-PTZ system in and out of resonance. They could show that the emission intensity goes through a maximum at the zero crossing of the energy difference. The concomitant quantum dynamics studies included spin–orbit interaction between the ^{3}LE and the CT states, vibronic interaction between the two triplet states, and hyperfine interaction between the singlet and triplet CT states as possible coupling terms. The outcome of these studies suggests that hyperfine interaction between the two CT states is by far too small and that the combination of spin–orbit and vibronic interaction is required to effectuate the rISC (compare also Chapter 12).

Also the theoretical study by Chen et al. [125] was motivated by the question why some butterfly-shaped blue D–A–D emitters show TADF properties although their lowest excited triplet state has LE character and the singlet–triplet splitting is substantial. Adachi and coworkers explained this behavior by reverse internal conversion (RIC) from the LE $^3(\pi\pi^*)$ to the ^{3}CT state that then undergoes rISC to the ^{1}CT state [116]. Although the investigated compounds are similar to those studied by Monkman and coworkers [111, 118, 120], Chen et al. [125] come to different conclusions. They computed rISC rates between the T_1 and S_1 states in harmonic oscillator approximation including Duschinsky effects using a Fermi's golden rule expression. Their ansatz is very similar to the one employed in our laboratory, except for the fact that they use two-component TDDFT [126] for computing the electronic coupling matrix elements. Not unexpectedly, the rate of rISC from the T_1 state with LE character to the S_1 state with ICT character is several orders of magnitude too small in Condon approximation. Scanning through the torsional angle between donor and acceptor subunits reveals an intersection of the two lowest triplet potential energy curves at 90° with a maximum for the lowest triplet state. Actually, the authors find a broken-symmetry minimum for T_1 in their D–A–D example. It is argued that the triplet states can approach the intersection by low-frequency motions and that nonadiabatic interactions are expected to play a significant role. Rates for vibration-assisted transitions have not been presented, though, in that work [125].

In the following, the photophysics of three metal-free TADF emitters shall be discussed in more detail, namely, of the green TADF emitter 1,2,3,5-tetrakis(carbazol-9-yl)-4,6-dicyanobenzene (4CzIPN) and of the assistant dopants 3-(9,9-dimethylacridin-10(9*H*)-*yl*)-9H-xanthen-9-one (ACRXTN) and 10-phenyl-10*H*,10'H-spiro[acridine-9,9'-anthracen]-10'-one (ACRSA). 4CzIPN has very high photoluminescence efficiency in apolar solvents and films [19, 112] and is considered a prototypical donor–acceptor multichromophoric system. ACRXTN, consisting of an acridine donor unit and a xanthone acceptor unit [24], is particularly interesting because of the already very involved photophysical properties of the heteroaromatic xanthone moiety [44]. ACRSA, finally, is one of the few spiro-compounds known to exhibit efficient TADF [122]. These compounds were investigated recently by means of high-level quantum chemical methods in our laboratory with the aim to shine light on some of the underlying mechanisms [127–129]. The (preliminary) results of these studies will be compared with experimental data and – where available – theoretical results from TDDFT calculations.

8.4.2.1 1,2,3,5-Tetrakis(carbazol-9-yl)-4,6-dicyanobenzene (4CzIPN)

Attaching carbazolyl donor units to dicyano-substituted benzene cores as acceptors, Uoyama et al. [19] presented a series of luminescence emitters, their color varying from turquoise to red depending on the number of carbazolyl units and the positions of the cyano substituents. The green emitter 1,2,3,5-tetrakis(carbazol-9-yl)-4,6-dicyanobenzene (4CzIPN, Figure 8.8) turned out to be a TADF emitter with excellent internal quantum efficiency in toluene and in 4,4′(bis-carbazol-9-yl)biphenyl (CBP) film [19]. OLEDs based on 4CzIPN show high luminance efficiencies and excellent operational stability [130].

4CzIPN is a CT system with small singlet–triplet energy gap, the magnitude of which has been estimated from Arrhenius plots assuming that only T_1 is

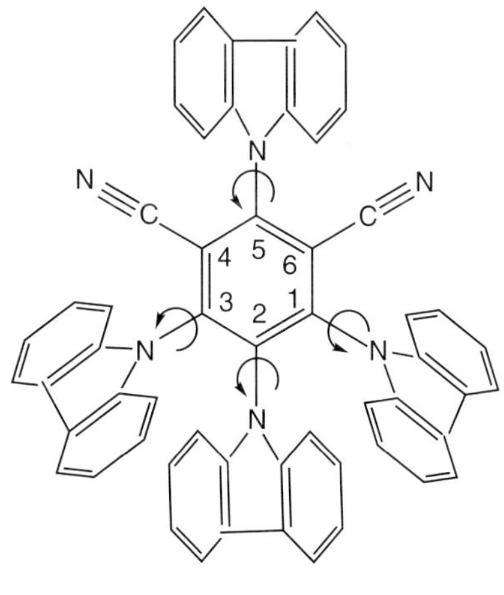

Figure 8.8 Chemical structure of the green TADF emitter 1,2,3,5-tetrakis(carbazol-9-yl)-4,6-dicyanobenzene (4CzIPN).

located energetically below S_1 (see, however, below). The estimates vary slightly depending on the conditions and the solvent. Uoyama et al. [19] report a value of $\Delta E_{ST} = 83$ meV in CBP film, whereas Ishimatsu et al. [112] give a somewhat larger value of 140 ± 19 meV in toluene. Also the measured luminescence lifetimes vary somewhat between the two studies, but the values are in the same ballpark. At room temperature, Uoyama et al. obtain time constants of 17.8 ns for prompt and 5.1 µs for delayed fluorescence in toluene solution under nitrogen atmosphere, while Ishimatsu et al. report 14.2 ns and 1.82 µs for these processes, respectively, in the same solvent. Uoyama et al. [19] also carried out quantum chemical calculations. The computed singlet–triplet energy gap depends strongly on the functional used in the TDDFT calculations, ranging from approximately 10 meV for the hybrid functionals B3LYP and PBE0 over 362 meV for M06-2X to 700 meV for the Coulomb attenuated CAM-B3LYP in the gas phase. Likewise, the computed vertical emission wavelengths vary substantially with the functional, ranging from 731 nm (B3LYP) to 430 nm (M06-2X) and 420 nm (ωB97X-D).

The energy gap law [34] states that in the weak coupling regime, that is, for small coordinate displacements, the rate of nonradiative transition between two electronic states decreases exponentially with their increasing energy separation (see also Section 8.2.1.2). This does not necessarily mean, however, that a smaller energy gap automatically leads to improved luminescence properties of a TADF emitter. A detailed experimental and theoretical study investigating the solvent effects on TADF in 4CzIPN revealed a surprising trend with regard to the efficiency of the delayed fluorescence. Ishimatsu et al. [112] observe larger Stokes shifts of the emission in polar solvents compared with toluene suggesting stronger ICT in the excited state. Concomitantly, the activation energy for rISC is lowered. However, despite the smaller magnitude of ΔE_{ST} in polar solvents, the photoluminescence quantum yield decreases from 94% in toluene solution over 54% in dichloromethane and 18% in acetonitrile to 14% in ethanol. Using TDDFT in conjunction with the M06-2X functional and modeling the solvent–solute interaction by a polarizable continuum model [131, 132], they also computed absorption and emission energies for 4CzIPN in these solvents. The authors consistently explain the experimentally observed trends for ΔE_{ST}, the emission wavelength, and the photoluminescence quantum yields by increasing ICT character of the excited states with increasing solvent polarity, leading in turn to weaker electric dipole transitions.

Ishimatsu et al. [112] report torsion angles between donors and acceptor of c. 50–65° for the electronic ground state, with the largest dihedral angle for the carbazolyl donor in 1-position and the smallest one for the donor in 3-position. In the DFT/B3LYP optimized structure obtained for the isolated molecule in our laboratory, the dihedral angles between the carbazolyl donors in 1-, 2-, and 3-positions and the isophthalonitrile core are very similar (+63 and +64°), while the carbazolyl donor in 5-position shows a stronger twist (−71°) [127]. For computing the spin-free properties of the electronically excited states, the parallelized version of the original DFT/MRCI [64, 67] method was employed in the preliminary calculations. The wave function of the first excited singlet state is dominated by the HOMO→LUMO ICT transition but with marked

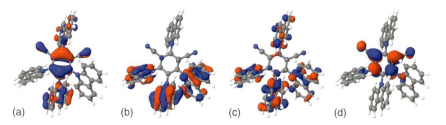

Figure 8.9 BHLYP molecular orbitals of 4CzIPN, important for the characterization of the lowest excited states according to Marian [127]. (a) HOMO-17, (b) HOMO-1, (c) HOMO, (d) LUMO.

contributions from local excitations. In particular, the local HOMO-17→LUMO excitation on the isophthalonitrile core (Figure 8.9) has a coefficient of nearly 0.1 in the S_1 wave function at the ground-state minimum. Local excitations lead to an enhancement of the oscillator strength for $S_0 \rightarrow S_1$. This transition is responsible for the shoulder at c. 450 nm in the absorption spectrum of 4CzIPN [19]. The vertical DFT/MRCI excitation energy of 2.94 eV (422 nm) in the gas phase is expected to be slightly redshifted due to solvent–solute interactions. The TDDFT excitation energy obtained by Ishimatsu et al. [112] for the M06-2X functional in a PCM (3.29 eV, 376 nm) is substantially higher. Interestingly, we find two triplet states below S_1 at the FC point, with the electronic structure of T_1 corresponding to S_2. The multiconfigurational T_1 wave function has the HOMO-1→LUMO configuration as leading term. T_2 is dominated by the HOMO → LUMO configuration. Both triplet excited states exhibit larger contributions from LE on the isophthalonitrile core than their singlet counterparts.

The T_1 nuclear arrangement was optimized with unrestricted DFT (UDFT). At the T_1 minimum, the donors in 1-, 2-, and 3-positions exhibit larger torsion angles with respect to the isophthalonitrile core (ranging between +68 and +73°) than in the electronic ground state, whereas the dihedral angle is flatter (−66°) for the donor in 5-position. This trend reflects the fact that electron density was mainly donated by the carbazolyl substituents in 1-, 2-, and 3-positions upon the ICT excitation to the T_1 state. At this point of the coordinate space, S_1 and T_1 result predominantly from the HOMO → LUMO excitation. Still, two triplets are found below the first excited singlet state in the DFT/MRCI calculations. S_1 and T_1 are separated by an energy gap of 86 meV here, in excellent agreement with experimental evidence. The T_2 state is located halfway between the T_1 and S_1 states at this geometry.

The SPOCK [101, 102] program was used to determine SOCMEs of the DFT/MRCI wave functions. The mutual SOCMEs of the three closely spaced electronic states are small. For the S_1 and T_1 DFT/MRCI wave functions, a sum over squared SOCMEs of 10^{-2} cm^{-2} was obtained. The electronic coupling of S_1 and T_2 is slightly larger (sum over squared SOMCEs 4×10^{-2} cm^{-2}). The presence of the intermediate triplet state might therefore accelerate the ISC and rISC processes. The calculation of ISC and rISC rate constants is in progress.

Fluorescence and phosphorescence rates were obtained at the MRSOCI level [76]. The MRSOCI calculations were performed at the UDFT-optimized T_1

minimum. Technically, they are at the limit of what can be handled by the current version of the SPOCK program. In the Davidson diagonalization of the MRSOCI matrix, 8 complex-valued eigenvectors with expansion lengths of $\approx 5 \times 10^8$ configuration state functions were determined. The sublevels of the T_1 states are degenerate for all practical purposes. Phosphorescence is not a competitive decay mechanism in 4CzIPN. The calculated rate constants for phosphorescence are of the order of merely 10^{-2}–10^{-1} s^{-1}.

Although the S_1 state wave function is dominated by ICT excitations ($\Delta\mu \approx 7$ D), the fluorescence exhibits substantial oscillator strength. The emission from the S_1 state gains intensity from small amounts of local excitations on the isophthalonitrile core. At the T_1 minimum geometry, a vertical emission wavelength of 482 nm is computed for the isolated molecule, in good agreement with the experimental peak maximum at 507 nm [19, 112] measured in toluene. For a comparison of the calculated (pure) radiative lifetime of 44 ns with measured time constants, quantum yields have to be taken into account. Using Eq. (8.15) and the experimentally determined quantum yield of 21.1% for prompt fluorescence at 300 K [19] yields a lifetime of $\tau \approx 9.3$ ns that compares well with the experimental values of 17.8 ns [19] and 14.2 ns [112] in toluene solution.

The magnitude of the local contributions to the $S_1 \rightarrow S_0$ transition density depends critically on the dihedral angles between the molecular planes of the carbazolyl and isophthalonitrile moieties. When this angle is constrained to 90°, the oscillator strength drops by four orders of magnitude, thus markedly decreasing the luminescence probability. In this case, only one triplet state of B_1 symmetry is located below the first excited singlet state, and their mutual SOCME vanishes by symmetry selection rules. This shows that the deviation from an orthogonal orientation of the donor and acceptor units is essential for the performance of the 4CzIPN TADF emitter.

8.4.2.2 Mechanism of the Triplet-to-Singlet Upconversion in the Assistant Dopants ACRXTN and ACRSA

Recently, one of us started investigating the photophysics of 3-(9,9-dimethylacridin-10(9H)-yl)-9H-xanthen-9-one (ACRXTN) and 10-phenyl-10H,10'H-spiro[acridine-9,9'-anthracen]-10'-one (ACRSA) (Figure 8.10) by quantum chemical methods [128, 129]. ACRXTN and ACRSA have been utilized as assistant dopants in OLEDs [24]. The idea behind this approach is to use triplet excitons for populating the S_1 state of the assistant dopant by rISC. Instead of radiatively decaying by fluorescence, the S_1 state transfers its excitation energy by FRET to a strongly fluorescent organic emitter. Nakanotani et al. [24] could show that the presence of the assistant dopant substantially improved the external electroluminescence quantum efficiency of the OLED, indicating an internal exciton production efficiency of nearly 100%.

In ACRXTN, acridine and xanthone are covalently linked, with the corresponding molecular planes arranged in a perpendicular fashion (see Figure 8.10). The HOMO of ACRXTN is a π-type orbital on the acridine moiety, whereas its LUMO is a π* orbital localized on xanthone. Hence, the lowest electronically excited state is expected to have ICT character. For computing electronically excited states, TDDFT [133] in conjunction with the B3LYP density functional,

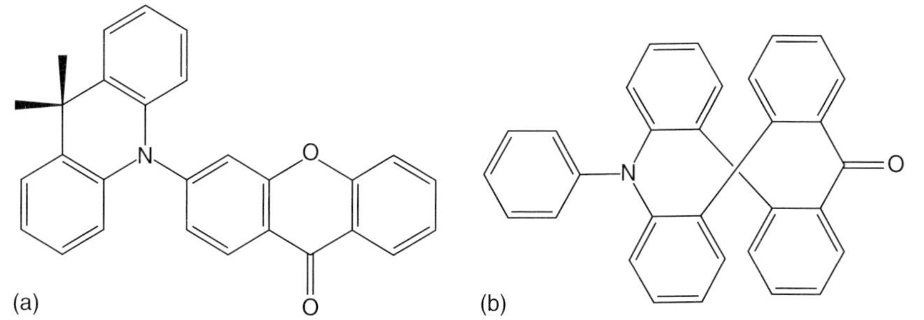

Figure 8.10 Chemical structures of the assistant dopants (a) ACRXTN and (b) ACRSA.

resolution-of-the-identity approximated coupled-cluster response methods (RI-CC2) [134, 135] as well as the redesigned DFT/MRCI-R [72] quantum chemical methods were employed. All theoretical methods agree that the lowest excited triplet and singlet states originate from an ICT excitation from acridine to xanthone [128]. Experimentally, the fluorescence (F) and phosphorescence (P) emissions in dichloromethane peak at 2.53 and 2.47 eV, respectively. The vertical DFT/MRCI-R emission energies in vacuum are only slightly larger (F: 2.77 eV, P: 2.71 eV) in the SV(P) basis, RI-CC2 yields an even higher value of 3.09 eV for both, whereas TDDFT/B3LYP gives 2.19 eV (F) and 2.18 eV (P). While the energetic separation between the LE and ICT states is nearly identical for DFT/MRCI-R and RI-CC2, this is not the case for TDDFT/B3LYP. Hence, it appears that RI-CC2 might be better suited for the optimization of excited-state geometries than TDDFT.

ACRXTN seems to have inherited some of the photophysical properties of the parent monochromophores [44]. In addition to the ICT states, two low-lying triplet states with $^3(n_O \pi_L^*)$ and $^3(\pi\pi^*)$ electronic structure as well as a $^1(n_O \pi_L^*)$ state are found that correspond to local excitations of the xanthone moeity [128]. So far, ISC and rISC rate constants have not been determined for this kind of complex. From the course of the potential energy curves and the knowledge of the coupling matrix elements, the following qualitative picture emerges.

In apolar media, the S_1 potential energy surface of ACRXTN exhibits at least two minima, the global minimum with ICT electronic structure and a local minimum originating from a local $(n_O \pi_L^*)$ excitation on xanthone. Two or three minima are expected on the lowest triplet excited-state surface, with the triplet ICT minimum being the global one. The second minimum on the T_1 surface exhibits $T_{\pi\pi^*}$ electronic structure. It is nearly degenerate with the $T_{n\pi^*}$ minimum in apolar media. In the ICT potential well, the singlet–triplet splitting is small enough (0.06 eV) to enable, at least in principle, the thermally activated rISC from the triplet to the corresponding singlet. However, the direct SOC between the states is too small (sum of squares $\approx 10^{-3}$ cm^{-2}) to make this process efficient. The carbonyl stretching vibration drives the system through a crossing with the $T_{\pi\pi^*}$ state that mediates the coupling of the ICT states and allows for an equilibration of the singlet and triplet ICT populations. In polar media, the $(n_O \pi^*)$ states are blueshifted, whereas the $T_{\pi\pi^*}$ and ICT states experience slight redshifts. Hence, a double minimum situation on the lowest triplet excited-state surface can be foreseen. In contrast, only one minimum with ICT character is expected on the lowest singlet excited-state potential energy surface. The $T_{\pi\pi^*}$ state continues to be the doorway state mediating the (r)ISC of the singlet and triplet ICT states in ARXCTN.

The spiro-compound ACRSA is very similar in that respect. Within an energy interval of 0.3 eV, five electronically excited states are found in the gas phase and in apolar solvents [129]. In ACRSA, the ICT from the phenylacridine to the anthracenone chromophores (HOMO → LUMO) is the lowest excitation in the vertical absorption region. Because of its low oscillator strength (4×10^{-5}), the singlet transition at 378 nm is barely visible in the calculated absorption spectrum. This is also true for the $^1(n_O\pi_L^*)$ excitation (346 nm) and the (HOMO-1 → LUMO) transition (318 nm). According to our calculations, the shoulder in the absorption spectrum around 310 nm stems from two local acridine transitions. The shoulder is slightly blueshifted with respect to experiment where a shoulder is observed around 320 nm in toluene solution [122]. This is also the case for the band maximum that is found at about 280 nm compared with the experimental value at approximately 300 nm. It arises from acridine to phenyl excitations. The overall shape of the computed absorption spectrum agrees very well with the experimental spectrum in toluene. Test calculations for the isolated system suggest that the computed excitation energies are lowered by about 0.1 eV when a larger basis set of valence triple-zeta plus polarization quality is used, thus improving the agreement with experiment.

With a relative permittivity of $\epsilon_r = 2.38$, toluene is an apolar solvent. Nevertheless, the solvent–solute interactions, modeled by conductor-like solvent model (COSMO) [136, 137], preferentially stabilize the ICT state by 0.09 eV, whereas the $n_O\pi_L^*$ state is destabilized by 0.10 eV with respect to the gas phase. Adiabatically, the lowest excited state does not stem from the (HOMO → LUMO) transition. At the DFT/MRCI-R level of theory, an LE state of the anthracenone moeity, $^3(n_O\pi_L^*)$, constitutes the global minimum on the T_1 potential energy surface. Close by, the ^{3}CT and the lowest $^3\pi\pi^*$ states are located. Likewise, the global minimum of the S_1 state has $^1(n_O\pi_L^*)$ character that is nearly degenerate with the ^{1}CT state. For this reason, strong nonadiabatic coupling is expected in addition to SOC. Similar to ACRXTN, the C=O stretching mode drives the low-lying states toward intersections of the potential energy surfaces (Figure 8.11). And indeed, rate constants for rISC of the order of 10^9 s^{-1} in toluene have been derived from quantum dynamics calculations including vibronic coupling and SOC simultaneously [129]. The efficient rISC in this compound is attributed to the presence of LE $n\pi^*$ and $\pi\pi^*$ states and their strong interaction with the CT states.

Vibronic interactions are also required to make fluorescence electric dipole allowed. The calculated vertical singlet emission energies of 2.93 eV (423 nm) for the ICT state and of 2.89 eV (429 nm) for the $n_O\pi_L^*$ state in toluene are substantially blueshifted with respect to the photoluminescence band maximum that is found at about 500 nm in the same solvent according to Nasu et al. [122]. At present, it is not clear where the discrepancy comes from. Two things are striking, however. Firstly, in DPEPO film the emission maximum is found experimentally at about 480 nm [122]. Due to the polarity of DPEPO, a bathochromic shift of the

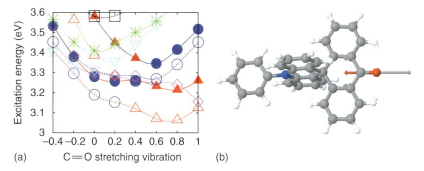

Figure 8.11 DFT/MRCI-R energy profiles of ACRSA along the C=O stretching normal coordinate (mode 138). Zero represents the ground-state equilibrium geometry (C=O bond length 122 pm), positive/negative distortions correspond to an elongation/a shortening of the C=O bond. Solid lines: singlets; dashed lines: triplets; triangles: ICT states with leading $\pi_H\pi_L^*$ configuration; circles: $n_O\pi_L^*$ states; diamonds: $\pi\pi^*$ states with leading $\pi_O\pi_L^*$ configuration; upside down triangles: $\pi\pi^*$ states with leading $\pi_H\pi_{L+1}^*$ configuration; squares: $\pi\pi^*$ states with leading $\pi_{H-1}\pi_L^*$ configuration; stars: $\pi\pi^*$ states with $\pi_{H-5}\pi_L^*$ leading configuration.

peak maximum with respect to its wavelength in toluene solution would have been expected. Instead, a hypsochromic shift of at least 20 nm is found.

Secondly, ACRSA is used as an assistant dopant for the blue fluorescence emitter tetra-ter-butylperylene (TBPe). Herein, it is assumed that ACRSA transfers its excitation energy by FRET to TBPe. Save for the proper orientation of the transition dipole moments, FRET is only efficient, however, if the emission spectrum of the FRET donor and the absorption spectrum of the FRET acceptor have substantial overlap [138]. As may be seen from Figure 1b of Ref. [24], there is barely any overlap between the emission spectrum ascribed to ACRSA and the absorption spectrum of TBPe that has its origin transition at wavelength shorter than 450 nm. As the enhancement of external quantum efficiency by the assistant dopant is undoubted, the shown emission spectrum can probably not be assigned to ACRSA.

Polar solvents such as acetonitrile ($\epsilon_r \approx 36$) shift the $(n_O\pi_L^*)$ states substantially toward higher excitation energies, whereas the lowest $^3\pi\pi^*$ state is nearly unaffected by the solvent. In contrast, the ICT states are significantly redshifted. Because of these trends, we expect the electronic states to be turned into and out of resonance depending on the particular environment and the temperature.

8.5 Outlook and Concluding Remarks

Insight into the factors that determine the probability of TADF is a key step toward the design and optimization of third-generation OLED emitters. Despite intensive research on this topic in the latest years, a complete and consistent rationalization of TADF is still missing. As outlined in this chapter, a small singlet–triplet splitting of the electronically excited emitter states is not sufficient for TADF to take place. Rather, the molecular parameters that steer the relative probabilities of excited-state processes such as intramolecular charge and energy transfer, ISC, rISC, fluorescence, phosphorescence, and nonradiative deactivation have to be understood. Computational chemistry can substantially contribute to this understanding. In particular, it can provide detailed information about spectroscopically dark states and their coupling to the luminescent ones, information that is difficult or even impossible to obtain from experimental data alone. Moreover, starting from a lead structure, quantum chemistry can easily assess the effects of chemical substitution.

With regard to internal quantum yields and rate constants, experimental and theoretical information is complimentary. Experimentally, (r)ISC rate constants are often determined indirectly from the quantum yields of the prompt and delayed components. Time-resolved spectra of such complexes, from which ISC rate constants could be retrieved directly, are still scarce. While suffering from uncertainties with regard to the underlying models and quantum chemical methods, theory can, in principle, determine rate constants and use them to derive internal TADF quantum yields that can be compared with experimental data.

The quantum chemical methods employed in our preliminary work are well suited for computing spectroscopic properties of the systems at hand. DFT/MRCI-R is the method of choice for obtaining reliable electronic excitation energies and excited-state properties at reasonable cost in the purely organic donor–acceptor systems. SPOCK is a powerful tool for computing electronic SOCME and phosphorescence rate constants. Less demanding single-reference linear response methods such as RI-CC2 or, where applicable, TDDFT can be used to obtain excited-state minimum geometries and vibrational frequencies. Because of the comparably low nonradiative transition rates, typical for (reverse) ISC in TADF emitters, a static Fourier transform formalism seems appropriate for calculating the rate constants. What is presently missing is an efficient way to compute vibronic SOC rates for larger systems. Vibronic SOC is considered essential in donor–acceptor systems because of the small magnitude of the direct electronic SOCMEs between singlet and triplet CT states.

The latest class of hyperfluorescent OLEDs combines high internal quantum efficiency and long operational stability by using assistant dopants for the harvesting of triplet and singlet excitons in addition to fluorescence emitters. So far, quantum chemical research in that direction is missing. Modeling the excitation energy transfer from the assistant dopant to the fluorescent acceptor beyond the ideal dipole approximation is a challenging task that could be worth considering.

References

1 Tang, C.W. and VanSlyke, S.A. (1987). Organic electroluminescent diodes. *Appl. Phys. Lett.* 51: 913–915.
2 Baldo, M.A., O'Brien, D.F., You, Y., Shoustikov, A., Sibley, S., Thompson, M.E., and Forrest, S.R. (1998). Highly efficient phosphorescent emission from organic electroluminescent devices. *Nature* 395: 151–154.
3 Adachi, C. (2014). Third-generation organic electroluminescence materials. *Jpn. J. Appl. Phys.* 53: 060101.

4 Adachi, C. and Adachi, J. (2015). High performance TADF for OLEDs. *Presentation at the 2015 SSL R&D WORKSHOP in San Francisco, CA*. http://energy.gov/sites/prod/files/2015/02/f19/adachi_oled-tadf_sanfrancisco2015.pdf (accessed 10 March 2018).

5 Thompson, M. (2007). The evolution of organometallic complexes in organic light-emitting devices. *MRS Bull.* 32: 694–701.

6 Kappaun, S., Slugovc, C., and List, E.J.W. (2008). Phosphorescent organic light-emitting devices: working principle and iridium based emitter materials. *Int. J. Mol. Sci.* 9: 1527–1547.

7 Yersin, H., Rausch, A.F., Czerwieniec, R., Hofbeck, T., and Fischer, T. (2011). The triplet state of organo-transition metal compounds. Triplet harvesting and singlet harvesting for efficient OLEDs. *Coord. Chem. Rev.* 255: 2622–2652.

8 Wagenknecht, P.S. and Ford, P.C. (2011). Metal centered ligand field excited states: their roles in the design and performance of transition metal based photochemical molecular devices. *Coord. Chem. Rev.* 255: 591–616.

9 Choy, W.C.H., Chan, W.K., and Yuan, Y. (2014). Recent advances in transition metal complexes and light-management engineering in organic optoelectronic devices. *Adv. Mater.* 26: 5368–5399.

10 Baldo, M.A., Lamansky, S., Burrows, P.E., Thompson, M.E., and Forrest, S.R. (1999). Very high-efficiency green organic light-emitting devices based on electrophosphorescence. *Appl. Phys. Lett.* 75: 4–6.

11 Sajoto, T., Djurovich, P.I., Tamayo, A.B., Oxgaard, J., Goddard, W.A. (2009). Temperature dependence of blue phosphorescent cyclometalated Ir(III) complexes. *J. Am. Chem. Soc.* 131: 9813–9822.

12 Hudson, Z.M., Sun, C., Helander, M.G., Chang, Y.L., Lu, Z.H., and Wang, S. (2012). Highly efficient blue phosphorescence from triarylboron-functionalized platinum(II) complexes of N-heterocyclic carbenes. *J. Am. Chem. Soc.* 134: 13930–13933.

13 Darmawan, N., Yang, C.H., Mauro, M., Raynal, M., Heun, S., Pan, J., Buchholz, H., Braunstein, P., and Cola, L.D. (2013). Efficient near-UV emitters based on cationic bis-pincer iridium(III) carbene complexes. *Inorg. Chem.* 52: 10756–10765.

14 Reineke, S., Thomschke, M., Lüssem, B., and Leo, K. (2013). White organic light-emitting diodes: status and perspective. *Rev. Mod. Phys.* 85: 1245–1293.

15 Cariati, E., Lucenti, E., Botta, C., Giovanella, U., Marinotto, D., and Righetto, S. (2016). Cu(I) hybrid inorganic-organic materials with intriguing stimuli responsive and optoelectronic properties. *Coord. Chem. Rev.* 306: 566–614.

16 Czerwieniec, R., Leitl, M.J., Homeier, H.H., and Yersin, H. (2016). Cu(I) complexes - thermally activated delayed fluorescence. Photophysical approach and material design. *Coord. Chem. Rev.* 325: 2–28.

17 Leitl, M.J., Zink, D.M., Schinabeck, A., Baumann, T., Volz, D., and Yersin, H. (2016). Copper(I) complexes for thermally activated delayed fluorescence: from photophysical to device properties. *Top. Curr. Chem.* 374: 25.

18 Parker, C.A. and Hatchard, C.G. (1961). Triplet-singlet emission in fluid solutions. phosphorescence of eosin. *Trans. Faraday Soc.* 57: 1894–1904.

19 Uoyoma, H., Goushi, K., Shizu, K., Nomura, H., and Adachi, C. (2012). Highly efficient organic light-emitting diodes from delayed fluorescence. *Nature* 492: 234–238.
20 Förster, T. (1959). 10th spiers memorial lecture. Transfer mechanisms of electronic excitation. *Discuss. Faraday Soc.* 27: 7–17.
21 Ermolaev, V.L. and Sveshnikova, E.B. (1963). Inductive-resonance transfer of energy from aromatic molecules in the triplet state. *Dokl. Akad. Nauk* 149: 1295–1298.
22 Baldo, M.A., Thompson, M.E., and Forrest, S.R. (2000). High-efficiency fluorescent organic light-emitting devices using a phosphorescent sensitizer. *Nature* 403: 750–753.
23 Fukagawa, H., Shimizu, T., Kamada, T., Yui, S., Hasegawa, M., Morii, K., and Yamamoto, T. (2015). Highly efficient and stable organic light-emitting diodes with a greatly reduced amount of phosphorescent emitter. *Sci. Rep.* 5: 9855/1–7.
24 Nakanotani, H., Higuchi, T., Furukawa, T., Masui, K., Morimoto, K., Numata, M., Tanaka, H., Sagara, Y., Yasuda, T., and Adachi, C. (2014). High-efficiency organic light-emitting diodes with fluorescent emitters. *Nat. Commun.* 5: 4016–4022.
25 Jou, J.H., Kumar, S., Agrawal, A., Li, T.H., and Sahoo, S. (2015). Approaches for fabricating high efficiency organic light emitting diodes. *J. Mater. Chem. C* 3: 2974–3002.
26 Jankus, V., Data, P., Graves, D., McGuinness, C., Santos, J., Bryce, M.R., Dias, F.B., and Monkman, A.P. (2014). Highly effcient TADF OLEDs: how the emitter-host interaction controls both the excited state species and electrical properties of the devices to achieve near 100% triplet harvesting and high efficiency. *Adv. Funct. Mater.* 24: 6178–6186.
27 Klán, P. and Wirz, J. (2009). *Photochemistry of Organic Compounds: From Concepts to Practice*. Wiley.
28 Vosskötter, S., Konieczny, P., Marian, C.M., and Weinkauf, R. (2015). Towards an understanding of the singlet-triplet splittings in conjugated hydrocarbons: azulene investigated by anion photoelectron spectroscopy and theoretical calculations. *Phys. Chem. Chem. Phys.* 17: 23573–23581.
29 Etinski, M., Petković, M., Ristić, M.M., and Marian, C.M. (2015). Electron-vibrational coupling and fluorescence spectra of tetra-, penta-, and hexacoordinated chlorophylls c_1 and c_2. *J. Phys. Chem. B* 119: 10156–10169.
30 Marian, C. (2001). Spin–orbit coupling in molecules. In: *Reviews in Computational Chemistry*, vol.17 (ed. K. Lipkowitz and D. Boyd), 99–204. Weinheim: Wiley-VCH.
31 Marian, C.M. (2012). Spin–orbit coupling and intersystem crossing in molecules. *WIREs Comput. Mol. Sci.* 2: 187–203.
32 Heß, B.A., Marian, C.M., Wahlgren, U., and Gropen, O. (1996). A mean-field spin–orbit method applicable to correlated wavefunctions. *Chem. Phys. Lett.* 251: 365–371.
33 El-Sayed, M.A. (1968). The triplet state: its radiative and nonradiative properties. *Acc. Chem. Res.* 1: 8–16.

34 Englman, R. and Jortner, J. (1970). Energy gap law for radiationless transitions in large molecules. *Mol. Phys.* 18: 145–164.

35 Freed, K.F. and Jortner, J. (1970). Multiphonon processes in the nonradiative decay of large molecules. *J. Chem. Phys.* 52: 6272–6291.

36 Marcus, R.A. (1984). Nonadiabatic processes involving quantum-like and classical-like coordinates with applications to nonadiabatic electron transfers. *J. Chem. Phys.* 81: 4494–4500.

37 Etinski, M. and Marian, C.M. (2010). Overruling the energy gap law: fast triplet formation in 6-azauracil. *Phys. Chem. Chem. Phys.* 12: 15665–15671.

38 Henry, B.R. and Siebrand, W. (1971). Spin–orbit coupling in aromatic hydrocarbons. Analysis of nonradiative transitions between singlet and triplet states in benzene and naphthalene. *J. Chem. Phys.* 54: 1072–1085.

39 Tatchen, J., Gilka, N., and Marian, C.M. (2007). Intersystem crossing driven by vibronic spin–orbit coupling: a case study on psoralen. *Phys. Chem. Chem. Phys.* 9: 5209–5221.

40 Perun, S., Tatchen, J., and Marian, C.M. (2008). Singlet and triplet excited states and intersystem crossing in free-base porphyrin: TDDFT and DFT/MRCI study. *ChemPhysChem* 9: 282–292.

41 Salzmann, S., Tatchen, J., and Marian, C.M. (2008). The photophysics of flavins: what makes the difference between gas phase and aqueous solution. *J. Photochem. Photobiol., A* 198: 221–231.

42 Rai-Constapel, V., Salzmann, S., and Marian, C.M. (2011). Isolated and solvated thioxanthone: a photophysical study. *J. Phys. Chem. A* 115: 8589–8596.

43 Peng, Q., Niu, Y., Shi, Q., Gao, X., and Shuai, Z. (2013). Correlation function formalism for triplet excited state decay: Combined spin–orbit and nonadiabatic couplings. *J. Chem. Theory Comput.* 9: 1132–1143.

44 Rai-Constapel, V., Etinski, M., and Marian, C.M. (2013). Photophysics of xanthone: a quantum chemical perusal. *J. Phys. Chem. A* 117: 3935–3944.

45 Rai-Constapel, V. and Marian, C.M. (2016). Solvent tunable photophysics of acridone: a quantum chemical perspective. *RSC Adv.* 6: 18530–18537.

46 Danovich, D., Marian, C.M., Neuheuser, T., Peyerimhoff, S.D., and Shaik, S. (1998). Spin–orbit coupling patterns induced by twist and pyramidalization modes in C_2H_4: a quantitative study and a qualitative analysis. *J. Phys. Chem. A* 102: 5923–5936.

47 Cogan, S., Haas, Y., and Zilberg, S. (2007). Intersystem crossing at singlet conical intersections. *J. Photochem. Photobiol., A* 190: 200–206.

48 Penfold, T.J. and Worth, G.A. (2010). The effect of molecular distortions on spin–orbit coupling in simple hydrocarbons. *Chem. Phys.* 375: 58–66.

49 Beljonne, D., Shuai, Z., Pourtois, G., and Bredas, J.L. (2001). Spin–orbit coupling and intersystem crossing in conjugated polymers: a configuration interaction description. *J. Phys. Chem. A* 105: 3899–3907.

50 Burin, A.L. and Ratner, M.A. (1998). Spin effects on the luminescence yield of organic light emitting diodes. *J. Chem. Phys.* 109: 6092–6102.

51 Etinski, M., Tatchen, J., and Marian, C.M. (2011). Time-dependent approaches for the calculation of intersystem crossing rates. *J. Chem. Phys.* 134: 154105.

52 Duschinsky, F. (1937). The importance of the electron spectrum in multi atomic molecules concerning the Franck-Condon principle. *Acta Physicochim.* 7: 551–566.

53 Etinski, M., Rai-Constapel, V., and Marian, C.M. (2014). Time-dependent approach to spin-vibronic coupling: implementation and assessment. *J. Chem. Phys.* 140: 114104.

54 Capano, G., Penfold, T.J., Röthlisberger, U., and Tavernelli, I. (2014). A vibronic coupling Hamiltonian to describe the ultrafast excited state dynamics of a Cu(I)-phenanthroline complex. *Chimia* 68: 227–230.

55 Capano, G., Chergui, M., Röthlisberger, U., Tavernelli, I., and Penfold, T.J. (2014). A quantum dynamics study of the ultrafast relaxation in a prototypical Cu(I) phenanthroline. *J. Phys. Chem. A* 118: 9861–9869.

56 Gibson, J., Monkman, A.P., and Penfold, T.J. (2016). The importance of vibronic coupling for efficient reverse intersystem crossing in thermally activated delayed fluorescence molecules. *ChemPhysChem* 17: 2956–2961.

57 Köppel, H., Domcke, W., and Cederbaum, L.S. (1984). Multi-mode molecular dynamics beyond the Born–Oppenheimer approximation. *Adv. Chem. Phys.* 57: 59–246.

58 Beck, M.H., Jäckle, A., Worth, G.A., and Meyer, H.D. (2000). The multiconfiguration time-dependent Hartree method: a highly efficient algorithm for propagating wavepackets. *Phys. Rep.* 324: 1–105.

59 Dreuw, A. and Head-Gordon, M. (2004). Failure of time-dependent density functional theory for long-range charge-transfer excited states: the zincbacteriochlorin–bacteriochlorin and bacteriochlorophyll–spheroidene complexes. *J. Am. Chem. Soc.* 126: 4007–4016.

60 Huang, S., Zhang, Q., Shiota, Y., Nakagawa, T., Kuwabara, K., Yoshizawa, K., and Adachi, C. (2013). Computational prediction for singlet- and triplet-transition energies of charge-transfer compounds. *J. Chem. Theory Comput.* 9: 3872–3877.

61 Moral, M., Muccioli, L., Son, W.J., Olivier, Y., and Sancho-García, J.C. (2015). Theoretical rationalization of the singlet-triplet gap in OLEDs materials: impact of charge-transfer character. *J. Chem. Theory Comput.* 11: 168–177.

62 Grimme, S. (2006). Semiempirical hybrid density functional with perturbative second-order correlation. *J. Chem. Phys.* 124: 034108.

63 Karton, A., Tarnopolsky, A., Lamère, J.F., Schatz, G.C., and Martin, J.M.L. (2008). Highly accurate first-principles benchmark data sets for the parametrization and validation of density functional and other approximate methods. Derivation of a robust, generally applicable, double-hybrid functional for thermochemistry and thermochemical kinetics. *J. Phys. Chem. A* 112: 12868–12886.

64 Grimme, S. and Waletzke, M. (1999). A combination of Kohn-Sham density functional theory and multi-reference configuration interaction methods. *J. Chem. Phys.* 111 (13): 5645–5655.

65 Becke, A.D. (1993). A new mixing of Hartree-Fock and local density-functional theories. *J. Chem. Phys.* 98: 1372–1377.

66 Silva-Junior, M.R., Schreiber, M., Sauer, S.P.A., and Thiel, W. (2008). Benchmarks for electronically excited states: time-dependent density functional theory and density functional theory based multireference configuration interaction. *J. Chem. Phys.* 129: 104103.

67 Kleinschmidt, M., Marian, C.M., Waletzke, M., and Grimme, S. (2009). Parallel multireference configuration interaction calculations on mini-β-carotenes and β-carotene. *J. Chem. Phys.* 130: 044708.

68 Escudero, D. and Thiel, W. (2014). Assessing the density functional theory-based multireference configuration interaction (DFT/MRCI) method for transition metal complexes. *J. Chem. Phys.* 140: 194105/1–8.

69 Kleinschmidt, M., van Wüllen, C., and Marian, C.M. (2015). Intersystem-crossing and phosphorescence rates in facIr^{III}(ppy)$_3$: A theoretical study involving multi-reference configuration interaction wavefunctions. *J. Chem. Phys.* 142: 094301.

70 Heil, A., Gollnisch, K., Kleinschmidt, M., and Marian, C.M. (2016). On the photophysics of four heteroleptic iridium(III) phenylpyridyl complexes investigated by relativistic multi-configuration methods. *Mol. Phys.* 114: 407–422.

71 Marian, C.M. and Gilka, N. (2008). Performance of the DFT/MRCI method on electronic excitation of extended π-systems. *J. Chem. Theory Comput.* 4: 1501–1515.

72 Lyskov, I., Kleinschmidt, M., and Marian, C.M. (2016). Redesign of the DFT/MRCI Hamiltonian. *J. Chem. Phys.* 144: 034104.

73 Jovanović, V., Lyskov, I., Kleinschmidt, M., and Marian, C.M. (2017). On the performance of DFT/MRCI-R and MR-MP2 in spin–orbit coupling calculations. *Mol. Phys.* 115: 109–137.

74 Grimme, S., Antony, J., Ehrlich, S., and Krieg, H. (2010). A consistent and accurate ab initio parametrization of density functional dispersion correction (DFT-D) for the 94 elements H-Pu. *J. Chem. Phys.* 132: 154104.

75 Helgaker, T., Jørgensen, P., and Olsen, J. (2000). *Molecular Electronic-Structure Theory*. Wiley.

76 Kleinschmidt, M., Tatchen, J., and Marian, C.M. (2006). SPOCK.CI: a multireference spin–orbit configuration interaction method for large molecules. *J. Chem. Phys.* 124: 124101.

77 Ågren, H., Vahtras, O., and Minaev, B. (1996). Response theory and calculations of spin–orbit coupling phenomena in molecules. *Adv. Quantum Chem.* 27: 71–162.

78 Christiansen, O., Gauss, J., and Schimmelpfennig, B. (2000). Spin–orbit coupling constants from coupled-cluster response theory. *Phys. Chem. Chem. Phys.* 2: 965–971.

79 Deaton, J.C., Switalski, S.C., Kondakov, D.Y., Young, R.H., Pawlik, T.D., Giesen, D.J., Harkins, S.B., Miller, A.J.M., Mickenberg, S.F., and Peters, J.C. (2010). E-type delayed fluorescence of a phosphine-supported Cu$_2$(μ-NAr$_2$)$_2$ diamond core: harvesting singlet and triplet excitons in OLEDs. *J. Am. Chem. Soc.* 132: 9499–9508.

80 Czerwieniec, R., Kowalski, K., and Yersin, H. (2013). Highly efficient thermally activated fluorescence of a new rigid Cu(I) complex [Cu(dmp)(phanephos)]$^+$. *Dalton Trans.* 42: 9826–9830.

81 Chen, X.L., Yu, R., Zhang, Q.K., Zhou, L.J., Wu, X.Y., Zhang, Q., and Lu, C.Z. (2013). Rational design of strongly blue-emitting cuprous complexes with thermally activated delayed fluorescence and application in solution-processed OLEDs. *Chem. Mater.* 25: 3910–3920.

82 Osawa, M., Kawata, I., Ishii, R., Igawa, S., Hashimoto, M., and Hoshino, M. (2013). Application of neutral d^{10} coinage metal complexes with an anionic bidentate ligand in delayed fluorescence-type organic light-emitting diodes. *J. Mater. Chem. C* 1: 4375–4383.

83 Linfoot, C.L., Leitl, M.J., Richardson, P., Rausch, A.F., Chepelin, O., White, F.J., Yersin, H., and Robertson, N. (2014). Thermally activated delayed fluorescence (TADF) and enhancing photoluminescence quantum yields of [Cu^{I}(diimine)(diphosphine)]$^{+}$ complexes–photophysical, structural, and computational studies. *Inorg. Chem.* 53: 10854–10861.

84 Czerwieniec, R. and Yersin, H. (2015). Diversity of copper(I) complexes showing thermally activated delayed fluorescence: basic photophysical analysis. *Inorg. Chem.* 54: 4322–4327.

85 Osawa, M., Hoshino, M., Hashimoto, M., Kawata, I., Igawa, S., and Yashima, M. (2015). Application of three-coordinate copper(I) complexes with halide ligands in organic light-emitting diodes that exhibit delayed fluorescence. *Dalton Trans.* 44: 8369–8378.

86 Zhang, Q., Chen, J., Wu, X.Y., Chen, X.L., Yu, R., and Lu, C.Z. (2015). Outstanding blue delayed fluorescence and significant processing stability of cuprous complexes with functional pyridine–pyrazolate diimine ligands. *Dalton Trans.* 44: 6706–6710.

87 Bergmann, L., Hedley, G.J., Baumann, T., Bräse, S., and Samuel, I.D.W. (2016). Direct observation of intersystem crossing in a thermally activated delayed fluorescence copper complex in the solid state. *Sci. Adv.* 2: e1500889.

88 Leitl, M.J., Küchle, F.R., Mayer, H.A., Wesemann, L., and Yersin, H. (2013). Brightly blue and green emitting Cu(I) dimers for singlet harvesting in oleds. *J. Phys. Chem. A* 117: 11823–11836.

89 Hofbeck, T., Monkowius, U., and Yersin, H. (2015). Highly efficient luminescence of Cu(I) compounds: thermally activated delayed fluorescence combined with short-lived phosphorescence. *J. Am. Chem. Soc.* 137: 399–404.

90 Krylova, V.A., Djurovich, P.I., Whited, M.T., and Thompson, M.E. (2010). Synthesis and characterization of phosphorescent three-coordinate Cu(I)–NHC complexes. *Chem. Commun.* 46: 6696–6698.

91 Krylova, V.A., Djurovich, P.I., Aronson, J.W., Haiges, R., Whited, M.T., and Thompson, M.E. (2012). Structural and photophysical studies of phosphorescent three-coordinate copper(I) complexes supported by an N-heterocyclic carbene ligand. *Organometallics* 31: 7983–7993.

92 Krylova, V.A., Djurovich, P.I., Conley, B.L., Haiges, R., Whited, M.T., Williams, T.J., and Thompson, M.E. (2014). Control of emission colour with N-heterocyclic carbene (NHC) ligands in phosphorescent three-coordinate Cu(I) complexes. *Chem. Commun.* 50: 7176–7179.

93 Leitl, M.J., Krylova, V.A., Djurovich, P.I., Thompson, M.E., and Yersin, H. (2014). Phosphorescence versus thermally activated delayed fluorescence. Controlling singlet–triplet splitting in brightly emitting and sublimable Cu(I) compounds. *J. Am. Chem. Soc.* 136: 16032–16038.

94 Osawa, M. (2014). Highly efficient blue-green delayed fluorescence from copper(I) thiolate complexes: luminescence color alteration by orientation change of the aryl ring. *Chem. Commun.* 50: 1801–1803.

95 Gernert, M., Müller, U., Haehnel, M., Pflaum, J., and Steffen A. (2017). A cyclic alkyl(amino)carbene as two-atom-pi-chromophore leading to the first phosphorescent linear Cu(I) complexes. *Chem. Eur. J.* 23: 2206–2216.

96 Gneuß, T., Leitl, M.J., Finger, L.H., Rau, N., Yersin, H., and Sundermeyer, J. (2015). A new class of luminescent Cu(I) complexes with tripodal ligands - TADF emitters for the yellow to red color range. *Dalton Trans.* 44: 8506–8520.

97 Cheng, G., So, G.K.M., To, W.P., Chen, Y., Kwok, C.C., Ma, C., Guan, X., Chang, X., Kwok, W.M., and Che, C.M. (2015). Luminescent zinc(II) and copper(I) complexes for high-performance solution-processed monochromic and white organic light-emitting devices. *Chem. Sci.* 6: 4623–4635.

98 Siddique, Z.A., Yamamoto, Y., Ohno, T., and Nozaki, K. (2003). Structure-dependent photophysical properties of singlet and triplet metal-to-ligand charge transfer states in copper(I) bis(diimine) compounds. *Inorg. Chem.* 42: 6366–6378.

99 Capano, G., Röthlisberger, R.U., Tavernelli, I., and Penfold, T.J. (2015). Theoretical rationalization of the emission properties of prototypical Cu(I)-phenanthroline complexes. *J. Phys. Chem. A* 119: 7026–7037.

100 Föller, J., Kleinschmidt, M., and Marian, C.M. (2016). Phosphorescence or thermally activated delayed fluorescence? Intersystem crossing and radiative rate constants of a three-coordinate Cu(I) complex determined by quantum chemical methods. *Inorg. Chem.* 55: 7508–7516.

101 Kleinschmidt, M., Tatchen, J., and Marian, C.M. (2002). Spin–orbit coupling of DFT/MRCI wavefunctions: method, test calculations, and application to thiophene. *J. Comput. Chem.* 23: 824–833.

102 Kleinschmidt, M. and Marian, C.M. (2005). Efficient generation of matrix elements of one-electron spin–orbit operators. *Chem. Phys.* 311: 71–79.

103 Etinski, M., Tatchen, J., and Marian, C.M. (2014). Thermal and solvent effects on the triplet formation in cinnoline. *Phys. Chem. Chem. Phys.* 16: 4740–4751.

104 Kirchhoff, J.R., Gamache, R.E., Blaskie, M.W., Paggio, A.A.D., Lengel, R.K., and McMillin, D.R. (1983). Temperature dependence of luminescence from $Cu(NN)^{2+}$ systems in fluid solution. Evidence for the participation of two excited states. *Inorg. Chem.* 22: 2380–2384.

105 Hirata, S., Sakai, Y., Masui, K., Tanaka, H., Lee, S.Y., Nomura, H., Nakamura, N., Yasumatsu, M., Nakanotani, H., Zhang, Q., Shizu, K., Miyazaki, H., and Adachi, C. (2015). Highly efficient blue electroluminescence based on thermally activated delayed fluorescence. *Nat. Mater.* 14: 330–336.

106 Zgierski, M.Z. (2003). Cu(I)-2,9-dimethyl-1,10-phenanthroline: density functional study of the structure, vibrational force-field, and excited electronic states. *J. Chem. Phys.* 118: 4045–4051.

107 Wang, F. and Ziegler, T. (2005). A simplified relativistic time-dependent density-functional theory formalism for the calculations of excitation energies including spin–orbit coupling effect. *J. Chem. Phys.* 123 (15): 154102.

108 Bergmann, L. (2016). *New Emitters for OLEDs: The Coordination- and Photo-Chemistry of Mononuclear Neutral Copper(I) Complexes (Beiträge zur Organischen Synthese)*. Logos Verlag Berlin.

109 Bernanose, A., Comte, M., and Vouaux, P. (1953). A new method of light emission by certain organic compounds. *J. Chim. Phys.* 50: 64–68.

110 Yersin, H., Czerwieniec, R., and Hupfer, A. (2012). Singlet harvesting with brightly emitting Cu(I) and metal-free organic compounds. Proceedings of SPIE, Volume 8435, p. 843508.

111 Dias, F.B., Bourdakos, K.N., Jankus, V., Moss, K.C., Kamtekar, K.T., Bhalla, V., Santos, J., Bryce, M.R. (2013). Triplet harvesting with 100% efficiency by way of thermally activated delayed fluorescence in charge transfer oled emitters. *Adv. Mater.* 25: 3707–3714.

112 Ishimatsu, R., Matsunami, S., Shizu, K., Adachi, C., Nakano, K., and Imato, T. (2013). Solvent effect on thermally activated delayed fluorescence by 1,2,3,5-tetrakis(carbazol-9-yl)-4,6-dicyanobenzene. *J. Phys. Chem. A* 117: 5607–5612.

113 Wu, S., Aonuma, M., Zhang, Q., Huang, S., Nakagawa, T., Kuwabara, K., and Adachi, C. (2014). High-efficiency deep–blue organic light-emitting diodes based on a thermally activated delayed fluorescence emitter. *J. Mater. Chem. C* 2: 421–424.

114 Tanaka, H., Shizu, K., Nakanotani, H., and Adachi, C. (2014). Dual intramolecular charge-transfer fluorescence derived from a phenothiazine–triphenyltriazine derivative. *J. Phys. Chem. C* 118: 15985–15994.

115 Wang, H., Xie, L., Peng, Q., Meng, L., Wang, Y., Yi, Y., and Wang, P. (2014). Novel thermally activated delayed fluorescence materials-thioxanthone derivatives and their applications for highly efficient OLEDs. *Adv. Mater.* 26: 5198–5204.

116 Zhang, Q., Li, B., Huang, S., Nomura, H., Tanaka, H., and Adachi, C. (2014). Efficient blue organic light-emitting diodes employing thermally activated delayed fluorescence. *Nat. Photonics* 8: 326–332.

117 Shizu, K., Tanaka, H., Uejima, M., Sato, T., Tanaka, K., Kaji, H., and Adachi, C. (2015). Strategy for designing electron donors for thermally activated delayed fluorescence emitters. *J. Phys. Chem. C* 119: 1291–1297.

118 Dias, F.B., Santos, J., Graves, D.R., Data, P., Nobuyasu, R.S., Fox, M.A., Batsanov, A.S., Palmeira, T., Berberan-Santos, M.N., Bryce, M.R., and Monkman, A.P. (2016). The role of local triplet excited states and D-A relative orientation in thermally activated delayed fluorescence: photophysics and devices. *Adv. Sci.* 3: 1600080.

119 Data, P., Pander, P., Okazaki, M., Takeda, Y., Minakata, S., and Monkman, A.P. (2016). Dibenzo[a,j]phenazine-cored donor-acceptor-donor compounds as green-to-red/NIR thermally activated delayed fluorescence organic light emitters. *Angew. Chem.* 128: 5833–5838.

120 Etherington, M., Gibson, J., Higginbotham, H.F., Penfold, T.J., and Monkman, A.P. (2016). Second-order spin-vibronic coupling mediates triplet harvesting and thermally-activated delayed fluorescence in charge transfer molecules. *Nat. Commun.* 7: 13680.

121 Méhes, G., Nomura, H., Zhang, Q., Nakagawa, T., and Adachi, C. (2012). Enhanced electroluminescence efficiency in a spiro-acridine derivative through thermally activated delayed fluorescence. *Angew. Chem. Int. Ed.* 51: 11311–11315.

122 Nasu, K., Nakagawa, T., Nomura, H., Lin, C.J., Cheng, C.H., Tseng, M.R., Yasuda, T., and Adachi, C. (2013). A highly luminescent spiro-anthracenone-based organic light-emitting diode exhibiting thermally activated delayed fluorescence. *Chem. Commun.* 49: 10385–10387.

123 Lu, J., Zheng, Y., and Zhang, J. (2015). Tuning the color of thermally activated delayed fluorescent properties for spiro-acridine derivatives by structural modification of the acceptor fragment: a DFT study. *RSC Adv.* 5: 18588–18592.

124 Shizu, K., Noda, H., Tanaka, H., Taneda, M., Uejima, M., Sato, T., Tanaka, K., Kaji, H., and Adachi, C. (2015). Highly efficient blue electroluminescence using delayed-fluorescence emitters with large overlap density between luminescent and ground states. *J. Phys. Chem. C* 119: 26283–26289.

125 Chen, X.K., Zhang, S.F., Fan, J.X., and Ren, A.M. (2015). Nature of highly efficient thermally activated delayed fluorescence in organic light-emitting diode emitters: nonadiabatic effect between excited states. *J. Phys. Chem. C* 119: 9728–9733.

126 Li, Z.D., Suo, B.B., Zhang, Y., Xiao, Y.L., and Liu, W.J. (2013). Combining spin-adapted open-shell TD-DFT with spin–orbit coupling. *Mol. Phys.* 111: 3741–3755.

127 Marian, C.M. (2017). Unpublished results.

128 Marian, C.M. (2016). Mechanism of the triplet-to-singlet upconversion in the assistant dopant ACRXTN. *J. Phys. Chem. C* 120: 3715–3721.

129 Lyskov, I. and Marian, C.M. (2017). Climbing up the ladder: intermediate triplet states promote the reverse intersystem crossing in the efficient TADF emitter ACRSA. *J. Phys. Chem. C* 121: 21145–21153.

130 Nakanotani, H., Masui, K., Nishide, J., Shibata, T., and Adachi, C. (2013). Promising operational stability of high-efficiency organic light-emitting diodes based on thermally activated delayed fluorescence. *Sci. Rep.* 3: 2127.

131 Improta, R., Barone, V., Scalmani, G., and Frisch, M.J. (2006). A state-specific polarizable continuum model time dependent density functional theory method for excited state calculations in solution. *J. Chem. Phys.* 125: 054103.

132 Improta, R., Scalmani, G., Frisch, M.J., and Barone, V. (2007). Toward effective and reliable fluorescence energies in solution by a new state specific polarizable continuum model time dependent density functional theory approach. *J. Chem. Phys.* 127: 074504.

133 Furche, F. and Ahlrichs, R. (2002). Adiabatic time-dependent density functional methods for excited state properties. *J. Chem. Phys.* 117: 7433–7447.

134 Hättig, C. and Weigend, F. (2000). CC2 excitation energy calculations on large molecules using the resolution of the identity approximation. *J. Chem. Phys.* 113: 5154–5161.

135 Hättig, C., Hellweg, A., and Köhn, A. (2006). Distributed memory parallel implementation of energies and gradients for second-order Møller-Plesset perturbation theory with the resolution-of-the-identity approximation. *Phys. Chem. Chem. Phys.* 8: 1159–1169.

136 Klamt, A. and Schüürmann, G. (1993). COSMO: A new approach to dielectric screening in solvents with explicit expressions for the screening energy and its gradient. *J. Chem. Soc., Perkin Trans. 2* 5: 799–805.

137 Schäfer, A., Klamt, A., Sattel, D., Lohrenz, J.C.W., and Eckert, F. (2000). COSMO implementation in turbomole: extension of an efficient quantum chemical code towards liquid systems. *Phys. Chem. Chem. Phys.* 2: 2187–2193.

138 May, V. and Kühn, O. (2004). *Charge and Energy Transfer Dynamics in Molecular Systems*. Wiley-VCH.

9

The Role of Vibronic Coupling for Intersystem Crossing and Reverse Intersystem Crossing Rates in TADF Molecules

Thomas J. Penfold and Jamie Gibson

Newcastle University, School of Chemistry, NE1 7RU Newcastle upon Tyne, UK

9.1 Introduction

Since the first report of electroluminescence (EL) in poly(*para*-phenylenevinylene) [1], there has been an enormous amount of effort dedicated to the development of organic light-emitting diodes (OLEDs). However, early attempts to implement OLEDs based solely upon organic materials were restricted by an intrinsically low EL efficiency. This arose from the limit imposed by the statistics of the spin states formed under electrical excitation. Indeed, following electrical excitation only 25% of the excitons exist in the singlet states, while the remaining 75% form triplet excitons [2]. For organic species the latter, the triplet states, are not (strongly) dipole coupled to the molecular ground state due to the absence of strong spin–orbit coupling (SOC). Consequently, this energy cannot be harvested and is dissipated nonradiatively.

This limitation was overcome by doping the emitting layer with metal–organic emitters [3] yielding the so-called phosphorescence-based organic light-emitting diodes (PhOLEDs). For these devices, the presence of a metal–organic complex containing heavy atoms opens the potential for significant SOC and gives rise to a radiative $T_1 \rightarrow S_0$ transition (*triplet harvesting*) [4]. The radiative rate of this transition, $k^r(T_1 \rightarrow S_0)$, crucial to device efficiency [5] is formally zero but gains intensity through spin–orbit mixing with the close-lying singlet states S_n:

$$k^r(T_1 \rightarrow S_0) = \frac{64\pi^2 \nu^3}{3hc^3} \left| \sum_n \frac{\langle T_1 | \hat{H}_{so} | S_n \rangle}{\Delta E_{T_1-S_n}} \cdot \langle S_n | e\mathbf{q} | S_0 \rangle \right.$$

$$\left. + \sum_m \frac{\langle T_m | \hat{H}_{so} | S_0 \rangle}{\Delta E_{S_0-T_m}} \cdot \langle T_1 | e\mathbf{q} | T_m \rangle \right|^2 \quad (9.1)$$

where $\langle S_0 | e\mathbf{q} | S_n \rangle$ is the transition dipole moment between the ground state (S_0) and excited singlet states, S_n. $\langle S_n | \hat{H}_{so} | T_1 \rangle / \Delta E_{S_n-T_1} = a$ describes the first-order mixing coefficient between triplet states ($T_1^{SOC} = T_1^{noSOC} + aS_n$) that besides stimulating $k^r(T_1 \rightarrow S_0)$ also promotes intersystem crossing (ISC) and reverse intersystem crossing (rISC). The second sum in the brackets describes the contribution to the phosphorescence rate owing to the perturbation of the electronic

Highly Efficient OLEDs: Materials Based on Thermally Activated Delayed Fluorescence, First Edition. Edited by Hartmut Yersin.
© 2019 Wiley-VCH Verlag GmbH & Co. KGaA. Published 2019 by Wiley-VCH Verlag GmbH & Co. KGaA.

ground state, i.e. singlet ground state gaining triplet character, in contrast to triplet excited states mixing with singlet states. It is often argued that the energy gap between the electronic ground state and the excited triplet states is so much larger than the energy difference between the T_1 state and the excited singlet states and that the second sum can be neglected. However, this disregards the fact that the sum over n includes $n = 0$, i.e. the electronic ground state, and that the term for $m = 1$ in the second perturbation sum partially cancels this contribution. To avoid artefacts in the calculation, both perturbation sums should be always considered.

However, although this approach can be extremely effective, until now the only phosphorescent materials found practically useful are iridium and platinum complexes, which are unappealing for commercial applications due to their high cost and low abundance [6]. Indeed, as iridium is the fourth most scarce naturally occurring element on the planet, it is unwise to base such large-scale technologies such as lighting and displays on it.

An exciting new avenue for effective harvesting of both the singlet and triplet excitons can be achieved by using materials that emit through thermally activated delayed fluorescence (TADF) [7]. Here, the triplet excitons (75%) are not emitted through the $T_1 \rightarrow S_0$ radiative transition, but use thermal energy to promote upconversion via rISC so that they can emit as singlet excitons. This means that effective TADF requires the emitters to have both a small energy gap $\Delta E_{S_1-T_1}$ between the emitting singlet and triplet states, preferably <0.1 eV, and a nonzero SOC that mixes their spin character and permits efficient rISC, as shown in Eq. (9.1). The size of the former ($\Delta E_{S_1-T_1}$) is two times the electron exchange interaction ($2J$) between the ground and excited states [8]:

$$J = \int \int \psi_1^*(\mathbf{q}_1)\psi_2(\mathbf{q}_2) \frac{e^2}{4\pi\epsilon_0|\mathbf{q}_1 - \mathbf{q}_2|} \psi_2(\mathbf{q}_1)\psi_1^*(\mathbf{q}_2) d\mathbf{q}_1 \, d\mathbf{q}_2 \qquad (9.2)$$

where ψ_1 and ψ_2 represent ground- and excited-state wave functions, respectively; e is the charge; and ϵ_0 is the dielectric constant. This can usually be well approximated using the spatial overlap between the highest occupied molecular orbital (HOMO) and lowest unoccupied molecular orbital (LUMO), and consequently, the most promising TADF emitters have exploited charge-transfer (CT) states that minimize the spatial overlap and therefore the exchange energy [9, 10]. Crucially, as shown in Eq. (9.1), this small energy gap is able to promote significant mixing between the singlet and triplet states, which depends inversely on the energy gap provided SOC is nonzero. This relationship reduces the emphasis upon the magnitude of SOC and opens the possibility for devices that contain exclusively organic materials [11, 12]. The challenge of this approach is that reducing the spatial overlap between the HOMO and LUMO orbitals also reduces the radiative rate of the emitting S_1 state. While high fluorescence efficiencies are important for high performing devices, it is more important that the radiative decay rate is much faster than all nonradiative rates (k_{nr}), and therefore during the design of TADF emitters, it is important to understand the processes contributing to k_{nr} and reduce them as much as possible.

As illustrated in the previous paragraph, for third-generation TADF OLED devices, the communication between low-lying singlet and triplet excited states is of great importance and plays a fundamental role in determining key molecular and material properties relevant in a device context. Consequently, a thorough understanding of the basic principles governing the interplay between these manifolds of spin states is of great importance if one is to achieve systematic material design. Computationally, the most common approach for elucidating mechanistic information about processes occurring in electronically excited states is to compute energy profiles along viable reaction and/or decay pathways. These are usually represented as one-dimensional curves, {Q} in Figure 9.1, containing displacements along multiple nuclear degrees of freedom. Important regions of the potential surface, such as crossings between states of the same or different multiplicities, potentially contributing to internal conversion (IC) and ISC, respectively (Figure 9.1), can also be identified. While useful, the resulting picture is static and lacks the dynamical information important for obtaining a complete understanding, especially given the complexity that can arise from dynamics occurring on a multidimensional potential energy surface (PES). This is especially pertinent in cases that exhibit a breakdown of the Born–Oppenheimer approximation, giving rise to nonadiabatic coupling [13]. Indeed, these nonadiabatic interactions arise from nuclear motion and occur via a dynamic effect in the sense that they become strong when a system traverses a region where coupled states are close in energy or degenerate, leading to a more efficient crossing.

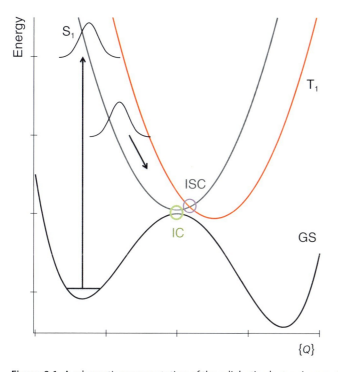

Figure 9.1 A schematic representation of the adiabatic electronic energies against an arbitrary nuclear configuration {Q}. Coupling between states of the same multiplicity originating from nonadiabatic coupling gives rise to internal conversion (IC). Transitions between states of different multiplicities, intersystem crossing (ISC), arise from spin–orbit coupling.

In this chapter an overview of dynamical methods used for understanding excited-state processes is provided. We then summarize our recent research using quantum nuclear dynamics to understand the role of vibronic coupling for ISC and rISC rates in two molecules with implications to TADF.

9.1.1 Background to Delayed Fluorescence

Molecular fluorescence is usually a two-step process, i.e. an initial absorption giving rise to an electronically excited state followed by the radiative decay of that state or another one of lower energy (Kasha's rule) into the electronic ground state. This is known as prompt fluorescence (PF). However, if the florescence and nonradiative decay rate from this state is a lot less than the rate of ISC, i.e. $k_F + k_{NR} \ll k_{ISC}$ (see Figure 9.2), then fluorescence may occur by a more complicated route, the triplet manifold. In this case, the S_1 state decays via ISC into the T_1 state. Subsequently, if the phosphorescence and nonradiative decay of the triplet state is slow, as expected, and the energy gap between the singlet and triplet states is small enough (normally ≤ 0.2 eV), then after vibrational thermalization, a second ISC, often called rISC, back to the S_1 can occur followed by emission. This is known as delayed fluorescence (DF). If all radiative and nonradiative processes are very small, this process can occur multiple times before DF actually occurs. The system can therefore be said to be cycling between the two states [15].

The first observation of DF was by Perrin and coworkers [16] who reported two long-lived emission bands in solid uranyl salts naming them *true phosphorescence* and *fluorescence of long duration*. This was later characterized in more detail by Lewis et al. [17] in rigid media and Parker and Hatchard [18] using Eosin. The latter study is responsible for the original name, E-type DF, which is now most commonly referred to as TADF.

The most common description of TADF is the equilibrium model, first proposed by Parker and Hatchard [18] and later used by Kirchhoff et al. [19] to describe the TADF of Cu(I) metal–organic complexes. This assumes that

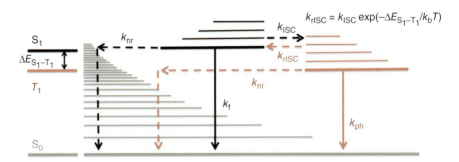

Figure 9.2 A simplified energy diagram representing a general schematic of the upconversion of triplet states to a higher-energy singlet state. k_{rISC} is approximately equal to the k_{ISC} multiplied by a Boltzmann term, which determines the number of states with sufficient energy to overcome the barrier ($\Delta E_{T_1-S_1}$). Deviations from this relationship arise from a higher density of final states expected for the direct intersystem crossing case. Source: Ref. [14]. Reproduced with permission of American Chemical Society.

$k_F \ll k_{rISC}$, and in this limit the steady-state populations of the emitting singlet and triplet states are determined by Boltzmann statistics, as the molecule spends sufficient time in the excited state for an equilibrium to form before emission eventually occurs. The relative population of the two states can be expressed using an equilibrium constant:

$$K = \frac{[S_1]}{[T_1]} = \frac{k_{rISC}}{k_{ISC}} = \frac{1}{3}\exp(-\Delta E_{S_1-T_1}/k_B T) \qquad (9.3)$$

making it possible to express the rate of the whole TADF process (k_{TADF}), i.e. rISC proceeded by fluorescence as the product of the amount of population in the S_1 state and the rate-limiting step, i.e. k_F:

$$k_{TADF} = \frac{1}{3}k_F \exp(-\Delta E_{S_1 T_1}/k_B T) \qquad (9.4)$$

Here, the prefactor of 1/3 is a consequence of the three triplet substates (i.e. $M_s = -1, 0, 1$). This approach, adopted by Adachi and coworkers [12, 20] for organic emitters, defines the TADF process as depending exclusively on the energy gap between the singlet and triplet states and crucially, as shown in Eq. (9.4), independent of the coupling between them, i.e. the rates of ISC (k_{ISC}) and rISC (k_{rISC}), provided they are not zero. This motives design procedures that focus upon minimizing this energy gap using CT states and suppressing larger amplitude molecular vibrations to reduce nonradiative pathways to ensure a maximum emission quantum yield, despite the potentially low radiative rates associated with the CT states.

While this represents a convenient approach for the analysis of photophysical data, the key assumption, i.e. $k_{rISC} \gg k_F$, must be and often is broken to support new emitters with stronger fluorescence yields [20, 21]. Within this regime, TADF is cast in terms of a kinetic process [22], for which the population of the emitting states in the longtime limit, assuming nonradiative and phosphorescence channels are small, can be expressed as

$$\frac{dS_1}{dt} = -k_F[S_1] + k_{rISC}[T_1] \qquad (9.5a)$$

$$\frac{dT_1}{dt} = -k_{rISC}[T_1] \qquad (9.5b)$$

Solving Eqs. (9.5a) and (9.5b) yields

$$T_1(t) = T_1(0)\exp(-k_{rISC}t) \qquad (9.6a)$$

$$S_1(t) = \left(\frac{k_{rISC}}{k_F - k_{rISC}}\right) T_1(0)\exp(-k_{rISC}t) \qquad (9.6b)$$

This illustrates the importance in not only tuning the energy gap, as is the focus from the equilibrium representations presented in Eq. (9.4), but also determining and optimizing other factors that influence k_{rISC}, for which a wide variety of values exist in the literature. This was recently affirmed by Inoue et al. [23] who demonstrated significant reduction in device roll-off efficiency for complexes exhibiting larger k_{rISC}, as the device is less susceptible the quenching effects, such as triplet–triplet annihilation.

9.1.2 The Mechanism of rISC

In Section 9.1.1 the core elements of TADF and the importance of k_{rISC} were established. However, despite the significant amount of research undertaken in this area, the mechanism for rISC and especially the efficient rISC recently reported in organic emitters ($k_{rISC} \sim 10^7$ s^{-1} [20, 21]) has not been fully established.

Transitions between two states of different spin multiplicities usually occur via the SOC interaction [24]. SOC between the lowest singlet and triplet metal–ligand CT states (^{1}MLCT and ^{3}MLCT) of Cu(I) complexes [25] is small but usually sufficient to explain the values of k_{ISC} and k_{rISC} reported [26]. For Cu(I) complexes that exhibit significant conformation freedom within the excited states, it can be important to also consider the SOC beyond the Franck–Condon geometry and incorporate how vibrational motion affects its magnitude, i.e. spin–vibronic coupling; see Refs. [27–32]. However, the situation is not quite so simple for donor–acceptor (D–A) complexes, which form the majority of organic TADF emitters. Indeed, Lim et al. [33] showed that SOC between ^{1}CT and ^{3}CT in these intramolecular charge transfer (ICT) complexes is formally zero. This is because spin–orbit coupling matrix elements (SOCMEs) between a singlet and triplet state with equal spatial orbital occupations are formally forbidden. The origin of this is analogous to El-Sayed's law [34] regarding conservation of total angular momentum during spin–flip transitions.

This issue, limiting our understanding of the mechanisms driving rISC, would appear reduced by the observation that the lowest triplet state for many complexes exhibiting efficient rISC is a $\pi\pi^*$ locally excited triplet state (^{3}LE) [21, 35]. In this case ^{3}LE $\rightarrow$ ^{1}CT conversion could have a finite, if small, SOCME permitting rISC. However, using a Fermi's golden rule (FGR) approach, Chen et al. [36] showed that the rate of ^{3}LE $\rightarrow$ ^{1}CT conversion in a series of TADF molecules was $\sim 10^2$ s^{-1}, four to five orders of magnitude less than the rates reported experimentally, and consequently a key component leading to complete understanding is clearly lacking.

To address this, Ogiwara et al. [37] used electron paramagnetic resonance (EPR) spectroscopy to probe the population of the ^{3}LE and ^{3}CT states. By fitting the transient experimental signals, they reported that complexes showing the largest rISC exhibited an EPR signal consistent with a mixture of both the ^{3}LE and ^{3}CT states. The authors used this to propose that efficient rISC includes not only the SOC pathway (^{3}LE $\rightarrow$ ^{1}CT) but also a hyperfine coupling-induced ISC pathway (^{3}CT $\rightarrow$ ^{1}CT). This mechanism arises from interactions between an electron's spin and the magnetic nuclei of its molecule. It is therefore completely local and not quenched by significant electron-hole separation experienced in CT states like SOC [38]. This conclusion is consistent with the proposal of Adachi and coworkers [20], who rationalized efficient rISC from ^{3}LE to ^{1}CT, as proceeding via reverse internal conversion (rIC) to the ^{3}CT and then using hyperfine coupling-induced ISC to cross to the ^{1}CT. However, crucially the hyperfine coupling constants are very small, usually in the range of 10^{-4} meV, and they therefore also appear highly unlikely that such coupling accounts for efficient rISC.

Recently D–A and D–A–D complexes reported by Ward et al. [39] point to a strong dynamical component to the rISC mechanism. Indeed, their work found that complexes including steric hindrance around the D–A dihedral angle switch off the main TADF pathway and make the molecule phosphorescent at room temperature. This *dynamical* aspect, i.e. motion along the D–A dihedral coordination, which appears to promote rISC is consistent with the recent simulations of Marian [40] who used multireference quantum chemistry methods to show, in agreement with Ref. [36], that direct SOC was indeed too small to explain efficient rISC. Marian proposed that it is mediated by mixing with an energetically close-lying ^{3}LE state along a carbonyl stretching mode, promoting spin–vibronic mixing between multiple excited states as being crucial to efficient rISC. Similar mechanisms have been widely reported for the ISC rate [27–32]. However while hugely informative, these calculations are inherently static and therefore do not provide mechanistic understanding of the TADF process. For this detail, simulations with temporal resolution are required and are the focus of the present work.

9.2 Beyond a Static Description

The rate of population transfer between two states can be described using a FGR approach as described in Ref. [28]. This was illustrated using high-level quantum chemistry calculations of the rISC rates for both Cu(I) and organic TADF molecules. This first-order perturbative approach is very effective, as demonstrated on numerous occasions by Marian and coworkers [27–29], describing the excited-state kinetics, provided that the coupling between the two states is small compared with their energy difference. If this is not the case, the validity of this approach, i.e. perturbation theory, becomes questionable, although it has still been used with some success [41].

Importantly, as described in the previous section, recent work demonstrates a dynamical mechanism for efficient rISC [39, 42]. This strongly motivates computations that go beyond inherently static quantum chemistry calculations and solves the time-dependent motion of the nuclei over a PES to elucidate a full understanding of the fine mechanistic details. In the following subsections we describe a variety of approaches for achieving this.

The starting point for any attempts aiming to elucidate fine details about excited-state dynamics is the time-dependent Schrödinger equation (TDSE):

$$i\hbar \frac{\partial}{\partial t} \Psi(q, Q, t) = \hat{H} \Psi(q, Q, t) \tag{9.7}$$

To derive working equations from this, the Born–Huang *ansatz* [43] for the total wave function is applied:

$$\Psi(q, Q, t) = \sum_{I}^{\infty} \Phi_I(q; Q) \Omega_I(Q, t) \tag{9.8}$$

where $\Phi(q, Q)$ describes a complete set of electronic states that are solutions of the time-independent electronic Schrödinger equation:

$$\hat{H}_{el}(q; Q) \Phi_I(q, Q) = E_{el}(Q) \Phi_I(q; Q) \tag{9.9}$$

In practice solutions for $E_{el}(Q)$ are achieved using a quantum chemistry method at various configurations in nuclear coordinate space, Q, which then provides the PES. Approaches for obtaining an accurate representation of the PES are discussed in Section 9.2.1. Inserting the Born–Huang *ansatz* into Eq. (9.7) and multiplying from the left by $\Phi(q, Q)$ yields, after integration over the electronic coordinates q, the equation describing the temporal evolution of the nuclear wave function $\Omega(Q, t)$:

$$i\hbar \frac{\partial \Omega_J(Q, t)}{\partial t} = -\sum_\gamma \frac{\hbar^2}{2M_\gamma} \nabla_\gamma^2 \Omega_J(Q, t) + \sum_I H_{JI}(Q)\Omega_I(Q, t)$$

$$\text{paying} + \sum_{\gamma I} \frac{\hbar^2}{2M_\gamma} D_{JI,\gamma}(Q)\Omega_I(Q, t)$$

$$- \sum_{\gamma, I \neq J} \frac{\hbar^2}{M_\gamma} d_{JI,\gamma}(Q)\nabla_\gamma \Omega_I(Q, t) \quad (9.10)$$

where

$$H_{JI}(Q) = \int \Phi_J^*(q; Q)\hat{H}_{el}\Phi_I(q; Q)dq \quad (9.11)$$

$d_{JI,\gamma}(Q)$ are the first-order nonadiabatic coupling vectors, defined

$$d_{JI,\gamma}(Q) = \int \Phi_J^*(q; Q)[\nabla_\gamma \Phi_I(q; Q)]dq \quad (9.12)$$

and $D_{JI,\gamma}(Q)$ are the second-order nonadiabatic coupling elements given

$$D_{JI,\gamma}(Q) = \int \Phi_J^*(q; Q)[\nabla_\gamma^2 \Phi_I(q; Q)]dq \quad (9.13)$$

$\{\Phi_J(q; Q)\}$ describes a complete set of electronic basis functions that are solutions of the time-independent Schrödinger equation, Eq. (9.9). Neglecting the latter two terms Eqs. (9.12) and (9.13) represents the Born–Oppenheimer approximation. This assumes complete decoupling between the electronic and nuclear degrees of freedom. In this scenario, a wave packet excited onto state I stays on state I indefinitely. To perform excited-state dynamics simulations, it is the solution of Eqs. (9.9) and (9.10) that is required. This is the subject of the following sections.

9.2.1 Obtaining the Potential Energy Surfaces

The Born–Oppenheimer approximation allows us to visualize the nuclei evolving over a PES generated by the electrons. Consequently, excited-state nuclear dynamics require an accurate potential. Given that the number of nuclear degrees of freedom in a molecule is $3N - 6$, calculating a full PES for a system containing more than five atoms is challenging. Indeed, if the potential is represented on a grid of points, the number of points, and therefore quantum chemistry calculations, is required to define the potential scales as N^f, where N is the number of grid points along each f degree of freedom.

This dimensionality problem has led to two distinct groups of excited-state dynamics simulations. The first retains a description of the potential and wave function upon a grid of points. Here the number of degrees of freedom that can be included is restricted, but significant effort is placed upon the development of the potential, i.e. the model Hamiltonian, with the aim of capturing the key physics of the problem understudy. Models adopting this approach are typically limited to a few tens of nuclear degrees of freedom, and therefore the excited-state dynamics occurring on this potential have traditionally been performed using grid-based quantum nuclear wavepacket dynamics [13]. Consequently, these simulations usually provide a rigorous description of the nuclear dynamics within the constraints imposed by the model PES. Provided the approximations are appropriate to capture the core elements of the dynamical process understudy, this approach can be extremely powerful. Indeed, such simulations have contributed to our understanding of a wide variety of fundamental photophysical and photochemical processes [32, 44–47].

The alternative is to adopt an *on-the-fly* approach, i.e. calculate the potential as and when it is required during the nuclear dynamics. The distinct advantage of this is that it removes the need to preparametrize a model PES and subsequently makes it possible, unlike the model approach, to follow the nuclear dynamics in unconstrained nuclear configuration space. This field was born from the development of *ab initio* molecular dynamics, which emerged following the pioneering work of Car and Parrinello [48], for studying dynamics and properties in the electronic ground state. However, more recently it has been extended to enable the study of excited-state dynamics [49–51].

Here the nuclear motion is described using either a Gaussian wavepacket basis or semiclassical trajectories [50, 52, 53], and does not require a grid of points to be defined. However, despite the advantages of full configuration space simulations, these simulations still require a large computational effort to reach statistical convergence, i.e. dynamics does not change when increasing either the number of trajectories or Gaussian functions. Indeed, a large number of these basis functions are usually required, and therefore the rigorous applicability of this approach remains limited to relatively small molecular systems, <100 atoms. In comparison with grid-based methods, one can roughly state that during such methods the approximations are shifted more to the nuclear dynamics themselves, rather than in the potential as is the case for grid-based methods. Of course in both cases, the accuracy of the potential depends upon the underlying accuracy of the quantum chemistry method used.

Independent of the method adopted for simulating the PES, computations of excited-state dynamics are complicated by the breakdown of the Born–Oppenheimer approximation and the subsequent presence of nonadiabatic coupling. The challenge of these terms can be seen by rewriting Eq. (9.12) as

$$d_{JI,\gamma}(Q) = \int \frac{\Phi_J^*(q; Q)(\nabla \hat{H}_{el})\Phi_I(q; Q)}{V_J - V_I} dq \quad (9.14)$$

where V_I and V_J represent the potential energy of states I and J, respectively. This clearly shows that within the adiabatic basis of Eqs. (9.10)–(9.13), which is

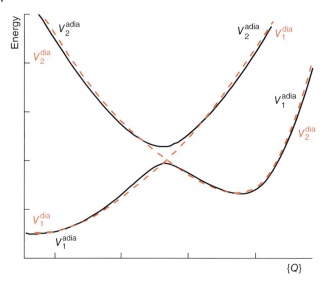

Figure 9.3 Schematic representation of the adiabatic (black solid) and diabatic (red dashed) representation of the electronic states along an arbitrary nuclear configuration {Q}.

standard for quantum chemistry programs, the nonadiabatic corrections necessary for describing the coupling between two different excited states depend inversely on the energy gap between surfaces. When this gap becomes small, the coupling increases induced coupling between the nuclear motions on different surfaces. If two surfaces become degenerate, the coupling becomes infinite, leading to a strong likelihood for numerical instabilities during the simulations. These singularities can be removed by switching to a diabatic electronic basis (Figure 9.3), which is therefore the method of choice for quantum dynamics simulations. In contrast to the adiabatic picture, which provides sets of energy-ordered PES and nonlocal coupling elements via nuclear momentum-like operators, Eqs. (9.12) and (9.13), the diabatic picture provides PES that are related to an electronic configuration, and the couplings are provided by local multiplicative Q-dependent potential-like operators. Importantly, as the surfaces in the diabatic picture are smooth, they can often be described by a low-order Taylor expansion, as described in the following section.

9.2.1.1 Vibronic Coupling Model Hamiltonian

One approach for obtaining an appropriate PES within the diabatic basis suitable for quantum dynamics simulations and the method of choice for the case studies presented below is within the framework vibronic coupling model [54]. This approach exploits two key aspects: Firstly, in contrast to adiabatic PES, the diabatic potential is usually smooth and can be expressed as a low-order Taylor expansion around a geometry of interest defined as $\mathbf{Q}_0$. Secondly, while transformations from the adiabatic to the diabatic representation are normally difficult to perform, the opposite way (i.e. diabatic to adiabatic) is relatively simple using a unitary transformation [55]. Consequently, starting from an initial guess, the diabatic potential can be refined using least-squares fit to the adiabatic potential computed using standard quantum chemical methods. At each iteration of the fit, the diabatic Hamiltonian is transformed into the adiabatic basis to assess the quality of the fit. This approach is often referred to as *diabatization by ansatz*. It is noted that this can also be achieved without fitting, using only a few point near $\mathbf{Q}_0$; see Refs. [44, 56]. While quick and computationally inexpensive, this approach can lack important information of the potential away from $\mathbf{Q}_0$ and can lead unrealistic potential curves at distorted geometries.

The starting point for obtaining the vibronic coupling Hamiltonian is a Taylor series expansion around $\mathbf{Q}_0$, using dimensionless (mass–frequency scaled) normal mode coordinates:

$$\mathbf{H} = \mathbf{H}^{(0)} + \mathbf{W}^{(0)} + \mathbf{W}^{(1)} + \cdots \tag{9.15}$$

Truncation at first order, as shown here, is referred to the linear vibronic coupling (LVC) model, while truncation at second order is referred to the quadratic vibronic coupling (QVC) model. The zeroth-order term is the ground-state harmonic oscillator approximation:

$$H^{(0)} = \sum_\alpha \frac{\omega_\alpha}{2} \left(\frac{\partial^2}{\partial Q_\alpha^2} + Q_\alpha^2 \right) \tag{9.16}$$

with the vibrational frequencies ω_α. The zeroth-order coupling matrix contains the adiabatic state energies at $\mathbf{Q}_0$. The adiabatic potential surfaces are equal to the diabatic surfaces at this point, so $\mathbf{W}^{(0)}$ is diagonal and is expressed as

$$W_{ij}^{(0)} = \sum_\alpha \langle \Phi_i(\mathbf{Q}_0) | \hat{H}_{\text{el}} | \Phi_j(\mathbf{Q}_0) \rangle \tag{9.17}$$

where $\mathbf{H}_{\text{el}}$ is the standard clamped nucleus electronic Hamiltonian and Φ is the diabatic electronic functions. The first-order linear coupling matrix elements are written as

$$W_{ij}^{(1)} = \sum_\alpha \langle \Phi_i(\mathbf{Q}_0) | \frac{\partial \hat{H}_{\text{el}}}{\partial Q_\alpha} | \Phi_j(\mathbf{Q}_0) \rangle Q_\alpha \tag{9.18}$$

where the on-diagonal and off-diagonal terms are written as

$$W_{ii}^{(1)} = \sum_\alpha \kappa_\alpha^{(i)} Q_\alpha \tag{9.19a}$$

$$W_{ij}^{(1)} = \sum_\alpha \lambda_\alpha^{(i,j)} Q_\alpha \tag{9.19b}$$

κ and λ are the expansion coefficients corresponding to the on- and off-diagonal matrix elements. The on-diagonal elements are the forces acting within an electronic surface and are responsible for structural changes of excited-state potentials compared with the ground state. The off-diagonal elements are the nonadiabatic couplings responsible for transferring wavepacket population between different excited states. The second-order terms, $\mathbf{W}^{(2)}$, are expressed as

$$W_{ij}^{(2)} = \sum_{\alpha\beta} \langle \Phi_i(\mathbf{Q}_0) | \frac{\partial^2 \hat{H}_{\text{el}}}{\partial Q_\alpha Q_\beta} | \Phi_j(\mathbf{Q}_0) \rangle Q_\alpha Q_\beta \tag{9.20}$$

where the specific on-diagonal and off-diagonal terms are written as

$$W_{ii}^{(2)} = \frac{1}{2} \sum_{\alpha,\beta} \gamma_{\alpha,\beta}^{(i)} Q_\alpha Q_\beta \qquad (9.21)$$

$$W_{ij}^{(2)} = \frac{1}{2} \sum_{\alpha,\beta} \mu_{\alpha,\beta}^{(i,j)} Q_\alpha Q_\beta \qquad (9.22)$$

The off-diagonal terms (μ) are rarely used; however the on-diagonal terms can often be very important. The quadratic terms (i.e. $Q_\alpha = Q_\beta$) are responsible for changes in frequency of the excited-state potential compared with the ground state, while the bilinear terms (i.e. $Q_\alpha \neq Q_\beta$) are responsible for mixing the normal modes in the excited state, i.e. the so-called Duschinsky rotation effect (DRE). Higher-order terms can be important and are included in the same manner [31, 57].

Figure 9.4 provides an illustration, up to second order, of the stepwise fitting procedure often used. The first two steps involve defining $H^{(0)}$ and $W^{(0)}$ and

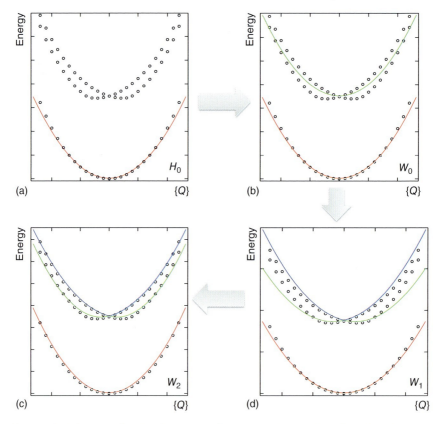

Figure 9.4 Schematic showing the stepwise fitting procedure used to obtain the vibronic coupling Hamiltonian. Initially, $H^{(0)}$ and $W^{(0)}$ are defined using a frequency and excited-state calculation at Q_0. Subsequently, the linear model and the quadratic terms are fit. This is done in a stepwise manner, always keeping parameters at lower orders fixed. It is noted that all states (red, green, and blue lines) are present in all of the fits but may not be observed as they overlap.

using the frequency of the normal modes and excited-state energies at Q_0, respectively, and obtained from quantum chemistry calculations. Subsequently, the linear model is refined (κ, λ), while the remaining parameters are set to zero. Lower-order terms, i.e. $H^{(0)}$ and $W^{(0)}$, are fixed to the value determined during the previous step. Next the second-order terms (γ) are refined. In this case only the quadratic terms are relevant, and the representative potentials shown only include displacements along one normal mode. The inclusion of bilinear terms requires calculation of the potential simultaneously along two normal modes (α and β). Again, parameters arising from lower-order terms are kept fixed.

An important consideration when setting up a model Hamiltonian for large polyatomic molecules is which nuclear degrees of freedom are required in the model to ensure an accurate description of the dynamics of interest. In addition, it is clear that for large systems the number of relevant terms, especially at second order, can rapidly become very large making it challenging to refine all of them. The former can be addressed by identifying important excited-state structures, such as energy minima and conical intersections, and expressing them as a function of the normal mode displacements from Q_0. This helps to clearly identify the modes responsible for significant structural changes in the excited state, but will not necessarily identify nonadiabatic coupling modes that do not always induce structural changes in the excited states. Here additional help is provided using, if present, molecular symmetry. Indeed, an LVC matrix element will only be nonzero if the product vibrational mode symmetry and that of the states is written as

$$\Gamma_i \otimes \Gamma_j \otimes \Gamma_\alpha \supset A \tag{9.23}$$

where Γ_i and Γ_j are symmetries of the two states and Γ_α is the symmetry of the mode. Therefore off-diagonal elements (nonadiabatic coupling elements) will only be nonzero if the product of the two states gives the symmetry of the specific mode. On-diagonal elements can only be nonzero for totally symmetric modes. While establishing an accurate Hamiltonian is still possible within low symmetry molecules, the use of such rules does significantly simplify the fitting process, and many of the coupling terms in the Hamiltonian can be excluded as they are known to be zero.

9.2.2 Solving for the Motion of the Nuclei

Motivated primarily by the desire to overcome the exponential scaling of computational effort with the number of degrees of freedom outlined in Section 9.2, a large number of approaches for addressing the nuclear dynamics over excited-state PES have been developed. These range from full quantum approaches, which explicitly address the nuclear motion according to the TDSE to mixed quantum–classical approaches, such as Tully's trajectory surface hopping (TSH). This latter approach replaces the wave packet with classical trajectories and incorporates nonadiabatic transitions using a stochastic algorithm [52, 53]. The appeal of mixed quantum–classical approach is driven largely by the relative simplicity in applying then within a framework of *on-the-fly* excited-state dynamics [49, 51, 58], as discussed above.

Developments of more sophisticated trajectory approaches based upon Gaussian wave packets instead of independent point trajectories have seen a revival in recent years and include multiple spawning [59, 60], coupled coherent states (CCS) [61], multiconfigurational Ehrenfest (MCE) [62], variational multiconfigurational Gaussian wave packet (vMCG) [50, 63], and most recently the multiple cloning method [64]. All of these can be applied to on-the-fly excited-state dynamics, but unlike TSH converged to the correct quantum description in the limit of a complete basis set. For larger molecules, >100 atoms, this limit will be very computationally expensive to reach, and consequently, despite significant recent growth in on-the-fly methodologies, there remains a strong place for grid-based quantum dynamics based upon model vibronic coupling Hamiltonians as described above. In the following two sections, we outline the details of one of the most efficient ways of doing this, the multiconfigurational time-dependent Hartree (MCTDH) method that we have focused upon and used for both of the case studies discussed below.

9.2.2.1 Multiconfigurational Time-Dependent Hartree Approach

The simplest way of solving the TDSE is to expand the nuclear wave function into a time-independent product basis set, with time-dependent coefficients:

$$\Psi(x_1, x_2, \ldots, x_f, t) = \sum_{j_1=1}^{n_1} \cdots \sum_{j_f=1}^{n_f} C_{j_1 \ldots j_f}(t) \prod_{k=1}^{f} \chi_{j_k}^{(k)}(x_k) \tag{9.24}$$

where χ_j is an orthonormal basis, such as the eigenfunctions of the harmonic oscillator, and x_k is the nuclear coordinate. The analogous approach in electronic structure theory is full configuration interaction (full CI). Consequently, while it rigorously describes the motion of a nuclear wave packet, like full CI, it suffers from a severe scaling problem. Indeed, it scales exponentially with the number of degrees of freedom, making it difficult to describe nuclear dynamics in systems containing more than five to six degrees of freedom; hence, as in exactly the same way as electronic structure theory, approximate methods are required.

One such approximate scheme is the time-dependent Hartree (TDH) method [65], also known as time-dependent self-consistent field (TDSCF). Here one assumes the total wave function of the system can be approximated as a single Hartree product of single-particle functions or orbitals (φ), and consequently the nuclear wave function is defined as

$$\Psi(x_1, x_2, \ldots, x_f, t) = C(t) \prod_{k=1}^{f} \varphi_k(x_k, t) \tag{9.25}$$

The effort required is clearly significantly reduced as the sum over j configurations and f degrees of freedom is absent. But obviously this comes at the cost that the correlation between the degrees of freedom is no longer treated correctly. In this case, in contrast to electronic structure theory, this is the correlation between the nuclear degrees of freedom. It can be shown that the error introduced by the TDH approach is small if the variation in the potential function over the width of the wave function is small. This is usually the case for dynamics around the equilibrium but clearly will be insufficient when considering excited-state dynamics due to the strong variations in the potential.

As the performance of the TDH method is usually rather poor, an obvious way of improving it is to move beyond a single configuration and take multiple configurations into account [66], i.e. multiconfigurational time-dependent self-consistent field (MC-TDSCF) approach. A particularly powerful variant of the MC-TDSCF family, which has become extensively used, is the MCTDH method [67]. In this approach the nuclear wave function *ansatz* takes the form

$$\Psi(Q_1, \ldots, Q_f, t) = \sum_{j_1=1}^{n_1} \cdots \sum_{j_f=1}^{n_f} A_{j_1 \ldots j_f}(t) \prod_{k=1}^{f} \varphi_{j_k}^{(k)}(Q_k, t) \tag{9.26}$$

where $Q_1, \ldots, Q_f$ are the nuclear coordinates, $A_{j_1 \ldots j_f}$ are the MCTDH expansion coefficients, and $\varphi_{j_k}^{(k)}$ are the n_k expansion functions for each degree of freedom k known as single-particle functions. Note that setting $n_1 = \cdots = n_f = 1$ retrieves the TDH formulation.

The *ansatz* for the MCTDH nuclear wave function appears rather similar to standard nuclear wavepacket approach Eq. (9.31); however crucially the basis functions are time dependent. This means that fewer basis functions are required to converge the calculations as they adapt to provide the best possible basis for the description of the evolving wave packet. In addition, the coordinate for each set of single-particle functions, Q_k, can be a composite coordinate of one or more system coordinates. Thus the basis functions are d-dimensional, where d is the number of system coordinates, usually between 1 and 4. This reduces the effective number of degrees of freedom for the purpose of the simulations. The memory required by the standard method is proportional to N^f, where N is the number of grid points for each f degree of freedom. In contrast, the memory needed by the MCTDH method scales as

$$\text{memory} = fnN + n^f \tag{9.27}$$

where the first term is due to the (single-mode) single-particle function representation and the second term is the wavefunction coefficient vector A. As $n < N$, often by a factor of 5 or more, the MCTDH method needs much less memory than the standard method, so allowing larger systems to be treated. Indeed the standard implementation of MCTDH can treat, depending on the exact details of the calculations, ~50 nuclear degrees of freedom. A detailed discussion on all of these aspects, beyond the remit of the present chapter, can be found in Ref. [68].

9.2.2.2 Density Matrix Formalism of MCTDH: ρMCTDH

The standard MCTDH wavefunction approach, outlined in the previous section, describes the evolution of a particular well-defined initial state, and therefore calculations are effectively performed at 0 K. However, given the importance of the thermal aspect in the present topic upon TADF, it clearly is important to explicitly address the temperature of the system, evoking the need to adopt a density matrix approach. In density operator form the 0 K state is expressed as

$$\rho = |\Psi\rangle\langle\Psi| \tag{9.28}$$

where it is a pure state. However a system at finite temperature is an incoherent mixture of very many thermally excited states, $|\Psi_n\rangle$, and therefore the correct description is in the form of a density matrix:

$$\rho = \sum_n p_n |\Psi_n\rangle\langle\Psi_n| \tag{9.29}$$

where p_n is a probability, not an amplitude of finding a system in state Ψ_n. So in addition to a probability of finding a particle described by a wave function at a specific location, there is now also a probability for being in a different state [69].

The equation of motion for the density matrix follows naturally from the definition of ρ and the TDSE, leading to the well-known Liouville–von Neumann equation:

$$i\hbar \frac{\partial \rho}{\partial t} = [H, \rho] \tag{9.30}$$

This applies to a so-called closed quantum system, i.e. the Hamiltonian is not in contact with a dissipative environment. When used to describe an open quantum system, the equation has an additional term used to describe dissipation or population decay; however this was not considered for the examples discussed below.

In common with the multiconfiguration wavefunction expansion used in the MCTDH scheme for wave functions, a similar expansion can be used in the case of density operators, in this case, expressed as

$$\rho(Q_1, \ldots, Q_f, Q'_1, \ldots, Q'_f, t) = \sum_{j_1=1}^{n_1} \cdots \sum_{j_f=1}^{n_f} B_{j_1 \ldots j_f}(t) \prod_{k=1}^{f} \sigma_{j_k}^{(k)}(Q_k, Q'_k, t) \tag{9.31}$$

where $B_{j_1 \ldots j_f}$ denotes the MCTDH coefficients and σ presents the single-particle density operators (SPDO's) analogous to the single-particle functions of the conventional scheme. This implementation of this approach is described in detailed in Refs. [70, 71]. It is important to bear in mind that compared with the wavefunction approaches described in the previous section, the numerical treatment of density operators is more difficult since the dimensionality of the system formally doubles [69].

9.3 Case Studies

In the following subsections two recent examples of using quantum wavepacket dynamics simulations to reveal information about the fine details of excited-state processes, in particular ISC and rISC, are discussed. The first is a study of the ultrafast relaxation in a prototypical Cu(I)–phenanthroline complex, $[Cu(dmp)_2]^+$ (dmp = 2,9-dimethyl-1,10-phenanthroline) [25, 32, 72, 73], as a prototypical example of the excited-state dynamics of Cu(I) complexes, which was one of the first to be applied as a TADF OLED device [74]. In the second example, we illustrate the critical contribution of vibronic coupling for achieving efficient rISC within organic D–A molecule [42], which have become the most popular area of research for TADF.

9.3.1 Ultrafast Dynamics of a Cu(I)–phenanthroline Complex

The necessity to harvest the 75% of the triplet excitons generated following electrical excitation has placed a great emphasis upon the heavier transition metals. Consequently, metal–organic complexes based upon first-row transition metal ions, such as Cu(I), were largely excluded as the absence of any significant heavy atom effect meant that the triplet lifetime of these complexes was often too long for devices [75]. However, while this is true when focusing upon the *triplet harvesting* mechanism, Cu(I) emitters have been known, since the pioneering work of McMillin and coworkers [19], to exhibit TADF, and therefore if one is able to optimize this *singlet harvesting* pathway, the Cu(I) emitters become an attractive, earth-abundant emitter. Indeed, as previously stated, Cu(I) complexes were the first to be used for TADF OLED devices [74].

Within the plethora of Cu(I) complexes that exist [76], a subset that has received significant attention, especially regarding their ultrafast dynamics and ISC rate, is the mononuclear Cu(I)–phenanthroline complexes [77, 78]; among which $[Cu(dmp)_2]^+$ is a prototypical example. In general, the ground state of these complexes adopts a pseudotetrahedral geometry with the two ligands being orthogonal; see Figure 9.5c [79]. Upon excitation into the singlet metal-to-ligand charge transfer (MLCT) states, the oxidation state of copper becomes Cu(II), and a flattening of the complex occurs, because the copper d^9 electronic configuration is susceptible to pseudo-Jahn–Teller (PJT) distortions [80], as shown in Figure 9.5c.

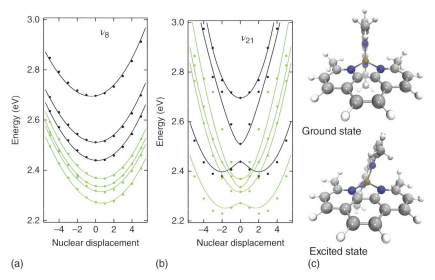

Figure 9.5 Cuts through the PES along (a) ν_8, a breathing mode acting on the Cu–N distances, and (b) ν_{21} responsible for the pseudo- Jahn–Teller distortion and therefore affecting the dihedral between the two ligands. The dots are results from the quantum chemistry calculations for the singlet (black) and triplet (green) states. The lines correspond to their fit from which the expansion coefficients are determined. (c) DFT-optimized geometry of the ground state (upper) and lowest triplet state (lower) of $[Cu(dmp)_2]^+$. The latter clearly shows the reduction of the dihedral angle between the two ligands arising from the pseudo-Jahn–Teller effect.

Over the past decade, numerous time-resolved spectroscopic studies of [Cu(dmp)$_2$]$^+$ have been performed to provide a detailed picture of kinetic processes occurring within the excited state [80–86]. Interestingly while these studies reach a consensus on the time scales observed, with three principal dynamical processes (50–100 fs, 500–900 fs, and 10–20 ps) consistently reported, the conclusions drawn from these kinetics are somewhat contrasting with some groups assigning subpicosecond ISC followed by a slow structural distortion, while others attribute the processes the other way around. To address this, we performed a quantum dynamics study, which is now described.

Importantly, as outlined in the previous section, we must first obtain a description for the PES. As [Cu(dmp)$_2$]$^+$ has 57 atoms, and therefore 165 nuclear degrees of freedom, calculating a full PES is unrealistic, and therefore we adopt the model vibronic coupling Hamiltonian approach outlined above. In total our model Hamiltonian is based upon a reduced subspace of the full configuration space, including eight vibrational degrees of freedom, most important for the first picosecond of excited-state dynamics [32, 73]. This identification was achieved using the magnitude of the linear coupling constants Eqs. (9.19a) and (9.19b) calculated for the lowest 60 frequency normal modes and by expressing the excited-state structural changes in normal mode coordinates. Although this clearly represents a significant reduction in the dimensionality of the potential, as shown in Figure 9.6, the modes included closely correspond to those identified from the femtosecond transient absorption study of Ref. [84]. Consequently, although the present Hamiltonian will be unable to capture longer time effects, such as vibrational cooling, this model Hamiltonian will provide accurate insight into the femtosecond dynamics.

Figure 9.5 shows the potential energy curves along two vibrational degrees of freedom identified to be most important during the excited dynamics. The first, (v_8), is a totally symmetric breathing mode acting on the Cu–N distances of the

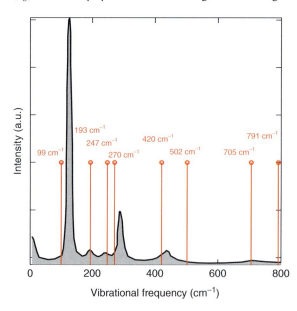

Figure 9.6 Reproduction of the Fourier transform power spectrum of the oscillatory components within the excited-state dynamics of [Cu(dmp)$_2$]$^+$ reported in Ref. [84] (filled with gray). This is overplayed with the vibrational frequencies (red) included in the model Hamiltonian. Source: Ref. [84]. Reproduced with permission of John Wiley & Sons.

first coordination sphere. The excited-state potentials are slightly shifted from the minimum energy geometry of the ground state (Q_0), and point to a Cu–N is slightly contracted in the excited state, consistent with previous experimental analysis [72]. The second potential (v_{21}) corresponds to the mode responsible for the PJT distortion, as illustrated by the double-well potential shape in both the lowest singlet and triplet states. This potential shape arises from strong nonadiabatic coupling between the low-lying excited states and therefore will be crucial for describing the population dynamics in the excited state.

Finally, in terms of generation of the model potential, it is noted that in Figure 9.5, the singlet and triplet states were calculated and fitted separately yielding a set of two spin-free PES. In order to probe the role of ISC, the singlet and triplet manifolds are coupled by SOC. The SOCMEs computed along the important normal modes were included to generate a complete Hamiltonian, enabling direct insight into the ultrafast spin–vibronic dynamics.

Using the Hamiltonian outlined above and described in detail in Refs. [32, 73], Figure 9.7 shows the relative diabatic state populations during the first picosecond after photoexcitation in the S_3 state for the model Hamiltonian, without (Figure 9.7a) and with (Figure 9.7b) SOC. In both cases their initial dynamics shows an initial ultrafast IC from the S_3 to the S_2 and S_1 states. The electronic energy of the wave packet on S_3 is then converted into kinetic energy on the lower states, resulting in an incoherent distribution of vibrationally hot levels. Initially (up to 200 fs) the two states (S_1 and S_2) can be said to be in equilibrium; however as the dynamics proceeds the population occurs increasingly in the S_1 state. When fitted with a biexponential, the first kinetic component of the S_3 state is ~100 fs, in good agreement experiment [82, 84].

Figure 9.7b shows that when SOCMEs between the low-lying singlet and triplet states, we observe ~80% of the wave packet populating the triplet states consistent with ultrafast ISC. Importantly, this cannot occur via SOC between the

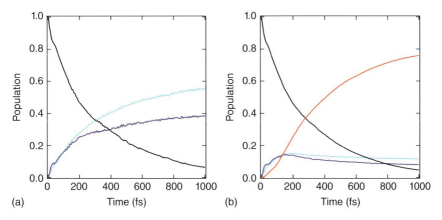

Figure 9.7 (a) Relative diabatic state populations of S_1 (cyan), S_2 (purple), and S_3 (black) for 1 ps following photoexcitation (no spin–orbit coupling). (b) Relative diabatic state populations of S_1 (cyan), S_2 (purple), S_3 (black), and the triplet (T_{1-4}) states (red), for 1 ps following photoexcitation (with spin–orbit coupling). *Source:* Ref. [32]. Reproduced with permission of American Chemical Society.

S_1 and T_1, as previously assumed in the literature, due to a relatively large energy gap (~0.2 eV) and small SOCMEs. Instead, ISC initially occurs via $S_1 \rightarrow T_2$ and $S_1 \rightarrow T_3$, due to the strong SOC and degeneracies of these three states along the PJT (v_{21}) mode (see Figure 9.5b) [73]. Here ISC occurs via a dynamical effect by traversing a region where the coupled singlet and triplet states are degenerate, leading to efficient and multiple ultrafast ISC channels. Therefore clearly the excited-state vibrational motion plays a crucial part in this.

In the context of TADF, and especially rISC, it is stressed that the present dynamics describing ultrafast ISC would not influence the k_{rISC} rate, as it derives from dynamics in the S_1 state. For rISC, the system would not only be in the triplet state but also in the relaxed geometry, and therefore the rate of rISC can be described using the SOCME between S_1 and T_1 at this geometry. However, these findings of ultrafast ISC have significant implications if one wishes to apply these complexes for solar energy conversion. Indeed, Huang et al. [87] demonstrated that when attached to a TiO_2, charge injection for a related Cu(I)–phenanthroline complex, from the ^{1}MLCT, is two orders of magnitude faster than from the ^{3}MLCT. Consequently for efficient charge injection one needs to restrict ISC.

9.3.2 The Contribution of Vibronic Coupling to the rISC of PTZ-DBTO2

In Section 9.1.2 the present understanding of the mechanism for efficient rISC in organic TADF emitters was discussed. Importantly, while low rISC rates of $\sim 10^{2-3} s^{-1}$ [23, 88] can be explained using an FGR description and simply invoking the SOCME between the lowest singlet and triplet states [36], this approach is clearly unable to explain the large $k_{rISC} \sim 10^7 s^{-1}$ reported [20, 21]. In addition, Ward et al. [39] showed a strong dynamic component to rISC. For different D–A–D molecules with very similar energy gaps ($\Delta E_{S_1-T_1}$), they reported very large variations in k_{rISC}. Indeed, by introducing steric hindrance to the motion of the D and A group, they reported that the emission could be switched from TADF to phosphorescence. Clearly, the restriction of molecular vibrations is crucial for reducing k_{rISC}. Such an effect cannot be explained within present models for TADF and is indicative of a mechanism, which is dynamic in nature, in the sense that it is promoted by molecular vibrations, which are thermally activated.

We have recently addressed these discrepancies using quantum dynamics simulations [42]. In this work we focused upon a D–A molecule composed of a phenothiazine donor and a dibenzothiophene-S,S-dioxide acceptor (PTZ-DBTO2), shown in Figure 9.8. This was chosen because it has been recently reported in Ref. [89], and this dimer analogue and the D–A–D trimer give identical photophysics and excellent OLED performance >19% EQE.

The normal modes and excited-state energies were computed using TDDFT(M06-2X) [90] within the Tamm–Dancoff approximation (TDA) [91]. In this context, it is important to bear in mind the limitations of TDDFT for describing CT excitations [92]. In this case the large fraction (54%) of nonlocal exact exchange incorporated within the M06-2X functional remedies this problem to a certain extent in a semiempirical manner [93] and gave good agreement with the experimental energetics, as shown in Figure 9.9. In addition, to ensure

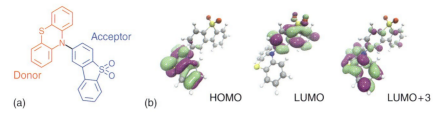

Figure 9.8 (a) Schematic representation of the donor–acceptor (D–A) molecule composed of a phenothiazine donor and a dibenzothiophene-S,S-dioxide acceptor (PTZ-DBTO2). (b) Molecular orbitals most involved in the low-lying excited states (^{3}LE, ^{3}CT, and ^{1}CT) considered herein. *Source:* Ref. [89]. Reproduced with permission of John Wiley & Sons.

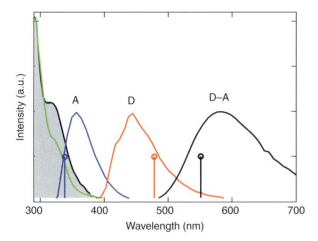

Figure 9.9 Experimental (filled gray) and TDDFT(M062X)-calculated (green) absorption spectra PTZ-DBTO2. The experimental spectrum was recorded in toluene. The calculated spectrum was shifted by 30 nm to match the experiment. Also shown is the experimental and calculated (stick) emission spectrum of the donor unit only, acceptor unit only, and PTZ-DBTO2, i.e. D–A.

the validity of the calculated Hamiltonian, we also studied the excited states with an optimally tuned LC-BLYP functional previously used to predict the energy gap and excitation energies of TADF emitters [94]. In both cases, while the absolute energies and the gaps between the low-lying states were slightly different, the vibronic couplings, one of the parameters for the Hamiltonian that, in contrast to the energetics, cannot be directly verified experimentally, were very similar. It is also emphasized for these simulations the TDA is critical for avoiding problems associated with the triplet instability [95].

The Hamiltonian used and detailed in Ref. [42] included the most relevant low-lying valence excited states of PTZ-DBTO2, which were found to be the lowest singlet CT state (^{1}CT), the lowest triplet CT state (^{3}CT), and the lowest triplet LE state (^{3}LE). All other excited states were sufficiently higher in energy, compared with the lowest ^{3}LE state, that their participation in thermally activated processes is known, from Boltzmann statistics, to be small. In agreement with recent experimental findings [39, 89], the TDDFT simulations showed that the lowest triplet state is the ^{3}LE on the donor group composed of an HOMO → LUMO+3 (Figure 9.8) transition. The two CT states (^{3}CT and ^{1}CT) are dominated by HOMO →LUMO transitions. The two spin manifolds (singlet and triplet) are coupled by SOC and the hyperfine interaction. The former couples the ^{3}LE and ^{1}CT states, and an SOCME with a magnitude of 2 cm^{-1} was calculated. In contrast, the two CT states were coupled using the hyperfine interaction. This is known to be small, and value of 0.2 cm^{-1} was used, which is consistent with typical experimentally recorded values used [37].

Given the C_s point group of PTZ-DBTO2, only vibrational mode of A″ symmetry could vibronically couple the two triplet states. Subsequently all A″ modes with a frequency <500 cm^{-1}, i.e. those which can reasonably be thermally activated (15 in total), were calculated and fitted to determine the magnitude of nonadiabatic coupling. From this three modes, with vibrational motion predominantly acting on the dihedral angle, were found to exhibit significant vibronic coupling and are shown in Figure 9.10.

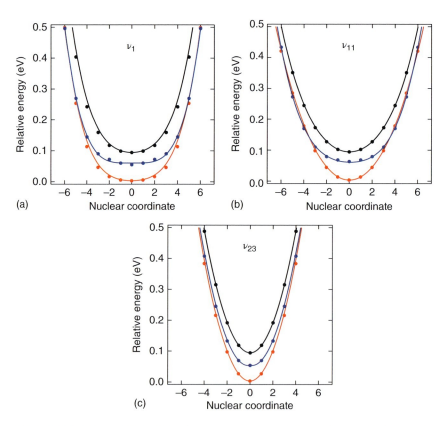

Figure 9.10 Potential energy curves along v_1, v_{11}, and v_{23} for the low-lying excited states of PTZ-DBTO2 (^{3}LE = red, ^{3}CT = blue, and ^{1}CT = black) relative to the ^{3}LE energy minimum and the corresponding fit of the model vibronic coupling Hamiltonian to these potentials.

Consequently, the $n \times n$ matrix Hamiltonian, where n is the number of electronic states, used to describe PTZ-DBTO2 is written as

$$\hat{H} = \begin{pmatrix} \frac{\omega}{2}\left(\frac{\partial^2}{\partial Q_i^2} + Q_i^2\right) + E_{^3\text{LE}}^{\text{rel}} & \lambda_{Q_i} & E_{\text{SOC}} \\ \lambda_{Q_i} & \frac{\omega}{2}\left(\frac{\partial^2}{\partial Q_i^2} + Q_i^2\right) + E_{^3\text{CT}}^{\text{rel}} & E_{\text{HFI}} \\ E_{\text{SOC}} & E_{\text{HFI}} & \frac{\omega}{2}\left(\frac{\partial^2}{\partial Q_i^2} + Q_i^2\right) + E_{^1\text{CT}}^{\text{rel}} \end{pmatrix}$$

(9.32)

where Q_i is the nuclear degrees of freedom, i.e. v_1, v_{11}, and v_{23}. E^{rel} is the excited-state energy, relative to the ^{3}LE state. This means that $E_{^3\text{LE}}^{\text{rel}}$ will be 0, while for $E_{^3\text{CT}}^{\text{rel}}$ and $E_{^1\text{CT}}^{\text{rel}}$, the energy represents the gap between this state and the ^{3}LE state.

Using this model, Figure 9.11a shows the relative population of the ^{3}LE state for 1 ns after excitation of the ^{1}CT state; therefore these dynamics follow the ISC kinetics, simulated using the standard wavefunction formalism of MCTDH. The black trace uses the model Hamiltonian as calculated for PTZ-DBTO2 and shows after 1 ns a population of the ^{3}LE state of ~0.075 corresponding to $k_{\text{ISC}} \sim 5 \times 10^5 \text{s}^{-1}$. However, if the vibronic coupling between the ^{3}LE and ^{3}CT states is removed (blue trace), k_{ISC} is significantly reduced, and within the time scale of the dynamics, population transfer to the ^{3}LE (T_1) state is negligible. In contrast removing the HFI, which couples the two CT states, has very little effect on the ISC kinetics. This indicates, contrary to the recent conclusions of Ogiwara et al. [37], that the HFI cannot be the dominant mechanism.

Figure 9.11b shows the same simulations, but in this case they are initiated in the lowest triplet state, ^{3}LE, to mimic the rISC dynamics. These simulations were performed at 300 K, using the density operator formalism of MCTDH described above. As expected, these indicate the same trends as observed in Figure 9.11a. Indeed, if the vibronic coupling between the ^{3}LE and ^{3}CT states is neglected (blue trace), the k_{rISC} is significantly reduced, while exclusion of HFI has almost no effect, despite the near degeneracy of the two CT states.

Importantly, the critical role of vibronic coupling between the two triplet states for the ^{3}LE $\to$ ^{1}CT conversion cannot be described within a similar first-order perturbation theory approach. Therefore following the work of Henry and Siebrand [96], a more general expression of the ISC/rISC rate derived from second-order perturbation theory must be adopted. Here the rISC rate is described as[1]

$$k_{T_1-S_1}^{\text{rISC}} = \frac{2\pi}{\hbar} \sum_f \left| \langle \Psi_f | \hat{H}_{\text{int}} | \Psi_i \rangle + \sum_n \frac{\langle \Psi_f | \hat{H}_{\text{int}} | \Psi_n \rangle \langle \Psi_n | \hat{H}_{\text{int}} | \Psi_i \rangle}{E_i - E_n} \right|^2 \delta(E_f - E_i)$$

(9.33)

---

1 It is noted that this equation equally holds in the case of ISC.

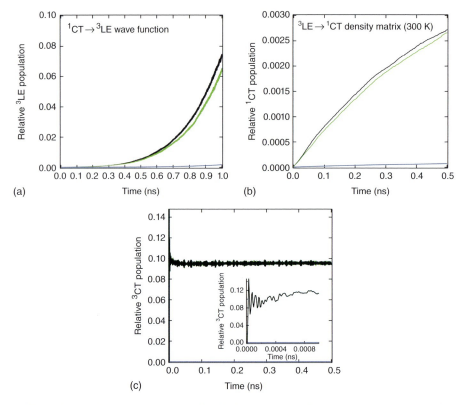

Figure 9.11 (a) Relative populations of the ^{3}LE state associated with intersystem crossing after excitation into the ^{1}CT state. (b) The relative populations of the ^{1}CT state associated with reverse intersystem crossing (rISC) after initially populating the ^{3}LE state (300 K). (c) Population kinetics of the ^{3}CT state during the rISC dynamics after initial population of the ^{3}LE state, i.e. (b) inset is a zoom into the early time kinetics. Black: full model Hamiltonian, green: no HFI, and blue: no vibronic coupling.

The first term is the normal first-order FGR describing the transition from some initial state (Ψ_i) to a final state (Ψ_f). In contrast coupling between the initial and final states in the second term, so-called second order, is mediated by an intermediate state (Ψ_n). In the present case, a direct second-order coupling would require population transfer between the two CT states, via the HFI, i.e. an initial ^{3}LE state populates the ^{3}CT via vibronic coupling, which decays into the ^{1}CT, via the HFI. However, as already demonstrated the HFI coupling plays an insignificant role. Consequently, the relevant terms for the present observations are

$$k_{\text{rIC}} = \frac{2\pi}{\hbar} |\langle \Psi_{^3\text{CT}} | \hat{H}_{\text{vib}} | \Psi_{^3\text{LE}} \rangle|^2 \delta(E_{^3\text{CT}} - E_{^3\text{LE}}) \qquad (9.34)$$

and

$$k_{\text{rISC}} = \frac{2\pi}{\hbar} \left| \frac{\langle \Psi_{^1\text{CT}} | \hat{H}_{\text{soc}} | \Psi_{^3\text{LE}} \rangle \langle \Psi_{^3\text{LE}} | \hat{H}_{\text{vib}} | \Psi_{^3\text{CT}} \rangle}{E_{^3\text{CT}} - E_{^3\text{LE}}} \right|^2 \delta(E_{^1\text{CT}} - E_{^3\text{LE}}) \qquad (9.35)$$

Eqs. (9.34) and (9.35) indicate a two-step mechanism. Firstly, the large vibronic coupling between ^{3}LE and ^{3}CT promotes, on a time scale much faster than the rISC (see Figure 9.11c), an equilibrium between the two states. Obviously, the position of this equilibrium depends both on the size of the vibronic coupling and the energy gap. Subsequently, the second-order term, Eq. (9.35), couples the ^{3}CT and the ^{1}CT, using the ^{3}LE as an intermediate. This latter second-order term is very efficient because of the good vibrational overlap between the almost degenerate initial and final states, ^{3}CT and ^{1}CT, respectively. Therefore, the two coupling terms driving this dynamics are the SOC and the vibronic coupling elements. This explains recent experimental results that demonstrated that steric hindrance of D–A dihedral angle switches the main pathway from TADF to phosphorescence [21, 39, 97]. This steric hindrance is equivalent to removing the vibronic coupling term, which is shown herein to be the strongest along modes exhibiting a distortion of the D–A dihedral angle.

It is also consistent with the time-resolved EPR study of Ogiwara et al. [37] who concluded rISC occurring in the spin–orbit and hyperfine pathways. However, the EPR only probes the population of the ^{3}LE and ^{3}CT states. For molecules where they found fast rISC, there was significant population of the ^{3}CT. In Ref. [22] they proposed that this population would undergo transfer via HFI interaction, and therefore the two routes for transfer into the ^{1}CT state made rISC faster. But as shown in our present work, an equilibrium between the ^{3}LE and ^{3}CT exists and depends on the strength of vibronic coupling. Therefore strong vibronic coupling would increase the population of the ^{3}CT state and provide rapid rISC, as they observe, via the second-order mechanism we demonstrated. Therefore again, these results are entirely consistent with our simulations.

In the context of recent results on similar molecules with the D–A–D setup, Santos et al. [97] reported a strong dependence on the relative energy levels according to the environment. This is not surprising given the interplay between CT and LE states, both of which will exhibit different responses to changes in the local embedding environment. This gives rise to three distinct scenarios, shown schematically in Figure 9.12: (i) the LE below the CT states, (ii) the LE degenerate with the CT states, and (iii) the LE above the CT states. Only the first case, the one studied here, requires the initial rIC step, but importantly, because the vibronic coupling is an order of magnitude larger than the other coupling mechanisms, this first step required in Figure 9.12a will be significantly faster than the k_{rISC} process. Indeed, as shown in Figure 9.11c, an equilibrium

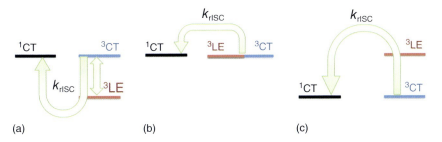

Figure 9.12 Schematic energy-level diagrams illustrating the effect to the environment of the relative splitting of the CT and LE states of PTZ-DBTO2. *Source:* Ref. [42]. Reproduced with permission of John Wiley & Sons.

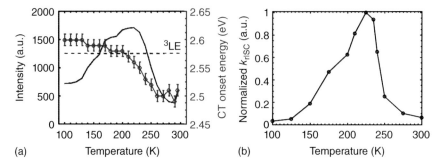

Figure 9.13 (a) The temperature dependence of the intensity (black line) and CT onset energy (purple circles). The change in CT onset energy plateaus below the glass temperature, representative of the PEO film becoming rigid. The black dashed line represents the energy of the ^{3}LE at 2.58 eV, with the peak in intensity apparent as the CT energy crosses resonance. (b) Relative rate of reverse intersystem crossing as a function of temperature for the D–A complex PTZ-DPTO2. The rates were extracted from the population of the 1CT state at 0.5 ns of dynamics simulations initiated from the lowest triplet state. *Source:* Ref. [98]. https://www.nature.com/articles/ncomms13680\LY1\textbackslash#rightslink.LicensedUnderCCBY4.0.

between the two triplet states is reached very quickly. It therefore is not the rate-determining step and consequently will not change the overall rate of rISC.

Importantly, this mechanism was recently demonstrated using a combination of photoinduced absorption and quantum dynamics, using the methods described above [98]. By placing the TADF molecule in a solid polar polyethylene oxide (PEO) host, it was shown that the rigidity and polarity of the host greatly affects the rISC rate and thus TADF. By exploiting the temperature-dependent polarity of the PEO host (Figure 9.13a), the CT states were brought into energetic resonance with the ^{3}LE state, and a significant increase in TADF was observed. These results allowed three distinct regimes of TADF to be categorized as TADF I (CT > LE), TADF II (CT = LE), and TADF III (CT < LE). These observations were simulated using quantum dynamics (Figure 9.13b) and the model Hamiltonian outlined above showing that vibronic coupling between the local and CT triplet states invokes an efficient second-order coupling from the triplet states to the singlet CT state. The model predicts the same resonant behavior of TADF as a function of temperature, confirming the proposed model for efficient TADF.

9.4 Conclusions and Outlook

The importance of a deeper first principles understanding of the properties of TADF emitters should not be underestimated, and in this regard theory and computations play an important role. It is therefore rather surprising that, despite the significant experimental interested in this area, the number of computational studies remains rather small. In this chapter, we have described computational approaches for simulating the excited-state dynamics of molecules, achieved using model PES and wavepacket dynamics simulations, which achieve a rigorous treatment of motion over this excited-state surface.

The two case studies used illustrate the importances of explicitly considering the role of excited-state nuclear motion when trying to establish a detailed understanding of ISC and rISC. As demonstrated in the second case study, this is critically important to describe the rate of efficient rISC in organic TADF molecules. These results, which are consistent with all recent experimental observations, demonstrate the dynamical mechanism of rISC. This also highlights the importance of not just tuning the $\Delta E_{S_1-T_1}$ gap but paying close attention to the gaps between the ^{3}LE and ^{1}CT and ^{3}CT and ^{1}CT states. There are now two energy gaps to consider! As the relative gaps between the CT and LE states are sensitive to the immediate embedding environment, this opens the ability to tune the performance of molecular emitters within the device by altering the rigidity or polarity of the host molecules. This effect has recently been demonstrated [98], and device tuning in this manner is likely to be crucial for achieving high external quantum efficiencies.

Finally, it is clear that for high performing devices, high fluorescence efficiencies are important. But more importantly, the radiative decay rate must be much faster than all nonradiative processes. Therefore, in terms of molecular vibrations, it will be important to identify which vibrational modes control radiative decay rates and rISC and the competition between them and with nonradiative processes, along with the role played by all excited states, including upper excited states, involved in TADF, as a function of molecular geometry and controlled degrees of torsional freedom. We must understand which vibrational modes couple ^{1}CT and ^{1}LE and each to the ground state, especially those mediating fast nonradiative decay of the ^{1}CT state and those mediating rISC. This leads to the key question: Can we block the former vibrational channels while retaining the latter and hence efficient TADF? This idea of dynamic molecular optimization also has the potential great ramifications in many different fields of molecular photophysics.

References

1 Burroughes, J.H., Bradley, D.D.C., Brown, A.R., Marks, R.N., Mackay, K., Friend, R.H, Burns, P.L., and Holmes, A.H. (1990). Light emitting diodes based on conjugated polymers. *Nature* 347: 539–541.
2 Wilson, J.S., Dhoot, A.S., Seeley, A.J.A.B., Khan, M.S. and Köhler, A., and Friend, R.H. (2001). Spin-dependent exciton formation in π-conjugated compounds. *Nature* 413: 828–831.
3 Baldo, M.A., O'Brien, D.F., You, Y., Shoustikov, A., Sibley, S., Thompson, M.E., and Forrest, S.R. (1998). Highly efficient phosphorescent emission from organic electroluminescent devices. *Nature* 395: 151–154.
4 Yersin, H. (2004). Triplet emitters for OLED applications. Mechanisms of exciton trapping and control of emission properties. In: *Transition Metal and Rare Earth Compounds*, Topics in Current Chemistry. 1–26. Berlin: Springer-Verlag.
5 Reineke, S., Walzer, K., and Leo, K. (2007). Triplet exciton quenching in organic phosphorescent light emitting diodes with Ir-based emitters. *Phys. Rev. B* 75: 125328.

6 Volz, D., Wallesch, M., Flechon, C., Danz, M., Verma, A., Navarro, J.M., Zink, D., Braese, S., and Baumann, T. (2015). From iridium and platinum to copper and carbon: new avenues for more sustainability in organic light-emitting diodes. *Green Chem.* 17: 1988–2011.

7 Monkowius, U. and Yersin, H. (2010). Komplexe mit kleinen Singulett-Triplett-Energie-Abständen zur Verwendung in opto-elektronischen Bauteilen (Singulett-Harvesting-Effekt) Complexes with small singlet-triplet energy separations for use in opto-electronic components (singlet harvesting effect). DE Patent Application DE200,810,033,563.

8 Endo, A., Ogasawara, A., Takahashi, A., Yokoyama, D., Kato, Y., and Adachi, C. (2009). Thermally activated delayed fluorescence from Sn^{4+}–porphyrin complexes and their application to organic light emitting diodes — a novel mechanism for electroluminescence. *Adv. Mater.* 21: 4802–4806.

9 Lee, J., Shizu, K., Tanaka, H., Nomura, H., Yasuda, T., and Adachi, C. (2013). Oxadiazole- and triazole-based highly efficient thermally activated delayed fluorescence emitters for organic light emitting diodes. *J. Mater. Chem. C* 1: 4599–4604.

10 Yersin, H. (2008). *Highly Efficient OLEDs with Phosphorescent Materials*. Wiley.

11 Uoyama, H., Goushi, K., Shizu, K., Nomura, H., and Adachi, C. (2012). Highly efficient organic light-emitting diodes from delayed fluorescence. *Nature* 492: 234–238.

12 Goushi, K., Yoshida, K., Sato, K., and Adachi, C. (2012). Organic light-emitting diodes employing efficient reverse intersystem crossing for triplet-to-singlet state conversion. *Nat. Photonics* 6: 253–258.

13 Worth, G.A. and Cederbaum, L.S. (2004). Beyond Born–Oppenheimer: molecular dynamics through a conical intersection. *Annu. Rev. Phys. Chem.* 55: 127–158.

14 Berberan-Santos, M.N. and Garcia, J.M.M. (1996). Unusually strong delayed fluorescence of C70. *J. Am. Chem. Soc.* 118: 9391–9394.

15 Baleizão, C. and Berberan-Santos, M.N. (2007). Thermally activated delayed fluorescence as a cycling process between excited singlet and triplet states: application to the fullerenes. *J. Chem. Phys.* 126: 204510.

16 Delorme, R. and Perrin, F. (1929). Duration of the fluorescence of solid uranyl salts and their solution. *J. Phys. Rad. Ser.* 10: 177–186.

17 Lewis, G.N., Lipkin, D., and Magel, T.T. (1941). Reversible photochemical processes in rigid media: a study of the phosphorescent state. *J. Am. Chem. Soc.* 63: 3005–3018.

18 Parker, C.A. and Hatchard, C.G. (1961). Triplet-singlet emission in fluid solutions: phosphorescence of Eosin. *Trans. Faraday Soc.* 57: 1894–1904.

19 Kirchhoff, J.R., Gamache, J.E., Blaskie, M.W., Del-Paggio, A.A., Lengel, R.K., and McMillin, D.R. (1983). Temperature dependence of luminescence from $Cu(NN)^{2+}$ systems in fluid solution. Evidence for the participation of two excited states. *Inorg. Chem.* 22: 2380–2384.

20 Zhang, Q., Li, J., Shizu, K., Huang, S., Hirata, S., Miyazaki, H., and Adachi, C. (2012). Design of efficient thermally activated delayed fluorescence materials for pure blue organic light emitting diodes. *J. Am. Chem. Soc.* 134: 14706–14709.

21 Dias, F.B., Bourdakos, K.N., Jankus, V., Moss K.C., Kamtekar, K.T., Bhalla, V., Santos, J., Bryce, M.R., and Monkman, A.P. (2013). Triplet harvesting with 100% efficiency by way of thermally activated delayed fluorescence in charge transfer OLED emitters. *Adv. Mater.* 25: 3707–3714.

22 Dias, F.B. (2015). Kinetics of thermal assisted delayed fluorescence in blue organic emitters with large singlet–triplet energy gap. *Philos. Trans. R. Soc. London, Ser. A* 373: 20140447.

23 Inoue, M., Serevičius, T., Nakanotani, H., Yoshida, K., Matsushima, T., Juršėnas, S., and Adachi, C. (2016). Effect of reverse intersystem crossing rate to suppress efficiency roll-off in organic light-emitting diodes with thermally activated delayed fluorescence emitters. *Chem. Phys. Lett.* 644: 62–67.

24 Marian, C.M. (2011). Spin–orbit coupling and intersystem crossing in molecules. *WIREs Comput. Mol. Sci.* 2: 187–203.

25 Capano, G., Rothlisberger, U., Tavernelli, I., and Penfold, T.J. (2015). Theoretical rationalization of the emission properties of prototypical Cu(I)–phenanthroline complexes. *J. Phys. Chem. A* 119: 7026–7037.

26 Leitl, M.J., Zink, D.M., Schinabeck, A.J., Baumann, T., Volz, D., and Yersin, H. (2016). Copper(I) complexes for thermally activated delayed fluorescence: from photophysical to device properties. *Top. Curr. Chem.* 374: 1–34.

27 Tatchen, J., Gilka, N., and Marian, C.M. (2007). Intersystem crossing driven by vibronic spin–orbit coupling: a case study on psoralen. *Phys. Chem. Chem. Phys.* 9: 5209–5221.

28 Etinski, M., Rai-Constapel, V., and Marian, C.M. (2014). Time-dependent approach to spin–vibronic coupling: implementation and assessment. *J. Chem. Phys.* 140: 114104.

29 Marian, C.M., Etinski, M., and Rai-Constapel, V. (2014). Reverse intersystem crossing in rhodamines by near-infrared laser excitation. *J. Phys. Chem. A* 118: 6985–6990.

30 Minns, R.S., Parker, D.S.N., Penfold, T.J., Worth, G.A., and Fielding, H.H. (2010). Competing ultrafast intersystem crossing and internal conversion in the "channel 3" region of benzene. *Phys. Chem. Chem. Phys.* 12: 15607–15615.

31 Penfold, T.J., Spesyvtsev, R., Kirkby, O.M., Minns, R.S., Parker, D.N.S., Fielding, H.H., and Worth, G.A. (2012). Quantum dynamics study of the competing ultrafast intersystem crossing and internal conversion in the "channel 3" region of benzene. *J. Chem. Phys.* 137: 204310.

32 Capano, G., Chergui, M., Rothlisberger, U., Tavernelli, I., and Penfold, T.J. (2014). A quantum dynamics study of the ultrafast relaxation in a prototypical Cu(I)–phenanthroline. *J. Phys. Chem. A* 118: 9861–9869.

33 Lim, B.T., Okajima, S., Chandra, A.K., and Lim, A.C. (1981). Radiationless transitions in electron donor-acceptor complexes: selection rules for S_1-T_1 intersystem crossing and efficiency of S_1-S_0 internal conversion. *Chem. Phys. Lett.* 79: 22–27.

34 El-Sayed, M.A. (1963). Spin–orbit coupling and the radiationless processes in nitrogen heterocyclics. *J. Chem. Phys.* 38: 2834–2838.

35 Dias, F.B., Santos, J., Graves, D., Data, P., Nobuyasu, R.S., Fox, M.S., Batsanov, A.S., Palmeira, T., Berberan-Santos, M.N., Bryce, M.R., and Monkman, A.P. (2016). The role of local triplet excited states and D-A relative orientation in

thermally activated delayed fluorescence: photophysics and devices. *Adv. Sci.* doi: 10.13140/RG.2.1.1821.1446.

36 Chen, X.K., Zhang, S.F., Fan, J.X., and Ren, A.M. (2015). Nature of highly efficient thermally activated delayed fluorescence in OLED emitters: non adiabatic effect between excited states. *J. Phys. Chem. C* 119: 9728–9733.

37 Ogiwara, T., Wakikawa, Y., and Ikoma, T. (2015). Mechanism of intersystem crossing of thermally activated delayed fluorescence molecules. *J. Phys. Chem. A* 119: 3415–3418.

38 Hontz, E., Chang, W., Congreve, D.N., Bulovic, V., Baldo, M.A., and Van-Voorhis, T. (2015). The role of electron–hole separation in thermally activated delayed fluorescence in donor–acceptor blends. *J. Phys. Chem. C* 119: 25591–25597.

39 Ward, J.S., Nobuyasu, R.S., Batsanov, A.S., Data, P., Monkman, A.P., Dias, F.B., and Bryce, M.R. (2016). The interplay of thermally activated delayed fluorescence (TADF) and room temperature organic phosphorescence in sterically-constrained donor–acceptor charge-transfer molecules. *Chem. Commun.* 52: 2612–2615.

40 Marian, C.M. (2016). On the mechanism of the triplet-to-singlet upconversion in the assistant dopant ACRXTN. *J. Phys. Chem. C* 120: 3715–3721.

41 Sousa, C., de Graaf, C., Rudavskyi, A., Broer, R., Tatchen, J., Etinski, M., and Marian, C.M. (2013). Ultrafast deactivation mechanism of the excited singlet in the light-induced spin crossover of $[Fe(2,2\text{-bipyridine})_3]^{2+}$. *Chem. Eur. J.* 19: 17541–17551.

42 Gibson, J., Monkman, A.P., and Penfold, T.J. (2016). The importance of vibronic coupling for efficient reverse intersystem crossing in TADF molecules. *ChemPhysChem* doi: 10.1002/cphc.201600662.

43 Born, M. and Huang, K. (1998). *Dynamical Theory of Crystal Lattices*. Oxford University Press.

44 Raab, A., Worth, G.A., Meyer, H.-D., and Cederbaum, L.S. (1999). Molecular dynamics of pyrazine after excitation to the S_2 electronic state using a realistic 24-mode model Hamiltonian. *J. Chem. Phys.* 110: 936–946.

45 Markmann, A., Worth, G.A., and Cederbaum, L.S. (2005). Allene and pentatetraene cations as models for intramolecular charge transfer: vibronic coupling Hamiltonian and conical intersections. *J. Chem. Phys.* 122: 144320.

46 Worth, G.A., Welch, G., and Paterson, M.J. (2006). Wavepacket dynamics study of $Cr(CO)_5$ after formation by photodissociation: relaxation through an (ExA)xe Jahn-Teller conical intersection. *Mol. Phys.* 104: 1095–1105.

47 Pápai, M., Rozgonyi, T., Vanko, G., and Penfold, T.J. (2016). High efficiency iron photosensitiser explained with quantum wavepacket dynamics. *J. Phys. Chem. Lett.* 7: 2009–2014.

48 Car, R. and Parrinello, M. (1985). Unified approach for molecular dynamics and density-functional theory. *Phys. Rev. Lett.* 55: 2471–2474.

49 Curchod, B.F.E., Rothlisberger, U., and Tavernelli, I. (2013). Trajectory-based nonadiabatic dynamics with time-dependent density functional theory. *ChemPhysChem* 14: 1314–1340.

50 Richings, G.W., Polyak, I., Spinlove, K.E., Worth, G.A., Burghardt, I., and Lasorne, B. (2015). Quantum dynamics simulations using gaussian wavepackets: the vMCG method. *Int. Rev. Phys. Chem.* 34: 269–308.
51 Wang, L., Akimov, A.V., and Prezhdo, O.V. (2016). Recent progress in surface hopping: 2011–2015. *J. Phys. Chem. Lett.* 7: 2100–2112.
52 Tully, J.C. and Preston, R.K. (1971). Trajectory surface hopping approach to nonadiabatic molecular collisions: the reaction of H$^+$ with D$_2$. *J. Chem. Phys.* 55: 562–572.
53 Tully, J.C. (1990). Molecular dynamics with electronic transitions. *J. Chem. Phys.* 93: 1061–1071.
54 Köppel, H., Domcke, W., and Cederbaum, L.S. (1984). Multimode molecular dynamics beyond the Born-Oppenheimer approximation. *Adv. Chem. Phys.* 57: 59–246.
55 Cattarius, C., Worth, G.A., Meyer, H.-D., and Cederbaum, L.S. (2001). All mode dynamics at the conical intersection of an octa-atomic molecule: multi-configuration time-dependent Hartree (MCTDH) investigation on the butatriene cation. *J. Chem. Phys.* 115: 2088–2100.
56 Mahapatra, S., Cederbaum, L.S., and Koppel, H. (1999). Theoretical investigation of Jahn-Teller and Pseudo-Jahn-Teller coupling effects on the photoelectron spectrum of allene. *J. Chem. Phys.* 111: 10452–10463.
57 Penfold, T.J. and Worth, G.A. (2009). A model Hamiltonian to simulate the complex photochemistry of benzene II. *J. Chem. Phys.* 131: 190–199.
58 Barbatti, M. (2011). Nonadiabatic dynamics with trajectory surface hopping method. *WIREs Comput. Mol. Sci.* 1: 620–633.
59 Martinez, T.J., Ben-Nun, M., and Levine, R.D. (1996). Multi-electronic-state molecular dynamics: a wave function approach with applications. *J. Phys. Chem.* 100: 7884–7895.
60 Ben-Nun, M. and Martınez, T.J. (1998). Nonadiabatic molecular dynamics: validation of the multiple spawning method for a multidimensional problem. *J. Phys. Chem.* 108: 7244–7257.
61 Shalashilin, D.V. and Child, M.S. (2004). The phase space CCS approach to quantum and semiclassical molecular dynamics for high-dimensional systems. *Chem. Phys.* 304: 103–120.
62 Shalashilin, D.V. (2009). Quantum mechanics with the basis set guided by Ehrenfest trajectories: theory and application to spin-boson model. *J. Chem. Phys.* 130: 244101.
63 Worth, G.A. and Burghardt, I. (2003). Full quantum mechanical molecular dynamics using Gaussian wavepackets. *Chem. Phys. Lett.* 368: 502–508.
64 Makhov, D.V., Glover, W.J., Martinez, T.J., and Shalashilin, D.V. (2014). Ab initio multiple cloning algorithm for quantum nonadiabatic molecular dynamics. *J. Chem. Phys.* 141: 054110.
65 McLachlan, A.D. (1964). A variational solution of the time-dependent Schrodinger equation. *Mol. Phys.* 8: 39–44.
66 Makri, N. and Miller, W.H. (1987). Time-dependent self-consistent field (TDSCF) approximation for a reaction coordinate coupled to a harmonic bath: single and multiple configuration treatments. *J. Chem. Phys.* 87: 5781–5787.

67 Beck, M.H., Jäckle, A., Worth, G.A., and Meyer, H.-D. (2000). The multiconfiguration time-dependent Hartree method: a highly efficient algorithm for propagating wavepackets. *Phys. Rep.* 324: 1–105.

68 Meyer, H.-D., Gatti, F., and Worth, G.A. (2008). *High Dimensional Quantum Dynamics: Basic Theory, Extensions, and Applications of the MCTDH Method*. Weinheim: VCH.

69 Tannor, D.J. (2007). *Introduction to Quantum Mechanics: A Time Dependent Perspective*. University Science Books.

70 Raab, A., Burghardt, I., and Meyer, H.-D. (1999). The multiconfiguration time-dependent Hartree method generalized to the propagation of density operators. *J. Chem. Phys.* 111: 8759–8772.

71 Raab, A. and Meyer, H.-D. (2000). Multiconfigurational expansions of density operators: equations of motion and their properties. *Theor. Chem. Acc.* 104: 358–369.

72 Penfold, T.J., Karlsson, S., Capano, G., Lima, F.A., Rittmann, J., Reinhard, M., Rittmann-Frank, M.H., Braem, O., Baranoff, E., Abela, R., Tavernelli, I., Rothlisberger, U., Milne, C.J., and Chergui, M. (2013). Solvent-induced luminescence quenching: static and time-resolved X-ray absorption spectroscopy of a copper(I) phenanthroline complex. *J. Phys. Chem. A* 117: 4591–4601.

73 Capano, G., Penfold, T.J., Rothlisberger, U., and Tavernelli, I. (2014). A vibronic coupling Hamiltonian to describe the ultrafast excited state dynamics of a Cu(I)–phenanthroline complex. *Chimia* 68: 227–230.

74 Deaton, J.C., Switalski, S.C., Kondakov, D.Y., Young, R.H., Pawlik, T.D., Giesen, D.J., Harkins, S.B., Miller, A.J.M., Mickenberg, S.F., and Peters, J.C. (2010). E-type delayed fluorescence of a phosphine-supported $Cu_2(\mu\text{-}NAr_2)_2$ diamond core: harvesting singlet and triplet excitons in OLEDs. *J. Am. Chem. Soc.* 132: 9499–9508.

75 Wallesch, M., Volz, D., Zink, D.M., Schepers, U., Nieger, M., Baumann, T., and Bräse, S. (2014). Bright coppertunities: multinuclear Cu(I) complexes with N-P ligands and their applications. *Chem. Eur. J.* 20: 6578–6590.

76 Dumur, F. (2015). Recent advances in organic light emitting devices comprising copper complexes: a realistic approach for low cost and highly emissive devices? *Org. Electron.* 21: 27–39.

77 McMillin, D.R., Kirchhoff, J.R., and Goodwin, K.V. (1985). Exciplex quenching of photo-excited copper complexes. *Coord. Chem. Rev.* 64: 83–92.

78 Armaroli, N. (2001). Photoactive mono- and polynuclear Cu(I)–phenanthrolines. A viable alternative to Ru(II)-polypyridines? *Chem. Soc. Rev.* 30: 113–124.

79 Zgierski, M.Z. (2003). Cu(I)-2,9-dimethyl-1,10-phenanthroline: density functional study of the structure, vibrational force-field, and excited electronic states. *J. Chem. Phys.* 118: 4045.

80 Siddique, Z.A., Yamamoto, Y., Ohno, T., and Nozaki, K. (2003). Structure-dependent photophysical properties of singlet and triplet metal-to-ligand charge transfer states in copper(I) bis(diimine) compounds. *Inorg. Chem.* 42: 6366–6378.

81 Chen, L.X., Shaw, G.B., Novozhilova, I., Liu, T., Jennings, G., Attenkofer, K., Meyer, G.J., and Coppens, P. (2003). MLCT state structure and dynamics of a copper(I) diimine complex characterized by pump-probe X-ray and laser spectroscopies and DFT calculations. *J. Am. Chem. Soc.* 125: 7022–7034.

82 Iwamura, M., Takeuchi, S., and Tahara, T. (2007). Real-time observation of the photoinduced structural change of bis(2,9-dimethyl-1,10-phenanthroline) copper(I) by femtosecond fluorescence fpectroscopy: a realistic potential curve of the Jahn-Teller distortion. *J. Am. Chem. Soc.* 129: 5248–5256.

83 Iwamura, M., Watanabe, H., Ishii, K., Takeuchi, S., and Tahara, T. (2011). Coherent nuclear dynamics in ultrafast photoinduced structural change of bis(diimine) copper(I) complex. *J. Am. Chem. Soc.* 133: 7728–7736.

84 Huang, J., Buyukcakir, O., Mara, M.W., Takeuchi, S., and Tahara, T. (2012). Highly efficient ultrafast electron injection from the singlet MLCT excited state of copper(I) diimine complexes to TiO_2 nanoparticles. *Angew. Chem. Int. Ed.* 51: 12711–12715.

85 Iwamura, M., Takeuchi, S., and Tahara, T. (2014). Substituent effect on the photoinduced structural change of Cu(I) complexes observed by femtosecond emission spectroscopy. *Phys. Chem. Chem. Phys.* 16: 4143–4154.

86 Shaw, G.B., Grant, C.D., Shirota, H., Castner, E.W., Meyer, G.J., and Chen, L.X. (2007). Ultrafast structural rearrangements in the MLCT excited state for copper(I) bis-phenanthrolines in solution. *J. Am. Chem. Soc.* 129: 2147–2160.

87 Huang, J., Buyukcakir, O., Mara, M.W., Coskun, A., Dimitrijevic, N.M., Barin, G., Kokhan, O., Stickrath, A.B., Ruppert, R., Tiede, D.M., Chen, L.X. (2012). Highly efficient ultrafast electron injection from the singlet MLCT excited state of copper(I) diimine complexes to TiO_2 nanoparticles. *Angew. Chem. Int. Ed.* 51: 12711–12715.

88 Masui, K., Nakanotani, H., and Adachi, C. (2013). Analysis of exciton annihilation in high-efficiency sky-blue organic light-emitting diodes with thermally activated delayed fluorescence. *Org. Electron.* 14: 2721–2726.

89 Nobuyasu, R.S., Ren, Z., Griffiths, G.C., Batsanov, A.S., Data, P., Yan, S., Monkman, A.P., Bryce, M.R., and Dias, F.B. (2016). Rational design of TADF polymers using donor–acceptor monomer with enhanced TADF efficiency induced by the energy alignment of charge transfer and local triplet excited states. *Adv. Opt. Mater.* 4: 597–607.

90 Zhao, Y. and Truhlar, D.G. (2008). The M06 suite of density functionals for main group thermochemistry, thermochemical kinetics, noncovalent interactions, excited states, and transition elements: two new functionals and systematic testing of four M06-class functionals and 12 other functionals. *Theor. Chem. Acc.* 120: 215–241.

91 Hirata, S. and Head-Gordon, M. (1999). Time-dependent density functional theory within the Tamm–Dancoff approximation. *Chem. Phys. Lett.* 314: 291–299.

92 Dreuw, A., Weisman, J.L., and Head-Gordon, M. (2003). Long-range charge-transfer excited states in time-dependent density functional theory require non-local exchange. *J. Chem. Phys.* 119: 2943–2946.

93 Li, R., Zheng, J., and Truhlar, D.G. (2010). Density functional approximations for charge transfer excitations with intermediate spatial overlap. *Phys. Chem. Chem. Phys.* 12: 12697–12701.

94 Penfold, T.J. (2015). On predicting the excited-state properties of thermally activated delayed fluorescence emitters. *J. Phys. Chem. C* 119: 13535–13544.

95 Peach, M.J.G., Williamson, M.J., and Tozer, D.J. (2011). Influence of triplet instabilities in TDDFT. *J. Comput. Theor. Chem.* 7: 3578–3585.

96 Henry, B.R. and Siebrand, W. (1971). Spin–Orbit coupling in aromatic hydrocarbons: analysis of nonradiative transitions between singlet and triplet states in benzene and naphthalene. *J. Chem. Phys.* 54: 1072–1085.

97 Santos, P.L., Ward, J.S., Data, P., Batsanov, A.S., Bryce, M.R., Dias, F.B., and Monkman, A.P. (2016). Engineering the singlet–triplet energy splitting in a TADF molecule. *J. Mater. Chem. C* 4: 3815–3824.

98 Etherington, M.K., Gibson, J., Higginbotham, H.F., Penfold, T.J., and Monkman, A.P. (2016). Revealing the spin–vibronic coupling mechanism of thermally-activated delayed fluorescence. *Nat. Commun.* 7: 13680.

10

Exciplex: Its Nature and Application to OLEDs

Hwang-Beom Kim[1], Dongwook Kim[2], and Jang-Joo Kim[3]

[1] *Seoul National University, Department of Materials Science and Engineering, 1 Gwanak-ro, Gwanak-gu, Seoul 151-744, South Korea*
[2] *Kyonggi University, Department of Chemistry, 154-42 Gwanggyosan-Ro, Yeongtong-Gu, Suwon, 16227, South Korea*
[3] *Seoul National University, Department of Materials Science and Engineering and Research Institute of Advanced Materials (RIAM), 1 Gwanak-ro, Gwanak-gu, Seoul 151-744, South Korea*

Exciplexes have been widely employed in organic light-emitting diodes (OLEDs) by virtue of the thermally activated delayed fluorescence (TADF) characteristics and the convenience of exploiting exciplexes to OLEDs. Exciplexes are readily formed when the blend films of donor molecules and acceptor molecules are excited. We treat the fundamentals of exciplexes and their application to OLEDs in this chapter. First, the electronic structures and properties of exciplexes and various transition processes associated with exciplex systems are covered. Later, OLEDs employing exciplexes as emitters as well as sensitizers are reviewed.

10.1 Introduction

Intermolecular interactions among various molecules have drawn a great deal of attention because they are ubiquitous not only in biology and supramolecular chemistry but also in organic electronics. In particular, the noncovalent weak intermolecular interactions among π-conjugated molecular units are of utmost interest for the organic electronics applications, e.g. OLEDs and organic photovoltaics (OPVs), because they affect the morphology of the active layer film, which has a close link to the device performance. Not surprisingly, these intermolecular interactions in the ground state are notably different from those in the excited state, generating new species. A dimer of the same kind bound in the excited state but repulsive in the ground state is defined as an excimer (**exci**ted + di**mer**); likewise, a dimer of a different kind is denoted as an exciplex (**exci**ted + com**plex**). Given that the fundamental processes in the organic electronics include the electronic excitations in the solid-state film of π-conjugated organic materials, the formation of exciplexes is supposed to be widespread. In OPVs, for example, the initial exciton of donor or acceptor materials, i.e. locally excited (LE) state of a donor or acceptor, dissociates at the donor–acceptor interface forming the

Highly Efficient OLEDs: Materials Based on Thermally Activated Delayed Fluorescence,
First Edition. Edited by Hartmut Yersin.
© 2019 Wiley-VCH Verlag GmbH & Co. KGaA. Published 2019 by Wiley-VCH Verlag GmbH & Co. KGaA.

charge-transfer (CT) exciplex state as an intermediate state, which is followed by the charge separation producing free charges. In OLEDs, reverse processes take place, i.e. electrons in acceptor molecules and holes in donor molecules come close by Coulomb interaction or by electric potential to form the CT exciplex states. Therefore, the electronic structure of such an exciplex state with regard to that of the LE state of constituent materials is one of the key factors to consider in the fabrication of efficient devices.

As a light-emitting guest in the emissive layer, the exciplex has remained out of interest due to their rather poor performances. Recently, however, the exciplex formation has been brought into researcher's attention. Goushi et al. and Park et al. demonstrated an enhancement with their devices of 10% external quantum efficiency (EQE) using a 4,4′,4″-tris[3-methylphenyl(phenyl)amino]triphenyl-amine (m-MTDATA):2,8-bis(diphenylphosphoryl)dibenzo[b,d]thiophene (PPT) and 4,4′,4″-tris(N-carbazolyl)-triphenylamine (TCTA) and bis-4,6-(3,5-di-3-pyridylphenyl)-2-methylpyrimidine (B3PYMPM) blend layers as emitting layers [1–3]. This work was followed by a lot of studies, including the ones of Hung et al. and Kim et al. where devices with an EQE of 11% and an internal quantum efficiency (IQE) of almost 100% were reported using exciplexes as emitters, respectively [4–6]. Provided that these materials are purely organic molecules, the strong spin–orbit coupling (SOC) and the phosphorescence from these materials are hardly expected. Therefore, these results strongly suggest that the exciplex formation plays a role in the conversion from the triplet excited state to the singlet one and hence improves the performance of OLEDs. In this regard, the fundamental understanding of various features of the exciplex is essential to fabricate more efficient OLEDs.

In this chapter, we begin our discussion with the fundamentals of exciplexes including their electronic structures, properties, and kinetics of various processes related to them. Later, we evolve our discussion to the applications of exciplexes to OLEDs.

10.2 Formation and Electronic Structures of Exciplexes

Excited CT complex – in short, exciplex – is formed by CT between an excited donor (D*) and an acceptor or a donor and an excited acceptor (A*) as shown in Figure 10.1. Generally, the energy levels of the highest occupied molecular orbital (HOMO) and the lowest unoccupied molecular orbital (LUMO) of the donor molecule are higher than those of the acceptor molecule. Since exciplex formation can be viewed as the CT from an LE molecule to its neighboring one, the wave function of the exciplex can be written as a linear combination of LE, CT, and neutral ground states: [7]

$$\psi([DA]^*) \approx c_3\phi(D^*A) + c_2\phi(DA^*) + c_1\phi(D^+A^-) + c_0\phi(DA) \qquad (10.1)$$

where the coefficients pertain to the relative contribution of each state and D and A represent electron donor and acceptor, respectively. The first two terms on the right-hand side of the equation correspond to the LE states of D and A, the third term is associated with the CT states, and the last one points to the ground

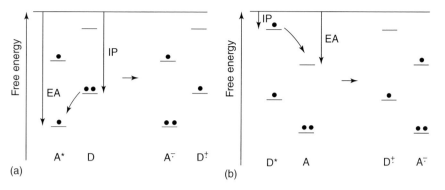

Figure 10.1 Schematic illustration of the exciplex formation via the charge transfer when either (a) the electron acceptor or (b) electron donor is excited [7]. Reproduced with permission of American Chemical Society.

state, respectively. If ground-state influences on the exciplex state are neglected, Eq. (10.1) further reduces to

$$\psi([DA]^*) \approx c_3\phi(D^*A) + c_2\phi(DA^*) + c_1\phi(D^+A^-) \tag{10.2}$$

The fraction of charge transfer f_{CT} in the exciplex is defined by the following equation:

$$f_{CT} = |c_1|^2 \tag{10.3}$$

The intermolecular CT state energy in the gas phase is represented by

$$E_{CT} = IP_D - EA_A - E_B = \Delta E_{DA} - E_B, \tag{10.4}$$

where IP_D is the ionization potential (IP) of the donor, EA_A is the electron affinity (EA) of the acceptor, E_B is the binding energy of the CT state, and $\Delta E_{DA} = IP_D - EA_A$.

From Figure 10.1, simple rules for determining the feasibility of electron transfer (ET) can be written as follows: [8]

$$D^* + A \rightarrow D^+ + A^-$$

$$IP_{D^*} - EA_A < 0 : \text{allowed} \tag{10.5}$$

$$IP_{D^*} - EA_A > 0 : \text{forbidden} \tag{10.6}$$

$$D + A^* \rightarrow D^+ + A^-$$

$$EA_{A^*} - IP_D < 0 : \text{forbidden} \tag{10.7}$$

$$EA_{A^*} - IP_D > 0 : \text{allowed} \tag{10.8}$$

For an exciplex, E_B can be defined as the difference between the exciplex energy and the sum of energies for the free radical ion pair:

$$E_B^{\text{exciplex}} = -\{E([DA]^*) - E(D^+) - E(A^-)\} \tag{10.9}$$

Therefore, the exciplex energy depends on the degree of CT, which will be described later in details. E_B stems mainly from the electrostatic interactions of a hole–electron pair. Because of the short distance between the hole and electron in the LE state, E_B for the LE state is in general larger than that for

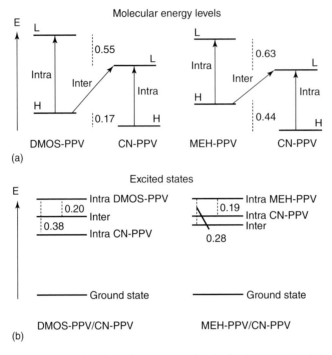

Figure 10.2 Molecular and state energy levels of DMOS-PPV/CN-PPV and MEH-PPV/CN-PPV systems [9]. Reproduced with permission of American Physical Society.

the CT state where the hole and electron are spatially separated. This indicates that although the above conditions are satisfied, the exciplex can be less stable than LE individual molecules. Upon excitation, for instance, a pair of poly[2-methoxy-5-(2′-ethylhexyloxy)-p-phenylenevinylene] (MEH-PPV) and poly(2,5,2′,5′-tetrahexyloxy-7,8′-dicyano-di-p-phenylenevinylene) (CN-PPV) undergoes an ET to form an exciplex. On the other hand, an energy transfer (ENT) takes place in the pair of poly(2-dimethyloctylsilyl-1,4-phenylenevinylene) (DMOS-PPV) and CN-PPV [9]. It was claimed that the transition from the LE state to the CT state requires the energy penalty on the order of 0.35 eV, leading to the higher CT state in energy for the latter. They proposed that the pair with the smaller ΔE_{DA} favors the ET, whereas the one with the larger ΔE_{DA} does the ENT; see Figure 10.2.

In the case of the triplet state, similar cases are more often found. Exchange energy, K, is often invoked to explain the difference between the singlet and triplet states; ignoring the electron correlation effect, and if the singlet and triplet excited states are of the same nature, then the energy difference between these states corresponds to $2K$ [10]. Coulomb integral, J, and K are defined as

$$J(\phi_i, \phi_j) = \left\langle \phi_i(1)\phi_j(2) \left| \frac{1}{r_{12}} \right| \phi_i(1)\phi_j(2) \right\rangle \tag{10.10}$$

$$K(\phi_i, \phi_j) = \left\langle \phi_i(1)\phi_j(2) \left| \frac{1}{r_{12}} \right| \phi_i(2)\phi_j(1) \right\rangle \tag{10.11}$$

where ϕ_i represents wave function for particles (an electron and a hole) and r_{12} is the distance between particles 1 and 2. Basically, K is a quantum mechanical correction for J between two charged particles. It arises from the fact that an electron is indistinguishable and subject to Pauli exclusion principle. As a result of the antisymmetry of the wave function, K of a hole–electron pair destabilizes the singlet excited state and stabilizes the triplet excited state, putting more emphasis on the role of E_B for the triplet excited state [8, 11, 12]. Provided that K is interpreted as the Coulomb interaction between the overlap of the hole and electron wave functions [13], K for the LE state is larger than that of CT state. Therefore, the binding energy for the triplet LE state is more enhanced, suggesting that the energy penalty for the transition from the triplet LE state to the triplet CT state be even larger than that for the singlet process. In order for the triplet CT state to be formed, ΔE_{DA} should be more significantly reduced [9, 11].

Since exciplex is formed by CT from an LE molecule to its neighboring ground-state molecule, the electronic couplings between frontier molecular orbitals of individual molecules play an important role in the exciplex formation. Given that such electron couplings are affected by the separation distance, orientation, and symmetry of molecules [14], the donor and acceptor molecules are supposed to have small interplanar distances and parallel alignments between their molecular planes to form the exciplexes [15, 16]. Inai et al., for example, investigated various structural isomers of polypeptides with a *p*-(dimethylamino)phenyl group as a donor unit and a naphthyl group as an acceptor unit [16]. They found that the exciplex formation is more effective in the face-to-face configurations than the head-to-tail ones. In addition, their exciplex is formed at the intermolecular spacing of 0.83 nm, but not at 1.20 nm. Thus, molecules with the twisted conformation [17] or bulky, three-dimensional groups [18] were designed to prevent the exciplex formation.

The nuclear reorganization energy also plays a role in the ET [7, 14]. The large geometric difference between the LE state and the exciplex state would require the large excess energy for the transition, impeding the exciplex formation [19]. The exciplex formation can also be inhibited by putting the electron-rich moiety right next to the electron-deficient moiety because they could weaken each other's electron-rich or electron-deficient properties [20]. On the other hand, the exciplex formation can also be controlled by the modulation of intermolecular interactions. The excited molecule with high dipole moment could induce a dipole in the neighboring molecule with high polarizability. This gives rise to an enhanced intermolecular interaction in the excited state and consequently facilitates the formation of exciplexes [21]. Small molecular pairs of donor and acceptor molecules are collected in the Appendix of this chapter.

Exciplexes in solid state are expected to have various dimer geometries with different distances between D and A molecules, resulting in a broad energy band [3, 22]. Exciplexes with different D–A distances have different degrees of CT between D and A. Exciplexes with shorter D–A distances have larger CT, lower energy due to the Coulomb interaction proportional to r^{-1}, and smaller ΔE_{ST} and slower decay rates due to the lower overlap between the frontier orbitals.

10.3 Optical Properties of Exciplexes

10.3.1 Photoluminescence of Exciplexes

Photoluminescence (PL) spectra of exciplexes are generally featureless, broad, and redshifted from the constituting molecules and depend on the degree of CT in the complex. Therefore, features of PL spectra, e.g. v_{0-0}, v_{av}, vibronic structure of the spectra, etc., provide important information regarding a given exciplex. Such characteristics are also influenced by the surrounding medium; as schematically displayed in Figure 10.3, the PL spectra of CT states tend to be more redshifted as the solvent becomes more polar. Because exciplexes, in general, have mixed LE and CT characters and spectral features of the pure LE and pure CT states are markedly different, PL spectra of exciplexes reflect the valuable information including its relative weight of CT character. Figure 10.4, for instance, shows the PL spectra of the exciplexes consisting of 9,10-dicyanoanthracene (DCA) and various alkylbenzene donors in cyclohexane [25]. Note that the excitation energy for DCA is significantly smaller than those of various alkylbenzene derivatives due to the longer π-conjugation length of the former, and hence the energies for the lowest LE state of DCA–alkylbenzene

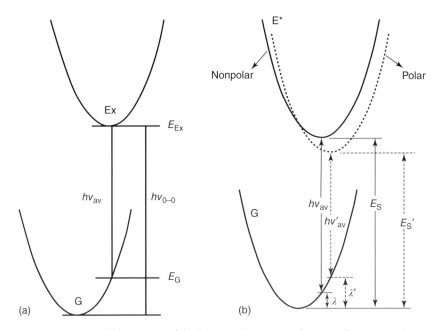

Figure 10.3 Pictorial description of the 0–0 transition energy, hv_{0-0}, and average emission energy, hv_{av} (a) [23]. Illustration of the effect of solvent polarity on the photoluminescence (PL) spectra (b) [24]. Ex and G denote the excited and ground states, respectively. Parabolas represent the respective electronic states of a given molecular system including the surround medium, e.g. solvent; λ is the reorganization energy for the emission. Here, we did not take into account of zero-point vibrational energy (ZPVE) levels for the sake of simplicity. The figure was modified from the original ones of the References. Reproduced with permission of American Chemical Society.

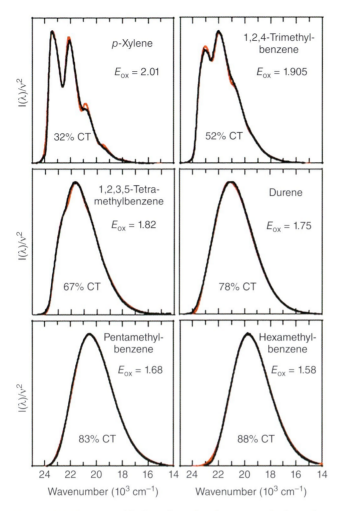

Figure 10.4 PL spectra (black) and simulated curves (red) of exciplexes with different degrees of CT consisting of DCA and various alkylbenzene donors in cyclohexane [25]. Reproduced with permission of John Wiley & Sons.

exciplexes remain constant. The decrease in the oxidation potentials of the donors (E_{ox} in the figure) corresponding to the elevation of the HOMO energy level of the donors stabilizes more the pure CT state with respect to the pure LE state. This increases the CT character of exciplexes, leading to the more redshifted PL spectra. In addition, such enhanced CT nature in the exciplexes gives rise to the less structured PL spectra because the vibrational progression is caused mainly by the pure LE state.

Recently, exciplex emission in the solid state was interpreted based on a broad energy band due to various dimer geometries, which have different distances between the donor (D) and acceptor (A) molecules (C.-K. Moon and J.-J. Kim, unpublished). Exciplexes with different D–A distances have different degrees of CT between D and A. The low-energy exciplex emission is largely from the CT state, but the LE state makes a larger contribution to high-energy exciplex emission.

10.3.2 Absorption Spectra of Exciplexes

In contrast to the PL spectra of exciplexes, the absorption peaks for the exciplex state are usually insignificant. Figure 10.5 shows the PL and absorption spectra of the TAPC and 6,6′-bis(2-4diphenylquinoline) (B1PPQ) neat films and TAPC:B1PPQ blend films as a function of their molar ratio, where TAPC serves as an electron donor and B1PPQ as an electron acceptor [26]. Only emissions from the exciplex state are observed in the blend films with 10–50 mol% load of B1PPQ. This indicates that almost all the excitons in the blend films end up with the exciplex. Nevertheless, absorption peaks corresponding to the exciplex state are not shown in such blend films as shown Figure 10.5b. The absorption spectra of the blend films are reproduced by the linear combination of the absorption spectra of the D and A molecules with no CT absorption. This indicates that the exciplexes are not directly formed from the ground state by photoexcitation, but formed by CT from the excitons of the constituent molecules. In other words, the exciton of the constituent molecule acts as a precursor to the exciplex.

However, recent elaborate studies identified new absorption peaks corresponding to the exciplex state in the donor–acceptor blend films [27–34]. Figure 10.6a shows the PL spectrum of the MDMO-PPV:PCBM (1:4) blend film [35]. In comparison with the PL spectra of the pristine films of the constituent molecules shown in the inset of Figure 10.6a, a broad peak around 1.35 eV was identified as an emission peak from the exciplex state to the ground state. Utilizing Fourier transform photocurrent spectroscopy (FTPS), an extremely sensitive technique for low-intensity peak via the photocurrent measurement, they found a new red-shifted, broad, and weak absorption peak around 1.65 eV as seen in Figure 10.6b [27]. This peak is lower in energy than the respective absorption of MDMO-PPV and PCBM and hence attributed to the singlet exciplex state.

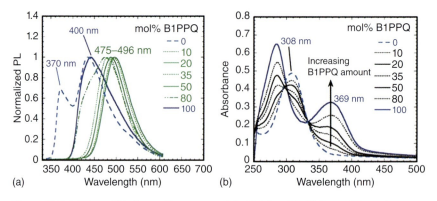

Figure 10.5 (a) PL and (b) absorption spectra of the TAPC and B1PPQ neat film and TAPC:B1PPQ blend films with various molar ratios [26]. Reproduced with permission of American Chemical Society.

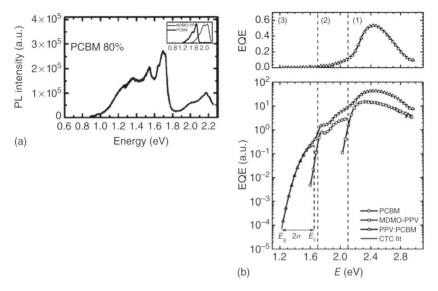

Figure 10.6 (a) PL [35] and (b) FTPS [27] spectra of the MDMP-PPV:PCBM (1 : 4) blend film. The new weak absorption below 1.65 eV is only observed in the blend film originating from the CT absorption. Reproduced with permission of John Wiley & Sons.

10.4 Decay Processes of the Exciplex in Solution

The fluorescence rate constant for a given exciplex is low compared with those of its constituent molecules, giving the exciplex the relatively long lifetime. Therefore, various processes involved in the decay of an exciplex are in competition as shown in Figure 10.7. Now, we turn to such processes, starting with the fluorescence rate of the exciplex.

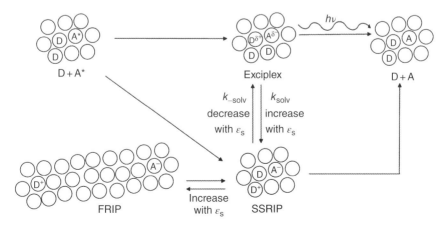

Figure 10.7 Illustration of various decay processes for the exciplex-forming system.

10.4.1 Fluorescence Rate Constant for the Exciplex State

Fluorescence rate constant, k_f, increases as the transition dipole moment, M, and the energy gap between the ground and the excited states increase. It is expressed as follows: [23]

$$k_f = \frac{16\pi^3}{3hc^3\varepsilon_0} f(n) v_{av}^3 |M|^2 \tag{10.12}$$

where h, c, and ε_0 is the Planck constant, the speed of light, and the dielectric constant for vacuum and M is transition dipole moment. $f(n)$ is a function of the refractive index, n, of surrounding medium and, assuming the spherical cavity of the emitting species in the solvent, written as

$$f(n) = n\left(\frac{n^2+2}{3}\right)^2 \tag{10.13}$$

and v_{av} is the average emission frequency, which is defined in Eq. (10.14):

$$v_{av} = \left(\frac{\int_0^\infty I_F(v)dv}{\int_0^\infty v^{-k}I_F(v)dv}\right)^{\frac{1}{3}} \tag{10.14}$$

where $I_F(v)$ is the emission intensity expressed in photons per unit time per unit spectral energy and k is 1 or 3 when the excited state has high CT character or high LE character, respectively.

The transition dipole moment, M, is defined as

$$M = \langle \psi_{Ex} | \hat{\mu} | \psi_G \rangle \tag{10.15}$$

where ψ_G and ψ_{Ex} represent the ground-state and excited-state wave functions, respectively, and $\hat{\mu}$ is the dipole moment operator. If the wave function for the exciplex state is plugged in Eq. (10.15), the expression for the transition dipole moment of the exciplex, $M_{exciplex}$, is obtained when LE states are originated from acceptors:

$$M_{exciplex} = c_2 M_{A^*} + c_1 M_{D^+A^-} \tag{10.16}$$

The transition dipole moment for the pure ion pair, i.e. $[D^+A^-]$, can be written as

$$M_{D^+A^-} = -\frac{H_{01}}{hv_{av}}\Delta\mu \tag{10.17}$$

where H_{01} is the matrix element for the electronic coupling between the ground and the pure ion pair states and $\Delta\mu$ is the dipole moment difference between such states. Note that $\Delta\mu$ and M_A are vectors and k_f is proportional to $|M|^2$, leading to the dependence of k_f on the angle, θ, between $\Delta\mu$ and M_A. Then, Eq. (10.12) can be modified to [23, 36–38]

$$\frac{k_f}{f(n)v_{av}^{ex}} = \frac{16\pi^3}{3h^3c^3\varepsilon_0}\left[(H_{01}|\Delta\mu|)^2 + 2(H_{01}H_{12}|\Delta\mu||M_{A^*}|\cos\theta)\left(\frac{hv_{av}^{ex}}{hv_{av}^A - hv_{av}^{ex}}\right) \right.$$
$$\left. + (H_{12}|M_{A^*}|)^2\left(\frac{hv_{av}^{ex}}{hv_{av}^A - hv_{av}^{ex}}\right)^2\right] \tag{10.18}$$

where $h\nu_{av}^{ex}$ and $h\nu_{av}^{A}$ denote the emission energies for the exciplex and acceptor (or the molecule associated with the lower LE state), respectively; H_{12} is the matrix element for the electronic coupling between the exciplex state and the LE state of the acceptor; and $\Delta\mu$ can be experimentally evaluated from the solvatochromism of the exciplex. The emission maximum in a solution, $\nu_{av}(\varepsilon)$, is influenced by the solvent polarity and expressed as [23]

$$\nu_{av}(\varepsilon) = \nu_{av}(0) - \frac{(\Delta\mu)^2}{hc\rho^3}\left\{2\left(\frac{\varepsilon-1}{2\varepsilon+1}\right) - \left(\frac{n^2-1}{2n^2+1}\right)\right\} \quad (10.19)$$

where ρ is the radius of the exciplex solvent cavity and ε and n are the solvent dielectric constant and refractive index, respectively, and thus $\nu_{av}(0)$ corresponds to the emission frequency in the vacuum. Here, the ground-state static dipole moment of the exciplex is ignored. However, such assumptions are not completely fulfilled, and the model strongly depends on ρ. Therefore, the $\Delta\mu$ value of this model should be taken with some caution.

Figure 10.8 shows the experimental value of $\frac{k_f}{(n^3 \nu_{av}^{ex})}$ for various exciplexes of TCA and alkylbenzene derivatives as a function of $h\nu_{av}/(h\nu_{av}^{A} - h\nu_{av})$ and fitted curve based on Eq. (10.18); note that $f(n)$ is replaced by n^3 here [38]. The experimental data were obtained from exciplexes consisting of TCA and alkylbenzene donors in various solvents. The accord between them is remarkable, suggesting that the model is highly useful.

These theoretical and experimental results show that the fluorescent rate constant of the exciplex decreases as the energy difference between the LE and the exciplex states increases – in other words, as the degree of CT in the exciplex increases (increased CT character).

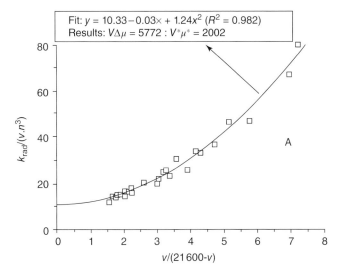

Figure 10.8 Experimental value of $k_f/(n^3 \nu_{av}^{ex})$ for exciplexes consisting of TCA and alkylbenzene as a function of $h\nu_{av}/(h\nu_{av}^{A} - h\nu_{av})$ where $h\nu_{av}^{A} = 21\,600$ cm^{-1} for TCA corresponding to the lower LE state in energy and the fit line of Eq. (10.18) [38]. Reproduced with permission of John Wiley & Sons.

10.4.2 Contact Radical Ion Pair (CRIP) Versus Solvent-separated Radical Ion Pair (SSRIP)

In going from the exciplex to the fully separated free ion pair in solution, contact radical ion pair (CRIP) and solvent-separated radical ion pair (SSRIP) are the intermediates of importance. In the presence of solvent molecules, free energies for such states include the solvation energy term as follows: [7]

$$G_{\text{CRIP}} = \text{IP}_D - \text{EA}_A - \frac{1}{4\pi\varepsilon_0} \frac{\mu^2}{\rho^3} \left(\frac{\varepsilon - 1}{2\varepsilon + 1}\right) - \frac{e^2}{4\pi\varepsilon_0 d_{\text{CRIP}}} \quad (10.20)$$

$$G_{\text{SSRIP}} = \text{IP}_D - \text{EA}_A$$
$$- \frac{e^2}{4\pi\varepsilon_0}\left(\frac{1}{2r_D} + \frac{1}{2r_A}\right)\left(1 - \frac{1}{\varepsilon}\right) - \frac{e^2}{4\pi\varepsilon_0 \varepsilon d_{\text{SSRIP}}} \quad (10.21)$$

where μ and ρ denote the static dipole moment of the CRIP and the radius of the CRIP solvent cavity; r_D and r_A are the radius of donor and that of acceptor, respectively; and d_{CRIP} and d_{SSRIP} refer to the distances between the radical ions in the CRIP and SSRIP, respectively. The first two terms correspond to the IP of the donor and EA of the acceptor, the third and last terms are associated with the solvation energy and Coulombic interactions of the CRIP/SSRIP.

Figure 10.9 shows the experimental data and the fitting lines of Δ_{RIP} as a function of the static dielectric constant of various solvents [39]. In the figure, $\Delta_{\text{CRIP}} \equiv G_{\text{CRIP}} - e(E_{\text{ox}} - E_{\text{red}})$ and $\Delta_{\text{SSRIP}} \equiv G_{\text{SSRIP}} - e(E_{\text{ox}} - E_{\text{red}})$. The open circles denote singlet CRIPs of 1,2,4,5-tetracyanobenzene (TCB), pyromellitic

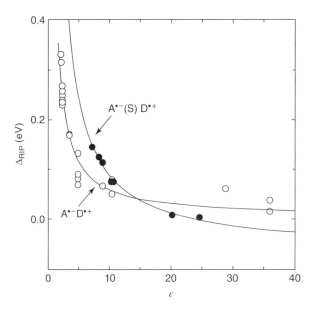

Figure 10.9 Experimental data and the fit lines of Δ_{RIP} for the CRIP (A$^{\bullet-}$D$^{\bullet+}$) and SSRIP (A$^{\bullet-}$(S)D$^{\bullet+}$) as a function of the static dielectric constant of various solvents [39]. Reproduced with permission of American Chemical Society.

dianhydride (PMDA), 1,4-dicyanobenzene (DCB), and TCA as acceptors with various alkylbenzenes as donors in various solvents. The closed circles denote SSRIPs of TCB with *p*-xylene. As demonstrated in Figure 10.9, in the case of nonpolar solvent, CRIP is more stable than SSRIP. However, the relative stability of SSRIP to CRIP is more enhanced as the solvent becomes more polar, and finally SSRIP becomes more stable than CRIP when $\varepsilon > 13$.

Both CRIP and SSRIP can return to the ground state via an ET. Therefore, the rate constant of such an ET influences the lifetimes of CRIP and SSRIP. By assuming that the solvent is characterized by a Debye relaxation spectrum, Jortner and Bixon obtained the ET rate constant [40]

$$k_{ET} = \frac{2\pi}{\hbar} \frac{1}{\sqrt{4\pi \lambda_s k_B T}} \sum_{n=0}^{\infty} F_n V^2 (1+H_n)^{-1} \exp\left[-\frac{(\Delta G + nh\nu_{eff} + \lambda_s)}{4\lambda_s k_B T}\right] \tag{10.22}$$

$$F_n = \exp(-S)\frac{S^n}{n!} \tag{10.23}$$

$$H_n = \frac{8\pi^2 F_n V^2 \tau_L}{h\lambda_s} \tag{10.24}$$

where k_B and T are the Boltzmann constant and the absolute temperature; λ_s is the reorganization energy for the ET regarding solvent relaxation and low-frequency vibration of the molecule that can be treated classically; V is the matrix element for the electronic coupling between initial and final electronic states; ΔG is the free energy difference in the ET; ν_{eff} is the frequency of the effective mode for the high-frequency molecular vibration; S is the Huang–Rhys factor defined as $S = \frac{\lambda_{eff}}{h\nu_{eff}}$; and τ_L is the Debye longitudinal relaxation time of the medium that corresponds to the dielectric relaxation time.

Figure 10.10 shows the radiationless recombination ET rate constant, k_{-et}, as a function of the associated free energy change, $-\Delta G_{-et}$. By fitting the experimental data using Eq. (10.22), the electronic coupling, V, and the reorganization energy, λ_s, regarding the ET were determined to be 0.12 and 0.55 eV, respectively, for CRIPs and 1.3 meV and 1.72 eV, respectively, for SSRIPs. The results demonstrate that the electronic coupling for the ET is much stronger for the CRIPs than SSRIPs, consistent with the shorter distance between the radical ion pair in the CRIPs. In addition, the much larger λ_s for SSRIPs can be attributed to the more solvent molecules involved in the solvation of SSRIPs.

10.4.3 Charge Separation Versus Charge Recombination

An exciplex can be generated by the charge-separation ET process from an LE complex. Figure 10.11 shows the comparison between the charge-separation (from the LE state to the CT state) and the charge-recombination rate constant (from the CT state to the ground state) as a function of the associated energy difference ($-\Delta E$) [43]. The charge-separation process to form the exciplex appears to be independent of the energy difference between the exciplex and the LE states as long as the $-\Delta E$ value is larger than approximately 0.3 eV, and hence a barrierless process. Considering that the charge-separation rate constant ($\sim$200 fs) matches

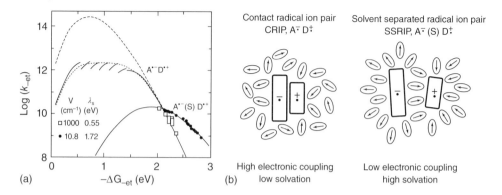

Figure 10.10 (a) Radiationless recombination ET rate constants as a function of the free energy of ET for experimental values for various exciplexes (CRIPs) of TCA and several alkylbenzene donors dissolved in various solvents (open rectangles) and various SSRIPs of TCA and DCA with several alkylbenzene donors dissolved in acetonitrile (closed circles) and fit lines (dotted lines overlapping with solid lines) by Eq. (10.22) with two sets of parameter of $H_{if} = 0.12$ eV, $\lambda_s = 0.55$ eV (CRIPs) and $H_{if} = 0.0013$ eV, $\lambda_s = 1.72$ eV (SSRIPs). (b) Schematic diagram of the CRIP (exciplex) and SSRIP [41]. Reproduced with permission of American Chemical Society.

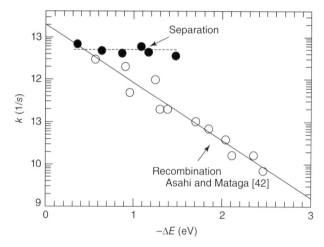

Figure 10.11 The charge-separation rate constant of six different exciplexes from CA, DCA, and TCA as acceptors and solutes and DMA and ANL as donors and solvents (closed circles) as a function of the energy gap between the locally excited complex state and the exciplex state; the charge-recombination rate constant of various exciplexes of TCNE, tetracyanoquinol-dimethane (TCNQ), PMDA, and phthalic anhydride (PA) with various polycyclic aromatic hydrocarbon as a function of the energy gap between the ground state and the exciplex state, which was reported by Asahi and Mataga (open circles) [43]. Reproduced with permission of AIP Publishing.

the intermolecular vibration frequency (100–200 cm^{-1}), it was suggested that the charge separation from the LE complex might be mediated via intermolecular stretching vibration. On the other hand, the rate constant of charge recombination, as discussed above regarding the CRIP, depends on the energy difference between the exciplex and the ground states. It was suggested that the charge-separation process would be the adiabatic barrierless reaction because the potential curves of the pure CT state and pure LE state cross each other.

10.4.4 Intersystem Crossing (ISC) in the Exciplex

In the case of a free ion pair, the correlation between the unpaired electrons of individual ions is negligible, and their spins are arbitrarily oriented, due to a long distance between them. As they get closer as in SSRIP or CRIP, however, such spin correlation between the radical ions becomes sizable, and hence they should be either in the singlet or the triplet states. Without the SOC, spin wave functions for the states with different multiplicities, e.g. the singlet and triplet states, remain orthogonal, and thus the interconversions between them, i.e. intersystem crossings (ISCs), are inherently forbidden. In other words, a nonzero SOC is essential to make the ISCs possible.

The SOC is caused by the magnetic torque generated by the electron orbit motion, which could change the electron spin [44]. Within the zeroth-order approximation and the central field approximation, the Hamiltonian for the SOC reads as

$$\hat{H}_{SO} \approx \alpha \hat{l} \cdot \hat{s} = \alpha \left[\frac{1}{2}(\hat{l}_- \cdot \hat{s}_+ + \hat{l}_+ \cdot \hat{s}_-) + \hat{l}_z \cdot \hat{s}_z \right] \quad (10.25)$$

where $\hat{l}_\pm$ and $\hat{s}_\pm$ denote raising(+)/lowering(−) operators for the orbital and spin angular momenta of a given electron, respectively, and α is the SOC constant; $\alpha \propto \frac{Z}{r^3}$ where Z is the nucleus charge and r is the distance of an electron from a given nucleus. Therefore, the SOC become stronger when the electron orbits closer around the nucleus with a higher charge; in comparison to the phosphors with heavy metal elements, e.g. Ir or Pt, the SOC for the most organic molecules are 2–3 orders of magnitudes smaller [24, 45]. As Eq. (10.25) implies, when the spin of an electron is flipped over, the angular momentum of the electron should also change. Consequently, the SOC between the singlet and triplet states of the same nature, e.g. the SOC between the singlet and triplet CT states, should remain negligible, as the so-called *El-Sayed rule* states [46]. Therefore, the ISC is expected to be faster between the singlet exciplex state with a dominant CT character and the triplet LE complex than that between the highly CT singlet and triplet exciplex states.

The ISC rate constant can be calculated from the equation analogous to Eq. (10.22) where SOC Hamiltonian matrix elements are used instead of the electronic coupling V [47]. Figure 10.12 shows the experimentally measured ISC rate constant from the singlet exciplex to the triplet states for the CP, DCA, and TCA complexes with alkylbenzene; the ISC rate constants were fitted using Eq. (10.22) and in fair accord with the experimental data [47]. It is noteworthy that the ISC in the organic molecular complexes can take place on a timescale of c. 100 ns or faster, despite the relatively small SOC. As the energy difference between the singlet and triplet state, ΔE_{ST}, decreases, the ISC rate increases, suggesting that such ISCs take place in the *inverted* regime. Figure 10.12b reflects that the degree of CT (z) is higher for the singlet exciplex but remains negligible for the triplet complex, supporting that the ISC for such systems occurs between the highly

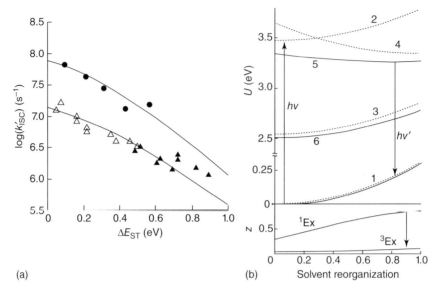

Figure 10.12 (a) Intersystem crossing rate, k_{ISC}, for CP (closed circle), DCA (close triangle), and TCA (open triangle) with alkylbenzene donors in various solvents as a function of ΔE_{ST}; the line is fitted using Eq. (10.22). (b) The energies of the ground state (1), the pure singlet LE state (2), the pure triplet LE state (3), the degenerated singlet and triplet CT state (4), the highly ionic singlet exciplex state (5), and the highly neutral triplet exciplex state (6) and the degree of charge transfer (z) of the singlet and triplet exciplex states as a function of solvent reorganization for exciplexes of CP with 1,3,5-trimethoxybenzenes in PrCN [47]. Reproduced with permission of Springer Nature.

CT singlet state and highly LE triplet state, for which the SOC is expected to be sizable.

The ISC between the highly CT singlet and triplet exciplexes has been considered to take place via hyperfine interactions between the nucleus spin and the electron spin. A few studies investigated the contribution from the ISC between the singlet and triplet CT states by the electron spin resonance (ESR) technique, implying the considerable ISC between these states [10, 48]. Recently, however, such interactions are expected to be even far smaller than the SOC between the singlet CT and the triplet LE states in intramolecular CT thermally activated delayed fluorescence (TADF) molecules [49, 50]. The strong vibronic coupling between the triplet CT state and the triplet LE state can facilitate the ISC between the singlet and triplet CT exciplexes unless the triplet CT and the triplet LE states are significantly apart in energy [49]. It is not clear at this moment that the same mechanism can be applied to intermolecular CT exciplexes. Further study is required to clarify the mechanism of ISC or reverse ISC in exciplex.

10.5 Exciplexes in Organic Solid Films

Organic films in organic electronics typically have low static dielectric constants of 3–4 [51–54]. Therefore, the Coulomb attraction in the radical ion pair is sufficiently strong to make the exciplex states more stable than the SSRIP states,

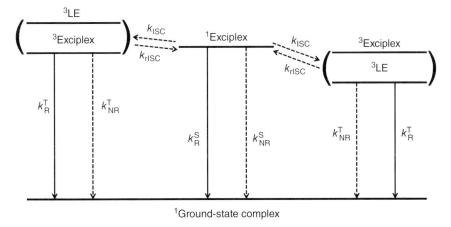

Figure 10.13 Schematic illustration of the energy diagram regarding the lowest excited states and various processes among them.

and the transition from the latter to the former is expected to be highly effective in organic films. In the case of OLEDs aiming for efficient light emission, the exciplex and the LE states are of high importance. In the case of OPVs, on the other hand, the transition from the exciplex state to the SSRIP, so-called charge-separated state, has long been of utmost interest because it is an essential process for the OPVs to be functional. Because our interests lie in the utilization of the exciplex in OLEDs in this chapter, we will limit our discussion within the processes regarding the exciplex and the LE states in the organic films hereafter.

It is useful at this stage to name two different systems as in Figure 10.13. If the energy of the triplet exciplex (^{3}Exciplex) is higher than the triplet LE state (^{3}LE) (right in Figure 10.13a), the transition from the ^{3}Exciplex to the ^{3}LE must be fast because it is the exothermic process. Therefore, the ^{3}Exciplex will end up in the ^{3}LE. On the other hand, if the ^{3}Exciplex is more stable (left in Figure 10.13a), the transition from the ^{3}Exciplex to the ^{3}LE is an energetically uphill process, and the complex remains in ^{3}CT. In the literature, the latter is called as a *confined* system but the former as a *non-confined* one.

10.5.1 Prompt Versus Delayed Fluorescence

When the ISC between the singlet and triplet states is sufficiently facilitated, the fluorescence of organic molecular systems including the exciplex should compete with ISCs. Therefore, the molecules in the S_1 state can fluoresce not only instantly (prompt fluorescence (PF)) but also after being back from the journey to the triplet state (delayed fluorescence (DF)). In the case of the electronic excitation in OLEDs, in particular, the light-emitting molecules are more prone to delay their emissions due to the dominant triplet state population.

Relying on the simplest two-model system (Figure 10.13), the rate equations for the concentrations of such a system in the S_1 and T_1 states are given in Eqs. (10.26) and (10.27):

$$\frac{d[S_1]}{dt} = G_S - (k_R^S + k_{NR}^S + k_{ISC})[S_1] + k_{rISC}[T_1] \tag{10.26}$$

$$\frac{d[T_1]}{dt} = G_T - (k_R^T + k_{NR}^T + k_{rISC})[T_1] + k_{ISC}[S_1] \qquad (10.27)$$

where $[S_1]$ and $[T_1]$ are the concentrations of the singlet and triplet states, respectively, and G_S and G_T are their respective generation rates. It is generally assumed that G_T is zero for photoexcitation and $G_S:G_T = 1:3$ for electrical excitation. The photoluminescence quantum yield (PLQY) (Φ_{PL}) of exciplexes can be obtained from the steady-state solution with $G_T = 0$ as follows when their phosphorescence is ignored:

$$\Phi_{PL} = \frac{k_R^S[S_1]}{G_S} = \Phi_p + \Phi_d \qquad (10.28)$$

where Φ_p and Φ_d are the PLQY of the prompt and delayed components, respectively, defined by

$$\Phi_p = \frac{k_R^S}{k_R^S + k_{NR}^S + k_{ISC}} \qquad (10.29)$$

$$\Phi_d = \frac{\Phi_{ISC}\Phi_{rISC}}{1 - \Phi_{ISC}\Phi_{rISC}}\Phi_p \qquad (10.30)$$

with $\Phi_{ISC} = k_{ISC}/(k_R^S + k_{NR}^S + k_{ISC})$ and $\Phi_{rISC} = k_{rISC}/(k_R^T + k_{NR}^T + k_{rISC})$.

In the transient PL experiments, generation rates of singlet and triplet exciplexes from the excitation source could be considered as the delta function at $t = 0$, and the concentration of the singlet exciplex state can be obtained by solving Eqs. (10.26) and (10.27) and described by

$$[S_1] = C_p \exp(-k_p t) + C_d \exp(-k_d t) \qquad (10.31)$$

where k_p and k_d and C_p and C_d are the decay rate constants and the intensities at $t = 0$ of the prompt and delayed emissions, respectively, which are expressed by

$$k_p, k_d = \frac{1}{2}(k_S + k_T)\left\{1 \pm \sqrt{1 - \frac{4k_S k_T - 4k_{ISC}k_{rISC}}{(k_S + k_T)^2}}\right\} \qquad (10.32)$$

$$C_p = \frac{k_S - k_d}{k_p - k_d}[S_1]_0 \qquad (10.33)$$

$$C_d = \frac{k_p - k_S}{k_p - k_d}[S_1]_0 \qquad (10.34)$$

where $[S_1]_0$ is the concentration of the singlet exciplex state at $t = 0$, $k_S = k_R^S + k_{NR}^S + k_{ISC}$, and $k_T = k_R^T + k_{NR}^T + k_{rISC}$. The diagram for transient PL profiles obeying two-exponential model of Eq. (10.31) is depicted in Figure 10.14 [2, 55–57].

To sum up, there are only four equations relating to the experimentally obtainable values (PLQY, k_p, k_d, and C_p/C_d) in spite of six unknowns ($k_R^S, k_{NR}^S, k_{ISC}, k_R^T, k_{NR}^T$, and k_{rISC}). Therefore, we need an approximation to get the rate constants, which depends on materials of interest. Generally, phosphorescence is hardly observed from organic molecules without heavy atoms at room temperature, and thus k_R^T of exciplexes could be ignored when heavy atoms are absent in the exciplex system at room temperature.

Figure 10.14 Schematic diagram of the transient PL profiles with prompt and delayed components obeying Eq. (10.31).

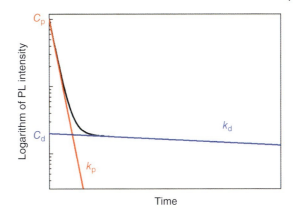

The ISC from the triplet to the singlet state plays an important role in improving the OLED efficiency:

$$k_{rISC} \approx A \exp\left(-\frac{\Delta E_{ST}}{k_B T}\right) \quad (10.35)$$

where A is the pre-exponential factor and ΔE_{ST} is the singlet–triplet energy splitting [2, 6, 22, 57, 58]. Therefore, when ΔE_{ST} is small, the rISC can occur effectively, and the exciplex emission is expected to show DF characteristics. Temperature-dependent DF variation, i.e. so-called thermally activated delayed fluorescence, for example, is observed in the 50 mol% m-MTDATA:3TPYMB blend film, as shown in Figure 10.15. As expected, the measured k_{rISC} does increase as the ΔE_{ST} decrease. ΔE_{ST} of the exciplex was computed to be about 50 meV [2], comparable to the thermal energy at the room temperature (~25 meV). Note that the LE states of the constituent molecules are higher in energy than the exciplex state (Figure 10.15c); the exchange energy, K, for the highly CT exciplex state and thus ΔE_{ST} between them are expectedly small. Assuming their complex is nonemissive in the triplet state, the increase of k_{rISC} could induce the increase of the population of the singlet exciplex state being an emissive state, resulting in higher PLQY of exciplexes as long as other transition rate constants remain the same.

One should note that the PLQY of DF does not solely depend on the ISC rate constant. As a typical example, Figure 10.16 shows PLQYs of PF, DF, and total fluorescence and the rate constants for various associated processes as a function of temperature for the TCTA:B4PYMPM exciplex. Their transient PL profiles are also shown in Figure 10.16a [6]. In the range of 150–300 K, the PLQYs of both total PL and DF counterintuitively increase as the temperature decreases. Detailed analyses, however, revealed that k_{rISC} did increase with temperature as expected. It was nonradiative decay processes (both for the singlet and triplet states) that become more effective and thus suppress the PLQY as the temperature increases [6]. ΔE_{ST} of this exciplex was estimated to be only 8.5 meV, smaller than the thermal energy even at 150 K (c. 13 meV).

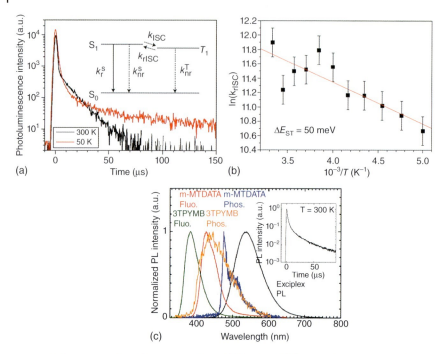

Figure 10.15 (a) Transient PL profiles of the m-MTDATA:3TPYMB exciplex in the 50 mol% m-MTDATA:PBD blend film. (b) k_{rISC} of the m-MTDATA: 3TPYMB exciplex in the 50 mol% m-MTDATA:PBD blend film at different temperature and a plot of Eq. (10.35). (c) Fluorescence (300 K) and phosphorescence (10 K) spectra of the constituent m-MTDATA and 3TPYMB molecules with the PL spectra (300 K) of their blend film [2]. Reproduced with permission of Springer Nature.

If $k_R^S, k_{NR}^S, k_R^T, k_{NR}^T \ll k_p$ and $4k_S k_T - 4k_{ISC} k_{rISC} \ll (k_S + k_T)^2$, ΔE_{ST} can be obtained from the transient PL profiles [55], which can be simulated using Eq. (10.36) from Eq. (10.31):

$$I(t) = I(0) \left\{ \frac{k_{ISC}}{k_{ISC} + k_{rISC}} \exp(-k_p t) + \frac{k_{rISC}}{k_{ISC} + k_{rISC}} \exp(-k_d t) \right\} \quad (10.36)$$

Then, k_{ISC}/k_{rISC} can be obtained from the ratio of the pre-exponential factors of two different exponential functions in Eq. (10.36). Considering the degeneracy for the triplet states (factor of 3), ΔE_{ST} is given in Eq. (10.37), which is the relation between the free energy change and the equilibrium constant of the reaction: [55, 59]

$$\Delta E_{ST} = RT \ln \frac{k_{ISC}}{3k_{rISC}} \quad (10.37)$$

where R is the gas constant.

10.5.2 Spectral Shift as a Function of Time

The time-resolved spectra of the exciplex emission generally redshift as time goes by. Figure 10.17 shows the time-resolved spectra of the TCTA:B3PYMPM

10.5 Exciplexes in Organic Solid Films | 351

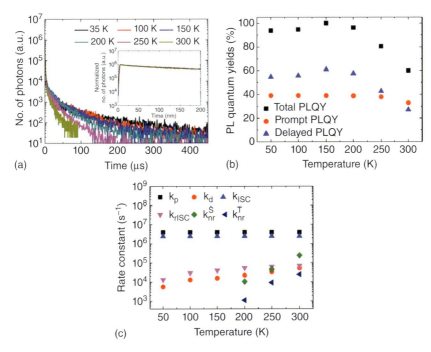

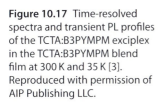

Figure 10.16 Transient PL profiles. (a), Total, prompt, and delayed PLQYs (b) and the prompt and delayed decay rate constants and four different radiationless transition rate constants (c) at various temperatures of the TCTA:B4PYMPM exciplex in the TCTA:B4PYMPM blend film [6]. Reproduced with permission of American Chemical Society.

Figure 10.17 Time-resolved spectra and transient PL profiles of the TCTA:B3PYMPM exciplex in the TCTA:B3PYMPM blend film at 300 K and 35 K [3]. Reproduced with permission of AIP Publishing LLC.

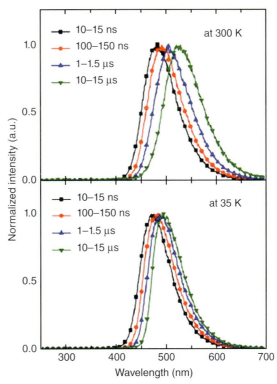

exciplex. The origin of the spectral shift is still under discussion. Adachi and coworkers suggested that the polarization of the medium be different when initially the S_1 state forms; it depends on the way it is generated, either via direct charge recombination or via the rISC from the T_1 state. Such medium polarization varies with time in the solid-state film and thus redshift with time is observed in transient emission spectra [2]. On the other hand, it was proposed that the broad distribution of the energy level of the exciplex state caused by the different geometries of the constituent molecules in the exciplexes and the relaxations to the stable geometries of the constituent molecules induce the transient redshift of the emission spectra of the exciplexes. They also explained that the transient redshift of the emission spectra of exciplexes is smaller at lower temperatures due to the reduced movements to the stable geometry of the molecules at lower temperatures, as shown in Figure 10.17. The transient PL profiles of exciplexes showing the multiexponential decays in Figure 10.17a could be explained by the broad distribution of ΔE_{ST} of exciplexes caused by the different geometries of the constituent molecules in the exciplexes [3]. Also the same group recently reported that different exciplexes with different dimer geometries have different degrees of CT and resultantly different decay rates. An exciplex with larger degree of CT gives emission at lower energy and longer decay time than an exciplex with smaller degree of CT. The spectral redshift with delay time originates from the superposition of the fast-decaying high-energy exciplex and slow-decaying low-energy exciplex (C.-K. Moon and J.-J. Kim, unpublished).

Baldo and coworkers proposed that the energy distribution of the HOMO of the donor and LUMO of the acceptor in the blend films causes the transient redshift as time flows. By means of the time-resolved spectrum measurement and the kinetic Monte Carlo simulation, they uncovered that the CT exciplex state migrates along the donor–acceptor grain boundary in the m-MTDATA:3TPYMB blend film to find more stable interfacial CT site, fluctuating the distance of the geminate pair of the hole and electron and thus their binding energy/CT state energy; see Figure 10.18 [60]. Therefore, the energy of the CT state gets lower, and the PL spectra redshift along with time. Further study is required for the clarification of the spectral shift with delay time.

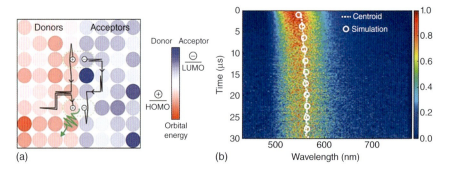

Figure 10.18 (a) Schematic diagram of exciplex diffusion by the geminate recombination of the exciplexes. (b) Streak camera measurement of the m-MTDATA:3TPYMB blend film with the centroid (dot) of the time-resolved PL spectra and the simulation (open circles) by the kinetic Monte Carlo simulation [60]. Reproduced with permission of Springer Nature.

10.6 OLEDs Using Exciplexes

10.6.1 Exciplexes as Emitters

Exciplex emissions at the interface between a charge transporting layer and an emitting layer in OLEDs were reported long time ago as an origin of low efficiency or as an active emission source for white OLEDs [61, 62]. The research activities on exciplex-based OLEDs increased rapidly after the demonstration of the potential to convert the triplet excited state to the singlet excited state without using the heavy atom effect, resulting in the larger singlet portion than 25% in OLEDs using fluorescent materials. Maximum achievable internal quantum efficiency (IQE_{max}) under the assumption of no electrical loss in the devices can be calculated based on the spin statistics of OLEDs. From the steady-state solution of Eqs. (10.26) and (10.27) with the $G_S : G_T = 0.25 : 0.75$, IQE_{max} is described by the following equation if the impacts of environments, e.g. Purcell factor, are not considered: [2, 6]

$$IQE_{max} = \frac{k_R^S [S_1]}{G_S + G_T} = \Phi_{PL}(0.25 + 0.75 \times \Phi_{rISC}) \qquad (10.38)$$

High Φ_{rISC} is required to achieve high efficiency, which in turn requires the reduction of the k_{NR}^T and ΔE_{ST} but the enhancement of the SOC.

Table 10.1 summarizes the device structures, maximum EQEs and PLQYs, EL peak wavelength, ΔE_{ST}, and whether the exciplex system is the triplet

Table 10.1 Characteristics of exciplex-emitting OLEDs.

No.	Device structure	Maximum EQE (%)/ PLQY (%)	EL peak (nm)	Confined system/ ΔE_{ST} (eV)	References (year)
1	ITO/NPB (50 nm)/PPSPP (50 nm)/ PyPySPyPy (10 nm)/Mg:Ag (100 nm)	3.4/62	495	—/—	[63] (2003)
2	ITO (110 nm)/m-MTDATA (20 nm)/ m-MTDATA:PBD (60 nm, m.r 5 : 5)/ PBD (20 nm)/LiF (0.8 nm)/Al (50 nm)	2/20	—	O/0.050	[2] (2012)
3	ITO (110 nm)/m-MTDATA (20 nm)/ m-MTDATA:3TPYMB (60 nm, m.r 5 : 5)/ 3TPYMB (20 nm)/LiF (0.8 nm)/Al	5.4/26	543	O/—	[2] (2012)
4	ITO/m-MTDATA (35 nm)/ m-MTDATA:PPT (30 nm, m.r 5 : 5)/ PPT (35 nm)/LiF (0.8 nm)/Al	10.0/28.5	—	—/—	[1] (2012)
5	ITO/NPB (30 nm)/NPB:TPBi (35 nm, 5 : 5)/ TPBi (35 nm)/LiF(1 nm)/Al(100 nm)	2.7/28	450	X/—	[64] (2013)
6	ITO/TAPC (30 nm)/TCTA (10 nm)/ TCTA:B3PYMPM (30 nm, m.r 5 : 5)/ B3PYMPM (30 nm)/LiF (1 nm)/Al (100 nm)	3.1/36	506	—/—	[3] (2013)
7	ITO/PEDOT:PSS (30 nm)/NPB (20 nm)/ TCTA (5 nm)/3P-T2T (75 nm)/Liq/Al	7.7/—	548	O/—	[5] (2013)

(Continued)

Table 10.1 (Continued)

No.	Device structure	Maximum EQE (%)/ PLQY (%)	EL peak (nm)	Confined system/ ΔE_{ST} (eV)	References (year)
8	ITO/PEDOT:PSS (30 nm)/NPB (20 nm)/ TCTA (5 nm)/TCTA:3P-T2T (25 nm, m.r 5 : 5)/3P-T2T (50 nm)/Liq/Al	7.8/—	548	O/—	[5] (2013)
9	ITO/m-MTDATA (27 nm)/m-MTDATA: BPhen (23 nm, m.r 3 : 7)/BPhen (20 nm)/ Alq$_3$ (13 nm)/LiF (1 nm)/Al (80 nm)	7.79/15	563	O/0.001	[65] (2014)
10	ITO/NPB (30 nm)/TCTA (10 nm)/mCP: HAP-3MF (20 nm, w.r 92 : 8)/DPEPO (10 nm)/ TPBi (40 nm)/LiF (0.8 nm)/Al (100 nm)	11.3/66.1	540	—/—	[66] (2014)
11	ITO/mCP (20 nm)/mCP:TPBi (20 nm, w.r 5 : 5)/TPBi (5 nm)/Bphen (30 nm)/LiF (1 nm)/Al (80 nm)	1.48/—	450	X/—	[67] (2015)
12	ITO/mCP (20 nm)/mCP:BPhen (20 nm, w.r 5 : 5)/ BPhen (35 nm)/LiF (1 nm)/Al (80 nm)	2.23/—	462	X/—	[67] (2015)
13	ITO/CuI (8 nm)/THCA (40 nm)/FIrpic (9 nm)/ TCz1 (10 nm)/Ca (50 nm)/Al (200 nm)	5/—	572	O/0.02	[68] (2015)
14	ITO (95 nm)/HATCN (5 nm)/TAPC (20 nm)/ m-MTDATA:26DCzPPy (20 nm, m.r 5 : 5)/ TmPyPB (50 nm)/LiF (1 nm)/Al (100 nm)	5.03/20.3	525	—/—	[69] (2015)
15	ITO/MoO$_3$ (3 nm)/TCTA (15 nm)/ TCTA:BPhen (12 nm, m.r 5 : 5)/BPhen (30 nm)/LiF/Al	2.65/—	461	—/—	[70] (2015)
16	ITO/MoO$_3$ (3 nm)/TAPC (15 nm)/ TAPC:3P-T2T (12 nm, m.r 5 : 5)/3P-T2T (25 nm)/LiF/Al	6.86/—	572	—/—	[70] (2015)
17	ITO/TPAPB (30 nm)/TPAPB:TPBi (30 nm, m.r 5 : 5)/TPBi (40 nm)/LiF (1 nm)/Al	7.0/44.1	464	X/—	[71] (2015)
18	ITO/MoO$_3$ (3 nm)/mCBP (20 nm)/ mCBP:PO-T2T (20 nm, m.r 5 : 5)/PO-T2T (40 nm)/LiF (0.8 nm)/Al	7.66/34	475	O/—	[72] (2015)
19	ITO/MoO$_3$ (3 nm)/TPD (20 nm)/ TPD:DMAC-DPS (25 nm, w.r 5 : 5)/ PO-T2T (45 nm)/LiF (0.8 nm)/Al	1.63/—	586	—/—	[73] (2015)
20	ITO/MoO$_3$ (3 nm)/mCBP (20 nm)/ DMAC-DPS:B4PyMPM (25 nm, w.r 5 : 5)/ B4PyMPM (40 nm)/LiF (0.8 nm)/Al	4.40/—	492	—/—	[73] (2015)
21	ITO/MoO$_3$ (3 nm)/mCBP (20 nm)/ DMAC-DPS:T2T (25 nm, w.r 5 : 5)/ T2T (40 nm)/LiF (0.8 nm)/Al	4.44/—	489	—/—	[73] (2015)
22	ITO/MoO$_3$ (3 nm)/mCBP (20 nm)/ DMAC-DPS:PO-T2T (25 nm, w.r 5 : 5)/ PO-T2T (40 nm)/LiF (0.8 nm)/Al	9.08/—	537	—/—	[73] (2015)
23	ITO (120 nm)/PEDOT:PSS (60 nm)/TAPC (30 nm)/TAPC:DTrz (25 nm, 5 : 5)/TSPO1 (5 nm)/TPBi (30 nm)/LiF (1 nm)/Al (200 nm)	3.27/—	525	—/0.05	[74] (2015)

(Continued)

Table 10.1 (Continued)

No.	Device structure	Maximum EQE (%)/ PLQY (%)	EL peak (nm)	Confined system/ ΔE_{ST} (eV)	References (year)
24	ITO (120 nm)/PEDOT:PSS (60 nm)/TAPC (30 nm)/TCTA:DTrz (25 nm, 5 : 5)/TSPO1 (5 nm)/TPBi (30 nm)/LiF (1 nm)/Al (200 nm)	5.52/—	512	—/0	[74] (2015)
25	ITO (120 nm)/PEDOT:PSS (60 nm)/TAPC (30 nm)/TAPC:CzTrz (25 nm, 5 : 5)/TSPO1 (5 nm)/TPBi (30 nm)/LiF (1 nm)/Al (200 nm)	8.88/—	512	—/0.04	[74] (2015)
26	ITO (120 nm)/PEDOT:PSS (60 nm)/TAPC (30 nm)/TCTA:CzTrz (25 nm, 5 : 5)/TSPO1 (5 nm)/TPBi (30 nm)/LiF (1 nm)/Al (200 nm)	12.62/—	505	—/0	[74] (2015)
27	ITO/TAPC (30 nm)/CDBP (10 nm)/ CDBP:PO-T2T (30 nm, w.r 5 : 5)/ PO-T2T (40 nm)/LiF (1 nm)/Al (100 nm)	13.0/51	480	O/0.03	[75] (2015)
28	ITO/HATCN (5 nm)/TAPC (55 nm)/ TCTA:Tm3PyBPZ (30 nm, 5 : 5)/Tm3PyBPZ (40 nm)/Liq (2 nm)/Al (100 nm)	13.1/—	528	O/—	[76] (2015)
29	ITO/TAPC (35 nm)/NPB (5 nm)/ NPB:DPTPCz (30 nm, w.r 5 : 5)/TmPyPB (40 nm)/LiF (1 nm)/Al (100 nm)	0.6/15	502	X/—	[75] (2015)
30	ITO/TAPC (35 nm)/TCTA (5 nm)/ TCTA:DPTPCz (30 nm, w.r 5 : 5)/TmPyPB (40 nm)/LiF (1 nm)/Al (100 nm)	11.9/55	510	O/0.062	[75] (2015)
31	ITO/TAPC (40 nm)/TAPC:DPTPCz (30 nm, w.r 5 : 5)/TmPyPB (40 nm)/ LiF (1 nm)/Al (100 nm)	15.4/68	520	O/0.047	[75] (2015)
32	ITO (70 nm)/TAPC (75 nm)/TCTA (10 nm)/ TCTA:B4PYMPM (30 nm, m.r 5 : 5)/ B4PYMPM (40 nm)/LiF (0.7 nm)/Al (100 nm).	11.0/60	530	—/0.009	[6] (2016)
33	ITO/4% ReO3:TCTA (60 nm)/TCTA (15 nm)/ TCTA:CNT2T (25 nm, m.r 5 : 5)/CN-T2T (50 nm)/Liq (0.5 nm)/Al (100 nm)	9.7/50.0	540	O/—	[55] (2016)
34	ITO/4% ReO$_3$:Tris-PCz (60 nm)/Tris-PCz (15 nm)/Tris-PCz:CN-T2T (25 nm, m.r 5 : 5)/ CN-T2T (50 nm)/Liq (0.5 nm)/Al (100 nm)	11.9/52.9	525	O/0.026	[55] (2016)
35	ITO/TCTA (85 nm)/TmPyTZ (85 nm)/ LiF (1 nm)/Al (150 nm)	10.1/—	540	O/0.030	[77] (2016)
36	ITO/TAPC (85 nm)/TmPyTZ (85 nm)/ LiF (1 nm)/Al (150 nm)	12.0/—	552	O/0.053	[77] (2016)
37	ITO/TAPC (40 nm)/mCP (10 nm)/ mCP:PO-T2T (20 nm, w.r 5 : 5)/PO-T2T (45 nm)/LiF (1 nm)/Al (150 nm)	8.6/—	476	O/0.010	[78] (2016)
38	ITO/TAPC (40 nm)/mCP (10 nm)/ MAC:PO-T2T (20 nm, w.r 5 : 5)/PO-T2T (45 nm)/LiF (1 nm)/Al (150 nm)	17.8/—	516	O/0.014	[78] (2016)

confined system or not for the exciplex-based OLEDs with the EQE over 1%. The underlined layers in device structures are the light-emitting layers. The EQEs have been increased rapidly from 5.4% to 17.8% during the last several years by properly developing new materials. Most of donors are amine- and carbazole-based hole-transporting materials such as TCTA, NPB, and m-MTDATA. A variety of electron-transporting materials such as Bphen, PO-T2T, and TPBi have been used as acceptors.

Confined exciplex systems appear to be useful to improve the device performance utilizing small ΔE_{ST} as exemplified in Devices 29–31 [79]. NPB:DPTPCz singlet exciplexes of Device 29 are higher in energy than the T_1 state of NPB (triplet non-confined system). The small EQE/PLQY of this device might be attributed to the less effective Φ_{rISC}. In contrast, TCTA:DPTPCz exciplexes of Device 30 and TAPC:DPTPCz exciplexes of Device 31 show large EQE/PLQY because of the triplet confined characteristics of them [79].

Observation of delayed emission in a mixed film does not necessarily mean that it is from TADF. Triplet–triplet annihilation (TTA) is another way to get singlet exciplexes from the triplet ones. The major difference between the TADF through the rISC and the TTA is molecularity of the reaction; the former is an unimolecular reaction, while the latter a bimolecular one [3, 22, 64]. Therefore, the intensity of the delayed emission arising from the TADF is linearly proportional to $[T_1]$, while that stemming from the TTA is linearly proportional to $[T_1]^2$. Therefore, the mechanism responsible for the delayed emission can be identified by investigating either the incident light intensity dependence or the current density dependence of the delayed emission. Figure 10.19a, b displays the delayed PL intensity of m-MTDATA:PBD exciplex and that of NPB:TPBi exciplex employed in Device 5 as a function of the pulse fluence, respectively [22, 64]. $[T_1]$ is linearly proportional to the pulse fluence for the m-MTDATA:PBD exciplex. Thus, it could be concluded that the delayed emission of m-MTDATA:PBD exciplexes arises from the TADF and that of NPB:TPBi exciplexes arises from the TTA inferred from its slope. The transition of the slope from 2 to 1 in Figure 10.19b may be because of the active quenching of the excitons caused by high concentration of excitons [80]. Device 17 with triplet non-confined exciplexes also shows higher EQE than the expected maximum EQE from PLQY of TPAPB:TPBi exciplexes, assuming they are the conventional fluorescent emitters, and TTA would be the dominant way for the delayed emission of TPAPB:TPBi exciplexes [71].

10.6.2 Exciplexes as Sensitizers

Exciplex OLEDs are interesting because of their capability to harvest triplet excited states leading to higher efficiency than the singlet limit. However, the PLQY of exciplexes is rather small up to now, below 70%, reflecting a rather small oscillator strength probably because of the small overlap between spatially separated HOMO (donor) and LUMO (acceptor). More importantly, the emission spectrum of exciplex is rather broader, limiting the exciplex application to the OLEDs in terms of color purity. Employing both exciplexes as the host and sensitizer and conventional fluorescent dopant with narrow emission spectra could enhance the device performance via the ENT from the singlet exciplex

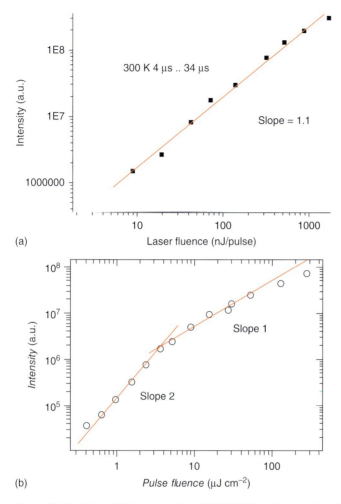

Figure 10.19 Delayed PL intensity of m-MTDATA:PBD exciplexes (a) and NPB:TPBi exciplexes [22] (b) as a function of the pulse fluence [64]. Reproduced with permission of John Wiley & Sons.

to the dopant. Assuming that the direct exciton formation and the ISC/rISC in dopants are negligible, the efficiency of generating the S_1-state dopant in this device, $\phi_{S_1(D)}$ can be written as Eq. (10.39):

$$\phi_{S_1(D)} = 0.25 \times \Phi^S_{ENT} \sum_{k=0}^{\infty} (\Phi'_{ISC}\Phi'_{rISC})^k + 0.75 \times \Phi^S_{ENT}\Phi'_{rISC} \sum_{k=0}^{\infty} (\Phi'_{ISC}\Phi'_{rISC})^k$$

(10.39)

where Φ^S_{ENT}, Φ'_{ISC}, and Φ'_{rISC} are the ENT efficiency from the singlet exciplex to the dopant and the ISC and rISC efficiencies of exciplex under the consideration of the ENT, respectively, and represented by

$$\Phi^S_{ENT} = \frac{k^S_{ENT}}{k^S_R + k^S_{NR} + k^S_{ISC} + k^S_{ENT}}$$

(10.40)

$$\Phi'_{ISC} = \frac{k_{ISC}}{k_R^S + k_{NR}^S + k_{ISC} + k_{ENT}^S} \quad (10.41)$$

$$\Phi'_{rISC} = \frac{k_{RISC}}{k_R^T + k_{NR}^T + k_{rISC} + k_{ENT}^T} \quad (10.42)$$

where k_{ENT}^S and k_{ENT}^T are the rate constants for the singlet and triplet ENT from the host to the dopant, respectively. Then, the EQE of OLEDs with exciplex sensitizers can be written as Eq. (10.43) if the Purcell factor is ignored:

$$EQE = \phi_{S_1(D)} \times \gamma \times PLQY_{(D)} \times \eta_{out} \quad (10.43)$$

where γ is the recombination efficiency of electrons and holes in OLEDs, $PLQY_{(D)}$ is the PLQY of the dopant, and η_{out} is the light-out coupling efficiency, which is generally considered to be 20–30% when the emitting transition dipole moment vectors are randomly distributed [64, 65].

Note here that the concentration ratio between the exciplex sensitizer and dopant must be carefully controlled to maximize the efficiency. If the concentration of exciplex sensitizer is too low, the conversion of triplet exciplexes to the singlet ones is not effective. Therefore, the exciplex sensitizing effect is hardly expected, undermining the device efficiency. In too high exciplex concentration, on the other hand, the average distance between the sensitizer and the dopant will increase, and the concentration of the ground-state dopant remains low. Therefore, the ENT from the sensitizer to the dopant would not be effective. In addition, at both extremes, the concentration quenching and triplet-polaron quenching of dopants or exciplexes are still expected to be active.

Regarding the ENT, both Förster and Dexter mechanisms are in action; the former stems from the transition dipole–transition dipole interactions between the energy donor and acceptor, hence proportional to r^{-6} where r is the distance between the exciplex and dopant molecule. To facilitate Förster ENT, the spectral overlap of the dopant absorption and the sensitizer emission should also be substantial. On the other hand, Dexter ENT occurs via the quantum electronic coupling, and thus the wave function overlap between molecules should be significant. As such, Dexter ENT decays exponentially as a function of distance. Given that the transition dipole between the singlet and triplet states is negligible for organic molecules, Dexter mechanism dominates the triplet ENT, whereas Förster ENT is more important for the singlet one. Therefore, the conditions for decreasing the Dexter ENT rate must be considered when an exciplex host and a dopant are selected, which warrants further studies.

Table 10.2 collects information on the exciplex-sensitized OLEDs: device structures, maximum EQEs, EL peak wavelengths, and PLQYs of the device, and singlet–triplet energy splitting and triplet confinement of the exciplex. The light-emitting layer in device structures is underlined. When both emissions from exciplexes and fluorescent dopants are observed, two EL maximum peaks are provided. Devices 5–8 performed better than the conventional fluorescent OLEDs, demonstrating the sensitizing effect of exciplex-forming hosts.

Table 10.2 Characteristics of OLEDs employing exciplexes as the host and conventional fluorescent emitters as the dopant.

No.	Device structure	Maximum EQE (%)/ PLQY (%)	EL peak (nm)	Confined system/ ΔE_{ST} (eV)	References (year)
1	ITO/MoO$_3$ (3 nm)/mCBP (20 nm)/ mCBP:PO-T2T (20 nm)/<u>mCBP:POT2T:1.0 wt.% Rubrene (5 nm)</u>/PO-T2T (40 nm)/LiF (0.8 nm)/Al	6.09/—	560	O/—	[72] (2015)
2	ITO/MoO$_3$ (3 nm)/mCBP (20 nm)/ <u>mCBP:PO-T2T:0.4 wt.% Rubrene (20 nm)</u>/ PO-T2T (40 nm)/LiF (0.8 nm)/Al	7.05/—	475, 550	O/—	[72] (2015)
3	ITO/MoO$_3$ (3 nm)/mCBP (20 nm)/ <u>mCBP:PO-T2T:0.2 wt.% DCJTB (20 nm)</u>/ PO-T2T (40 nm)/LiF (0.8 nm)/Al	5.75/—	475, 582	O/—	[72] (2015)
4	ITO/MoO$_3$ (3 nm)/mCBP (20 nm)/ <u>mCBP:PO-T2T:0.4 wt.% DCJTB (20 nm)</u>/ PO-T2T (40 nm)/LiF (0.8 nm)/Al	6.16/—	480, 600	O/—	[72] (2015)
5	ITO/MoO$_3$ (3 nm)/NPB (20 nm)/TCTA (8 nm)/<u>TCTA:3P-T2T:1 wt.% DCJTB (15 nm)</u>/ 3P-T2T (45 nm)/LiF (1 nm)/Al	10.15/68.5	544, 605	—/—	[81] (2015)
6	ITO (100 nm)/TAPC (75 nm)/TCTA (10 nm)/ TCTA:B4PYMPM:0.5 wt.% DCJTB (30 nm)/ B4PYMPM (50 nm)/LiF (0.7 nm)/Al (100 nm)	10.6/73	600	—/—	[82] (2015)
7	ITO/TAPC (40 nm)/ <u>TAPC:DPTPCz:1 wt.% C545T (30 nm)</u>/TmPyPB (50 nm)/ LiF (1 nm)/Al (100 nm)	7.5/64	496	—/—	[4] (2015)
8	ITO/TAPC (40 nm)/<u>TAPC:DPTPCz: 0.2 wt% C545T (30 nm)</u>/TmPyPB (50 nm)/ LiF (1 nm)/Al (100 nm)	14.5/68	496	—/—	[4] (2015)
9	ITO/4% ReO$_3$:Tris-PCz (60 nm)/Tris-PCz (15 nm)/ <u>Tris-PCz:CN-T2T:1 wt.% Rubrene (25 nm)</u>/ CN-T2T (50 nm)/Liq (0.5 nm)/Al (100 nm)	6.9/—	510, 566	O/0.026	[55] (2016)
10	ITO/4% ReO$_3$:Tris-PCz (60 nm)/Tris-PCz (15 nm)/ <u>Tris-PCz:CN-T2T:1 wt.% DCJTB (25 nm)</u>/ CN-T2T (50 nm)/Liq (0.5 nm)/Al (100 nm)	9.7/—	510, 610	O/0.026	[55] (2016)

As discussed before, the concentration of dopants does influence the properties of OLEDs. OLEDs using small concentration of dopants show the exciplex emission as seen in Devices 2–4, 9, and 10, indicating less effective ENT from the exciplex to dopants. At high concentration of dopants, on the other hand, devices do not show the exciplex emission, but EQE could decrease as seen in Devices 7 and 8 [4, 55, 72]. The concentration quenching of dopants might also play a role in the reduction in PLQY [82, 83]. Thus far, the highest efficiency of the conventional fluorescent dye-doped OLEDs is reportedly about 15%, and there must be a room for further enhancement by developing proper exciplex hosts and dopants molecules, tuning the concentration ratio, and so on.

10.7 Summary and Outlook

In this chapter, we covered the fundamentals of exciplexes from their electronic structures and optical properties to various transition processes associated with them. Due to its dominant CT character, the exciplexes are not such effective in light emission. Nevertheless, inherently small ΔE_{ST} in *confined* systems, they have potentials to effectively convert the triplet excited states to the singlet ones. This paves the road toward OLEDs with the IQE of unity using organic molecules and their significant fabrication cost reduction. As we discussed, the application of exciplexes will remain as one of the important topics for OLED applications.

Apart from their utilization as emitters or sensitizers in OLEDs, interfaces between different molecules are ubiquitous in organic electronics. In addition, it is noteworthy that the electronic structures and properties of exciplexes are also influenced by their configurations as mentioned earlier, and thus their configuration control is considered to have fundamental importance; the intermolecular interactions in the exciplex state, however, are markedly different than that in the ground state and are relatively weak, leading to various configurations. As such, further studies of exciplexes are highly warranted.

Appendix

10.A.1 Small Molecular Pairs of Donors and Acceptors Forming Exciplexes

The exciplex consists of a pair of an electron donor molecule and an electron acceptor molecule. Here, electron donor molecules generally contain electron-donating moieties such as carbazoles, triarylamines, and 9,10-dihydroacridines [84–88]. Electron acceptor molecules typically have electron-withdrawing moieties such as 1,3,5-triazines, quinoxalines, pyrazines, pyrimidines, quinolines, phenanthrolines, pyridines, 1,3,4-oxadiazoles, benzimidazoles, benzotriazoles, triarylboranes, siloles, phosphine oxides, sulfones, cyano groups, fluoro groups, and carbonyl groups [84, 86, 87, 89–96]. The HOMO/LUMO energy levels of electron donor molecules and electron acceptor molecules and exciplex emission peak energies are summarized in Table 10.A.1.

10.A.2 Small Molecules with Electron-donating Moieties Forming Exciplexes

Figure 10.A.1 shows the molecular structures of electron donors employed in OLEDs utilizing small molecule-based exciplexes. Figure 10.A.1a depicts

Table 10.A.1 Energy levels of HOMO and LUMO of molecules forming exciplexes and PL-spectra peaks of exciplexes in the blended films.

Donor	HOMO/LUMO (eV)	Acceptor	HOMO/LUMO (eV)	Exciplex emission peak (eV)	References	Donor	HOMO/LUMO (eV)	Acceptor	HOMO/LUMO (eV)	Exciplex emission peak (eV)	References
CBP	−6.1/−	B3PYMPM	−6.77/−3.2	2.92	[97]	NPB	−5.5/−2.4	PPSPP	−5.9/−3.1	2.50	[15]
CBP	−6.1/−2.5	B4PYMPM	−7.1/−3.5	2.82	[98]	NPB	−5.5/−2.4	PyPySPyPy	−5.9/−3.2	2.25	[15]
CBP	−5.9/−2.5	B4PyPPM	−7.2/−3.4	2.78	[99]	NPB	−5.5/−2.4	TPBi	−6.4/−2.7	2.73	[64]
CBP	−5.5/−2.0	CPP	−6.2/−3.3	2.40	[100]	TPD	−5.4/−2.3	BCP	−6.4/−2.8	2.78	[101]
CBP	−5.7/−2.2	CPQ	−6.2/−3.3	2.26	[93]	TPD	−5.4/−2.4	BPhen	−6.4/−2.8	2.64	[102]
CDBP	−5.86/−2.41	PO-T2T	−7.10/−3.30	2.61	[75]	TPD		Gd(DBM)$_3$bath	−/−	2.12	[103]
mCBP	−6.1/−2.4	PO-T2T	−6.8/−2.8	2.70	[72]	TPD	−5.4/−2.4	STO	−7.7/−3.0	2.23	[104]
mCP	−6.1/−2.4	3P-T2T	−6.5/−3.0	2.83	[67]	TPD	−5.4/−2.3	Tb(PMIP)$_3$(Phen)	−5.9/−2.7	2.30	[105]
mCP	−6.1/−2.4	B3PYMPM	−6.77/−3.2	2.97	[106]	TPD		TBOB	−/−	2.41	[107]
mCP	−6.1/−2.4	BPhen	−6.4/−2.5	3.06	[67]	FL2	−5.4/−2.4	Alq$_3$	−5.9/−3.2	2.38	[108]
mCP	−5.9/−2.4	HAP-3MF	−6.0/−3.4	2.23	[66]	FL3	−5.3/−2.3	Alq$_3$	−5.9/−3.2	2.36	[108]
mCP	−6.1/−2.4	POPH	−6.7/−2.5	3.22	[109]	m-MTDATA	−5.1/−2.0	3TPYMB	−6.8/−3.3	2.34	[110]
mCP	−6.1/−2.4	PO-T2T	−7.5/−3.5	2.64	[111]	m-MTDATA	−5.1/−2.1	Alq$_3$	−5.9/−3.2	2.16	[108]
mCP	−6.1/−2.4	TPBi	−6.4/−2.7	3.16	[67]	m-MTDATA		BCP	−6.0/−2.5	2.34	[112]
THCA	−4.63/−1.98	Alq$_3$	−5.7/−3.1	2.21	[113]	m-MTDATA	−5.1/−1.9	BPhen	−6.0/−2.5	2.36	[114]
THCA	−4.63/−1.98	BPhen	−6.4/−3.0	2.25	[115]	m-MTDATA	−5.1/−2.0	PBD	−6.1/−2.4	2.25	[22]
THCA	−4.63/−1.98	FIrpic	−5.8/−2.9	2.21	[68]	m-MTDATA	−5.1/−2.0	PPT	−6.7/−3.0	2.38	[1]
TCTA	−5.8/−2.5	3P-T2T	−6.4/−3.0	2.31	[81]	m-MTDATA	−5.1/−2.0	PtL^{2}Cl	−5.6/−3.1	2.07	[61]
TCTA	−5.83/−2.43	B3PYMPM	−6.77/−3.2	2.55	[3]	m-MTDATA	−5.1/−2.0	T2T	−6.5/−2.8	2.14	[116]
TCTA	−5.83/−2.43	B4PYMPM	−7.30/−3.71	2.44	[6]	m-MTDATA	−5.10/−1.92	TmPyPb	−6.63/−2.54	2.59	[117]
TCTA	−5.7/−2.3	BPhen	−6.1/−2.8	2.68	[118]	m-MTDATA	−5.1/−1.9	TPBi	−6.2/−2.7	2.36	[119]
TCTA	−5.70/−	DTrz	−7.01/−3.34	2.41	[74]	MTDATA	−5.0/−1.9	B1PPQ	−5.7/−2.7	2.11	[26]
TCTA	−5.8/−2.4	POPH	−6.7/−2.5	2.77	[109]	TDDP	−4.7/−1.6	BCP	−6.2/−2.6	2.46	[120]

(Continued)

Table 10.A.1 (Continued)

Donor	HOMO/LUMO (eV)	Acceptor	HOMO/LUMO (eV)	Exciplex emission peak (eV)	References	Donor	HOMO/LUMO (eV)	Acceptor	HOMO/LUMO (eV)	Exciplex emission peak (eV)	References
TCTA	−5.7/−2.5	TBOB	−6.6/−2.9	2.48	[121]	TPB	−5.2/−2.0	Alq$_3$	−5.9/−3.2	2.30	[108]
TCTA	−5.8/−2.4	Tm3PyBPZ	−6.6/−3.1	2.41	[76]	SiDMAC	−5.60/−1.97	BPSPF	−6.60/−2.69	2.84	[122]
TCTA	−5.83/−2.5	TmPyTZ	−6.63/−3.09	2.34	[77]	MAC	−5.64/−2.97	PO-T2T	−7.04/−3.22	2.42	[78]
TCTA	−5.7/−2.4	TPBi	−6.1/−2.8	2.85	[123]	TPAPB	−5.36/−2.35	TPBi	−6.20/−2.70	2.65	[71]
TCTA	−5.8/−2.4	TPO	−7.1/−2.75	2.77	[124]	DMAC-DPS	−5.9/−2.9	B4PyMPM	−7.3/−3.7	2.52	[73]
TCTA	−5.8/−2.4	TPOB	−6.7/−3.0	2.67	[125]	DMAC-DPS	−5.9/−2.9	PO-T2T	−7.5/−3.5	2.31	[73]
Cz3d	−5.64/−2.68	Oxa3d	−5.91/−2.78	2.62	[126]	DMAC-DPS	−5.9/−2.9	T2T	−6.5/−3.0	2.62	[73]
Tris-PCz	−5.6/−2.1	CN-T2T	−6.7/−2.78	2.34	[55]	TPD	−5.5/−2.4	DMAC-DPS	−5.9/−2.9	2.25	[73]
TAPC	−5.5/−	3P-T2T	−/−	2.28	[70]	m-MTDATA	−5.1/−1.9	26DCzPPy	−6.05/−2.56	2.41	[69]
TAPC	−5.3/−1.8	B1PPQ	−5.7/−2.7	2.50	[26]	TAPC	−5.3/−1.8	26DCzPPy	−6.05/−2.56	2.77	[69]
TAPC	−5.6/−2.0	BTPS	−6.7/−2.9	2.62	[127]	TAPC	−5.50/−	CzTrz	−6.08/−3.28	2.41	[74]
TAPC	−5.50/−	DTrz	−7.01/−3.34	2.36	[74]	TCTA	−5.70/−	CzTrz	−6.08/−3.28	2.43	[74]
TAPC	−5.6/−2.0	POPH	−6.7/−2.5	2.70	[109]	NPB	−5.50/−2.30	CzPhB	−5.92/−2.91	2.67	[20]
TAPC	−5.39/−1.8	TmPyTZ	−6.63/−3.09	2.24	[77]	m-MTDATA	−4.97/−1.88	2d	−6.18/−3.19	2.13	[128]
TAPC	−5.5/−2.1	TPBi	−6.1/−2.8	2.71	[129]	NPB	−5.43/−2.40	2d	−6.18/−3.19	2.38	[128]
2-TNATA	−5.1/−	Alq$_3$	−/−	2.15	[130]	TAPC	−5.47/−2.00	2d	−6.18/−3.19	2.38	[128]
2-TNATA	−5.15/−2.13	Spiro-DPVBi	−5.50/−2.50	2.36	[131]	TPD	−5.6/−2.5	CzOxa	−6.22/−3.13	2.62	[132]
TTA	−5.3/−1.8	B1PPQ	−5.7/−2.7	2.50	[26]	NPB	−5.40/−2.81	DPTPCz	−6.03/−3.08	2.53	[79]
NPB	−5.5/−2.5	Alq$_3$	−5.9/−3.2	2.38	[108]	TAPC	−5.43/−2.67	DPTPCz	−6.03/−3.08	2.47	[79]
NPB	−5.4/−2.4	B4PyPPM	−7.15/−3.44	2.25	[133]	TCTA	−5.51/−2.71	DPTPCz	−6.03/−3.08	2.47	[79]
NPB	−5.2/−2.2	CPP	−6.2/−3.3	1.94	[100]	THCA	−4.63/−1.98	IC2	−5.56/−2.73	1.97	[132]
NPB	−5.2/−	CPQ	−6.2/−3.3	1.98	[93]	TPD	−5.40/−2.20	ANHC	−5.60/−2.63	2.39	[21]
NPB	−5.2/−2.1	F2Py	−6.19/−2.92	2.27	[134]	TPD	−5.40/−2.20	CNHC	−5.65/−2.70	2.37	[21]
NPB	−5.4/−2.4	PBD	−6.1/−2.7	2.57	[135]						

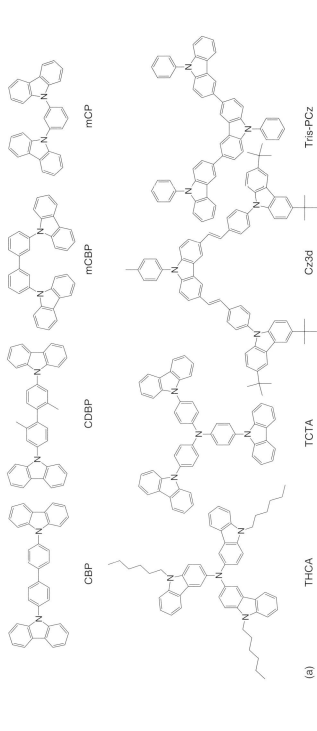

Figure 10.A.1 Electron donors containing carbazoles (a) and triphenylamines (b).

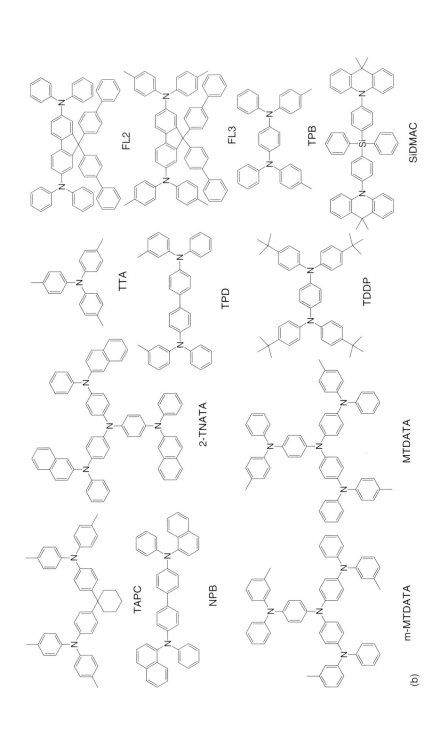

Figure 10.A.1 (*Continued*)

the molecular structures of electron donors containing carbazoles, which are 4,4′-N,N′-dicarbazole-biphenyl (CBP) [93, 97–100], 4,4′-bis(9-carbazolyl)-2,2′-dimethylbiphenyl (CDBP) [75], 9,9′-biphenyl-3,3′-diylbis-9H-carbazole (mCBP) [72], N,N′-dicarbazolyl-3,5-benzene (mCP) [66, 67, 78, 106, 109, 111, 129, 136, 137], tri(9-hexylcarbazol-3-yl)amine (THCA) [68, 113, 115], TCTA [3, 5, 6, 70, 74, 76, 77, 81, 82, 109, 118, 121, 123–125, 129, 138, 139], 3,6-bis(2-{4-[3,6-bis(4-*tert*-butylphenyl)carbazole-9-yl]phenyl}-vinyl)-9-*p*-tolyl-9H-carbazole (Cz3d) [126], and 9,9′,9″-triphenyl-9H,9′H,9″H-3,3′:6′,3″-tercarbazole (Tris-PCz) [55].

Figure 10.A.1b shows the molecular structures of the electron donors including triphenylamines. TAPC [26, 70, 74, 77, 109, 127, 129], 4,4′,4″-tris[2-naphthyl (phenyl)amino]triphenylamine (2-TNATA) [130], tris(*p*-tolyl)amine (TTA) [26], N,N′-diphenyl-N,N′-bis(1-naphthylphenyl)-1,1′-biphenyl-4,4′-diamine (NPB) [15, 63, 64, 93, 100, 108, 133–135], N,N′-diphenyl-N,N′-bis-(3-methylphenyl)-1,1′-biphenyl-4,4′-diamine) (TPD) [101–105, 107, 140–142], 2,7-bis(diphenylamino)-9,9-bis(1,1′-biphenyl-4-yl)-9H-fluorene (FL2) [108], 2,7-bis[di(4-methylphenyl)amino]-9,9-bis(1,1′-biphenyl-4-yl)-9H-fluorene (FL3) [108], m-MTDATA [1, 2, 22, 61, 65, 108, 110, 112, 114, 116, 117, 119, 143–147], 4,4′,4″-tris (3-methylphenylamino)triphenylamine (MTDATA) [26], 1,4-bis(di-4-*tert*-butylphenylamino)benzene (TDDP) [120], 1,4-bis[4-methylphenyl(phenyl)amino] benzene (TPB) [108], and bis[4-(9,9-dimethyl-9,10-dihydroacridine)phenyl] diphenylsilane (SiDMAC) [122] are used for OLEDs employing small molecular exciplexes.

10.A.3 Small Molecules with Electron-accepting Moieties Forming Exciplexes

Figure 10.A.2 shows the molecular structures of electron acceptors employed in OLEDs utilizing small molecule-based exciplexes. Electron acceptors containing 1,3,5-triazines for small molecular OLEDs using exciplexes are 3P-T2T [5, 67, 70, 81], 3′,3‴,3‴‴-(1,3,5-triazine-2,4,6-triyl)tris(([1,1′-biphenyl]-3-carbonitrile)) (CN-T2T) [55], 3,3′-bis(4,6-diphenyl-1,3,5-triazin-2-yl)-1,1′-biphenyl (DTrz) [74], 2,4,6-tris(3-(pyridin-3-yl)phenyl)-1,3,5-triazine (TmPyTZ) [77], 2,4,6-tris(biphenyl-3-yl)-1,3,5-triazine (T2T) [116], (1,3,5-triazine-2,4,6-triyl)tris(benzene-3,1-diyl))tris(diphenylphosphine oxide) (PO-T2T) [72, 75, 78, 111], 2,4,6-tris(3′-(pyridin-3-yl)biphenyl-3-yl)-1,3,5-triazine (Tm3PyBPZ) [76], and 2,5,8-tris(4-fluoro-3-methylphenyl)-1,3,4,6,7,9,9b-heptaazaphenalene (HAP-3MF) [66] shown in Figure 10.A.2a.

Figure 10.A.2b shows the molecular structures of electron acceptors having quinoxalines such as 6,7-dicyano-2,3-di-[4-(2,3,4,5-tetraphenylphenyl)phenyl] quinoxaline (CPQ) [93] and 2,3-bis(9,9-dihexyl-9H-fluoren-2-yl)-6,7-difluoroquinoxaline (F2Py) [134], having pyrazines such as 2,3-dicyano-5,6-di-(4-(2,3,4,5-tetraphenylphenyl)phenyl)pyrazine (CPP) [100], and having pyrimidines such as bis-4,6-(3,5-di-3-pyridylphenyl)-2-methylpyrimidine (B3PYMPM) [3, 97, 106, 136–139], B4PYMPM [6, 82, 98], 2-phenyl-4,6-bis(3,5-di-4-pyridylphenyl) pyrimidine (B4PyPPM) [99, 133].

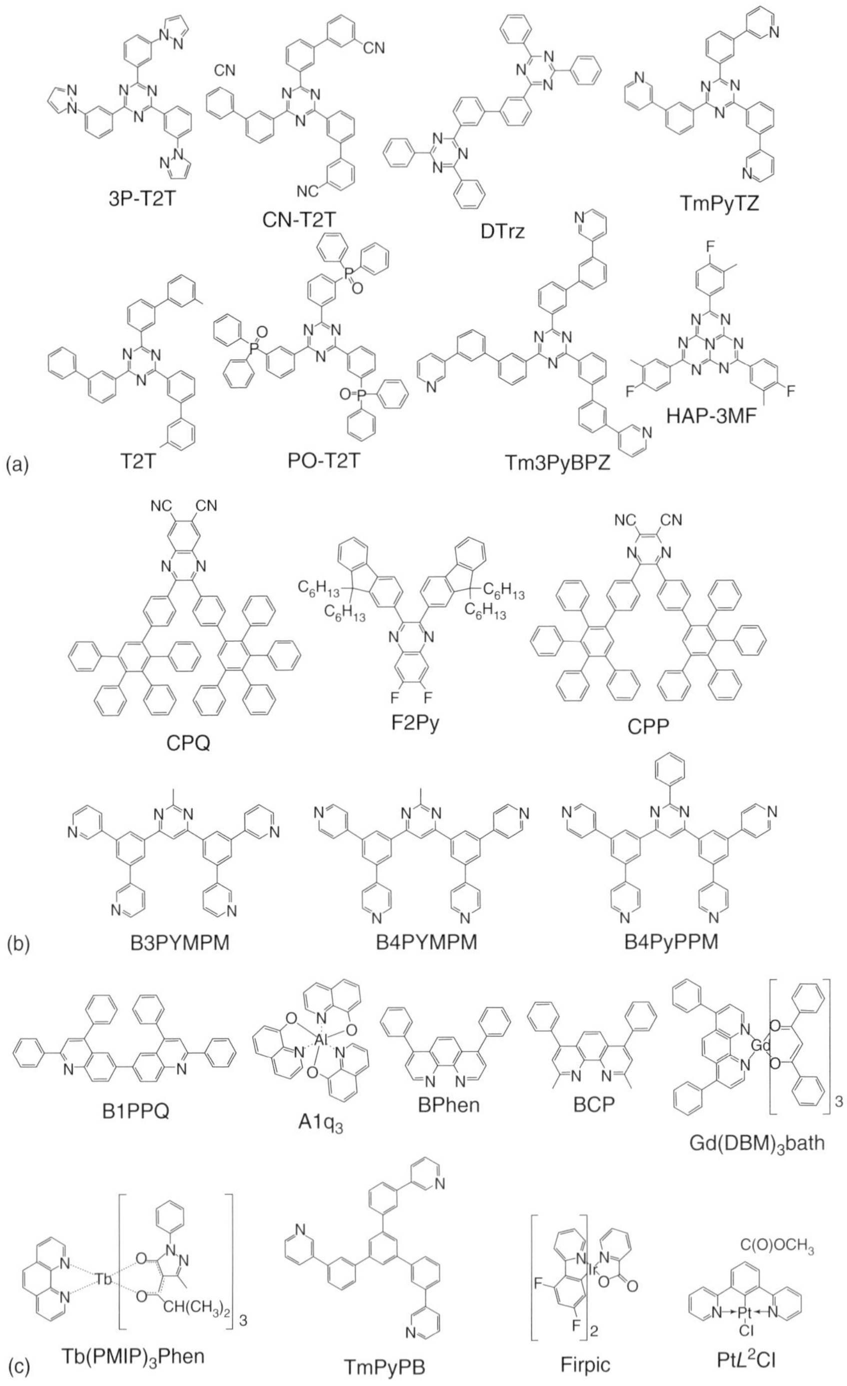

Figure 10.A.2 Electron acceptors containing 1,3,5-triazines (a), quinoxalines (b), quinolones (c), 1,3,4-oxadiazoles (d), and triarylboranes or siloles or phosphine oxides or sulfones (e).

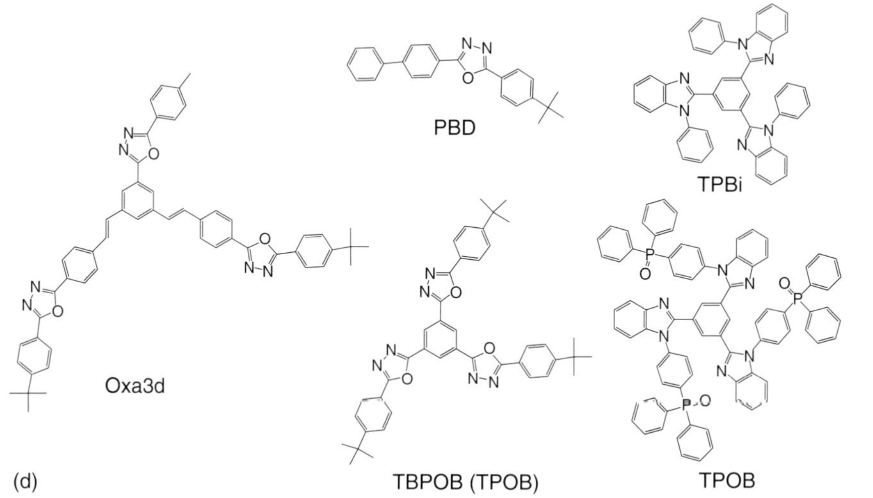

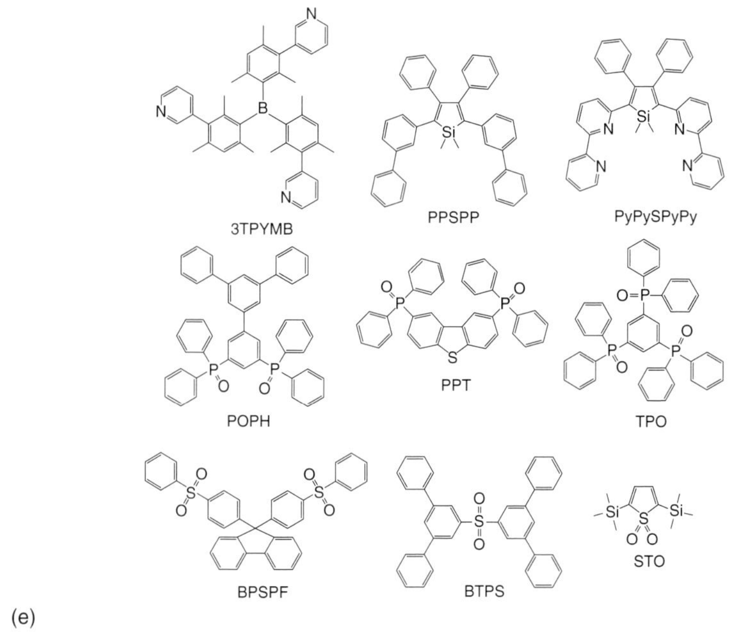

10.A.3 Small Molecules with Electron-accepting Moieties Forming Exciplexes | 367

(d)

Oxa3d   PBD   TPBi

TBPOB (TPOB)   TPOB

(e)

3TPYMB   PPSPP   PyPySPyPy

POPH   PPT   TPO

BPSPF   BTPS   STO

Figure 10.A.2 (*Continued*)

Figure 10.A.2c depicts the molecular structures of electron acceptors containing quinolines such as B1PPQ [26] and tris(8-hydoxyquinolate)aluminum (Alq_3) [108, 113, 130]; containing phenanthrolines such as 4,7-diphenyl-1,10-phenanthroline (Bphen) [65, 67, 70, 114, 115, 118, 144–146], 2,9-dimethyl-4,7-diphenyl-1,10-phenanthroline (BCP) [101, 112, 120], gadolinium(dibenzoylmethanato)$_3$ (bathophenanthroline) (Gd(DBM)$_3$bath) [103, 140], and tris-(1-phenyl-3-methyl-4-isobutyryl-5-pyrazolone)-1,10-phenanthroline-terbium (Tb(PMIP)3(Phen)) [105]; and containing pyridines such as 1,3,5-tri(m-pyrid-3-yl-phenyl)benzene

(TmPyPB) [117], iridium(III) bis-[4,6-difluorophenyl]-pyridinato-*N*,C2']-picolinate (FIrpic) [68], and platinum [methyl-3,5-di-(2-pyridyl) benzoate] chloride (PtL^2Cl) [61].

Molecular structures of electron acceptors with 1,3,4-oxadiazoles such as 2-[3,5-bis(2-{4-[5-(4-*tert*-butylphenyl)-1,3,4-oxadiazole-2-yl]phenyl}vinyl) phenyl]-5-(4-methylphenyl)-1,3,4-oxadiazole (Oxa3d) [126], 2-(biphenyl-4-yl)-5-(4-*tert*-butylphenyl)-1,3,4-oxadiazole (PBD) [2, 22, 135], and 1,3,5-tris(4-*tert*-butylphenyl-1,3,4-oxadiazolyl)benzene (TBPOB (TPOB)) [107, 121] and electron acceptors with benzimidazoles such as 1,3,5-tris(*N*-phenyl-benzimidazol-2-yl) benzene (TPBi) [64, 65, 67, 119, 123, 125, 129, 147] and 1,3,5-tris(1-(4-(diphenylphosphoryl)phenyl)-1H-benzo[*d*]imidazol-2-yl)benzene (TPOB) [125] are shown in Figure 10.A.2d.

Figure 10.A.2e shows the electron acceptors involving triarylboranes or siloles or phosphine oxides or sulfones. 3TPYMB [2, 110, 143] contains a triarylborane. 2,5-Di-(3-biphenyl)-1,1-dimethyl-3,4-diphenylsilacyclopentadiene (PPSPP) [63] and 2,5-bis(2′,2″-bipyridin-6-yl)-1,1-dimethyl-3,4-diphenylsilacyclopentadiene (PyPySPyPy) [15] contain siloles. (5-Terphenyl-1,3-phenylene)bis(diphenylphosphine oxide) (POPH) [109], 2,8-bis(diphenylphosphoryl)dibenzo[*b*,*d*] thiophene (PPT) [1], and 1,3,5-tris(diphenylphosphoryl)benzene (TPO) [124] have phosphine oxides. 9,9-Bis[4-(phenylsulfonyl)phenyl]fluorene (BPSPF) [122], 5′,5″″-sulfonyl-di-1,1′:3′,1″-terphenyl (BTPS) [127], and 2,5-bis(trimethylsilyl)-thiophene-1,1-dioxide (STO) [104, 141, 142] involve sulfones.

10.A.4 Small Molecules with Electron-donating and Electron-accepting Moieties Forming Exciplexes

Not only could typical pairs of donor molecules and acceptor molecules form exciplexes, but bipolar molecules could form exciplexes with typical donors or acceptors. The examples of the bipolar small molecules that were employed for OLEDs as exciplexes are shown in Figure 10.A.3. 6-(9,9-Dimethylacridin-10(9H)-yl)-3-methyl-1H-isochromen-1-one (MAC) [78] and bis[4-(9,9-dimethyl-9,10-dihydroacridine)phenyl]sulfone (DMAC-DPS) [73] contain 9,10-dihydroacridines as donor units. (4-Dimesitylboryl)phenyltriphenylamine (TPAPB) [71] has a triarylamine as a donor unit. (3′-(4,6-Diphenyl-1,3,5-triazin-2-yl)-(1,1′-biphenyl)-3-yl)-9-carbazole (CzTrz) [74, 148], 3-{4-(1,1-dimesitylboryl)-phenyl}-9-ethyl-9H-carbazole (CzPhB) [20], 9-[2,8]-9-carbazole-[dibenzothiophene-S,S-dioxide]-carbazole (2d) [128], iridium(III)-bis{5-(9-carbazolyl)-2-phenyl-1, 2,3-benzotriazolato-N,C2′}acetyl-acetonate complex (IC2) [132], 2,6-bis(3-(carbazol-9-yl)phenyl)pyridine (26DCzPPy) [69], 3-(4,6-diphenyl-1,3,5-triazin-2 -yl)-9-phenyl-9H-carbazole (DPTPCz) [4, 79], 2-(4-biphenylyl)-5-(4-carbazole-9-yl)phenyl-1,3,4-oxadiazole (CzOxa) [149], 3,6-di-(p-acetylphenyl)-*N*-hexylcarbazole (ANHC) [21], and 3,6-di-(p-cyanophenyl)-*N*-hexylcarbazole (CNHC) [21] have carbazoles as donor moieties. Meanwhile, MAC and ANHC have carbonyl groups as acceptor units. TPAPB and CzPhB have triarylboranes as electron acceptor moieties. CzTrz and DPTPCz contain 1,3,5-triazines as

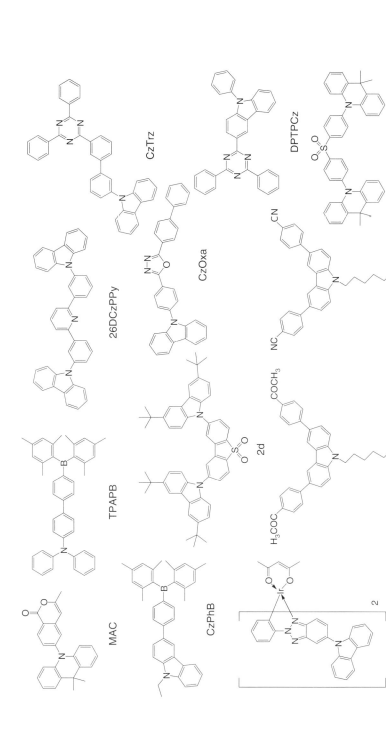

Figure 10.A.3 Bipolar molecules employed for OLEDs as exciplexes.

acceptor units. IC2, 26DCzPPy, CzOxa, and CNHC involve acceptor moieties that are a benzotriazole, a pyridine, a 1,3,4-oxadiazole, and a cyano group, respectively.

References

1. Goushi, K. and Adachi, C. (2012). *Appl. Phys. Lett.* 101: 23306.
2. Goushi, K., Yoshida, K., Sato, K., and Adachi, C. (2012). *Nat. Photonics* 6: 253.
3. Park, Y.S., Kim, K.H., and Kim, J.J. (2013). *Appl. Phys. Lett.* 102: 153306.
4. Liu, X.K., Chen, Z., Zheng, C.J., Chen, M., Liu, W., Zhang, X.H., and Lee, C.S. (2015). *Adv. Mater.* 27: 2025.
5. Hung, W., Fang, G., Chang, Y., Kuo, T., Chou, P., Lin, S., and Wong, K. (2013). *ACS Appl. Mater. Interfaces* 5: 6826.
6. Kim, K.H., Yoo, S.J., and Kim, J.J. (2016). *Chem. Mater.* 28: 1936.
7. Kavarnos, G.J. and Turro, N.J. (1986). *Chem. Rev.* 86: 401.
8. Kim, D. (2015). *J. Phys. Chem. C* 119: 12690.
9. Halls, J., Cornil, J., dos Santos, D., Silbey, R., Hwang, D.-H., Holmes, A., Brédas, J., and Friend, R. (1999). *Phys. Rev. B* 60: 5721.
10. Hontz, E., Chang, W., Congreve, D.N., Bulović, V., Baldo, M.A., and Van Voorhis, T. (2015). *J. Phys. Chem. C* 119: 25591.
11. Lee, K. and Kim, D. (2016). *J. Phys. Chem. C* 120: 28330.
12. Samanta, P.K., Kim, D., Coropceanu, V., and Brédas, J.L. (2017). *J. Am. Chem. Soc.* 139: 4042.
13. Dexter, D.L. (1953). *J. Chem. Phys.* 21: 836.
14. Brédas, J.-L., Beljonne, D., Coropceanu, V., and Cornil, J. (2004). *Chem. Rev.* 104: 4971.
15. Palilis, L.C., Murata, H., Uchida, M., and Kafafi, Z.H. (2003). *Org. Electron.* 4: 113.
16. Inai, Y., Sisido, M., and Imanishi, Y. (1990). *J. Phys. Chem.* 94: 6237.
17. Wang, H., Xie, L., Peng, Q., Meng, L., Wang, Y., Yi, Y., and Wang, P. (2014). *Adv. Mater.* 26: 5198.
18. Lee, J., Yuan, Y.Y., Kang, Y., Jia, W.L., Lu, Z.H., and Wang, S. (2006). *Adv. Funct. Mater.* 16: 681.
19. Yip, W.T. and Levy, D.H. (1996). *J. Phys. Chem.* 100: 11539.
20. Lin, S.L., Chan, L.H., Lee, R.H., Yen, M.Y., Kuo, W.J., Chen, C.T., and Jeng, R.J. (2008). *Adv. Mater.* 20: 3947.
21. Nayak, P.K., Agarwal, N., Periasamy, N., Patankar, M.P., and Narasimhan, K.L. (2010). *Synth. Met.* 160: 722.
22. Graves, D., Jankus, V., Dias, F.B., and Monkman, A. (2014). *Adv. Funct. Mater.* 24: 2343.
23. Gould, I.R., Young, R.H., Mueller, L.J., Albrecht, A.C., and Farid, S. (1994). *J. Am. Chem. Soc.* 116: 8188.
24. Gould, I.R., Boiani, J.A., Gaillard, E.B., Goodman, J.L., and Farid, S. (2003). *J. Phys. Chem. A* 107: 3515.

25 Young, R.H., Feinberg, A.M., Dinnocenzo, J.P., and Farid, S. (2015). *Photochem. Photobiol.* 91: 624.
26 Kulkarni, A.P. and Jenekhe, S.A. (2008). *J. Phys. Chem. C* 112: 5174.
27 Vandewal, K., Gadisa, A., Oosterbaan, W.D., Bertho, S., Banishoeib, F., Van Severen, I., Lutsen, L., Cleij, T.J., Vanderzande, D., and Manca, J.V. (2008). *Adv. Funct. Mater.* 18: 2064.
28 Itaya, A., Kitagawa, T., Moriyama, T., Matsushita, T., and Miyasaka, H. (1997). *J. Phys. Chem. B* 101: 524.
29 Goris, L., Haenen, K., Nesládek, M., Wagner, P., Vanderzande, D., De Schepper, L., D'haen, J., Luisen, L., and Manca, J.V. (2005). *J. Mater. Sci.* 40: 1413.
30 Goris, L., Poruba, A., Hod'Ákova, L., Vaněček, M., Haenen, K., Nesládek, M., Wagner, P., Vanderzande, D., De Schepper, L., and Manca, J.V. (2006) *Appl. Phys. Lett.* 88: 52113.
31 Vandewal, K., Goris, L., Haeldermans, I., Nesládek, M., Haenen, K., Wagner, P., and Manca, J.V. (2008). *Thin Solid Films* 516: 7135.
32 Vandewal, K., Tvingstedt, K., Gadisa, A., Inganäs, O., and Manca, J.V. (2009). *Nat. Mater.* 8: 904.
33 Clarke, T.M. and Durrant, J.R. (2010). *Chem. Rev.* 110: 6736.
34 Vandewal, K., Albrecht, S., Hoke, E.T., Graham, K.R., Widmer, J., Douglas, J.D., Schubert, M., Mateker, W.R., Bloking, J.T., Burkhard, G.F., Sellinger, A., Frechet, J.M., Amassian, A., Riede, M.K., McGehee, M.D., Neher, D., and Salleo, A. (2014). *Nat. Mater.* 13: 63.
35 Hallermann, M., Kriegel, I., Da Como, E., Berger, J.M., Von Hauff, E., and Feldmann, J. (2009). *Adv. Funct. Mater.* 19: 3662.
36 Gould, I.R., Noukalcis, D., Gomez-Jahn, L., Young, R.H., Goodman, J.L., and Farid, S. (1993). *Chem. Phys.* 176: 439.
37 Bixon, M., Jortner, J., and Verhoeven, J.W. (1994). *J. Am. Chem. Soc.* 116: 7349.
38 Verhoeven, J.W., Scherer, T., Wegewijs, B., Hermant, R.M., Jortner, J., Bixon, M., Depaemelaere, S., and de Schryver, F.C. (1995). *Rec. Trav. Chim. Pays Bas* 114: 443.
39 Arnold, B.R., Farid, S., Goodman, J.L., and Gould, I.R. (1996). *J. Am. Chem. Soc.* 118: 5482.
40 Jortner, J. and Bixon, M. (1988). *J. Chem. Phys.* 88: 167.
41 Gould, I.R., Young, R.H., Moody, R.E., and Farid, S. (1991). *J. Phys. Chem.* 95: 2068.
42 Asahi, T. and Mataga, N. (1991). *J. Phys. Chem.* 95: 1956.
43 Iwai, S., Murata, S., Katoh, R., Tachiya, M., Kikuchi, K., and Takahashi, Y. (2000). *J. Chem. Phys.* 112: 7111.
44 Turro, N.J. (1991). *Modern Molecular Photochemistry*. University Science Books.
45 Morais, J., Hung, R.R., Grabowski, J.J., and Zimmt, M.B. (1993). *J. Phys. Chem.* 97: 13138.
46 El-Sayed, M.A. (1963). *J. Chem. Phys.* 38: 2834.
47 Kuzmin, M.G., Soboleva, I.V., Dolotova, E.V., and Dogadkin, D.N. (2005). *High Energy Chem.* 39: 86.

48 Ogiwara, T., Wakikawa, Y., and Ikoma, T. (2015). *J. Phys. Chem. A* 119: 3415.
49 Gibson, J., Monkman, A.P., and Penfold, T.J. (2016). *Chem. Phys. Chem.* 17: 2956.
50 Marian, C.M. (2016). *J. Phys. Chem. C* 120: 3715.
51 Rao, A., Chow, P.C.Y., Gélinas, S., Schlenker, C.W., Li, C.-Z., Yip, H.-L., Jen, A.K.Y., Ginger, D.S., and Friend, R.H. (2013). *Nature* 500: 435.
52 Giebink, N.C., D'Andrade, B.W., Weaver, M.S., MacKenzie, P.B., Brown, J.J., Thompson, M.E., and Forrest, S.R. (2008). *J. Appl. Phys.* 103: 44509.
53 Berleb, S., Brütting, W., and Paasch, G. (2000). *Org. Electron.* 1: 41.
54 Noguchi, Y., Miyazaki, Y., Tanaka, Y., Sato, N., Nakayama, Y., Schmidt, T.D., Brütting, W., and Ishii, H. (2012). *J. Appl. Phys.* 111: 114508.
55 Hung, W.Y., Chiang, P.Y., Lin, S.W., Tang, W.C., Chen, Y.T., Liu, S.H., Chou, P.T., Hung, Y.T., and Wong, K.T. (2016). *ACS Appl. Mater. Interfaces* 8: 4811.
56 Baldo, M., Adachi, C., and Forrest, S.R. (2000). *Phys. Rev. B* 62: 10967.
57 Baleizão, C. and Berberan-Santos, M.N. (2007). *J. Chem. Phys.* 126: 204510.
58 Frederichs, B. and Staerk, H. (2008). *Chem. Phys. Lett.* 460: 116.
59 Bachilo, S.M., Benedetto, A.F., Weisman, R.B., Nossal, J.R., and Billups, W.E. (2000). *J. Phys. Chem. A* 104: 11265.
60 Deotare, P.B., Chang, W., Hontz, E., Congreve, D.N., Shi, L., Reusswig, P.D., Modtland, B., Bahlke, M.E., Lee, C.K., Willard, A.P., Bulović, V., Van Voorhis, T., and Baldo, M.A. (2015). *Nat. Mater.* 14: 1130.
61 Kalinowski, J., Cocchi, M., Virgili, D., Fattori, V., and Williams, J.A.G. (2007). *Adv. Mater.* 19: 4000.
62 Tamoto, N., Adachi, C., and Nagai, K. (1997). *Chem. Mater.* 9: 1077.
63 Palilis, L.C., Makinen, A.J., Uchida, M., and Kafafi, Z.H. (2003). *Appl. Phys. Lett.* 82: 2209.
64 Jankus, V., Chiang, C.J., Dias, F., and Monkman, A.P. (2013). *Adv. Mater.* 25: 1455.
65 Zhang, T., Chu, B., Li, W., Su, Z., Peng, Q.M., Zhao, B., Luo, Y., Jin, F., Yan, X., Gao, Y., Wu, H., Zhang, F., Fan, D., and Wang, J. (2014). *ACS Appl. Mater. Interfaces* 6: 11907.
66 Li, J., Nomura, H., Miyazaki, H., and Adachi, C. (2014). *Chem. Commun. (Camb).* 50: 6174.
67 Zhang, T., Zhao, B., Chu, B., Li, W., Su, Z., Wang, L., Wang, J., Jin, F., Yan, X., Gao, Y., Wu, H., Liu, C., Lin, T., and Hou, F. (2015). *Org. Electron.* 24: 1.
68 Cherpak, V., Stakhira, P., Minaev, B., Baryshnikov, G., Stromylo, E., Helzhynskyy, I., Chapran, M., Volyniuk, D., Hotra, Z., Dabuliene, A., Tomkeviciene, A., Voznyak, L., and Grazulevicius, J.V. (2015). *ACS Appl. Mater. Interfaces* 7: 1219.
69 Chen, D., Wang, Z., Wang, D., Wu, Y.C., Lo, C.C., Lien, A., Cao, Y., and Su, S.J. (2015). *Org. Electron. physics, Mater. Appl.* 25: 79.
70 Zhao, B., Zhang, T., Chu, B., Li, W., Su, Z., Luo, Y., Li, R., Yan, X., Jin, F., Gao, Y., and Wu, H. (2015). *Org. Electron. physics, Mater. Appl.* 17: 15.

71 Chen, Z., Liu, X.-K., Zheng, C.-J., Ye, J., Liu, C.-L., Li, F., Ou, X.-M., Lee, C.-S., and Zhang, X.-H. (2015). *Chem. Mater.* 27: 5206.
72 Zhang, T., Zhao, B., Chu, B., Li, W., Su, Z., Yan, X., Liu, C., Wu, H., Gao, Y., Jin, F., and Hou, F. (2015). *Sci. Rep.* 5: 10234.
73 Zhang, T., Zhao, B., Chu, B., Li, W., Su, Z., Yan, X., Liu, C., Wu, H., Jin, F., and Gao, Y. (2015). *Org. Electron. physics, Mater. Appl.* 25: 6.
74 Oh, C.S., Kang, Y.J., Jeon, S.K., and Lee, J.Y. (2015). *J. Phys. Chem. C* 119: 22618.
75 Liu, X.K., Chen, Z., Qing, J., Zhang, W.J., Wu, B., Tam, H.L., Zhu, F., Zhang, X.H., and Lee, C.S. (2015). *Adv. Mater.* 27: 7079.
76 Zhang, L., Cai, C., Li, K.F., Tam, H.L., Chan, K.L., and Cheah, K.W. (2015). *ACS Appl. Mater. Interfaces* 7: 24983.
77 Chen, D., Xie, G., Cai, X., Liu, M., Cao, Y., and Su, S.J. (2016). *Adv. Mater.* 28: 239.
78 Liu, W., Chen, J.-X., Zheng, C.-J., Wang, K., Chen, D.-Y., Li, F., Dong, Y.-P., Lee, C.-S., Ou, X.-M., and Zhang, X.-H. (2016). *Adv. Funct. Mater.* 26: 2002.
79 Liu, X.K., Chen, Z., Zheng, C.J., Wang, K., Chen, D.-Y., Li, F., Dong, Y.-P., Lee, C.-S., Ou, X.-M., and Zhang, X.-H. (2015). *Adv. Mater.* 27: 2378.
80 Mikhnenko, O.V., Blom, P.W.M., and Nguyen, T.-Q. (2015). *Energy Environ. Sci.* 8: 1867.
81 Zhao, B., Zhang, T., Chu, B., Li, W., Su, Z., Wu, H., Yan, X., Jin, F., Gao, Y., and Liu, C. (2015). *Sci. Rep.* 5: 10697.
82 Kim, K.H., Moon, C.K., Sun, J.W., Sim, B., and Kim, J.J. (2015). *Adv. Opt. Mater.* 3: 895.
83 Kawamura, Y., Brooks, J., Brown, J.J., Sasabe, H., and Adachi, C. (2006). *Phys. Rev. Lett.* 96: 11.
84 Shirota, Y. and Kageyama, H. (2007). *Chem. Rev.* 107: 953.
85 Zhigang, R.L. (2015). *Organic Light-Emitting Materials and Devices*. CRC Press.
86 Ge, Z., Hayakawa, T., Ando, S., Ueda, M., Akiike, T., Miyamoto, H., Kajita, T., and Kakimoto, M. (2008). *Chem. Lett.* 37: 262.
87 Chen, Y.-M., Hung, W.-Y., You, H.-W., Chaskar, A., Ting, H.-C., Chen, H.-F., Wong, K.-T., and Liu, Y.-H. (2011). *J. Mater. Chem.* 21: 14971.
88 Zhang, X., Shen, W., Zhang, D., Zheng, Y., He, R., and Li, M. (2015). *RSC Adv.* 5: 51586.
89 Gaspar, D.J. and Polikarpov, E. (2015). *OLED Fundamentals: Materials, Devices, and Processing of Organic Light-Emitting Diodes*. CRC Press.
90 Hughes, G. and Bryce, M.R. (2005). *J. Mater. Chem.* 15: 94.
91 Jeon, S.O. and Lee, J.Y. (2012). *J. Mater. Chem.* 22: 4233.
92 Sasabe, H., Seino, Y., Kimura, M., and Kido, J. (2012). *Chem. Mater.* 24: 1404.
93 Xu, X., Yu, G., Chen, S., Di, C., and Liu, Y. (2008). *J. Mater. Chem.* 18: 299.
94 Murata, H., Malliaras, G.G., Uchida, M., Shen, Y., and Kafafi, Z.H. (2001). *Chem. Phys. Lett.* 339: 161.
95 Banal, J.L., Subbiah, J., Graham, H., Lee, J.-K., Ghiggino, K.P., and Wong, W.W.H. (2013). *Polym. Chem.* 4: 1077.

96 Tekarli, S.M., Cundari, T.R., and Omary, M.A. (2008). *J. Am. Chem. Soc.* 130: 1669.
97 Park, Y.S., Jeong, W.I., and Kim, J.J. (2011). *J. Appl. Phys.* 110: 124519.
98 Komatsu, R., Sasabe, H., Inomata, S., Pu, Y.J., and Kido, J. (2015). *Synth. Met.* 202: 165.
99 Seino, Y., Inomata, S., Sasabe, H., Pu, Y.J., and Kido, J. (2016). *Adv. Mater.* 28: 2638.
100 Xu, X., Chen, S., Yu, G., Di, C., You, H., Ma, D., and Liu, Y. (2007). *Adv. Mater.* 19: 1281.
101 Zhu, H., Xu, Z., Zhang, F., Zhao, S., and Song, D. (2008). *Appl. Surf. Sci.* 254: 5511.
102 Kalinowski, J., Cocchi, M., Virgili, D., and Sabatini, C. (2006). *Appl. Phys. Lett.* 89: 11105.
103 Wang, D.Y., Li, W.L., Chu, B., Liang, C.J., Hong, Z.R., Li, M.T., Wei, H.Z., Xin, Q., Niu, J.H., and Xu, J.B. (2006). *J. Appl. Phys.* 100: 24506.
104 Mazzeo, M., Pisignano, D., Favaretto, L., Sotgiu, G., Barbarella, G., Cingolani, R., and Gigli, G. (2003). *Synth. Met.* 139: 675.
105 Xin, H., Guang, M., Li, F.Y., Bian, Z.Q., Huang, C.H., Ibrahim, K., and Liu, F.Q. (2002). *Phys. Chem. Chem. Phys.* 4: 5895.
106 Shin, H., Lee, S., Kim, K.H., Moon, C.K., Yoo, S.J., Lee, J.H., and Kim, J.J. (2014). *Adv. Mater.* 26: 4730.
107 Ogawa, H., Okuda, R., and Shirota, Y. (1998). *Mol. Cryst. Liq. Cryst. Sci. Technol. Sect. A.* 315: 187.
108 Matsumoto, N., Nishiyama, M., and Adachi, C. (2008). *J. Phys. Chem. C* 112: 7735.
109 Ban, X., Sun, K., Sun, Y., Huang, B., Ye, S., Yang, M., and Jiang, W. (2015). *ACS Appl. Mater. Interfaces* 7: 25129.
110 Huang, Q., Zhao, S., Xu, Z., Fan, X., Shen, C., and Yang, Q. (2014). *Appl. Phys. Lett.* 104: 161112.
111 Lee, J.H., Cheng, S.H., Yoo, S.J., Shin, H., Chang, J.H., Wu, C.I., Wong, K.T., and Kim, J.J. (2015). *Adv. Funct. Mater.* 25: 361.
112 Chen, L.L., Li, W.L., Li, M.T., and Chu, B. (2007). *J. Lumin.* 122–123: 667.
113 Michaleviciute, A., Gurskyte, E., Volyniuk, D.Y., Cherpak, V.V., Sini, G., Stakhira, P.Y., and Grazulevicius, J.V. (2012). *J. Phys. Chem. C* 116: 20769.
114 Wang, D., Li, W., Chu, B., Su, Z., Bi, D., Zhang, D., Zhu, J., Yan, F., Chen, Y., and Tsuboi, T. (2008). *Appl. Phys. Lett.* 92: 53304.
115 Chapran, M., Ivaniuk, K., Stakhira, P., Cherpak, V., Hotra, Z., Volyniuk, D., Michaleviciute, A., Tomkeviciene, A., Voznyak, L., and Grazulevicius, J.V. (2015). *Synth. Met.* 209: 173.
116 Nakanotani, H., Furukawa, T., Morimoto, K., and Adachi, C. (2016). *Sci. Adv.* 2: e1501470.
117 Wang, S., Wang, X., Yao, B., Zhang, B., Ding, J., Xie, Z., and Wang, L. (2015). *Sci. Rep.* 5: 12487.
118 Duan, Y., Sun, F., Yang, D., Yang, Y., Chen, P., and Duan, Y. (2014). *Appl. Phys. Express* 7: 52102.

119 Zhang, G., Li, W., Chu, B., Su, Z., Yang, D., Yan, F., Chen, Y., Zhang, D., Han, L., Wang, J., Liu, H., Che, G., Zhang, Z., and Hu, Z. (2009). *Org. Electron. physics, Mater. Appl.* 10: 352.

120 Wu, C., Djurovich, P.I., and Thompson, M.E. (2009). *Adv. Funct. Mater.* 19: 3157.

121 Ogawa, H., Okuda, R., and Shirota, Y. (1998). *Appl. Phys. A Mater. Sci. Process.* 67: 599.

122 Seino, Y., Sasabe, H., Kimura, M., Inomata, S., Nakao, K., and Kido, J. (2016). *Chem. Lett.* 45: 283.

123 Wang, X., Wang, R., Zhou, D., and Yu, J. (2016). *Synth. Met.* 214: 50.

124 Ban, X., Sun, K., Sun, Y., Huang, B., and Jiang, W. (2016). *Org. Electron.* 33: 9.

125 Ban, X., Sun, K., Sun, Y., Huang, B., and Jiang, W. (2016). *ACS Appl. Mater. Interfaces* 8: 2010.

126 Cha, S.W. and Jin, J.-I. (2003). *J. Mater. Chem.* 13: 479.

127 Seino, Y., Sasabe, H., Pu, Y.J., and Kido, J. (2014). *Adv. Mater.* 26: 1612.

128 Jankus, V., Data, P., Graves, D., McGuinness, C., Santos, J., Bryce, M.R., Dias, F.B., and Monkman, A.P. (2014). *Adv. Funct. Mater.* 24: 6178.

129 Kim, B.S. and Lee, J.Y. (2014). *Adv. Funct. Mater.* 24: 3970.

130 Okumoto, K. and Shirota, Y. (2000). *J. Lumin.* 87: 1171.

131 Li, G., Kim, C.H., Zhou, Z., Shinar, J., Okumoto, K., and Shirota, Y. (2006). *Appl. Phys. Lett.* 88: 10.

132 Cherpak, V., Stakhira, P., Minaev, B., Baryshnikov, G., Stromylo, E., Helzhynskyy, I., Chapran, M., Volyniuk, D., Tomkuté-Lukšiené, D., Malinauskas, T., Getautis, V., Tomkeviciene, A., Simokaitiene, J., and Grazulevicius, J.V. (2014). *J. Phys. Chem. C* 118: 11271.

133 He, S.J., Wang, D.K., Jiang, N., Tse, J.S., and Lu, Z.H. (2016). *Adv. Mater.* 28: 649.

134 Zhang, W., Yu, J.S., Huang, J., Jiang, Y.D., Zhang, Q., and Cao, K.L. (2010). *Chinese Phys. B* 19: 47802.

135 Zhao, D., Zhang, F., Xu, C., Sun, J., Song, S., Xu, Z., and Sun, X. (2008). *Appl. Surf. Sci.* 254: 3548.

136 Sun, J.W., Lee, J.H., Moon, C.K., Kim, K.H., Shin, H., and Kim, J.J. (2014). *Adv. Mater.* 26: 5684.

137 Lee, S., Shin, H., and Kim, J.J. (2014). *Adv. Mater.* 26: 5864.

138 Lee, S., Kim, K.H., Limbach, D., Park, Y.S., and Kim, J.J. (2013). *Adv. Funct. Mater.* 23: 4105.

139 Park, Y.S., Lee, S., Kim, K.H., Kim, S.Y., Lee, J.H., and Kim, J.J. (2013). *Adv. Funct. Mater.* 23: 4914.

140 Liang, C.J., Zhao, D., Hong, Z.R., Li, R.G., Li, W.L., and Yu, J.Q. (2000). *Thin Solid Films* 371: 207.

141 Mazzeo, M., Pisignano, D., Della Sala, F., Thompson, J., Blyth, R.I.R., Gigli, G., Cingolani, R., Sotgiu, G., and Barbarella, G. (2003). *Appl. Phys. Lett.* 82: 334.

142 Mazzeo, M., Thompson, J., Blyth, R.I.R., Anni, M., Gigli, G., and Cingolani, R. (2002). *Phys. E Low-Dimensional Syst. Nanostructures* 13: 1243.

143 Lei, Y., Zhang, Q., Chen, L., Ling, Y., Chen, P., Song, Q., and Xiong, Z. (2016). *Adv. Opt. Mater.* 4: 694.
144 Zhu, L., Xu, K., Wang, Y., Chen, J., and Ma, D. (2015). *Front. Optoelectron.* 8: 439.
145 Yan, F., Chen, R., Sun, H., and Sun, X.W. (2014). *Appl. Phys. Lett.* 104: 2012.
146 Kalinowski, J. (2008). *J. Non. Cryst. Solids* 354: 4170.
147 Zhou, D.Y., Cui, L.S., Zhang, Y.J., Liao, L.S., and Aziz, H. (2014). *Appl. Phys. Lett.* 105: 2012.
148 Zhang, D., Cai, M., Zhang, Y., Bin, Z., Zhang, D., and Duan, L. (2016). *ACS Appl. Mater. Interfaces* 8: 3825.
149 Guan, M., Bian, Z.Q., Zhou, Y.F., Li, F.Y., Li, Z.J., and Huang, C.H. (2003). *Chem. Commun. (Camb).* 21: 2708.

11

Thermally Activated Delayed Fluorescence Materials Based on Donor–Acceptor Molecular Systems

Ye Tao[1,2], Runfeng Chen[1], Huanhuan Li[1,2], Chao Zheng[1], and Wei Huang[1,2]

[1] *Nanjing University of Posts and Telecommunications, Key Laboratory for Organic Electronics and Information Displays & Institute of Advanced Materials, Jiangsu National Synergistic Innovation Center for Advanced Materials, 9 Wenyuan Road, Nanjing, 210023, PR China*
[2] *Nanjing Tech University, Key Laboratory of Flexible Electronics & Institute of Advanced Materials, National Synergistic Innovation Center for Advanced Materials, 30 South Puzhu Road, Nanjing 211816, PR China*

Thermally activated delayed fluorescence (TADF) materials with efficient transition and interconversion between the lowest singlet (S_1) and triplet (T_1) excited states offer unique optical and electronic properties for organic light-emitting diode (OLED) applications. In this chapter, we present an overview of the quick development in the molecular structure engineering of TADF materials in donor–acceptor (D–A) molecular architecture. Fundamental design principles and the common relations between the molecular structures and optoelectronic properties for the diversified device applications as emitters, sensitizers, or hosts in OLEDs have been discussed. Especially, a survey of recent progress in the studies of the D–A type TADF materials, with a particular emphasis on the different molecular building blocks for TADF phenomenon, is highlighted.

11.1 Introduction

TADF materials, developed to harvest triplet exciton for luminescence and to avoid the use of expensive and resource-limited noble metals, can realize an internal quantum efficiency of up to 100% in OLEDs through the efficient utilization of both singlet and triplet excitons assisted by efficient reverse intersystem crossing (rISC) process for triplet-to-singlet transformation (Figure 11.1) [1–5]. Importantly, TADF can be facilely designed in purely organic donor–acceptor (D–A) molecular architecture; based on the abundant donors and acceptors already developed, a large number of D–A type TADF materials with varied molecular structures and optoelectronic properties have been reported in the recent years, leading to significant progress in TADF-related OLED applications.

Important milestones in the development of TADF materials are illustrated in Figure 11.2. Firstly discovered in 1961 by Parker and Hatchard in eosin dye, TADF was previously named as *E*-type delayed fluorescence (DF) [6]. Then,

Highly Efficient OLEDs: Materials Based on Thermally Activated Delayed Fluorescence,
First Edition. Edited by Hartmut Yersin.
© 2019 Wiley-VCH Verlag GmbH & Co. KGaA. Published 2019 by Wiley-VCH Verlag GmbH & Co. KGaA.

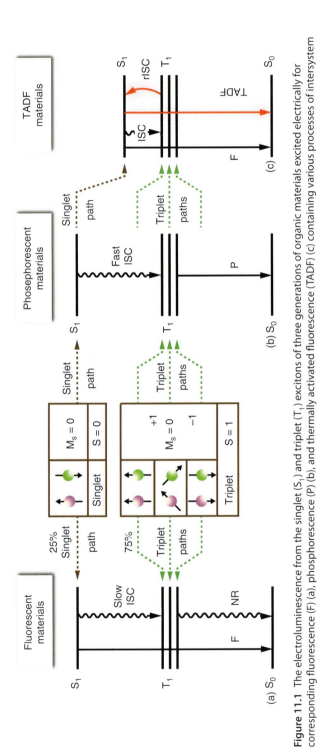

Figure 11.1 The electroluminescence from the singlet (S_1) and triplet (T_1) excitons of three generations of organic materials excited electrically for corresponding fluorescence (F) (a), phosphorescence (P) (b), and thermally activated fluorescence (TADF) (c) containing various processes of intersystem crossing (ISC), nonradiative relaxation (NR), and reverse intersystem crossing (rISC). Source: Ref. [1]. Reproduced with permission of Elsevier.

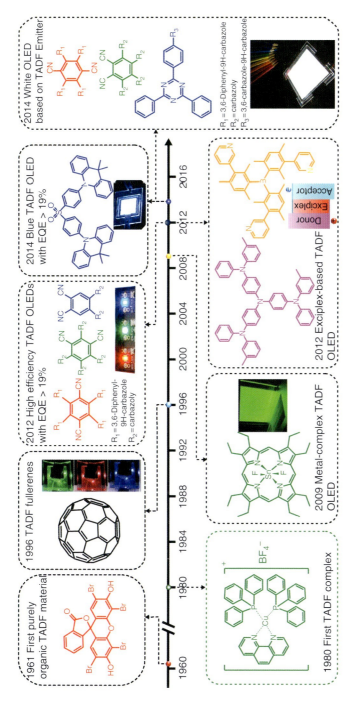

Figure 11.2 Milestones in the development of TADF materials. The TADF phenomenon in eosin was firstly discovered by Parker and Hatchard [6]. Blasse and McMillin reported the first TADF Cu-complex in 1980 [7]. The TADF emission in fullerenes was found by Berberan-Santos and Garcia [8]. Endo et al. demonstrated a TADF OLED using a Sn(IV)-complex in 2009 [9]. In 2012, OLED using the TADF exciplex was reported by Goushi et al. [10]. The remarkable progress of TADF OLEDs was achieved by Adachi and coworkers with EQE of up to 19.3% [11]. In 2014, Adachi and coworkers reported a blue and white TADF OLED with EQE of 19.5% and 17%, respectively [12, 13].

efficient TADF in metal-containing materials (Cu(I)-complex) was found in 1980 [7], followed by the observation of TADF in fullerenes in 1996, when Berberan-Santos and Garcia firstly applied the triplet-state-related TADF in the detection of the oxygen and temperature [8]. The modern research of TADF in OLEDs began in 2009, although the devices required a rather high onset voltage of around 10 V [9]. Researches on TADF OLEDs culminated after the work of Adachi and coworkers in 2012 by designing TADF molecules in D–A molecular structure [11]. Thanks to the large varieties of organic donors and acceptors that have been already developed in organic electronics, the D–A type TADF materials with varied solubility, stability, emitting color, charge transport property, etc. can be rationally obtained with relatively straightforward molecular design strategy and facile synthetic route; the red, green, blue, and white TADF OLEDs with corresponding external quantum efficiency (EQE) of up to ~13% [14], 31.2% [15], 36.7% [16], and 25.5% [17] have been reported, which undoubtedly break the efficiency limitations of fluorescent OLEDs and become comparable with PhOLEDs based on rare metal complexes. The distinguished developments of the noble metal-free TADF materials have truly revolutionized our understandings of organic semiconductors and optoelectronics, and the replacement of source-limited rare metal complexes, at least in OLEDs, is highly expected in the near future.

In this chapter, we will summarize the recent progress on the molecular designs and properties of TADF materials based on D–A molecular systems as well as their recent advances in OLED applications. We begin by describing the TADF OLEDs, focusing on the device structures and operation mechanism. Next, we will describe the basic considerations in molecular design of D–A type TADF materials involving fundamental design principle, the control of the singlet–triplet energy splitting and the modulation of luminescent quantum yield. Subsequently, the emphasis will be placed on different types of TADF molecular systems with varied donor and acceptor building blocks, accompanied by thorough coverage of the recent efforts and latest research progresses on high efficient OLEDs to take advantage of TADF effects. A major goal of this chapter is to provide illustrative accounts on recent progress and to systematize our knowledge of the subject, extracting fundamental principles on design strategies of D–A type TADF materials and the common relationship between the molecular structures of TADF materials and their optoelectronic properties for OLEDs applications.

11.2 TADF OLEDs

11.2.1 Device Structures and Operation Mechanisms of TADF OLED

The device structure of TADF OLEDs has no difference from the conventional sandwich configuration of fluorescence (nondoped) and phosphorescence (doped) OLEDs, containing a series of functional layers including electrodes of metal cathode and indium tin oxide (ITO) anode, hole and electron injection and transport layers, an emissive layer (EML), and sometime exciton-blocking layers.

The device operation mechanisms of TADF OLEDs are also very similar to the traditional OLEDs, which require efficient charge injection and balanced charge transport for high device performance. However, significant difference begins when singlet and triplet excitons are electronically excited. For fluorescent OLEDs, only 25% singlet excitons can be harvested for luminescence, and 75% triplet excitons are wasted through nonradiative decays. For phosphorescence OLEDs, both singlet and triplet excitons are emissive with 100% internal electroluminescence quantum efficiency due to singlet–triplet mixing via the strong spin–orbit coupling of the heavy metal atoms. In the case of TADF OLEDs, although only singlet excitons are emissive, triplet excitons can be easily transformed to singlet excitons via facile rISC processes, leading to 100% exciton harvesting for electroluminescence.

Generally, there are four important processes for TADF emission in TADF OLEDs (Figure 11.3):

1) The singlet and triplet excitons are formed after electron and hole recombination in a singlet-to-triplet ratio of 1 : 3 according to spin statistics.
2) The higher-lying exciton states are relaxed to the lowest singlet (S_1) and/or triplet (T_1) states via fast vibrational relaxation (VR), internal conversion (IC), and intersystem crossing (ISC).
3) The formed triplet excitons at T_1 are efficiently back transferred to S_1 through rISC process with the aid of thermal activation.
4) The singlet excitons at S_1 formed either initially or back transferred from T_1 are radiatively deactivated to S_0 for prompt fluorescence (PF) and long-lived DF, respectively.

In these processes, the efficient rISC process from T_1 to S_1 is very critical to utilize triplet excitons to improve the efficiency of the device. In order to facilitate the rISC process, small singlet–triplet splitting (ΔE_{ST}) is the key to enhance the $T_1 \rightarrow S_1$ transition. Of course, when maintaining such an efficient rISC, the ISC is even more efficient than rISC due to the higher energy level of S_1 than T_1; the majority of the excitons are located at T_1 with lower energy level in

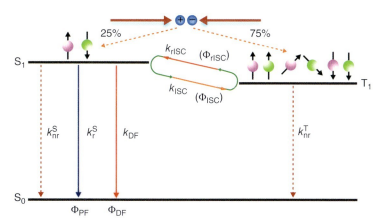

Figure 11.3 Electroluminescence processes of TADF molecules.

TADF materials. Therefore, T_1 should be stable enough with slow nonradiative deactivation rate (k_{nr}^T) to support the rISC process for DF emission.

11.2.2 TADF Molecules as Emitters for OLEDs

With apparent advantages in harvesting both singlet and triplet excitons for electroluminescence and theoretically 100% internal quantum efficiency, TADF molecules were widely used as emitters in OLEDs [2]. Upon electrical excitation, the theoretical maximum EQE of the TADF OLEDs can be estimated by Eq. (11.1) [2]:

$$\begin{aligned} EQE &= \eta_{int}\eta_{out} = \gamma\eta_r\eta_{PL}\eta_{out} \\ &= \gamma\left[\sum_{k=0}^{\infty}(0.75\Phi_{PF}\Phi_{rISC}(\Phi_{ISC}\Phi_{rISC})^k + 0.25\Phi_{PF}(\Phi_{ISC}\Phi_{rISC})^k)\right]\eta_{out} \\ &= \gamma[0.25(\Phi_{PF}+\Phi_{DF})+0.75\Phi_{rISC}(\Phi_{PF}+\Phi_{DF})]\eta_{out} \\ &= \gamma(0.25+0.75\Phi_{rISC})\eta_{PL}\eta_{out} \end{aligned} \quad (11.1)$$

where η_{int} is the internal quantum efficiency of the device, η_{out} is the out-coupling constant, γ is the ratio of the charge combination to the electron and hole transportation, η_r is the excitation–production ratio, η_{PL} is the photoluminescence efficiency, Φ_{PF} is the photoluminescence quantum yield (PLQY) of the PF, Φ_{DF} is the PLQY of the DF, k is the number of ISC and rISC transitions, and Φ_{ISC} and Φ_{rISC} are the ISC and rISC efficiencies, respectively. If the triplet exciton is very stable with very low nonradiative decay rate ($k_{nr}^T \to 0$), the Φ_{rISC} will approach 1 and EQE = $\gamma \times \eta_{PL} \times \eta_{out}$, suggesting that 100% of the excitons can be used for electroluminescence. When $\eta_{PL} = 1$, the internal quantum efficiency will be 100%. Therefore, in order to obtain the high efficiency of TADF OLEDs, TADF emitters should have high PLQY ($\eta_{PL} = 1$), small ΔE_{ST}, and stable triplet state for total transformation of triplet excitons to singlet ones through efficient reverse ISC ($\Phi_{rISC} = 1$) and properly aligned the highest occupied molecular orbital (HOMO) and the lowest unoccupied molecular orbital (LUMO) energy levels for efficient charge injection as well as good bipolar mobility for balanced and efficient charge transportation to increase the charge recombination ratio ($\gamma \to 1$). In addition, the TADF emitters should also possess the other characteristics, such as good film-forming properties, sufficiently high glass transition temperature to avoid crystallization within the desired operation lifetime of the device, and appropriate solubility and stability for fabrication procedures, for example, spin coating, inkjet printing, and vacuum deposition [16, 18–20].

11.2.3 TADF Molecules as Host Materials and Sensitizers for OLEDs

A large number of TADF molecules, especially in D–A molecular architecture, are found to have low PLQY due to their too much separated HOMO and LUMO to efficiently support radiative decay for luminescence, eliminating their applications as efficient OLED emitters according to Eq. (11.1). Fortunately, these D–A type TADF molecules have balanced and bipolar charge transporting/injection properties, which are very attractive to function as high-performance host materials and singlet exciton sensitizers in OLED applications [2, 21].

Host materials are important in OLEDs to disperse emitters, especially for the long-lived triplet ones, to alleviate the quenching effects, such as concentration quenching, triplet–triplet annihilation (TTA), triplet exciton-polaron quenching (TPQ), and electric field-induced exciton dissociation. An efficient host material should possess the following features [22–24]:

i) Sufficiently higher triplet energy (E_T) than the doped emitters (guests) to prevent reverse energy transfer from the guest to the host.
ii) Properly aligned HOMO and LUMO levels to match with those of the neighboring layers to ensure effective charge injection from the adjacent layers.
iii) Larger HOMO–LUMO energy gap (E_g) than that of the guest to facilitate direct charge trapping on the doped emitter.
iv) Balanced charge transport properties for hole and electron recombination process and confinement of the excitons formation zone in the EML.
v) Good morphological stability, film-forming ability, and structure compatibility to make the dispersion of dopant in host uniform during the device fabrication and operation. Moreover, the orientation of the host matrix may also have great impact on the device performance [25–27].

Singlet exciton sensitizers can harvest triplet excitons generated in OLEDs for fluorescent emitters via energy transfer [28]. To guarantee efficient triplet exciton harvesting using a TADF sensitizer, two critical requirements should be satisfied [29]:

i) The excitons should be formed in TADF materials, and the direct exciton trapping by fluorescent dopant should be avoided.
ii) The energy transfer from TADF sensitizer to fluorescent materials should only be mainly the singlet–singlet Förster energy transfer, and triplet–triplet Dexter energy transfer should be suppressed.

Therefore, the HOMO and LUMO energy levels of the TADF host/sensitizer should be close to that of fluorescent dopant to prevent the direct trapping of the guest/emitter; the doping concentration of fluorescent emitter should be very low to minimize the Dexter energy transfer to alleviate the triplet exciton formation in fluorescent materials.

11.2.4 Host-free TADF OLEDs

The TADF emission is closely related to the long-lived triplet excited states via rISC process. Therefore, TADF emitters are generally needed to be dispersed in a host and protected for high OLED performance. However, the host–guest systems are susceptible to crystallize and aggregate for phase separation due to the different structural compatibility of host and guest molecules, leading to a poor device performance especially in long-term device operation. Moreover, the doping of emitters into the host requires precise control of the concentrations of both guest and host, which is a complicated process, causing a high cost of fabrication. And the host molecules may influence the photoluminescent (PL) and electroluminescent (EL) properties of the guests, inducing large variations in the emitting color, PLQY, and lifetime of the EML due to strong host and guest interactions [30]. Therefore, there is a trend, recently, to eliminate the use of host materials in both metal complex-based PhOLEDs and TADF OLEDs. Thanks to the large structural diversity of purely organic molecules, TADF materials can be designed relatively easier to be not as sensitive as metal complexes toward host materials and doping levels; the doping concentrations of the metal-free TADF emitters can be varied in a wide range from 1 to 40 wt% and even 100% (host free) [31, 32]. Of course, with the aforementioned advantages without using host materials, TADF molecules for the nondoped devices should have the following additional characteristics:

i) The host-free film of TADF materials should show a high PLQY.
ii) Special attentions should be paid during molecular design of TADF molecules to reduce quenching effects of the long-lived triplet excitons excited electrically in OLEDs.
iii) The general requirements of OLED EMLs, including matched frontier orbital energy levels toward adjacent charge transport layers for efficient and balanced charge injection and high quality nanoscale thin film with smooth and stable morphology, should be satisfied.

11.3 Basic Considerations in Molecular Design of TADF Molecules

11.3.1 Design Principles of Donor–Acceptor Molecular Systems for TADF Emission

The key process in TADF emission is the rISC ($T_1 \rightarrow S_1$) that upconverts excitons from the low-lying T_1 to S_1 at a higher energy level, when the ΔE_{ST} between S_1 and T_1 is small enough (<0.37 eV) that can be overcome by the thermal vibration with the aid of surrounding temperature [33, 34]. The small ΔE_{ST} can be achieved through separated HOMO and LUMO distributions or, more specifically, through separated frontier orbital distributions [33]. Another important factor for TADF emission is the high luminescent efficiency, which requires overlapped HOMO and LUMO for high efficient luminescence with small transition dipole moment [4, 35]. Therefore, these two contradictory sides in simultaneously realizing small ΔE_{ST} for efficient rISC (separated HOMO and LUMO) and high luminescent efficiency for strong emission (overlapped HOMO and LUMO) should be taken into consideration to an optimized extent in the molecular design of the metal-free organic materials for efficient TADF emissions.

The majority of the reported TADF materials based on D–A molecular system (Figure 11.4) were designed via intramolecular charge transfer (ICT) state with the following characteristics:

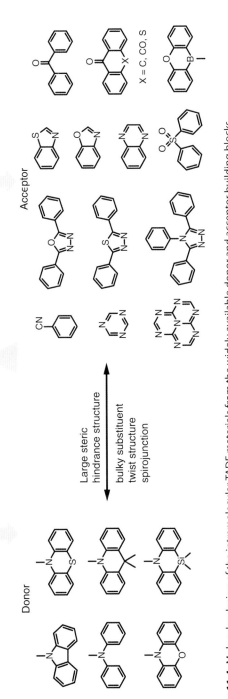

Figure 11.4 Molecular design of the intramolecular TADF materials from the widely available donor and acceptor building blocks.

i) A small ΔE_{ST} in donor–acceptor structure is realized by separating HOMO and LUMO on donor and acceptor groups, respectively [36].
ii) The steric hindrance structures, such as twisty, bulky, or spiro junctions, were introduced to connect donor and acceptor units to effectively separate the spatial overlaps between HOMO and LUMO [37, 38].
iii) The π-conjugation length and redox potentials of donor and acceptor moieties along with the interruption extent of the conjugation between them should also be taken into consideration entirely to achieve an emission with desired color and high PLQY [39].
iv) Densely combined donors and acceptors are used to aggrandize the overlap between HOMO and LUMO wave functions and enhance molecular structural rigidity, hence to increase radiative luminescent efficiency for high PLQY [11].

In addition to the intramolecular D–A type TADF emitters, the ICT between two electron-donating and electron-accepting molecules can also produce TADF emissions from the in situ formed exciplexes, showing broad and redshifted emissions compared with individual donor and acceptor molecules [40]. The electron transition from the LUMO of an acceptor to the HOMO of a donor in a large electron-hole separation distance for exciplex emission should have a small ΔE_{ST} for TADF emission via efficient rISC from the triplet charge transfer (CT) state to the singlet CT state (Figure 11.5). However, this CT emission of exciplex is generally very low in efficiency. To enhance the TADF emission in exciplexes, appropriate donor and acceptor molecules should be carefully selected with the following considerations:

i) Electron-donating and accepting molecules should have high triplet energy levels to confine the triplet exciplex state by preventing the quenching of the triplet state via the triplet energy back transfer to the donor or acceptor [10].
ii) Both the shallow HOMO levels in the electron-donating molecules and deep LUMO levels in the electron-accepting molecules, as well as high PL efficiency, are important for exciplex-based OLEDs (Figure 11.5).
iii) Electron-donating and electron-accepting molecules should have a planar structure for a flat-on orientation in solid states to avoid the formation of excimer in the EML [41].

11.3.2 Control of Singlet–Triplet Energy Splitting (ΔE_{ST})

A small ΔE_{ST} is of fundamental importance for TADF materials featured by a fast rISC process [2]. Theoretically, the molecular energy of the lowest singlet (E_S) and triplet (E_T) excited state is determined by orbital energy (E), electron repulsion energy (K), and exchange energy (J) of the two unpaired electrons at the excited states, as illustrated in Eqs. (11.2) and (11.3). By definition, ΔE_{ST} is the difference between E_S and E_T and is then equal to the twice of J (Eq. (11.4)) by subtracting Eq. (11.3) from Eq. (11.2), suggesting J is the most decisive factor for the ΔE_{ST} [42]:

$$E_S = E + K + J \tag{11.2}$$

11.3 Basic Considerations in Molecular Design of TADF Molecules

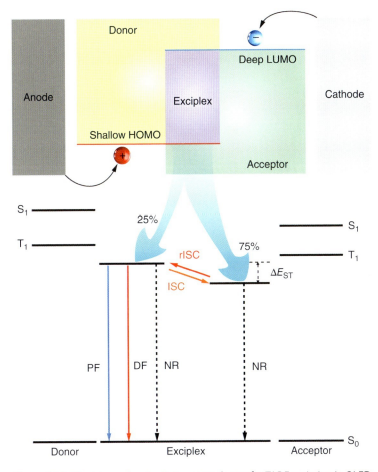

Figure 11.5 The intermolecular D–A system design for TADF emission in OLEDs.

$$E_T = E + K - J \tag{11.3}$$

$$\Delta E_{ST} = E_S - E_T = 2J \tag{11.4}$$

At S_1 and T_1, the unpaired two electrons are mainly distributed on the frontier orbitals of the HOMO and the LUMO, respectively, leading to the same J values regardless of the different spin states. Therefore, the exchange energy (J) of these two electrons at HOMO and LUMO can be determined by Eq. (11.5):

$$J = \iint \varphi_L(1)\varphi_H(2) \left(\frac{e^2}{r_1 - r_2} \right) \varphi_L(2)\varphi_H(1) dr_1 \, dr_2 \tag{11.5}$$

where φ_H and φ_L are the HOMO and LUMO wave functions, respectively, and e represents the electron charge. From Eq. (11.5), J is determined by spatial separation ($r_1 - r_2$) and overlap integral of φ_H and φ_L, i.e. spatial wave function separation of frontier orbitals. Thus, a small ΔE_{ST} can be expected when there is a small overlap and/or a large separation between HOMO and LUMO, and, in contrast, a large overlap and/or a small separation will lead to a large ΔE_{ST} (Eq. (11.6)):

$$\Delta E_{ST} = \iint \varphi_L(1)\varphi_H(1)\left(\frac{2e^2}{r_1 - r_2}\right)\varphi_L(2)\varphi_H(2)dr_1\,dr_2 \quad (11.6)$$

The calculation of ΔE_{ST} in Eq. (11.6) can be simplified by using the orbital overlap integral ($I_{H/L}$) and the mean separation distance ($<r_{H/L}>$) of HOMO and LUMO. $I_{H/L}$ represents the extent of overlap between HOMO (H) and LUMO (L), which can be calculated using the overlap integral function of Multiwfn (Eq. (11.7)) [33, 43]:

$$I_{H/L} = \int |\varphi_H(r)||\varphi_L(r)|dr \quad (11.7)$$

$<r_{H/L}>$ is obtained from the barycenter (r_{tot}) of the absolute value of the corresponding molecular orbitals to summarize the effects of the spatial separation ($r_1 - r_2$) (Eqs. (11.8)–(11.11)). Further adopting the function outputting statistic data of the points in specific spatial and value range, the r_{tot} of the absolute value of the molecular orbital can be computed as in Eq. (11.8):

$$r_{tot} = \sum_k^{all} r_k f(r_k) \bigg/ \sum_k^{all} f(r_k) \quad (11.8)$$

where r denotes coordinate vector, f represents the data value, and k runs over all grid points including positive and negative points, respectively. Thus, the barycenter of HOMO and LUMO (r_H and r_L, respectively) are

$$r_H = \sum_k^{all} r_k f_H(r_k) \bigg/ \sum_k^{all} f_H(r_k) \quad (11.9)$$

$$r_L = \sum_k^{all} r_k f_L(r_k) \bigg/ \sum_k^{all} f_L(r_k) \quad (11.10)$$

Thus, the mean distance between HOMO and LUMO is

$$\langle r_{H/L} \rangle = |r_H - r_L| \quad (11.11)$$

Assuming that the variation of the separation distance between HOMO and LUMO ($r_1 - r_2$) is in a small range around $<r_{H/L}>$, ΔE_{ST} could be related to $I_{H/L}$ and $<r_{H/L}>$ in Eq. (11.12):

$$\Delta E_{ST} = 28.8 \frac{I_{H/L}^2}{\langle r_{H/L} \rangle} \quad (11.12)$$

where ΔE_{ST} is in eV and $<r_{H/L}>$ is in Å.

However, using the simple HOMO→LUMO transition to picture the transition nature of S_1 or T_1 states is not accurate. To give a whole picture of the excited states, natural transition orbital (NTO) analysis was performed to offer a compact orbital representation for the electronic transition density matrix to consider all the possible transitions [44, 45]. Similarly, the overlap extent (I_S and I_T in Eqs. (11.13)–(11.14)) and mean separation distance ($<r_S>$ and $<r_T>$ in Eqs. (11.15)–(11.20)) of the highest occupied natural transition orbitals (HONTOs) (φ_H') and the lowest unoccupied natural transition orbitals (LUNTOs) (φ_L') at both S_1 and T_1 states were calculated to give a full-picture analysis of ΔE_{ST} using Multiwfn [33, 43]:

$$I_S = \int |\varphi_{H'}^S(r)||\varphi_{L'}^S(r)|\,dr \tag{11.13}$$

$$I_T = \int |\varphi_{H'}^T(r)||\varphi_{L'}^T(r)|\,dr \tag{11.14}$$

$$r_{H'}^S = \sum_k^{all} r_k f_{H'}^S(r_k) \Big/ \sum_k^{all} f_{H'}^S(r_k) \tag{11.15}$$

$$r_{L'}^S = \sum_k^{all} r_k f_{L'}^S(r_k) \Big/ \sum_k^{all} f_{L'}^S(r_k) \tag{11.16}$$

$$r_{H'}^T = \sum_k^{all} r_k f_{H'}^T(r_k) \Big/ \sum_k^{all} f_{H'}^T(r_k) \tag{11.17}$$

$$r_{L'}^T = \sum_k^{all} r_k f_{L'}^T(r_k) \Big/ \sum_k^{all} f_{L'}^T(r_k) \tag{11.18}$$

$$\langle r_S \rangle = |r_{H'}^S - r_{L'}^S| \tag{11.19}$$

$$\langle r_T \rangle = |r_{H'}^T - r_{L'}^T| \tag{11.20}$$

With the aid of NTO analysis, the description of ΔE_{ST} (Eq. (11.12)) can be updated to Eq. (11.21) using above-defined parameters of I_S, I_T, $\langle r_S \rangle$, and $\langle r_T \rangle$ to consider the whole picture of the electron interactions of the corresponding excited states:

$$\Delta E_{ST} = C_S \frac{I_S^2}{\langle r_S \rangle} + C_T \frac{I_T^2}{\langle r_T \rangle} \tag{11.21}$$

where C_S and C_T are the combination constants of S_1 and T_1 states, respectively. Consequently, the ultralow ΔE_{ST} can be obtained with both separated HONTO and LUNTO at S_1 and T_1 states (low I_S and I_T, long $\langle r_S \rangle$ and $\langle r_T \rangle$); small ΔE_{ST} can be resulted with separated HONTO and LUNTO at S_1 state but small overlapped at T_1 state (low I_S but high I_T, long $\langle r_S \rangle$ but short $\langle r_T \rangle$); and theoretically, ideal TADF molecules can be designed with small ΔE_{ST} from separated HONTO and LUNTO at T_1 state (low I_T, long $\langle r_T \rangle$) and high luminescent efficiency from overlapped HONTO and LUNTO at S_1 state (high I_S, short $\langle r_S \rangle$) [33].

11.3.3 Modulation of Luminescent Efficiency of TADF Emission

According to Eq. (11.1), the theoretical maximum EQE of the TADF OLED is proportion to the PLQY (η_{PL}) of TADF emitters [46]. In order to improve the device performance of TADF OLEDs to compete with that of PhOLEDs, the PLQY of TADF molecules should be very high. Compared with the relatively sophisticated control of ΔE_{ST} to obtain TADF by separating HOMO and LUMO of a molecule, the concurrently achieved small ΔE_{ST} and high PLQY are still a challenge. Considering important processes for both PF and DF in OLEDs, internal electroluminescence quantum efficiency (η_{int}) of TADF compounds can be expressed as follows (Eq. 11.22)):

$$\eta_{int} = \left[\sum_{k=0}^{\infty} (0.75\Phi_{PF}\Phi_{rISC}(\Phi_{ISC}\Phi_{rISC})^k + 0.25\Phi_{PF}(\Phi_{ISC}\Phi_{rISC})^k) \right]$$

$$= \left[\frac{0.75\Phi_{PF}\Phi_{rISC}}{1-\Phi_{ISC}\Phi_{rISC}} + \frac{0.25\Phi_{PF}}{1-\Phi_{ISC}\Phi_{rISC}} \right]$$

$$= \left[\frac{0.25 + 0.75\Phi_{rISC}}{1-\Phi_{ISC}\Phi_{rISC}} \right] \Phi_{PF} \tag{11.22}$$

Here, Φ_{PF} is typically expressed as

$$\Phi_{PF} = k_f \tau_{PF} \tag{11.23}$$

where k_f is the PF decay rate and τ_{PF} is the PF lifetime that can be obtained by fitting the decay curve of the time-resolved PL spectrum. The relationship between k_f and the absorption coefficient in fluorescent molecules can be expressed as [47, 48]

$$k_f = 2.88 \times 10^{-9} n^2 \langle \overline{v}_f^{-3} \rangle^{-1} \int \varepsilon(v_a) \mathrm{d}\ln v_a \tag{11.24}$$

$$\langle \overline{v}_f^{-3} \rangle^{-1} = \frac{\int f(v_f) \mathrm{d}v_f}{\int f(v_f) v_f^{-3} \mathrm{d}v_f} \tag{11.25}$$

where v_f and $f(v_f)$ represent the fluorescence wave number and the fluorescence spectrum, v_a and $\varepsilon(v_a)$ are the absorption wave number and the molar absorption coefficient at v_a, and n is the refractive index. From Eq. (11.24), k_f is closely related to $\varepsilon(v_a)$, which can be experimentally measured by absorption spectrum. To obtain high $\varepsilon(v_a)$ for high k_f, the molecules can be designed according to the relationship between $\varepsilon(v_a)$ and oscillator strength (F) and transition dipole moment (Q) for absorption as [49]

$$F = 4.32 \times 10^{-9} n^{-1} \int \varepsilon(v_a) \mathrm{d}v_a = \left(\frac{8\pi^2 m_e c \langle v_f \rangle}{3he^2} \right) |Q|^2 \tag{11.26}$$

where m_e, c, h, and e are the electron mass, light speed, Planck's constant, and elementary charge, respectively, and $<v_f>$ represents the average wave number of fluorescence. The values of F and Q can be either experimentally determined or theoretically predicted.

High η_{int} needs high Φ_{PF}, which can be achieved through large k_f, based on Eqs. (11.22)–(11.23); large k_f in turn requires large F and Q values, according to Eqs. (11.24)–(11.26). Therefore, for the development of TADF materials with high η_{int}, those key characteristics involving high k_f, strong oscillator strength for absorption (F), and large transition dipole moment (Q) should be satisfied. From the molecular design point of view, one can extend the molecular orbitals to suppress a decrease in F and k_f meanwhile limit the overlap between HOMO and LUMO for a low ΔE_{ST}. Following this strategy, Adachi and coworkers obtained nearly 100% of η_{int} in TADF molecules by inducing a large oscillator strength (for high PL efficiency) at even a small overlap between the two wave functions of the frontier orbitals (for small ΔE_{ST}) [48].

11.4 Typical Donor–Acceptor Molecular Systems with High TADF Performance

Generally, it is convenient to construct TADF molecules using donors and acceptors in D–A molecular systems in either intramolecular or intermolecular architectures for CT state with separated HOMO (distributed on donor) and LUMO (distributed on acceptor) and small ΔE_{ST}. However, a suitable bridge is needed to connect donor and acceptor units together properly to build an integral molecular system for TADF emissions. Therefore, it is important to smartly choose not only suitable donor and acceptor building blocks but also the connecting bridge to realize simultaneously small ΔE_{ST}, stable T_1, and highly luminescent S_1 with desired emission color, efficiency, and charge injection and transport properties for OLED applications. A large variety of organic donors and acceptors are available for the selection, but the major choice of donor units is concentrated on the N-containing aromatics of carbazole, diphenyl amine, phenoxazine, and their derivatives, due to their strong electron-donating ability, facile synthesis, stable and high triplet states, etc. For intermolecular D–A type TADF systems, the bridge between donors and acceptors are not needed, but interactions between them should be strong enough to form a large number of stable exciplexes with strong CT feature for small ΔE_{ST} and TADF emission. As a result, the suitable donor and acceptor pairs for efficient TADF emission in intermolecular system is relatively difficult to be satisfied, and there remains plenty of challenges and opportunities in developing intermolecular TADF materials for full-color and high efficient TADF OLEDs. Besides these intra- and intermolecular TADF small molecules, TADF polymers with apparent advantages in solution-processable devices are also developed recently. Such a wide range of available rare metal-free D–A type TADF materials will certainly promote the investigations of organic electronics and boost the practical applications in the near future.

11.4.1 Cyano-based TADF Molecules

Due to the strong electron-withdrawing ability of cyano group, cyano-substituted aromatophors exhibit promising electron affinities to act as electron-accepting building blocks in the molecular scaffold for constructing high-performance D–A type TADF molecules [50, 51]. Adachi and coworkers prepared a TADF emitter (**1**, Scheme 11.1) with two cyano electron-accepting moieties and two di-p-tolylamino electron-donating moieties, which were orthogonally connected through a spiro bridge to provide large steric hindrance and bulkiness between the donor and acceptor moieties for the effective separation of HOMO and LUMO with increased thermal and morphological stabilities and reduced aggregation tendency in the solid state [52]. Compound **1** shows yellow emission with low PL quantum efficiency of 27% and

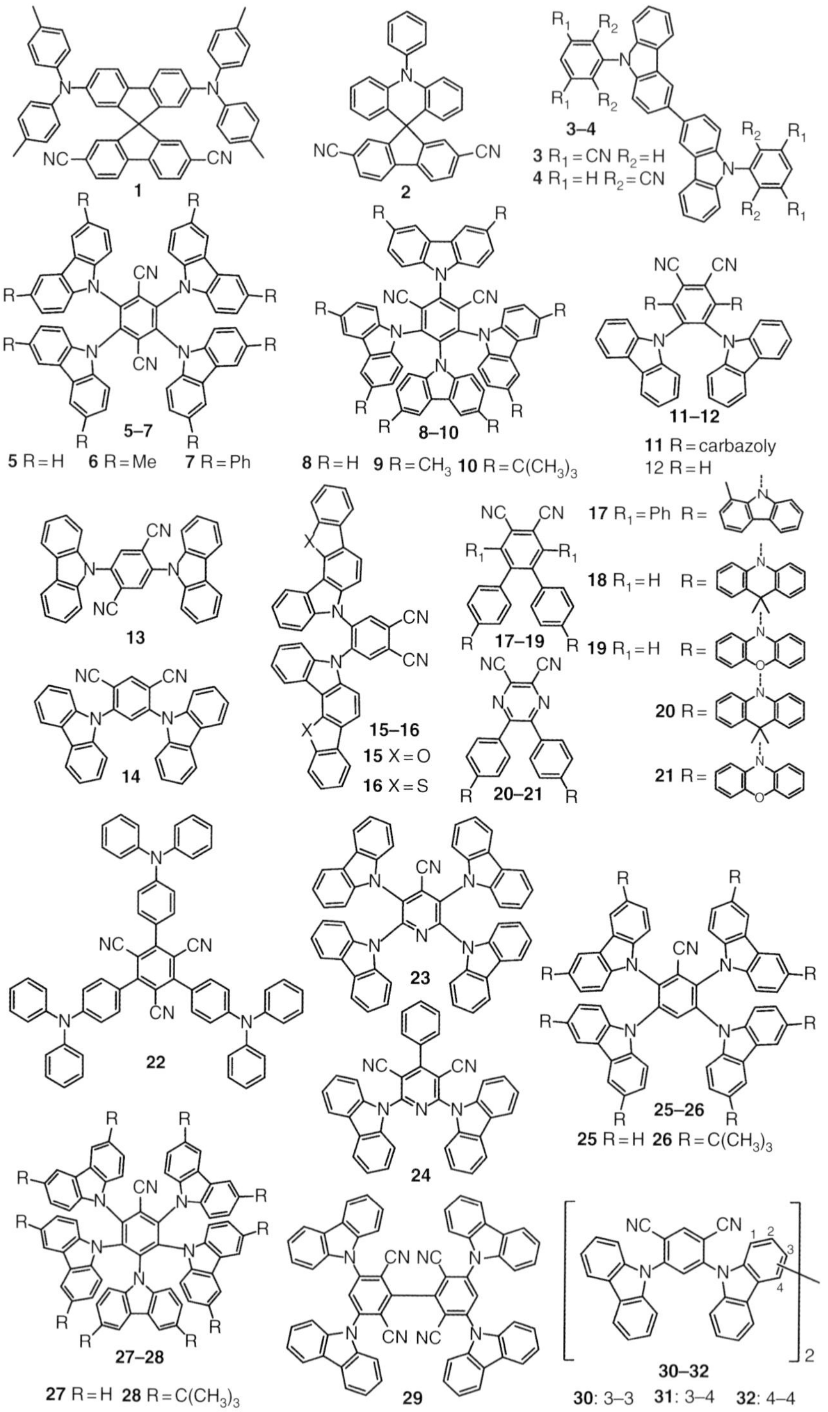

Scheme 11.1 Cyano-based TADF molecules of **1–32**.

a small ΔE_{ST} of 0.057 eV. The TADF OLEDs doped with **1** as the emitter (ITO/α-NPD/6 wt% **1**: mCP/Bphen/Mg Ag/Ag) exhibited an EQE of 4.4%, maximum current efficiency (CE) of 13.5 cd A^{-1}, and power efficiencies (PE) of 13.0 lm W^{-1}, with the aid of N,N′-4,4′-dicarbazole-3,5-benzene (mCP) as the host materials and 4,4-bis[N-(1-naphthyl)-N-phenylamino] biphenyl (a-NPD) and 4,7-diphenyl-1,10-phenanthroline (Bphen) as hole-transport and electron-transport layers (ETL), respectively. To improve the TADF OLED device performance, a modified spiroacridine derivative (**2**, $\Delta E_{ST} = 0.028$ eV, Scheme 11.1) was also synthesized by the same group [53]. The efficient TADF emission with a high PLQY of 67.3% endowed the OLED devices of **2** (ITO/TAPC/mCP/6 wt% **2**: TPSiF/TmPyPB/LiF/Al) with a maximum EQE of 10.1%, which undoubtedly breaks through the theoretical EQE limitation of fluorescent OLEDs (5%).

To understand the influence of the twisting angle between donor and acceptor units on the TADF performance, Adachi and coworkers investigated two bicarbazolyl dicyanobenzene derivatives in A–D–A structure with different steric hindrance (**3–4**, Scheme 11.1) [54]. The two TADF molecules exhibit small ΔE_{ST} of 0.14 and 0.06 eV, bright blue photoluminescence at 470 and 488 nm, high PL quantum yields of 50% and 72%, and efficient devices performance with maximum EQE of 9.2% and 9.6%, respectively. The larger twisting angle induced by ortho-substitution of CN because of steric hindrance in **4** results in more efficient separation of HOMO and LUMO for smaller ΔE_{ST}, shorter fluorescence lifetime, and reduced efficiency roll-off at high luminance.

Further improvement of the TADF device efficiency with EQE of up to 19.3% was realized in 2012 by employing a series of highly efficient TADF materials based on carbazolyl dicyanobenzene, where the carbazoles and dicyanobenzenes are used as electron-donating and electron-accepting groups, respectively (Scheme 11.1) [11]. The emission color was easily tuned from sky blue to orange by changing the number of carbazolyl groups or by introducing different substituents on carbazole (Figure 11.6). The cyano group efficiently suppresses both nonradiative deactivation and geometries change at the excited states of those materials, and therefore their PL efficiencies are largely improved. The steric hindrance induced by the densely substituted donors results in the distorted carbazolyl units from the dicyanobenzene plane, leading to the breakage of π-conjugation with localized HOMO and LUMO on the donor and acceptor, respectively, for small ΔE_{ST} (0.083–0.15 eV). The blue, green, and orange TADF OLEDs reached high maximum EQEs of 8.0 ± 1.0%, 19.3 ± 1.5%, and 11.2 ± 1.0%, respectively, which are comparable with that achieved in PhOLEDs based on noble metal–organic complexes. These TADF OLEDs suffer from high-efficiency roll-offs due to serious exciton quenching effects including singlet–triplet annihilation (STA) [55, 56] and TTA. In order to improve the device stability, a more stable device structure for **8** was fabricated through controlling the recombination zone position by carefully selecting the exciton-blocking layers, tuning the interfaces of the EML, and optimizing the concentration of the emitter [57]. These **8**-doped TADF OLEDs showed excellent device stability with operation lifetime more than 2500 h at an initial luminance of 1000 cd m^{-2} and over 10 000 h at 500 cd m^{-2}, demonstrating that the TADF material is intrinsically

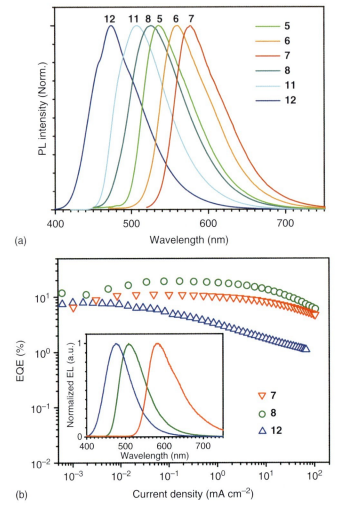

Figure 11.6 Normalized photoluminescence spectra measured in toluene solution (a) and EQE as a function of current density for TADF OLEDs. Inset, normalized electroluminescence spectra of the OLEDs. The compounds of **5–12** are shown in Scheme 11.1.

stable under electrical excitation. Furthermore, the performance of TADF OLEDs can be also influenced by the molecular orientation and the deposition temperature of the host materials [26]; when **8** is dispersed in a randomly orientated host of CBP, the efficiency roll-off of the device was suppressed by 30% at a current density of 100 mA cm^{-2}. Further studies show that the host materials are very important in realizing high-performance TADF OLEDs based on **8**; the EQE can be improved to 21.2% [58], 24.2% [59], 26.7% [60], 28.6% [61], 29.6% [62], and 31.2% [15] with various hosts. Using a carrier-/exciton-confining structure and energy transfer from an exciplex, the **8**-based TADF OLEDs showed a low turn-on voltage of 2.33 V and maximum PE of 107 lm W^{-1} with a record of 79.4 lm W^{-1} at 1000 cd m^{-2}, which is 1.6 times higher than that of state-of-the-art TADF OLEDs [63].

To realize the ultimate potential of organic electronics in large-area, low-cost, and high-efficiency display or lighting products, intense and growing interests were paid to solution-based processes, such as spin-coating or inkjet printing [64–66]. Roll-to-roll solution processing techniques were also applied to the low-cost fabrication of TADF OLEDs based on **8**; high EQEs (19.1%) comparable with those achieved by conventional vacuum deposition were demonstrated [67]. Moreover, TADF molecules (**9–10**, Scheme 11.1) with improved solubility for the fabrication of high-performance solution-processed TADF OLEDs were further designed by modifying **8** with highly soluble methyl or *tert*-butyl groups by Lee's group [18]. The resulting molecules show good solubility in common solvent and green TADF emissions with high PLQY of 67% for **9** and 78% for **10**, and maximum EQEs of 8.2% and 18.3% were achieved in their solution processed TADF OLEDs, respectively.

Efficient TADF can be also observed when reducing the number of carbazole substituents on the dicyanobenzene core. The two carbazole-substituted **12** shows blue TADF for OLEDs with EQE of up to 13.6% [68]. By changing the substitution site on the dicyanobenzene core, analogues of **13** (PLQY = 85%, ΔE_{ST} = 0.08 eV) and **14** (PLQY = 35%, ΔE_{ST} = 0.05 eV) are also efficient in blue TADF emission, exhibiting turn-on voltages of 4.2 V and 3.5 V and EQEs of 15% [24] and 16.4% [69], respectively, in TADF OLEDs. Furthermore, **14** is an excellent host material for yellow PhOLEDs. The PhOLED of iridium(III) bis(4-phenylthieno[3,2-c]pyridinato-N,C2′)acetylacetonate (PO-01) hosted by **14** achieved a turn-on voltage of 2.0 V and a maximum EQE of 24.9%, which are among the best results of yellow PhOLEDs reported so far. Still, **14** is excellent as a host for white OLEDs with a maximum EQE of up to 22.9% and a warm white color coordinate of (0.39, 0.43).

Besides carbazole, other electron donors are also effective in constructing TADF molecules when they were connected on the dicyanobenzenes. Lee et al. successfully developed two TADF molecules (**15–16**, Scheme 11.1) based on benzofurocarbazole and benzothienocarbazole as the donor [70]. These two TADF emitters exhibit high thermal property with T_g of 204 °C and 219 °C, T_d of 397 °C and 434 °C, and small ΔE_{ST} of 0.13 eV and 0.17 eV; blue photoluminescence with high PLQY of 94.6% and 94.0% in solution; and high TADF OLED performance with maximum EQE of 12.1% and 11.8%, respectively. Yasuda and coworkers prepared a series of wedge-shaped TADF molecules using various donor units (1-methylcarbazole, 9,9-dimethylacridan, and phenoxazine). These molecules (**17–21**) show small ΔE_{ST} of 0.04–0.36 eV, efficient TADF emissions that cover the entire visible range from blue to red and high-performance TADF OLEDs with high EQE of up to 18.9% [71].

Besides the dicyanobenzene acceptors, tricyanobenzene (**22**) and mono-/dicyanopyridines (**23**, **24**) are also applicable in designing cyano-based TADF molecules. The tricyanobenzene-based TADF molecule (**22**) has six substitutions with two different types of alternating substituents of three cyano groups and three triphenylamines, resulting in reduced vibrational deactivation processes because of the high steric hindrance [72]. The molecule shows green emission with a peak at 533 nm, small ΔE_{ST} of 0.103 eV, and PLQY of 100% in doped film, demonstrating the almost complete suppression of nonradiative decay for a maximum EQE of 21.4% in its TADF OLEDs. Based on cyanopyridine, Tao and coworkers also synthesized a soluble TADF molecule (**23**, Scheme 11.1) with carbazole as the donor [19]. The compound shows green emission peaked at 536 nm in $CHCl_3$ and yellow emission peaked at 560 nm in doped film with PLQY of 55%, ΔE_{ST} of 0.07 eV, and thermal property of 408 °C for T_d and 159 °C for T_g. Its solution-processed TADF OLED devices with the configuration of ITO/PEDOT: PSS/8 wt% 23: mCP)/TmPyPB/LiF/Al achieved a maximum EQE of 11.3%, CE of 38.9 cd A^{-1}, and PE of 14.8 lm W^{-1}. Zhang and coworkers prepared a blue TADF emitter with dicyanopyridine (pyridine-3,5-dicarbonitrile) as the acceptor and carbazole as the donor (**24**, Scheme 11.1). Separated HOMO and LUMO distribution for extremely low ΔE_{ST} (0.04 eV), high thermal property with T_d of 350 °C, blue emission with high PLQY (~50%) in doped film, and a maximum EQE of 21.2%, CE of 47.7 cd A^{-1}, and PE of 42.8 lm W^{-1} in the TADF OLEDs were achieved [73]. Duan and coworkers developed a new series of blue TADF molecules based on cyanobenzene (**25–28**, Scheme 11.1). The four or five carbazole-substituted cyanobenzenes show small ΔE_{ST} of 0.17–0.30 eV, high PLQY of 0.49–0.86, and high thermal properties with T_d of 436–454 °C and T_g of 316–325 °C. The blue TADF OLEDs achieved high efficiencies with a maximum EQE of up to 21.2% and excellent stability with a record long T50 of 770 h at an initial luminance of 500 cd m^{-2}, showing improved efficiency and stability by protecting luminance core with the steric shielding effects of the *tert*-butyl substituents [74].

For the further development of TADF molecules, Lee and coworkers proposed a new design concept via a dual TADF-emitting core containing four cyano substituents [75]. The molecule of **29** was designed and synthesized through a simple coupling of two TADF cores of **14**. The resulted dual-core TADF emitter shows small ΔE_{ST} of 0.13 eV, blue emission with PLQY of 91%, and a high TADF device performance with a low turn-on voltage of 3.5 V, maximum EQE of 18.9%, and CIE coordinate of (0.22, 0.46). Following this design strategy, they further prepared three other twin TADF emitters (**30–32**, Scheme 11.1), which exhibit blue and green colors with high PLQY of 61–87%, small ΔE_{ST} of 0.11–0.21 eV, and high efficient OLEDs with EQE of up to ~20% [76].

11.4.2 Nitrogen Heterocycle-based TADF Molecules

Due to the electron-deficient nature of aromatic systems containing electronegative nitrogen (N) atoms, N-incorporated arenes are typical electron-accepting moieties with promising electron affinities for the molecular design of D–A type TADF molecules.

Triazine derivatives. The highly electron-deficient triazine containing three N atoms with three easily modifiable sites is highly attractive for the construction of TADF molecules by connecting electron-donating substituents to the electron-accepting triazine core either symmetrically or asymmetrically [77]. For example, Adachi and coworkers prepared an asymmetric triazine-based TADF emitter (**33**, Scheme 11.2) by introducing two bulky indolocarbazoles and one biphenyl onto the triazine core; the strong donor of indolocarbazole with large steric hindrance results in a separated HOMO and LUMO with small ΔE_{ST} (0.11 eV) for blue-green TADF emission [78]. The TADF device with the configuration of ITO/α-NPD/mCP/6 wt% **33**: mCP/BP4mPy/LiF/Al, where BP4mPy (3,3′,5,5′-tetra[(M-pyridyl)-phen-3-yl]biphenyl) acts as ETL, achieved an maximum EQE of 5.3%. This TADF molecule of **33** is also a high-performance host material for PhOLED due to its balanced charge transport and good injection properties; the green PhOLEDs (using phosphorescent guest of tris(2-phenylpyridinato) iridium (III), Ir(ppy)$_3$) hosted by **33** exhibited a very low onset voltage of 2.19 V, maximum CE of 68 cd A^{-1}, and PE of 60 lm W^{-1} [21]. TADF-sensitized fluorescent OLEDs using **33** as a sensitizer for a yellow fluorescent dopant was also reported to show a maximum EQE of 4.5% and a PE of 12.3 lm W^{-1} at 10 cd m^{-2}. Using another TADF sensitizer of **34** with one bulky indolocarbazole and two biphenyl substituents on triazine, the EQE is further improved to 11.7%, which apparently exceeds the theoretical EQE upper limit (5%) of fluorescent OLEDs [79]. Similar to **33**, **34** is also an excellent host for green PhOLED [80], showing a maximum EQE of 23.9% and a PE of 77.0 lm W^{-1} of the devices. After a minor modification of the number and structure of donors in **33** and **34**, new TADF molecules of **35**, **36**, and **37** can be conveniently obtained due to the existence of strong ICT state when directly connecting carbazoles to the triazine framework through N atoms. Compared with **33**, the TADF molecule of **35** has a smaller ΔE_{ST} of 0.02 eV, higher PLQY (45 ± 1%), and much better device performance of the TADF OLEDs with significantly improved EQE of 14% [81]. When the donors of indolocarbazoles were replaced with bicarbazoles, the ΔE_{ST} of the resulted molecules of **36** [82] and **37** [83] are 0.06 and 0.09 eV, respectively, leading to EQEs of up to 11% and 6% in their corresponding TADF OLEDs. Additionally, the yellowish-green PhOLEDs employing **37** as the host materials reached a maximum EQE of 20.1% [84].

Besides the direct connecting of donor substituents and acceptor core for triazine-based TADF molecules, aromatic bridges are also useful in controlling the interactions between the donor and acceptor for TADF emission. When a phenyl bridge is adopted as in **38–39** (Scheme 11.2) [85], relatively larger ΔE_{ST} of ~0.25 eV due to slightly overlapped HOMO and LUMO on the bridge were observed. These blue TADF materials show high thermal stability with T_d of ~500 °C and T_g of ~220 °C, and their TADF OLEDs achieved high maximum EQE of 17.5% for **38** and 18.9% for **39**, respectively. More importantly, the **39**-based TADF OLED showed a lifetime of 52 h of up to 80% of initial luminance at 500 cd m^{-2}, which is almost three times as long as that of blue phosphorescent OLED under the same device structure using tris[1-(2,4-diisopropyldibenzo[b,d]furan-3-yl)-2-phenylimidazole]iridium (Ir(dbi)$_3$) as the emitter. A subsequent study by Lee and coworkers reported the synthesis of triazine-based TADF emitters (**40–42**, Scheme 11.2) following a similar design rule using dense substitution of the phenyl bridge [16]. These materials exhibit ΔE_{ST} in a range of 0.07–0.23 eV but have high PLQY of up to 100% and tunable emission color from blue to green. High device performance with maximum efficiency of 25% was achieved in those TADF emitters doped OLEDs. Dendritic diphenylamine and carbazole with large molecular size and electron-donating ability were also connected to the phenyl bridge to prepare new TADF molecules with high PLQY by controlling the spatial overlap between the frontier orbitals to suppress nonradiative decay. The resulting blue (**43**) [48]

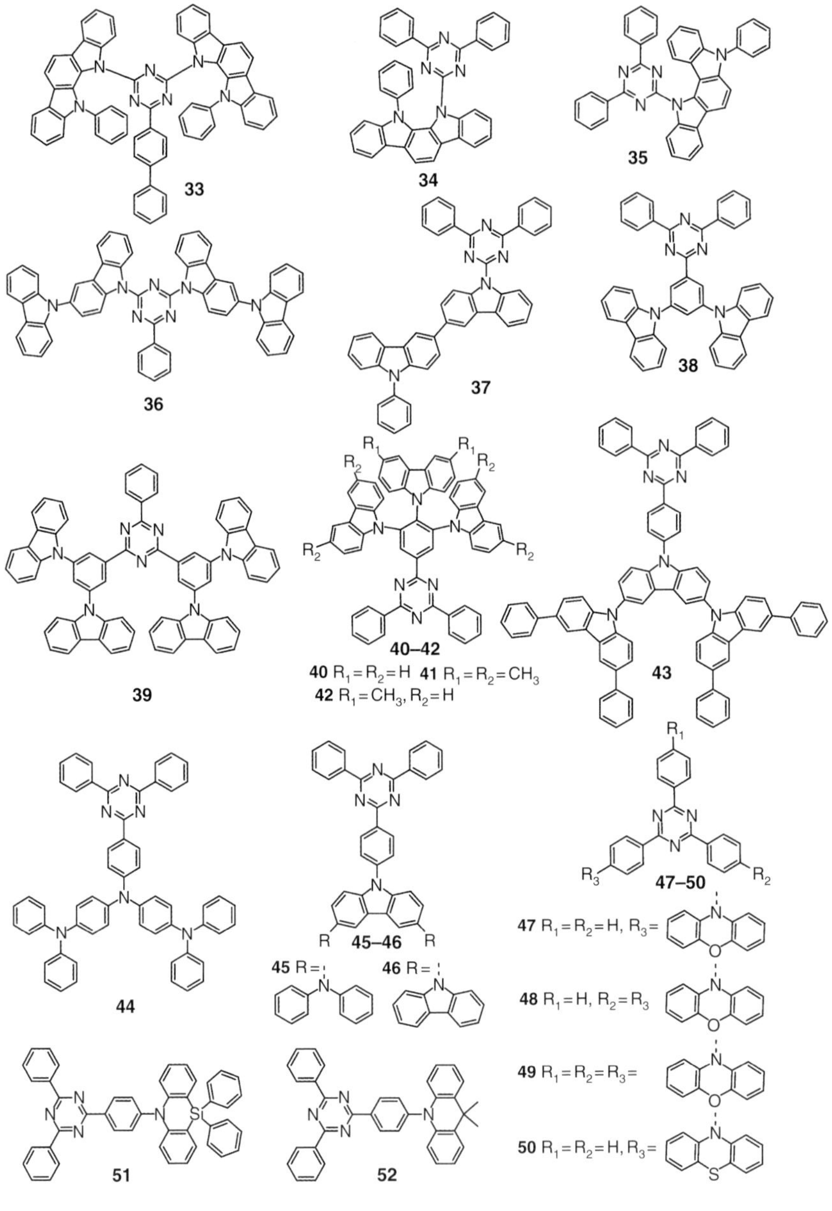

Scheme 11.2 Triazine-based TADF molecules of **33–52**.

and green (**44**) [86] TADF molecules show small ΔE_{ST} (<0.1 eV) and high PLQY (80% for blue and 100% for green) for high-performance TADF OLEDs with maximum EQE of 20.6% and 13.8% for blue and green, respectively. Especially, using a carbazole and diphenylamino hybrid dendritic donor, the obtained green TADF emitter (**45**) has a small ΔE_{ST} of 0.026 eV and 100% PLQY and 100% reverse ISC efficiency; the TADF OLED with **45** as emitter showed a maximum EQE of 29.6% [87]. A white-light TADF OLED based on red (**7**), green (**8**), and blue (**46**) triazine-based TADF materials as the emissive dopants was also achieved, exhibiting a high EQE over 17% with CIE coordinate of (0.30, 0.38) [13].

Benefited from the phenyl bridge, which separates the donor substituents and triazine acceptor core and modulates their interactions effectively, various donors are applicable to prepare high-performance triazine-based TADF molecules with partially overlapped frontier orbitals on the bridge for small ΔE_{ST} and high PLQY simultaneously. Tanaka et al. introduced the electron donor of phenoxazine into the triphenyltriazine to construct TADF material (**47**, Scheme 11.2) [46]. Due to the high steric repulsion with large dihedral angle (74.9°) between the phenoxazine moiety and the phenyl ring, the twisted D–A structure facilitates a spatially separation of HOMO and LUMO, leading to a small ΔE_{ST} (0.07 eV) for efficient TADF emission with high PLQY of 65.7% in doped film. The TADF device with the structure of ITO/α-NPD/6 wt% **47**: CBP/TPBi/LiF/Al achieved an outstanding EL performance with a maximum EQE of 12.5% and high luminance of 10000 cd m^{-2}. Further investigations on orientation order of **47** regulated by tuning the temperature during fabrication of the thin films revealed that the film deposited at 200 K showed a horizontal orientation and the EQE of the device can be enhanced by 24% compared with corresponding OLEDs with the vertically orientated film deposited at 300 K [27]. By increasing the numbers of the phenoxazine unit connected to the triphenyltriazine core, emission colors of the resulted TADF emitters (**48**, **49**, Scheme 11.2) can be tuned from 545 to 568 nm, and the HOMO and LUMO are much more separated in **48** and **49** than that in **47** due to the increased steric hindrance, leading to a smaller ΔE_{ST} of 0.054 eV for **48** and 0.065 eV for **49** [88]. High PLQYs of 64% and 58% were observed in **48** and **49** doped mCBP (3,3′-bis(N-carbazolyl)-1,1′-biphenyl) films. The TADF OLEDs based on the two emitters with the configuration of ITO/α-NPD/6 wt% **48** or **49**: mCBP/TPBi/LiF/Al exhibited maximum EQE of 9.1 ± 0.5% and 13.3 ± 0.5%, respectively. In addition, the TADF-sensitized fluorescent OLEDs with **47** and **49** as sensitizers were also studied by Adachi and coworkers. The resulting OLEDs showed high EQEs of 18% and 17.5% for yellow and red fluorescent dopants and a relatively long operational lifetime of up to 194 h [28]. They continued to modify the structure of **47** using phenothiazine to replace phenoxazine [89]. The resulted molecule (**50**, Scheme 11.2) demonstrates a dual ICT fluorescence with TADF characteristics peaked at 409 and 562 nm, which is composed of the direct $^1CT_1^A \rightarrow S_0^A$ transition with a large ΔE_{ST} between the $^1CT_1^A$ and $^3CT_1^A$ states and an indirect $^1CT_1^E \rightarrow S_0^E$ transition allowed through the successive $^3CT_1^E \rightarrow {}^1CT_1^E$ upconversion followed via rISC with a small ΔE_{ST} between the $^1CT_1^E$ and $^3CT_1^E$ states, due to the presence of two different quasi-axial (CT_1^A) and quasi-equatorial (CT_1^E) conformers at excited states, where the superscript E and A represent equatorial and axial, respectively. A maximum EQE of 10.8 ± 0.5% was observed in the **50**-doped TADF OLEDs with dual ICT fluorescence.

Kim and coworkers firstly introduced a highly rigid, bulky, and electron-donating azasiline to the phenyl bridge of triazine to prepare TADF molecules [90]. The resulting **51** shows deep blue TADF emission with high PLQY of 74 ± 2% in doped film and small ΔE_{ST} of 0.14 eV. The TADF device achieved a maximum EQE of 22.3% with a CIE coordinate of (0.149, 0.197). Tsai et al. reported a TADF

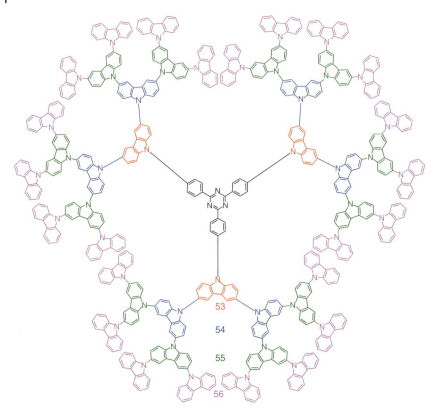

Scheme 11.3 Triazine-based TADF dendrimers of **53–56**.

emitter (**52**, Scheme 11.2) by connecting 9,9-dimethyl-9,10-dihydroacridine as the electron-donating unit to the phenyl bridge of 1,3,5-triazine [91]. The resulted molecule exhibits a small ΔE_{ST} of ~0.05 eV, high glass transition of 91 °C, and high PLQY of 90% in a doped film and 83% in a nondoped (neat) films. Both the nondoped and doped TADF OLEDs achieved outstanding device performance with maximum EQE of 20% and 26.5%, respectively.

In light of the great potential of the nondoped solution-processable TADF molecules for large-scale and easy fabrication, Yamamoto and coworkers prepared a series of TADF dendrimers based on dendritic carbazole (D) and 2,4,6-triphenyl-1,3,5-triazine (A) [92]. These dendritic TADF materials (**53–56**, Scheme 11.3) show similar emission profiles with high PLQY of ~100% in the solution and small ΔE_{ST} of 0.03–0.06 eV. The solution-processed undoped TADF OLEDs in a device structure of ITO/PEDOT: PSS/EML/TPBI/Ca/Al achieved EQEs of 2.4% (**54**), 3.4% (**55**), and 1.5% (**56**).

Pyrimidine derivatives. Pyrimidine is a highly π-deficient acceptor that can be easily modified on 2,4,6-positions. Kido and coworkers synthesized a series of TADF molecules (**57–59**, Scheme 11.4) using pyrimidine as the acceptor and phenylacridine as the donor [93]. These molecules doped in bis-(2(diphenylphosphino)phenyl)ether oxide) (DPEPO) films exhibit light blue

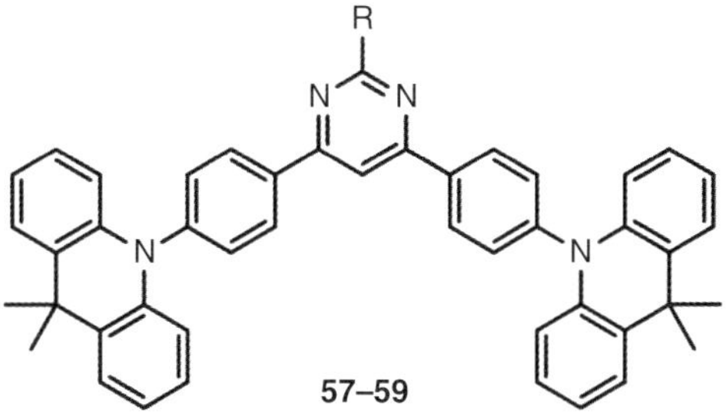

Scheme 11.4 Pyrimidine-based TADF molecules of **57–59**.

emission with high PLQY of ~80%, small ΔE_{ST} of <0.20 eV, and high device performance with low turn-on voltage of 2.8 V, high EQE of 24.5%, and high PE of 61.6 lm W^{-1}.

Triazole, oxadiazole, and thiadiazole derivatives. The N-containing five-membered aromatic rings of oxadiazole, triazole, and thiadiazole are widely used in building the electron-transporting materials for OLEDs due to their good electron transportation and injection abilities [94, 95]. Adachi and coworkers developed a series of TADF emitters (**60–63**, Scheme 11.5) based on oxadiazole and triazole derivatives in D–A or D–A–D architecture to realize separated HOMO and LUMO distribution for small ΔE_{ST} [96]. The films with **61** and **63** as the dopants hosted by DPEPO show high PLQY of 52 ± 3% and 87 ± 3%, respectively. Employing these doped films as the EMLs, maximum EQE of 6.4% (sky blue) and 14.9% (green) with corresponding CIE color coordinates of (0.16, 0.15) and (0.25, 0.45) at 10 mA cm^{-2} were observed in the TADF OLEDs with the device configuration of ITO/α-NPD/mCP/6 wt% **61** or **63**: DPEPO/DPEPO/TPBi/LiF/Al. Adachi and coworkers also used 1,3,4-thiadiazole to prepare TADF molecules (**64**, Scheme 11.5) [97]. They found that the vacant 3d orbitals of divalent sulfur in the thiadiazole heteroring can cause electron-pair-accepting conjugative effect for narrowed bandgap, enhanced $S_1 \to T_1$ ISC, increased contribution of the DF component, and reduced ΔE_{ST} for enhancing reverse ISC. The maximum EQE of its TADF OLEDs reached 10.0%.

Heptazine derivatives. Heptazine, which possesses a rigid and planar heterocyclic system of six C=N bonds surrounding a central sp^2-hybridized N-atom, was also applied as an acceptor building block in constructing the TADF materials in intramolecular D–A molecular structure [98]. The orange-red TADF emitter (**65**, Figure 11.7) containing a heptazine core and three electron-donating *tert*-butyl substituents of triphenylamine shows high decomposition temperature at 519.9 °C, relatively small ΔE_{ST} of 0.17 eV due to slight overlap of HOMO and LUMO, and high PLQY of 91 ± 0.9% in the doped film using 26mCPy as the host matrix. Although the component of DF of the **65** doped film is only 6% in PL, the OLED with the structure of ITO/α-NPD/6 wt% **65**: 26mCPy/Bphen/MgAg/Ag exhibited a high device performance with a turn-on voltage of 4.4 V, a maximum luminance of 17000 cd m^{-2}, and a maximum EQE of 17.5%, CE of 25.9 ± 1.6 cd A^{-1}, and PE of 22.1 lm W^{-1}, suggesting that rISC process is highly efficient under

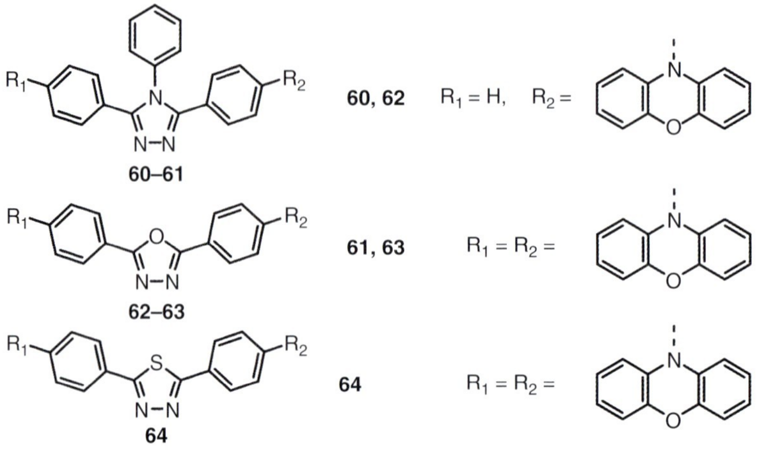

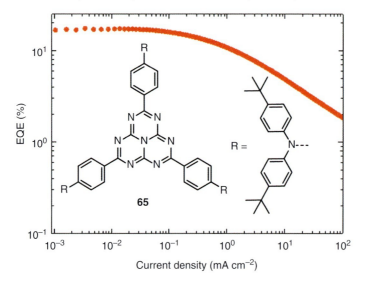

Scheme 11.5 N-containing five-membered heteroring-based TADF molecules of **60–64**, where triazole, oxadiazole, and thiadiazole are acceptors and phenoxazine is donor.

Figure 11.7 EQE-current density curve of **65**. (inset, the molecular structure of **65**).

electrical excitation for high OLED efficiencies, even though **65** demonstrates quite weak TADF in the PL process.

1,4-Diazatriphenylene derivatives. The 1,4-diazatriphenylenes (ATP), which has a diaza-heterocyclic and three fused benzene rings, are well known as a strong electron-withdrawing unit for wide applications in OLEDs due to their large π-conjugated framework and high triplet energy. A series of interesting TADF emitters (**66–70**, Scheme 11.6) based on ATP were developed by combining phenoxazine, 9,9-dimethylacridane, and 3-(diphenylamino)carbazole as the donor units, respectively, in a D–A–D molecular architecture [99]. These molecules show typical TADF characteristics with small ΔE_{ST} (0.04–0.26 eV), high PLQY of up to 81%, and high EQE of the doped TADF OLED of up

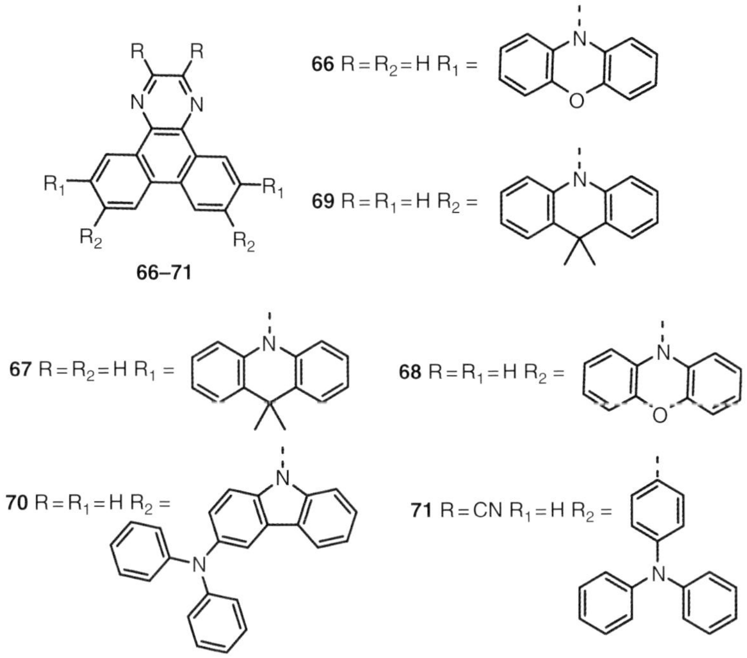

Scheme 11.6 1,4-Diazatriphenylenes-based TADF molecules of **66–71**.

to 12%. A near-infrared (NIR)TADF molecule (**71**, Scheme 11.6) based on cyano-substitution diazatriphenylenes and triphenylamine in V-shaped D–A–D configuration was reported to have a relatively small ΔE_{ST} of 0.13 eV, a large k_F value of 9.0×10^7 s^{-1}, and strong aggregation-induced emission (AIE) effect with high PLQY of 14% in solid film. The nondoped and doped TADF devices using **71** as the emitter achieved the maximum EQE of 2.1% and 9.8% with electroluminescence peaks at 668 nm and 708 nm, respectively. Those data are among the best results of the most efficient deep red/NIR OLEDs, including the PhOLEDs, reported so far [100].

Benzothiazole and benzoxazole derivatives. Benzothiazole and benzoxazole derivatives are efficient electron acceptors for charge transport materials in organic electronics due to their strong electron-deficient nature. Therefore, they are promising building blocks for constructing TADF materials [101]. Adachi and coworkers reported a series of TADF emitters (**72–76**, Scheme 11.7) containing benzothiazole and benzoxazole as acceptor in D–A or D–A–D architecture to realize separated HOMO and LUMO distribution for small ΔE_{ST} (0.033–0.071 eV). These D–A–D type TADF materials exhibited higher PLQY of 98% (**75**), 81% (**76**), and 80% (**74**) compared with the D–A type molecules of **72** (72%) and **73** (75%). Maximum EQE of 16.6%, 14.4%, and 14.0% for the D–A–D type emitters and 9.1% and 12.1% for the D–A type emitters were observed in the TADF OLEDs employing the doped films as the EMLs. Due to the introduction of an exciton-blocking layer between the EML and ETL, the maximum EQE of **75**-based OLED can be increased to 17.6%.

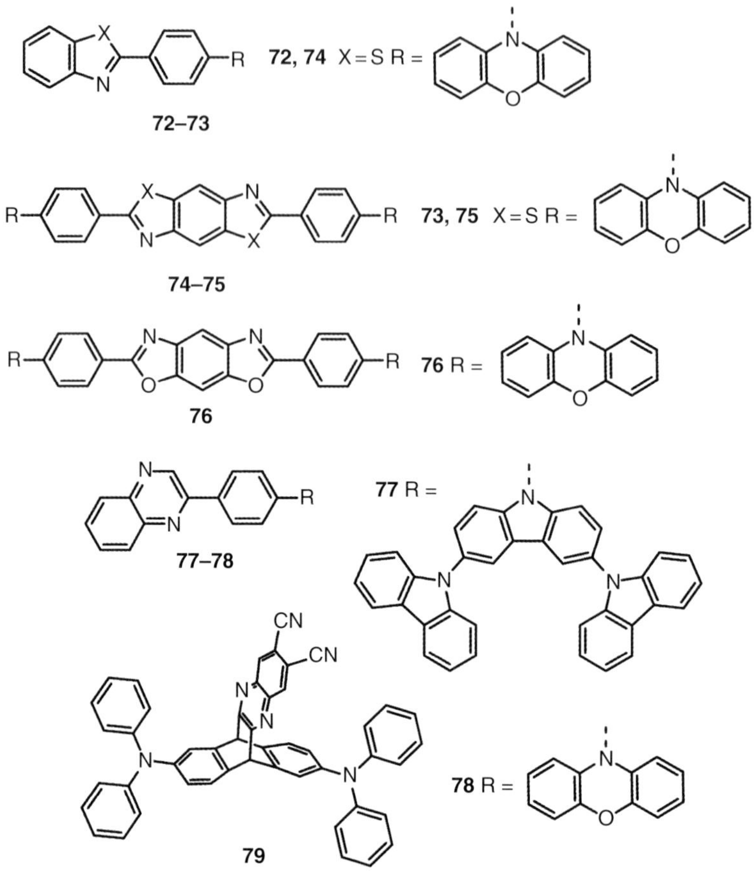

Scheme 11.7 Benzothiazole-, benzoxazole-, and quinoxaline-based TADF derivatives of 72–79.

Quinoxaline derivatives. In light of the high electron-withdrawing ability and good thermal stability of quinoxaline, Adachi and coworkers developed two TADF molecules (**77–78**, Scheme 11.7) using quinoxaline as the acceptor unit and different arylamines as the donor [102]. They found that the TADF properties can be controlled through tuning the twisting angle between the electron-donating and accepting units. The **77** with small twisting angle of 50° exhibits higher PLQY of 74% and electroluminescence EQE of 12.8% compared with PLQY of 66% and EQE of 10.4% of **78** with large twisting angle of 77.9°. Swager and coworkers developed a TADF molecule (**79**, Scheme 11.7) using cyano-substitution quinoxaline as the acceptor, triptycene as the scaffold, and diphenylamine as the donor [103]. The prepared molecule shows small ΔE_{ST} of 0.11 eV, high thermal stability ($T_d > 380\,°C$), and high PLQY of 44% in oxygen-free cyclohexane solutions. The OLEDs using **79** as TADF emitters achieved an emission peak at 573 nm and a maximum EQE of 9.4%.

Diazafluorene derivatives. With orthogonally connected π-conjugated system and strong electron-accepting ability, diazafluorene would be a promising

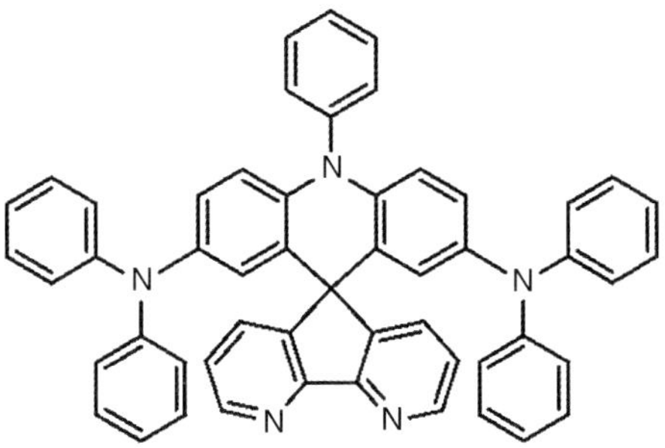

Scheme 11.8 Diazafluorene-based TADF molecule of **80**.

building block for TADF molecules. Adachi and coworkers reported a TADF molecule (**80**, Scheme 11.8) using diazafluorene as an electron-accepting unit and bis(diphenylamino)acridane as an electron-donating unit [104]. The prepared compound **80** exhibits a small ΔE_{ST} of 0.021 eV, blue-greenish TADF emission with a PLQY of 70% in a mCP doped film at room temperature, and high EQE of the doped TADF OLEDs of up to 9.6%.

11.4.3 Diphenyl Sulfoxide-based TADF Molecules

Diphenyl sulfoxide, containing a powerful electron-withdrawing group with a twisted angle in the center, is one of the most famous and typical electron-acceptor components for TADF molecule constructions [105]. A series of diphenyl sulfoxide-based TADF molecules (**81–83**, Scheme 11.9) using diphenylsulfonyl and arylamines as acceptor and donor, respectively, have been developed [39]. These molecules show broad and structureless emission bands from the ICT singlet state (^{1}CT) and well resolved and characteristic phosphorescence from $^3\pi-\pi^*$ state localized at the donor. Therefore, the mechanism of the observed TADF emission involves a reverse IC from T$_1$ ($^3\pi-\pi^*$) to ^{3}CT and a successive rISC from ^{3}CT to ^{1}CT (S$_1$). In order to enhance the energy interchange between T$_1$ and S$_1$, the $^3\pi-\pi^*$ state should be close to or even higher than the ^{3}CT state. Due to the multiple energy transfer from T$_1$ to S$_1$, those TADF materials of **81–83** show relatively large ΔE_{ST} of 0.54, 0.45, and 0.32 eV, respectively. By controlling the redox potential and π-conjugation length of donor and acceptor moieties, bright pure blue TADF emission peaked at 421, 430, and 423 nm with high PLQY of 60, 66, and 80% in the doped film hosted by DPEPO, and efficient TADF devices with maximum EQEs of 2.9, 5.6, and 9.9% are achieved in **81–83**, respectively. The best device performance was observed in the **83**-based TADF OLEDs, showing a standard blue emission with EQE of 9.9% and CIE of (0.15, 0.07) due to the lowest ΔE_{ST} induced by slightly raised ^{1}CT but greatly raised $^3\pi-\pi^*$ after the replacement of diphenylamine with carbazole. However, serious EQE roll-off was generally observed in these devices, because of the longer triplet exciton lifetime induced by the relatively larger ΔE_{ST}. In order to reduce the ΔE_{ST} for low-efficiency roll-off, the more electron-rich methoxy groups were introduced to replace the *tert*-butyl

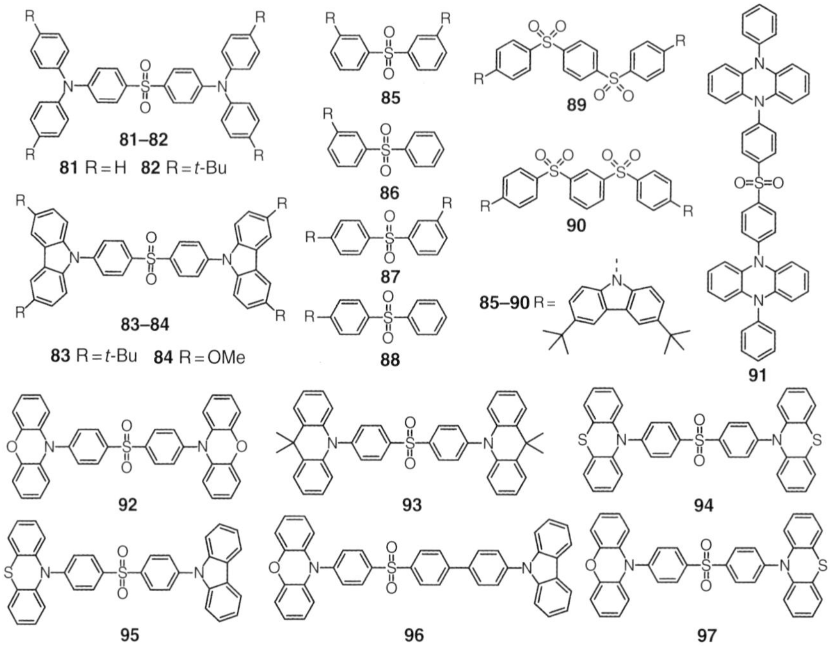

Scheme 11.9 Diphenylsulfoxide-based TADF molecules of **81–97**.

substituents on the carbazole units [106]. The resulting TADF emitter of **84** (Scheme 11.9) shows deep blue emission with small $\Delta E_{ST} = 0.21$ eV, a relatively short triplet exciton lifetime, a high PLQY of 80% in the doped DPEPO film, and a maximum EQE of 14.5% with significantly reduced efficiency roll-off in its TADF OLEDs. Even at a bright emission of 100 cd m^{-2}, the EQE of the **84**-doped OLEDs maintains above 9%. The effects of the number and the linking position of the electron-donating units on the ΔE_{ST} of the diphenylsulfoxide-based TADF molecules (**83**, **85–88**, Scheme 11.9) were also investigated [107]; the results show that ΔE_{ST} can be tuned from 0.39 to 0.22 eV. To further reduce the ΔE_{ST} of the diphenylsulfoxide-based TADF emitters, Kido and coworkers enhanced the electron-accepting diphenylsulfoxide by inserting one more phenyl sulfone into the diphenylsulfoxide core; two TADF molecules based on bis(phenylsulfonyl)benzene (**89–90,** Scheme 11.9) [108] show smaller ΔE_{ST} of 0.05 and 0.24 eV, blue emission peak at 446 and 422 nm, high PLQY of 66.6% and 71.0% in doped film, and maximum EQE of 5.5% and 10% in TADF OLEDs were observed, respectively.

Besides diphenylamines and carbazoles, other electron-donating structures have been also employed as donors to interact with diphenylsulfone acceptor to build efficient TADF molecules. Zhang et al. reported a series of TADF molecules using different donors of 5-phenyl-5,10-dihydrophenazine (**91**), phenoxazine (**92**), and 9,9-dimethyl-9,10-dihydroacridine (**93**) [12]. These molecules with ΔE_{ST} around 0.08 eV and high k_{ISC} at the order of $\sim 10^7$ s^{-1} show PL peaks at 577, 507, and 460 nm with high PLQY of 3, 80, and 80%, respectively. Blue TADF OLED employing **93** as the dopant achieved a maximum EQE of

19.5%, low turn-on voltage of 3.7 V, and blue emission with CIE coordinate of (0.16, 0.20). Importantly, the EQEs of these TADF OLEDs still maintained 16.0% at a brightness of 1000 cd m^{-2}, showing very low-efficiency roll-off and great potential in replacing the noble metal-based phosphorescent complexes. Furthermore, a nondoped blue TADF OLED based on **93** without the help of host molecules still exhibits a high device performance with maximum EQE of 19.5% [32], mainly attributing to the special molecular structure composed of large twist structure of diphenyl sulfoxide and steric hindrance of methyl groups on the 9,9-dimethyl-9,10-dihydroacridine that greatly suppress the aggregation-caused quenching effect in film state. Furthermore, a hybrid white OLED using the blue emission TADF molecules of **93** as the triplet harvester and green and red fluorescent emitters as singlet harvesters realized a high EQE over 12% and CIE coordinates of (0.25, 0.31) [109].

Chi and coworkers designed two AIE TADF molecules (**94–95**, Scheme 11.9) with strong AIE property through regulation of molecular interactions by introducing phenothiazine to diphenylsulfoxide moieties [110]. The mono- (**95**) and bi-phenothiazine (**94**) substituted TADF molecules show green emission peak with high PLQY of 93.3% and 52.8% and small ΔE_{ST} of 0.2 and 0.03 eV, respectively. Interestingly, the compound **95** also exhibits a significant mechanoluminescent induced by asymmetric molecular structure and formation of a noncentrosymmetric arrangement in crystals. When applying an asymmetric molecular structure design of the diphenylsulfone with two different donor substituents (**96**), dual-emissive TADF compounds that fully inherit the photophysical properties of the parent molecules can be produced [111], and the asymmetric TADF emitters (**97**) are more efficient than the symmetric ones with enhanced EQE of the host-free OLEDs of up to 17% [112].

11.4.4 X-bridged Diphenyl Sulfoxide-based TADF Molecules

For the further development of TADF molecules, X-bridges are introduced to connect the diphenyl of the high-performance TADF building block of diphenyl sulfoxide to construct a more rigid molecular structure with improved optical and electronic properties. With C-bridge and acridine as donors, Lee and coworkers developed a deep blue TADF emitter (**98**, Scheme 11.10), which shows a zero ΔE_{ST}, high PLQY of 100% in doped film, and high device performance with a maximum EQE of 19.8% and deep blue CIE coordinate of (0.15, 0.13) [113]. Using C-bridge of diphenyl sulfoxide and phenoxazine as donor, **99** (Scheme 11.10) was prepared and found to have an environment-dependent TADF with a tunable ΔE_{ST} and a maximum EQE of 13.5% in OLEDs [114]. An additional sulfoxide bridge was also introduced into the diphenyl sulfoxide, providing a stronger acceptor of doubly O=S=O bridged diphenyl for the design of D–A type TADF molecules. Combining with donors of 9,9-dimethyl-9,10-dihydroacridin and phenoxazine, Su and coworkers prepared two highly efficiently TADF molecules (**100–101**, Scheme 11.10), which exhibit small ΔE_{ST} of 0.058 and 0.048 eV and high PLQY of 71% and 62%, respectively. Their TADF OLEDs showed low turn-on voltage of 3.5 and 3.7 V, maximum EQE of 19.2% and 16.7% for evaporation device and turn-on voltage of 3.7

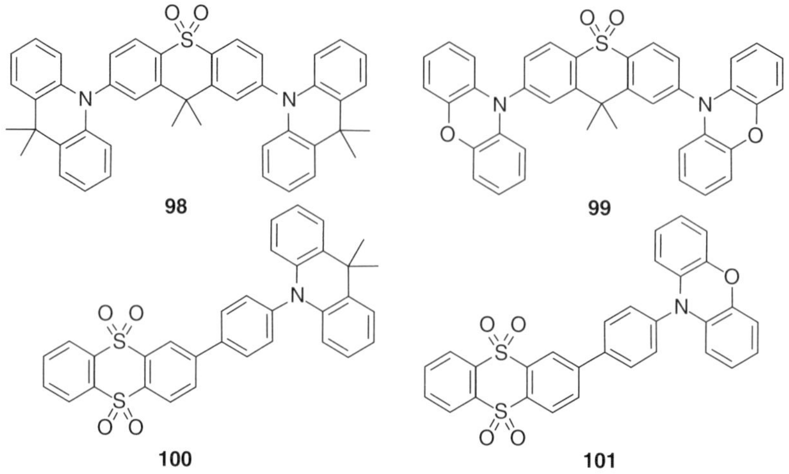

Scheme 11.10 X-bridged diphenyl sulfoxide-based TADF molecules of **98–101**.

and 4.1 V, and maximum EQE of 17.5% and 15.2% for solution-process device, respectively [115]. Besides X-bridge, the direct linking of the diphenyl to produce the sulfurafluorene dioxide was also reported to be applicable to produce TADF molecules [116, 117].

11.4.5 Diphenyl Ketone-based TADF Molecules

Containing a highly electron-withdrawing C=O group with a twist angle in the center, diphenyl ketone is known as a special purely organic phosphor with efficient ISC ($k_{ISC} = 10^{11}$ s^{-1}) that is highly attractive for building TADF materials [118–120]. Recently, Adachi and coworkers developed a series of butterfly-shaped D–A–D type TADF emitters (**102–106**, Scheme 11.11) using diphenyl ketone as the acceptor moiety and carbazoles or phenoxazines as the donor moiety [121]. These molecules show small ΔE_{ST} (0.03–0.21 eV) and highly efficient TADF emissions in full color with emission peaks at 444, 475, 538, 541, and 555 nm and PLQY of 55, 73, 70, 71, and 36%, respectively. High-performance monochromatic TADF OLEDs of **102–106** with maximum EQEs of 8.1, 14.3 10.7, 6.9, and 4.2% (Figure 11.8); turn-on voltages at 4.3, 4.4, 3.2, 3.6, and 2.8 V; and maximum luminance of 510, 3900, 86 100, 57 120, and 50 820 cd m^{-2} were observed, respectively. Notably, a white TADF OLED using blue (**103**) and red (**106**) dopants was also successfully realized, showing a maximum EQE of 6.7% with CIE coordinate of (0.32, 0.39). To obtain a white-light-emitting TADF molecule, Chi and coworkers designed an asymmetric TADF molecule (**107**, Scheme 11.11) based on **102** by asymmetrically connecting carbazolyl and phenothiazinyl substituents to the diphenyl ketone core [122]. A white TADF emission from two complementary colors emission induced by carbazole (blue) and phenothiazine (yellow) with ΔE_{ST} of 0.04 eV and obvious AIE phenomenon was observed. Researches on subsequent structural modification of **103** by Lee to cross the donor and acceptor in an X-shaped molecular structure resulted in two new TADF molecules (**108–109**, Scheme 11.11), showing separated

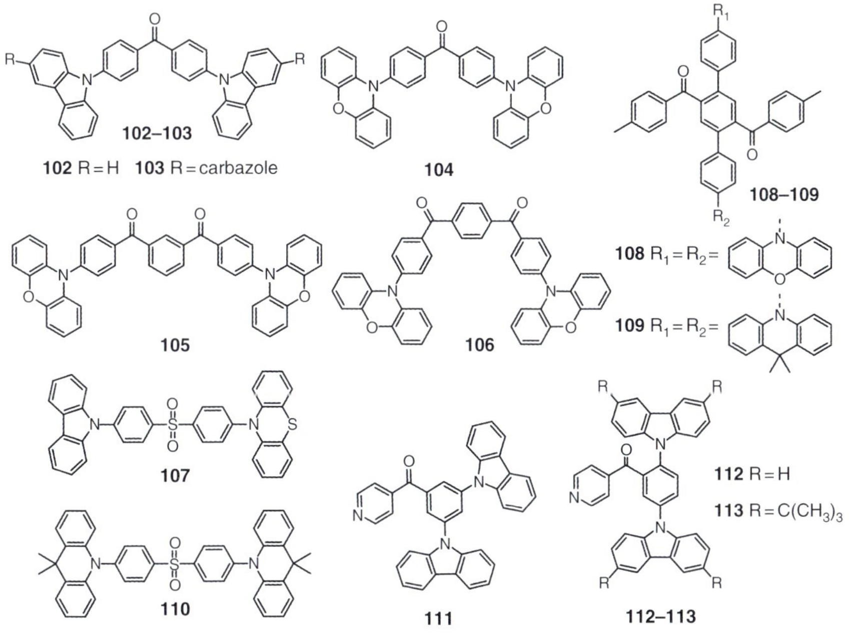

11.4 Typical Donor–Acceptor Molecular Systems with High TADF Performance

Scheme 11.11 Diphenyl ketone- and phenyl(pyridin-4-yl)methanone-based TADF molecules of **102–113**.

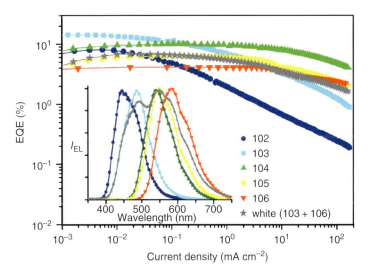

Figure 11.8 EQE-current density curves and electroluminescence spectra (inset) of TADF OLEDs based on **102–106**.

HOMO and LUMO with small ΔE_{ST} of 0.02 and 0.05 eV, green and sky blue emission band at 510 and 496 nm in the toluene, high PLQY in doped films of 57% and 46% with high k_f of ~7.0×10^6 s^{-1}, and EQEs of 11.3% and 10.0% in the green TADF OLEDs, respectively [123]. Zhang et al. introduced the large and twisty 9,9-dimethyl-9,10-dihydroacridine to diphenyl ketone [32]. The resulting molecule (**110**, Scheme 11.11) shows a green TADF emission with high PLQY of 90%, small ΔE_{ST} of ~0.07 eV, and a maximum EQE of 18.9% with small efficiency roll-off in an undoped TADF OLED.

Besides diphenyl ketone acceptors, phenyl(pyridin-4-yl)methanone is also applicable in preparing D–A type TADF molecules. Cheng and coworkers prepared **111** by connecting two *meta* carbazolyl groups on the phenyl ring of phenyl(pyridin-4-yl)methanone [124]. The TADF molecule (**111**) shows a small ΔE_{ST} of 0.06 eV, reversibly switchable emission via external stimuli, and high OLED performance with a low turn-on voltage of 3.6 V, high EQE of 18.4%, and CIE coordinate of (0.18, 0.25). By tuning the substitution position of the carbazolyl units, two new TADF molecules (**112** and **113**) were resulted; those materials show blue and green emission with high PLQY of 88% and 91.4% (doped film), small ΔE_{ST} of 0.03 eV and 0.04 eV, and high efficient OLEDs with maximum EQE of 24.0% and 27.2%, respectively [125].

11.4.6 X-bridged Diphenyl Ketone TADF Molecules

To improve the TADF performance of diphenyl ketone-based molecules, various atoms including oxygen, carbon, and sulfur have been introduced into diphenyl ketone as an X-bridge to promote the optical and electronic properties by constructing rigid molecular structure. Through the C-bridge of diphenyl ketone (spiro-anthracenone), Adachi and coworkers developed a blue-greenish TADF molecule (**114**, Scheme 11.12) that shows a small ΔE_{ST} of 0.04 eV, a high PLQY of 81%, and a maximum EQE of 16.5% in the TADF OLEDs [30]. Also, high-performance TADF-sensitized OLEDs employing **114** and its O-bridged analogue (**115**, Scheme 11.12) as the sensitizers to transfer ~100% electrically excited excitons to conventional fluorescence molecules were found to show high EQEs of 13.5% and 15.8%, respectively [28].

Another ketone bridge was also introduced to diphenyl ketone to enhance its electron deficiency and fasten ISC process; the formed C=O bridge diphenyl ketones (anthraquinone) is highly effective as an acceptor to cooperate with different donors in D–A–D configuration for a series of TADF molecules (**116–123**, Scheme 11.12) [14]. These compounds show large fluorescence rate (k_F) of up to 6.1×10^7 s^{-1}, small ΔE_{ST} of 0.07–0.29 eV, tunable emission color from yellow to red with PLQY of up to 80%, and high-performance TADF OLEDs with low turn-on voltages around 3 V; high maximum EQEs of 12.5, 9.0, 9.0, and 6.9%; and varied emission bands at 624, 637, 574, and 584 nm, respectively.

The S-bridged diphenyl ketones (thioxanthones) can benefit from both ketone and sulfoxide bridges for high rate ISC (k_{ISC}) and high quantum yield of triplet formation, showing great potential in constructing high-performance TADF molecules. Recently, two novel TADF emitters (**124–125**, Scheme 11.12) with small ΔE_{ST} (0.052 and 0.073 eV) and high k_{ISC} of 10^7 s^{-1} using thioxanthone as

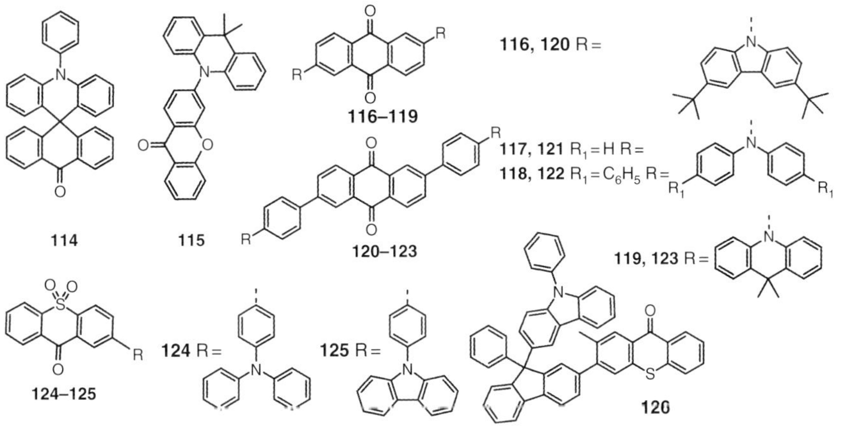

Scheme 11.12 X-bridged diphenyl ketone-based TADF molecules of **114–126**.

the acceptor and arylamine as the donor were reported by Wang et al. [126] **124** and **125** show PL peaks at 625 and 570 nm with pronounced AIE characteristic, high PLQY of 36 ± 2% and 96 ± 2% in film, and highly efficient TADF OLEDs with turn-on voltages at 5.3 and 4.7 V and maximum EQEs of 18.5% and 21.5%, respectively. Wang et al. also introduced a sulfur bridge to diphenyl ketone to prepare TADF molecule (**126**, Scheme 11.12). The resulting **126** containing a bulky donor shows a small ΔE_{ST} of 0.19 eV for efficient rISC to harvest the triplet excitons and high thermal property with T_d of 456 °C and T_g of 127.3 °C for device operation. When **126** was used as a host material for the orange and red PhOLEDs, high EQEs of 11.8% and 15.6% with low-efficiency roll-off was observed, due to significantly reduced triplet concentration via efficient rISC from triplet to singlet and thus subdued TTA and TPA processes on host [127].

11.5 Organoboron-based TADF Molecules

Boron-containing fragment, which contains a boron-incorporated aromatic moiety with strong electron-accepting property due to the high accepting ability of the boron atom, has emerged as a new acceptor unit for TADF molecule design [128, 129]. Four TADF molecules (**127–130**, Scheme 11.13) containing 10*H*-phenoxaborin and acridan as acceptor and donor were prepared by Adachi and coworkers [130]. These compounds show blue TADF emissions with high PLQY (56–100%) and small ΔE_{ST} of 0.06–0.12 eV; their TADF OLEDs exhibited a high EQE of 21.7, 13.3, 19.0, and 20.1%, respectively. Kitamoto et al. also developed two 10*H*-phenoxaborin-based TADF molecules (**131–132**, Scheme 11.13) [131], which show blue and green emissions with high PLQYs of 98% and 99% in the doped films, small ΔE_{ST} of 0.013 eV and 0.028 eV, and excellent blue (peaked at 466 nm) and green (peaked at 503 nm) TADF OLEDs with maximum EQEs of 15.1% and 22.1%, respectively. Recently, Wu and coworkers used boron atom as a hub for the spiro linker of light-emitting molecules, leading to

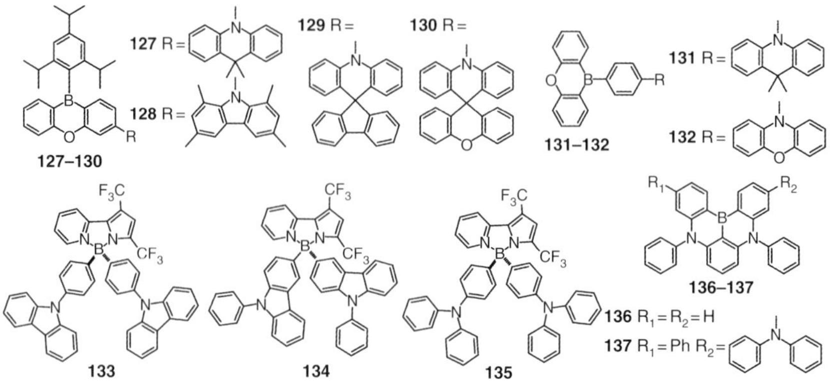

Scheme 11.13 Organoboron-based TADF molecules of **127–137**.

spatially separated HOMO and LUMO with orthogonal orientation in boron complexes (**133–135**, Scheme 11.13). The electron-deficient pyridyl pyrrolide and electron-donating phenylcarbazolyl fragments or triphenylamine connected by boron leads to intense green TADF, and the OLEDs showed an EQE of up to 13.5% [132]. Hatakeyama et al. constructed two new TADF molecules (**136–137**, Scheme 11.13) based on a rigid polycyclic aromatic framework containing triphenylboron and two nitrogen atoms; the multiple resonance effect of boron and nitrogen atoms results in separated HOMO and LUMO for TADF emission with small ΔE_{ST} of 0.14 eV and 0.18 eV, high PLQY of 84% and 90%, and ultrapure blue emission with emission peak at 462 and 470 nm, respectively [133]. The blue TADF OLEDs using **136–137** as emitters achieved high performance with EL peaks at 459 nm and 467 nm, FWHM of 28 nm and 28 nm, CIE coordinates of (0.13, 0.09) and (0.12, 0.13), and high EQE of 13.5% and 20.2%, respectively.

11.6 TADF Polymers

Currently, the majority of the materials capable of TADF emission are limited to small molecules; it is a great challenge to prepare TADF polymers for solution-processable polymeric OLEDs (P-OLEDs) with apparent advantages in reducing the fabrication costs and increasing the device size. Recently, Nikolaenko et al. proposed a strategy, which they called "intermonomer TADF," to provide a promising route for molecular design of TADF polymers [134]. This concept involves of spatial separation of the HOMO and LUMO on donor and acceptor units, respectively, for the formation of intermonomer CT states to realize small ΔE_{ST} and act as the charge transporting units, and an additional comonomer with high triplet energy level to act as high T_1 spacer units to prevent aggregation of the TADF emissive centers (intermonomer D–A structure) and to fulfill solution processability. The synthesized TADF polymer (Scheme 11.14) shows a small ΔE_{ST} of 0.22 eV, a green emission peaked at ~535 nm with PLQY of ~41%, and a high performance in solution-processed TADF P-OLEDs with EQE of up to 10%.

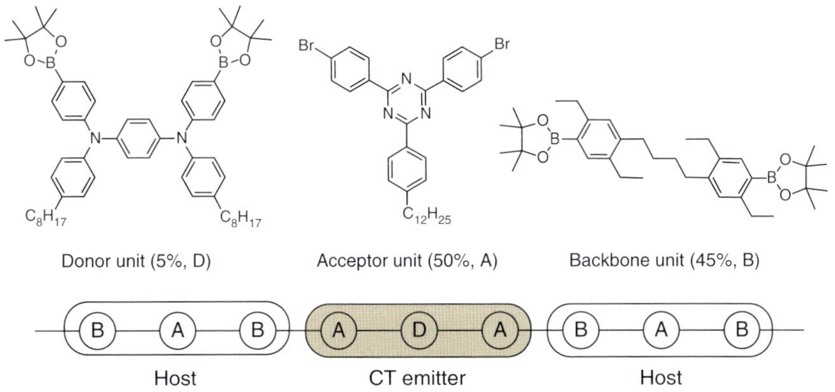

Scheme 11.14 The design of TADF polymers based on "intermonomer TADF" with donor (D), acceptor (A), and backbone (B) units.

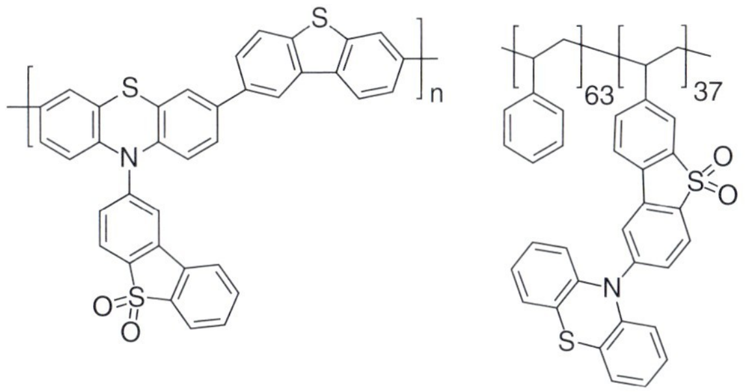

Scheme 11.15 Backbone and side-chain type TADF polymer.

In other efforts to prepare TADF polymers, Dias and coworkers connected efficient TADF small molecules to the side chain of a flexible polystyrene and incorporated into a conjugated backbone (Scheme 11.15); an acceptable EQE of 3.5% at 100 cd m^{-2} was observed [135]. Yang and coworkers presented a series of TADF polymers through grafting the TADF emitter onto the side chain of the polymer backbone following side-chain engineering strategy. The OLEDs using these TADF grafted polymers as the emitter exhibited a maximum EQE of 4.3% [136].

11.7 Intermolecular D–A System for TADF Emission

Besides intramolecular D–A molecules, intermolecular D–A systems are also applicable to compose TADF emissions. The well-known exciplex is an intermolecular D–A structure formed between a donor (D) and an acceptor (A) via CT under electrical excitation in OLED devices. Exciplex emission, which is low in efficiency and broad in emission spectrum, was generally not favored in previous OLEDs studies [40]. However, recent studies show that for the exciplex emission of the intermolecular excited states, the electron transition from the LUMO of an acceptor to the HOMO of a donor over a large electron-hole separation distance can provide a smaller exchange energy compared with that of intramolecular D–A systems, resulting in a small ΔE_{ST} for TADF emission [10]. The exciplex-based TADF OLEDs can be fabricated either in bilayer with two layers of donor and acceptor molecules or in single layer with their mixture. A slight redshift between the PF and DF components was observed in exciplex-type TADF, which is different from the TADF emission from single D–A molecules. The reason for the spectral shift has not been clarified at this moment. Adachi and coworkers explained that DF is generated immediately after the rISC process, so that the nuclear configuration of S_1 formed through rISC is affected by that of T_1. The Franck–Condon factor (including polarization in the host medium) of DF would be different from that of PF, leading to slight different PL spectra of PF and DF [10]. The existence of a broad distribution of the energy level of the exciplex originating from different geometric arrangements between the two molecules in the exciplex has been regarded as another plausible reason [137].

Blended single-layer exciplex. The significant performance breakthrough of exciplex-based OLEDs to utilize TADF effects was recently demonstrated in m-MTDATA (D): 3TPYMB (A) intermolecular D–A system [10] (Scheme 11.16). The device based on single-layer m-MTDATA: 3TPYMB (1 : 1) exciplex exhibited a high EQE of 5.4% that is higher than the limit of conventional fluorescence-based OLEDs even at a rather low PLQY (26%) of the exciplex; the efficient rISC (86.5%) greatly contributes to increase EQE of the device. Yang and coworkers [138] changed the ETL thickness of the same intermolecular D–A system and achieved a lower turn-on voltage of 2.1 V, a higher luminescence of up to 17100 cd m^{-2}, a higher CE of 36.79 cd A^{-1}, and a lower efficiency roll-off. To further improve the performance of OLEDs based on TADF emissions of exciplexes, a new electron-accepting molecule 2, 8-bis(diphenylphosphoryl) dibenzo-[b,d]thiophene (PPT), which has higher triplet energy to confine the triplet exciplex, was adopted [139]. After carefully optimizing the ratios of the donor (m-MTDATA) and the acceptor (PPT) for high PLQY and suitable recombination zone, the green OLEDs showed a high PE of 47.0 lm W^{-1} and an EQE of up to 10%. Peng and coworkers selected m-MTDATA and BPhen as donor and acceptor components, respectively; a maximum EQE of 7.79% was achieved in the TADF OLEDs containing m-MTDATA: 70 mol% BPhen as the EML with an extremely low ΔE_{ST} close to zero [140].

Kim and coworkers reported a new D–A material pair for exciplex-type TADF emission using a donor of 4,4′,4″-tris(N-carbazolyl)-triphenylamine (TCTA) and an acceptor of bis-4,6-(3,5-di-3-pyridylphenyl)-2-methylpyrimidine (B3PYMPM). The PLQY of the exciplex emission is only 10% at 195 K but almost 100% at 35 K [137]. The EQE of the OLEDs increased from 3.1% to 10% when the temperature dropped from 300 to 195 K. In-depth studies of this exciplex system indicated that k_{ISC} (1.1×10^7 s^{-1}) is faster than k_r^S (7×10^6 s^{-1}) and k_{rISC} has a distribution in the range from 3×10^6 to 3×10^4 s^{-1} depending on the delay time and is almost independent of temperature. The temperature-independent rISC rate indicates that the rISC takes place without thermal activation, which is very different from that in single molecular D–A type TADF materials. The broad range of the k_{rISC} was suggested to result from the broad energy level

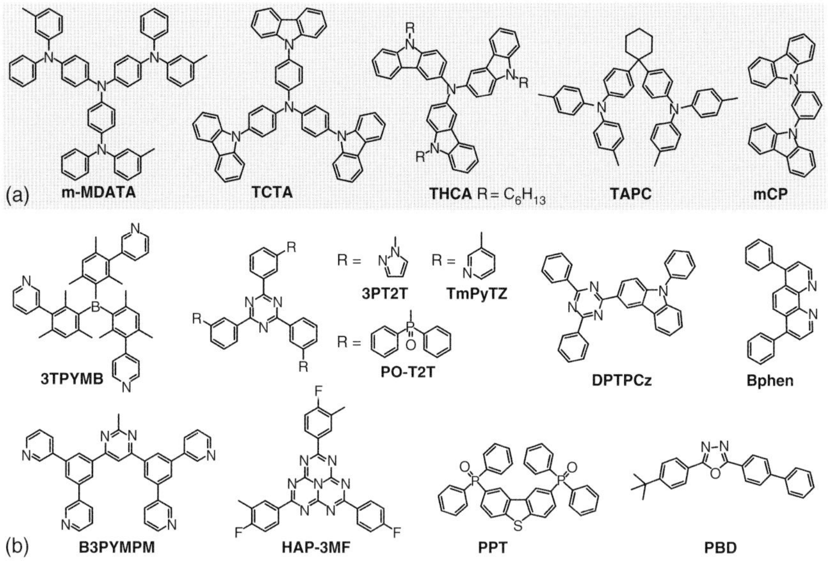

Scheme 11.16 Representative donors (a) and acceptors (b) applicable for intermolecular D–A type TADF emission.

distributions of the singlet and the triplet exciplexes. Further, the singlet CT states may lie energetically below the triplet CT states, because the kinetic exchange dominates the interactions of the donor and acceptor in the exciplex, which favors the singlet state with short intermolecular distance and coulombic interaction at CT states [141, 142]. Therefore, the singlet and triplet exciplex have broad energy level distributions, and energy difference between them (ΔE_{ST}) is extremely small that can be either positive or negative. And the rISC from the triplet to singlet exciplex can occur even without thermal activation in this system. The especially low ΔE_{ST} (estimated to be 5 meV) was also observed by Graves et al. in a green-emitting exciplex (525–550 nm) between m-MTDATA and PBD in a 50 : 50 blended film [143]. Recently, a new electron-transporting material of ((1,3,5-triazine-2,4,6-triyl)tris(benzene-3,1-diyl))tris (diphenylphosphine oxide) (PO-T2T) that has very low LUMO and HOMO (−2.83 and −6.83 eV, respectively) was synthesized by Wong and coworkers [144]. Based on PO-T2T, a panchromatic range of exciplex emission from blue to red was realized by systematically tuning the HOMO of the hole-transporting materials. And a tandem, all-exciplex-based white OLED was also fabricated, showing excellent EQE (11.6%), CE (27.7 cd A^{-1}), PE (15.8 lm W^{-1}), CIE (0.29, 0.35), and color-rendering indices (CRI) of 70.6. Zhang and coworkers also reported a new exciplex system with PO-T2T as the acceptor and a new TADF emitter of MAC (6-(9,9-dimethylacridin-10(9H)-yl)-3-methyl-1H-isochromen-1-one) as the donor. The new exciplex D–A system has two rISC routes on both the single-molecule TADF emitter of MAC and the exciplex emitter; this TADF molecular design can greatly enhance the utilization of the triplet excitons, increase the device efficiency, and decrease the efficiency roll-off, achieving a maximum EQE of 17.8% and low-efficiency roll-off (12.3% at 1000 cd m^{-2}) [145].

The widely used commercial host material of mCP was also found to be effective in constructing intermolecular D–A system for TADF emission when combined with a new electron-accepting molecule of 2,5,8-tris(4-fluoro-3-methylphenyl)-1,3,4,6,7,9,9b-heptaazaphenalene (HAP-3MF) [146]. The OLEDs containing 8 wt% HAP-3MF: mCP as the EML exhibited a high EQE of up to 11.3%, which is far above the EQE limit (5%) of fluorescent OLEDs but is still far below the theoretical limit (20%) of TADF OLEDs due to the low PLQY. More recently, Liu et al. reported an exciplex pair of TAPC: DPTPCz, which shows the highest PLQY (68%) and EQE (15.4%) in the exciplex OLEDs reported so far [147]. Similar to the intramolecular D–A type TADF emitters, exciplexes were also used as singlet exciton sensitizers to harvest triplet excitons in OLEDs for fluorescent emitters. Liu et al. fabricated a TADF-sensitized OLED by using TAPC:DPTPCz as singlet exciton sensitizer; a low turn-on voltage of 2.8 V and high CE, PE, and EQE of 44.0 cd A^{-1}, 46.1 lm W^{-1}, and 14.5% were observed, respectively [29].

Bilayer-type exciplex : Alternatively, the exciplex-based TADF OLEDs can also function in a bilayer structure using an appropriate combination of donor and acceptor layers. The TADF OLED efficiencies of a bilayer-type exciplex can be significantly enhanced according to the following criteria:

i) Both the donor and acceptor molecules should have a planar structure for the flat-on orientation of the molecules in solids to avoid the appearance of excimer emission in the EML.
ii) High hole and electron mobilities should be satisfied to allow sufficient charge carriers to the interface for exciplex generation.
iii) The charge carriers can be accumulated at the interfacial region by the large differences between the frontier energy levels of donor and acceptor, giving a high propensity of electron-hole recombination to generate S_1 and T_1 states.
iv) More importantly, the triplet energies of donor and acceptor need to be higher than that of the exciplex to confine the electro-generated triplet states within the interfacial region.

Consequently, the nonemissive T_1 state can predominantly shuttle back to emissive S_1 state, leading to a high-efficiency bilayer-type exciplex OLED.

In 2013, Hung et al. reported a simple bilayer-type exciplex TADF OLED based on a carbazole-centered electron-donating material (TCTA) and a triazine-based electron-accepting material (3P-T2T) [148]. The yellow OLED exhibited a high CE of 22.5 cd A^{-1} and an EQE of up to 7.7%. Su et al. reported a highly efficient bilayer TADF OLED containing TAPC as the donor and TmPyTZ as the acceptor, which exhibited an extremely low turn-on voltage of 2.14 V, high maximum CE of 37.8 cd A^{-1}, EQE of up to 12.02%, and PE of 52.8 lm W^{-1} [149]. Recently, Cherpak et al. used a metal complex (FIrpic) as an acceptor to form interface exciplex with THCA [41]. This new bilayer OLED emits yellow light consisted of a combination of the blue phosphorescent emission from FIrpic and a broad efficient DF from the exciplex, exhibiting a high maximum CE of 15 cd A^{-1}, brightness of 38 000 cd m^{-2}, and EQE of ~5%.

11.8 Summary and Outlook

The design and characterization of TADF materials for optoelectronic applications represents an active area of recent research in organic electronics. Widely considered as the third generation OLED materials, noble metal-free TADF materials either in small molecules, dendrimers, polymers, or exciplex pairs are promising candidates as emitters, sensitizers, and hosts with small ΔE_{ST} for theoretically 100% excitons utilization in D–A molecular structure for balanced charge transport and injection properties, which are highly attractive for a wide variety of high-performance optoelectronic devices. TADF OLEDs using TADF emitters have shown significant improvements in PLQYs, color varieties, and device EQEs, since their first report in 2009. Up till now, high PLQY of ~100%, maximum EQEs over 30%, low efficiency roll-off, blue-to-red and white TADF OLEDs, and long operation lifetimes have been evidenced in many TADF molecular systems. In addition, the low sensitivity of TADF molecules toward concentrations allow the fabrication of nondoped TADF OLEDs with EQEs of up to 20%; solution-processable TADF small molecules, dendrimers, and polymers have also been successfully developed, demonstrating the ultimate potentials of organic materials in large-area, low-cost, and high-efficiency display and lighting products. TADF-sensitized fluorescent OLEDs using TADF sensitizers can also harvest 100% excitons through the efficient rISC of the TADF molecules, exhibiting excellent device performance with full-color emission, promising operation stability, and EQEs of up to 15% that obviously exceeds 5% limitation of traditional fluorescent OLEDs. PhOLEDs using TADF molecules as host materials show EQEs over 20%, extremely low driving voltages of ~2.0 V, and low-efficiency roll-off (<10%) due to the balanced and bipolar charge transport/injection properties induced by the D–A configuration and small ΔE_{ST}. The outstanding performance in certain aspects of these judiciously designed D–A type TADF materials has even transcended that of noble metal-containing phosphorescent complexes, in addition to their distinct relevant competitive advantages of low-cost, rich resources, and easy material preparation. The success in the breakthrough of the theoretical and technical challenges and the revolution of the understandings of organic optoelectronics that arise in developing high-performance TADF materials may pave the way to shape the future of organoelectronics.

References

1 Yersin, H., Rausch, A.F., Czerwieniec, R., Hofbeck, T., and Fischer, T. (2011). *Coord. Chem. Rev.* 255: 2622.
2 Tao, Y., Yuan, K., Chen, T., Xu, P., Li, H., Chen, R., Zheng, C., Zhang, L., and Huang, W. (2014). *Adv. Mater.* 26: 7931.
3 Czerwieniec, R., Yu, J., and Yersin, H. (2011). *Inorg. Chem.* 50: 8293.
4 Czerwieniec, R., Leitl, M.J., Homeier, H.H.H., and Yersin, H. (2016). *Coord. Chem. Rev.* 325: 2.
5 Yersin, H. and Monkowius, U. (2008). GER Patent DE20081033563.

6 Parker, C.A. and Hatchard, C.G. (1961). *Trans. Faraday Soc.* 57: 1894.
7 Blasse, G. and McMillin, D.R. (1980). *Chem. Phys. Lett.* 70: 1.
8 Berberan-Santos, M.N. and Garcia, J.M.M. (1996). *J. Am. Chem. Soc.* 118: 9391.
9 Endo, A., Ogasawara, M., Takahashi, A., Yokoyama, D., Kato, Y., and Adachi, C. (2009). *Adv. Mater.* 21: 4802.
10 Goushi, K., Yoshida, K., Sato, K., and Adachi, C. (2012). *Nat. Photonics* 6: 253.
11 Uoyama, H., Goushi, K., Shizu, K., Nomura, H., and Adachi, C. (2012). *Nature* 492: 234.
12 Zhang, Q., Li, B., Huang, S., Nomura, H., Tanaka, H., and Adachi, C. (2014). *Nat. Photonics* 8: 326.
13 Nishide, J., Nakanotani, H., Hiraga, Y., and Adachi, C. (2014). *Appl. Phys. Lett.* 104: 233304.
14 Zhang, Q., Kuwabara, H., Potscavage, W.J., Huang, S., Hatae, Y., Shibata, T., and Adachi, C. (2014). *J. Am. Chem. Soc.* 136: 18070.
15 Lee, D.R., Kim, B.S., Lee, C.W., Im, Y., Yook, K.S., Hwang, S., and Lee, J.Y. (2015). *ACS Appl. Mater. Interfaces* 7: 9625.
16 Lin, T.-A., Chatterjee, T., Tsai, W.-L., Lee, W.-K., Wu, M.-J., Jiao, M., Pan, K.-C., Yi, C.-L., Chung, C.-L., Wong, K.-T., and Wu, C.-C. (2016). *Adv. Mater.* 28: 6976.
17 Liu, X.K., Chen, Z., Qing, J., Zhang, W.J., Wu, B., Tam, H.L., Zhu, F., Zhang, X.H., and Lee, C.S. (2015). *Adv. Mater.* 27: 7079.
18 Cho, Y.J., Yook, K.S., and Lee, J.Y. (2014). *Adv. Mater.* 26: 6642.
19 Tang, C., Yang, T., Cao, X., Tao, Y., Wang, F., Zhong, C., Qian, Y., Zhang, X., and Huang, W. (2015). *Adv. Opt. Mater.* 3: 786.
20 Tao, Y., Guo, X., Hao, L., Chen, R., Li, H., Chen, Y., Zhang, X., Lai, W., and Huang, W. (2015). *Adv. Mater.* 27: 6939.
21 Zhang, D., Duan, L., Zhang, D., Qiao, J., Dong, G., Wang, L., and Qiu, Y. (2013). *Org. Electron.* 14: 260.
22 Chaskar, A., Chen, H., and Wong, K. (2011). *Adv. Mater.* 23: 3876.
23 Tao, Y.T., Yang, C.L., and Qin, J.G. (2011). *Chem. Soc. Rev.* 40: 2943.
24 Nishimoto, T., Yasuda, T., Lee, S.Y., Kondo, R., and Adachi, C. (2014). *Mater. Horiz.* 1: 264.
25 Kim, K., Moon, C., Lee, J., Kim, S., and Kim, J. (2014). *Adv. Mater.* 26: 3844.
26 Komino, T., Nomura, H., Koyanagi, T., and Adachi, C. (2013). *Chem. Mater.* 25: 3038.
27 Komino, T., Tanaka, H., and Adachi, C. (2014). *Chem. Mater.* 26: 3665.
28 Nakanotani, H., Higuchi, T., Furukawa, T., Masui, K., Morimoto, K., Numata, M., Tanaka, H., Sagara, Y., Yasuda, T., and Adachi, C. (2014). *Nat. Commun.* 5: 4011.
29 Liu, X., Chen, Z., Zheng, C., Chen, M., Liu, W., Zhang, X., and Lee, C. (2015). *Adv. Mater.* 27: 2025.
30 Nasu, K., Nakagawa, T., Nomura, H., Lin, C., Cheng, C., Tseng, M., Yasuda, T., and Adachi, C. (2013). *Chem. Commun.* 49: 10385.
31 Song, W., Lee, I.H., Hwang, S., and Lee, J.Y. (2015). *Org. Electron.* 23: 138.

32 Zhang, Q., Tsang, D., Kuwabara, H., Hatae, Y., Li, B., Takahashi, T., Lee, S.Y., Yasuda, T., and Adachi, C. (2015). *Adv. Mater.* 27: 2096.
33 Chen, T., Zheng, L., Yuan, J., An, Z., Chen, R., Tao, Y., Li, H., Xie, X., and Huang, W. (2015). *Sci. Rep.* 5: 10923.
34 Leitl, M.J., Krylova, V.A., Djurovich, P.I., Thompson, M.E., and Yersin, H. (2014). *J. Am. Chem. Soc.* 136: 16032.
35 Yao, L., Yang, B., and Ma, Y. (2014). *Sci. China Chem.* 57: 335.
36 Klessinger, M. (1995). *Angew. Chem. Int. Ed.* 34: 549.
37 Dias, F.B., Jankus, V., Kamtekar, K.T., Santos, J., and Monkman, A.P. (2013). *Adv. Mater.* 25: 3707.
38 Saragi, T., Spehr, T., Siebert, A., Fuhrmann-Lieker, T., and Salbeck, J. (2007). *Chem. Rev.* 107: 1011.
39 Zhang, Q.S., Li, J., Shizu, K., Huang, S.P., Hirata, S., Miyazaki, H., and Adachi, C. (2012). *J. Am. Chem. Soc.* 134: 14706.
40 Li, G., Kim, C.H., Zhou, Z., Shinar, J., Okumoto, K., and Shirota, Y. (2006). *Appl. Phys. Lett.* 88: 253505.
41 Cherpak, V., Stakhira, P., Minaev, B., Baryshnikov, G., Stromylo, E., Helzhynskyy, I., Chapran, M., Volyniuk, D., Hotra, Z., Dabuliene, A., Tomkeviciene, A., Voznyak, L., and Grazulevicius, J.V. (2015). *ACS Appl. Mater. Interfaces* 7: 1219.
42 Fan, M.G. and Yao, J.N. (2009). *Photochemistry and Optical Function Materials*. Science Press.
43 Lu, T. and Chen, F. (2012). *J. Comput. Chem.* 33: 580.
44 Salman, S., Kim, D., Coropceanu, V., and Brédas, J. (2011). *Chem. Mater.* 23: 5223.
45 Kim, D., Coropceanu, V., and Bredas, J.L. (2011). *J. Am. Chem. Soc.* 133: 17895.
46 Tanaka, H., Shizu, K., Miyazaki, H., and Adachi, C. (2012). *Chem. Commun.* 48: 11392.
47 Strickler, S.J. and Berg, R.A. (1962). *J. Chem. Phys.* 37: 814.
48 Hirata, S., Sakai, Y., Masui, K., Tanaka, H., Lee, S.Y., Nomura, H., Nakamura, N., Yasumatsu, M., Nakanotani, H., Zhang, Q., Shizu, K., Miyazaki, H., and Adachi, C. (2014). *Nat. Mater.* 14: 330.
49 Mulliken, R.S. (1939). *J. Chem. Phys.* 7: 14.
50 Thomas, K., Velusamy, M., Lin, J.T., Tao, Y.T., and Cheun, C.H. (2004). *Adv. Funct. Mater.* 14: 387.
51 Yassar, A., Demanze, F., Jaafari, A., El Idrissi, M., and Coupry, C. (2002). *Adv. Funct. Mater.* 12: 699.
52 Nakagawa, T., Ku, S.Y., Wong, K.T., and Adachi, C. (2012). *Chem. Commun.* 48: 9580.
53 Mehes, G., Nomura, H., Zhang, Q.S., Nakagawa, T., and Adachi, C. (2012). *Angew. Chem. Int. Ed.* 51: 11311.
54 Li, B., Nomura, H., Miyazaki, H., Zhang, Q., Yoshida, K., Suzuma, Y., Orita, A., Otera, J., and Adachi, C. (2014). *Chem. Lett.* 43: 319.
55 Higgins, R.W.T., Monkman, A.P., Nothofer, H.G., and Scherf, U. (2001). *Appl. Phys. Lett.* 79: 857.
56 Zhang, Y. and Forrest, S.R. (2012). *Phys. Rev. Lett.* 108: 267404.

57 Nakanotani, H., Masui, K., Nishide, J., Shibata, T., and Adachi, C. (2013). *Sci. Rep.* 3: 2127.

58 Im, Y. and Lee, J.Y. (2014). *Chem. Mater.* 26: 1413.

59 Kim, B.S. and Lee, J.Y. (2014). *ACS Appl. Mater. Interfaces* 6: 8396.

60 Cho, Y.J., Yook, K.S., and Lee, J.Y. (2014). *Adv. Mater.* 26: 4050.

61 Kim, B.S. and Lee, J.Y. (2014). *Adv. Funct. Mater.* 24: 3970.

62 Sun, J.W., Lee, J., Moon, C., Kim, K., Shin, H., and Kim, J. (2014). *Adv. Mater.* 26: 5684.

63 Seino, Y., Inomata, S., Sasabe, H., Pu, Y., and Kido, J. (2016). *Adv. Mater.* 28: 2638.

64 Liu, H., Bai, Q., Yao, L., Hu, D., Tang, X., Shen, F., Zhang, H., Gao, Y., Lu, P., Yang, B., and Ma, Y. (2014). *Adv. Funct. Mater.* 24: 5881.

65 Ding, J., Zhang, B., Lu, J., Xie, Z., Wang, L., Jing, X., and Wang, F. (2009). *Adv. Mater.* 21: 4983.

66 Yook, K.S. and Lee, J.Y. (2014). *Adv. Mater.* 26: 4218.

67 Kawano, K., Nagayoshi, K., Yamaki, T., and Adachi, C. (2014). *Org. Electron.* 15: 1695.

68 Masui, K., Nakanotani, H., and Adachi, C. (2013). *Org. Electron.* 14: 2721.

69 Cho, Y.J., Yook, K.S., and Lee, J.Y. (2015). *Sci. Rep.* 5: 7859.

70 Lee, D.R., Hwang, S., Jeon, S.K., Lee, C.W., and Lee, J.Y. (2015). *Chem. Commun.* 51: 8105.

71 Park, I.S., Lee, S.Y., Adachi, C., and Yasuda, T. (2016). *Adv. Funct. Mater.* 26: 1813.

72 Taneda, M., Shizu, K., Tanaka, H., and Adachi, C. (2015). *Chem. Commun.* 51: 5028.

73 Liu, W., Zheng, C., Wang, K., Chen, Z., Chen, D., Li, F., Ou, X., Dong, Y., and Zhang, X. (2015). *ACS Appl. Mater. Interfaces* 7: 18930.

74 Zhang, D., Cai, M., Zhang, Y., Zhang, D., and Duan, L. (2016). *Mater. Horiz.* 3: 145.

75 Cho, Y.J., Jeon, S.K., Chin, B.D., Yu, E., and Lee, J.Y. (2015). *Angew. Chem. Int. Ed.* 54: 5201.

76 Kim, M., Jeon, S.K., Hwang, S., Lee, S., Yu, E., and Lee, J.Y. (2016). *Chem. Commun.* 52: 339.

77 An, Z.F., Chen, R.F., Yin, J., Xie, G.H., Shi, H.F., Tsuboi, T., and Huang, W. (2011). *Chem. Eur. J.* 17: 10871.

78 Endo, A., Sato, K., Yoshimura, K., Kai, T., Kawada, A., Miyazaki, H., and Adachi, C. (2011). *Appl. Phys. Lett.* 98: 83302.

79 Zhang, D., Duan, L., Li, C., Li, Y., Li, H., Zhang, D., and Qiu, Y. (2014). *Adv. Mater.* 26: 5050.

80 Zhang, D., Duan, L., Zhang, D., and Qiu, Y. (2014). *J. Mater. Chem. C* 2: 8983.

81 Sato, K., Shizu, K., Yoshimura, K., Kawada, A., Miyazaki, H., and Adachi, C. (2013). *Phys. Rev. Lett.* 110: 247401.

82 Lee, S.Y., Yasuda, T., Nomura, H., and Adachi, C. (2012). *Appl. Phys. Lett.* 101: 93306.

83 Serevicius, T., Nakagawa, T., Kuo, M.C., Cheng, S.H., Wong, K.T., Chang, C.H., Kwong, R.C., Xia, S., and Adachi, C. (2013). *Phys. Chem. Chem. Phys.* 15: 15850.

84 Chang, C.H., Kuo, M.C., Lin, W.C., Chen, Y.T., Wong, K.T., Chou, S.H., Mondal, E., Kwong, R.C., Xia, S.A., Nakagawa, T., and Adachi, C. (2012). *J. Mater. Chem.* 22: 3832.
85 Kim, M., Jeon, S.K., Hwang, S., and Lee, J.Y. (2015). *Adv. Mater.* 27: 2515.
86 Uejima, M., Nomura, H., Sato, T., Tanaka, K., Kaji, H., Adachi, C., and Shizu, K. (2015). *Phys. Rev. Appl.* 3: 14001.
87 Kaji, H., Suzuki, H., Fukushima, T., Shizu, K., Suzuki, K., Kubo, S., Komino, T., Oiwa, H., Suzuki, F., Wakamiya, A., Murata, Y., and Adachi, C. (2015). *Nat. Commun.* 6: 8476.
88 Tanaka, H., Shizu, K., Nakanotani, H., and Adachi, C. (2013). *Chem. Mater.* 25: 3766.
89 Tanaka, H., Shizu, K., Nakanotani, H., and Adachi, C. (2014). *J. Phys. Chem. C* 118: 15985.
90 Sun, J.W., Baek, J.Y., Kim, K., Moon, C., Lee, J., Kwon, S., Kim, Y., and Kim, J. (2015). *Chem. Mater.* 27: 6675.
91 Tsai, W., Huang, M., Lee, W., Hsu, Y., Pan, K., Huang, Y., Ting, H., Sarma, M., Ho, Y., Hu, H., Chen, C., Lee, M., Wong, K., and Wu, C. (2015). *Chem. Commun.* 51: 13662.
92 Albrecht, K., Matsuoka, K., Fujita, K., and Yamamoto, K. (2015). *Angew. Chem. Int. Ed.* 54: 5677.
93 Komatsu, R., Sasabe, H., Seino, Y., Nakao, K., and Kido, J. (2016). *J. Mater. Chem. C* 4: 2274.
94 Kulkarni, A.P., Tonzola, C.J., Babel, A., and Jenekhe, S.A. (2004). *Chem. Mater.* 16: 4556.
95 Zhu, R., Wen, G.A., Feng, J.C., Chen, R.F., Zhao, L., Yao, H.P., Fan, Q.L., Wei, W., Peng, B., and Huang, W. (2005). *Macromol. Rapid Commun.* 26: 1729.
96 Lee, J., Shizu, K., Tanaka, H., Nomura, H., Yasuda, T., and Adachi, C. (2013). *J. Mater. Chem. C* 1: 4599.
97 Tanaka, H., Shizu, K., Lee, J., and Adachi, C. (2015). *J. Phys. Chem. C* 119: 2948.
98 Li, J., Nakagawa, T., MacDonald, J., Zhang, Q., Nomura, H., Miyazaki, H., and Adachi, C. (2013). *Adv. Mater.* 25: 3319.
99 Takahashi, T., Shizu, K., Yasuda, T., Togashi, K., and Adachi, C. (2014). *Sci. Technol. Adv. Mater.* 15: 34202.
100 Wang, S., Yan, X., Cheng, Z., Zhang, H., Liu, Y., and Wang, Y. (2015). *Angew. Chem. Int. Ed.* 54: 13068.
101 Sagara, Y., Shizu, K., Tanaka, H., Miyazaki, H., Goushi, K., Kaji, H., and Adachi, C. (2015). *Chem. Lett.* 44: 360.
102 Shizu, K., Tanaka, H., Uejima, M., Sato, T., Tanaka, K., Kaji, H., and Adachi, C. (2015). *J. Phys. Chem. C* 119: 1291.
103 Kawasumi, K., Wu, T., Zhu, T., Chae, H.S., Van Voorhis, T., Baldo, M.A., and Swager, T.M. (2015). *J. Am. Chem. Soc.* 137: 11908.
104 Ohkuma, H., Nakagawa, T., Shizu, K., Yasuda, T., and Adachi, C. (2014). *Chem. Lett.* 43: 1017.
105 Grabowski, Z.R., Rotkiewicz, K., and Rettig, W. (2003). *Chem. Rev.* 103: 3899.

106 Wu, S., Aonuma, M., Zhang, Q., Huang, S., Nakagawa, T., Kuwabara, K., and Adachi, C. (2014). *J. Mater. Chem. C* 2: 421.
107 Huang, B., Qi, Q., Jiang, W., Tang, J., Liu, Y., Fan, W., Yin, Z., Shi, F., Ban, X., Xu, H., and Sun, Y. (2014). *Dyes Pigm.* 111: 135.
108 Liu, M., Seino, Y., Chen, D., Inomata, S., Su, S., Sasabe, H., and Kido, J. (2015). *Chem. Commun.* 51: 16353.
109 Higuchi, T., Nakanotani, H., and Adachi, C. (2015). *Adv. Mater.* 27: 2019.
110 Xu, S., Liu, T., Mu, Y., Wang, Y., Chi, Z., Lo, C., Liu, S., Zhang, Y., Lien, A., and Xu, J. (2015). *Angew. Chem. Int. Ed.* 54: 874.
111 Xu, B., Mu, Y., Mao, Z., Xie, Z., Wu, H., Zhang, Y., Jin, C., Chi, Z., Liu, S., Xu, J., Wu, Y., Lu, P., Lien, A., and Bryce, M.R. (2016). *Chem. Sci.* 7: 2201.
112 Lee, I.H., Song, W., and Lee, J.Y. (2016). *Org. Electron.* 29: 22.
113 Lee, I. and Lee, J.Y. (2016). *Org. Electron.* 29: 160.
114 Santos, P.L., Ward, J.S., Data, P., Batsanov, A.S., Bryce, M.R., Dias, F.B., and Monkman, A.P. (2016). *J. Mater. Chem. C* 4: 3815.
115 Xie, G., Li, X., Chen, D., Wang, Z., Cai, X., Chen, D., Li, Y., Liu, K., Cao, Y., and Su, S. (2016). *Adv. Mater.* 28: 181.
116 Gan, S., Luo, W., He, B., Chen, L., Nie, H., Hu, R., Qin, A., Zhao, Z., and Tang, B.Z. (2016). *J. Mater. Chem. C* 4: 3705.
117 Ward, J.S., Nobuyasu, R.S., Batsanov, A.S., Data, P., Monkman, A.P., Dias, F.B., and Bryce, M.R. (2016). *Chem. Commun.* 52: 2612.
118 Hoshino, S. and Suzuki, H. (1996). *Appl. Phys. Lett.* 69: 224.
119 Wolf, M.W., Legg, K.D., Brown, R.E., Singer, L.A., and Parks, J.H. (1975). *J. Am. Chem. Soc.* 97: 4490.
120 Yuan, W.Z., Shen, X.Y., Zhao, H., Lam, J.W.Y., Tang, L., Lu, P., Wang, C., Liu, Y., Wang, Z., Zheng, Q., Sun, J.Z., Ma, Y., and Tang, B.Z. (2010). *J. Phys. Chem. C* 114: 6090.
121 Lee, S.Y., Yasuda, T., Yang, Y.S., Zhang, Q., and Adachi, C. (2014). *Angew. Chem. Int. Ed.* 53: 6402.
122 Xie, Z., Chen, C., Xu, S., Li, J., Zhang, Y., Liu, S., Xu, J., and Chi, Z. (2015). *Angew. Chem. Int. Ed.* 54: 7181.
123 Lee, S.Y., Yasuda, T., Park, I.S., and Adachi, C. (2015). *Dalton Trans.* 44: 8356.
124 Rajamalli, P., Senthilkumar, N., Gandeepan, P., Ren-Wu, C., Lin, H., and Cheng, C. (2016). *J. Mater. Chem. C* 4: 900.
125 Rajamalli, P., Senthilkumar, N., Gandeepan, P., Huang, P., Huang, M., Ren-Wu, C., Yang, C., Chiu, M., Chu, L., Lin, H., and Cheng, C. (2016). *J. Am. Chem. Soc.* 138: 628.
126 Wang, H., Xie, L., Peng, Q., Meng, L., Wang, Y., Yi, Y., and Wang, P. (2014). *Adv. Mater.* 26: 5198.
127 Wang, H., Meng, L., Shen, X., Wei, X., Zheng, X., Lv, X., Yi, Y., Wang, Y., and Wang, P. (2015). *Adv. Mater.* 27: 4041.
128 Kinoshita, M., Kita, H., and Shirota, Y. (2002). *Adv. Funct. Mater.* 12: 780.
129 Nagai, A., Kobayashi, S., Nagata, Y., Kokado, K., Taka, H., Kita, H., Suzuri, Y., and Chujo, Y. (2010). *J. Mater. Chem.* 20: 5196.
130 Numata, M., Yasuda, T., and Adachi, C. (2015). *Chem. Commun.* 51: 9443.

131 Kitamoto, Y., Namikawa, T., Ikemizu, D., Miyata, Y., Suzuki, T., Kita, H., Sato, T., and Oi, S. (2015). *J. Mater. Chem. C* 3: 9122.
132 Shiu, Y., Cheng, Y., Tsai, W., Wu, C., Chao, C., Lu, C., Chi, Y., Chen, Y., Liu, S., and Chou, P. (2016). *Angew. Chem. Int. Ed.* 55: 3017.
133 Hatakeyama, T., Shiren, K., Nakajima, K., Nomura, S., Nakatsuka, S., Kinoshita, K., Ni, J., Ono, Y., and Ikuta, T. (2016). *Adv. Mater.* 28: 2777.
134 Nikolaenko, A.E., Cass, M., Bourcet, F., Mohamad, D., and Roberts, M. (2015). *Adv. Mater.* 27: 7236.
135 Nobuyasu, R.S., Ren, Z., Griffiths, G.C., Batsanov, A.S., Data, P., Yan, S., Monkman, A.P., Bryce, M.R., and Dias, F.B. (2016). *Adv. Opt. Mater.* 4: 597.
136 Luo, J., Xie, G., Gong, S., Chen, T., and Yang, C. (2016). *Chem. Commun.* 52: 2292.
137 Park, Y., Kim, K., and Kim, J. (2013). *Appl. Phys. Lett.* 102: 153306.
138 Huang, Q., Zhao, S., Xu, Z., Fan, X., Shen, C., and Yang, Q. (2014). *Appl. Phys. Lett.* 104: 161112.
139 Goushi, K. and Adachi, C. (2012). *Appl. Phys. Lett.* 101: 23306.
140 Zhang, T., Chu, B., Li, W., Su, Z., Peng, Q.M., Zhao, B., Luo, Y., Jin, F., Yan, X., Gao, Y., Wu, H., Zhang, F., Fan, D., and Wang, J. (2014). *ACS Appl. Mater. Interfaces* 6: 11907.
141 Difley, S., Beljonne, D., and Van Voorhis, T. (2008). *J. Am. Chem. Soc.* 130: 3420.
142 Segal, M., Singh, M., Rivoire, K., Difley, S., Van Voorhis, T., and Baldo, M.A. (2007). *Nat. Mater.* 6: 374.
143 Graves, D., Jankus, V., Dias, F.B., and Monkman, A. (2014). *Adv. Funct. Mater.* 24: 2343.
144 Hung, W., Fang, G., Lin, S., Cheng, S., Wong, K., Kuo, T., and Chou, P. (2014). *Sci. Rep.* 4: 5161.
145 Liu, W., Chen, J., Zheng, C., Wang, K., Chen, D., Li, F., Dong, Y., Lee, C., Ou, X., and Zhang, X. (2016). *Adv. Funct. Mater.* 26: 2002.
146 Li, J., Nomura, H., Miyazaki, H., and Adachi, C. (2014). *Chem. Commun.* 50: 6174.
147 Liu, X., Chen, Z., Zheng, C., Liu, C., Lee, C., Li, F., Ou, X., and Zhang, X. (2015). *Adv. Mater.* 27: 2378.
148 Hung, W., Fang, G., Chang, Y., Kuo, T., Chou, P., Lin, S., and Wong, K. (2013). *ACS Appl. Mater. Interfaces* 5: 6826.
149 Chen, D., Xie, G., Cai, X., Liu, M., Cao, Y., and Su, S. (2016). *Adv. Mater.* 28: 239.

12

Photophysics of Thermally Activated Delayed Fluorescence

Andrew Monkman

University of Durham, OEM Research Group, Department of Physics, South Road, Durham, DH1 3LF, England

12.1 Introduction

One of the major successes of OLED research has been the discovery of methods to overcome the limitation imposed by charge recombination spin statistics [1]. As we know, upon charge recombination three times more triplet excited states are generated than singlet states; therefore the maximum internal efficiency (IQE) of an OLED is limited to 25% [2], unless methods to "harvest" the triplets, i.e. convert them to emissive singlet states, are used. Saying this, in many fluorescent OLEDs, the 25% limit is exceeded by the "accidental" use of triplet fusion (TF), one decay channel from the process of triplet–triplet annihilation (TTA) [3]. In this case, the maximum theoretical total singlet yield can reach 62.5% with devices showing >40% TF contribution having been demonstrated [4, 5]; however this approach is unsatisfactory as it falls far short of what can be achieved using phosphorescent emitters and also in practice only gives rise to best blue fluorescent OLEDs with around 11% IQE.

Phosphorescent materials containing Ir(III), Pt(II), or other heavy metals can harvest both singlet and triplet excited states by means of enhanced intersystem crossing (ISC), via the heavy atom effect, and have been used widely to increase the internal emission yield in an OLED to ~100% [6]. Currently, these provide the state-of-the-art green and red emitting complexes but have to be dispersed in charge transporting matrices to avoid aggregation-induced quenching, and obtaining stable long lifetime deep blue emitting phosphors has proved difficult and requires suitable high triplet level hosts [7]. Thus long lifetime deep blue phosphorescent OLEDs have not been demonstrated because of a variety of reasons [8, 9]. This has led to new searches for methods by which triplets can be converted to singlets efficiently to yield near 100% efficient OLEDs, especially in the elusive deep blue.

What in fact has been known for over 90 years is that there is just such another triplet harvesting method that uses thermal activation to enable the upconversion of triplet into singlet states [10, 11]. Perrin in 1929 [12] first proposed that thermal activation of a "dark" state to a singlet state gave rise to delayed fluorescence. It was not until the mid-1940s that such a dark state was proposed to be a

Highly Efficient OLEDs: Materials Based on Thermally Activated Delayed Fluorescence,
First Edition. Edited by Hartmut Yersin.
© 2019 Wiley-VCH Verlag GmbH & Co. KGaA. Published 2019 by Wiley-VCH Verlag GmbH & Co. KGaA.

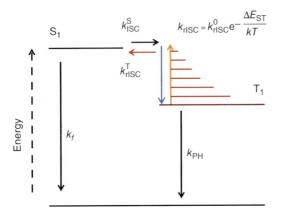

Scheme 12.1 Simplified schematic energy level diagram for an E-type process with transition rates shown.

triplet excited state [13, 14]. E-type delayed fluorescence was described by Parker et al. [15] as the thermal activation of a triplet state to a higher-lying vibronic state followed by "reverse" intersystem crossing (rISC) to a resonant singlet vibronic state, giving the first real experimental proof of the existence of unique triplet excited states in organic materials. It was termed E-type delayed fluorescence having been first observed in eosin dyes. It is distinct from P-type delayed fluorescence that arises from triplet annihilation, often incorrectly ascribed as E-type [16], and is directly proportional to the energy gap between the triplet and singlet states, ΔE_{ST}, as depicted in Scheme 12.1. As will be shown later, this is a very simplified overview of the physics involved in the rISC step. However, the term "TADF," or thermally activated delayed fluorescence, now used throughout the OLED literature was first used by Wilkinson and Horrocks as far back as 1968. In order to maximize E-type delayed fluorescence, ΔE_{ST} should be minimized. As predicted by Beens and Weller [17], intramolecular charge-transfer (ICT) states where the electron and hole are decoupled on different orbitals of the donor and acceptor fragments of the system have zero exchange energy to first order, giving rise to E-type delayed fluorescence [18]. This is because they are separated spatially to a large extent and do not interact, or in some cases the donor and acceptor fragments are orthogonal and again the two electrons do not interact [19]. E-type delayed fluorescence remained a scientific "curiosity" until very recently because no molecules with low enough ΔE_{ST} were discovered and almost all work was carried out in solution, not solid state.

The two-step nature of the rISC is shown in this scheme, whereby thermal activation is required to increase the triplet state to a vibrational level isoenergetic with one in the singlet manifold such that an adiabatic spin flip can proceed. This is required because the energy change during the spin flip is negligible, $\ll \mu J$. The overall rate of rISC is thus dictated by both the Boltzmann term and the rate of the spin flip, k^0_{rISC}. In the case of charge-transfer (CT) molecules however, this is not the case, as explained here and in the Chapter 9 by Tom Penfold.

The recent change in fortune for E-type delayed fluorescence is told in the opening chapters of this book, and the term *thermally activated delayed fluorescence* or "TADF" reused by Adachi et al. [20] ICT molecules based on electron- and hole-transporting molecules used in OLEDs were found to have sufficiently small ΔE_{ST} such that the TADF effect greatly enhanced OLED efficiency [11, 21], up to a point where near 100% internal efficiency has been reported [22, 23]. Similar progress was also made in Cu(I) metalorganic complexes as well (see Chapter 1 by Yersin et al.), although devices are more difficult to fabricate with these systems. Surprisingly, the ΔE_{ST} gap measured for most of these early systems was still rather large, much greater than the expected difference between the singlet and triplet CT states in a simple ICT system (<100 meV), but could yield efficient TADF. For example, in a series of materials containing electron donor (**D**) and electron acceptor (**A**) units, and **D–A–D** structures, showing singlet–triplet energy gaps (ΔE_{ST}) from 0.32 to 0.54 eV and efficient TADF in the blue spectral region [23], the authors tentatively describe the origin of TADF to a mechanism that involves reverse internal conversion (rIC), from the donor-centered $^3\pi\pi^*$ lowest triplet state to the ^{3}CT, followed by rISC to the ^{1}CT emissive state. However, the total energy gap between the lowest singlet ^{1}CT and triplet $^3\pi\pi^*$ states is still surprisingly high in order to give such efficient TADF, when compared with the 25 meV, kT value at room temperature (RT).

From these initial promising findings using TADF to harvest triplets, we set out to use a range of different spectroscopic measurements to try to unravel the mechanisms that drive and control both rISC and hence TADF. We established that the energy levels involved are more complex than previously appreciated, with heteroatom lone pair orbitals playing a key role in this process, and that the key excited states involved in rISC are the lowest-energy local triplet state along with the singlet and triplet CT states. In this chapter we show the types of measurements we have employed and how the analysis of these results has led us to a detailed understanding of rISC and TADF, as well as factors that greatly influence TADF efficiency in devices.

In addition to ICT molecules, the CT states formed in an exciplex (excited-state complex) between donor and acceptor molecules also give rise to small ΔE_{ST} gap, and exciplex TADF devices with enhanced efficiency have been subsequently demonstrated [11, 21]. However, as our spectroscopic studies have shown, the picture in exciplexes is made more complex by structural inhomogeneity [24]. This leads on to very complex, mixed photophysics with contributions from dimer states [25] as well as exciplex delayed fluorescence and strongly competing TTA [16]. Thus, for clarity, in this chapter, only measurements on ICT systems are given so that a clearer picture can emerge. However, from a large-scale investigation of exciplex systems, we have been able to conclude that the same basic underlying photophysics and mechanisms control TADF in exciplexes as we find in the ICT molecules but the dispersion caused by morphological inhomogeneity smears out all responses and the competition with TTA is far more prevalent than in ICT systems. The reader is pointed to the work of dos Santos for a clearer understanding of the photophysics of exciplexes [26].

12.2 Comments on the Techniques Used in Our Studies

Basic absorption and fluorescence measurements using standard laboratory spectrometers can give a good idea about the potential efficiency of TADF in a candidate molecule, especially using slightly nonstandard measurements, described in subsequent sections. In all cases we extract excited-state energies from spectra by calculating the onset of emission. This is obvious for local excited states, but in the case of CT states, there is a tendency to take the energy from the peak position of the broad emission band. This we believe is incorrect because as with local bands, the onset gives pure electronic energy, whereas the lower-energy part of the band arises from transitions to vibronic levels of the ground state. The band is structureless because of inhomogeneity and vibronic and phonon coupling within the manifold of emitting molecules. However, these basic measurements can only go so far, and it is time-resolved measurements that are the real workhorse here. The main problem though is the time regime of most interest, the microsecond, because this is the most neglected in time-resolved spectroscopy. Whereas both picosecond and nanosecond fluorescence lifetime measurements are now rather routine, especially using time-correlated single-photon counting techniques [27], measuring emission at longer times is not common, and moreover, to study delayed emission, the (usually) far stronger prompt emission must be gated out to give sufficient signal to noise to allow proper kinetic measurement and lifetime determination of the delayed component to be made. In Durham we have been developing time-resolved gated emission measurements to study both delayed fluorescence [27, 28] and phosphorescence (PH) [29, 30] from OLED materials for over a decade. Our methodology is based on the use of intensified CCD cameras and 100 ps pulse Nd:YAG (355 nm) and 1 ns pulse nitrogen laser (337 nm) sources. The iCCDs can be gated within 100 ps of the excitation pulse and detection delays ranging from 200 ps to 100's of milliseconds programmed. This makes them ideal tools to study delayed emission, and with careful sliding gate and delay protocols, single measurements can be made with very high dynamic range, up to 9 decades of time and 13 decades of intensity [31, 32]. These measurements give not only the lifetime of the various decay channels seen in these ICT materials but also the emission spectrum at each delay time so that contributions from various species can be differentiated and assigned. Further, and very importantly, the intensity dependence of the delay emission can be measured to unambiguously differentiate between TADF and TTA and geminate pair recombination [33].

12.3 Basic Absorption and Emission Properties

Photophysical measurements in isolation on one material tell you a lot about that material but little about the general photophysics and mechanisms underpinning the class of molecular responses. Thus it is vital for the spectroscopist to work hand in hand with chemistry colleagues to generate series of materials and next-generation materials designed from what is learned from photophysics. In

Durham we are very fortunate to have Prof. Martin Bryce and his team in chemistry whom we have worked with for a very long time; without their input we could not have made much progress in this field. Our first ICT materials were a family based around our then newly designed electron-transport building block (the acceptor unit, **A**) dibenzothiophene-*S,S*-dioxide, with weak donor groups (**D**) based on fluorene units [34, 35]. These systems are more classical ICT systems and display the characteristic responses of CT excited states but little hint of TADF as the ΔE_{ST} is large, but from them though the basic **D–A–D** motif was born and a new generation of materials synthesized. A systematic series of **D–A–D** materials comprising five different core **A** units substituted with five different **D** units was subsequently synthesized [23, 36, 37]. These then led on to new generations of materials with very large TADF efficiencies. Various evolutions of the **D–A–D** (and analogous **D–A** molecules) motif are shown throughout the chapter, where possible X-ray crystal structures are also given to show the orientation of **D** and **A**, which is very informative, even if somewhat of a single "snapshot" of the molecular structure. We have found through careful study of pure compounds very little difference in the photophysics and optical properties of **D–A–D** and **D–A** analogues [38]. What is important for all efficient TADF molecules we have found so far is the bonding pattern between **D** and **A**; they should be bound through a nitrogen atom at the *para* position (with respect to the A sulfur heteroatom) of the **A** unit, which in most cases leads to near orthogonality between **D** and **A** unit. For a very typical TADF ICT molecule, phenothiazine-dibenzothiophene-*S,S*-dioxide (PTZ-DBTO2), this can be seen in the X-ray structure of the molecule (Figure 12.1).

Consistent with the single-crystal X-ray data, showing potentially very weak ground-state electronic coupling between the **D** and **A** units, the absorption spectrum of PTZ-DBTO2 is almost entirely formed by the superposition of the **D** and **A** fragment absorptions (Figure 12.2). This confirms the very limited conjugation

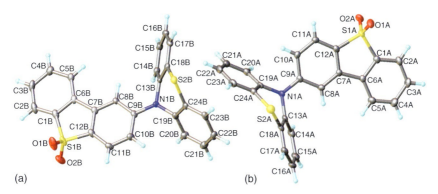

Figure 12.1 X-ray crystal structure of phenothiazine-dibenzothiophene-*S,S*-dioxide. The asymmetric unit contains two molecules (a and b) of similar conformation. In both, the dibenzothiophene system is planar, while the phenothiazine moiety is folded along the N…S vector by 47.0° (a) and 52.6° (b). The N atom is nearly, but not fully, planarized (sum of bond angles 356.0° and 358.1°); its bonding plane is inclined by 10.2° (a) and 5.2° (b) to the dibenzothiophene plane. The angle between dibenzothiophene and the C(9)N(1)S(2) plane equals 83.2° (a) and 87.1° (b).

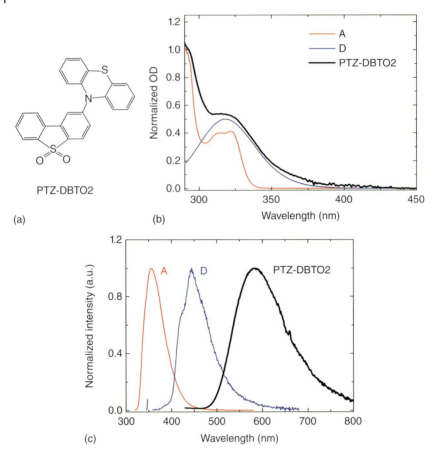

Figure 12.2 (a) Chemical structures of PTZ-DBTO2. (b) The absorption spectra of the **D** and **A** fragments as well as that of PTZ-DBTO2 in dilute toluene solution. (c) Fluorescence spectra of the same in toluene solution.

between the **D** and **A** units, i.e. electronically decoupled nature of the **D** and **A** fragments in the molecule, such that the **D** and **A** are effectively isolated electronically from one another, which is imposed by the nearly perpendicular **D–A** orientation. The emission of the **D–A** molecule is strongly redshifted compared with the emission of either **D** or **A** fragments. The fluorescence from PTZ-DBTO2 appears broad and featureless, as is typically observed from excited states with strong CT character, in both solution and solid state [20]. This Gaussian band shape is not dependent on the polarity of the medium in which the PTZ-DBTO2 is dissolved, unlike ICT molecules with weaker CT or nonperpendicular **D–A** structure. In these cases structured emission is usually seen in nonpolar solvents. The emission Stokes shift is large and is dependent on solvent polarity. The emissive singlet state in PTZ-DBTO2 is thereafter identified as ^{1}CT.

More careful study of the PTZ-DBTO2 absorption spectrum compared with its **D** and **A** fragments reveals a very small "tail" absorption to the red, 400–500 nm region ($\varepsilon \sim 100$ M^{-1} cm^{-1}), indicative of a weak n–π* transition, but which is

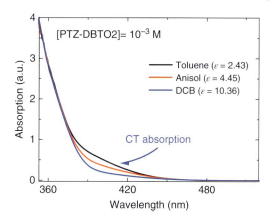

Figure 12.3 Absorption of PTZ-DBTO2 obtained at high concentration (10^{-3} M) in solvents of increasing polarity (DCB = 1,2-dichlorobenzene). A weak direct CT absorption is identified at the onset of the main absorption band.

found to be solvent dependent with a weak redshift on increasing solvent polarity, consistent with a mixed n–π^*/π–π^* transition [19, 39–42] (Figure 12.3). Further, excitation into this very weak band still gives rise to ^{1}CT emission and is thus identified as a direct CT absorption: ^{1}CT ← S_0.

However, the exact nature of this absorption feature is hard to reconcile. Take 2,7-bis(phenoxazin-10-yl)-9,9-dimethylthioxanthene-S,S-dioxide (DPO-TXO2) as another example, a **D–A–D** ICT molecule, again with near-perpendicular **D** and **A** units, but with a more deformed **A** structure and **D** with a different heteroatom opposite the bonding N (Figure 12.4). The normalized optical absorption spectra of the **A** and **D** units as well as DPO-TXO2 molecule, measured in dilute methylcyclohexane (MCH), are also shown in Figure 12.4. The inset of Figure 12.4 shows the absorption peak around 390 nm in different solvents. In the DPO-TXO2 absorption spectrum, the first two absorption peaks (higher energy) match well with a linear combination of the **A** and **D** absorption. The third peak, around 390 nm, does not appear in neither the **D** nor the **A** absorption spectra and is much stronger when compared with the tail emission seen in PTZ-DBTO2. This band shows a slight redshift on increasing solvent polarity, which again suggests a n–π^*/π–π^* mixed-state transition. We again can assign this peak to the direct ^{1}CT absorption. This is a relatively strong transition confirming a more pronounced π–π^* character [41]. The exact nature of these "CT" absorption bands, the degree of state mixing, and the transfer of oscillator strength from A to D or vice versa are as yet not well understood.

It is interesting to note that PTZ-DBTO2 has more efficient TADF (as defined by the measure DF/PF ratio; see Section 12.5.1) than DPO-TXO2 and makes much more efficient OLEDs, and correlation between this direct CT absorption band and the efficiency of TADF is not trivial. The reason for this we believe arises from the nature of the wave function of this ground state. Marian [40] has shown that many of the states in these **D–A** systems have highly mixed n–π^* and π–π^* character and as such cannot be simply interpreted. More work is required to fully understand these low-energy low oscillator strength bands in the highly twisted ICT molecules. However, for weaker ICT molecules, we have observed strong competition between planarity and twisted ground-state configuration as

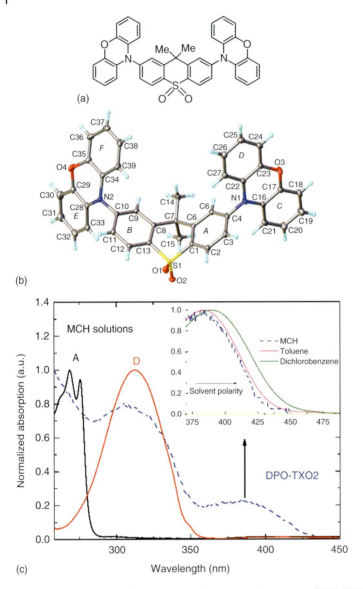

Figure 12.4 (a) and (b): The structure and X-ray crystal structure of DPO-TXO2. The optical absorption spectra of the **A** and **D** fragments and DPO-TXO2 molecule, all measured in diluted in methylcyclohexane (MCH). The inset shows the absorption spectra peak around 390 nm of DPO-TXO2 in different solvents clearly showing the band to redshift with increasing polarity.

a function of polarity driven by the competition between π conjugation and steric hindrance [42].

As we have seen from the data so far, the emission spectra of the ICT molecules are a broad Gaussian profile, redshifted with respect to the individual **D** and **A** fragments. In the systems with large dihedral angle between **D** and **A**, the CT excited state is stabilized in all environments, nonpolar as well as polar. This

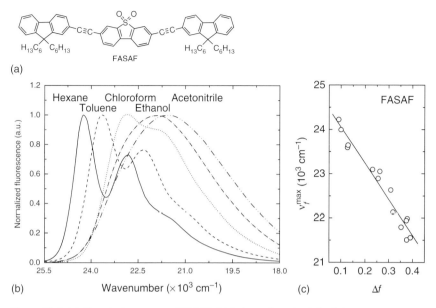

Figure 12.5 (a) Chemical structure of FASAF. (b) Solvatochromic shift of the emission spectrum of FASAF in solvents with different orientation polarizability Δf. (c) Solvatochromic shift of the emission maximum, ν_{max}, FASAF against the solvent polarity parameter.

is indicative of strong CT in these systems, that is to say, near full transfer of an electron from **D** to **A**. In other ICT systems where the geometric structure is not so well defined, or the CT is weaker, this is not the case, and the CT band is only observed in polar solvents, which help to stabilize the CT. Strong solvatochromism is observed however in all CT bands [34, 43]. In the cases where the CT state is not stabilized in nonpolar solvents or host materials, only local **D** (or **A**) emission is observed. This we term ^{1}LE emission arising from the lowest-energy "local" excited singlet state. Such behavior is exemplified by FASAF, Figure 12.5.

The plots of the emission maxima, ν_{max}, of FASAF as a function of the solvent orientation polarizability parameter (Lippert–Mataga Eq. (12.1) [44, 45]) (Figure 12.5)

$$\Delta f = \frac{\varepsilon - 1}{2\varepsilon + 1} - \frac{n^2 - 1}{2n^2 + 1} \qquad (12.1)$$

showed a good linear relationship (correlation coefficient of 0.98), with a decrease in the energy of the emission maximum with an increase in Δf, showing that the excited-state dipole moment is larger [46] for FASAF, compared with its ground-state dipole moment. Note that no significant differences were found between protic and nonprotic solvents, showing that hydrogen bonding is not responsible for the observable redshift. The Lippert–Mataga plot is also linear with no inflection showing a simple evolution from one state, the ^{1}LE state, given by the well-structured blue emission in weakly polar solvents, to the ^{1}CT state, highly redshifted Gaussian band in highly polar solvents. The redshift and loss of structure is due to the increasing stabilization of the CT character of the excited state:

$$\Psi_{\text{Exc}} = c_1 \,|\, D^*A >_{\text{Loc}} + c_2 \,|\, DA^* >_{\text{Loc}} + c_3 \,|\, D^+A^- >_{\text{CT}} \quad (12.2)$$

As the polarity of the surrounding medium increases, the charge separation increases, and thus the CT character of the excited state increases. Equation (12.2) also explains how one can observe dual fluorescence, i.e. emission from both ^{1}LE and ^{1}CT states at the same time.

The energy of the CT state is given by the following equation to first order:

$$h\nu_{\text{EX}}^{\text{max}} \approx\sim I_D - A_A - E_C \quad (12.3)$$

where I_D is the ionization potential of the **D**, A_A the electron affinity of the **A**, and E_C the Coulomb energy term from the charge separation. Thus, as the charges separate further or the charges are screened by the surrounding (polar) environment, Ec increases, and the energy of the CT state decreases, i.e. redshifts. In weak CT states, much of the relaxation process is a "twisting mechanism," TICT [44], assuming that the molecule is capable of internal rotation, leading to a 90° twist between the **D** and **A** units in the final equilibrated CT state.

One can study this relaxation mechanism in further detail using temperature. If a rotation about the **D–A** bond is active in stabilizing the CT state, then it can be frozen out as the solvent freezes and a marked change in the emission should be observed, as is seen in **D–A–D** molecules where the **D** and **A** and bonded through a C—C link [42] (Figure 12.6).

For ICT molecules such as PTZ-DBTO2, where **D** and **A** are bounded through a nitrogen atom, which is already highly twisted in the ground state, we observe stabilization of the CT state even in nonpolar solvents, thus favoring the TICT picture, but with an already ideal molecular structure in the ground state. However,

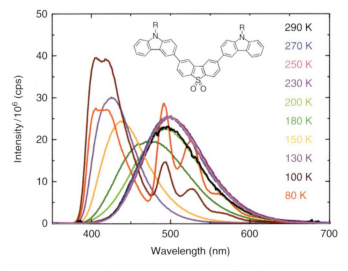

Figure 12.6 Temperature-dependent emission of DCz-DBTO2 (structure given) dissolved in ethanol. At room temperature the emission is strongly CT in character. On cooling one observes the onset of dual fluorescence (^{1}LE + ^{1}CT). Upon freezing of the solvent, ^{1}LE emission dominates accompanied by phosphorescence from the ^{3}LE state.

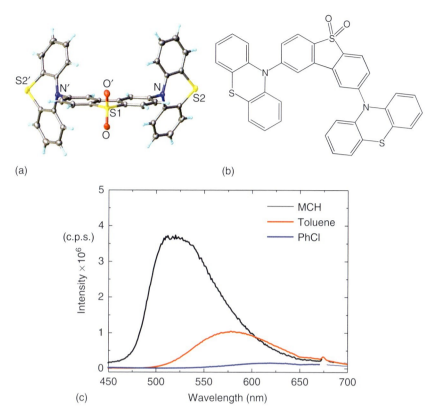

Figure 12.7 Chemical structure, X-ray crystal structure, and solvatochromic shift observed in DPTZ-DBTO2. Strong ^{1}CT emission is observed in nonpolar MCH solution that redshifts with increasing solvent polarity, accompanied by an increasing loss of emission yield.

again strong solvatochromism is observed, identifying the role of the polarity shielding the separated charges (Figure 12.7). This is one of the reasons why only the **D–A** and **D–A–D** molecules having such a highly twisted ground state give efficient TADF, especially in devices where weakly polar host materials are used.

For these **D–A–D** twisted molecules, one observes strong CT emission in nonpolar solvents such as MCH, with further redshift of the CT emission in more polar solvents such as toluene and chlorobenzene, as shown in Figure 12.7. Along with this redshift one also observes strong loss in the emission yield. For the typical molecule DPTZ-DBTO2, as we see from the X-ray structure having almost perfect orthogonality between **D** and **A** units, [47] the increasing polarity cannot further relax the molecule by increasing the twist angle between the **D** and **A** units. However, from Eq. (12.3) we see that part of the energy of the CT state is governed by the Coulomb term between the two separated charges and here the solvent polarity acts to shield the charges so that they can separate further on the molecule, increasing the Coulomb energy term and thus reducing the energy of the CT state. This accounts the redshift on increasing polarity in these highly twisted molecules. The loss in emission yield is most probably accounted for by the energy gap law [48], combined with the fact that the lifetime of the more relaxed ^{1}CT state increases and thus competition with internal conversion back to the ground state also increases. We shall return to this observation later.

As we have seen in Figures 12.3 and 12.4, a weak direct CT absorption band is observed on the red edge of the main absorption band of many ICT molecules. Excitation into this band excites the ^{1}CT, the emission from which is readily detected. For DPTZ-DBTO2 we explored this transition in more detail using highly concentrated solutions, c. 10^{-2} M. The calculated extinction coefficient for this transition is of order 10^2 M^{-1} cm^{-1}. Unlike DPO-TXO2, this red edge absorption shows weak but measureable blueshift with increasing solvent polarity (Figure 12.8) indicative of a transition with n–π* character. However following Marian [40] we believe it is again a transition having considerable mixed character.

Because this red edge band is so weak compared with the neighboring transitions, there is considerable overlap between it and the higher-lying band associated with the ^{1}LE ← S$_0$ transition. Therefore, it proved necessary to excite very low down in this band to observe the pure nature of the transition. A series of measurements were made exciting at lower energies, as indicated in Figure 12.8,

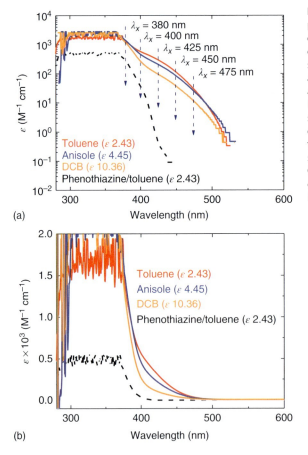

Figure 12.8 Effect of solvent polarity on the red edge, direct CT absorption spectra of DBTZ-DBTO2. On the left shows the weak red edge absorption band measured at high concentration on a log extinction coefficient scale, as well as the various excitation wavelengths used to obtain the emission spectra shown in Figure 12.9. On the right the absorption data plotted on a linear scale is shown.

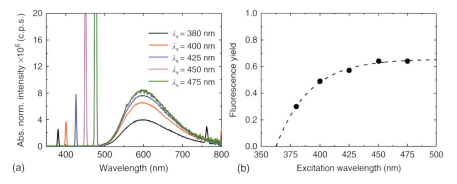

Figure 12.9 (a) Emission spectra of DPTZ-DBTO2 in toluene with varying excitation wavelengths shown in Figure 12.9. The spikes show the corresponding excitation wavelength. (b) The variation of the DPTZ-DBTO2 fluorescence yield with excitation wavelength. All measurements at room temperature.

and the emission spectra recorded (Figure 12.9). These were normalized to the absorptivity and excitation power at each excitation wavelength so that the effect on the ^{1}CT emission quantum yield could be observed. In this way we could then probe the mechanistic difference between direct and indirect CT formation. These will be discussed fully in the following section once time-resolved measurements have been introduced. At this juncture, the main observation, which is rather surprising at first glance, is the fact that exciting the "pure" direct CT absorption leads to a near doubling in the measured emission quantum yield from the ^{1}CT state as compared with excitation via the ^{1}LE state (Figure 12.9).

This near doubling of the emission yield is however only observed in the **D–A–D** molecule, and exciting the **D** unit through the red edge of its absorption band shows no increase in emission yield, unambiguously showing that this is a phenomenon related to the channel through which the CT state is directly photocreated [47]. This difference we hypothesize is due to the competition between slow electron transfer (ET) from ^{1}LE to ^{1}CT and "fast" ISC from ^{1}LE to ^{3}LE. When exciting directly into the ^{1}CT state, the competition is avoided and subsequently gives a higher emission yield. This result has direct ramifications on device performance. As is well known in an OLED device within the emitter to region, electrons are injected into the LUMO of the emitter and holes into the HOMO. For an ICT molecule this means that the electrons are injected into the acceptor unit, which possesses the lowest-energy LUMO, and the holes are injected into the donor unit, which has the highest-energy HOMO. Electrochemical cyclic voltammetry measurements confirm this to be the case for all ICT molecules measured to date [38]. This of course means that the CT state is directly created on charge recombination on the ICT molecule, avoiding the excited donor species, and from our observations above, this implies that the ICT molecule when acting as the emitter in the OLED will have a much higher emission quantum yield than measured in a standard integrating sphere measurement on the sample film. In the case of DPTZ-DBTO2, this has been directly measured, devices made using this material as the emitter give greater than 19% EQE requiring a photoluminescence quantum yield of the emitter of order 65%, whereas a standard measurement of the photoluminescence quantum yield of DPTZ-DBTO2 returns a value of 30%, clearly showing that recombination in the device does not form excited donor units but instead directly creates the CT excited state, which as we have seen above then has a photoluminescence quantum yield of the required 65%. This observation gives great insight into the charge recombination and excited-state formation step within an OLED.

12.4 Phosphorescence and Triplet State Measurements

As seen in Figure 12.6, strong PH can be observed in many ICT molecules, even at RT in some cases [23, 38, 39]. Being able to accurately measure PH from the spectra to extract the energy of the lowest triplet state is one of the most important measurements to make on TADF systems.

In Figure 12.10 a typical PH spectrum of a **D–A–D** molecule (DPTZ-DBTO2) is shown, compared with the PH of the individual **D** and **A** units. Sample films were prepared by spin coating with sample–zeonex ratio of (1 : 20 w/w) onto sapphire substrates. Zeonex is a commercial branched polyolefin that is primarily used to make CDs and makes an ideal inert host material with excellent film-forming properties and optical transparency. The PH spectra are measured using the gated iCCD method, with a delay time between 1 and 10 ms and integration typically 50–100 ms. These measurements are explained further in the next section. Zeonex is also close to MCH in polarity, so a nonpolar host. For DPTZ-DBTO2 a well-structured PH emission is observed very close in energy to the ^{1}CT emission. However, on careful inspection, when compared with the PH from its constituent **D** and **A** units, it is clear that the DPTZ-DBTO2 PH is a linear combination of both **D** and **A** PH. Again this reflects the electronically decoupled nature of the **D** and **A** fragments in the molecule. Moreover, this leads to potential errors in estimating the triplet energy of the molecule. Clearly,

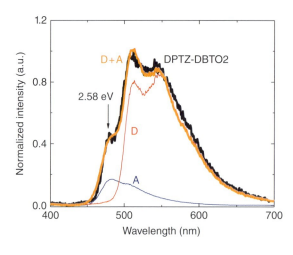

Figure 12.10 Phosphorescence spectra of DPTZ-DBTO2 (see Figure 12.5) in zeonex at 80 K (black), compared with the phosphorescence of the individual **D** (red) and **A** (blue) fragments.

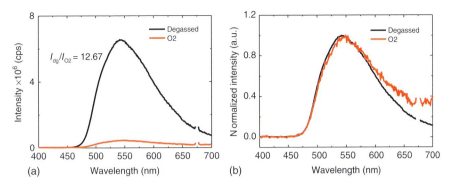

Figure 12.11 (a) A comparison is shown between aerated (red) and degassed (black) steady-state fluorescence spectra of DPTZ-DBTO2 in MCH solution at RT. Upon degassing the solution, the integrated fluorescence signal increases by a factor of c. 13. (b) The normalized emissions obtained in aerated and degassed solutions very closely match each other, confirming that the same CT state gives rise to the emission in both cases.

the lowest-energy triplet state of the molecule is the ^{3}LE state of its **D** fragment and cannot simply be taken from the onset of the PH spectrum. This is a very critical observation dictating that measurements of triplet states in TADF molecules must be made with care and good resolution and be compared with **D** and **A** units for completeness. We observe such dual PH in many TADF materials.

One of the most simplest spectroscopic methods for determining the nature of the states involved in the generation of delayed fluorescence in TADF molecules is to use oxygen quenching to quench all triplet states from the system, and from this, one can determine what proportion of the delay fluorescence arises from an excited state, which sometime in its lifetime has been a triplet state [49]. In Figure 12.11 the result for DPTZ-DBTO2 is shown; this material shows a particularly strong oxygen dependence on its delayed fluorescence. Comparing the integrated fluorescence from degassed and oxygenated solutions, we find a ratio of 12.67, equating to circa 95% of all ^{1}CT excited-state emission arising through a channel in which the excited state has passed through a triplet state, thus the triplet formation quantum yield, and only 5% of ^{1}CT states emit promptly, i.e. before they have time to cross to the triplet manifold. These experiments require rigorous degassing. Typically we use five freeze–thaw cycles to remove all traces of oxygen from the solution before measurement [47]. Although this experiment clearly shows that the vast majority of delayed fluorescence comes from states that had formal triplet character at some time in its life, it does not tell you which triplet state it was, i.e. CT or local excited, nor how the triplet was originally formed, i.e. ^{1}LE → ^{3}LE or ^{1}CT → ^{3}LE or ^{1}CT → ^{3}CT. This simple experiment is an excellent first step in the identification of potentially efficient TADF molecules. We shall subsequently show that this measurement correlates well with the measured ratio of prompt to delayed emission in time-resolved spectroscopic measurements where rISC is moderately slow.

12.5 Characteristics of the Delayed Fluorescence

12.5.1 Time-resolved Emission in Solution

The evolution of emission from any typical TADF **D–A** system can be measured using the gated time-resolved emission measurements as outlined in the experiment section of this chapter [50]. Depending on the absorption spectrum of the molecule to be investigated, we typically use a 355 nm or 337 nm pulsed laser as the excitation source and the gated iCCD detection system. Measurements can be made either in solution or in solid state – solution measurements as a function of temperature are more difficult because of the fact that many solvents do not form stable glass at low temperatures, limiting the temperature range over which the measurements can be made. However, it is very informative to make such time-resolved measurements on molecules in different solvents of different polarities to see how the various components of emission change as a function of polarity as well as to make the measurement as a function of temperature. To start, a typical set of data for a good TADF material in solution is discussed before going on to solid-state measurements where the interpretation of the decay curves is a little more involved. For all good TADF materials, the same characteristic decay curves are measured; however the decay times of various different decaying emissions change markedly.

The delayed fluorescence in nonpolar MCH and more polar toluene solutions of DPO-TXO2, from the early prompt emission (time delay, TD = 1.1 ns) to the end of the DF (TD = 0.14 ms), at different temperatures is shown in Figure 12.12a. DPO-TXO2 has weak ion pair character in the CT state in nonpolar environments, i.e. the state retains some local excited-state character [41]. The contribution of triplet excited states to the overall emission was determined by comparing the emission intensity in aerated and degassed solutions. The time-integrated emission spectra in degassed and nondegassed solutions matched each other, showing that DF and prompt fluorescence come from the same ^{1}CT state. The CT emission intensity in MCH increases by a factor of 3.10 when oxygen is removed, indicating a "through triplet" contribution to delayed fluorescence of 52%.

The decay curves for DPO-TXO2 are rather complex, and prompt and delayed emission regions are not well defined, but the DF emission has clearly higher intensity at high temperatures, indicating TADF. The analyses of the normalized spectra (Figure 12.12b) taken over the entire measurement region show that at c. 1.1 ns (and earlier) the emission spectra observed match that of the ^{1}LE$_D$ emission of the donor unit. In DPO-TXO2 it is possible to identify the two types of the donor emission, from separate measurements of the donor unit photophysics [51]: a peak around 375 nm identified in steady-state measurements as the phenoxazine singlet emission and a dimer emission, peak around 450 nm, that can be seen only in the time-gated measurements [41]. However, in the ICT molecule, the ^{1}LE$_D$ is sufficiently quenched by ET that the dimer emission appears as the significant contribution to the very early time prompt emission (first few nanoseconds). This is a rather unique system but highlights the possible complexities these ICT systems can show. Increasing the time delay (TD), the emission spectra progressively shift to longer wavelengths, moving from the ^{1}LE$_D$ dimer emission

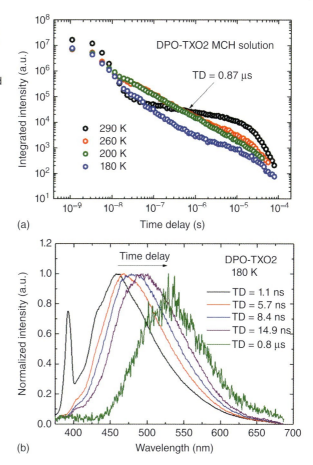

Figure 12.12 (a) Time-resolved fluorescence decay of DPO-TXO2 in MCH solution. The curves were obtained with 355 nm excitation. (b) Time-resolved normalized emission spectra in the entire region of analyses.

to that of stabilized ^{1}CT emission, passing through a region where dual emission is observed, indicating very slow ET, with a typical rate, $K_{ET} \sim 10^8$ s^{-1}. This is very common in the highly twisted D–A molecules. At TD = 14.9 ns, the CT emission is clearly observed and has onset at (2.94 ± 0.02) eV. However, the emission shows a slight redshift at very late times, indicating a contribution from donor PH. After TD = 0.8 μs, the PH emission is stabilized with onset at (2.78 ± 0.02) eV and was collected until 0.14 ms. The observed triplet emission could come from the localized triplet state of the donor and/or the acceptor. We measured the PH of the donor and acceptor units separately to identify this behavior. The ^{3}LE$_D$ PH of the donor unit was found to have onset at (2.79 ± 0.02) eV in toluene solution, corroborating with the PH spectra obtained for DPO-TXO2 and also with the triplet levels of phenoxazine reported previously in other solvents. [51] The PH spectra of the acceptor unit were found to have onset at (3.39 ± 0.02) eV, higher energy than the PH of DPO-TXO2. In this way we identify the triplet emission observed in DPO-TXO2 to come from the donor units. Thus, the ^{1}CT state and the ^{3}LE$_D$ state were identified in a MCH solution, and the splitting between these states was found to be $\Delta E_{ST} = (0.16 \pm 0.03)$ eV.

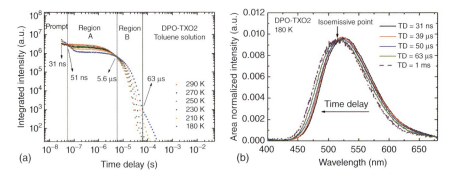

Figure 12.13 (a) Time-resolved fluorescence decay of DPO-TXO2 molecules in toluene solution in different temperatures. The curves were obtained with 337 nm excitation and data collected from 31 ns onward. (b) Time-resolved area normalized emission spectra in the entire region of analyses.

The delayed fluorescence in slightly more polar toluene solution was also measured and clearly demonstrates how sensitive the CT state is to environment as the CT state is strongly stabilized (Figure 12.13). The contribution of DF to the overall emission in DPO-TXO2 was found to be 4.8% or 82% from degassing studies, indicating a much strong ion pair character to the CT state. Figure 12.13a shows the decay curve of DPO-TXO2 in toluene solution. There is a clear biexponential decay with a fast component of $\tau_{PF} = 24.8$ ns, assigned to a prompt fluorescence ^{1}CT component, and a longer decay of $\tau_{DF} = 4.61$ μs, assigned to the delayed fluorescence ^{1}CT component. The ratio between I_{DF} and I_{PF}, the intensity integrals of DF and PF regions, respectively, was calculated and found to be $\overline{n} = 4.81$, where $\overline{n}$ is emission contribution of TADF. This is in excellent agreement with the value calculated by the degassing test in toluene solution ($\overline{n} = 4.80$). From this, the rISC rate constant was determined using Eq. (12.4) and to be 1.04×10^6 s^{-1} [5]:

$$k_{rISC} = \frac{\int I_{DF}(t)dt}{\int I_{PF}(t)dt} \cdot \frac{1}{\tau_{DF}} \quad (12.4)$$

The decay curves of DPO-TXO2 in toluene solution at different temperatures clearly exhibit four distinct regions, delineated by three well-defined crossing points, i.e. isoemissive points: 51 ns, 5.6 μs, and 63 μs. The decay below 51 ns is dominated by ^{1}CT → S$_0$ prompt fluorescence, and between 51 ns and 63 μs is the ^{1}CT → S$_0$ delayed fluorescence. ^{3}LE$_D$ → S$_0$ PH is also clearly observed in the region from 63 μs to 1 ms. The isoemissive points show that these are separate well-defined species that are independent of each other. As a function of temperature, the prompt emission increases in intensity on cooling, and at 180 K the onset of DF occurs at later times, showing increased prompt emission dominating the decreasing DF component as less thermal activation occurs. In region A, the DF emission is clearly observed to increase in intensity with increasing temperature, indicating the thermally activated nature of the DF mechanism. As can also be seen, the DF intensity is nearly time independent, a characteristic of efficient TADF and a dispersion in the rISC rate that gives rise to a more power law

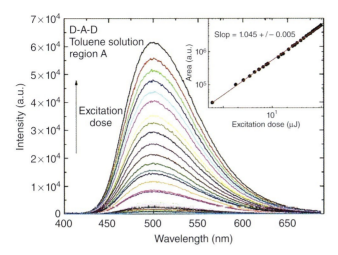

Figure 12.14 Delayed fluorescence spectra of DPO-TXO2 at different laser excitation doses. Inset graph shows the area of each spectrum as a function of excitation dose. Spectra collected with TD = 100 ns.

decay. [24] Again, this shows the great sensitivity of the rISC rate of each emitter to local environment. The intensity dependence of the DF emission in this region was analyzed as a function of the laser excitation dose, and a linear gradient of 1.045 ± 0.005 was found (Figure 12.14).

This result confirms the thermally assisted mechanism as opposed to TTA and is a very important measurement to make for every material to confirm that the temperature-dependent DF component is TADF, not TTA or a combination of both [52]. In region B the intensity of the emission increases as the system temperature decreases, mirroring the behavior of the prompt emission. The DF intensity dependence spectra as a function of excitation dose in this region also show a linear gradient (1.025 ± 0.004), confirming that this region also has monomolecular thermally dependent character. The behavior of the temperature dependence in this region is unexpected and can be understood by analyses of the triplet lifetime (τ_T), given by Eq. (12.5):

$$\tau_T = \frac{1}{k_{rISC} + k_T} \tag{12.5}$$

where k_{rISC} is the rate of the rISC process and k^T is the sum of nonradiative (NR) and radiative triplet decays. At high temperature, $k_{rISC} \gg k^T$, and the triplet lifetime is τ_T. At low temperature, k_{rISC} and k_T decrease due to their temperature dependence, and consequently, τ_T increases, making the decay curve of DPO-TXO2 longer lived, so PH occurs with higher efficiency at longer times where TADF no longer occurs. Hence, we can see a crossing point at TD = 5.6 μs in the decay curves in different temperatures indicating the turning point in this TADF/PH competition. Also, the dielectric constant of toluene slightly increases with decreasing temperature [53]. Hence, at lower temperature, the CT state has slightly lower energy, leading to an increased ΔE_{ST} and consequently longer

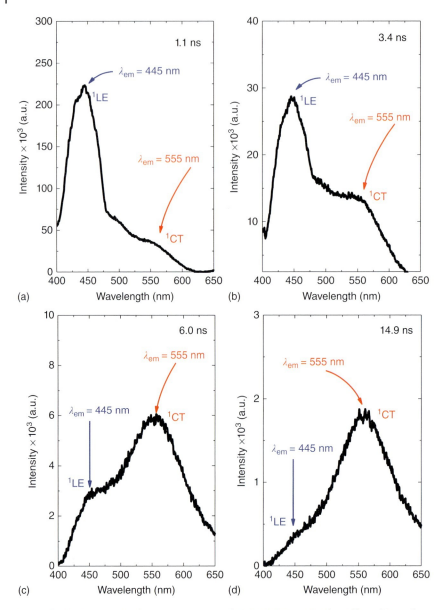

Figure 12.15 Time-resolved emission spectra of DPTZ-DBTO2 in CBP host film, obtained over the first 15 ns following excitation. Evolution of the decay of the local donor excited-state ^{1}LE by electron transfer to form the singlet charge-transfer excited-state ^{1}CT at (a) 1.1 ns, (b) 3.4 ns, (c) 6 ns, and (d) 14.9 ns.

triplet lifetime. Beyond region B the longest-lived component of the decay curve is observed. This appears just at low temperatures, vanishing at RT, and is assigned to the long-lived ^{3}LE$_D$ PH.

Figure 12.13b shows the area normalized emission spectra at 180 K, giving the peak position of each spectrum. From TD = 31 ns to 39 μs, all the spectra have the same position, onset at (2.70 ± 0.02) eV, and same spectral shape. From TD = 39 to 63 μs, a very slight and continuous blueshift is observed, due to the growing contribution of ^{3}LE$_D$ PH. The prompt and DF emission, to TD = 63 μs, have almost identical spectral shape, showing that both originate from the same singlet excited state, again as one must expect for TADF. After 63 μs the spectra reach an onset at (2.77 ± 0.02) eV, and the same emission was collected until TD = 1 ms. The isoemissive point (labeled on the figures) gives a clear indication for the presence of two excited-state species, firstly the ^{1}CT state and then a triplet state (^{3}LE$_D$). The PH emission has onset very close to that found in MCH solution (2.78 ± 0.02) eV, showing that the PH of the donor does not change in different solvents, as expected for a neutral local excited state. Thus, the ^{1}CT state and the ^{3}LE$_D$ state are identified in DPO-TXO2 in toluene solution, and the splitting between these states was found to be ΔE_{ST} = (0.07 ± 0.03) eV. Additionally, the activation energy (E_A) required for the DF process was calculated from an Arrhenius fit of the temperature-dependent DF emission intensity and found to be E_A = (0.031 ± 0.005) eV. The difference between these two values is rather insightful [54]. From modeling of the rISC process, described in Chapter 9 (Penfold), these two independent measurements actually measure two different properties of the rISC process: The optical one measures the ^{1}CT (fluorescence) ^{3}LE (PH) ΔE_{ST}, whereas thermal measurement gives the energy of the main molecular vibration that couples the singlet ^{3}LE and CT states, wherein 31 meV (c. 250 cm^{-1}) is indicative of a torsional rocking mode between the D and A units about the N—C bond.

Comparing both environments, we observed that in toluene solution the contribution of delayed fluorescence is higher than in MCH solution, 82% and 52% respectively. This is a direct consequence of the larger ΔE_{ST} in MCH solution. Also, the lifetime of the DF in MCH is clearly longer than in toluene, due to the larger ΔE_{ST}, corroborating measurements from other TADF emitters [55]. It is important to notice that the ^{3}LE$_D$ energy level in MCH and in toluene solutions differs by just (0.01 ± 0.02) eV less than the uncertainty of the measurement, whereas the ^{1}CT energy levels change considerably with solvent polarity, causing a big difference in the energy splitting according to the environment. Therefore, the environment has a strong influence on the ΔE_{ST} energy value through the extreme sensitivity of the CT energy to polarity and the very small energy gaps that differentially change as the CT energies change.

One may ask how the situation changes if the energy ordering of the triplet states such that the ^{3}CT state lies below the ^{3}LE state. In this situation dynamic modeling indicates that the mechanism for rISC is still the same and that vibronic coupling between these two triplet states is still required to mediate the spin flip back to the singlet CT manifold [56]. Spectroscopically, the situation is a little different with the ^{3}CT state being the lowest-energy triplet state, which becomes the reservoir for all triplet excitations. The coupling of the ^{3}CT state to the ground state will be effectively zero given that PH is forbidden between the ^{3}CT and singlet ground state and the electronic coupling between the CT states and the ground state is negligible in these highly twisted molecules. Therefore, one will not observe PH at long times in the decay of the molecule. Instead one will see a much longer DF emission as the triplet excitations slowly reverse intersystem across back to the singlet manifold. Again any dispersion in environment of the system will lead to a dispersion in rISC rates, and the decay of the DF will take on a more power-like slope as opposed to an exponential decay.

Throughout we have only seen **D–A–D** systems where it is the donor fragment of the molecule that has the lowest-energy local singlet and triplet energy levels. This is not a prerequisite, and recently unambiguous spectroscopic data has been presented for a **D–A–D** system with a rather different accept unit whose singlet and triplet local energy levels are below that of the donor units [57]. Even though it is the acceptor unit that is photoexcited and has the lowest-energy local triplet state, the photophysics of the **D–A–D** molecule and the strong TADF that it shows is identical to the other TADF molecules presented here with a donor having the lowest local energy levels.

12.5.2 Time-resolved Emission in Solid State

Solid-state measurements here refer mainly to the study of films of an emitter in a host material, either the inert, nonpolar polymer zeonex or as a coevaporated film in a typical OLED host material, not a pure film of the emitter. This avoids self-quenching issues. The initial emission in the first few nanoseconds after excitation from DPTZ-DBTO2 in CBP host (Figure 12.15) clearly shows ^{1}LE fluorescence characteristic of the donor phenothiazine unit, as confirmed from the integrated fluorescence spectra of phenothiazine. This decays with an average lifetime of c. 3 ns and is effectively quenched by ET to form the CT state, as seen by the rise of the "prompt" ^{1}CT emission. This shows how slow the ET step is when the **D** and **A** fragments are near perpendicular to one another. Compared with other simple **DA** molecules, the rate of ET in DPTZ-DBTO2, $K_{ET} \sim 10^8$ s^{-1}, estimated from the lifetime of the prompt ^{1}LE emission, is 3–4 orders of magnitude slower than in a molecule free to reorganize its geometry. [58] Because the ET rate is slow, it is in competition with both radiative decay of the initial local excited state and ISC of ^{1}LE to ^{3}LE. Given the time resolution of our iCCD camera is around 1 ns, we cannot always resolve this first ET step. This initial emission is not usually seen in more polar environments.

This early time behavior of DPTZ-DBTO2 in solvent-free rigid matrix can be understood from the basic equation governing ET, $K_{ET} = 4\pi^2/h V_{DA}^2$ (FCWD) [59, 60], where V_{DA} is the electronic **D–A** coupling, which is minimized in the ICT molecules having the near orthogonality of the **D** and **A** fragments, and the FCWD is the Franck–Condon weighted density of states, the measure of the total spatial overlap of all **D** and **A** vibrational modes. In the case where **D** and **A** are nearly perpendicular, only out-of-plane vibrational modes can thus couple to enable state crossing, also causing the FCWD to be rather small. Thus, in rigid matrix, initial intramolecular ET is extremely slow.

Over the full decay period, lasting into 10's of ms, the decay is very similar to what is observed in solution, except for the added complication of the appearance of PH at longer times (Figure 12.16).

In a nonpolar zeonex matrix, during the formation of the ^{1}CT state, over the first 10–15 ns, the prompt emission progressively shifts to longer wavelengths (from 512 nm to 560 nm) due to the growing contribution of the ^{1}CT emission

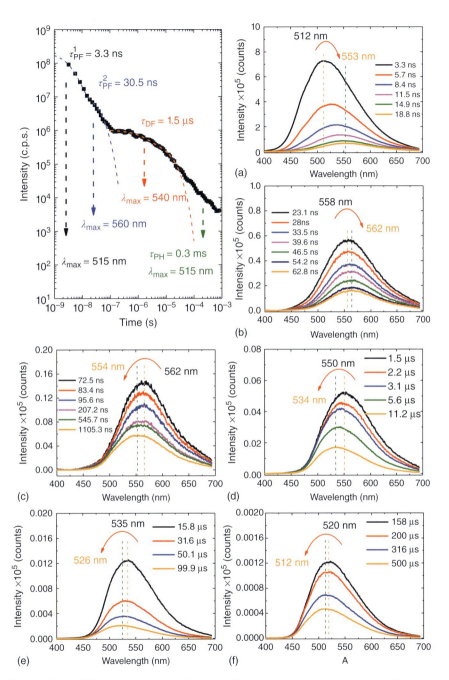

Figure 12.16 A full series of emission decay and time-resolved spectra obtained at different delay times following the intensity decay curve (left) from (a) 1 ns to (f) 1 ms, for DPTZ-DBTO2 dispersed in zeonex at RT, showing the complex evolution of the spectral decay components over many decades of time. Clearly, measuring over a limited time window can cause difficulties if interpreting the photophysics of the molecule under investigation.

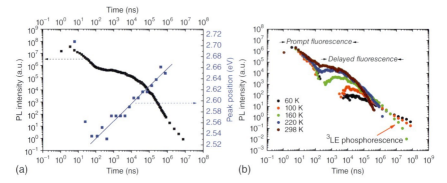

Figure 12.17 (a) Plot of the DPTZ-DBTO2 emission intensity decay in CBP, obtained at RT with the change in emission peak position also observed as a function of time. (b) Temperature dependence of the DPTZ-DBTO2 emission decay in CBP matrix clearly identifying that the DF increases strongly with increasing temperature, whereas the PF is unaffected.

(Figure 12.15). The prompt ^{1}CT state (those singlet CT states that never cycle through the triplet manifold) decays with a 30 ns time constant, measured by an exponential fit of the data (Figure 12.16). The delayed fluorescence, also from the ^{1}CT state, as seen from the time-resolved spectra, decays with a time constant of 1.5 μs, the emission peak being observed at 560 nm up to a delay time of approximately 400 ns. It then progressively shifts back to shorter wavelengths, due to the growing contribution of the underlying ^{3}LE PH. For delay times in the μs time range, the delayed emission maximum is at 540 nm (2.29 eV), and at late times, into the ms time range, the emission peaks at 512 nm, closely matching the peak of the ^{3}LE PH.

The temperature dependence of the DPTZ-DBTO2 emission decay in nonpolar CBP matrix is given in Figure 12.17. At RT the decay of the emission is very similar to that measured in zeonex because both hosts are similar in polarity and the rISC energy barriers are very similar in both, on the order of 20 meV. From 298 to 220 K, the emission decay appears almost independent of temperature, in line with this small energy barrier. Below 220 K, the ^{1}CT delayed fluorescence is progressively quenched, and at later times, the long-lived ^{3}LE PH becomes dominant. As the PH is of higher energy than ^{1}CT, this gives rise to a blueshift of emission at later times. In all cases, only the delayed ^{1}CT emission shows thermally activated behavior. These decay curves are typical for all ICT molecules that have a significant TADF contribution.

The intramolecular TADF origin of the delayed fluorescence observed in DPTZ-DBTO2 films in zeonex and CBP was again confirmed by the strictly linear dependence of the integrated delayed fluorescence intensity with excitation intensity. Moreover, the temperature dependence of the DPTZ-DBTO2 delayed fluorescence (Figure 12.17b) shows a pronounced increase of the emission intensity with temperature from 20 to 250 K, but remains practically constant above 250 K, due to the very small ^{3}CT–^{3}LE energy barrier, determined as 0.02 eV.

12.5.3 Kinetics of the ^{1}CT Prompt State

As we have seen, photoexcitation of the local singlet excited-state ^{1}LE leads on to ET to form the ^{1}CT state within the first few nanoseconds. In a polar host environment, one then observes an energetic relaxation of the ^{1}CT state over the next 50–100 ns. After this time there is no further observable energetic relaxation of ^{1}CT state (Figure 12.18). From close inspection of the measured emission spectra during this first 100 ns, we see a more pronounced loss of the blue edge of the emission band, which we ascribe to dispersion in the lifetimes of ^{1}CT states of different energy. Higher-energy ^{1}CT states have shorter lifetimes as they have more local exciton character (see Eq. (12.2)), while lower-energy ^{1}CT states that are more energetically relaxed have longer lifetimes as the excited state has more CT character. Thus, during this initial period, we see the decay of prompt CT

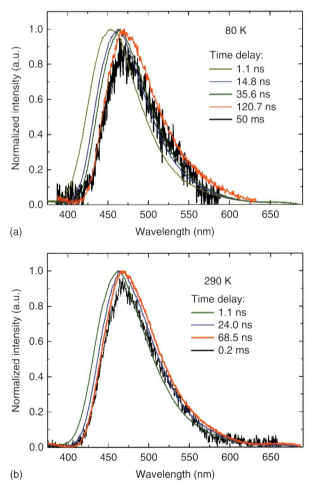

Figure 12.18 Time-resolved normalized emission spectra of DDMA-TXO2 in bis[2-(diphenylphosphino)phenyl] ether oxide (DPEPO) matrix at (a) 80 K and (b) 290 K. The first 100 ns evolution of the ^{1}CT state and the DF in the millisecond are highlighted.

states having a dispersion in lifetime. This means that it is difficult to ascribe a single well-defined energy to the ^{1}CT states. If we assume this dispersion in lifetime arises from an energetic relaxation of the state, then energy of the ^{1}CT state is the energy on initial creation, and so one should measure the CT energy from its emission spectrum at the earliest times.

12.6 Understanding Which Excited States are Involved

Taking DPTZ-DBTO2 as our model ICT-TADF molecule, in nonpolar MCH and zeonex, the CT states lie above ^{3}LE, as we have already seen in Figure 12.10. To investigate what happens to the rISC efficiency as this energy ordering changes, we can make use of the differential polarity dependence of these excited states. 20 µM solutions of DPTZ-DBTO2 in toluene ($\varepsilon = 2.38$) and 2-methyltetrahydrofuran (MeTHF) ($\varepsilon = 6.97$) were studied alongside films using zeonex and polyethylene oxide (PEO). Figure 12.19 shows the resulting steady-state emission spectra at RT. Zeonex has a similar polarity to nonpolar solvents such as MCH ($\varepsilon = 2.02$) and also has a very high glass transition temperature (375 K), meaning that there is no "solvent" reorientation over standard measurement temperatures and the CT energy and the molecular structure of the guest are fixed for all temperatures. Although PEO has a dielectric constant of $\varepsilon = 5$, this polarity is measured at microwave frequencies [61], and at optical frequencies PEO is seen to have a similar polarity to toluene at RT ($\varepsilon = 2.38$). It also has a low glass transition temperature (T_g) of 220 K, much lower than that of zeonex [62], which gives PEO a temperature-dependent polarity, whereas zeonex is a polarity inert reference. The emission peak of DPTZ-DBTO2 in toluene and MeTHF is redshifted compared with that in zeonex by 0.2 and 0.4 eV, respectively, in line with the polarity of the solvents. In PEO the emission is at about the same energy as toluene, but DPTZ-DBTO2 has a higher emission quantum yield probably due to reduced NR quenching in the very viscous polymer host. In MeTHF with high polarity, we see the characteristic high redshift through Coulomb relaxation and strong loss of emission yield as previously described. Also the relative energy splitting between ^{1}CT and ^{3}LE will be large, with ^{1}CT, and by inference ^{3}CT, shifting some 0.3–0.4 eV below ^{3}LE. Such a large energy

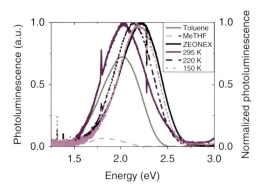

Figure 12.19 Photoluminescence of DPTZ-DBTO2 in a variety of hosts. The transparent curves of toluene and MeTHF solution (2×10^{-5} M) correspond to the left scale, and the bold lines ($\lambda_{ex} = 400$ nm) of zeonex and PEO (at 295, 220, and 150 K) correspond to the right scale ($\lambda_{ex} = 375$ nm). The reduction in magnitude for MeTHF is polarity related, while the shift of emission in PEO is related to the glass transition temperature.

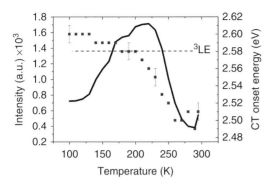

Figure 12.20 The temperature dependence of the intensity (black line) and CT onset energy (purple squares). The change in CT onset energy plateaus below the T_g, representative of the PEO film becoming rigid. The black dashed line represents the energy of the ^{3}LE at 2.58 eV, with the peak in intensity apparent as the CT energy crosses resonance. The error bars are indicative of the error in the fit of 0.01 eV for all points; the same error is expected on the ^{3}LE energy.

gap between these states will greatly reduce the efficiency of rISC and thus the contribution of TADF to total emission. For DPTZ-DBTO2 we have seen that in MCH this accounts for 95% of the total emission!

Using the two polymer hosts enables temperature-dependent measurements to be made using the key difference between the two polymers, the temperature-dependent polarity of PEO and its low T_g. By passing through T_g we can restrict vibrational motion of the guest and further affect the energetics of the system.

Figure 12.20 shows how the optical properties of DPTZ-DBTO2 in PEO change as a function of temperature, especially around the glass transition temperature. The ^{1}CT energy blueshifts as T_g is approached and then stabilizes when the PEO becomes rigid below T_g. With this increase in ^{1}CT energy, the intensity of the emission also increases up to T_g but then reduces at lower temperatures. Given that the energy of the localized ^{3}LE state is little affected by temperature and polarity, energetically this means that the ^{1}CT energy level shifts from below the ^{3}LE state, passing through resonance at about 220 K and then increasing further and stabilizing above the ^{3}LE state at low temperature. The total shift in the ^{1}CT energy onset is from 2.50 to 2.60 eV, which is a large differential shift in terms of the effect it has on the rISC rate, and is directly reflected in the shape of the temperature-dependent intensity curve.

This behavior clearly demonstrates and identifies the key role played by the relative energies of the CT and ^{3}LE states on rISC, and hence the TADF efficiency and fully supports the second-order spin–orbit coupling (SOC) mechanism [54] (Chapter 9, Penfold) where the ^{3}LE state mediates SOC by vibronic coupling, a more quantum mechanically correct picture of the SOC mechanism in CT systems where direct SOC between ^{1}CT and ^{3}CT is negligible [63, 64]. These results show the critical differences between rISC in an ICT molecule with orthogonal D and A units and simple molecular systems where ^{1}LE and ^{3}LE are energetically close enough to allow weak rISC.

12.7 Excited-state Properties

So far we have discussed the characterization of excited states that emit light, which does however preclude any understanding of dark states, i.e. those excited states that have very weak or zero radiative decay rates or metastable excited states that decay by nonradioactive transitions to lower-lying excited states. These dark states can however be characterized optically using the technique of photoinduced absorption (PIA) [65]. The most simple PIA measurement is that of quasi-steady-state PIA [66]; here the sample is excited using an electrically modulated (chopped) laser, typically a diode laser at 375 nm, and at the same time an intense white light source is used to probe the sample. After passing the sample the white light passes through a monochromator and is detected using various photodiodes and a lock-in amplifier. The lock-in amplifier provides the master oscillator (73 Hz), which electrically modulates the diode laser. Lock-in detection measures the difference in optical absorption of the sample with and without the laser excitation, at a set wavelength. This gives a normalized measure of the induced absorption, that is, the absorption of the excited state created by the laser excitation normalized to the absolute absorption of the sample at the measured wavelength. Using a scanning monochromator one can then build up an absorption spectrum of the excited states of the molecule and from this identify potentially different excited-state species that coexist. To aid identification of the different excited states, we again use the **D** unit of the ICT molecule as a reference. The lock-in technique also enables signals that are in phase with the chopping frequency and those that are out of phase to be separated; in-phase is related to species that decay on a timescale faster than the inverse chopping frequency. Here this would be the prompt ^{1}CT emission, for example, and the out-of-phase component is generally related to species that live longer than the inverse chopping frequency, such as a triplet state-induced absorption.

Figure 12.21 shows the PIA for DPTZ-DBTPO2 and the pure phenothiazine donor unit (PTZ) at RT and 20 K in zeonex. It is apparent, through comparison with the PTZ spectra, that the PIA of DPTZ-DBTO2 is dominated by the triplet

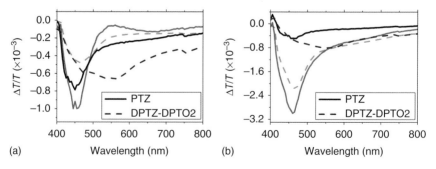

Figure 12.21 (a) The in-phase component of the PIA of phenothiazine (black line) and DPTZ-DBTO2 (purple dashes) in a zeonex host matrix, measured at room temperature (bold) and 20 K (transparent). (b) The out-of-phase component of the PIA of the same materials. (PTZ = phenothiazine donor unit).

absorption of the local phenothiazine triplet, ^{3}LE$_D$, at 20 K [67, 68]. The PIA of the acceptor dibenzothiophene has been reported at higher energies than phenothiazine in the literature [69] and thus not relevant for these spectra. At RT DPTZ-DBTO2 deviates from the PTZ spectra with the appearance of a broad absorption between 500 and 800 nm. This absorption is attributed to the CT states. A broad band similar to this has been observed in PTZ solution and assigned to solvated electrons, created from absorption of the stabilized phenothiazine radical anion in solution analogous to the CT states here. [67, 68] This broad CT absorption is less prominent at lower temperatures in accordance with the temperature dependence of rISC, and so this low-energy induced absorption must be dominated by the ^{1}CT state. Thus, the ^{3}LE and ^{1}CT excited states can be differentiated in the **D–A–D** system. As we use a slow chop, 73 Hz, the ^{1}CT states are the long-lived states that give rise to DF and thus directly linked to the TADF signal; however it is clear from the out-of-phase signal that the ^{1}CT state has a much shorter lifetime than the local triplet, which is as we expect from time-resolved emission experiments. The residual CT PIA observed out of phase may well arise from a small ^{3}CT population.

We can again use polarity effects in the PIA investigations. DPTZ-DBTO2 was dissolved in toluene and MeTHF at concentrations of 10 mM (Figure 12.22). The in-phase spectra show that the CT emission dominates and that in toluene it is stronger than in MeTHF, a result mirrored in the steady-state PL. However, the out-of-phase spectra for both solvent systems are dominated by a broad, long-lived CT state-induced absorption. This is attributed to the fact that the energy of the CT states is stabilized in the more polar environments such that they now lie below the ^{3}LE (unaffected by host polarity). The ^{3}CT state now becomes the reservoir for the triplet excitations. This PIA is identified as the ^{3}CT state due to it being dominant in the out-of-phase PIA signal, concomitant with it having a much longer lifetime than the emissive ^{1}CT state (in-phase signal). This shows that the PIA spectra can help to identify the energetic arrangements in the excited states in a TADF system, which gives the first experimental evidence that there is a well-defined ^{3}CT state.

The temperature dependence of the PIA of DPTZ-DBTO2 was also measured in PEO films (Figure 12.23). The in-phase component (Figure 12.23a) shows that the intensity of the ^{1}CT emission peaks at around 210–220 K, the T_g of PEO. The polarity of the PEO decreases with decreasing temperature, causing the ^{1}CT energy to blueshift and close the gap between the CT states and the unaffected

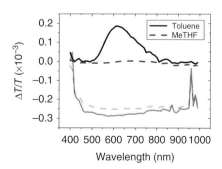

Figure 12.22 The PIA of DPTZ-DPTO2 in toluene and MeTHF at concentrations of 10 mM. The in-phase spectra are the bold lines and the out-of-phase spectra the transparent lines. The broad absorption between 500 nm and 1000 nm is related to the ^{3}CT states.

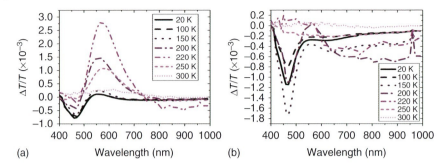

Figure 12.23 (a) The in-phase component of the PIA of DPTZ-DPTO2 in PEO over a temperature range from 300 K to 20 K. The resonance peak occurs near the T_g at 220 K. (b) The out-of-phase component of the PIA of DPTZ-DPTO2 in PEO. Between 200 and 220 K there is a crossover between triplet dominated and CT dominated absorption, highlighting the resonant point of the system.

^3LE_D state. The increasing intensity of emission indicates that TADF becomes more efficient because the rISC barrier reduces, which has a more important effect than reducing the amount of thermal energy available to vibronically activate the rISC. Below T_g the system is dominated by the ^3LE_D triplet-induced absorption, similar to that observed in zeonex. This is attributed to the fact that at lower temperatures the ^1CT now lies above the ^3LE state and TADF becomes less efficient and there is less thermal activation energy available, so rISC efficiency reduces and the long-lived triplet states dominate.

In the out-of-phase signal (see Figure 12.23b) near T_g, especially 200 K, both triplet absorption and ^1CT absorption are seen. This is again due to the polarity of the PEO host bringing the ^1CT into resonance with ^3LE_D. At 220 K, the ^1CT absorption is the only signal, a result of rISC being extremely efficient so that a large number of triplet states are rapidly recycled into emissive ^1CT states. As the temperature is further increased above T_g, the ^1CT state moves below the local triplet and rISC efficiency starts to reduce, in line with the reduction in in-phase emission. These temperature-dependent effects are not observed in zeonex as it is nonpolar and its T_g is so high that the zeonex is a rigid matrix at all measured temperatures. The reasons why we observe a kind of phase transition around 220 K have not been explored so far, but we assume that once the host matrix becomes "solid," then the polarity effects weaken as the host cannot rearrange around the ICT molecule to accommodate the changing electron distribution on the ICT molecule [42, 52].

The PIA measurements show that the energy separation between the local triplet state, the ^3LE state, and the **D–A–D** molecule CT states critically controls the efficiency of the TADF process and fully supports the mechanism of rISC where ^1LE mediates second-order SOC. Furthermore, the rigidity of the host and its temperature-dependent polarity have a major effect on rISC and hence TADF, which could be a crucial consideration in host materials for devices. It also indicates that intramolecular motion of the **D–A–D** molecule is critical in the TADF process, especially radiative emission. This is further explored in the next section. Three distinct regimes for TADF can thus be identified

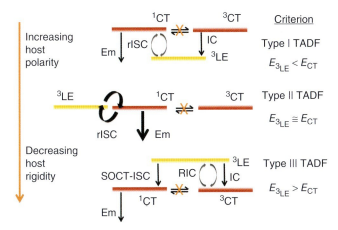

Scheme 12.2 Energy level schemes identifying three types of TADF dependent on the relative energy levels of the three key states involved in rISC: ^{1}CT, ^{3}CT, and ^{3}LE.

dependent on the relative energetic positions of the ^{1}CT, ^{3}CT, and ^{3}LE states (Scheme 12.2). From these results it is clear that both careful molecular design and host environment are critical in controlling the efficiency of TADF and both must be correct to achieve the desired type II TADF regime.

12.8 Dynamical Processes

As stressed throughout this chapter, one of the most important criteria for an efficient TADF emitter is the near-perpendicular orientation of the **D** and **A** such that they are electrically decoupled in the ground state and strong CT ensues in the excited state. Further, this brings the lowest ^{3}LE state into resonance with both ^{1}CT and ^{3}CT to facilitate a high rate of rISC. We assumed that to enhance the rate of rISC and at the same time further reduce the rate of IC, making the **D–A–D** molecule as rigidly orthogonal as possible would be a way forward.

A series of sterically hindered **D–A–D** molecules were synthesized based on the DPTZ-DBTO2 motif (Figure 12.24) [39]. Using the DPTZ-DBTO2 building blocks gave us a well-characterized **D–A–D** core, and by incorporating addition bulky groups on the 1- and/or 9-positions of the phenothiazine **D** units, the near orthogonality of each **D** and **A** unit was locked in. This strategy was confirmed by NMR measurements were for the bulkiest substituents used; at RT the torsional rocking of the **D** groups about the C—N bridging bond between each **D** and **A** pair was slowed down and could be resolved on the NMR timescale. Only heating the solutions up to 373 K could well-resolved NMR spectra of a single species be recorded. X-ray crystal structure determination confirmed near 90° dihedral angles between **D** and **A**.

From the time-resolved emission decays, we can see that no ^{1}CT emission can be measured, either prompt or delayed, in these highly sterically hindered **D–A–D** molecules. Similarly, in the integrated emission, only ^{1}LE and ^{3}LE

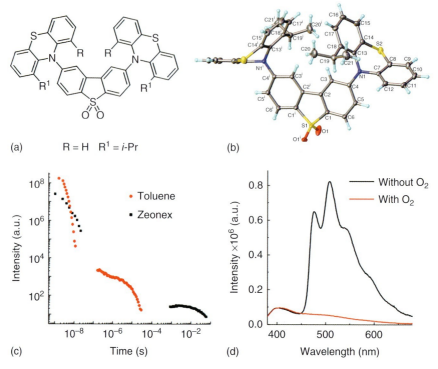

Figure 12.24 Chemical and X-ray molecular structure of *i*Pr (propyl) substituted DPTZ-DBTO2, displayed with thermal ellipsoids at 50% probability. Time-resolved emission decay at 290 K (355 nm excitation) and oxygen-dependent emission at 290 K in Zeonex (excited at 355 nm). Almost all ^{1}CT emission, prompt and delayed, is absent, but strong room temperature phosphorescence is observed. All measurements at room temperature. An absence of data points in the time-resolved emission decays indicates signal at the noise flaw of the detector.

emission is observed. From this we conclude that excitation into the **D** ^{1}LE excited state does not then undergo ET to form the ^{1}CT state as seen in the parent DPTZ-DBTO2 molecule, with a rate $k_{ET} \sim 10^8$ s^{-1}. Because of the bulky *i*Pr substituent preventing torsional rocking about the **D–A** bonds, the rate of ET is slowed from the already slow rate of the parent. ET is then outcompeted by both IC and ISC of the ^{1}LE state back to S$_0$ or ^{3}LE, respectively. Clearly a large **D** triplet population is created, leading to the strong RT PH that is observed. This also suggests that IC also has a very slow rate. With such a slow ET rate, the coupling between **D** and **A** must be extremely weak or negligible, which would also imply that the ^{1}CT would have negligible coupling to the ground state and so very weak radiative decay as well. From the X-ray structure we see that the **D–A** dihedral angles are very similar in both the parent and substituted molecules and the major difference between them is the ability to torsionally rock about the **D–A** bonds in the parent as revealed by the NMR results, i.e. in the substituted cases this motion is greatly slowed down. DPTZ-DBTO2 shows very strong TADF, whereas the *i*PR substitution effectively kills all CT formation and emission. Thus we conclude that the dynamic torsional rocking is very important in facilitating both ET and ^{1}CT radiative decay. From our vibronically coupled model of second-order SOC, we find that both ISC (^{1}CT to either ^{3}CT/^{3}LE) and rISC are controlled by such molecular vibrations, and this added complexity of TADF shows yet another fascinating side of a very complex but highly efficient (in some cases) mechanism.

12.9 Emitter–host Interactions

As we have seen in the preceding sections, the critical rISC step that harvests the triplet excitations requires a near degeneracy between one of the local triplet states, ^{3}LE, and both of the CT excited states, ^{1}CT and ^{3}CT. Because the energies of the CT states are very sensitive to environment, whereas the ^{3}LE is more robust, we have seen that it is possible to tune the CT energies in and out of resonance with ^{3}LE using the polarity of the host environment. This fact has a major impact on the efficiency of TADF, and because a TADF emitter has to be codeposited into a (ambipolar) charge transporting host to make an OLED, the polarity of that host will perturb the CT energies of the TADF emitter. Thus, to optimize the TADF efficiency, both the emitter and host must be considered as a combined system. This "host tuning" can be dramatic given that we aim to minimize the ^{3}LE–^{1}CT gap to well below 25 meV. This is readily demonstrated with the blue emitting **D–A–D** molecule 2,7-bis(9,9-dimethyl-acridin-10-yl)-9,9-dimethylthioxanthene-S,S-dioxide (DDMA-TXO2) [70, 71] (Figure 12.25).

Through the introduction of the carbon bridge in both the **D** and **A** units, the triplet energy of both is increased, as well as the CT energies of the molecule. In nonpolar zeonex the energy splitting between ^{1}CT and ^{3}LE was found to be $\Delta E_{ST} = (0.15 \pm 0.03)$ eV, which is rather large, and one would expect that the material would not be such an efficient TADF emitter.

Time-resolved emission measured in zeonex shows a characteristic decay curve of a TADF system with a well-resolved temperature-dependent DF in the

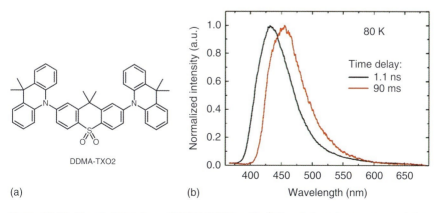

Figure 12.25 Chemical structure of DDMA-TXO2 and its ^{1}CT and phosphorescence emission measured in zeonex.

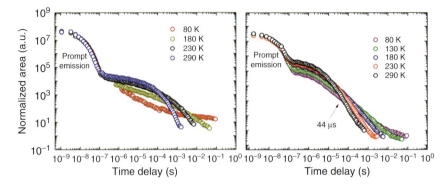

Figure 12.26 Temperature-dependent time-resolved emission decay of DDMA-TXO2 measured in (a) zeonex and (b) DPEPO.

microsecond region, which is quenched by 80 K, indicative of the large ΔE_{ST} energy barrier (Figure 12.26). The intensity-dependent DF curve has a slope of 0.96, and emission is from ^{1}CT from the earliest times until the end of the decay, giving purely TADF, apart from at 80 K when **D** PH is observed. However codepositing DDMA-TXO2 in DPEPO, a high triplet host material [72] with higher polarity than zeonex, we observed large changes in the emission decay (Figure 12.26b). We see the onset of the temperature-dependent DF component at early times and an order of magnitude increase in intensity compared with the prompt CT emission, indicating a faster ISC rate. Moreover, the DF decays much faster, indicating a faster rISC rate as well. At 80 K the DF component is still very strong, indicating that the ΔE_{ST} in DPEPO must be smaller in magnitude than in zeonex. The polar nature of the DPEPO host has reduced the CT energy to bring it into resonance with the ^{3}LE state. This is clearly seen in Figure 12.27, where the prompt ^{1}CT emission is seen to be redshift as compared with that in zeonex, the onset now being in resonance with the PH. The EL from a device using DDMA-TXO2:DPEPO as an emitter layer has onset at the same energy. The EL peak is redshifted, but this was found to be caused by an exciplex emission contribution (610 nm) arising at the interface with the electron-transport layer.

We calculate that the ΔE_{ST} in DPEPO is on the order of (0.01 ± 0.03) eV at RT. Such a small energy barrier is the reason why DF is still strongly observed at 80 K and concomitantly yields near 100% triplet harvesting in an OLED with device EQE > 22% in the deep blue. This host tuning of the rISC energy gap confirms the mechanism of rISC that requires the ^{3}LE state to be in near resonance with ^{1}CT and act as intermediary vibronic coupling state with ^{3}CT. Further, one would expect there be a difference in the PLQY of DDMA-TXO2 in DPEPO if DF contributes much more strongly to the total luminescence signal. Careful O_2-free measurements give PLQY of 0.3 ± 0.03 in zeonex and 0.95 ± 0.03 in DPEPO, clearly verifying the model of rISC described.

It is also clear that this effect has to be taken into account very carefully when optimizing a TADF OLED because of the sensitivity of the rISC rate to the ΔE_{ST} energy barrier and the rather small magnitude of kT and RT. Even small induced

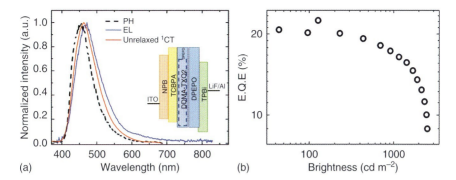

Figure 12.27 A comparison of the emission of DDMA-TXO2:DPEPO, prompt 1CT, El, and phosphorescence. The inset depicts the device structure used. On the right is the EQE curve for the device showing better than 22% EQE from the device.

shifts in the energies of the CT states can be enough to reduce the rISC rate by several orders of magnitude and thus greatly affect the overall TADF efficiency.

12.10 Energy Diagram for TADF

D–A–D ICT molecules that have efficient TADF, as we have seen from the polarity-dependent measurements and PIA, show a subtle interplay between the relative energy separations of the ^{1}CT, ^{3}CT, and ^{3}LE energy levels, which dictate the absolute efficiency of the rISC rate. From dynamic quantum chemistry studies (Chapter 9, Penfold), we find that the vibronic coupling mechanism is rather independent of the energy ordering, the rate being dominated by the energy gap between the ^{3}CT and ^{3}LE states that are vibronically coupled. Thus, bearing this in mind, we can give a somewhat generic Jablonski diagram for the TADF process (Scheme 12.3).

12.11 Final Comments

As we have seen throughout this chapter, the key mechanism that underlies rISC and thus the efficiency of TADF is the electronic coupling between ^{3}LE and ^{3}CT that mediates SOC to enable the conversion of triplet to singlet states. The experimental data clearly shows the importance of bringing the ^{3}LE state into resonance with ^{1}CT and hence ^{3}CT; this is a very clear and important difference between our original understanding of TADF where it was envisaged that the gap between ^{1}CT and ^{3}CT was the barrier that controlled rISC and TADF. From this new understanding most of the photophysics of the **D–A** molecules showing TADF can then be understood. It is also important to understand that we are minimizing the gap between two dislike states, a local excited state and a CT state, and these states behave very differently to their environment. From this, we have shown that the thermal activation gap is readily tuned by environment just as much as molecular structure. It is important to start from a molecule where the donor and acceptor

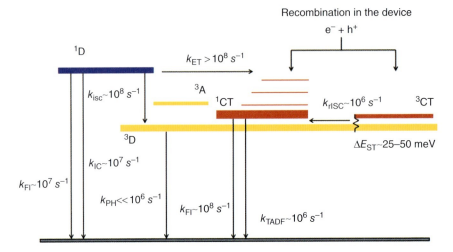

Scheme 12.3 Generic Jablonski diagram for TADF. Optical excitation of the ^{1}D state leads to lower PLQY because of the competition between slow ET and ISC to the ^{3}D state. rISC is mediated through a vibronically coupled ^{3}LE–^{3}CT state. Direct CT formation either by direct CT absorption or charge recombination gives higher PLQY.

units are near perpendicular to each other and this is controlled by steric effects, so that the CT state is stabilized in any environment and ^{1}CT and ^{3}CT are effectively isoenergetic; this is still an important criterion. Fine-tuning then comes from the environment in which molecule is coevaporated, for example. However, as we have shown, if we take this to its logical conclusion and synthesize a molecule that is rigidly orthogonal, then the electronic coupling between donor and acceptor becomes so weak that CT is outcompeted by other NR processes, i.e. internal conversion and ISC. This indicates that a certain degree of dynamic torsional rocking is required to both enable efficient ET and subsequent radiative decay of the one CT state. One also has to remember that in a device the creation mechanism of the excited state through charge recombination is different to that of photocreation through photoexcitation of the donor or acceptor unit leading on to CT. As we have shown, when one photogenerates CT states through a direct CT absorption, a strong increase in PLQY is observed by avoiding NR losses, i.e. internal conversion and ISC associated with the initially excited ^{1}LE local state, and in the device we also avoid these losses so that the TADF emitter has an effective higher PLQY than expected. Effective electronic coupling between the CT state and the ground state is still required so that radiative recombination has a faster rate to outcompete the NR decay from the CT state. We now have a new set of guide rules for producing efficient TADF molecules that go much further to explaining how to design and optimize the TADF molecule to work in an efficient OLED. Using this new knowledge we have shown, for example, with DDMA-TXO2:DPEPO, efficient deep blue TADF OLEDs having EQE > 22% can be demonstrated. With improved device efficiency we hope that improved device lifetime also follows on although this is the next great challenge, but these initial results go a long way to giving the holy grail of the hundred percent efficient red, green, and blue emitters for displays, although as yet it is impossible to say if the technology will go full TADF or mixed blue TADF, red and green phosphorescent, only time will tell.

Acknowledgments

I wish to thank everyone who has contributed to the work described in this chapter; Paloma Lays dos Santos, Heather Higginbotham, Fernando Dias, Marc Etherington, Jose Santos, David Graves, Przemyslaw Data, Roberto Nobuyasu, Hameed Al'Attar, Jonathan Ward, Mark Fox, Andrei Batsanov, and Martin Bryce. I thank the EPSRC and the European Union for funding our work.

References

1 Forrest, S.R. (2004). *Nature* 428: 911.
2 Baldo, M.A., O'Brien, D.F., Thompson, M.E., and Forrest, S.R. (1999). *Phys. Rev. B: Condens Matter* 60: 14422.
3 Monkman, A.P., Rothe, C., and King, S.M. (2009). *Proc. IEEE* 97: 1597.
4 Kondakov, D.Y., Pawlik, T.D., Hatwar, T.K., and Spindler, J.P. (2009). *J. Appl. Phys.* 106: 124510.
5 King, S.M., Cass, M., Pintani, M., Coward, C., Dias, F.B., Monkman, A.P., and Roberts, M. (2011). *J. Appl. Phys.* 109: 074502.
6 Adachi, C., Baldo, M.A., Thompson, M.E., and Forrest, S.R. (2001). *J. Appl. Phys.* 90: 5048.
7 Baldo, M.A., Forrest, S.R. and Thompson, M.E. (2005). *Organic Electroluminescence* (ed. Z.H. Kafafi), 267. CRC Press.
8 Sivasubramaniam, V., Brodkorb, F., Hanning, S., Loebl, H.P., van Elsbergen, V., Boerner, H., Scherf, U., and Kreyenschmidt, M. (2009). *Cent. Eur. J. Chem* 7: 836.
9 Sivasubramaniam, V., Brodkorb, F., Hanning, S., Loebl, H.P., van Elsbergen, V., Boerner, H., Scherf, U., and Kreyenschmidt, M. (2009). *J. Fluor. Chem.* 130: 640.
10 Wilkinson, F. and Horrocks, A. (1968). Phosphorescence and delayed fluorescence of organic substances. In: *Luminescence in Chemistry* (ed. E.J. Bowen), 116–153. London: Van Nostrand.
11 Goushi, K., Yoshida, K., Sato, K., and Adachi, C. (2012). *Nat. Photonics* 6: 253.
12 Perrin, F. (1929). *Ann. Phys. (Paris)* 12: 169.
13 Terenin, A.N. (1943). *Acta Phys. Chim. USSR* 18: 210.
14 Lewis, G.N. and Kasha, M. (1944). *J. Am. Chem. Soc.* 66: 2100.
15 Parker, C.A. and Hatchard, C.G. (1961). *Trans. Faraday Soc.* 57: 1894.
16 Jankus, V., Chiang, C.-J., Dias, F., and Monkman, A.P. (2013). *Adv. Mater.* 25: 1455.
17 Beens, H. and Weller, A. (1968). *Acta Phys. Pol.* 34: 539.
18 Frederichs, B. and Staerk, H. (2008). *Chem. Phys. Lett.* 460: 116.
19 Marian, C.M. (2012). *WIREs Comput. Mol. Sci.* 2: 187.

20 Endo, A., Ogasawara, M., Takahashi, A., Yokoyama, D., Kato, Y., and Adachi, C. (2009). *Adv. Mater.* 21: 4802.
21 Goushi, K. and Adachi, C. (2012). *Appl. Phys. Lett.* 101: 023306.
22 Uoyama, H., Goushi, K., Shizu, K., Nomura, H., and Adachi, C. (2012). *Nature* 492: 234.
23 Dias, F.B., Bourdakos, K.N., Jankus, V., Moss, K.C., Kamtekar, K.T., Bhalla, V., Santos, J., Bryce, M.R., and Monkman, A.P. (2013). *Adv. Mater.* 25: 3707.
24 Graves, D., Jankus, V., Dias, F.B., and Monkman, A. (2014). *Adv. Funct. Mater.* 24: 2343.
25 Jankus, V., Data, P., Graves, D., McGuinness, C., Santos, J., Bryce, M.R., Dias, F.B., and Monkman, A.P. (2014). *Adv. Funct. Mater.* 24: 6178.
26 dos Santos, P.L., Dias, F.B., and Monkman, A.P. (2016). *J. Phys. Chem. C* 120: 18259.
27 Rothe, C., Guentner, R., Scherf, U., and Monkman, A.P. (2001). *J. Chem. Phys.* 115: 9557.
28 Rothe, C. and Monkman, A. (2002). *Phys. Rev. B* 65: 073201.
29 Sinha, S., Rothe, C., Güntner, R., Scherf, U., and Monkman, A.P. (2003). *Phys. Rev. Lett.* 90: 127402.
30 Rothe, C., Pålsson, L.O., and Monkman, A.P. (2002). *Chem. Phys.* 285: 95.
31 Rothe, C., Hintschich, S.I., and Monkman, A.P. (2006). *Phys. Rev. Lett.* 96: 163601.
32 Rothe, C., King, S.M., and Monkman, A.P. (2006). *Phys Rev Lett.* 97: 076602.
33 Aydemir, M., Jankus, V., Dias, F.B., and Monkman, A. (2014). *Phys. Chem. Chem. Phys.* 16: 21543.
34 Dias, F., Pollock, S., Hedley, G., Pålsson, L.-O., Monkman, A., Perepichka, I.I., Perepichka, I.F., Tavasli, M., and Bryce, M.R. (2006). *J. Phys. Chem. B* 110: 19329.
35 King, S.M., Matheson, R., Dias, F.B., and Monkman, A.P. (2008). *J. Phys. Chem. B* 112: 8010.
36 King, S.M., Perepichka, I.I., Perepichka, I.F., Dias, F.B., Bryce, M.R., and Monkman, A.P. (2009). *Adv. Funct. Mater.* 19: 586.
37 Moss, K.C., Bourdakos, K.N., Bhalla, V., Kamtekar, K.T., Bryce, M.R., Fox, M.A., Vaughan, H.L., Dias, F.B., and Monkman, A.P. (2010). *J. Org. Chem.* 75: 6771.
38 Nobuyashu, R.S., Ren, Z., Griffiths, G.C., Batsanov, A.S., Data, P., Yan, S., Monkman, A.P., Bryce, M.R., and Dias, F.B. (2016). *Adv. Opt. Mater.* 4: 597.
39 Ward, J.S., Nobuyasu, R.S., Batsanov, A.S., Data, P., Monkman, A.P., Dias, F.B., and Bryce, M.R. (2016). *Chem. Commun.* 52: 2612.
40 Marian, C.M. (2016). *J. Phys. Chem. C* 120: 3715.
41 dos Santos, P.L., Ward, J.S., Data, P., Batsanov, A.S., Bryce, M.R., Dias, F.B., and Monkman, A.P. (2016). *J. Mater. Chem. C* 4: 3815.
42 Aydemir, M., Haykír, G., Türksoy, F., Gümüş, S., Dias, F.B., and Monkman, A.P. (2015). *Phys. Chem. Chem. Phys.* 17: 25572.
43 Dias, F.B., King, S., Monkman, A.P., Perepichka, I.I., Kryuchkov, M.A., Perepichka, I.F., and Bryce, M.R. (2008). *J. Phys. Chem. B* 112: 6557.
44 Grabowski, Z.R., Rotkiewicz, K., and Rettig, W. (2003). *Chem. Rev.* 103: 3899.
45 Herbich, J. and Kapturkiewicz, A. (1998). *J. Am. Chem. Soc.* 120: 1014.

46 Yoshihara, T., Druzhinin, S.I., Demeter, A., Kocher, N., Stalke, D., and Zachariasse, K.A. (2005). *J. Phys. Chem. A* 109: 1497.

47 Dias, F.B., Santos, J., Graves, D., Data, P., Nobuyasu, R.S., Fox, M.A., Batsanov, A.S., Palmeira, T., Berberan-Santos, M.N., Bryce, M.R., and Monkman, A.P. (2016). *Adv. Sci.* 3: doi: 10.1002/advs.201600080.

48 Bixon, M., Jortner, J., Cortes, J., Heitele, H., and Michel-Beyerle, M.E. (1994). *J. Phys. Chem.* 98: 7289.

49 Grewer, C. and Brauer, H.-D. (1994). *J. Phys. Chem.* 98: 4230.

50 Rothe, C., Al Attar, H.A., and Monkman, A.P. (2005). *Phys. Rev. B* 72: 155330.

51 Huber, J.R. and Mantulin, W.W. (1972). *J. Am. Chem. Soc.* 94: 3755.

52 Aydemir, M., Haykır, G., Battal, A., Jankus, V., Sugunan, S.K., Dias, F.B., Al-Attar, H., Türksoy, F., Tavaslı, M., and Monkman, A.P. (2016). *Org. Electron.* 30: 149.

53 Mopsik, F.I. (1969). *J. Chem. Phys.* 50: 2559.

54 Gibson, J., Monkman, A.P., and Penfold, T.J. (2016). *ChemPhysChem* 17: 2956.

55 Lee, J., Shizu, K., Tanaka, H., Nakanotani, H., Yasuda, T., Kaji, H., and Adachi, C. (2015). *J Mater Chem C* 3: 2175.

56 Gibson, J. and Penfold, T.J. (2017). *PhysChemChemPhys* 19: 8428.

57 Data, P., Pander, P., Okazaki, M., Takeda, Y., Minakata, S., and Monkman, A.P. (2016). *Angew Chem Int Edit* 55: 5739.

58 Galievsky, V.A., Druzhinin, S.I., Demeter, A., Mayer, P., Kovalenko, S.A., Senyushkina, T.A., and Zachariasse, K.A. (2010). *J. Phys. Chem. A* 114: 12622.

59 Gould, I.R., Noukakis, D., Gomez-Jahn, L., Young, R.H., Goodman, J.L., and Farid, S. (1993). *Chem. Phys.* 176: 439.

60 Gould, I.R., Noukakis, D., Gomez-Jahn, L., Goodman, J.L., and Farid, S. (1993). *J. Am. Chem. Soc.* 115: 4405.

61 Fanggao, C., Saunders, G.A., Lambson, E.F., Hampton, R.N., Carini, G., Di Marco, G., and Lanza, M. (1996). *J. Polym. Sci., Part B: Polym. Phys.* 34: 425.

62 Fontanella, J.J., Wintersgill, M.C., Welcher, P.J., Calame, J.P., and Andeen, C.G. (1985). *IEEE Trans. Electr. Insul.* 20: 943.

63 Lim, B.T., Okajima, S., Chandra, A.K., and Lim, E.C. (1981). *Chem. Phys. Lett.* 79: 22.

64 vanWilligen, H., Jones, G., and Farahat, M.S. (1996). *J. Phys. Chem.* 100: 3312.

65 Etherington, M.K., Gibson, J., Higginbotham, H.F., Penfold, T.J., and Monkman, A.P. (2016). *Nat. Commun.* 7: 13680.

66 Vardeny, Z. and Tauc, J. (1985). *Phys. Rev. Lett.* 54: 1844.

67 Nath, S., Pal, H., Palit, D.K., Sapre, A.V., and Mittal, J.P. (1998). *J. Phys. Chem. A* 102: 5822.

68 Ghosh, H.N., Sapre, A.V., Palit, D.K., and Mittal, J.P. (1997). *J. Phys. Chem. B* 101: 2315.

69 Henry, B.R. and Lawler, E.A. (1974). *J. Mol. Spectrosc.* 51: 385.

70 dos Santos, P.L., Ward, J.S., Bryce, M.R., and Monkman, A.P. (2016). *J. Phys. Chem. Lett.* 7: 3341.

71 Lee, I. and Lee, J.Y. (2016). *Org. Electron.* 29: 160.

72 Han, C.M., Zhao, Y., Xu, H., Chen, J., Deng, Z., Ma, D., Li, Q., and Yan, P. (2011). *Chem. Eur. J.* 17: 5800.

13

Thioxanthone (TX) Derivatives and Their Application in Organic Light-emitting Diodes

Xiaofang Wei, Ying Wang, and Pengfei Wang

Technical Institute of Physics and Chemistry, Chinese Academy of Sciences, Beijing, 100190, China

This chapter summarizes our recent research on the thermally activated delayed fluorescence (TADF) materials based on thioxanthone (TX) derivatives and their application in organic light-emitting diodes (OLEDs). Synthesis strategies that lead to a series of TX-based TADF materials are reported, and the tunability of photoluminescent (PL) properties through modifying the molecular structure character of the TX unit is also discussed. Transient and steady-state photophysical measurement combined with theoretical analyses provides significant insight into their excited-state properties and intersystem crossing (ISC) process. Efforts have also been devoted to the exploitation of the materials as the emissive materials and hosts for preparation of high-efficiency OLEDs.

13.1 Organic Light-emitting Diodes

OLEDs refer to the light-emitting diodes in which the emissive electroluminescent (EL) layer is composed of a film of organic emissive compounds that emits light in response to electric current. The simplest structure of OLEDs has the emissive layer of organic emitters sandwiched between two electrodes, the anode and cathode, and at least one of the electrodes is transparent. André Bernanose et al. first observed EL in organic materials under high alternating voltage in air in the early 1950s [1]. In 1987, Tang and Slyke of Eastman Kodak Company reported a novel diode device with low operating voltage and improved efficiency, in which two separate hole-transporting and electron-transporting layers were incorporated and light emission occurs in the middle of the organic layer [2]. These findings opened the new era of OLEDs. In these devices, organic semiconductors are indispensable as they can be used as the carrier-transporting/injection materials and emissive materials. These semiconductors are small molecular or polymer materials. Their structures and properties can be tuned to obtain ideal emissive materials with excellent electrical properties. They can also be processed by solution methods, such as inject printing [3], spin-coating [4], and screen printing [5]. Theoretically, production cost is lower when mass production methods lower the cost. OLEDs can be fabricated on flexible substrates [6], such as polyethylene terephthalate, thereby eliminating fragile and heavy glass substrates and leading to lightweight roll-up displays. Moreover, OLEDs show low power dissipation, high efficiencies, and excellent color qualities without inherent pronounced directionality, which are superior to those of liquid crystal displays (LCDs). These advantages endow OLEDs with great opportunity for full-color displays and lighting, and OLEDs have already replayed LCDs in many portable applications [7].

OLEDs are composed of separate carrier-transporting and carrier-emitting layers sandwiched between the anode and the cathode. When OLEDs are working, electrons are injected from the cathode, and holes are injected from the anode. Holes and electrons meet in the emissive layer and recombine, leading to the formation of excitons. These excitons, which are located on the emissive molecules, radiatively relax to the ground state and emit the observed EL. Holes and electrons are odd electron species that have a spin of +1/2 or −1/2. Thus, the excitons formed by the recombination of holes and electrons can have either a singlet or triplet configuration. According to the spin statistics, charge carriers injected in OLED recombine to form singlet and triplet excitons at a 1 : 3 ratio. The singlet exciton decays rapidly from S_1 to S_0, thereby yielding fluorescence, whereas the radiative decay of triplet excitons can lead to phosphorescence. For fluorescence emitters, the first-generation emitters, the theoretical maximum EL quantum yield of fluorescent emitters is limited to 25% because the radiative transition from the triplet excited state to the singlet ground state is prohibited. Thus, all triplet excitons are lost for the generation of light and transferred into heat dissipation. Therefore, the performance of OLEDs based on these fluorescence emitters is low. For phosphorescent emitters, the second-generation emitters, strong spin–orbit coupling (SOC) facilitates the radiative path for the emission from triplet to ground state. Efficient transfer from the populated singlet state to the emitting triplet state can also be expected due to the strong SOC. Thus, all the excitons can be transferred to the triplet state for light harvesting, reaching the 100% internal quantum yield. The performance of OLEDs based on them is four times higher efficiency than OLEDs based on fluorescent emitters. However, transition metal ions, such as Ir and Pt, are generally introduced to enhance the SOC in phosphorescent emitters. These noble metals are expensive and indispensable. Their reserves are limited, which increase the cost for the emitters. Phosphorescent emitters require the population of the triplet excitons, and high triplet exciton densities will be produced to achieve high brightness at high current density. The high triplet exciton densities will result in exciton quenching processes, leading to the efficiency roll-off. Moreover, blue phosphorescent materials often suffer from poor stability and low emission quantum yield. These may limit the implementation of OLEDs in bright display and lighting. Yersin et al. first proposed the application of molecular TADF effect for singlet harvesting in OLEDs [8]. In 2009, Adachi's group reported pure organic aromatic compounds with efficient TADF to attain high EL efficiency (Figure 13.1) [9]. These emitters exhibit a sufficiently small energy gap between the singlet and triplet (ΔE_{ST}) to enable upconversion of the triplet excitons to singlet excitons and achieve 100% internal quantum efficiency. High external quantum efficiencies up to 25% have been reported for OLEDs employing TADF emitters, and the TADF emitters have now been accepted as the third generation

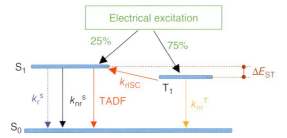

Figure 13.1 Energy diagram of TADF mechanism is the radiative decay rate from singlet state; k_r^S is the radiative decay rate from singlet state (fluorescence); k_{nr}^S is the irradiative decay rate from singlet state; k_{nr}^T is the irradiative decay rate from triplet state; k_{rISC} is the rate of the reverse intersystem crossing (rISC) from triplet state to singlet state; TADF is referred to the thermally activated delayed fluorescence; ΔE_{ST} Is the energy gap between singlet state and triplet state (phosphorescence); S_1, S_0, and T_1 are the energy level of singlet state, ground state, and triplet state.

of OLEDs emitter. Also these TADF materials have been applied to the host of OLEDs.

13.2 Pure Organic TADF Materials in OLEDs

Two distinct mechanisms exist for molecular delayed fluorescence (DF): P-type DF and E-type DF. In the P-type DF mechanism, a DF emission can be caused by a triplet–triplet annihilation. Parker and Hatchard observed P-type DF for the deoxygenated solutions of pyrene in ethanol [10]. Two molecules in the T_1 state can collide with each other in a concentrated solution. The collision provides enough energy for one of the molecules to return to the S_1 state and results in emission with the same spectral distribution as normal fluorescence. The decay time constant of the P-type DF process is half the lifetime of the triplet state in a dilute solution, and the DF intensity shows a quadratic dependence with the excitation light intensity [10]. In the E-type DF mechanism, the DF fluorescence emission with same spectral distribution as the prompt fluorescence occurs via triplet manifold and then exhibits a much longer decay time constant than the normal fluorescence. Once molecules are excited and achieve the singlet state, ISC to the triplet manifold occurs, followed by reverse ISC (rISC) from the triplet to the singlet state when ΔE_{ST} is small and fluorescent emission from the singlet. The cycle of ISC and rISC may repeat several times before fluorescence emission is achieved [11]. Owing to ΔE_{ST} between triplet and singlet, the rISC process always corresponds to a thermally activated process, viz the rate constant of the rISC process is strongly temperature dependent. Thus, the fluorescence emission is thermally activated and the E-type DF is called TADF. TADF is still a rare phenomenon since the first observation of TADF with eosin [12], and for the most fluorophores, TADF is usually much weaker than its prompt fluorescence. In 2011, Prof. Adachi opened up the new pathway for highly efficient EL by harvesting triplet excitons for light emission with aromatic TADF molecules [13]. Enormous efforts have been endeavored to the development of aromatic TADF molecules and their application to the high efficient OLEDs. Versatile molecular systems with TADF have been reported, including spiroacridine, triazine, spirobifluorene, phthalonitrile, diphenyl sulfone derivative, and so on [14]. High external quantum efficiency (EQE) up to 25% has been achieved for OLEDs using TADF emitters. TADF molecules have also been employed as the host for phosphorescent OLEDs to reduce the efficiency roll-off and increase the device stability [15]. Moreover, TADF materials have been used as assistant dopants for OLEDs based on traditional fluorescent emitters [16, 17]. This chapter focuses on the development and study of TX-based TADF materials.

13.3 TX Derivatives for OLED

TX (molecular structure shown in Figure 13.2) is a commonly used triplet sensitizer and an archetype of widely used photoinitiators of polymerization reactions, with a radiative decay rate of 5×10^7 s^{-1} from the singlet state and a lowest triplet energy of 2.80 eV [18]. The fluorescence quantum yield and the maximum fluorescence wavelength of TX increase with the increasing solvent polarity. This is attributed to the proximity effect, that is, the proximity of the S_2 ($^1n\pi^*$) and S_1 ($^1\pi\pi^*$) singlet excited states of TX leads to the change in the singlet internal conversion [19]. In the nonpolar solvents, S_2 and S_1 are very close together, and the rate of internal conversion from S_1 to S_0 is high, leading to a small fluorescence quantum yield and lifetime. In polar hydroxylic solvents, the ($^1\pi\pi^*$) states are stablilized, while the ($^1n\pi^*$) states are destabilized by solute–solvent interactions [20]. Thus, the energy gap between S_2 and S_1 increases, and the rate of internal conversion from S_1 to S_0 decreases, leading to a higher fluorescence quantum yield and a longer fluorescence lifetime [20, 21]. The energy gap between the first singlet and triplet excited state of TX had been reported to be lower than 0.3 eV, which is comparable with that of the molecule with efficient TADF [22]. The small ΔE_{ST} will enhance the rISC from T_1 to S_1. ΔE_{ST} can also be further reduced by the careful design of organic molecular structures. Thus, TX unit can be regarded as the excellent building blocks for the efficient TADF materials.

13.3.1 High Efficient OLEDs Based on TX-based TADF Materials

13.3.1.1 Design and Characterization of TX-based TADF Emitters

For the efficient TADF emitters, the emitters need to show both a small ΔE_{ST} and a reasonable radiative decay rate (>10^6 s^{-1}) to overcome competitive nonradiative decay pathways and get highly luminescent efficiency. TX unit exhibits

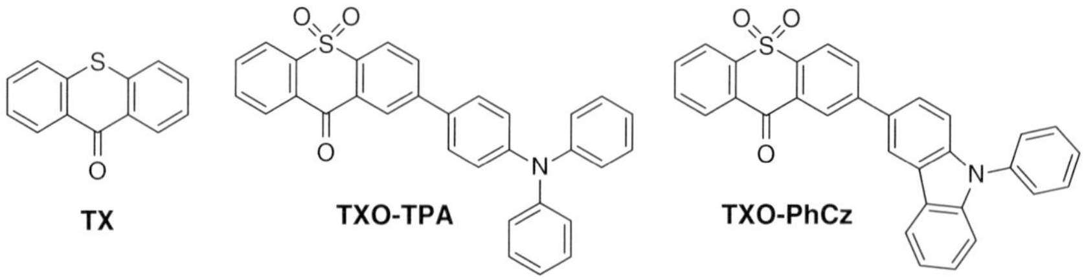

Figure 13.2 Molecular structures of TX, TXO-TPA, and TXO-PhCz.

a radiative decay rate of 5×10^7 s^{-1}. Thus, highly efficient TADF emitters can be obtained by optimizing the molecular structure of TX derivatives to reduce ΔE_{ST} and maintain the high radiative decay rate. ΔE_{ST} is given by the following equation [13]:

$$\Delta E_{ST} = E_S - E_T = 2J_{LU}$$

where E_S and E_T are the energy levels of singlet and triplet excited state; and J_{LU} is the exchange integral between the spatial wave functions of the ground level (the highest occupied molecular orbital (HOMO)) and an excited level (the lowest unoccupied molecular orbital (LUMO)). The exchange integral can be expressed as

$$J_{LU} = \int \psi_H(r_1)\psi_L(r_1)\frac{1}{r_{12}}\psi_H(r_2)\psi_L(r_2)dr_1\,dr_2$$

where ψ_H and ψ_L are the electron wave function of HOMO and LUMO, r_1 and r_2 are the coordinate of electrons, and r_{12} is the distance between electrons [23]. Thus, a small ΔE_{ST} can be achieved by reducing the overlap of HOMO and LUMO of the molecules. Intramolecular charge-transfer (ICT) molecules with a donor–acceptor (D–A) structure attain large spatial separation of the wave functions of HOMO and LUMO. Thus, ICT molecules with a D–A structure are constructed based on TX unit and the molecular structures as shown in Figure 13.2. The conventional hole-transporting moieties of the triphenylamine (TPA) unit or N-phenylcarbazole (N-PhCz) unit are introduced as the electron donor unit, and the S atom in TX unit is oxidated to enhance the electron-accepting ability. The twisted D–A structure facilitates the effective separation of electron densities of the HOMO and LUMO in a single molecule. TXO-TPA and TXO-PhCz were synthesized by Suzuki coupling reaction of the 9-H-thioxanthen-9-one-10,10-dioxide (TXO) and the electron donor unit of TPA or PhCz.

In order to qualitatively understand the nature of the excited state of TXO-TPA and TXO-PhCz, density functional theory (DFT) calculation was performed at the B3LYP/6-31G** level with the Gaussian 09 package to investigate their spatial electronic distribution, and the results of the HOMO–LUMO calculation are shown in Figure 13.3. The HOMOs of TXO-TPA and TXO-PhCz distribute over the TPA and PhCz moiety, whereas their LUMOs are dispersed on TXO moiety. Only a small overlap was observed between their HOMO and LUMO on the phenyl ring of the TXO moiety with the substituted donor group. These findings indicate the strong charge-transfer (CT) characteristic of TXO-TPA and TXO-PhCz, and a small electron exchange energy and the resulting small ΔE_{ST} can also be expected. Time-dependent DFT calculations were further performed to optimize the geometries for the lowest singlet (S_1) and triplet (T_1). The effective HOMO–LUMO separation of TXO-TPA and TXO-PhCz induced a strong ICT transition. Both compounds exhibited very small ΔE_{ST}, 0.03 eV for TXO-TPA and 0.14 eV for TXO–PhCz, which are much smaller than that of TX and comparable with those reported TADF emitters [24–26].

Table 13.1 summarizes the absorption and PL peaks of TXO-TPA and TXO-PhCz in different solvents and their corresponding quantum yields.

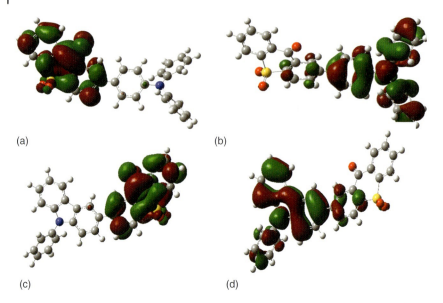

Figure 13.3 Calculated spatial distributions of the highest occupied molecular orbital (HOMO) and lowest unoccupied molecular orbital (LUMO) electron densities: (a) LUMO and (b) HOMO of TXO-TPA and (c) LUMO and (d) HOMO of TXO-PhCz.

Table 13.1 Absorption and emission characteristics of **TXO-PhCz** and **TXO-TPA**.

Solvent	TXO-TPA			TXO-PhCz		
	$\lambda_{ab}^{a)}$ (nm)	$\lambda_{em}^{b)}$ (nm)	Φ_f (10^{-3})	$\lambda_{ab}^{a)}$ (nm)	$\lambda_{em}^{b)}$ (nm)	Φ_f (10^{-3})
Hexane	414	555	32.3[c]	385	491	26.8[c]
Toluene	415	586	23.6[c]	396	522	25.4[c]
1,4-Dioxane	406	613	15.1[c]	390	540	22.6[c]
Chloroform	424	665	3.6[c]	400	587	18.0[c]
Ethyl acetate	405	663	2.1[c]	390	573	14.9[c]
THF	414	660	2.0[c]	393	579	5.1[c]
Ethanol	420	n.d.	—	395	n.d.	—
Acetonitrile	406	n.d.	—	402	n.d.	—
Methanol	410	n.d.	—	413	n.d.	—
Thin film	450	630	360[d]	420	580	930[d]

a) Absorption maximum in the long wavelength.
b) Emission maximum.
c) Quantum yield (±10%) was estimated using quinine sulfate (Φ_f 54.9% in 1 N H_2SO_4) as the standard and obtained after nitrogen degassing.
d) Absolute quantum yield, determined with an integrating sphere.

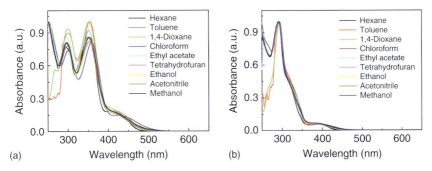

Figure 13.4 Absorption spectra of (a) TXO-TPA and (b) TXO-PhCz in different solvents.

TXO-TPA in all solutions exhibited a CT absorption band associated with the electron transfer from the TPA moiety to TXO moiety at around 415 nm (as shown in Figure 13.4). Similar CT absorption band centered at 400 nm can also be observed for TXO-PhCz in solutions. All the solutions of TXO-TPA and TXO-PhCz emit weak light, when excited by 365 nm, and even no emission can be detected in solvents with high polarity. For TXO-PhCz and TXO-TPA, the electronic excitation is also associated with an ICT coupled with rotational relaxation toward a twisted conformation, twisted intramolecular charge transfer (TICT). The nonemittive TICT state provides that additional nonradiative pathway for them, leading to the low quantum efficiency of them in polar solvents. Similar to other TICT compounds [22, 27], a pronounced positive solvatochromism of emission can also be observed: The emission peak shifted to the long wavelength in the solvent with high polarity, and the quantum yield of the solutions decreased with the solvent polarity. We also investigated the phosphorescent emission of TXO-TPA and TXO-PhCz at 77 K in oxygen-free 2-MeTHF (Figure 13.5). Both compounds exhibit unstructured phosphorescent emission spectra with peaks of 546 and 503 nm for TXO-TPA and TXO-PhCz, respectively. The triplet energy levels can be calculated to 2.46 and 2.27 eV for TXO-TPA and TXO-PhCz, respectively, from the emission peaks. To avoid the temperature and solvent effect on the PL emission, the PL emission of TXO-TPA and TXO-PhCz in 2-MeTHF at 77 K was also studied to get their

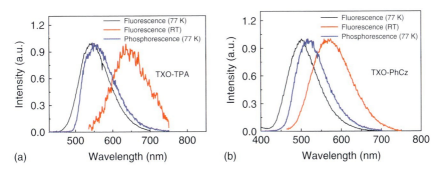

Figure 13.5 Phosphorescent and photoluminescence (PL) spectra of (a) TXO-TPA and (b) TXO-PhCz in oxygen-free 2-Me THF.

ΔE_{ST}. TXO-TPA and TXO-PhCz in 2-MeTHF at 77 K showed a PL emission peaks at 555 and 524 nm, respectively. Thus, ΔE_{ST} can be estimated to 0.04 eV for TXO-TPA and 0.09 eV for TXO-PhCz from the emission peaks of the PL and phosphorescence at 77 K. The large shift of the fluorescence spectra with the temperature can be ascribed to the orientation of the solvent molecules in the reaction field of the electronically excited solute molecules in different temperatures [19]. The ΔE_{ST} are comparable with those of TADF emitters of spiroacridine [28], azine derivatives [29–31], spirobifluorene [25], phthalonitrile [9], diphenyl sulfone derivative [32, 33], and anthraquinone [34], and much lower than that of TX derivatives. The small ΔE_{ST} possibly endows them TADF properties.

The materials and formed films should be stable during the OLEDs fabrication and working. Therefore, we carefully investigated the thermal stability of TXO-TPA and TXO-PhCz through the thermal gravimetric analysis (TGA) and differential scanning calorimetry (DSC) under a nitrogen atmosphere. TXO-TPA and TXO-PhCz showed good thermal stability: The decomposition temperatures are 380.9 and 392.6 °C for TXO-TPA and TXO-PhCz, and the glass transition temperatures are 87.4 and 115.14 °C for TXO-TPA and TXO-PhCz. Such excellent thermal stability of them is very convenient for making OLEDs devices and readily endows them with high morphological stability, which is desirable for high-performance OLEDs during operation.

Figure 13.6 shows the UV–Vis absorption and PL spectra of TXO-TPA and TXO-PhCz in thin films. TXO-PhCz exhibited a broad emission at 570 nm, and TXO-TPA exhibited an emission peak at 625 nm. The absolute fluorescence quantum yields (Φ_{PL}) of TXO-PhCz and TXO-TPA films were 0.93 and 0.36, respectively, which are much higher than those in solutions. Such remarkable PL enhancement in solid states can also be observed in aggregation-induced emission (AIE) materials. The AIE activity of TXO-TPA and TXO-PhCz can be further demonstrated by the emission behavior in acetonitrile/water mixtures with the quantitative change of the water content. Interestingly, the small change in the molecular structure of TXO-TPA and TXO-PhCz also leads to the sharp

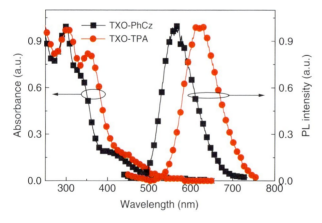

Figure 13.6 UV–Vis absorption and photoluminescence (PL) spectra of TXO-TPA and TXO-PhCz in film.

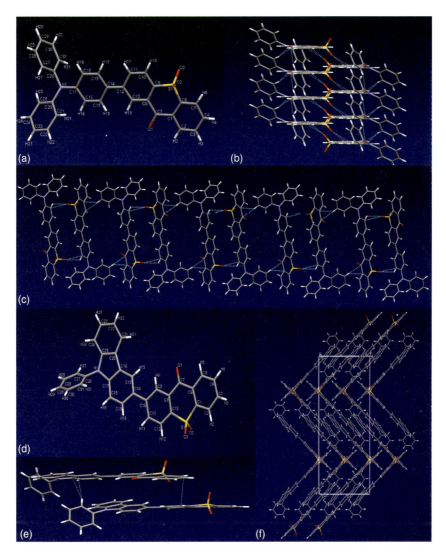

Figure 13.7 Single-crystal structures of TXO-TPA and TXO-PhCz.

difference in their Φ_{PL}. Thus, the single crystals of TXO-TPA and TXO-PhCz were grown, and their structures (as shown in Figure 13.7) were explored by X-ray single-crystal analysis. Both molecules showed an asymmetric molecular geometry. The TXO unit was almost in one plane, except the two oxygen atoms connected to the sulfur atom, and the TPA and PhCz units adopted a highly twisted conformation. When in solution, the twisted donor units will undergo intramolecular rotations, leading to the quenching of their emission and the low Φ_{PL} in solution. While in the aggregated state, there are intermolecular short contacts and local interactions between the neighboring molecules. These interactions impeded such intramolecular rotation and rigidified the molecular conformation, leading to the reduction of the loss via nonradiative relaxation.

These afforded the high emission of TXO-TPA and TXO-PhCz in the solid states [35]. Although only one single bond is changed from TXO-TPA to TXO-PhCz, the packing arrangements in their single crystals are totally different. TXO-TPA molecules in the solid state are held together by the intermolecular π–π interaction and short C—H···π contact, and two TXO-TPA molecules formed a flatly spread dimer with anticonformation with nearly parallel TXO units. Along the a-axis, TXO-TPA molecules formed a columnar stacking with an interplanar separation of 3.49 Å, and this columnar stacking affords a large intermolecular π overlap. For TXO-PhCz, two adjacent molecules with parallel configurations were held together by weaker intermolecular π–π interaction and C—H···π short contact and formed a TXO-PhCz dimer. The C—H···π short contact held the dimers in a face-to-edge manner, and the dimers were packed tightly in the crystal with the sandwiched herringbone structure. The stronger intermolecular π–π interaction of TXO-TPA facilitated the transport and thus annihilation of the excitons, leading to the lower Φ_{PL} in solid film.

To demonstrate the TADF properties of TXO-TPA and TXO-PhCz, the transient PL decay of the doped films in the 1,3-bis(9H-carbazol-9-yl) benzene (mCP) host and their temperature dependence were investigated. The doped films were selected here because TADF mechanism takes place via the triplet manifold. mCP was used as the host due to its high T_1 and S_1 states, which facilitate the confinement of the triplet excitons within the guest molecules. The emission of the 5 wt% doped TXO-TPA:mCP film blueshifted to 580 nm with a PL efficiency of 85%. The increased PL efficiency can be ascribed to the suppression of exciton annihilation by doping. The doped film showed a clear second-order exponential decay at room temperature, and the delay component decayed completely within 78 μs. The delayed component can be significantly suppressed by the oxygen blowing. The presence of the delayed component and the oxygen sensitivity demonstrated that the triplet exciton formed after excitation. The rISC from the triplet to the singlet can be further demonstrated by the overlap of the PL spectra of the prompt and delayed components (inset in Figure 13.8a and b). Figure 13.8 shows the temperature dependence of the transient PL decay from 100 to 300 K. Remarkable temperature dependence can be observed for the delayed component: The delayed component increased from 42% to 82% when the temperature increased from 150 to 300 K. Similar photophysical characteristics were observed for the 5 wt% doped film based on TXO-PhCz. All these affirm that both TXO-TPA and TXO-PhCz exhibit excellent TADF properties. Given that the prompt component showed a very low ratio and no pronounced temperature dependence was observed, the triplet formation efficiency (Φ_T) and ΔE_{ST} can be derived using the Berberan-Santos relation from the temperature dependence results [36]:

$$\ln\left[\frac{I_{prompt}}{I_{delayed}} - \left(\frac{1}{\Phi_T} - 1\right)\right] = \ln\left(\frac{k_p + k_{nr}^T}{k_{rISC}}\right) + \frac{\Delta E_{ST}}{RT}$$

where I_{prompt} and $I_{delayed}$ are the intensity of the prompt and delayed components, respectively. k_p is the phosphorescence rate constant, k_{nr}^T is the nonradiative rate constant from T_1, k_{rISC} is the $T_1 \rightarrow S_1$ rISC rate constant, and R is the gas constant. The Φ_T of TXO-TPA and TXO-PhCz were 99 and 98%, respectively, indicating

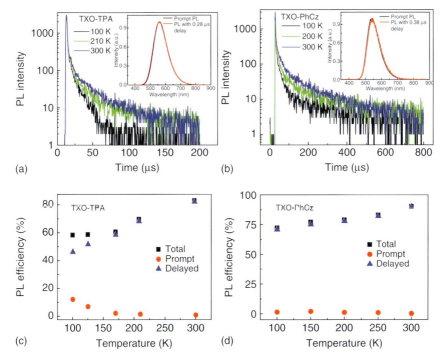

Figure 13.8 Transient photoluminescence (PL) characteristics of TXO-TPA and TXO-PhCz films. (a) Temperature dependence of the transient photoluminescence (PL) spectra for 5 ± 1 wt% TXO-TPA doped in 1,3-bis(9H-carbazol-9-yl) benzene (mCP) (inset is the PL emission for prompt and for 0.28 μs delay). (b) Temperature dependence of the transient PL spectra for 5 ± 1 wt% TXO-PhCz doped in mCP (inset is the PL emission for prompt and for 0.38 μs delay). (c) Temperature dependence of the total photoluminescence (black squares), prompt fluorescence (red circles), and delayed fluorescence (blue triangles) for 5 ± 1 wt% TXO-TPA doped in mCP. (d) Temperature dependence of the total photoluminescence (black squares), prompt fluorescence (red circles), and delayed fluorescence (blue triangles) for 5 ± 1wt%TXO-PhCz doped in mCP. The determination of the total, prompt, and delayed fluorescence can be referred to [13]. *Source:* Ref. [13]. Reproduced with the permission of AIP Publishing LLC.

that most of the singlet excitons by optical excitation were converted into the triplet excitons. These are consistent with the results that k_{ISC} of both compounds are two magnitudes higher than their k_r (see Table 13.2). ΔE_{ST} of 52 meV for TXO-TPA and 73 meV for TXO-PhCz were obtained from the slop of the fitting, which are comparable with the results calculating from the emission spectra and TDDFT.

The excellent TADF property of the doped films and their thermal stability enabled them to be the emissive molecules for the high-efficiency OLEDs. The OLED structure was ITO/PEDOT (30 nm)/TAPC (20 nm)/EML (35 nm)/TmPyPB (55 nm)/LiF (0.9 nm)/Al (100 nm), where poly(3,4-ethylene-dioxythiophene) (PEDOT) was used as the hole-injection layer (HIL) and 1,1-bis[4-[N,N′-di(p-tolyl)amino]phenyl] cyclohexane (TAPC) was used as the hole-transporting layer (HTL), 1,3,5-tri(m-pyrid-3-yl-phenyl)benzene

Table 13.2 Data extracted from the transient characterization of 5 ± 1 wt% TXO-TPA:mCP film and 5 ± 1 wt% TXO-PhCz:mCP film.

Compounds	Φ_{total} (%)	Φ_{prompt} (%)	Φ_{TADF} (%)	ΔE_{ST} (meV)	$k_s(s^{-1})$	$k_{ISC}(s^{-1})$	Φ_T	$k_{nr}(s^{-1})$	$k_r(s^{-1})$
TXO-TPA	83.00	1.19	81.81	52	3.87×10^7	3.79×10^7	98	5.60×10^4	7.44×10^5
TXO-PhCz	90.20	0.46	89.74	73	5.18×10^7	5.13×10^7	99	4.90×10^4	4.51×10^5

k_s is the rate constant of prompt fluorescence components; k_{ISC} is the intersystem crossing rate from singlet to triplet; Φ_T is the triplet formation efficiency; and k_r and k_{nr} are the radiative and nonradiative rate constants from S_1 state to S_0 state.

(TmPyPB) was used as electron-transporting layer (ETL) and hole-blocking layer (HBL), and 5 ± 1 wt% TXO-TPA:mCP or 5 ± 1 wt% TXO-PhCz:mCP is used as the emitting layer (EML). The molecular structures of the materials in the devices, their energy levels, and the EL characteristics of the devices are shown in Figure 13.9. Both devices showed excellent performances. The devices based on TXO-TPA emitted yellow light with color coordinates of CIE (0.45, 0.53) and can be turned on at 5.3 V with a maximum current efficiency of 43.3 cd A^{-1}, a maximum power efficiency of 47.4 lm W^{-1}, and a maximum EQE of 18.5%. The devices incorporating TXO-PhCz emitted green light with a color coordinates of CIE (0.31, 0.56). The devices were turned on at 4.7 V and afforded a maximum

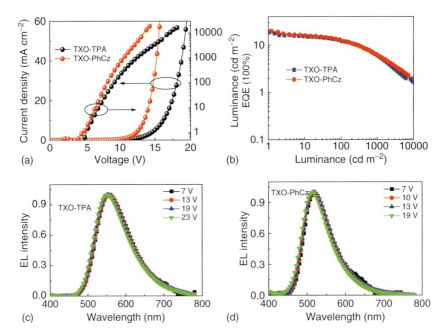

Figure 13.9 Electroluminescent (EL) characteristics of the OLEDs based on TXO-TPA and TXO-PhCz. (a) External quantum efficiency (EQE)-current density characteristics; (b) current density–voltage–luminance characteristics; (c) EL spectra operated at different voltages of the device incorporating TXO-TPA; and (d) EL spectra operated at different voltages of the device incorporating TXO-PhCz.

current efficiency of 76 cd A^{-1}, a maximum power efficiency of 70 lm W^{-1}, and a maximum EQE of 21.5%. The EQEs of those devices were beyond the theoretical limit of fluorescent OLEDs.

TADF emitters are generally doped into the host to avoid the concentration quenching. Since the host molecules will interact with the TADF emitter molecules, the photophysical properties of the emitters are sensitive to the surroundings. Careful consideration of the intermolecular interactions is promising to construct high-efficiency OLEDs based on TADF emitters. TXO-PhCz is the CT molecule with both donor and acceptor units. High-efficiency OLEDs have been constructed with the hole-dominating material of mCP as the host. Adachi's group [37] reported a carbazole-based TADF emitter, 2-biphenyl-4,6-bis(12-peylindolo[2,3-a]-cabazole-11-yl)-1,3,5-triazine (PIC-TRZ), with a ΔE_{ST} of almost 0 meV. The electron-dominating material of 9,9'-(2,6-pyridinediyl)-bis-9H-carbazole (PYD2) was used as the host of PIC-TRZ and afforded a high EQE of 14%, higher than that of the compartment with mCP as host. To further enhance the performance of OLEDs based on TXO-PhCz, the electron-dominating material of 1,3,5-Tris(N-phenylbenzimidazole-2-yl)benzene (TPBI) was introduced as the host of TXO-PhCz [37].

Figure 13.10 showed the absorption of the TXO-PhCz and TPBI films and PL spectra of TXO-PhCz:TPBI and pure TPBI films. Large spectra overlap can be observed between the emission of TPBI film and the absorption of TXO-PhCz film, indicating the efficient energy transfer from TPBI to TXO-PhCz. The PL spectrum of TXO-PhCz:TPBI 5 wt% film exhibited only a broad, unstructured emission of TXO-PhCz countered at 522 nm. No PL emission of TPBI or exciplex of TPBI-TXO-PhCz was observed. The PLQY of TXO:TPBI film was 88%, slightly lower than that of TXO-PhCz in mCP. The transient PL decay of TXO-PhCz in TPBI host also showed clear second-order exponential decays (as shown in Figure 13.11). The prompt component was estimated to be 23 ns, corresponding the radiative decay rate constant of 3.8×10^7 s^{-1} in TPBI. The delayed component was estimated to be 48 μs, shorter than that in mCP (87 μs). The lifetime of the delayed component (τ_D) can be expressed as [24]

$$\frac{1}{\tau_D} = k_{nr}^T + (1 - \Phi_T)k_{rISC}$$

As the PLQYs of TXO-PhCz in TPBI and mCP are similar, the shorter τ_D cannot be ascribed to the higher nonradiative decay of triplet excitons. Thus, the temperature dependence of the transient PL decays of the TXO-PhCz:TPBI film was further investigated. Figure 13.12 shows the dependence of the total PL (Φ_{Total}), prompt fluorescence (Φ_P), and DF (Φ_D) of the film on the temperature. As can be observed, Φ_P varied very slightly around 13–17% as the temperature between 100 and 250 K, and then, Φ_P decreases with the increase of the temperature and reached about 5% at 300 K. Φ_D increases monotonically as the temperature increased and saturated to around 82% above 250 K. The synergistic result was that Φ_{Total} increases monotonically from 100 to 250 K and then decreases above 250 K, leading to the maximum peak of 99% at 250 K. The decrease of the prompt component above was ascribed to the enhanced nonradiative decay from S_1 at high temperature. The triplet formation efficiency

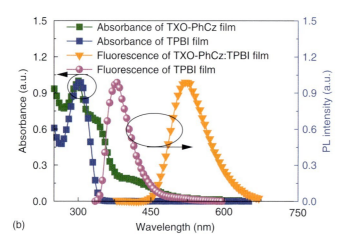

Figure 13.10 (a) Molecular structure of TXO-PhCz, 1,3,5-tris(N-phenylbenzimidazole-2-yl) benzene (TPBI), and 1,3-bis(9H-carbazol-9-yl) benzene (mCP) and (b) absorption of TXO-PhCz and TPBI films and fluorescence of TXO-PhCz:TPBI-doped film and pure TPBI film.

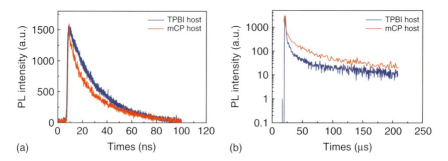

Figure 13.11 Transient photoluminescence (PL) decay of TXO-PhCz:TPBI-doped film at 300 K in the time range of (a) 200 ns and (b) 200 μs.

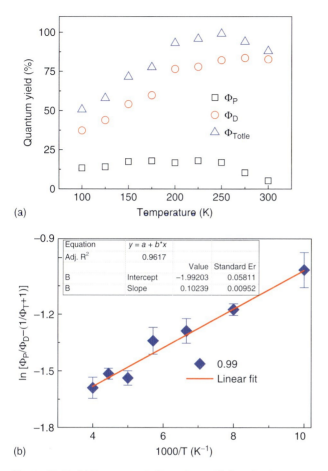

Figure 13.12 (a) Temperature dependence of photoluminescence quantum efficiencies for totle (Φ_{Totle}), prompt (Φ_P), and delayed (Φ_D) component; and (b) log plot of the intensity ratio of prompt fluorescence to delayed fluorescence versus 1/T.

was estimated to be 99%, and ΔE_{ST} is 8.8 meV using the Berberan-Santos plot. ΔE_{ST} was much lower than that of TXO-PhCz in mCP, which was ascribed to the shift of the singlet and triplet energy levels due to the interaction between host molecule and guest molecule.

Multilayer OLEDs were fabricated with TPBI as the host, and the device structure is the following: indium tin oxide (ITO) (150 nm)/poly(3,4-ethylenedioxythiophene) (PEDOT) (20 nm)/1,1-bis[4-[N,N'-di(p-tolyl) amino]phenyl] cyclohexane (TAPC) (20 nm)/mCP (10 nm)/TXO-PhCz:TPBI (5 wt%) (30 nm)/ 1,3,5-tri(m-pyrid-3-yl-phenyl)benzene (TmPyPB) (65 nm)/LiF (0.9 nm)/Al (100 nm). The energy level of the materials in the device and the performance of the device are shown in Figure 13.13. The device was turned on at 5 V with a green emission peak centered at 520 nm. The EL spectra of the device were almost identical to the PL spectra of the TXO-PhCz:TPBI-doped film, demonstrating the main contribution of the radiative decay of the singlet of TXO-PhCz. The device afforded a maximum current efficiency of 71.9 cd A^{-1}, a maximum

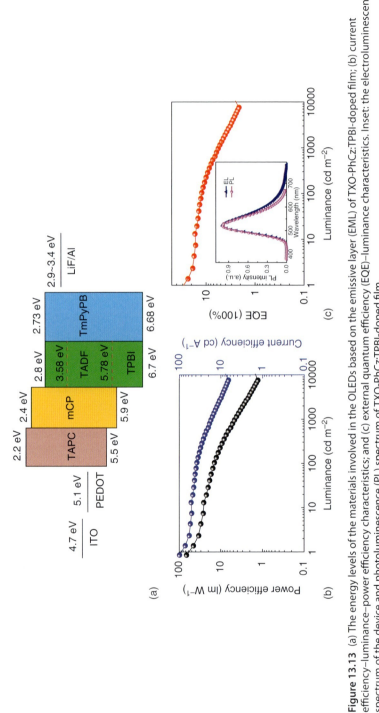

Figure 13.13 (a) The energy levels of the materials involved in the OLEDs based on the emissive layer (EML) of TXO-PhCz:TPBI-doped film; (b) current efficiency–luminance–power efficiency characteristics; and (c) external quantum efficiency (EQE)–luminance characteristics. Inset: the electroluminescent (EL) spectrum of the device and photoluminescence (PL) spectrum of TXO-PhCz:TPBI-doped film.

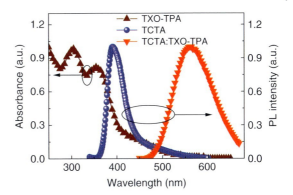

Figure 13.14 Absorbance of TXO-TPA and photoluminescence (PL) spectra of 4,4′,4″-tris(carbazole-9-yl)triphenylamine (TCTA) film and 5 wt% TXO-TPA:TCTA-doped film.

power efficiency of 45.2 lm W^{-1}, and a maximum EQE of 23.2%, which is higher than those of the device with mCP as the host. These results demonstrated that the higher performance of the device based on TADF emitters can be achieved by the dedicate consideration of the interaction between TADF emitter and host. The novel hosts matching the TADF emitters also are another main path to facilitate the future application of TADF emitter.

It is also significant to construct high-efficiency OLEDs based on TADF emitters by solution methods [38, 39]. 4,4′,4″-Tri(9-carbazoyl)triphenylamine (TCTA) can be easily deposited by the solution method to form amorphous film, and its high first-triplet energy level (2.78 eV) and wide bandgap can meet the requirement for the host materials for TXO–TPA [40–42]. Figure 13.14 shows the absorption spectrum of TXO-TPA and the PL spectra of the TCTA film and the 5 wt% TXO-TPA:TCTA film. A large overlap between the absorption spectrum of TXO-TPA and the PL spectrum of TCTA was observed, indicating the efficient energy transfer from TCTA to TXO-TPA. The PL emission of 5 wt% TXO-TPA:TCTA film showed only a broad emission centered at 570 nm, and there was no other shoulder or emission from TCTA. These findings further demonstrated the efficient energy transfer from TCTA to TXO-TPA, and no exciplex formed between TCTA and TXO-TPA. The device structure fabricated here was ITO/PEDOT (30 nm)/TXO-TPA:TCTA (5 wt%) (30 nm)/ TPBI (50 nm)/LiF (0.9 nm)/Al (100 nm). The emission layer of 5 wt% TXO-TPA:TCTA was spin-coated on top of PEDOT layer, and TPBI, LiF, and Al were subsequently deposited by vacuum evaporation to function as the electron-transport layer, electron-injection layer, and cathode, respectively. The device can be turned on at 3.5 V with a maximum current efficiency of 15.2 cd A^{-1}, a maximum power efficiency of 13.8 lm W^{-1}, and a maximum EQE of 5.97% (Figure 13.15). The performance of the devices can be further improved by the optimization of the device structure.

13.3.1.2 Nondoped OLEDs Based on TADF Emitters with Quantum Well Structure

For the TADF emitters, the triplet excitons can be upconverted into singlet excitons for light harvesting. The triplet excitons can cause the concentration quenching, depressing the efficiency of the devices. Thus, the emitting layer of

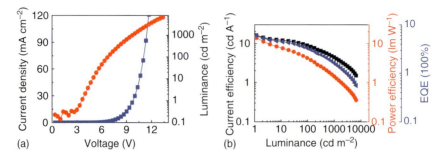

Figure 13.15 Electroluminescence (EL) characteristics: (a) current density–luminance–voltage curve and (b) current efficiency–power efficiency–external quantum efficiency (EQE)–luminance curves.

the host–guest doping systems is the requisite for high-efficiency OLEDs based on TADF emitters. However, TADF emitters are CT materials, and their photophysical properties are generally sensitive to the molecular atmosphere around the TADF emitters [43, 44]. The doping method with coevaporation is a complicated process and needs the precise control of the concentration of dopant [45]. These will cause the device reproducibility problem, impeding the large-scale commercial application of TADF OLEDs. Thus, undoped OLEDs based on TADF emitters are highly desirable and promising. Quantum well (QW) structures can effectively confine the charge carriers and excitons inside the emitting layer, affording high-efficiency light-emitting diodes [46–48]. These structures have been regarded as an efficient approach to construct nondoped OLEDs to achieve high efficiency. Yang et al. reported the multi-QW OLEDs based on blue phosphorescent emitter with a peak EQE of 20.31%, current efficiency of 40.31 cd A^{-1}, and power efficiency of 30.14 lm W^{-1}. These excellent results motivated our exploration of undoped TADF OLEDs with QW structure [48].

TXO-PhCz was chosen as the TADF emitters, and mCP was used as the potential barrier layer for the undoped TADF OLEDs with QW structure. First we optimized the width of the QW. The structures and the performance of the devices with single QW are shown in Figure 13.16, in which a thin layer of TXO-PhCz was inserted at the center of mCP as the emitting layer. The device with $x = 0.5$ nm affords a current efficiency of 2.9 cd A^{-1}, a power efficiency of 1.3 lm W^{-1}, and an EQE of 1.2%. Similar performances were obtained for the devices with $x = 1$ and 1.5 nm, and no prominent thickness dependence of TXO-PhCz can be observed. The EL spectra of the devices exhibited both blue emission band (370–450 nm) and green emission band. Under the electrical excitation, the radiative decay of mCP and the energy transfer from mCP to TXO-PhCz for light harvesting will compete each other. The obvious blue band indicated the inefficiency energy transfer from the mCP to TXO-PhCz. The blue emission bands of the devices were suppressed by the increase of the TXO-PhCz thickness. Combining with the similar device results, it can be concluded that the energy transfer cannot be enhanced by the increased width of the QW. However, with the increase of the well width, the emission peak redshifted, and the full wave at half maximum (FWHM) of the EL spectra increased. The shift of the emission

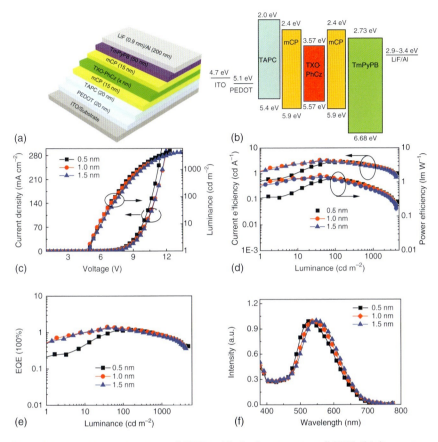

Figure 13.16 (a) Device structures of OLEDs with single quantum well (QW); (b) the energy-level diagram of OLEDs with single QW; and the performance of the device with the single QW: (c) current density–voltage–luminance curves; (d) power efficiency–current efficiency–luminance curves; (e) external quantum efficiency (EQE)–luminance curves; and (f) EL spectra.

peak and the change of the FWHM of the emission band for OLEDs coincided with the reported organic single QW devices [49]. Figure 13.17 shows the AFM image of TXO-PhCz on the multilayer films: ITO/PEDOT (20 nm)/TAPC (20 nm)/TXO-PhCz (x nm). The mCP film had the very smooth surface with a root-mean-square (RMS) of 0.24 nm. With the deposition of TXO-PhCz, similar smooth surface of TXO-PhCz film with RMS of 0.243 nm was achieved for the thickness of 0.5 nm. Further increasing the TXO-PhCz thickness, some bulges formed, and the RMS of TXO-PhCz films increased. These suggested that TXO-PhCz film grows on the mCP film via a layer-by-layer mode at the beginning stage. Then, the TXO-PhCz molecules aggregated, leading to the rough surface of the TXO-PhCz film. The transient PL decay of the TAXO-PhCz film with different thickness also supported this molecular aggregation (Figure 13.18). All these films exhibited both the prompt and the delayed components for their transient PL decay. For the TXO-PhCz film with 0.5 nm, the lifetime of the delayed component was about 137 μs. The thicker the TXO-PhCz film was, the

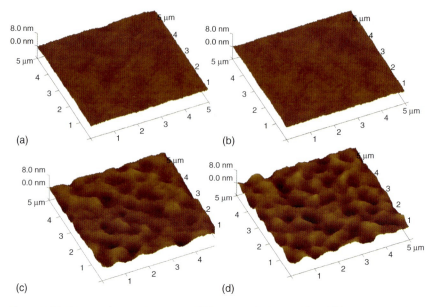

Figure 13.17 Atomic force microscope (AFM) images of the multilayer films of indium tin oxide (ITO)/poly(3,4-ethylenedioxythiophene)-poly(styrenesulfonate) (PEDOT) (20 nm)/di-[4-(N,N'-di-p-tolyl-amino)-phenyl]cyclohexane (TAPC) (20 nm)/1,3-bis(9H-carbazol-9-yl)benzene (mCP) (15 nm)/TXO-PhCz (x nm): (a) 0 nm, (b) 0.5 nm, (c) 1 nm, and (d) 1.5 nm.

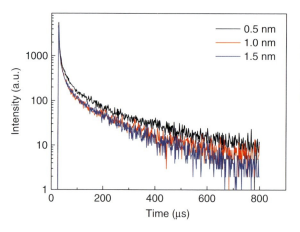

Figure 13.18 Transient photoluminescence (PL) decay curves of the multilayer films with the structure of 1,3-bis(9H-carbazol-9-yl)benzene (mCP) (15 nm)/TXO-PhCz ($x = 0.5$, 1, and 1.5 nm)/mCP (15 nm).

shorter the lifetime of the delayed component of the film was (127 μs for $x = 1$ nm and 112 μs for $x = 1.5$ nm). The shorter lifetimes of the delayed component of the films were ascribed to the serious singlet/triplet–triplet annihilation due to the self-aggregation of TXO-PhCz molecules. Thus, the width of the QW was better to be 0.5 nm to avoid the singlet/triplet–triplet annihilation.

To enhance the energy transfer from mCP to TXO-PhCz, OLEDs with multi-QWs were constructed. The thickness of the emission layer was fixed to 30 nm, and the 0.5-nm-thick TXO-PhCz layers were symmetrically inserted in mCP layer (as shown in Figure 13.19a). Figure 13.19b and c show the performance

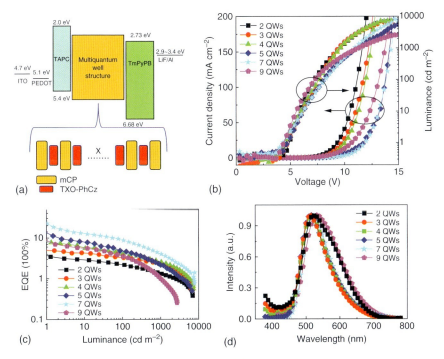

Figure 13.19 Device structures and energy-level diagram of the multiquantum wells. The thickness of TXO-PhCz of the quantum wells is fixed to be 0.5 nm (a). X is the number of quantum well, and Y is the thickness of the 1,3-bis(9H-carbazol-9-yl) benzene (mCP) block layer in the emitting layers: for $X = 2$, $Y = 10$ nm; for $X = 3$, $Y = 7.5$ nm; for $X = 4$, $Y = 6$ nm; for $X = 5$, $Y = 5$ nm; for $X = 7$, $Y = 3.75$ nm; and for $X = 9$, $Y = 3$ nm. Electronic properties of the devices with multiquantum well: (b) current density–voltage–luminance plots, (c) external quantum efficiency (EQE)–luminance plots, and (d) electroluminescence (EL) spectra at 9 V. ITO is referred to indium tin oxide; PEDOT is referred to poly(3,4-ethylenedioxythiophene)-poly(styrenesulfonate); TAPC is referred to di-[4-(N,N'-di-p-tolyl-amino)-phenyl]cyclohexane; LiF is referred to lithium fluoride; Al is referred to aluminum.

of the devices with multi-QWs. All the devices showed similar turn-on voltage. With the increase of the QW number, the current densities of the devices at the same voltage first decreased and then increased. The increase of the QW number caused much more defects or traps in TXO-PhCz films or at mCP/TXO-PhCz interface, leading to the low current density of the devices. With the increase of the QW number, the thickness of the mCP barrier layer in the EML will decrease, facilitating the direct trapping and thus transporting of the carriers in TXO-PhCz. These synergistic effects caused the lowest current density of the devices with seven QWs. Similar trend can also be observed for the performance of the devices. The device with seven QWs afforded the highest performance with a current efficiency of 69 cd A^{-1}, a power efficiency of 50 lm W^{-1}, and a maximum EQE of 22.6%. The EL spectra of the devices with QW structure are shown in Figure 13.19d. All the devices exhibited one main emission peak at 520 nm corresponding the emission from TXO-PhCz. For the two QWs, weak emission at the wavelength lower than 450 nm and an emission shoulder at around 580 nm were also observed, which were ascribed to the inefficient energy transfer from mCP to TXO-PhCz and the leakage of the excitons into the TAPC layer. The intensity of the emission at 570 nm and wavelengths lower than 450 nm were suppressed as the increase of the number of QWs. This emission disappeared for the device with five or seven QWs. TXO-PhCz is a bipolar molecular with both donor and acceptor units, and the increased number of QWs shifted the recombination zone to the TAPC/EML interface, leading to the leakage of the excitons to the TAPC layer. All these results demonstrated that the EML with QW structure is a cost-effective method for the construction of nondoped TADF OLEDs with high efficiency.

13.3.1.3 White OLEDs Based on Blue Fluorescent Emitter and Yellow TX-based TADF Emitters

Lighting by OLEDs also attracted the commercial interests, and efficient and thin large-area lighting source is now available in the market [50–52]. Generally, these white organic light-emitting diodes (WOLEDs) exhibit stable CIE and a color rendition index (CRI) over 75. Highly efficient WOLEDs with a power efficiency of 59.9 lm W^{-1} and CRI over 80 have been reported using a yellow phosphorescent emitter [53, 54]. As the efficient participation of both singlet and triplet excitons of TADF emitters, it is promising to producing high efficient WOLEDs with nearly 100% internal quantum efficiency [55, 56]. TXO-TPA emitted yellow EL that was centered at 552 nm with color coordinate of CIE (0.45, 0.53). Thus, it is rational to construct pure fluorescent OLED incorporating blue fluorescent emitter and TXO-TPA emitter with double-layer structure.

4P-NPB is a typical blue fluorescent emitter with a high fluorescent quantum yield of 92% and has been used for highly efficient WOLEDs with phosphorescent emitters [54]. It can be arranged with TXO-TPA as the two primary-color emitters for WOLEDs. WOLEDs can be arranged with two emitting layers, and a common host was chosen for the two emitters to reduce structural heterogeneity and facilitate the charge transport in the two emitting layers. mCP was chosen for its high triplet energy level to confine the triplet excitons on the emitters. Its appropriate HOMO and LUMO also facilitated the adjustment of the carrier balance in the emission layer and the construction of highly efficient WOLEDs. The absorbance and PL spectra of 4P-NPB and TXO-TPA in thin film and the PL spectra of mCP film are shown in Figure 13.20. mCP showed two sharp emission peaks of 364 and 350 nm with a long tail. 4P-NPB showed a broad absorption with a peak of 362 nm. Large spectra overlap between the emission of mCP and the absorbance of 4P-NPB, indicating the efficient energy transfer of the singlet from mCP to 4P-NPB. Similar overlap can also be observed for the absorbance of TXO-TPA and the emission of mCP. Therefore, the singlet excitons that formed on the mCP host can be effectively transferred to the two doped emitters through a Förster resonant process. The triplet excitons typically have long diffusion lengths of about 100 nm [57, 58]. The triplet excitons on mCP host can migrate to the TXO-TPA:mCP layer and transfer to the TXO-TPA by Dexter transfer process, even if the excitons were generated in the TXO-TPA:mCP layer. TXO-TPA can emit yellow light by the efficient upconversion from triplet to singlet. The device structure of WOLED design and the chemical structure and

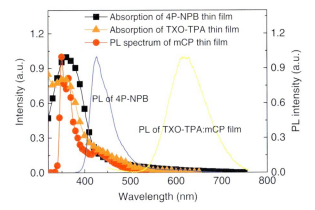

Figure 13.20 The absorption spectrum of N,N'-di-(1-naphthalenyl)-N,N'-diphenyl-[1,1':4',1'':4''',1'''-quaterphenyl]-4-4'''-diamine (4P-NPB) and TXO-TPA thin films and photoluminescence (PL) spectra of 1,3-bis(9H-carbazol-9-yl) benzene (mCP), 4P-NPB, and 5 wt% TXO-TPA:di-[4-(N,N'-di-p-tolyl-amino)-phenyl]cyclohexane (TAPC) film.

energy level of the used materials are shown in Figure 13.21. mCP shows a high hole mobility of 1.2×10^{-4} cm^2 V^{-1} s^{-1}, which was three times higher than that of electron mobility [59]. Similarly, the hole mobility of 4P-NPB is significantly higher than that of the electron mobility [60]. Thus, the exciton generation zone will be located at the interface of 4P-NPB:mCP and TmPyPB layer. Although some exciton losses in the transfer processes cannot be avoided, unity internal quantum efficiency can be achieved for the high efficient WOLEDs.

To get high efficient WOLED, the balance of yellow emission and blue emission can be tuned by the delicate management of singlet and triplet excitons in the two emitting layers. The doped concentration of TXO-TPA was maintained at 5 wt% as highly efficient yellow OLEDs have been achieved. The doping concentration of 4P-NPB was optimized from 0.5 to 20 wt%. The performances of the devices are summarized in Table 13.3. The devices showed similar turn-on voltages of about 4 V. Yellow light can only be emitted for the device with a 4P-NPB doping concentration of 0.5 wt%. By increasing the doping concentration of 4P-NPB, the blue emission peak appeared, and its intensity increased. The color coordinate shifts from the yellow light region into the white region (Figure 13.22). When the doping concentration was increased to 20 wt%, only blue light was observed, and no emission from TXO-TPA was observed. At low doping concentration of 4P-NPB, the singlet harvested for blue light, thereby leading to the yellow emission alone. The loss of singlet and triplet excitons during diffusion led to low efficiency, whereas the nonradiative loss of triplet excitons on 4P-NPB trapped by Dexter energy transfer diminished the efficiency of the devices at high doping concentration of 4P-NPB. Moreover, the back Dexter transfer from TXO-TPA to 4P-NPB at the interface of the two emission layers is another possible reason for triplet exciton loss (Figure 13.21c). Thus, the optimized concentration of 4P-NPB is 5 wt%, and the device affords a current efficiency of 10.9 cd A^{-1}, a power efficiency of 8.5 lm W^{-1}, and an EQE of 4.4%. The maximum total efficiencies of the devices can be up to 18.5 cd A^{-1}, 14.5 lm W^{-1}, and 7.5%, which are comparable

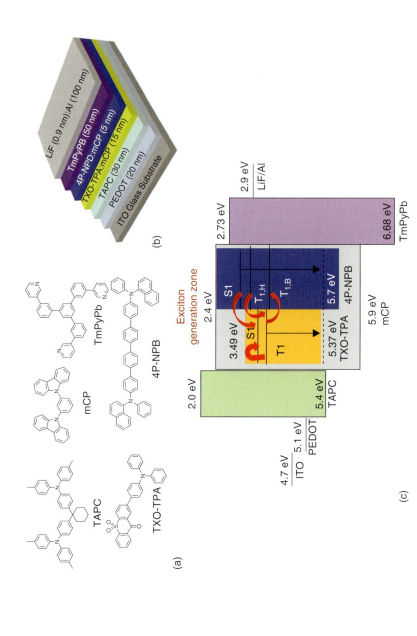

Figure 13.21 (a) The molecular structure of the materials used for the construction of white OLEDs (WOLEDs); (b) device structure and molecule arrangement of the WOLEDs; and (c) energy-level scheme for materials used in the WOLEDs and exciton energy diagram of the emitting layers.

Table 13.3 Electroluminescence characteristics of the OLEDs with different doping concentrations of 4P-NPB.

Doping concentration of 4P-NPB (%)	Turn-on voltage (V)	Maximum values			Color coordinate at 6 V
		Current efficiency (cd A^{-1})	Power efficiency (lm W^{-1})	EQE (%)	
0.5	4	5.6	2.9	2.0	(0.39, 0.51)
2	4	10.9	8.5	4.4	(0.37, 0.42)
5	4	7.2	5.7	3.6	(0.31, 0.35)
10	4.1	4.9	2.4	2.5	(0.30, 0.33)
20	4	3.2	1.5	1.9	(0.27, 0.27)

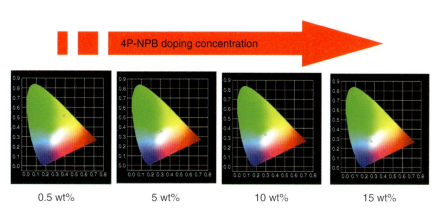

Figure 13.22 Color coordinate of International Commission on Illumination (CIE) at 6 V for the devices with different doping concentrations. 4P-NPB is referred to N,N'-di-(1-naphthalenyl)-N,N'-diphenyl-[1,1':4',1'':4'',1'''-quaterphenyl]-4-4'''-diamine.

with those of the devices based on fluorescent and phosphorescent emitters with similar structures [58]. Notably, the EL spectra of the device were very stable, and no derivation of the peak intensity and new peaks was observed even at high voltage up to 10 V.

To optimize the CIE and the performance of WOLEDs, device I, II, and III were fabricated, and the device structures were as follows: device I: ITO/PEDOT (20 nm)/TAPC (40 nm)/TPA:mCP (5%) (5 nm)/4P-NPB:mCP (2%) (5 nm)/TmPyPB (50 nm)/LiF(0.9 nm)/Al (100 nm); device II: ITO/PEDOT (20 nm)/TAPC (40 nm)/TPA:mCP (5%) (10 nm)/4P-NPB:mCP (2%) (10 nm)/TmPyPB (50 nm)/LiF (0.9 nm)/Al (100 nm); device III: ITO/PEDOT (20 nm)/TAPC (40 nm)/TPA:mCP (5%) (10 nm)/mCP (5 nm)/4P-NPB:mCP (2%) (10 nm)/TmPyPB (50 nm)/LiF (0.9 nm)/Al (100 nm). All these devices showed white EL emission. Device II and III with broad yellow-emitting layers showed higher performance due to the diminished quenching effect of the triplet excitons. As shown in Table 13.4, device II showed the highest performance:

Table 13.4 Electroluminescence characteristics of the devices I, II, and III.

Device structure	Turn-on voltage (V)	Maximum values			
		Current efficiency (cd A^{-1})	Power efficiency (lm W^{-1})	EQE (%)	Color coordinate at 6 V
Device I	3.6	3.2	2.4	1.8	(0.35, 0.32)
Device II	4.0	8.1	6.4	4.7	(0.34, 0.34)
Device III	4.3	8	5.8	4.4	(0.37, 0.36)

The device can be turned on at 4 V with a maximum EQE of 4.7%, a current efficiency of 8.1 cd A^{-1}, and a power efficiency of 6.4 lm W^{-1}. Impressively, the device showed low-efficiency roll-off. The critical current density of the device where EQE declines to half of its peak was approximately 112 mA cm^{-2}, which is among the best of reported multilayer WOLED based on phosphorescent emitters. No interlayer is needed to prevent the mutual exciton transfer and quenching [54].

13.3.2 TADF Host for Phosphorescent Emitters

TADF materials can also be applied as the host of phosphorescent OLEDs, which is inspired by their fast rISC from triplet to singlet. Generally, the efficiency of phosphorescent OLEDs tends to decrease with increasing brightness, the so-called efficiency "roll-off." The serious efficiency roll-off impedes them from the applications requiring high brightness [61]. There are two main annihilation processes on host molecules that are most relevant to the EQE roll-off of phosphorescent OLEDs: triplet–triplet and triplet-polaron annihilations. Thus, the efficiency roll-off can be effectively alleviated by the reduction of the triplet exciton density on the host materials. Lee's group [62] reported the exciplex with TADF properties as the host for the fluorescent emitter to sensitize the triplet excitons in the fluorescent OLEDs through the fast rISC process of the exciplex, and high-performance fluorescent OLEDs with EQE of 14.5% were achieved, which surpassed the efficiency limitation of fluorescent OLEDs. The exciplex host was the main carrier recombination center, and the triplet excitons were upconverted into the singlet excitons and then resonantly transferred to the singlet of fluorescent molecules for light emission. Thus, reducing the triplet exciton density by using TADF molecules as the host of phosphorescent emitter [16, 17] and alleviating the efficiency roll-off of phosphorescent OLEDs are rational.

MTXSFCz (Figure 13.23) was synthesized by Suzuki coupling reaction, in which the hole-transporting PhCz and electron-withdrawing TX unit were bridged with unconjugated fluorene. The HOMO of MTXSFCz was mainly localized on the PhCz unit, and the LUMO was dispersed over the electron-withdrawing TX unit. No obvious overlap between HOMO and LUMO was observed, which is similar to the reported bipolar phosphorescent host [63, 64].

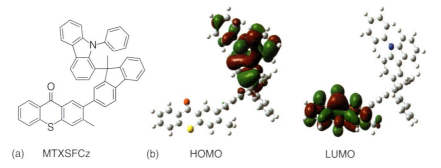

(a) MTXSFCz   (b) HOMO   LUMO

Figure 13.23 (a) Molecular structure of MTXSFCz and (b) calculated spatial distributions of the highest occupied molecular orbital (HOMO) and lowest unoccupied molecular orbital (LUMO) electron densities of MTXSFCz by Gaussian 03 at the B3LYP/6-31G(d) level.

The unconjugated fluorine unit caused the effective separation of the electron densities of HOMO and LUMO, leading to a small ΔE_{ST} of 0.06 eV calculated on the ground-state geometries. The small ΔE_{ST} facilitated the upconversion process from triplet to singlet. MTXSFCz showed a high decomposition temperature of 456 °C and a high glass transition temperature of 127.3 °C. The excellent stability and the twisted structure were beneficial to form homogeneous and amorphous films with good stability [65]. MTXSFCz exhibited bipolar charge transport properties with a hole and electron mobilities of 1.45×10^{-5} and 1.75×10^{-8} cm^2 (V s)$^{-1}$, respectively.

There was an absorption peak around 380 nm arising from the n–π* transition of TX core in the UV–Vis absorption spectrum [19, 21, 66], and no prominent CT absorption was observed over 400 nm [21] (see Figure 13.24). The HOMO energy level of −5.57 eV can be obtained from the cyclic voltammetry, and the LUMO energy level can be estimated to −2.83 eV, combining the energy gap estimated from the onset of the absorption spectrum. MTXSFCz in the DCM solution emitted blue light with a peak centered at 420 nm, and the emission peak in neat film centered at 451 nm, redshifted approximately 30 nm. The PL quantum yield of the MTXSFCz neat film measured by the integrating sphere was only 5.1%, and the low PL quantum yield was beneficial for the construction

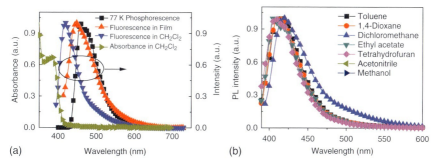

Figure 13.24 (a) Absorbance, photoluminescence (PL) of MTXSFCz in CH$_2$Cl$_2$ and in film, and spectrum measured in oxygen-free 2-methyl tetrahydrofuran (2-MeTHF) at 77 K and (b) PL emission of MTXSFCz in different solvents.

of highly efficient phosphorescent OLEDs. The triplet energy of 2.71 eV was estimated from its phosphorescent spectrum measured in oxygen-free 2-methyl THF at 77 K. Thus, ΔE_{ST} was estimated to be 0.19 eV, which is comparable with that from the DFT calculation and those of the reported TADF materials [14, 64].

The transient PL decay of the doped film of MTXSFCz in mCP and the temperature dependence were performed to demonstrate the TADF properties of MTXSFCz further (as shown in Figure 13.25). The doped film emitted blue light with the emission peak at 455 nm, and no emission from mCP was observed. The emission spectra of the film blueshifted due to the temperature dependence of the relaxation process, and the relative intensity decreased when the film is cool down to lower temperature [67, 68]. The PL decay curves were resolved into two components: the prompt and delayed components. The transient decay times of the prompt and delayed components were 3.1 ns and 45.3 μs from second-order exponential decay fitting. The PL spectra of these two components were identical. Thus, the DF can be assigned to the TADF that occurs through rISC. This finding can be further demonstrated by the monotonically increase of the delayed component as the temperature increases [24].

Efficient Förster energy transfer from MTXSFCz to Ir(piq)$_2$acac and Ir(2-phq)$_3$ (molecular structure as shown in Figure 13.26) can be expected due to the overlap between the PL emission of MTXSFCz and the absorption of Ir(piq)$_2$acac and

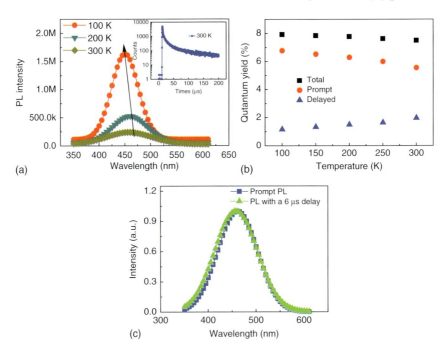

Figure 13.25 (a) Temperature dependence of the photoluminescence (PL) intensity for 5 ± 1 wt% MTXSFCz doped in 1,3-bis(9H-carbazol-9-yl) benzene (mCP) and inset is the PL decay curve of the doped film at 300 K. (b) Temperature dependence of prompt proportion (red circles) and delayed proportion (blue triangles) for 5 ± 1 wt% MTXSFCz doped in mCP based on lifetime fitting results. (c) The photoluminescence spectrum of the same film without delay time and 6 μs delay time. The determination of the total, prompt, and delayed fluorescence can be referred to [24]. Source: Ref. [24]. Reproduced with the permission of Springer Nature.

Ir(2-phq)$_3$. Thus, phosphorescent OLEDs based on MTXSFCz were fabricated, and the device structure was ITO/PEDOT (30 nm)/TAPC (20 nm)/10 wt% Ir(pig)$_2$acac: MTXSFCz or Ir(2-phq)$_3$: MTXSFCz (35 nm)/TmPyPB (55 nm)/LiF (0.9 nm)/Al (100 nm). The energy-level diagram was shown in Figure 13.26, and the device results were shown in Table 13.5. All devices were turned on at a voltage of about 4.0 V, and the EL spectra were independent of the applied

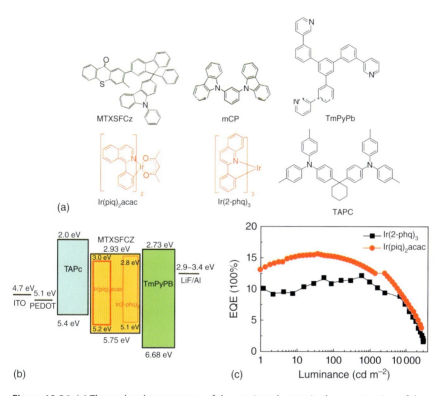

Figure 13.26 (a) The molecular structures of the semiconductors in the construction of the phosphorescent OLEDs; (b) the energy-level scheme for the phosphorescent OLEDs based on MTXSFCz; and (c) external quantum efficiency (EQE)–luminance curves of the phosphorescent OLEDs based on MTXSFCz. The doping concentration of the phosphorescent guest is 10 wt%. ITO is referred to indium tin oxide; PEDOT is referred to poly(3,4-ethylenedioxythiophene)-poly(styrenesulfonate); LiF is referred to lithium fluoride; Al is referred to aluminum.

Table 13.5 The performance summary of orange and red PhOLEDs based on reported hosts and MTXSFCz.

Device	Voltage (V)	Dopant concentration (wt%)	Current efficiency (cd A^{-1})	Power efficiency (lm W^{-1})	EQE$_{Max}$ (%)	EQE (%) at 1 000 cd m^{-2}	EQE (%) at 10 000 cd m^{-2}
Ir(2-phq)$_3$	4.0	10	26.8	18.0	11.8	11.2	7.6
	4.0	5	19.6	15.4	8.6	7.9	5.8
Ir(piq)$_2$acac	4.4	10	13.8	8.2	15.6	13.1	8.0
	3.8	5	10.5	8.6	12.0	9.4	—

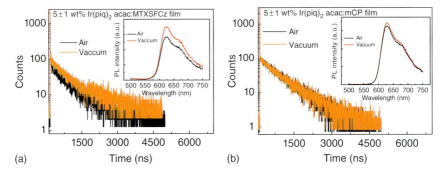

Figure 13.27 (a) The lifetime of 5 ± 1 wt% Ir(piq)$_2$acac:MTXSFCz film in vacuum and in air and inset is the photoluminescence (PL) intensity of 5 ± 1 wt% Ir(piq)$_2$acac:MTXSFCz film in vacuum and in air. (b) The lifetime of 5 ± 1 wt% Ir(piq)$_2$acac:1,3-bis(9H-carbazol-9-yl) benzene (mCP) film in vacuum and in air and the inset is the PL intensity of 5 ± 1 wt% Ir(piq)$_2$acac:mCP film in vacuum and in air.

voltage. The orange device with CIE (0.56, 0.43) showed a maximum current efficiency of 26.8 cd A^{-1}, a maximum power efficiency of 18.0 lm W^{-1}, and a maximum EQE of 11.8%. The red device with CIE (0.68, 0.32) afforded a current efficiency of 13.8 cd A^{-1}, a power efficiency of 8.2 lm W^{-1}, and an EQE of 15.6%. Interestingly, the efficiency roll-offs of both devices were low. J_0, the current density at which the EQE drops to half of its maximum value, was 80 mA cm^{-2} for the orange devices and 174 mA cm^{-2} for the red devices. These values are much larger than those of reported phosphorescent OLEDs and that of the reference devices based on mCP [68–73].

To confirm that the TADF properties of the MTXSFCz host can alleviate the efficiency roll-off, the decay curves of 5 wt% Ir(pig)$_2$acac-doped films with MTXSFCz and mCP were investigated (as shown in Figure 13.27). For the low doping concentration, the main emission of the doped film harvested from the triplet state of the guest through the FRET from the host to the guest [74, 75]. The decay time of the Ir(piq)$_2$acac emission in vacuum was about 864.4 ns. The decay time and the intensity of the emission were suppressed upon exposure to air. Meanwhile, no such effect was observed for the emission of Ir(piq)$_2$acac in mCP. Thus, the quenching effect of oxygen on Ir(piq)$_2$acac emission in MTXSFCz was attributed to the storage of the triplet excitons on MTXSFCz [76]. Both mCP and MTXSFCz had high triplet energy, at least 0.7 eV higher than that of Ir(piq)$_2$acac. All triplet excitons were confined on Ir(piq)$_2$acac guest, and thus, similar decay dynamics and lifetime should be observed [77]. However, the lifetime of Ir(piq)$_2$acac in MTXSFCz in vacuum was 864.4 ns, about 100 ns longer than that of Ir(piq)$_2$acac in mCP (764.3 ns). The longer lifetime can be attributed to the presence of the ISC and rISC on photoexcited MTXSFCz (as shown in Figure 13.28). The rate constant of FRET from host to guest can be estimated from the following equation [78]:

$$k_{ET} = \left(\frac{1}{\tau_D}\right)\left(\frac{R_0^6}{R^6}\right)$$

where τ_D is the radiative decay time of donor molecular, R_0 is the Förster transfer radius, and R is the average distance between donor and acceptor molecules.

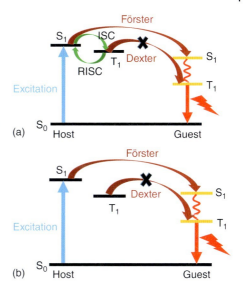

Figure 13.28 The mechanism of the energy transfer from the host to Ir(piq)$_2$acac in (a) MTXSFCz and (b) 1,3-bis(9H-carbazol-9-yl) benzene (mCP). Since the doping concentration of both host–guest system is low, the Dexter energy transfer from MTXSFCz to Ir(piq)$_2$acac can be omitted. The cycling of the exciton between singlet and triplet will contribute to the longer lifetime of Ir(piq)$_2$acac emission in MTXSFCz.

R_0 of the Ir(piq)$_2$acac:MTXSFCz film was 2.6 nm using a molecular modeling program, and the rate constant was $>10^{10}$ s^{-1}, which was nearly two orders of magnitude faster than those of ISC and rISC of TADF molecules [79]. The fast FRET converted the electrically generated singlet excitons of the host to the singlet and triplet excitons of guest, thereby facilitating the rISC of the MTXSFCz host from triplet to singlet. These decreased the triplet concentration on the MTXSFCz host, leading to the low-efficiency roll-off of the devices based on the MTXSFCz host.

13.4 Concluding Remarks and Outlook

TADF materials have attracted much attention in recent years due to their wide application in OLEDs, such as emitter, host, sensitizer of fluorescent emitters, and so on. TX unit, which benefited from both ketone and sulfur connections, showed a very small singlet–triplet energy gap, indicating its potential application in the construction of TADF materials. We reported high green and yellow TADF emitters, namely, TXO-TPA and TXO-PhCz, by the incorporation of donor units of TPA or PhCz and oxidation of sulfur atoms. The photophysical results of the emitters demonstrated their very small ΔE_{ST} (0.052 and 0.073 eV), which afforded fast rISC from triplet to singlet. The high-efficiency OLEDs based on TXO-TPA and TXO-PhCz showed a maximum EQE of 18.5 and 21.5%, respectively. These values surpassed the limitation of the OLEDs based on traditional fluorescent emitters. The high-efficiency multilayer WOLEDs incorporating 4P-NPB and TXO-TPA showed a maximum EQE of 4.7% with a CIE coordinate of (0.34, 0.34). We also reported the phosphorescent host of MTXSFCz with TADF, which had a TX and PhCz unit linked by spirofluorene. The orange and red PHOLEDs based on the host showed the high maximum EQE of 11.8 and 15.6%, respectively. The efficient rISC of the host suppressed the subdued annihilation of the triplet on the host, resulting in the low-efficiency roll-off. These works will be helpful in the development of new TADF materials for highly efficient OLEDs.

Our studies are only the primary step in the exploration of the TX-based TADF materials. Thus, the relationship between the molecular structures of the TX system and their photophysical properties should be studied for the rational design of TADF materials. Blue TADFs based on TX unit are particularly indispensable for the construction of OLEDs for the display and lightings. Solution-processed TADF emitters are also appealing for the low-cost OLEDs. Furthermore, new TADF hosts based on TX unit are also an interesting topic. Further studies should focus on the rational design of TADF hosts and their application in blue phosphorescent OLEDs. Moreover, the use of TADF materials is not limited to the emitters or hosts. OLEDs with TADF emitters as the sensitizer of fluorescent emitters had been reported. Thus, the new applications of TADF materials based on TX, such as sensitizer or other applications, are another further step toward the realization of highly efficient OLEDs.

Acknowledgments

This work was financially supported by the National Natural Science Foundation of China (Grant No. 61420106002, No.51373189, No. 61178061, and No.61227008), the "Hundred Talents Program" of the Chinese Academy of Sciences, and the Start-Up Fund of the Technical Institute of Physics and Chemistry of the Chinese Academy of Sciences.

References

1 Bernaose, A., Comte, M., and Vouaux, P. (1953). *J. Chim. Phys.* 50: 261.
2 Tang, C.W. and Vanslyke, S.A. (1987). *Appl. Phys. Lett.* 51: 913.
3 Seki, S., Uchida, M., Sonoyama, T., Ito, M., Watanabe, S., Sakai, S., and Miyashita, S. (2012). *SID Symposium Digest of Technical Papers* 40: 593.
4 Cai, M., Xiao, T., Hellerich, E., Chen, Y., Shinar, R., and Shinar, J. (2011). *Adv. Mater.* 23: 3590.
5 Lee, D.-H., Choi, J.S., Chae, H., Chung, C.-H., and Cho, S.M. (2008). *Display* 29: 436.
6 Li, Y., Tan, L.W., Hao, X.T., Ong, K.S., and Zhu, F. (2005). *Appl. Phys. Lett.* 86: 153508.
7 Gaspar, D.J. and Polikarpov, E. (2015). *OLED Fundamentals-Materials, Device, and Processing of Organic Light-Emitting Diodes*. Boca Raton: CRC Press.
8 Yersin, H. and Monkowius, U. (2006). *DE 10 2008 033 563*. Germany: Internal filing at the University of Regensburg.
9 Endo, A., Ogasawara, M., Takahashi, A., Yokoyama, D., Kato, Y., and Adachi, C. (2009). *Adv. Mater.* 21: 4802.
10 Parker, C.A. and Hatchard, C.G. (1961). *Trans. Faraday Soc.* 57: 1894.
11 Baleizão, C. and Berberan-Santos, M.N. (2007). *J. Chem. Phys.* 126: 204510.
12 Valeur, B. (ed.) (2002). *Molecular Fluorescence: Principles and Applications*. Weinheim: Wiley-VCH.

13 Endo, A., Sato, K., Yoshimura, K., Kai, T., Kawada, A., Miyazaki, H., and Adachi, C. (2011). *Appl. Phys. Lett.* 98: 083302.
14 Tao, Y., Yuan, K., Chen, T., Xu, P., Li, H., Chen, R., Zheng, C., Zhang, L., and Huang, W. (2014). *Adv. Mater.* 26: 7931.
15 Wang, H., Meng, L., Shen, X., Wei, X., Zheng, X., Lv, X., Yi, Y., Wang, Y., and Wang, P. (2015). *Adv. Mater.* 27: 5861.
16 Nakanotani, H., Higuchi, T., Furukawa, T., Masui, K., Morimoto, K., Numata, M., Tanaka, H., Sagara, Y., Yasuda, T., and Adachi, C. (2014). *Nat. Commun.* 5: 4016.
17 Zhang, D., Duan, L., Li, C., Li, H., Zhang, D., and Qiu, Y. (2014). *Adv. Mater.* 26: 5050.
18 Fouassier, J.P. and Ruhlmann, D. (1993). *Eur. Polym. J.* 29: 505.
19 Lai, T. and Lim, E.C. (1980). *Chem. Phys. Lett.* 73: 244.
20 Cavaleri, J.J., Prater, K., and Bowman, R.M. (1996). *Chem. Phys. Lett.* 259: 495.
21 Ley, C., Morlet-Savary, F., Jacques, P., and Fouassier, J.P. (2000). *Chem. Phys.* 255: 335.
22 Allonas, X., Ley, C., Bibaut, C., Jacques, P., and Fouassier, J.P. (2000). *Chem. Phys. Lett.* 322: 483.
23 Shuai, Z. and Peng, Q. (2016). *Nat. Sci. Rev.* doi: 10.1093/nsr/nww024.
24 Uoyama, H., Goushi, K., Shizu, K., Nomura, H., and Adachi, C. (2012). *Nature* 492: 234.
25 Nakagawa, T., Ku, S.-Y., Wong, K.-T., and Adachi, C. (2012). *Chem. Commun.* 48: 9580.
26 Lee, C.W. and Lee, J.Y. (2014). *Chem. Mater.* 26: 1616.
27 Soujanya, T., Fessenden, R.W., and Samanta, A. (1996). *J. Phys. Chem.* 100: 3507.
28 Méhes, G., Nomura, H., Zhang, Q., Nakagawa, T., and Adachi, C. (2012). *Angew. Chem. Int. Ed.* 51: 11311.
29 Li, J., Nakagawa, T., MacDonald, J., Zhang, Q., Nomura, H., Miyazaki, H., and Adachi, C. (2013). *Adv. Mater.* 25: 3319.
30 Serevičius, T., Nakagawa, T., Kuo, M.-C., Cheng, S.-H., Wong, K.-T., Chang, C.-H., Kwong, R.C., Xia, S., and Adachi, C. (2013). *Phys. Chem. Chem. Phys.* 15: 15850.
31 Hirata, S., Sakai, Y., Masui, K., Tanaka, H., Lee, S.Y., Hiroko, N., Nakamura, N., Yasumatsu, M., Nakanotani, H., Zhang, Q., Shizu, K., Miyazaki, H., and Adachi, C. (2015). *Nat. Mater.* 14: 330.
32 Zhang, Q., Li, B., Huang, S., Nomura, H., Tanaka, H., and Adachi, C. (2014). *Nat. Photon.* 8: 326.
33 Zhang, Q., Li, J., Shizu, K., Huang, S., Hirata, S., Miyazaki, H., and Adachi, C. (2012). *J. Am. Chem. Soc.* 134: 14706.
34 Zhang, Q., Kuwabara, H., Postscavage, W.J., Huang, S., Hatae, Y., Shibata, T., and Adachi, C. (2014). *J. Am. Chem. Soc.* 136: 18070.
35 Hong, Y., Lam, J.W., and Tang, B.Z. (2009). *Chem. Commun.* 4332.
36 Berberan-Santos, M.N. and Garcia, J.M.M. (1996). *J. Am. Chem. Soc.* 118: 9391.
37 Liu, Z., Helander, M.G., Wang, Z., and Lu, Z. (2013). *Org. Electron.* 14: 852.

38 Friend, R.H., Gymer, R.W., Holmes, A.B., Burroughes, J.H., Marks, R.N., Taliani, C., Bradley, D.D.C., Dos Santos, D.A., Bredas, J.L., Logdlund, M., and Salaneck, W.R. (1999). *Nature* 397: 121.

39 Kraft, A., Grimsdale, A.C., and Holmes, A.B. (1998). *Angew. Chem. Int. Ed.* 37: 402.

40 Chen, J.S., Shi, C.S., Fu, Q., Zhao, F.C., Hu, Y., Feng, Y.L., and Ma, D.G. (2012). *J. Mater. Chem.* 22: 5164.

41 Park, J.J., Park, T.J., Jeon, W.S., Pode, R., Jang, J., Kwon, J.H., Yu, E.S., and Chae, M.Y. (2009). *Org. Electron.* 10: 4115.

42 Doh, Y.J., Park, J.S., Jeon, W.S., Pode, R., and Kwon, J.H. (2012). *Org. Electron.* 13: 586.

43 Sato, K., Shizu, K., Yoshimura, K., Kawada, A., Miyazaki, H., and Adachi, C. (2013). *Phys. Rev. Lett.* 110: 247401.

44 Ishimatsu, R., Matsunami, S., Shizu, K., Adachi, C., Nakano, K., and Imato, T. (2013). *J. Phys. Chem. A* 117: 5607.

45 Wang, Q., Osawald, I.W.H., Perez, M.R., Jia, H., Shahub, A.A., Qiao, Q., Gnade, B.E., and Omary, M.A. (2014). *Adv. Funct. Mater.* 24: 4746.

46 Yang, X., Wu, F.-I., Haverinen, H., Cheng, C.-H., and Jabbour, G.E. (2011). *Appl. Phys. Lett.* 98: 033302.

47 Zhao, Y., Chen, J., and Ma, D. (2013). *ACS Appl. Mater. Interfaces* 5: 965–971.

48 Yang, X., Zhang, S., Qiao, X., Mu, G., Wang, L., Chen, J., and Ma, D. (2012). *Opt. Exp.* 20: 24411.

49 Huang, J., Xie, Z., Yang, K., Jiang, H., and Liu, S. (1999). *J. Phys. D: Appl. Phys.* 32: 2841.

50 Chang, Y.L., Song, Y., Wang, Z., Helander, M.G., Qiu, J., Chai, L., Liu, Z., Scholes, G.D., and Lu, Z. (2013). *Adv. Funct. Mater.* 23: 705.

51 Ying, L., Ho, C.-L., Wu, H., Cao, Y., and Wong, W.-Y. (2014). *Adv. Mater.* 26: 2459.

52 Kido, J., Kimura, M., and Nagai, K. (1995). *Science* 267: 1332.

53 Sasabe, H., Takamatsu, J., Motoyama, T., Watanabe, S., Wagenblast, G., Langer, N., Molt, O., Fuchs, E., Lennartz, C., and Kido, J. (2010). *Adv. Mater.* 22: 5003.

54 Sun, N., Wang, Q., Zhao, Y., Chen, Y., Yang, D., Zhao, F., Chen, J., and Ma, D. (2014). *Adv. Mater.* 26: 1617.

55 Higuchi, T., Nakanotani, H., and Adachi, C. (2015). *Adv. Mater.* 27: 2019.

56 Kim, B.S., Yook, K.S., and Lee, J.Y. (2014). *Sci. Rep.* 4: 6019.

57 Baldo, M.A. (1999). *Phys. Rev. B* 66: 14422.

58 Sun, Y., Giebink, N.C., Kanno, H., Ma, B., Thompson, M.E., and Forrest, S.R. (2006). *Nature* 440: 908.

59 Lan, Y.-H., Hsiao, C.-H., Lee, P.-Y., Bai, Y.-C., Lee, C.-C., Yang, C.-C., Leung, M.-Y., Wei, M.-K., Chiu, T.-L., and Lee, J.-H. (2011). *Org. Electron.* 12: 756.

60 Hofmann, S., Rosenow, T.C., Gather, M.C., Lössem, B., and Leo, K. (2012). *Phys. Rev. B: Condens. Matter Mater. Phys.* 85: 245209.

61 Murawski, C., Leo, K., and Gather, M.C. (2013). *Adv. Mater.* 25: 6801.

62 Liu, X.-K., Chen, Z., Zheng, C.-J., Chen, M., Liu, W., Zhang, X.-H., and Lee, C.-S. (2015). *Adv. Mater.* 27: 2025.

63 Yook, K.S. and Lee, J.Y. (2012). *Adv. Mater.* 24: 3169.
64 Wang, H., Xie, L., Peng, Q., Meng, L., Wang, Y., Yi, Y., and Wang (2014). *Adv. Mater.* 26: 5198.
65 Ye, S., Liu, Y., Chen, J., Lu, K., Wu, W., Du, C., Liu, Y., Wu, T., Shuai, Z., and Yu, G. (2010). *Adv. Mater.* 22: 4167.
66 Ishijima, S., Higashi, M., and Yamaguchi, H. (1994). *J. Phys. Chem.* 98: 10432.
67 Guha, S., Rice, J.D., Yau, Y.T., Martin, C.M., Chandrasekhar, M., Chandrasekhar, H.R., Guentner, R., Scanduicci de Freitas, P., and Scherf, U. (2003). *Phys. Rev. B* 67: 125204.
68 Platt, A.D., Kendrick, M.J., Loth, M., Anthony, J.E., and Ostroverkhova, O. (2011). *Phys. Rev. B* 84: 235209.
69 Orselli, E., Maunoury, J., Bascour, D., and Catinat, J.-P. (2012). *Org. Electron.* 13: 1506.
70 Wan, J., Zheng, C.-J., Fung, M.-K., Liu, X.-K., Lee, C.-S., and Zhang, X.-H. (2012). *J. Mater. Chem.* 22: 4502.
71 Jou, J.-H., Wu, P.-H., Lin, C.-H., Wu, M.-H., Chou, Y.-C., Wang, H.-C., and Shen, S.-M. (2010). *J. Mater. Chem.* 20: 8464.
72 Shih, P.I., Chien, C.H., Wu, F.I., and Shu, C.F. (2007). *Adv. Funct. Mater.* 17: 3514.
73 Su, Y.J., Huang, H.L., Li, C.L., Chien, C.H., Tao, Y.T., Chou, P.T., Datta, S., and Liu, R.S. (2003). *Adv. Mater.* 15: 884.
74 Jeon, W.S., Park, T.J., Kim, S.Y., Pode, R., Jang, J., and Kwon, J.H. (2009). *Org. Electron.* 10: 240.
75 Jen, Y.D., Yang, J.P., Heremans, P.L., der Auweraer, M.V., Rousseau, E., Geise, H.J., and Borghs, G. (2000). *Chem. Phys. Lett.* 320: 387.
76 Yang, X.H., Jaiser, F., Klinger, S., and Neher, D. (2006). *Appl. Phys. Lett.* 88: 021107.
77 Zhang, Q., Komino, T., Huang, S., Matsunami, S., Goushi, K., and Adachi, C. (2012). *Adv. Funct. Mater.* 22: 2327.
78 Kawamura, Y., Brooks, J., Brown, J.J., Sasabe, H., and Adachi, C. (2006). *Phys. Rev. Lett.* 96: 017404.
79 Furukawa, T., Nakanotani, H., Inoue, M., and Adachi, C. (2015). *Sci. Rep.* 5: 8429.

14

Solution-Processed TADF Materials and Devices Based on Organic Emitters

Nidhi Sharma[1,2], Michael Yin Wong[1], Ifor D.W. Samuel[2], and Eli Zysman-Colman[1]

[1] University of St Andrews, Organic Semiconductor Centre, EaStCHEM School of Chemistry, St Andrews, KY16 9ST, Fife, UK
[2] University of St Andrews, Organic Semiconductor Centre, SUPA School of Physics and Astronomy, St Andrews, KY16 9SS, Fife, UK

14.1 Introduction

Since the seminal discovery of electroluminescence (EL) in a bilayer organic light-emitting diode (OLED) by Tang and VanSlyke [1], OLEDs have blossomed over the last three decades due to their enormous potential in flat-panel display and solid-state lighting (SSL) applications. Compared with existing liquid crystal displays (LCDs), OLED displays give high contrast and viewing angle, together with high brightness and vivid colors. In addition, OLED displays provide low power usage, the capacity to display true black, a more robust, and environmentally friendly design than LCD screens. An extensive worldwide research and development effort has brought OLED technology to global adoption for displays, and the first commercial lighting devices are available.

OLEDs produce light when a voltage is applied to them, leading to the injection of charges. The injected electrons and holes can combine to form an exciton (bound electron-hole pair). The spin of the two charges can be combined in four ways, leading to 25% of the excitons formed under electrical excitation being in the singlet spin state and 75% triplets [2]. In fluorescent materials, electrically generated triplet excitons dissipate their energy in the form of heat resulting in a theoretical limit of 25% internal quantum efficiency (IQE). Taking into account a light-outcoupling efficiency of ca. 20% in a typical OLED, this results in a maximum external quantum efficiency (EQE) of 5% for the device. Even though a higher proportion of singlets has been suggested in conjugated polymers [3], triplet formation is nevertheless a major loss mechanism that needs to be overcome for highly efficient devices.

To increase the IQE limit of the OLED, phosphorescent materials were developed to harvest the nonemissive triplet excitons [4]. Phosphorescence is achieved through the use of organometallic complexes that incorporate a heavy metal. Its presence facilitates a rapid intersystem crossing (ISC) as a result of the large

Highly Efficient OLEDs: Materials Based on Thermally Activated Delayed Fluorescence,
First Edition. Edited by Hartmut Yersin.
© 2019 Wiley-VCH Verlag GmbH & Co. KGaA. Published 2019 by Wiley-VCH Verlag GmbH & Co. KGaA.

spin–orbit coupling (SOC) constant associated with the metal. Likewise, SOC induced by the heavy metal enhances radiative decay from the excited triplet state to the singlet ground state. Using this approach, both singlet and triplet excitons are harvested, which in turn leads to 100% IQE. Indeed, OLEDs employing phosphorescent metal complex emitters have been shown to exhibit EQEs of over 30% [5]. The majority of these highly efficient phosphorescent materials are based on scarce and expensive iridium(III) and platinum(II) complexes. In order to achieve a truly inexpensive OLED technology, a paradigm shift in materials development is required. Innovative design strategies relying on triplet–triplet annihilation (TTA) [6] or thermally activated delayed fluorescence (TADF) [7] have emerged in recent years. The maximum IQE for TTA is 62.5%, whereas for TADF it is a tantalizing 100%. TADF-based materials development has progressed rapidly since the first report of a TADF material used in an OLED in 2009 [8]. Recruitment of triplet excitons in TADF materials occurs through their conversion to singlet excitons through a rapid reverse intersystem crossing (rISC) made possible by a vanishingly small singlet–triplet excited-state energy difference (ΔE_{ST}). Small ΔE_{ST} values (usually defined as <200 meV) are achieved when the HOMO–LUMO exchange integral is small, which occurs when the HOMO and LUMO are spatially separated. Examples of OLEDs using noble metal-free TADF emitters have already achieved EQEs of over 30% in the blue [5b] and green [5a] and close to 19% in red [9] devices, demonstrating their great potential as replacements for high-efficiency phosphorescent OLEDs [10].

Generally, the fabrication of OLEDs has been categorized into one of two approaches: Thermal evaporation under vacuum is predominantly used for small molecule-based OLEDs [11]; for polymeric [12] and dendrimeric materials [13], solution processing techniques such as spin-coating and inkjet printing are employed [14]. Most OLEDs based on TADF emitters possess a vacuum-evaporated multilayer architecture. Careful choice of these layers permits optimal charge injection and transport and exciton recombination kinetics leading to high-efficiency devices. However, the complexity, inefficiency of materials use, and cost of vacuum deposition limit the mass production of cost-effective OLEDs, which is particularly germane for SSL applications [15]. Solution-processed OLEDs on the other hand hold great promise for reducing in dramatic fashion the fabrication cost of the device, particularly when it comes to pixel patterning and large area OLEDs [16]. Inkjet and roll-to-roll printing can effectively be used to fabricate large area high-resolution full-color flat-panel displays; these fabricated onto a wide range of substrates including flexible plastic and metal foil [17].

In general, for a solution-processed OLED material, it is necessary to have a uniform, amorphous film. Light-emitting polymers have been the most intensively studied on this platform [14a, 18], though the performance metrics of OLEDs employing polymers are typically lower than those using thermally deposited small molecules. Over the past few years, the device efficiency and lifetime of polymer-based devices has improved to the point of being attractive candidates for commercial OLED devices [14a]. There remain still some intrinsic challenges when using polymeric materials such as batch-to-batch variation and purification of the polymers. Light-emitting dendrimers were developed as an alternative class of solution-processed material, although macromolecular are also monodisperse. In this approach the core of the dendrimer defines the key electronic properties (e.g. color of light emission), while conjugated dendrons keep the cores apart and surface groups confer solubility [19]. In this way a very wide range of chromophores can be made solution processable, and their aggregation controlled [20]. Solution-processed small molecules are also attractive because they too have well-defined molecular structures, reproducibility of their synthesis, and facile purification. Thus, it is a good strategy to develop OLEDs based on solution-processable small molecules. Unfortunately, the majority of the small molecule emitters designed for vacuum deposition are not suitable for the solution processing due to their poor film morphology and easy crystallization upon spin casting. Device performance of such small molecule solution-processed OLEDs remains inferior to that of vacuum-deposited OLEDs [16].

The evolution of the development of solution-processed OLEDs mirrors that of vacuum-deposited OLEDs and started with polymers [12] and dendrimers [13, 19a], and then systems based on fluorescent [1] small molecules [21] were developed. Phosphorescent materials replaced fluorescent analogues in vacuum-deposited devices in order to achieve higher device efficiencies [4]. In the phosphorescent-based solution-processed OLEDs, device performance was enhanced by using dendrimers to confer solubility and control aggregation, leading to an EQE of 16% and green emission for the device. Soluble hosts and phosphorescent light-emitting materials [15a, 20, 22] have also achieved highly efficient solution-processed devices – a small molecule host and dopant materials for vacuum-deposited phosphorescent OLEDs were modified or used without modifications. Using this approach, EQEs close to 20% were reported for red [15a, 22a, 23], green [22b, 24], and blue [22c, d, 25] phosphorescent OLEDs, which were comparable with their vacuum-deposited counterparts.

Although most of the studies for achieving highly efficient devices have been focused on the development of solution-processable phosphorescent materials, recent progress in TADF materials has opened up the possibility for the development of high-efficiency solution-processed OLEDs using this class of materials. Figure 14.1 shows the comparison of the maximum EQEs achieved for green-emitting TADF OLEDs fabricated by vacuum deposition and solution processing over the past 5 years, the data of which were obtained at their turn-on voltages ($\sim$1–5 cd m^{-2} brightness). Although it is well established in OLED development to quote the efficiencies at display brightnesses (e.g. 100 cd m^{-2}), surprisingly in the field of TADF materials, many papers do not do this. Since the first report of a solution-processed TADF emitter in late 2014 [26], the field has progressed rapidly, and already EQEs comparable to those of highly efficient vacuum-deposited TADF emitters have been achieved.

In this review, we discuss the development of purely organic (i.e. metal-free) small molecule organic solution-processed TADF materials and their use in OLED devices and compare their performance with the state-of-the-art vacuum-deposited TADF materials and devices. Solution-processed TADF polymers and dendrimers are also discussed, and their device performance is contrasted with small molecule solution-processed TADF OLEDs.

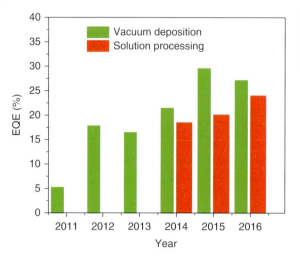

Figure 14.1 Chronological comparison of maximum EQEs achieved in OLEDs using green TADF emitters.

14.1.1 Solution-Processed Blue TADF Materials and Devices

The color of an OLED is defined by its Commission Internationale de l'Éclairage (CIE) coordinates. For a pure blue emitter, standard CIE coordinates defined by the National Television System Committee (NTSC) are (0.14, 0.08), whereas Phase Alternating Line (PAL) systems define pure blue color by CIE coordinates of (0.15, 0.06). Not all reports in the literature document CIE coordinates, and so in this chapter we have taken a simplified definition of blue emitters to be those materials with EL maxima (λ_{EL}) shorter than 490 nm. The photophysical properties of blue emitters **1–8** and EL characteristics of the corresponding OLEDs are summarized in Tables 14.1 and 14.2, respectively. The first report of a solution-processed blue TADF emitter was published by Mei et al. in 2015 [27]. They employed an inductively electron-withdrawing trifluoromethyl (-CF$_3$) group as the acceptor with weakly donating carbazole donors in compounds **1 (4CzCF$_3$Ph)** and **2 (5CzCF$_3$Ph)**. Compound **1** exhibited blue emission with emission maxima, λ_{PL}, at 442 nm in toluene and 445 nm in a thin film consisting of 10 wt% **1** doped in mCP. With an additional carbazole donor, **2** displayed a redshifted emission with λ_{PL} at 475 nm in toluene and 495 nm in a thin film consisting of 10 wt% **2** doped in mCP. Both the compounds exhibited microsecond delayed emission decays, τ_d, of 9.31 μs for **1** and 15.33 μs for **2** in degassed toluene solution, indicating the emission was delayed fluorescence. The photoluminescence quantum yields, Φ_{PL}, for **1** and **2** in toluene are 13% and 24%, respectively, under ambient conditions, which increased to 24% and 43%, respectively, under oxygen-free conditions. The ΔE_{ST}, determined from the energy difference of the highest energy vibronic subbands of the 77 K fluorescence and phosphorescence spectra, was found to be 0.24 eV for **1** and 0.02 eV for **2**, suggesting more efficient rISC in **2** than in **1**. Solution-processed devices were fabricated using the device stack: indium tin oxide (ITO)/poly(3,4-ethylenedioxythiophene)polystyrene sulfonate (PEDOT:PSS) (40 nm)/mCP:10 wt% **1 or 2** (40 nm)/TmPyPB (60 nm)/LiF (0.8 nm)/Al (100 nm). Compounds **1** and **2** exhibited λ_{EL} of 464 and 484 nm with CIE coordinates of (0.17, 0.18) and (0.21, 0.33), respectively. Devices

Table 14.1 Photophysical properties of emitters **1–40** of interest.

Material	Solution $\lambda_{PL}/\Phi_{PL}/\tau_d$ (medium) nm/%/μs	Solid state $\lambda_{PL}/\Phi_{PL}/\tau_d$ (medium) nm/%/μs	ΔE_{ST} (eV)	HOMO (eV)	LUMO (eV)	References
4CzCF₃Ph (**1**)	440/24/9.31(PhMe)	445/–/– (10 wt% in mCP)	0.24	−5.60	−2.62	[27]
4CzCF₃Ph (**2**)	481/43/15.33(PhMe)	495/–/– (10 wt% in mCP)	0.02	−5.57	−2.75	[27]
3CzFCN (**3**)	–/–/28(PhMe)	443/76/– (10 wt% in CzSi)	0.06	−6.38	−3.56	[28]
4CzFCN (**4**)	–/–/17(PhMe)	453.100/– (10 wt% in CzSi)	0.06	−6.31	−3.49	[28]
TB-1PXZ (**5**)	–/–/–	478/12/5.9(10 wt% in CzSi)	0.12	−5.08	−2.24	[29]
TB-2PXZ (**6**)	–/–/–	484/47/2.9(10 wt% in CzSi)	0.05	−5.08	−2.29	[29]
DV-MOC-DPS (**8**)	–/–/–	430/71/50 (9 wt% DV-CDBP)	0.31	−5.29	−2.03	[30]
4CzIPN (**9**)	507/94/5.1 (PhMe)	507/94/3.1 (1 wt% CzSi)	0.08	—	—	[31]
m4CzIPN (**10**)	–/–/–	527/67/2.6 (1 wt% CzSi)	0.02	—	—	[32]
t4CzIPN (**11**)	–/–/–	530/78/2.9 (1 wt% CzSi)	0.01	—	—	[32]
4CzCNPy (**12**)	536/–/– (CHCl₃)	560/54.9/8.4 (8 wt% in 4CzPy)	0.07	−5.72	−3.26	[33]
13	572/2.1/– (MeCN)	500/90/6.34 (10 wt% PMMA)	—	−5.87	−2.99	[34]
3 ACR-TRZ (**14**)	–/–/–	504/98/6.7 (16 wt% CBP)	0.015	—	—	[35]
DMAC-TRZ (**15**)	–/–/–	500/83/3.6 (Neat)	—	—	—	[36d]
TB-3PXZ (**16**)	–/–/–	509/95/1.3 (10 wt% CzSi)	—	−5.08	−2.38	[29]
ACRDSO₂ (**17**)	~520/34/– (PhMe)	520/71/8.3 (6 wt% CBP)	0.01	−5.26	−2.65	[37]
Red-1b (**18**)	593/–/– (PhMe)	644/28/0.82 (Neat)	0.40	—	—	[38]
FDQPXZ (**19**)	–/–/–	600/74/3.2 (5 wt% Bepp₂)	0.05	−5.06	−2.91	[39]
PXZDSO₂ (**20**)	–/–/–	540/62/5.0 (6 wt% CBP)	0.048	−5.06	−2.67	[37]
21	–/–/–	~540/43.6/1–50 000 (neat)	0.22	—	—	[40]
P12 (**22**)	491/34/4.06 (PhMe)	494/34/2.36 (neat)	—	—	—	[41]
pCzBP (**23**)	472/28/– (PhMe)	508/23/74 (10 wt% TCTA:TAPC)	0.16	−5.64	−2.76	[42]

(Continued)

Table 14.1 (Continued)

Material	Solution $\lambda_{PL}/\Phi_{PL}/\tau_d$ (medium) nm/%/μs	Solid state $\lambda_{PL}/\Phi_{PL}/\tau_d$ (medium) nm/%/μs	ΔE_{ST} (eV)	HOMO (eV)	LUMO (eV)	References
pAcBP (24)	550/26/– (PhMe)	540/46/24 (10 wt% TCTA:TAPC)	0.004	−5.41	−2.78	[42]
PAPCC (25)	–/–/–	472/9/0.32 (neat)	0.37	−5.38	−2.57	[43]
PAPTC (26)	–/–/–	510/44/0.23 (neat)	0.13	−5.33	−2.77	[43]
27	–/–/–	511/–/1–100 (neat)	0.28	—	—	[44b]
28	–/–/–	535/–/1–100 (neat)	0.35	−5.31	−3.10	[44a]
29	–/–/–	546/–/1–100 (neat)	0.46	−5.33	−3.06	[44a]
30	–/–/–	553/–/1–100 (neat)	0.42	−5.42	−3.06	[44a]
31	–/–/–	556/–/1–100 (neat)	0.40	−5.43	−3.05	[44a]
PCzDP-5, 10, 15 and 20 (32)	425, 510/50–67/– (PhMe)	497–509/39–74/∼1.9–2.3 (neat)	0.02	—	—	[45]
33	–/–/–	∼496/71/∼296 (neat)	0.12	−5.14	−1.82	[46]
G2TAZ (34)	473/94/– (PhMe)	500/52/3.1 (neat)	0.03	−5.76	−3.01	[26]
G3TAZ (35)	473/100/– (PhMe)	500/31/1.9 (neat)	0.06	−5.72	−2.97	[26]
G4TAZ (36)	473/94/– (PhMe)	500/8.4/2.4 (neat)	0.06	−5.68	−2.80	[26]
CDE (37)	–/–/0.52 (PhMe)	520/77/– (neat)	0.11	−5.12	−2.54	[47]
CDE1 (38)	–/–/0.62 (PhMe)	499/75/– (neat)	0.15	−5.25	−2.69	[47]
Tz-Cz (39)	∼480/–/– (PhMe)	490/58/∼3–4 (neat)	0.20	−5.20	−2.31	[48]
Tz-3Cz (40)	∼490/–/– (PhMe)	487/76/∼3–4 (neat)	0.20	−5.00	−2.11	[48]

based on **2** showed a maximum EQE of 5.2% with luminance and current efficiency of 2436 cd m^{-2} and 11.6 cd A^{-1}, respectively, whereas for **1** these values decreased to 1032 cd m^{-2} and 1.03 cd A^{-1} with an EQE of 0.67%. The significantly higher device efficiency for **2** was attributed to its significantly smaller ΔE_{ST} (Figure 14.2).

One of the key factors in achieving highly efficient solution-processed TADF OLEDs is to improve the solubility of the emitters in organic solvents. To address this issue, Lee and coworkers [28] reported fluorine-substituted benzonitrile acceptor-based TADF emitters **3** (3CzFCN) and **4** (4CzFCN) with improved solubility and blue emission color. The design principles for **3** and **4** were almost identical to those of **1** and **2**. Here, the authors replaced the trifluoromethylbenzene acceptor with difluorobenzonitrile/fluorobenzonitrile acceptors. Improvement in the solubility of **3** and **4** in dichloromethane and blue

Table 14.2 Electroluminescence properties of emitters 1–40 of interest.

Material	λ_{EL} (nm)	ΔE_{ST} (eV)	Host	Device structure	η_{ext} (%)	V_{on} (V)	L_{max} (cd m^{-2})	PE_{max} (lm W^{-1})	CE_{max} (cd A^{-1})	CIE/(x, y)	References
1	464	0.24	mCP	ITO/PEDOT: PSS/mCP:10 wt% **1**/TmPyPB/LiF/Al	0.67	4.8	1032	—	1.03	(0.17, 0.18)	[27]
2	484	0.02	mCP	ITO/PEDOT: PSS/mCP:10 wt% **2**/TmPyPB/LiF/Al	5.2	3.9	2436	—	11.8	(0.21, 0.33)	[27]
3	463	0.06	CzSi	ITO/PEDOT: PSS/PVK/CzSi:10 wt% **3**/TSPO1/TPBi/LiF/Al	17.8	—	—	—	26.9	(0.16, 0.19)	[28]
4	471	0.06	CzSi	ITO/PEDOT: PSS/PVK/CzSi:15 wt% **4**/TSPO1/TPBi/LiF/Al	20.0	—	—	—	36.1	(0.16, 0.25)	[28]
5	478	0.12	CzSi	ITO/PEDOT: PSS/CzSi:10 wt% **5**/TmPyPB/LiF/Al	1.0	—	—	0.7	1.7	(0.19, 0.29)	[29]
6	484	0.05	CzSi	ITO/PEDOT: PSS/CzSi:10 wt% **6**/TmPyPB/LiF/Al	8.9	—	—	13.8	21.0	(0.18, 0.40)	[29]
8(7)	444	0.31	DV-CDBP(7)	ITO/PEDOT: PSS/**5** : 9 wt% **8**/TPBi/LiF/Al	2.0	5.3	899	0.9	1.6	(0.12, 0.13)	[30]
9	507	0.08	CzSi	ITO/PEDOT: PSS/PVK/CzSi:2 wt% **9**/TSPO1/LiF/Al	8.1	—	—	8.9	—	(0.20, 0.43)	[32]
9	520	0.08	CBP	ITO/Buf-HIL/CBP:1 wt% **9**/TPBi/LiF/Al	24	3.8	—	58	73	(0.20, 0.43)	[49]
9	520	0.08	CBP	ITO/PEDOT: PSS/CBP:6 wt% **9**/TPBi/LiF/Al	18.5	3.4	—	—	—	(0.20, 0.43)	[50]
9	520	0.08	CBP	ITO/PEDOT: PSS/CBP:6 wt% **9**/T2T/Bpy-TP2/LiF/Al	11.9	3.8	—	—	—	(0.20, 0.43)	[50]
9	520	0.08	CPCB	ITO/PEDOT: PSS/CPCB:6 wt% **9**/T2T/Bpy-TP2/LiF/Al	9.9	3.4	—	—	—	(0.20, 0.43)	[50]
9	520	0.08	CBP	ITO/PEDOT: PSS/CBP:5 wt% **9**/TPBi/LiF/Al	8.4	3.4	—	20	26	(0.20, 0.43)	[51]
9	520	0.08	CBP	ITO/PEDOT: PSS/CBP:15 wt% **9**/TPBi/LiF/Al	9.9	3.3	—	26	33	(0.20, 0.43)	[51]
9	520	0.08	CBP	ITO/PEDOT: PSS/CBP:5 wt% **9**/B3PyPB/LiF/Al	15	3.1	—	51	50	(0.20, 0.43)	[51]

(Continued)

Table 14.2 (Continued)

Material	λ_{EL} (nm)	ΔE_{ST} (eV)	Host	Device structure	η_{ext} (%)	V_{on} (V)	L_{max} (cd m^{-2})	PE_{max} (lm W^{-1})	CE_{max} (cd A^{-1})	CIE/(x, y)	References
9	520	0.08	CBP	ITO/PEDOT: PSS/CBP:5 wt% **9**/B4PyMPM/LiF/Al	12	2.5	—	48	39	(0.20, 0.43)	[51]
10	518	0.02	CzSi	ITO/PEDOT: PSS/PVK/CzSi:2 wt% **10**/TSPO1/LiF/Al	8.2	—	—	13.2	—	(0.29, 0.57)	[32]
11	520	0.01	CzSi	ITO/PEDOT: PSS/PVK/CzSi:2 wt% **11**/TSPO1/LiF/Al	18.3	—	—	42.7	—	(0.31, 0.59)	[32]
12	524	0.07	mCP	ITO/PEDOT: PSS/mCP:8 wt% **12**/TmPyPB/LiF/Al	11.3	6.2	—	14.8	38.9	(0.35, 0.59)	[33]
12	524	0.07	4CZPy	ITO/PEDOT: PSS/4CZPy:8 wt% **12**/TmPyPB/LiF/Al	10.4	4.7	—	17.9	35.4	(0.34, 0.59)	[33]
12	516	0.07	pCNBCzmMe	ITO/PEDOT: PSS/pCNBCzmMe:8 wt% **12**/TmPyPB/LiF/Al	10.8	3.2	16 100	29.9	34.5	(0.31, 0.60)	[52]
12	520	0.07	pCNBCzoCF$_3$	ITO/PEDOT: PSS/pCNBCzoCF$_3$: 8 wt% **12**/TmPyPB/LiF/Al	9.7	3.7	14 370	22.7	32.2	(0.32, 0.61)	[52]
12	520	0.07	pCNBCzmCF$_3$	ITO/PEDOT: PSS/pCNBCzmCF$_3$: 8 wt% **12**/TmPyPB/LiF/Al	10.9	3.3	19 200	30.9	35.7	(0.33, 0.60)	[52]
13	546	—	Neat	ITO/PEDOT: PSS/PVK/**13**/B3PYMPM/LiF/Al	5.1	3.6	—	10.3	14.9	(0.41, 0.53)	[34]
14	504	0.015	CBP	ITO/PEDOT: PSS/CP:16 wt% **14**/BmPyPhB/Liq/A)	18.6	—	—	36.3	—	—	[35]
15	498	—	Neat	ITO/PEDOT: PSS/**15**/BmPyPhB/Liq/Al	17.6	4.8	6000	—	29.5	—	[53]
15	498	—	Neat	ITO/PEDOT: PSS/**15**/B4PyMPM/Liq/Al	16.8	4.0	10 000	—	—	—	[53]
16	510	0.01	CzSi	ITO/PEDOT: PSS/CzSi:10 wt% **16**/TmPyPB/LiF/Al	13.9	5.2	—	32.6	41.5	(0.23, 0.54)	[29]
17	534	0.05	CBP	ITO/Buf-HIL/CBP:6 wt% **17**/TmPyPB/LiF/Al	17.5	3.7	—	53.3	—	(0.32, 0.58)	[37]
18	644	0.40	Neat	ITO/PEDOT: PSS/PVK/**18**/TPBi/CsF/Al	1.75	5.0	7800	—	1.22	(0.65, 0.33)	[38]
19	600	0.05	CBP:PBD	ITO/PEDOT: PSS/CBP: PBD:10 wt% **19**/TmPyPB/LiF/Al	9.0	3.0	—	—	—	(0.53, 0.46)	[39]
20	552	0.04	CBP	ITO/Buf-HIL/CBP:6 wt% **20**/TmPyPB/LiF/Al	15.2	4.1	—	45.1	—	(0.42, 0.5)	[37]

21	530	0.22	Neat	ITO/PEDOT: PSS/Interlayer/**21**/NaF/Al/Ag	10.0	—	—	—	—	(0.32, 0.58)	[40]
22	506		Neat	ITO/PEDOT:PSS/poly-TPD/neat **22**/TmPyPB/Ba/Al	4.3	3.1	—	11.2	10.7	(0.24, 0.43)	[41]
23	500	0.16	TAPC:TCTA	ITO/PEDOT: PSS/TAPC:TCTA:10 wt% **22**/TmPyPB/LiF/Al	8.1	6.0	5100	9.0	24.9	(0.28, 0.43)	[42]
24	548	0.004	TAPC:TCTA	ITO/PEDOT: PSS/TAPC:TCTA:10 wt% **23**/TmPyPB/LiF/Al	9.3	4.3	30800	20.3	31.8	(0.38, 0.57)	[42]
25	508	0.37	Neat	ITO/PEDOT: PSS/Neat **25**/TmPyPB/LiF/Al	1.34	3.0	10251	37.1	41.8	(0.25, 0.47)	[43]
26	521	0.13	Neat	ITO/PEDOT: PSS/Neat **26**/TmPyPB/LiF/Al	12.36	2.6	556	3.67	3.6	(0.30, 0.59)	[43]
27	512	0.28	PBD:PVK	ITO/PEDOT: PSS/PVK:PBD:10 wt% **27**/TmPyPB/LiF/Al	11.5	2.1	—	1.5	5.7	—	[44b]
28	533	0.35	mCP	ITO/PEDOT: PSS/mCP:10 wt% **28**/TPBi/LiF/Al	20.1	5.8	—	40.1	61.3	(0.36, 0.55)	[44a]
29	536	0.46	mCP	ITO/PEDOT: PSS/mCP:10 wt% **29**/TPBi/LiF/Al	15.2	6.8	—	32.6	52.5	(0.37, 0.53)	[43]
30	539	0.42	mCP	ITO/PEDOT: PSS/mCP:10 wt% **30**/TPBi/LiF/Al	1.8	6.9	—	1.9	5.0	(0.39, 0.53)	[43]
31	556	0.40	mCP	ITO/PEDOT: PSS/mCP:10 wt% **31**/TPBi/LiF/Al	1.4	7.4	—	1.6	3.3	(0.43, 0.51)	[43]
32	496	0.02	Neat	ITO/PEDOT:PSS/**32**/TmPyPB/Liq/Al	16.1	—	—	14.3	38.6	(0.24, 0.40)	[45]
34	510	0.03	Neat	ITO/PEDOT: PSS/**34**/TPBi/LiF/Al	2.4	3.3	—	—	—	(0.25, 0.49)	[26]
35	500	0.06	Neat	ITO/PEDOT: PSS/**35**/TPBi/LiF/Al	3.4	3.5	—	—	—	(0.26, 0.48)	[26]
36	498	0.06	Neat	ITO/PEDOT: PSS/**36**/TPBi/LiF/Al	1.5	3.5	—	—	—	(0.23, 0.36)	[26]
37	546	0.11	Neat	ITO/PEDOT: PSS/**37**/TPBi/Liq/Al	12	4.8	>10000	17.3	38.9	(0.38, 0.56)	[47]
38	522	0.15	Neat	ITO/PEDOT: PSS/**38**/TPBi/Liq/Al	5.2	7.7	2512	8.7	13.9	(0.32, 0.52)	[47]
39	520	0.20	Neat	ITO/PEDOT: PSS/**39**/TPBi/CS$_2$CO$_3$/Al	10.1	4.0	18200	—	20.0	(0.24, 0.51)	[48]
40	520	0.20	Neat	ITO/PEDOT: PSS/**40**/TPBi/CS$_2$CO$_3$/Al	6.5	3.6	22950	—	30.5	(0.24, 0.51)	[48]

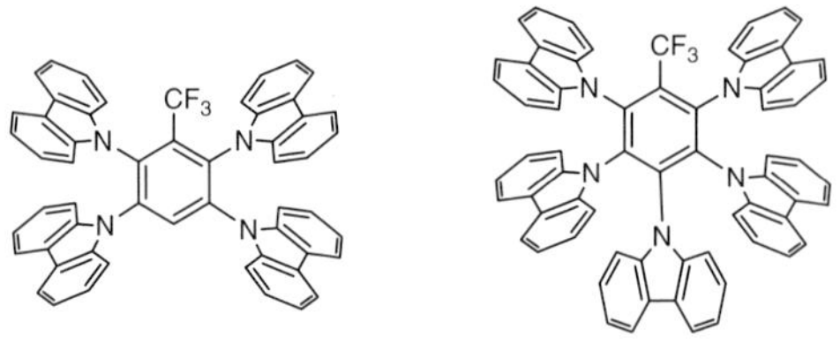

Figure 14.2 Chemical structures of emitters **1** and **2**.

emission was attributed to the hydrophobic and weak electron-withdrawing nature of the F atom, respectively. Both **3** and **4** show a remarkably small ΔE_{ST} of 0.06 eV. Compound **3** emits at 443 nm with a Φ_{PL} of 76% in 10 wt% CzSi thin film, while the λ_{PL} for **4** is 453 nm but with remarkable 100% Φ_{PL} under oxygen-free conditions. Their solution-processed devices were fabricated using the following device structure: ITO/PEDOT:PSS (60 nm)/PVK (15 nm)/CzSi:10 wt% **3** or 15 wt% **4** (25 nm)/TSPO1 (5 nm)/TPBi (30 nm)/LiF (1 nm)/Al (100 nm). OLED devices showed blue emission with λ_{EL} of 463 and 471 nm and CIE coordinates of (0.16, 0.19) and (0.16, 0.25) for **3** and **4**, respectively. At the same time, vacuum-deposited devices for **3** and **4** were also fabricated for comparison with device structures of ITO/NPB(35 nm)/CzSi:10 wt% **3** or **4** 15 wt% (20 nm)/TSPO1 (5 nm)/TPBi (30 nm)/LiF(1 nm)/Al (100 nm). The solution-processed devices were highly efficient with excellent maximum EQEs of 17.8% (8% at 100 cd m^{-2}) for **3** and 20% (16.3% at 100 cd m^{-2}) for **4**, which were higher than their vacuum-deposited counterparts (maximum EQE of 12.9% (10.9% at 100 cd m^{-2}) for **3**, EQE of 17.3% (12.3% at 100 cd m^{-2}) for **4**). These examples represent the highest reported EQE values for solution-processed blue TADF devices to date (Figure 14.3).

Liu et al. [29] employed triarylborane-based acceptors and decorated their materials with phenoxazine donors. Boron-based groups are weak electron-acceptor materials, thanks to the vacant p-orbital present on the boron atom. Using this approach, they reported solution-processed TADF emitters based on phenoxazine as donor units and a sterically hindered triarylborane acceptor **5 (TB-1PXZ)** and **6 (TB-2PXZ)**. The rISC was deliberately modulated by controlling the number of donor units. Compounds **5** and **6** showed λ_{PL} at 478 and 484 nm with Φ_{PL} of 12% and 47%, respectively, in 10 wt% CzSi films under nitrogen-saturated conditions. Small ΔE_{ST} of 0.12 and 0.05 eV and τ_d of 5.9 and 2.9 μs, respectively, confirmed the TADF nature of their emission.

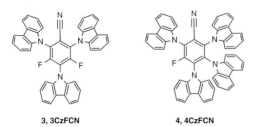

Figure 14.3 Chemical structures of emitters **3** and **4**.

Figure 14.4 Chemical structures of emitters **5** and **6**.

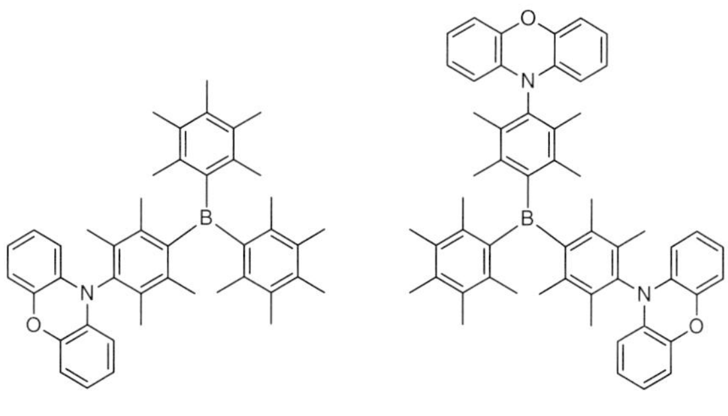

5, TB-1PXZ   6, TB-2PXZ

Solution-processed devices based on the architecture ITO/PEDOT:PSS (30 nm)/CzSi:**5**, **6** (10 wt%, 40 nm)/TmPyPB(40 nm)/Liq(2 nm)/Al(100 nm) exhibited λ_{EL} of 480 and 490 nm with CIE coordinates of (0.19, 0.29) and (0.18, 0.40) for **5** and **6**, respectively. Maximum EQEs of 1.0% and 8.9% were realized for **5** and **6**, which reduced to 0.9% and 8.8% at a luminance of 100 cd m^{-2}, respectively. Although the reported EQEs are significantly lower than the blue solution-processed TADF emitters of Lee and coworkers [28], a careful further optimization of the device stacks and molecular design of the emitter could improve the efficiency of the OLEDs (Figure 14.4).

To achieve high efficiencies, TADF emitters are usually employed as dopants in a host material to prevent concentration quenching. The choice of host material becomes critical, especially for blue emitters since a wide bandgap material with high triplet energy is required. Solution processability of such materials further limits the choice of host materials available. To overcome this limitation, Sun et al. [30] reported a host and emitter design **7 (DV-CDBP)** and **8 (DV-MOC-DPS)**, respectively, with thermally cross-linkable pendant units designed to improve the film-forming capabilities of emissive layer. Compound **7** was chosen as the host because of its high triplet energy of 2.95 eV. Compound **8** exhibited a ΔE_{ST} of 0.31 eV but with a λ_{PL} of 430 nm and Φ_{PL} of 71% in 9 wt% doped film of **8** in **7**. The transient photoluminescence spectrum revealed a prompt emission decay, τ_p, of 11 ns and a τ_d of 50 μs in degassed toluene, a typical profile for a TADF emitter. A solution-processed device using the device stack ITO/PEDOT:PSS (30 nm)/**8** doped in **7** in the mass ratio 1 :0.09 (50 nm)/TPBI (40 nm)/Cs$_2$CO$_3$ (2 nm)/Al (100 nm) exhibited a maximum EQE of 2.0% (1.3% at 100 cd m^{-2}) with current and power efficiencies of 1.6 cd A^{-1} and 0.9 lm W^{-1}, respectively. A deep blue emission with λ_{EL} of 444 nm and CIE coordinates of (0.12, 0.13) was realized for the device. Although the efficiencies of the devices were below the 5% threshold to unequivocally demonstrate that triplet excitons are being harvested in the device, these materials do demonstrate a promising strategy to transform emitters designed for vacuum-deposited OLEDs [54] into ones compatible for solution-processed OLEDs.

As described at the beginning of this section, a deep blue emitter is defined by its CIE coordinates ranging from (0.14, 0.08) to (0.15, 0.06).

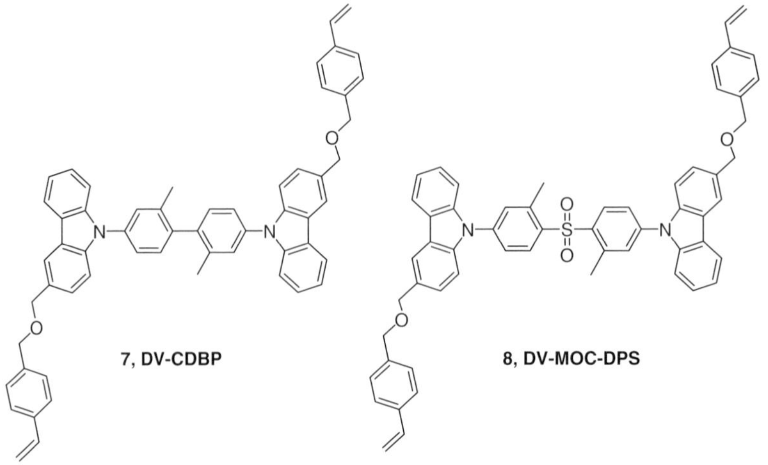

Figure 14.5 Chemical structures of emitters **7** and **8**.

The solution-processed TADF OLED with CIE coordinates most closely approaching the blue standard so far is based on emitter **4**, exhibiting an EQE of 20% but with CIE coordinates of (0.16, 0.25), which are still fairly far from the standard. The "bluest" solution-processed OLED showed better CIE coordinates of (0.12, 0.13) using **8** as the emitter; however the maximum EQE for this device was considerably lower at 2%. Thus, the development of high-performance deep blue TADF OLEDs remains a challenge for both solution-processed and vacuum-deposited devices (Figure 14.5).

14.1.2 Solution-Processed Green TADF Materials and Devices

The NTSC standard defines green color by CIE coordinates of (0.20, 0.71), and CIE coordinates of a pure green emission defined by the PAL standard are (0.29, 0.60). In this section, we have defined green emitters as those with λ_{EL} from 490 to 550 nm. Although a pure 490 nm emission would not be green, the broad emission spectrum of the organic semiconductors means that materials with emission peaking at 490 nm are typically closer to green than blue. The photophysical properties of emitters **9–17** and EL characteristics of the corresponding OLEDs are summarized in Tables 14.1 and 14.2, respectively.

The era of development of high-efficiency green TADF OLEDs began in 2012 when Uoyama et al. [31] reported the emitter 2,4,5,6-tetra(carbazol-9-yl)-1,3-dicyanobenzene (**4CzIPN**), **9**. Compound **9** exhibited a λ_{PL} at 507 nm and a very high Φ_{PL} of 94% in degassed toluene. The relatively short τ_d of 5.1 μs is indicative of TADF behavior, which was further confirmed by a small ΔE_{ST} of 83 meV. The device structure used was based on the following architecture: ITO/NPB (35 nm)/**9**: CBP (6 wt%, 15 nm)/TPBI (65 nm)/LiF (0.8 nm)/Al (100 nm). An EQE close to 19.3% was obtained with λ_{EL} at 520 nm for the OLED fabricated by vacuum deposition. This high device efficiency was comparable with those of highly efficient green phosphorescent OLEDs. Inspired by these results, Cho et al. [32] developed high-efficiency solution-processed OLEDs by modifying the structure of **9** with either methyl (**10; m4CzIPN**) or *tert*-butyl (**11; t4CzIPN**) groups decorating the carbazole donors to improve the solubility of the materials. In their study, **9** displayed λ_{PL} at 507 nm in 1 wt% doped CzSi film, while the emission of **10** and **11** were redshifted by about 20 nm due to the inductively electron-donating nature of the alkyl groups that strengthen the donor character. The Φ_{PL} of 1 wt% doped films of **10** and **11** in CzSi were 67% and 78%, respectively, where CzSi serves as a high bandgap host, and were lower than that of **9** (Φ_{PL} = 94%). Compounds **9–11** in 1 wt% CzSi spin-coated films showed τ_d of 3.1, 2.6, and 2.9 μs, respectively. Uniform film morphology and efficient energy transfer from the host to the dopant translated to high-performance solution-processed devices with the following architecture: ITO/PEDOT:PSS (60 nm)/PVK (15 nm)/**9, 10, 11**: CzSi (1 wt%, 30 nm)/TSPO1 (35 nm)/LiF(1 nm)/Al(200 nm). Devices exhibited green emission with λ_{EL} of 520, 540, and 545 nm and CIE coordinates of (0.20, 0.43), (0.29, 0.57), and (0.31, 0.59) for **9**, **10**, and **11**, respectively. The maximum EQE reported for these devices using **9–11** were of 8.1%, 8.2%, and 18.3%, which reduced to 7.2%, 5.4%, and 12.0% at a luminance of 1000 cd m^{-2}, respectively (Figure 14.6).

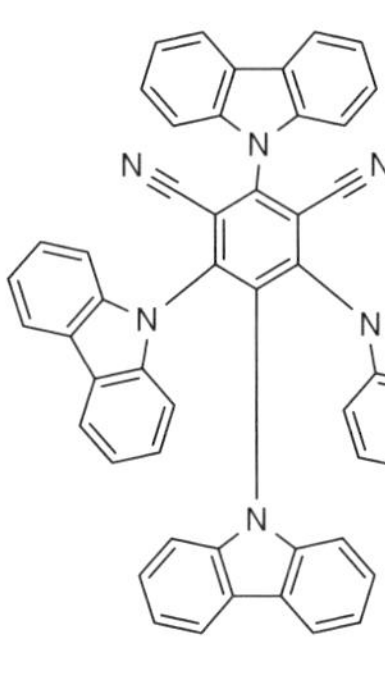

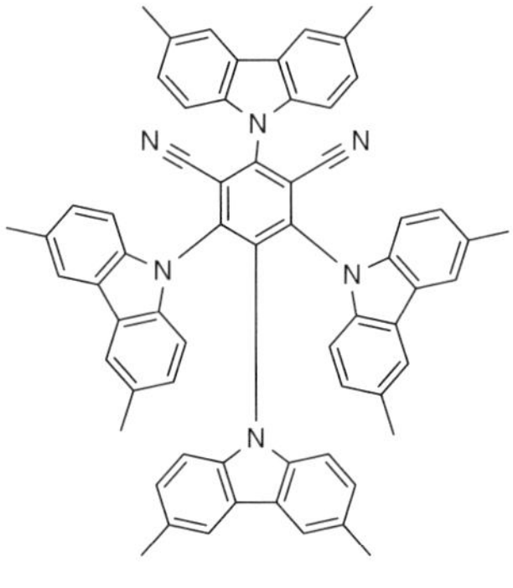

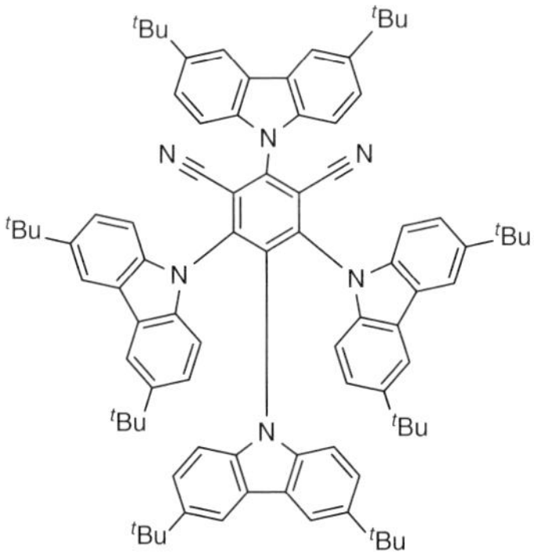

Figure 14.6 Chemical structures of emitters **9**, **10**, and **11**.

Although the solution-processed OLED using **9** reported by Cho et al. [32] did not perform close to its vacuum-deposited analogue, a much improved OLED using **9** was realized later by Suzuki et al. [50] They developed a new host material, 1,3-bis{3-[3-(9-carbazolyl)phenyl]-9-carbazolyl}benzene (CPCB), possessing a high triplet energy of 2.79 eV, and compared the device performance using this bespoke host to that with CBP as the host. Conjugation length and high triplet energy were maintained by linking four carbazole units to three phenyl bridges via the metapositions. The host CPCB showed good solubility (>5 wt%) in many polar solvents, which are typically used in solution processing; 6 wt% doped films of **9** in CPCB exhibited a high Φ_{PL} of 86% under nitrogen-saturated conditions. A solution-processed device based on the architecture ITO/PEDOT:PSS(40 nm)/**9**:CPCB 6 wt% (30 nm)/T2T(10 nm)/Bpy-TP2(40 nm)/LiF(1 nm)/Al (100 nm) showed an EQE of 9.9% (9.8 at 1000 cd m^{-2}). They also fabricated solution-processed devices using CBP as the host with the same device structure and realized a slightly improved EQE of 11.9% (11.4% at 1000 cd m^{-2}). Although devices doped in CPCB have lower efficiencies than their CBP-based counterparts, nonetheless CPCB was proved to be an effective host material for solution-processed TADF devices. Moreover, replacing the Bpy-TP2 electron-transporting layer (ETL) with TPBI resulted in a dramatic improvement of the EQE to 18.5% (17.2% at 1000 cd m^{-2}), which is close to the 19.3% realized for the corresponding vacuum-deposited OLED [31]. The improvement in efficiency was attributed to the increased quenching of excitons by the PEDOT:PSS layer due to its lower singlet energy (1.6 eV) in the case where Bpy-TP2 was used as the ETL. On using T2T/Bpy-TP2 as ETLs, the recombination zone in the EML was shifted toward (host/hole-injection layer) HIL/EML interface (i.e. PEDOT:PSS/EML) due to improved electron transport in ETL and injection into the EML from T2T/Bpy-TP2 layer.

After optimizing the device performance with **9** as the emitter by varying the host material, Komatsu et al. [51] fabricated solution-processed devices and evaluated their performance as a function of ETL material. Specifically, a comparison of device performance was drawn using B3PyPB and B4PyMPM as ETL versus TPBI as ETL in devices with the following architecture: ITO/PEDOT:PSS (30 nm)/**9**:CBP (35 nm)/ETL (65 nm)/Liq (3 nm)/Al (100 nm), with **9** doped at 5 or 15 wt% in CBP. TPBI-based OLEDs with doping concentration of **9** at 5 wt% showed a maximum EQE of 9.4% at a luminance of 1000 cd m^{-2}, which varied little upon increasing the doping concentration to 15 wt% (maximum EQE = 9.9%; 9.3% at a luminance of 1000 cd m^{-2}). Compound **9** has a deep ionization potential/electron affinity of 6.1/3.7 eV. Since TPBI has a shallow ionization potential of 6.3 eV, it does not have sufficient hole-blocking ability and therefore efficiency suffers. To solve this problem, B3PyPB and B4PyMPM were used as ETL materials due to their deep HOMO levels of 6.6 eV [55] and 7.1 eV [56], respectively, and high electron mobility of ∼10^{-4} cm^2 V^{-1} s^{-1}. OLEDs with 5 wt% **9** in CBP and using B3PyPB or B4PYMPM showed much improved maximum EQEs of 15% (13% at 1000 cd m^{-2}) and 16% (13% at 1000 cd m^{-2}), respectively, due to more effective hole blocking by the ETL. These maximum EQEs are among the highest reported efficiencies for solution-processed TADF devices and are not only close to highly efficient vacuum-deposited green TADF materials but also comparable to the best solution-processed green phosphorescent materials [20, 24b]. Further improvement in solution-processed TADF green devices using **9** as the emitter was realized when Kim et al. [49] simplified the device structure by eliminating the PVK HTL (hole transporting layer). TADF devices without HTL can suffer from severe exciton quenching at the PEDOT:PSS/EML interface, which can limit the efficiency to a great extent. To overcome this problem, a self-organized buffer hole-injection layer (Buf-HIL) was employed consisting of PEDOT:PSS and tetrafluoroethylene-perfluoro-3,6-dioxa-4-methyl-7-octene-sulfonic acid copolymer (PFI) [57]. The lower surface energy of PFI resulted in its rise to the top surface during spin-coating of Buf-HIL. This process led to the formation of a self-organized Buf-HIL with gradient PFI concentration, [PFI], which increased gradually from bottom to top of the HIL. Due to the deep ionization potential of PFI, this gradient [PFI] resulted in a gradual increase of the work function (WF) of Buf-HIL from 5.20 eV at the bottom to 5.95 eV at the top surface of the HIL. This gradual increase in WF facilitated the hole injection from ITO to the bottom surface and hole transporting to the host from the top surface. CBP was chosen as the host as a good match of the Buf-HIL WF with the HOMO (6.0 eV) of CBP. This not only improved the hole/electron balance in the EML but also avoided exciton quenching at the interface of the HIL/EML by eliminating the accumulation of holes at that interface. Solution-processed devices (ITO/Buf-HIL (40 nm)/**9**: CBP (40 nm)/ TPBI(50 nm)/ LiF(1 nm)/ Al(100 nm)) attained a maximum EQE of 24%, PE of 58 lm W^{-1}, which were much greater than the PEDOT:PSS-based OLEDs (maximum EQE = 15%, PE = 17.5 lm W^{-1}) and the vacuum-deposited OLEDs (maximum EQE = 19.3%). This device is the most efficient reported to date for a solution-processed OLED employing a pure organic TADF emitter.

Inspired by the structure of **9**, Tang et al. [33] employed a related acceptor in 4-cyanopyridine, producing the emitter 2,3,5,6-tetracarbazolyl-4-cyano-pyridine (**4CzCNPy**), **12**, and simultaneously developed a cyano-free new host material 4CzPy. Compound **12** showed strong green-yellow emission with λ_{PL} at 536 nm in chloroform solution and a yellow emission (λ_{PL} = 560 nm) in 8 wt% doped in 4CZPy thin film. The Φ_{PL} of the film was 24.7% under ambient conditions, which increased to 54.7% under oxygen-free conditions. A τ_d of 8.4 µs and ΔE_{ST} of 0.07 eV confirmed the TADF nature of the emission. By using **12** as the dopant in either the high triplet energy 4CZPy (2.8 eV) or mCP as hosts in a simple three layer solution-processed device of ITO/PEDOT:PSS (40 nm)/**12**:host (8 wt%, 30 nm)/TmPyPB (60 nm)/LiF (1 nm)/Al (100 nm), maximum EQEs of 10.4% (9.6% at 1000 cd m^{-2}) and 11.3% (10.6% at 1000 cd m^{-2}) were achieved, respectively, with λ_{EL} of 524 nm and CIE coordinates of (0.35, 0.59) obtained for both the devices. A maximum luminance of 21 431 cd m^{-2} and current density of 273 mA cm^{-2} were obtained at an operational voltage of 13.9 V using 4CZPy as the host. On the other hand, these values were 16 305 cd m^{-2}, 225 mA cm^{-2}, and 14.4 V for mCP-based devices. The low cost, high luminance, and high efficiency demonstrated the promise of emitter **12** in EL devices. To further study the effect of the host material on the device performance, the same group developed three new hybrid TADF host materials based on alkyl-substituted bicarbazole/cyanobenzene [52]. Three simple bicarbazole derivatives, namely, pCNBCzmMe (**a**), pCNBCzoCF$_3$ (**b**), and pCNBCzmCF3 (**c**), were synthesized by decorating the 9-positions of bicarbazole with cyanobenzene to improve the solubility and electron-transporting capabilities of the materials, while methyl and trifluoromethyl moieties were introduced to fine-tune their optoelectronic properties. The chemical structures of these three host materials are shown in Figure 14.7, and the data collected in Table 14.3. These three materials exhibited TADF properties with ΔE_{ST} of 0.30, 0.19, and 0.14 eV for **a**–**c**, respectively, in pristine neat films. They fabricated solution-processed devices for double TADF host/dopant system using these three TADF materials as host and **12** as TADF dopant with the configuration of ITO/PEDOT:PSS(40 nm)/**12**:Host (8 wt%, 40 nm)/TmPyPB (60 nm)/LiF (1 nm)/Al (100 nm). A maximum brightness >14 000 cd m^{-2} and EQEs ranging from 10% to 11% that remained almost identical at a luminance of 100 cd m^{-2} were achieved containing these hosts. Although devices with double TADF host–guest system did not much improve the EQE of the OLED with **12**, it did reveal an interesting approach of designing host–guest systems that may improve charge transport.

A distinct approach of using charged emitters in nondoped solution-processed TADF OLEDs was first reported by our group [34]. We developed a charged

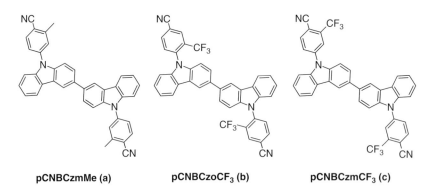

Figure 14.7 Chemical structures of emitter **12** and host materials **a**–**c**.

Table 14.3 Chemical structures of host, hole transporting, and electron-transporting materials.

Material	Structure	Type
PEDOT:PSS		Hole transporting layer (HTL)
PVK	Poly-9-vinylcarbazole	Host/hole-injection layer (HIL)
mCP	1,3-Bis(carbazol-9-yl)benzene	Host
CzSi	9-(4-tert-butylphenyl)-3,6-bis(triphenylsilyl)-9H-carbazole	Host
CBP	4,4′-Bis(carbazol-9-yl)biphenyl	Host

(Continued)

Table 14.3 (Continued)

Material	Structure	Type
DV-CDBP	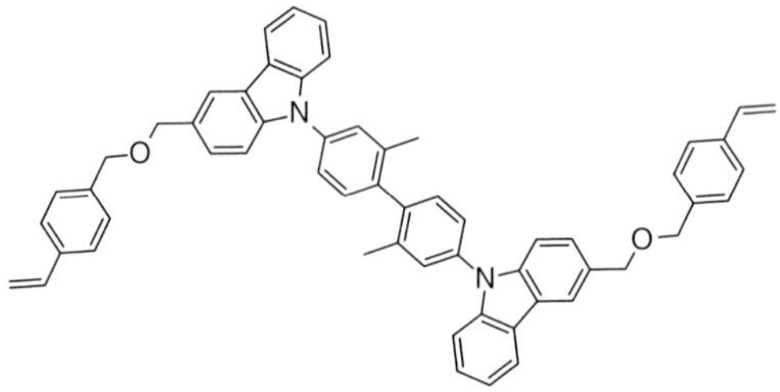 9,9′-(2,2′-dimethyl-[1,1′-biphenyl]-4,4′-diyl)bis(3-(((4-vinylbenzyl)oxy)methyl)-9H-carbazole)	Host
CPCB	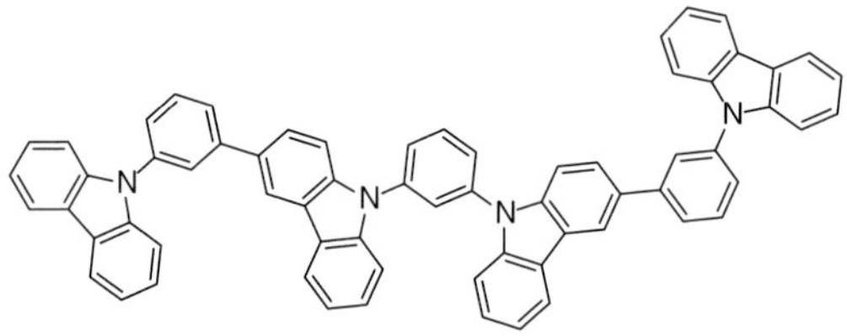 1,3-bis(3-(3-(9H-carbazol-9-yl)phenyl)-9H-carbazol-9-yl)benzene	Host
4CZPy	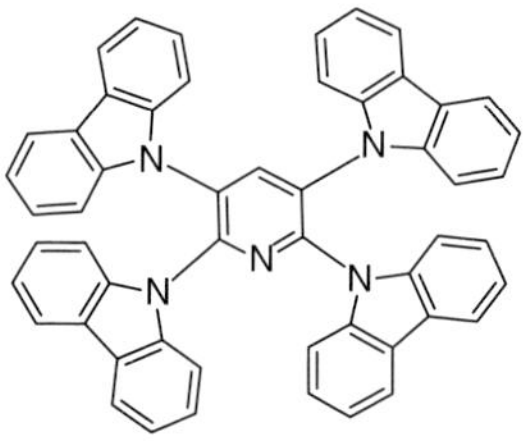 9,9′,9″,9‴-(pyridine-2,3,5,6-tetrayl)tetrakis(9H-carbazole)	Host
pCNBCzmMe	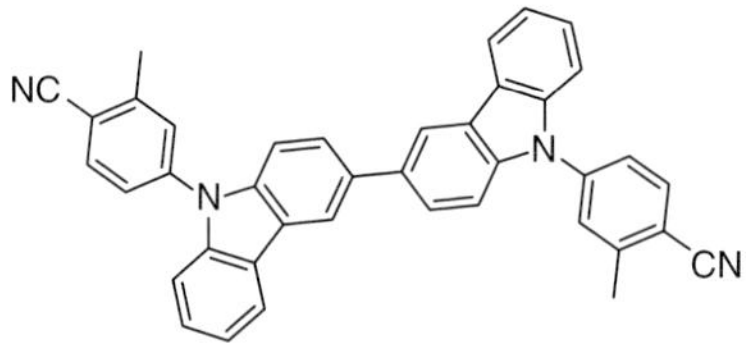 4,4′-(9H,9′H-[3,3′-bicarbazole]-9,9′-diyl)bis(2-methylbenzonitrile)	Host
pCNCZoCF$_3$	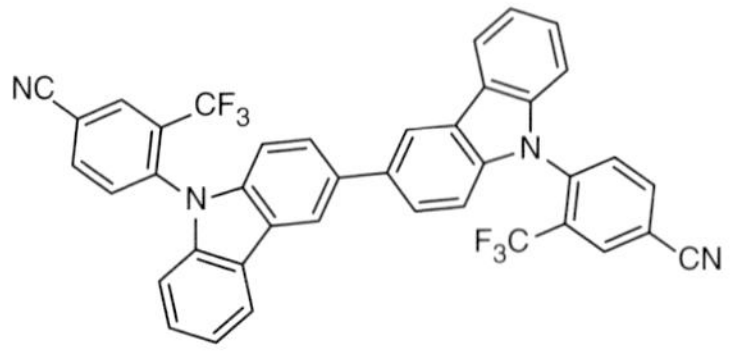 4,4′-(9H,9′H-[3,3′-bicarbazole]-9,9′-diyl)bis(3-(trifluoromethyl)benzonitrile)	Host

(Continued)

Table 14.3 (Continued)

Material	Structure	Type
pCNCZmCF$_3$	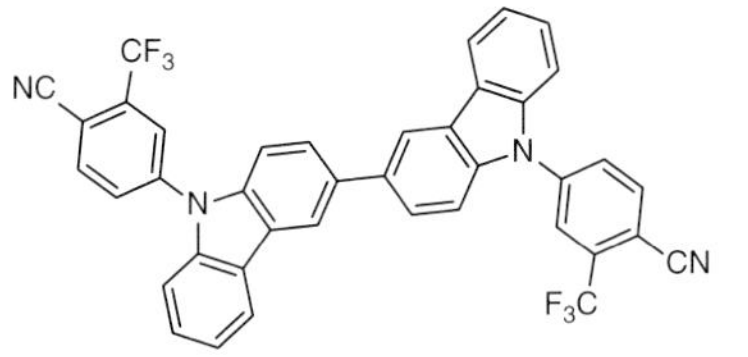 4,4′-(9H,9′H-[3,3′-bicarbazole]-9,9′-diyl)bis(2-(trifluoromethyl)benzonitrile)	Host
PBD	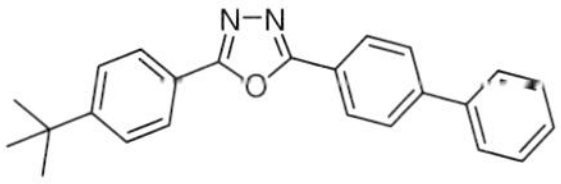 2-(4-Biphenyl)-5-(4-tert-butylphenyl)-1,3,4-oxadiazole	Host
TAPC	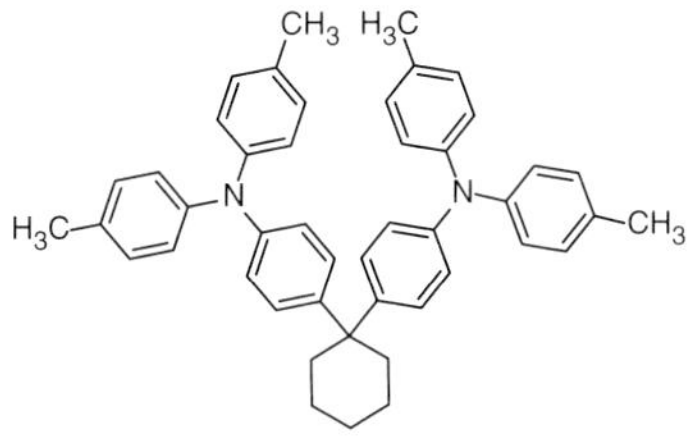 4,4′-(cyclohexane-1,1-diyl)bis(N,N-di-p-tolylaniline)	Host
TCTA	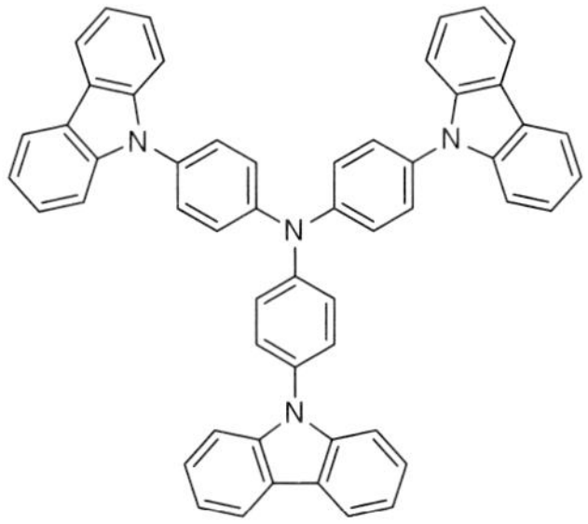 2-(4-Biphenyl)-5-(4-tert-butylphenyl)-1,3,4-oxadiazole	Host
TPBI	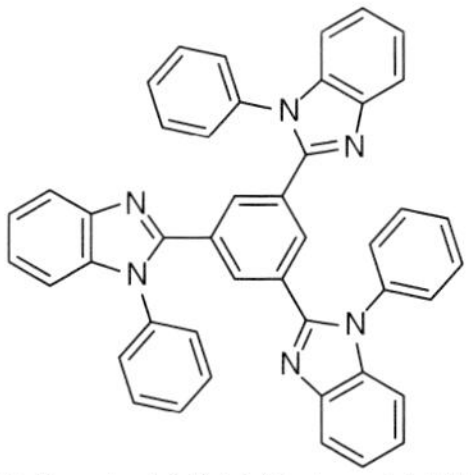 2,2′,2″-(1,3,5-Benzinetriyl)-tris(1-phenyl-1-H-benzimidazole)	Electron-transporting layer (ETL)

(Continued)

Table 14.3 (Continued)

Material	Structure	Type
TmPyPB	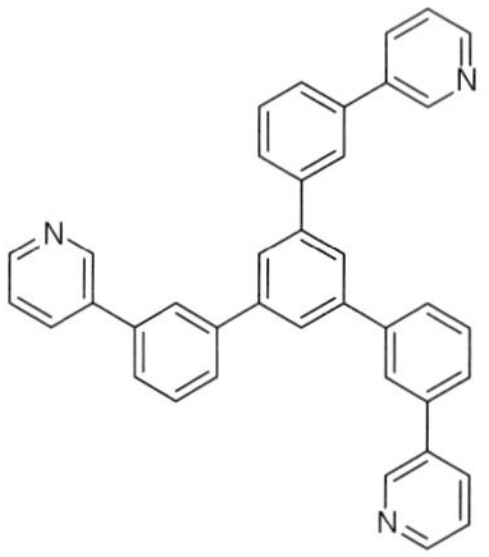 1,3,5-Tri[(3-pyridyl)-phen-3-yl]benzene	Electron transporting layer (ETL)
TSPO1	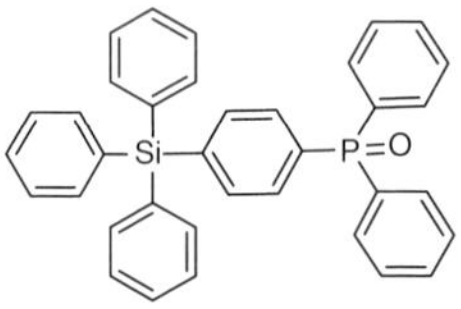 Diphenyl-4-triphenylsilylphenyl-phosphine oxide	Hole-blocking layer (HBL)
T2T	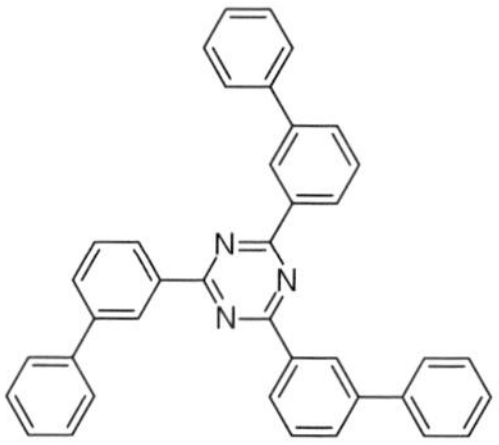 2,4,6-tri([1,1′-biphenyl]-3-yl)-1,3,5-triazine	Hole-blocking layer (HBL)
Bpy-TP2	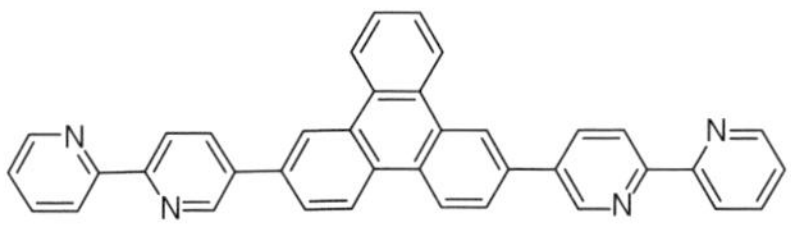 2,7-di([2,2′-bipyridin]-5-yl)triphenylene	Electron-transporting layer (ETL)
B3PyPB	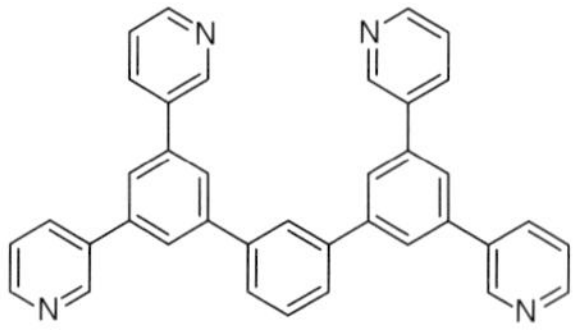 3,3″,5,5″-tetra(pyridin-3-yl)-1,1′: 3′,1″-terphenyl	Electron-transporting layer (ETL)

(Continued)

Table 14.3 (Continued)

Material	Structure	Type
B3PyMPM	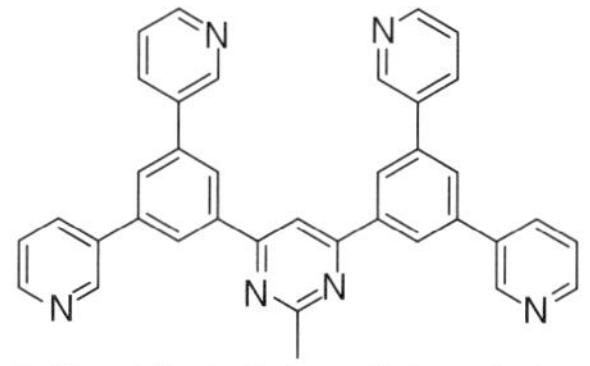 4,6-Bis(3,5-di(pyridin-3-yl)phenyl)-2-methylpyrimidine	Electron-transporting layer (ETL)
B4PyMPM	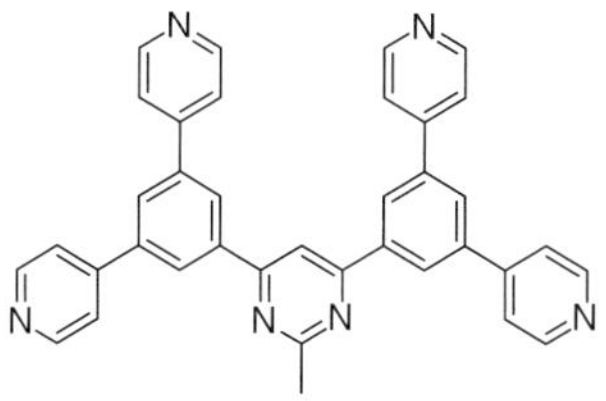 4,6-Bis(3,5-di(pyridin-4-yl)phenyl)-2-methylpyrimidine	Electron-transporting layer (ETL)

PEDOT and PSS, Poly(3,4-ethylenedioxythiophene)polystyrene sulfonate.

emitter **13** by tethering methylimidazolium units onto 2CzPN [31]. The Φ_{PL} of **13** was 73% doped in 10 wt% PMMA film under ambient conditions, which increased to 90% under nitrogen-saturated conditions. The τ_d of 6.34 μs confirmed its TADF nature. Solution-processed OLEDs using the device structure of ITO/PEDOT:PSS(30 nm)/PVK(30 nm)/**13**:mCP:OXD7 or neat **13** (10 wt%, 30 nm)/B3PyMPM(60 nm)/Ca(20 nm)/Al(100 nm) were fabricated using a double host system with mCP and OXD7 to balance and confine the hole ratio in the EML. The device with neat **13** as the EML produced a green OLED with $\lambda_{EL} = 546$ nm and CIE coordinates of (0.41, 0.53) with a maximum EQE of 5.1% with low turn-on voltage of 3.6 V, which outperformed the device with the doped EML (maximum EQE = 2.7%, turn-on voltage of 5.8 V). Although the neat-film device did not exceed the previous reported EQEs of green solution-processed OLEDs, its performance metrics were still superior to the few reports of nondoped TADF OLEDs in the literature [26, 38] (Figure 14.8).

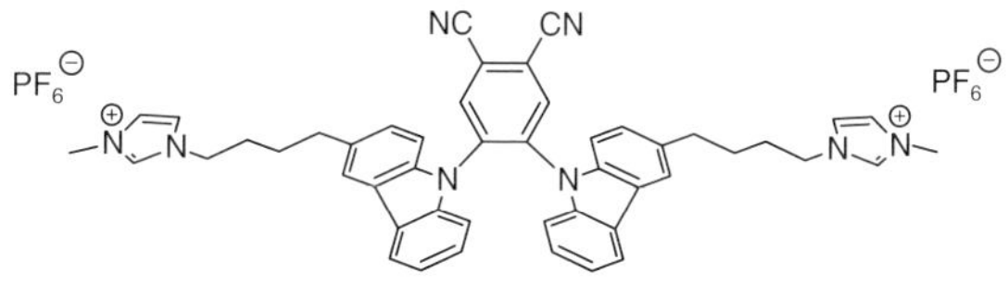

Figure 14.8 Chemical structure of emitter **13**.

So far, most of the research have focused on developing and modifying the D–A materials that are mainly based on carbazole as the donor. A handful of reports of vacuum-deposited highly efficient TADF emitters exist in the literature where disubstituted acridan units are used as donors [36]. Acridan was first employed as a donor in a solution-processable TADF emitter **3ACR-TRZ, 14**, which contains a triazine acceptor unit [35]. They doped 16 wt% **14** into CBP, which resulted in an emission with λ_{PL} at 504 nm with an outstanding Φ_{PL} of 98% under nitrogen-saturated atmosphere. The τ_d was measured to be 6.7 μs, while a ΔE_{ST} of only 15 meV confirmed the TADF nature of the emitter. Solution-processed OLEDs based on the device structure ITO/PEDOT:PSS (35 nm)/16 wt% **14**:CBP (55 nm)/B3PyPB (30 nm)/Liq (1 nm)/Al(100 nm) were fabricated. The OLEDs produced blue-green emission with λ_{EL} = 504 nm, a maximum EQE of 18.6% and a power efficiency of 36.3 lm W^{-1}. Motivated by the excellent device performance of **14**, Wada et al. [53] developed emitter **DMAC-TRZ, 15**, based on the same electron donor and acceptor fragments as **14** and reported a maximum EQE of 20% by vacuum depositing a host-free neat film of **15**. These encouraging results led the same group to fabricate a solution-processed device with **15** as the emitter [53]. A neat film of **15** prepared by spin-coating showed a Φ_{PL} of 84% that was almost identical to that found for its vacuum-deposited film (Φ_{PL} = 83%). Solution-processed devices based on ITO/PEDOT:PSS (40 nm)/**15** (30 nm)/BmPyPB (60 nm)/Liq (1 nm)/Al (100 nm) exhibited a maximum EQE of 17.5% with λ_{EL} of 498 nm, which was not only comparable to the nondoped vacuum-deposited counterpart (maximum EQE = 20%) but is also the highest value reported to date for a nondoped TADF solution-processed OLED (Figure 14.9).

After employing acridan to develop green high-performance TADF emitters, Liu et al. [29] reported solution-processed TADF emitters based on phenoxazine as donor units. They developed a series of triarylborane-based TADF emitters, namely, **TB-1PXZ (5)**, **TB-2PXZ (6)**, and **TB-3PXZ (16)** by varying the number of donor units and studied their optoelectronic properties. Emitters **5** and **6** produce blue light and were discussed in the previous section. Compound **16** showed a λ_{PL} at 509 nm with a near-unity Φ_{PL} 95% in 10 wt% doped CzSi films under nitrogen-saturated conditions. A barely resolved ΔE_{ST} of 0.01 eV and a very

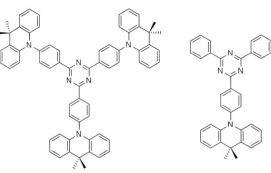

Figure 14.9 Chemical structures of emitters **14** and **15**.

14, 3ACR-TRZ

15, DMAC-TRZ

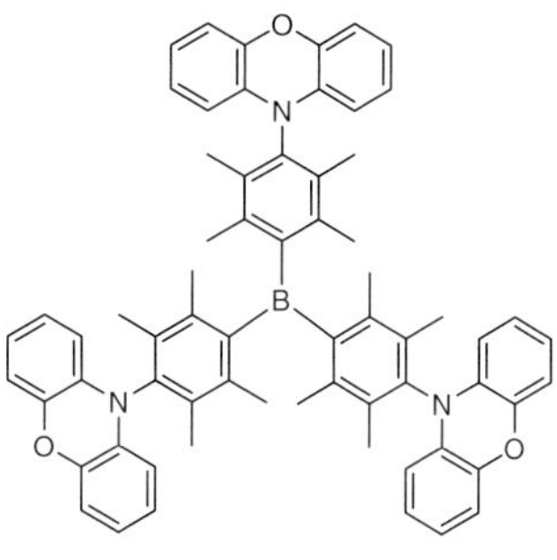

Figure 14.10 Chemical structure of emitter **16**.

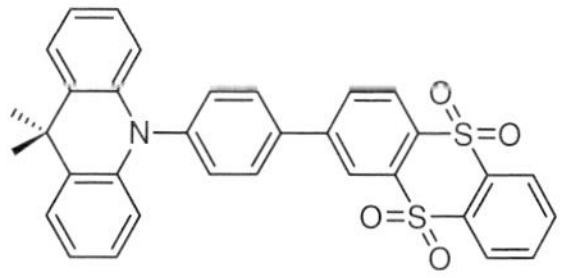

16, TB-3PXZ

Figure 14.11 Chemical structure of emitter **17**.

17, ACRDSO2

short τ_d of 1.3 μs confirmed its TADF nature. Solution-processed devices based on ITO/PEDOT:PSS (30 nm)/**16**:CzSi: (10 wt%, 40 nm)/TmPyPB(40 nm)/Liq (2 nm)/Al (100 nm) exhibited a maximum EQE of 13.6% (12.3% at luminance of 100 cd m^{-2}) with an λ_{EL} of 510 nm and CIE coordinates of (0.23, 0.54). The high efficiency observed for the OLED with **16** was attributed to its high Φ_{PL} and efficient rISC process due to the very small ΔE_{ST} (Figure 14.10).

To further enhance the device performance, another TADF emitter **17** (**ACRDSO$_2$**) was reported by Xie et al. [37] In their design a novel acceptor, thianthrene-9,9′,10,10′-tetraoxide, was coupled to a dimethylacridan, DMAC, donor through a phenyl bridge. Compound **17** exhibited a λ_{PL} at 520 nm with Φ_{PL} of 71% in 6 wt% doped CBP films. Moreover, a ΔE_{ST} of 0.058 eV and a short τ_d of 8.3 μs are indications of efficient rISC. Solution-processed devices with the device structure ITO/Buf-HIL (40 nm)/6 wt% **17**:CBP (35 nm)/TmPyPB (55 nm)/LiF (1 nm)/Al (100 nm) showed a maximum EQE of 17.5% at a luminance of 100 cd m^{-2} with an λ_{EL} at 534 nm (Figure 14.11).

14.1.3 Solution-Processed Yellow-to-Red TADF Materials and Devices

NTSC standard defines red color by CIE coordinates of (0.67, 0.33), and CIE coordinates of a pure red emission defined by PAL standard are (0.64, 0.33). In this section, we have defined the yellow-to-red emitters as those whose λ_{EL} lie between 550 nm and 644 nm. The photophysical properties of **18–20** and the EL characteristics of corresponding OLEDs are summarized in Tables 14.1 and 14.2, respectively.

Far less effort has been devoted to the design of efficient long wavelength TADF emitters compared with the focus on blue and green emitters. Due to a trade-off between singlet–triplet energy splitting and high fluorescence radiative rates [58], it is difficult to design and realize highly efficient TADF emitters, especially orange/red emitters, which tend to exhibit low Φ_{PL} values because nonradiative

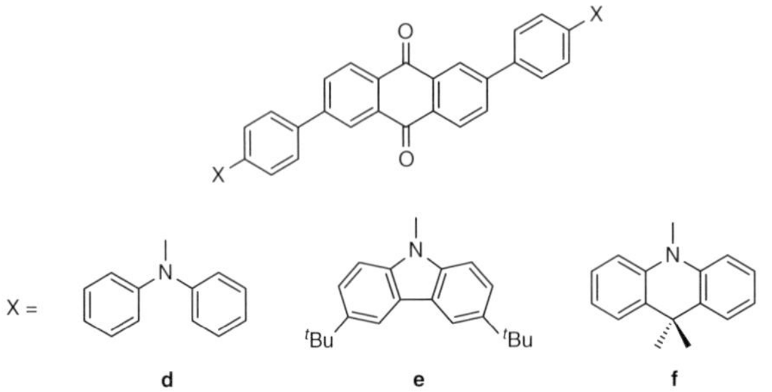

Figure 14.12 Chemical structures of emitters **d–f**.

decay rates increase exponentially with the emission wavelength following the well-known energy gap law [59]. Uoyama et al. [31] reported an orange emitter 4CzTPN where the vacuum-deposited OLED produced orange light with a λ_{EL} of 585 nm in EL and maximum EQE of 11.2%. In order to improve the efficiency of red-emitting devices, the same group developed emitters employing a central anthraquinone acceptor and two donors separated by a phenyl bridge from the acceptor. The introduction of the bridge was designed to enhance the radiative decay rates. Employing emitters **d–f** (Figure 14.12) in OLEDs led to devices with EQEs of 12.5%, 9.0%, and 6.9%, respectively, with corresponding λ_{EL} values of 624, 637 and 584 nm, respectively [60]. These efficiencies are among the highest reported to date for vacuum-deposited red TADF OLEDs.

To realize solution-processed red TADF OLEDs, Chen et al. [38] reported a dithienylbenzothiadiazole-based TADF emitter, **18 (red-1b)**, which exhibited λ_{PL} at 644 nm in neat film with a Φ_{PL} of 28% under nitrogen-saturated conditions. The ΔE_{ST} of 0.40 eV is unusually large for a TADF material, and the short τ_d of 828 ns is probably reflective of increased nonradiative decay. A solution-processed non-doped device based on ITO/PEDOT:PSS (35 nm)/PVK (30 nm)/**18** (30 nm)/TPBI (60 nm)/CsF(1 nm)/Al (100 nm) realized a maximum EQE of 1.75% (1.27% at a luminance of 100 cd m^{-2}) with pure red emission (λ_{EL} = 644 nm). The authors explained and confirmed the delayed fluorescence in the EL device by performing magneto-electroluminescence (MEL) measurements and attributed the emission mechanism to the coexistence of TADF and TTA. MEL is a tool to study rISC and TTA by the relative variation of EL intensity versus the strength of an external magnetic field. For an emitter showing prompt and delayed fluorescence, prompt fluorescence is a magnetic field independent because the direct radiative decay of the singlets is not affected by magnetic field. In the absence of an external magnetic field, rISC is very efficient between three degenerate triplet substates (T_{-1}, T_0, and T_{+1}) and the singlet state (S_1). In the presence of a strong external magnetic field, Zeeman splitting of the triplet states lifts the degeneracy between T_{-1} and T_{+1} states, leaving only one channel for rISC from T_0 to S; partly blocking the rISC and as a result leading to a decreased TADF. The MEL of emitter **18** consisted of a rapid decrease in the presence of external magnetic field (Figure 14.13).

To overcome the existing limited EQE observed for red TADF OLEDs, Yu et al. [39] reported an asymmetric D–A TADF molecule (**FDQPXZ**), **19**

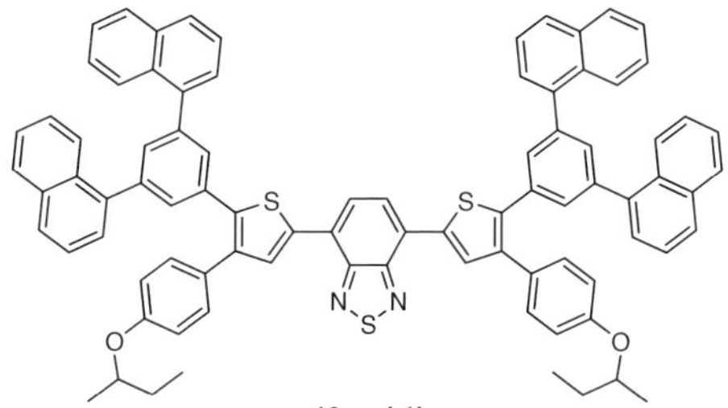

Figure 14.13 Chemical structure of emitter **18**.

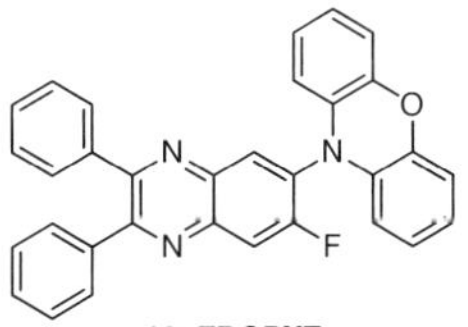

18, red-1b

Figure 14.14 Chemical structure of emitter **19**.

19, FDQPXZ

containing fluorine-substituted quinoxaline acceptor and phenoxazine donor units. Fluorine was introduced to increase the acceptor strength and redshift the emission further. Compound **19** showed a very small ΔE_{ST} of 0.04 eV, favorable for efficient rISC with a λ_{PL} of 606 nm in 5 wt% doped Bepp$_2$ films with a Φ_{PL} of 74%. The prompt and delayed lifetimes are 29 ns and 3.2 μs, respectively, demonstrating the TADF nature of the emission. Solution-processed devices based on ITO/PEDOT:PSS(40 nm)/10 wt% **19**:CBP:PBD (40 nm)/TmPyPB (50 nm)/LiF (1 nm)/Al (100 nm) exhibited a maximum EQE of 9.0% (7.8% at a luminance of 100 cd m^{-2}) with a λ_{EL} of 600 nm and CIE coordinates of (0.53, 0.46), which was the first example of a high-performance solution-processed orange-red TADF OLED (Figure 14.14).

To further enhance the device performance, red TADF emitter **20** was reported by Xie et al. [37] They developed a novel acceptor thianthrene-9,9′,10,10′-tetraoxide and coupled it with phenoxazine (**20; PXZDSO$_2$**). In many TADF emitter designs effective separation between the HOMO and LUMO, and hence, small ΔE_{ST} is achieved as a result of a large torsional angle between donor and acceptor. However, an overly large dihedral angle prevents efficient radiative decay from S$_1$ to S$_0$ by reducing the transition dipole moment according to Fermi's golden rule [61]. Thus, the introduction of the phenyl spacer group was designed to weaken the nonradiative internal conversion by elongating the distance between donor and acceptor moieties, and hence, to increase the Φ_{PL} while maintaining a small ΔE_{ST}. Compound **20** exhibited λ_{PL} at 540 nm with Φ_{PL} of 62% in 6 wt% doped CBP films. A ΔE_{ST} of 48 meV and a short τ_d of 5.0 μs for **20** favored an efficient rISC. Solution-processed devices with the device structure ITO/Buf-HIL (40 nm)/6 wt% **20**:CBP (35 nm)/TmPyPB (55 nm)/LiF (1 nm)/Al (100 nm) showed a maximum EQE 15.2% at a luminance of 100 cd m^{-2} for **20**. The OLED with **20** exhibited yellow emission with λ_{EL} at 552 nm and CIE coordinates of (0.42, 0.55). The OLED with the yellow emitter **20** proved to

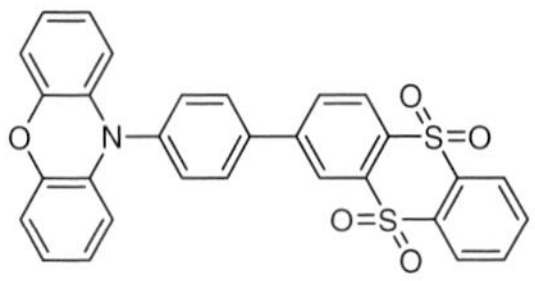

Figure 14.15 Chemical structures of emitter **20**.

20, PXZDSO2

be the first example of a solution-processed device that exhibited a comparable performance with a vacuum-deposited OLED of the same approximate energy (Figure 14.15).

14.1.4 Comparison of State-of-the-Art Solution-Processed OLEDs to Vacuum-Deposited Counterparts

To further optimize the device performance of solution-processed TADF OLEDs, it is crucial to understand the performance difference between state-of-the-art OLEDs fabricated by solution processing and vacuum deposition. Since the first report of solution-processed TADF OLEDs in 2014, this field has progressed remarkably, and EQEs comparable to state-of-the-art vacuum-deposited TADF OLEDs have already been achieved for each of blue, green, and red colors. Achieving a deep blue emission yet maintaining high efficiency is quite challenging. The deepest blue vacuum-deposited TADF emitter reported to date is based on a sulfone acceptor and *tert*-butylcarbazole donor (emitter **g**, Figure 14.16) moieties and possessed an EQE of 9.9% (λ_{EL} 423 nm) while exhibiting impressive CIE coordinates of (0.15, 0.07) [54]. A redshift in the EL spectrum usually accompanies a more efficient OLED in the blue. For example, the OLED with emitter **h** has a λ_{EL} of 468 nm yet retains deep blue CIE coordinates of (0.12, 0.13) along with an EQE of 20.2% [62]. A further shift in λ_{EL} to 490 nm for emitter **i** coincided with an impressive maximum EQE of 36.7% (35% at 100 cd m^{-2}) emitting with CIE coordinates of (0.18, 0.43) [63]. In fact, this device remains the most efficient reported for small molecule-based TADF OLEDs to date. By comparison, the best solution-processed TADF blue OLED in terms of maximum EQE reported thus far is based on emitter **4** with the OLED exhibiting a maximum EQE of 20% with CIE coordinates of (0.16, 0.25). Vacuum-deposited TADF OLEDs based on emitter **j** showed an EQE of 19.5% with CIE coordinates (0.16, 0.23); almost identical in emission color to solution-processed TADF OLED of emitter **4** [64]. From this comparison, there is still room for improved solution-processed blue TADF OLEDs. For the green OLEDs, the most efficient vacuum-deposited device is based on emitter **k** and exhibited maximum EQE of 29.6% (26% at 100 cd m^{-2}) with λ_{EL} at 520 nm [65]. Solution-processed devices provide comparable performance with the most efficient based on emitter **9** showing an EQE of 24% with an identical EL emission maximum and CIE at (0.20, 0.43). When it comes to the comparison of OLEDs emitting at lower energy, EQEs of 15.2% and 9.0% have been achieved for emitters **20** and **19** with λ_{EL} of 552 and 600 nm, respectively, for solution-processed devices, whereas for the most efficient vacuum-deposited OLEDs, these values are higher at 18.5% for yellow with λ_{EL} at 552 nm (emitter **l**) [66] and 17.5% for red with λ_{EL} at 610 nm (emitter **m**) [67].

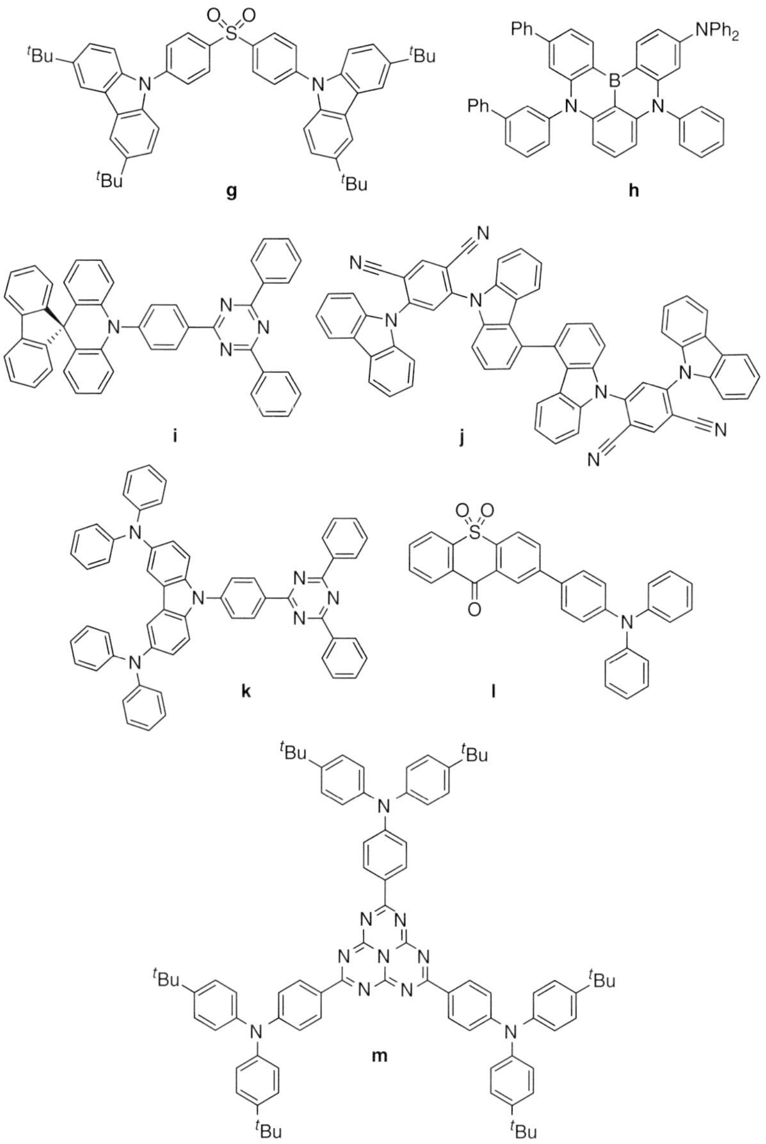

Figure 14.16 Chemical structures of emitters **g–m**.

14.1.5 Solution-Processed TADF Polymers and Dendrimers

We have so far only discussed the transfer of TADF technology from vacuum-deposited OLEDs to solution-processed OLEDs using small molecule TADF emitters and hosts. In the solution processing of small molecules, the tactic is to blend these functionalized small molecule TADF dopants into various hosts in order to maintain good charge transport and film morphology characteristics. However, the host–guest blend approach has significant challenges including dopant aggregation, diffusion of dopants into the clusters of hosts and into neighboring layers, and the propensity for the dopant to form microcrystalline domains, leading to poor film morphology [68]. Polymers, on the other hand, are particularly suitable for solution processing to make high quality films, e.g. by spin-coating or blade-coating. Furthermore the viscosity of polymer solutions can be readily adjusted to be ideal for inkjet printing. Light-emitting dendrimers have also proved a very effective strategy for conferring solution processability and controlling aggregation [13, 20, 69]. However, it is not trivial to achieve TADF in polymers and dendrimers given the challenge of simultaneously achieving small ΔE_{ST} and suppression of internal conversion in molecules containing many atoms. Additionally, intermolecular and intramolecular TTA quenches the triplet populations in polymers easily. Despite these challenges, TADF polymers and dendrimers would be attractive materials for solution-processed OLEDs. To date through, there have only been a handful of reports of TADF polymers and dendrimers. In this section we discuss these materials (emitters **21–40**) in terms of their photophysical properties and EL characteristics of the corresponding OLEDs (Tables 14.1 and 14.2, respectively).

Keeping the design strategies of TADF in mind, π-conjugated polymers have been utilized in realizing solution-processed TADF OLEDs. Nikolaenko et al. [40] reported the first example of a TADF polymer based on a strategy, which they called "intermonomer TADF." They proposed an approach where donor and acceptor centers were distinct monomers, which upon polymerization, formed the charge transfer (CT) emitter. The polymer, **21**, exhibited a small ΔE_{ST} due to the spatial separation of HOMO and LUMO, enabling efficient thermal upconversion of triplet excitons (**Figure** 14.17). They used a bis(diarylamine) donor and a 1,3,5-triazine acceptor as monomers and added a high T_1 energy backbone monomer 1,4-bis(4-(4,4,5,5-tetramethyl-1,3,2-dioxaborolan-2-yl)phenyl)butane to break the conjugation to avoid the TTA in polymeric structure. The optoelectronic

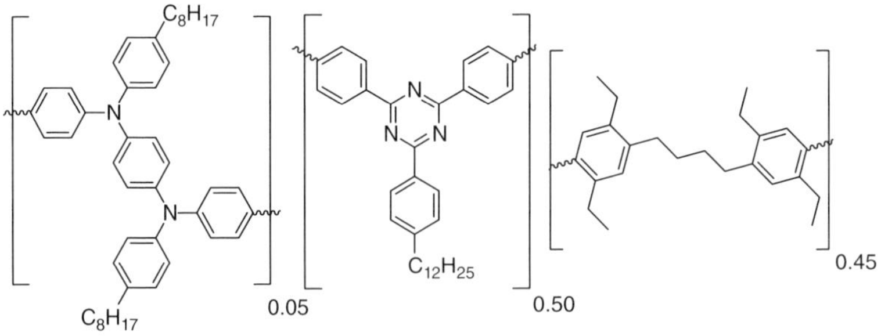

Figure 14.17 Chemical structure of emitters **21**.

characterization of the neat film of **21** revealed TADF character with the transient PL decay consisting of two components: prompt (15–50 ns) and delayed (1 μs to 50 ms), the Φ_{PL} in air of 41% increasing slightly to 44% under nitrogen-saturated conditions, and a ΔE_{ST} of 220 meV, determined from the peak position of the fluorescence and phosphorescence spectra. Polymer OLEDs were fabricated using device structure ITO/PEDOT:PSS (65 nm)/interlayer (40 nm)/**21** (80 nm)/NaF (2 nm)/Al (100 nm)/Ag (100 nm). They used a polymeric interlayer (poly(2,5-dihexylbenzene-N-(4-(2-butyl)phenyl)-diphenylamine)) possessing a reasonably high triplet level ($T_1 = 2.52$ eV) to minimize the triplet quenching in the TADF emissive layer and to prevent the aggregation of the TADF emissive cores. Devices showed a green emission with λ_{EL} at 530 nm and a maximum EQE of 10%.

Luo et al. [41] prepared a series of copolymers (**P0–P12**) using a known TADF chromophore (**PXZ-OXD**) [70] unit tethered to a poly(carbazole) backbone. The best device employed **P12 (22)**. Polymer **22** exhibited a λ_{PL} at 494 nm with a Φ_{PL} of 34% under nitrogen-saturated conditions and a τ_d of 2.36 μs as a neat thin film. The solution-processed OLED with the architecture (ITO/PEDOT:PSS (30 nm)/poly-TPD (40 nm)/neat **22** (50 nm)/TmPyPB (35 nm)/Ba (20 nm)/Al (100 nm)) showed a maximum EQE of 4.3% (2.4% at luminance of 100 cd m^{-2}) with λ_{EL} at 506 nm (Figure 14.18).

Using a similar strategy to that of Nikolaenko et al. Lee et al. [42] developed a class of π-conjugated polymers **23 (pCzBP)** and **24 (pAcBP)** based on alternating benzophenone acceptor and carbazole/acridan donor cores. A 10 wt% doped blend film of these polymer in a mixed host matrix (TCTA:TAPC(65 : 25)) showed λ_{PL} at 508 and 540 nm with Φ_{PL} of 23 and 46% under nitrogen-saturated conditions for **23** and **24**, respectively. Transient PL decays of **23** and **24** exhibited both prompt and delayed components where $\tau_d = 74$ and 10 μs, respectively. The ΔE_{ST} extracted from the onset wavelengths of the doped film emission

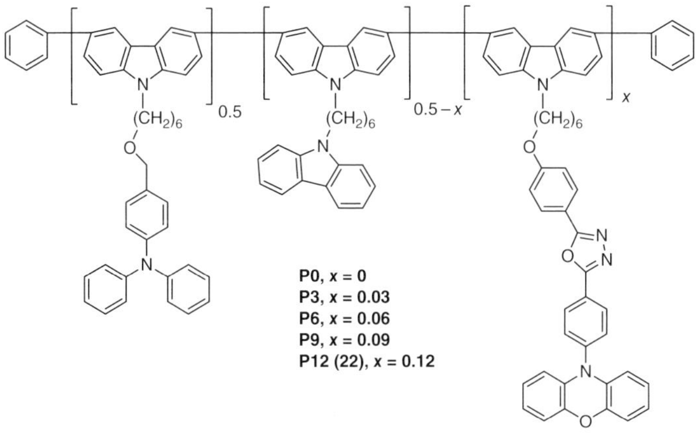

Figure 14.18 Chemical structure of emitter **22**.

spectra at 300 and 5 K were exceedingly small at 16 and 4 meV for **23** and **24**, respectively. OLED devices with the structure of ITO/PEDOT:PSS (40 nm)/**23, 24**:TCTA:TAPC (10 : 65 : 25, 40 nm)/TmPyPB (40 nm)/LiF (0.8 nm)/Al (80 nm) showed green emission for **23** and yellow emission for **24**. The device using **23** achieved an EQE of 8.1% with λ_{EL} of 500 nm, while that of **24** showed an EQE of 9.3% with λ_{EL} of 548 nm; EQE of **24** remained as high as 8.1% at a luminance of 1000 cd m^{-2}. These devices greatly exceeded the theoretical limit for conventional fluorescent OLEDs and provided a molecular design strategy for the further development of solution-processed TADF polymer OLEDs (Figure 14.19).

Zhu et al. [43] proposed an alternative strategy, exemplified in polymers **25 (PAPCC)** and **26 (PAPTC)**, wherein the donor units are contained within

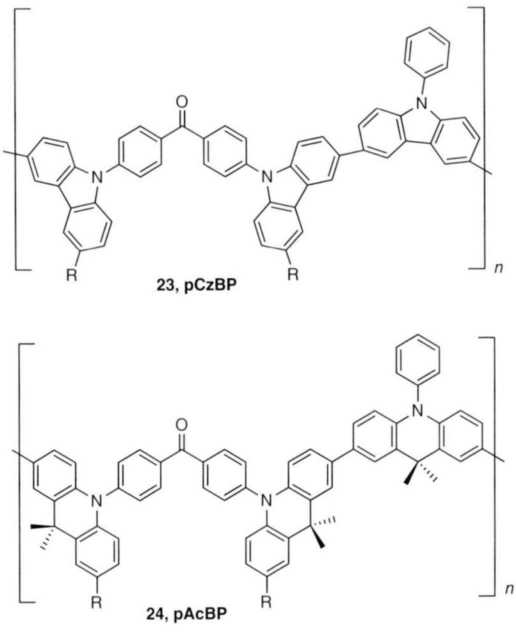

Figure 14.19 Chemical structure of emitters **23** and **24**.

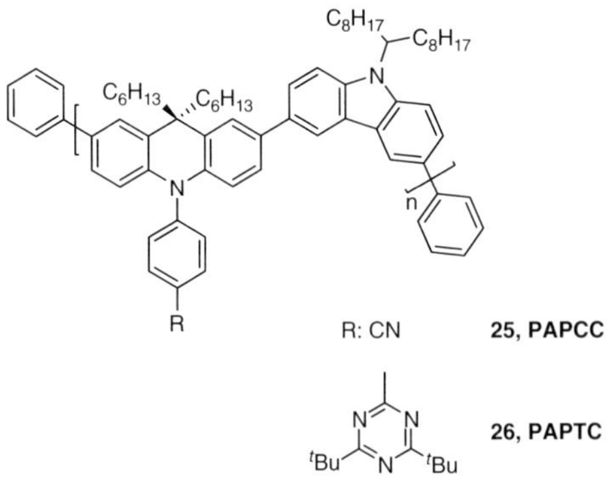

Figure 14.20 Chemical structure of emitters **25** and **26**.

the polymer backbone and the acceptor units grafted as the side chains (**Figure** 14.20). Pristine thin films of **25** and **26** displayed a broad, structureless emission with λ_{PL} at 472 and 510 nm and Φ_{PL} of 9% and 44%, respectively, under nitrogen-saturated conditions. Polymers **25** and **26** exhibited ΔE_{ST} of 0.37 and 0.13 eV, respectively, with very short τ_d of 0.32 and 0.23 μs, respectively. Solution-processed devices based on ITO/PEDOT:PSS (50 nm)/neat **25** or **26** (40 nm)/TmPyPB (50 nm)/LiF (1 nm)/Al(100 nm) were fabricated and showed EQEs of 1.3% at a luminance of 6 cd m^{-2} with a λ_{EL} of 508 nm and 12.3% at a luminance of 180 cd m^{-2} with a λ_{EL} of 521 nm, respectively. The relatively high EQE for the device with **26** was attributed to its smaller ΔE_{ST} of 0.13 eV, which facilitated the effective upconversion of triplet excitons.

Employing similar strategies as those illustrated previously, a variety of highly efficient polymers **27–31** based on phenothiazine donors and a dibenzothiophene-S,S-dioxide acceptors were reported jointly by the groups of Bryce and coworkers [44]. Polymer **27** contained the phenothiazine donor in the backbone along with dibenzothiophene, whereas, a polystyrene (PS) backbone was used as a spacer dispersing unit in polymers **28–31** to suppress the internal conversion and TTA. Polymers **28–30** comprised of different molecular ratios of donor–acceptor with respect to styrene units. All the polymers showed broad and featureless emission spectra as pristine thin films with emission maxima ranging from 535 (**28**) to 556 nm (**31**). Characteristic PL decay kinetics were observed with τ_d ranging from 1 to 100 μs. ΔE_{ST} values for **27–31** were calculated to be 0.28, 0.35, 0.46, 0.42, and 0.40 eV, respectively. Solution-processed polymer OLEDs were fabricated using the device structures ITO/PEDOT:PSS (40 nm)/**27**:PBD:PVK (10 : 40 : 50, 45 nm)/TPBI (50 nm)/LiF (1 nm)/Al (100 nm) and ITO/PEDOT:PSS (40 nm)/**28–31**:mCP (10 wt%, 45 nm)/TPBI (50 nm)/LiF (1 nm)/Al (100 nm) and showed maximum EQEs of 11.5% at a luminance of 1000 cd m^{-2} (λ_{EL} = 512 nm), 20.1% (λ_{EL} = 533 nm, EQE at 100 cd m^{-2} = 5.3%), 15.2% (λ_{EL} = 536 nm, EQE at 100 cd m^{-2} = 2.8%), 1.8% (λ_{EL} = 539 nm, EQE at 100 cd m^{-2} = 1.8%), and 1.4% (λ_{EL} = 556 nm, EQE at 100 cd m^{-2} = 0.6%) for **27–31**, respectively. A sharp drop in EQE was observed with decreasing styrene content in the family **28–31** owing to the corresponding increase in TTA. A device of **28** was also fabricated using CBP as a host, and an EQE of 2.5% was

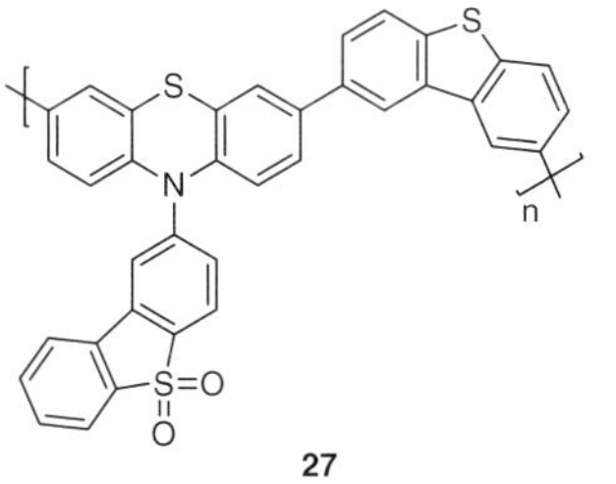

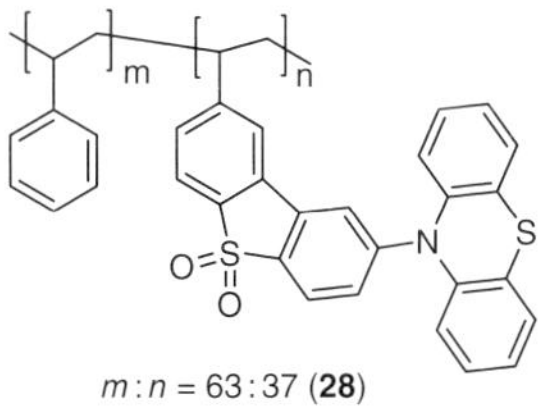

$m:n = 63:37$ (**28**)
$m:n = 54:46$ (**29**)
$m:n = 33:67$ (**30**)
$m:n = 0:100$ (**31**)

Figure 14.21 Chemical structures of emitters **27–31**.

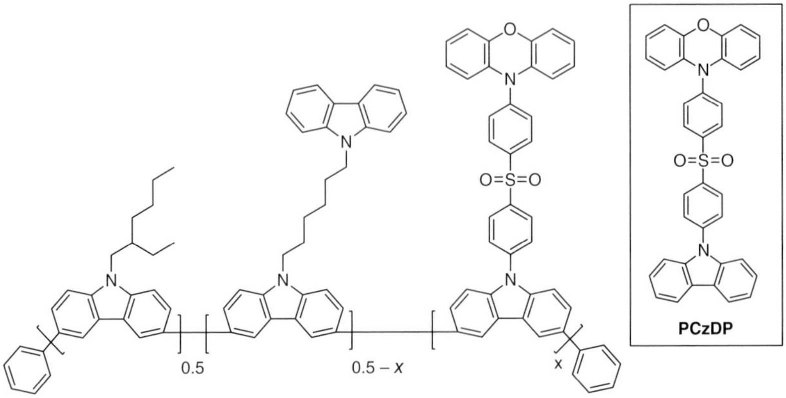

Figure 14.22 Chemical structure of emitter **32** ($x = 5, 10, 15$, or 20). Also shown in the inset is the reference TADF emitter (**PCzDP**).

achieved, while the EQE of the device with **28** using mCP as the host was much higher at 20.1%, demonstrating the importance of the host materials to the device performance (Figure 14.21).

Recently, Xie et al. [45] reported a series of copolymers incorporating the TADF emitter 10-(4-((4-(9H-carbazol-9-yl)phenyl)-sulfonyl)phenyl)-10H-phenoxazine (**PCzDP**). They synthesized four copolymers comprised of different molar ratios of **PCzDP** along with carbazole comonomers (**32**, Figure 14.22). In deaerated toluene solution, **PCzDP-5** (i.e. 5 mol% of **PCzDP** incorporated in the polymer) showed emission from both the backbone carbazole moiety ($\lambda_{PL} = \sim 425$ nm) and **PCzDP** ($\lambda_{PL} = \sim 510$ nm) with a Φ_{PL} of 50%, which clearly indicated that the concentration of **PCzDP** was insufficient for complete energy transfer from the comonomer. Polymers with higher concentrations of **PCzDP** (**PCzDP-10**, **PCzDP-15**, and **PCzDP-20**) showed much less emission from the carbazole backbone and higher Φ_{PL} (67%, 66%, and 60%, respectively). On the other hand, in neat films all displayed emission solely from **PCzDP** ($\lambda_{PL} = 497–509$ nm) due to the presence of both intrachain and interchain energy transfer as a function of the greater packing density of the polymer in the solid state. **PCzDP-5** neat film had the lowest Φ_{PL} of 39%, whereas **PCzDP-10**, **PCzDP-15**, and **PCzDP-20** showed higher values ($\Phi_{PL} = 74\%$, 70%, and 59%, respectively). The TADF nature of the polymers was confirmed by the presence of a delayed fluorescence with very similar emission lifetimes ($\tau_d = 1.9–2.3$ μs). The best device (ITO/PEDOT:PSS (50 nm)/**PCzDP-10** (70 nm)/TmPyPB (40 nm)/Liq (1 nm)/Al (100 nm)) showed a bluish-green emission with λ_{EL} and CIE at 496 nm and (0.24, 0.40), respectively. The maximum EQE achieved was 16.1%, which dropped to 11.3% at a brightness of 100 cd m^{-2}.

Wei et al. [46] very recently demonstrated how a non-TADF monomer (Figure 14.23) can be turned into a TADF macromolecule (macrocycle or polymer **33**). In the monomer, the dihedral angle between the donor and the acceptor is as small as 50.6°, and hence a large ΔE_{ST} of 0.37 eV results. However, though this same dihedral angle is 50.7° in both the polymer and macrocycle,

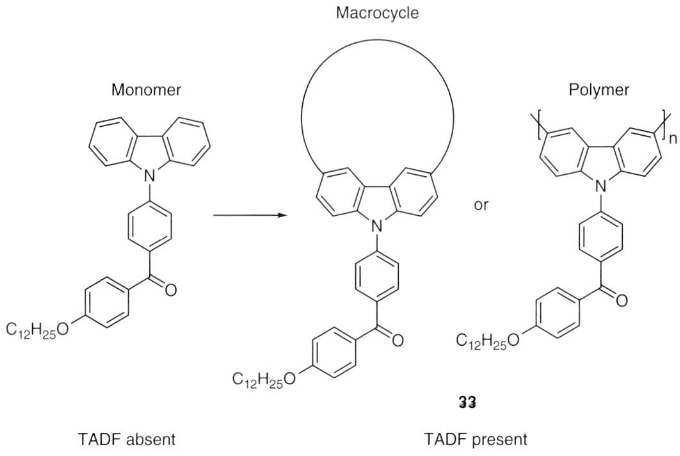

Figure 14.23 Chemical structures of A TADF-inactive monomer (left) that can be transformed to be TADF-active macrocycle or polymer (**33**). Source: Ref. [51]. Reproduced with permission of Elsevier.

each has a much higher HOMO level of −5.14 eV compared with −5.59 eV for the monomer and hence the CT character of the excited state is greatly enhanced, which results in a significantly smaller ΔE_{ST} of 0.12 eV. Compound **33** (both polymer and macrocycle) exhibited a λ_{PL} of 496 nm, Φ_{PL} of 71% with a τ_d of 296 μs in 2 wt% doped film of **33** in a PS host matrix, confirming its TADF character. This is the first report of turning on TADF activity by polymerization of a TADF-inactive monomer. However, no device data are reported.

The previous examples of solution-processed TADF OLEDs based on both small molecules and polymers demonstrate the potential of this family of devices. Another class of solution-processable materials are dendrimers, and phosphorescent dendrimers have been widely explored as efficient solution-processable OLED materials [13, 71] but only recently has attention turned to adapting this class of compounds into TADF emitters.

Albrecht et al. [26] reported the first example of a TADF dendrimer and its use in a nondoped, solution-processed OLED. They developed three carbazole dendrimer generations with a triazine core (**34 (G2TAZ)**, **35 (G3TAZ)**, and **36 (G4TAZ)**). All three dendrons showed almost identical pristine thin film emission spectra with λ_{PL} at 500 nm. Compounds **34–36** exhibited miniscule ΔE_{ST} of 0.03, 0.06, and 0.06 eV, respectively, with Φ_{PL} of 52%, 31%, and 8.4% under nitrogen-saturated conditions. Their τ_d were measured to be 3.1, 1.9, and 2.4 μs for **34**, **35**, and **36**, respectively. Nondoped solution-processed devices were fabricated with the structure ITO/PEDOT:PSS (30 nm)/**34**, **35**, and **36** (35 nm)/TPBI (40 nm)/Ca (10 nm)/Al(100 nm). Maximum EQEs reached 2.4% (λ_{EL} = 510 nm), 3.4% (λ_{EL} = 500 nm), and 1.5% (λ_{EL} = 498 nm) for **34–36**, respectively. These efficiencies were higher than the EQEs expected assuming that only singlet emission was contributing to the EL emission. Although the EQEs were less than 5%, this report nevertheless demonstrated the utility of employing dendrimers in solution-processable, nondoped OLED devices (Figure 14.24).

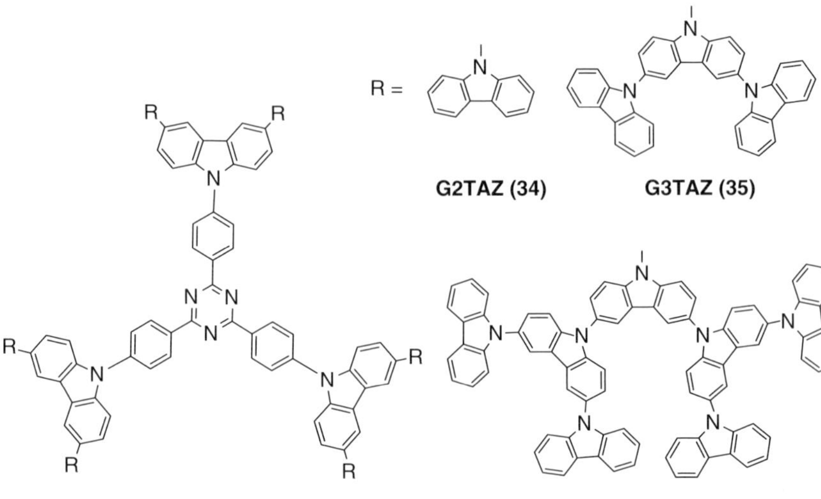

Figure 14.24 Chemical structure of emitters **34**, **35**, and **36**.

To achieve high EQE in dendrimer-based OLEDs, Li et al. [47] developed two TADF dendrons, **37** (**CDE1**) and **38** (**CDE2**). These dendrons were based on a TADF emissive core, DMAC–BP [72], which was previously shown to be an excellent TADF emitter with a high Φ_{PL} of 85% in thin film. They revealed that these dendrimer-based OLEDs showed a dual channel emissive feature where both TADF and interfacial exciplex emission were observed. Compounds **37** (λ_{PL} = 520 nm) and **38** (λ_{PL} = 499 nm) showed a blueshifted emission in neat film compared with DMAC-BP (λ_{PL} = 506 nm), suggesting that the appended dendrons completely suppressed the intermolecular interactions among the emissive cores. Compounds **37** and **38** exhibited pristine thin film Φ_{PL} of 77% and 75%, respectively. The ΔE_{ST} were 0.11 and 0.15 eV for **37** and **38**, respectively. Both dendrimers showed both prompt and delayed fluorescence in oxygen-free toluene with τ_d of 523 ns and 627 ns for **37** and **38**, respectively. Solution-processed double layer devices were fabricated using the device architecture of ITO/PEDOT:PSS (30 nm)/**37** or **38** (70 nm)/TPBI (40 nm)/Liq (2 nm)/Al (100 nm) and showed maximum EQEs of 12% (11.9% at 1000 cd m^{-2}) and 5.2% (4.1% at 1000 cd m^{-2}) for **37** and **38** with λ_{EL} of 546 and 522 nm, respectively. They observed a significant redshift in the EL spectra compared with the PL spectra for **37** and **38** (λ_{PL} = 520 nm; λ_{EL} = 546 nm for **37**, λ_{PL} = 499 nm; λ_{EL} = 522 nm for **38**). This redshift was attributed to an exciplex type of emission at the interface of the EML and ETL. To verify this hypothesis, solution-processed devices with three different ETLs including BPhen, B3PyMPM, and TmPyPB, in place of TPBI were fabricated for **37**, and all the devices with different ETLs exhibited redshifts in the EL spectra compared with the single layer (i.e. without ETL) device, which confirmed

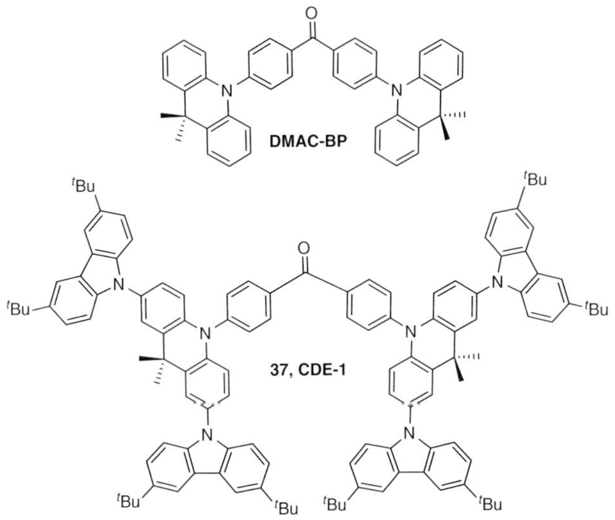

Figure 14.25 Chemical structures of emitter DMAC-BP and **37**.

the interfacial exciplex formation between **37** and the ETLs. A maximum EQE of 13.3% was observed upon using TmPyPB as the ETL. This increase in efficiency upon changing the ETL to TmPyPB suggested a dual emission pathway of TADF from the emitter **37** and exciplex emission originating from the emitter **37** and ETL TmPyPB. This work presented a judicious design tactic for nondoped solution-processable TADF emitters and demonstrated how a single material could show both TADF and exciplex emission (Figures 14.25 and 14.26).

Ban et al. [48] reported two generations of dendrimers, **39 (TZ-Cz)** and **40 (TZ-3Cz)**, based on a triazine core and carbazole donors. They modified the emissive core **TZ** with different generations of carbazole to produce **39** and **40** and compared their photophysical and EL properties with the emissive core **TZ**. Both the materials showed an expected blueshift in emission maxima than **TZ** (502 nm) in neat films (λ_{PL} = 490 nm for **39**; 487 nm for **40**). The thin film Φ_{PL} for **39** and **40** were 58% and 76% under nitrogen-saturated conditions, both higher than that of **TZ** (Φ_{PL} = 32%). The thin film ΔE_{ST} of **TZ**, **39**, and **40** were 0.28, 0.20, and 0.20 eV, respectively. Solution-processed, nondoped devices were fabricated using the device structure ITO/PEDOT:PSS (30 nm)/**TZ**, **39** or **40** (70 nm)/TPBI (40 nm)/Cs$_2$CO$_3$ (2 nm)/Al (100 nm). All the devices showed green emission with λ_{EL} of 520 nm. The maximum EQE of the device with **39** was 10.1%, which remained unchanged at a luminance of 1000 cd m^{-2} and was ten times higher than that of **TZ** (1.9%). Compound **40** showed an EQE of 6.5% at a luminance of 1000 cd m^{-2}, which was likewise higher than **TZ** but was inferior to that of **39** indicating the importance of the choice of peripheral dendrons (Figure 14.27).

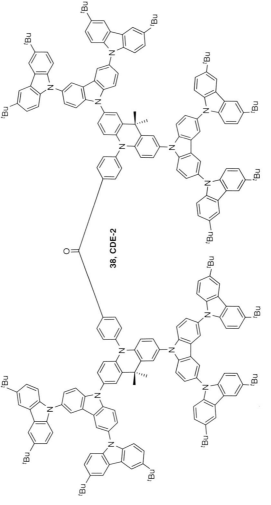

Figure 14.26 Chemical structure of emitter **38**.

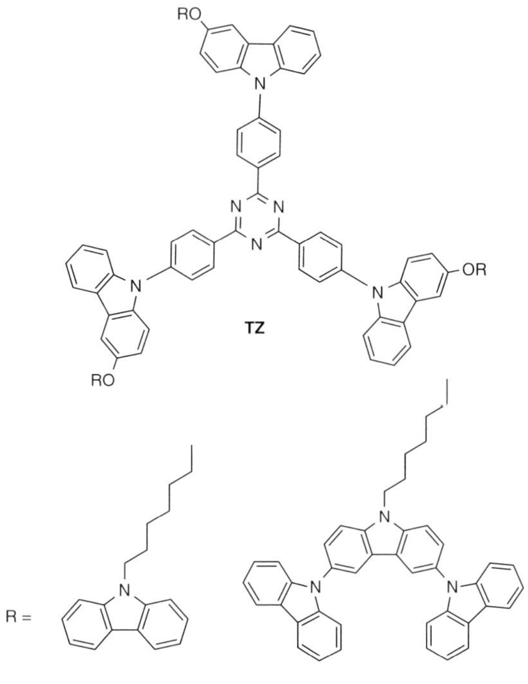

Figure 14.27 Chemical structure of emitter **39** and **40**.

TZ-Cz (39)   **TZ-3Cz (40)**

14.2 Summary and Outlook

The development of solution-processed TADF small molecules and macromolecules, including polymers and dendrimers, will continue to draw increasing attention since the initial reports, as summarized in this chapter, demonstrate the viability of these devices, which combine high efficiency with low-cost fabrication processes. Indeed, the most efficient solution-processed OLEDs are already approaching their vacuum-evaporated counterparts. However, their device lifetimes, which is a key performance metric, need considerable further work.

Alongside the development of new soluble materials, an understanding of the structure–property relationships is of equal importance. Thus far, the majority of the solution-processed TADF emitters have been developed as a function of the incorporation of alkyl or alkoxy groups to an already-established TADF structure whose efficiencies have already been optimized in the context of a vacuum-deposited device. At this point, no design rule exists on how to specifically design TADF emitters for solution processing nor are there design rules for which solubilizing groups and how many should be incorporated to control the aggregation-caused quenching to produce the best films.

Thus, a further improvement in the molecular design, device configuration, and fabrication techniques to tackle device stability issues with solution-processed

TADF emitters is still required. Nonetheless, the tremendous improvement in device performance over the past 3 years augurs well for a future in which low-cost, high-performance solution-processed TADF OLEDs dominate the marketplace.

References

1 Tang, C.W. and VanSlyke, S.A. (1987). *Appl. Phys. Lett.* 51: 913.
2 Baldo, M.A., O'Brien, D.F., Thompson, M.E., and Forrest, S.R. (1999). *Phys. Rev. B* 60: 14422.
3 (a) Wilson, J.S., Dhoot, A.S., Seeley, A.J.A.B., Khan, M.S., Kohler, A., and Friend, R.H. (2001). *Nature* 413: 828. (b) Wohlgenannt, M., Jiang, X.M., Vardeny, Z.V., and Janssen, R.A.J. (2002). *Phys. Rev. Lett.* 88: 197401. (c) Beljonne, D., Ye, A., Shuai, Z., and Brédas, J.L. (2004). *Adv. Funct. Mater.* 14: 684.
4 (a) Baldo, M.A., O'Brien, D.F., You, Y., Shoustikov, A., Sibley, S., Thompson, M.E., and Forrest, S.R. (1998). *Nature* 395: 151. (b) Adachi, C., Baldo, M.A., Thompson, M.E., and Forrest, S.R. (2001). *J. Appl. Phys.* 90: 5048.
5 (a) Kim, K.H., Moon, C.K., Lee, J.H., Kim, S.Y., and Kim, J.J. (2014). *Adv. Mater.* 26: 3844. (b) Lee, C.W. and Lee, J.Y. (2013). *Adv. Mater.* 25: 5450.
6 Partee, J., Frankevich, E.L., Uhlhorn, B., Shinar, J., Ding, Y., and Barton, T.J. (1999). *Phys. Rev. Lett.* 82: 3673.
7 (a) Parker, C.A. and Hatchard, C.G. (1961). *Trans. Faraday Soc.* 57: 1894. (b) Endo, A., Sato, K., Yoshimura, K., Kai, T., Kawada, A., Miyazaki, H., and Adachi, C. (2011). *Appl. Phys. Lett.* 98: 083302. (c) Yersin, H. and Monkowius, U. (2008). DE 10 2008 033 563 A1, Germany.
8 (a) Endo, A., Ogasawara, M., Takahashi, A., Yokoyama, D., Kato, Y., and Adachi, C. (2009). *Adv. Mater.* 21: 4802. (b) Deaton, J.C., Switalski, S.C., Kondakov, D.Y., Young, R.H., Pawlik, T.D., Giesen, D.J., Harkins, S.B., Miller, A.J.M., Mickenberg, S.F., and Peters, J.C. (2010). *J. Am. Chem. Soc.* 132: 9499.
9 Cho, Y.J. and Lee, J.Y. (2011). *Adv. Mater.* 23: 4568.
10 Wong, M.Y. and Zysman-Colman, E. (2017). *Adv. Mater.* 29: 1605444.
11 (a) Yang, X., Xu, X., and Zhou, G. (2015). *J. Mater. Chem. C* 3: 913. (b) Yang, X., Zhou, G., and Wong, W.Y. (2015). *Chem. Soc. Rev.* 44: 8484.
12 Burroughes, J.H., Bradley, D.D.C., Brown, A.R., Marks, R.N., Mackay, K., Friend, R.H., Burns, P.L., and Holmes, A.B. (1990). *Nature* 347: 539.
13 Burn, P.L., Lo, S.C., and Samuel, I.D.W. (2007). *Adv. Mater.* 19: 1675.
14 (a) Sekine, C., Tsubata, Y., Yamada, T., Kitano, M., and Doi, S. (2014). *Sci. Technol. Adv. Mater.* 15: 034203. (b) Wu, H.B., Zou, J.H., Liu, F., Wang, L., Mikhailovsky, A., Bazan, G.C., Yang, W., and Cao, Y. (2008). *Adv. Mater.* 20: 696. (c) Gong, S., Yang, C., and Qin, J. (2012). *Chem. Soc. Rev.* 41: 4797.
15 (a) Kim, H., Byun, Y., Das, R.R., Choi, B.-K., and Ahn, P.-S. (2007). *Appl. Phys. Lett.* 91: 093512. (b) Sonoyama, T., Ito, M., Seki, S., Miyashita, S., Xia,

S., Brooks, J., Cheon, K.-O., Kwong, R.C., Inbasekaran, M., and Brown, J.J. (2008). *J. Soc. Inf. Disp.* 16: 1229.
16 Duan, L., Hou, L., Lee, T.-W., Qiao, J., Zhang, D., Dong, G., Wang, L., and Qiu, Y. (2010). *J. Mater. Chem.* 20: 6392.
17 Xia, S., Cheon, K.-O., Brooks, J.J., Rothman, M., Ngo, T., Hett, P., Kwong, R.C., Inbasekaran, M., Brown, J.J., Sonoyama, T., Ito, M., Seki, S., and Miyashita, S. (2009). *J. Soc. Inf. Disp.* 17: 167.
18 Zhong, C., Duan, C., Huang, F., Wu, H., and Cao, Y. (2011). *Chem. Mater.* 23: 326.
19 (a) Halim, M., Pillow, J.N.G., Samuel, I.D.W., and Burn, P.L. (1999). *Adv. Mater.* 11: 371. (b) Halim, M., Pillow, J.N.G., Samuel, I.D.W., and Burn, P.L. (1999). *Synth. Met.* 102: 922.
20 Lo, S.C., Male, N.A.H., Markham, J.P.J., Magennis, S.W., Burn, P.L., Salata, O.V., and Samuel, I.D.W. (2002). *Adv. Mater.* 14: 975.
21 Kim, K.-H., Huh, S.-Y., Seo, S.-M., and Lee, H.H. (2008). *Appl. Phys. Lett.* 92: 093307.
22 (a) Chen, J., Shi, C., Fu, Q., Zhao, F., Hu, Y., Feng, Y., and Ma, D. (2012). *J. Mater. Chem.* 22: 5164. (b) Cai, M., Xiao, T., Hellerich, E., Chen, Y., Shinar, R., and Shinar, J. (2011). *Adv. Mater.* 23: 3590. (c) Jou, J.-H., Wang, W.-B., Chen, S.-Z., Shyue, J.-J., Hsu, M.-F., Lin, C.-W., Shen, S.-M., Wang, C.-J., Liu, C.-P., Chen, C.-T., Wu, M.-F., and Liu, S.-W. (2010). *J. Mater. Chem.* 20: 8411. (d) Lee, C.W. and Lee, J.Y. (2013). *Adv. Mater.* 25: 596.
23 Hou, L., Duan, L., Qiao, J., Zhang, D., Dong, G., Wang, L., and Qiu, Y. (2010). *Org. Electron.* 11: 1344.
24 (a) Zhu, M., Ye, T., He, X., Cao, X., Zhong, C., Ma, D., Qin, J., and Yang, C. (2011). *J. Mater. Chem.* 21: 9326. (b) Rehmann, N., Hertel, D., Meerholz, K., Becker, H., and Heun, S. (2007). *Appl. Phys. Lett.* 91: 103507.
25 (a) Yook, K.S. and Lee, J.Y. (2011). *Org. Electron.* 12: 1711. (b) Gong, S., Fu, Q., Wang, Q., Yang, C., Zhong, C., Qin, J., and Ma, D. (2011). *Adv. Mater.* 23: 4956.
26 Albrecht, K., Matsuoka, K., Fujita, K., and Yamamoto, K. (2015). *Angew. Chem. Int. Ed.* 54: 5677.
27 Mei, L., Hu, J., Cao, X., Wang, F., Zheng, C., Tao, Y., Zhang, X., and Huang, W. (2015). *Chem. Commun.* 51: 13024.
28 Cho, Y.J., Chin, B.D., Jeon, S.K., and Lee, J.Y. (2015). *Adv. Funct. Mater.* 25: 6786.
29 Liu, Y., Xie, G., Wu, K., Luo, Z., Zhou, T., Zeng, X., Yu, J., Gong, S., and Yang, C. (2016). *J. Mater. Chem. C* 4: 4402.
30 Sun, K., Xie, X., Liu, Y., Jiang, W., Ban, X., Huang, B., and Sun, Y. (2016). *J. Mater. Chem. C* 4: 8973.
31 Uoyama, H., Goushi, K., Shizu, K., Nomura, H., and Adachi, C. (2012). *Nature* 492: 234.
32 Cho, Y.J., Yook, K.S., and Lee, J.Y. (2014). *Adv. Mater.* 26: 6642.
33 Tang, C., Yang, T., Cao, X., Tao, Y., Wang, F., Zhong, C., Qian, Y., Zhang, X., and Huang, W. (2015). *Adv. Opt. Mater.* 3: 786.
34 Wong, M.Y., Hedley, G.J., Xie, G., Kölln, L.S., Samuel, I.D.W., Pertegás, A., Bolink, H.J., and Zysman-Colman, E. (2015). *Chem. Mater.* 27: 6535.

35 Wada, Y., Shizu, K., Kubo, S., Suzuki, K., Tanaka, H., Adachi, C., and Kaji, H. (2015). *Appl. Phys. Lett.* 107: 183303.

36 (a) Takahashi, T., Shizu, K., Yasuda, T., Togashi, K., and Adachi, C. (2016). *Sci. Technol. Adv. Mater.* 15: 034202. (b) Jang, M.E., Yasuda, T., Lee, J., Lee, S.Y., and Adachi, C. (2015). *Chem. Lett.* 44: 1248. (c) Zhang, Q., Li, B., Huang, S., Nomura, H., Tanaka, H., and Adachi, C. (2014). *Nat. Photonics* 8: 326. (d) Tsai, W.-L., Huang, M.-H., Lee, W.-K., Hsu, Y.-J., Pan, K.-C., Huang, Y.-H., Ting, H.-C., Sarma, M., Ho, Y.-Y., Hu, H.-C., Chen, C.-C., Lee, M.-T., Wong, K.-T., and Wu, C.-C. (2015). *Chem. Commun.* 51: 13662.

37 Xie, G., Li, X., Chen, D., Wang, Z., Cai, X., Chen, D., Li, Y., Liu, K., Cao, Y., and Su, S.J. (2016). *Adv. Mater.* 28: 181.

38 Chen, P., Wang, L.P., Tan, W.Y., Peng, Q.M., Zhang, S.T., Zhu, X.H., and Li, F. (2015). *ACS Appl. Mater. Interfaces* 7: 2972.

39 Yu, L., Wu, Z., Xie, G., Zhong, C., Zhu, Z., Cong, H., Ma, D., and Yang, C. (2016). *Chem. Commun.* 52: 11012.

40 Nikolaenko, A.E., Cass, M., Bourcet, F., Mohamad, D., and Roberts, M. (2015). *Adv. Mater.* 27: 7236.

41 Luo, J., Xie, G., Gong, S., Chen, T., and Yang, C. (2016). *Chem. Commun.* 52: 2292.

42 Lee, S.Y., Yasuda, T., Komiyama, H., Lee, J., and Adachi, C. (2016). *Adv. Mater.* 28: 4019.

43 Zhu, Y., Zhang, Y., Yao, B., Wang, Y., Zhang, Z., Zhan, H., Zhang, B., Xie, Z., Wang, Y., and Cheng, Y. (2016). *Macromolecules* 49: 4373.

44 (a) Ren, Z., Nobuyasu, R.S., Dias, F.B., Monkman, A.P., Yan, S., and Bryce, M.R. (2016). *Macromolecules* 49: 5452. (b) Nobuyasu, R.S., Ren, Z., Griffiths, G.C., Batsanov, A.S., Data, P., Yan, S., Monkman, A.P., Bryce, M.R., and Dias, F.B. (2016). *Adv. Opt. Mater.* 4: 597.

45 Xie, G., Luo, J., Huang, M., Chen, T., Wu, K., Gong, S., and Yang, C. (2017). *Adv. Mater.* 29: 1604223.

46 Wei, Q., Kleine, P., Karpov, Y., Qiu, X., Komber, H., Sahre, K., Kiriy, A., Lygaitis, R., Lenk, S., Reineke, S., and Voit, B. (2017). *Adv. Funct. Mater.* 27: 1605051.

47 Li, Y., Xie, G., Gong, S., Wu, K., and Yang, C. (2016). *Chem. Sci.* 7: 5441.

48 Ban, X., Jiang, W., Lu, T., Jing, X., Tang, Q., Huang, S., Sun, K., Huang, B., Lin, B., and Sun, Y. (2016). *J. Mater. Chem. C* 4: 8810.

49 Kim, Y.H., Wolf, C., Cho, H., Jeong, S.H., and Lee, T.W. (2016). *Adv. Mater.* 28: 734.

50 Suzuki, Y., Zhang, Q., and Adachi, C. (2015). *J. Mater. Chem. C* 3: 1700.

51 Komatsu, R., Sasabe, H., Inomata, S., Pu, Y.-J., and Kido, J. (2015). *Synth. Met.* 202: 165.

52 Cao, X., Hu, J., Tao, Y., Yuan, W., Jin, J., Ma, X., Zhang, X., and Huang, W. (2017). *Dyes Pigments* 136: 543.

53 Wada, Y., Shizu, K., Kubo, S., Fukushima, T., Miwa, T., Tanaka, H., Adachi, C., and Kaji, H. (2016). *Appl. Phys. Express* 9: 032102.

54 Zhang, Q., Li, J., Shizu, K., Huang, S., Hirata, S., Miyazaki, H., and Adachi, C. (2012). *J. Am. Chem. Soc.* 134: 14706.

55 Sasabe, H., Gonmori, E., Chiba, T., Li, Y.-J., Tanaka, D., Su, S.-J., Takeda, T., Pu, Y.-J., Nakayama, K.-i., and Kido, J. (2008). *Chem. Mater.* 20: 5951.
56 Sasabe, H., Tanaka, D., Yokoyama, D., Chiba, T., Pu, Y.-J., Nakayama, K.-I., Yokoyama, M., and Kido, J. (2011). *Adv. Funct. Mater.* 21: 336.
57 Lee, T.W., Chung, Y., Kwon, O., and Park, J.J. (2007). *Adv. Funct. Mater.* 17: 390.
58 Tao, Y., Yuan, K., Chen, T., Xu, P., Li, H., Chen, R., Zheng, C., Zhang, L., and Huang, W. (2014). *Adv. Mater.* 26: 7931.
59 Bixon, M., Jortner, J., Cortes, J., Heitele, H., and Michel-Beyerle, M.E. (1994). *J. Phys. Chem.* 98: 7289.
60 Zhang, Q., Kuwabara, H., Potscavage, W.J. Jr., Huang, S., Hatae, Y., Shibata, T., and Adachi, C. (2014). *J. Am. Chem. Soc.* 136: 18070.
61 Matsushita, T., Asada, T., and Koseki, S. (2007). *J. Phys. Chem. C* 111: 6897.
62 Hatakeyama, T., Shiren, K., Nakajima, K., Nomura, S., Nakatsuka, S., Kinoshita, K., Ni, J., Ono, Y., and Ikuta, T. (2016). *Adv. Mater.* 28: 2777.
63 Lin, T.-A., Chatterjee, T., Tsai, W.-L., Lee, W.-K., Wu, M.-J., Jiao, M., Pan, K.-C., Yi, C.-L., Chung, C.-L., Wong, K.-T., and Wu, C.-C. (2016). *Adv. Mater.* 28: 6976.
64 Kim, M., Jeon, S.K., Hwang, S.H., Lee, S.S., Yu, E., and Lee, J.Y. (2016). *Chem. Commun.* 52: 339.
65 Kaji, H., Suzuki, H., Fukushima, T., Shizu, K., Suzuki, K., Kubo, S., Komino, T., Oiwa, H., Suzuki, F., Wakamiya, A., Murata, Y., and Adachi, C. (2015). *Nat. Commun.* 6: 8476.
66 Wang, H., Xie, L., Peng, Q., Meng, L., Wang, Y., Yi, Y., and Wang, P. (2014). *Adv. Mater.* 26: 5198.
67 Li, J., Nakagawa, T., MacDonald, J., Zhang, Q., Nomura, H., Miyazaki, H., and Adachi, C. (2013). *Adv. Mater.* 25: 3319.
68 Shibata, M., Sakai, Y., and Yokoyama, D. (2015). *J. Mater. Chem. C* 3: 11178.
69 Lo, S.C., Richards, G.J., Markham, J.P.J., Namdas, E.B., Sharma, S., Burn, P.L., and Samuel, I.D.W. (2005). *Adv. Funct. Mater.* 15: 1451.
70 Lee, J., Shizu, K., Tanaka, H., Nomura, H., Yasuda, T., and Adachi, C. (2013). *J. Mater. Chem. C* 1: 4599.
71 (a) Furuta, P., Brooks, J., Thompson, M.E., and Fréchet, J.M.J. (2003). *J. Am. Chem. Soc.* 125: 13165. (b) Kwon, T.W., Alam, M.M., and Jenekhe, S.A. (2004). *Chem. Mater.* 16: 4657.
72 Zhang, Q., Tsang, D., Kuwabara, H., Hatae, Y., Li, B., Takahashi, T., Lee, S.Y., Yasuda, T., and Adachi, C. (2015). *Adv. Mater.* 27: 2096.

15

Status and Next Steps of TADF Technology: An Industrial Perspective

Alhama Arjona-Esteban[1] and Daniel Volz[2]

[1] cynora GmbH, Werner-von-Siemens-Straße 2-6, Gebäude 5110, 76646 Bruchsal, Germany
[2] McKinsey & Company, Inc., Taunustor 1, 60310 Frankfurt, Germany

The scope of this chapter is a discussion of the thermally activated delayed fluorescence (TADF) technology from an industrial point of view, focusing on potential motivating factors favoring the use of TADF in commercial products, as well as current roadblocks.

In the last few years, the development of new materials has had a significant impact on the industrial application of organic light-emitting diodes (OLEDs). Thanks to new functional materials, OLED displays with superior performance are now being used in smart watches, smartphones, and TVs. However, there is still room for improvement in areas like display resolution and energy efficiency.

With TADF, a new concept has been established to tackle these issues and will – once again through new, advanced materials – lead to superior products.

This chapter is divided in four sections. First, the key arguments for the use of TADF materials from the market's perspective will be discussed. Second, a status update for blue TADF technology as of the beginning of the year 2017 is presented. Third, we point out new trends concerning the use of TADF materials as hosts and finally conclude this chapter with an outlook.

15.1 What Does the Market Want?

Despite their first development in the late 1980s [1], OLEDs began their rise in commercial products only some years ago. These flat and flexible large-area light sources offer unique benefits compared with other electroluminescent devices such as color tunability, very fast switching timer, easy processing of the devices, and no need of color filters (even though LCD technology and OLED technology may be fused). The materials are nontoxic and in principle can be sustainable [2], offering the possibility to address mass markets.

Many applications are imaginable for OLED technology, ranging from printed light sources to flexible displays. While people dream of realizing printed OLEDs, which was proven to be possible on lab scale, OLEDs made their commercial debut in vacuum-processed, high-performance displays that were employed in

Highly Efficient OLEDs: Materials Based on Thermally Activated Delayed Fluorescence,
First Edition. Edited by Hartmut Yersin.
© 2019 Wiley-VCH Verlag GmbH & Co. KGaA. Published 2019 by Wiley-VCH Verlag GmbH & Co. KGaA.

smartphones and smart watches, manufactured by companies such as Samsung or LG in Korea.

Unlike LEDs, OLEDs consist of several very thin stacked layers [3, 4]. Because of their form factor, they can serve as two-dimensional light sources, thus enabling new applications: Recently realized smartphones demonstrated bendable displays using OLED technology. Furthermore, product designers dream of incorporating luminescent foils made from OLEDs into the design of consumer packaging as "smart labels," on semitransparent light-emitting windows and even on the front of buildings as decorative elements. While it is still early to predict if we will ever see such products, it is likely that the use of OLED technology for high-end and potentially mass-produced low-cost applications will drastically increase.

15.1.1 The Emitter Materials: Heart of the OLED

Herein, we will focus on the emitting materials (see Table 15.1 and Figure 15.1), which are responsible for transforming excitons into visible light. The excitons formed in the device by recombination of positive and negative charge carriers are divided into singlet (25%) and triplet (75%) excitons due to spin statistics. In order to fabricate efficient OLEDs, emitter molecules that are able to transform both kinds of excitons into light are preferred.

Looking at the emitter materials in more detail, three main radiative decay mechanisms are used to transform electrical energy into light: fluorescence (FLUO), phosphorescence (PHOS), and TADF. In the first OLEDs, fluorescent materials were used [5]. Around 1998, it was found that phosphorescent materials could also be applied in OLEDs, which led to a drastic increase in efficiency [6, 7]. Recently, emitters based on TADF were also shown to be suitable for very efficient OLED devices [8–12]. For a more detailed description of the technical background of TADF, we refer the reader to the other chapters in this book.

Table 15.1 Main difference between the main emission mechanisms used in OLEDs.

	Performance	Key feature
FLUO	25% of the energy can be used for light generation	Focus on efficient emission from the S_1 to S_0 level to use singlet exciton emission. Generally being used for blue and often green pixels
PHOS	100% of the energy can be used for light generation	Heavy-metal-based materials that feature large spin–orbit coupling to harvest both singlet and triplet excitons. Metals can be iridium, platinum, or osmium, for example.
		Generally used for red pixels and in some cases also in green pixels. Blue PHOS emitters do not show enough stability
TADF	100% of the energy can be used for light generation	Comparable efficiencies to PHOS but a higher stability is expected
		Materials feature carefully adjusted energy levels, S_1 and T_1, to allow for a small $\Delta E(S–T)$ to harvest both singlet and triplet excitons

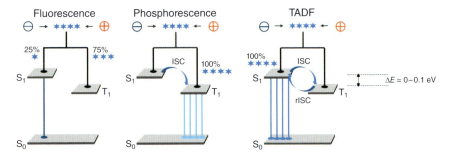

Figure 15.1 Exciton statistics in OLEDs and their emissive decay paths. Fluorescence (FLUO) often relates to the first emitter material generation. Phosphorescence (PHOS) is the second material generation. TADF marks the latest conceptual development. *Source*: Figure courtesy of CYNORA.

A potential TADF emitter material for industrial OLED applications has to satisfy several criteria:

1. *Performance*: OLEDs are ultimately manufactured as display components in devices such as TVs and smartphones; further applications are lighting units for general lighting application and for the automotive industry (taillights). For both fields, the key performance indicators, *device efficiency, color point of the device*, and its *operational stability*, are critical. While the assessment of a material's potential is possible based on photoluminescence quantum efficiency (PLQY), a high PLQY is ultimately irrelevant when the proven performance in a device is lower and cannot be raised.

2. *Compatibility*: Apart from providing the expected performance, a TADF material also needs to be compatible with the basic OLED layout. If the reduction or oxidation potential is far away from the other materials, it cannot be implemented in a working OLED architecture (compare with Figure 15.2). The energy levels (highest occupied molecular orbital (HOMO)

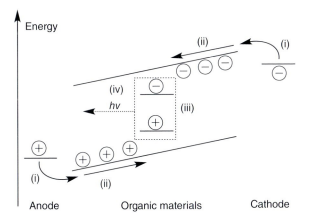

Figure 15.2 Elemental processes during the operation of an OLED. Basic steps are (i) injection of charge carriers in the organic materials, (ii) transport of the charge carriers, which is driven by Coulomb attraction, (iii) formation of excitons, and (iv) emission of light. In this complex system, the emitter (dashed line) has to match all other layers to function in an efficient manner. *Source*: Ref. [2]. Reproduced with permission of Royal Society of Chemistry.

and lowest unoccupied molecular orbital (LUMO) energy) have to match the other materials in the OLED stack in order to facilitate barrier-free charge transport, a low operating voltage, high brightness, and good stability. Another compatibility aspect, which is discussed in Section 1.2, is processing. At this point, vacuum processing still dominates industrial OLED manufacturing, meaning that nonsublimable compounds are likely not to be suitable for implementation in a commercial product in the foreseeable future.

3. *Availability*. To be implemented in a commercial product, TADF materials need to be available in a sufficient amount. This is somewhat connected to the number of synthetic steps and the complexity of the molecular structure, but there are no fixed guidelines since (looking at many commercial specialty chemicals and pharmaceuticals) optimization of synthetic procedures often opens straightforward routes to seemingly incredible complex structures, while relatively simple-looking structures need to be sorted out because they cannot be made in large quantities. Availability can also encompass considerations of intellectual property and having a so-called freedom to operate (FTO), which goes beyond the scope of this chapter [13]. What is partly connected to availability of a material is its sustainability, an aspect we previously discussed in detail [2] and will also be covered in Section 1.3.

The drastic impact of using different emitter materials on commercial products can be seen in Figure 15.3, where selected key performance indicators are plotted for various generations of Samsung Galaxy smartphones. The initial commercialization of OLED displays was possible only after the development

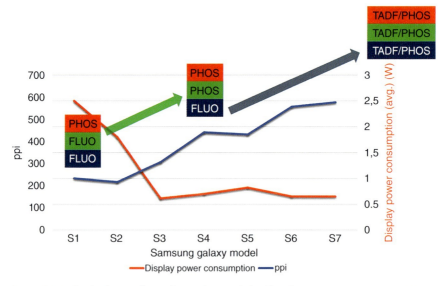

Figure 15.3 The display resolution (in pixel per inch (ppi)) and average power consumption (in W) of several generation of the Samsung Galaxy smartphone, which features OLED displays. Significant improvement was connected to the use of new material generations. In the future, all three colors may be generated by phosphorescent or TADF emitters. *Source*: Figure courtesy of CYNORA.

of phosphorescent red emitters with suitable stability and efficiency [6]. Consequently, the first generation, the S1, featured fluorescent materials for green and blue pixels, while red pixels contained phosphorescent materials. After the introduction of green phosphorescent materials, the average power consumption significantly dropped from 2.5 to well below 1 W. On the other hand, the display's resolution was doubled to roughly 400 pixels per inch. Using the more efficient phosphorescent green materials allowed the display manufacturers to significantly decrease the size of green pixels, which allowed for an increase of the resolution and an improved customer experience.

15.1.2 Processing Aspects

The large majority of the OLED products available on the market are based on small molecules deposited via physical vapor deposition. The current deposition technique of choice for industrial OLED production is vacuum thermal evaporation (VTE). The organic molecules or metals are heated in a vacuum chamber with pressures below 5×10^{-5} mbar; they evaporate and then condense on a cooled substrate. The temperature needed for sublimation is most typically up to 300 °C for organics and between 500 and 1000 °C for metals. The patterning of the layers is generally done with a shadow mask.

These techniques allow for the fabrication of multilayer OLEDs with a good control of the layer thicknesses [3]. It is also convenient to employ vacuum equipment that is already used in the semiconductor industry. Moreover, the purity of the films reflects the purity of the materials; therefore pure and stable materials lead to long-lived OLEDs. However, it is difficult to obtain homogenous layers on very large substrates, and it is very "material consuming" since the vast majority of the material end up on the walls of the vacuum chamber rather than on the substrate itself (see Figure 15.4).

The large waste of material during vacuum deposition prompted academic and industrial researchers to develop solution processing methodologies to allow for OLEDs fabricated by printing and coating techniques. The advantages of this concept would be evident: They could be performed at atmospheric pressure, are better suited to large-area devices, and are much less material consuming. With inkjet printing, for example, one can deposit a droplet exactly where it is needed instead of depositing material everywhere and using a shadow mask. Both coating and printing are suitable for large-scale fabrication and incorporation into roll-to-roll fabrication lines [14]. Several challenges need to be addressed prior to reaching market readiness for this approach:

1. *Blending of the functional layers during or after deposition* [15, 16], a phenomenon also called interlayer diffusion, may occur when a solvent used for the deposition of further layers is able to dissolve already deposited layers. This issue can be prevented with orthogonal solvent strategies, i.e. using a solvent that is not able to dissolve previously cast layers. This strategy requires tuning of the material properties to achieve sufficient solubility in the desired solvent while preventing solvation in other solvents. Often, the hole-transport layer is processed by this approach.

Figure 15.4 Photo of an evaporation chamber that has been used to deposit silver electrodes after several hours of operation. The silver is peeling off the chambers' walls. Only a small fraction of the used materials is actually deposited on the substrate, while large amounts are lost. Because of the high level of purity that is necessary to manufacture long-living OLED devices, reusing this excess material is not straightforward. Source: Ref. [2]. Reproduced with permission of Royal Society of Chemistry.

2. *Morphology and process parameter impact on device properties*, which is more challenging for solution processing [17, 18]. Morphological defects like crystalline grains in functional layers may act as charge traps [19], while aggregation causes emission quenching, which decreases the efficiency and may also affect the lifetime of the devices [11]. Such morphology defects can be avoided by using materials with a low crystallization tendency, which corresponds to a low lattice energy and a good solubility, or by immobilizing the relevant molecules, e.g. by attaching them to a polymeric backbone [20–27].
3. *Potential necessity of crosslinking*. Both morphological problems and troubles related to blending of layers during processing may be addressed with crosslinking [16, 28–35] of each layer after deposition, which prevents the dissolving of the layer during or after processing and stabilizes it against diffusion during device operation. Known approaches cover photochemical [36, 37], chemical [31, 35], thermal [32], and electrochemical techniques [33, 34]. Among various thermal reactions, simple click-chemistry-based processes are of particular interest [11, 21, 38–40]. A covalently anchored emitter prevents interlayer diffusion and promises to enhance the device lifetime drastically [30, 31].

From an industrial perspective, it is unclear if and when solution processing methodologies will be mature enough to be considered for the industrial manufacturing of OLEDs in general. Comparing solution processing with vacuum processing, in most cases both efficiency and device lifetime are lower, while the turn-on voltage is higher for the former technique, even when identical materials are used.

15.1.3 Sustainability Aspects

Looking at potential supply bottlenecks as a result of the incorporation of rare metals, we concluded in an earlier study [2] that the growth of the market share of OLED technology for flat-screen displays will not be hindered due to the low amount of material incorporated per device and the comparably low sales numbers. However, we pointed out that future industrial mass market applications such as lighting panels and especially the hypothetical use as smart labels could not be possible without finding alternatives for elements such as iridium, which is being used in PHOLED displays now (see Figure 15.3). The replacement of phosphorescent materials with TADF materials could solve this problem.

However, sustainability goes beyond avoiding supply bottlenecks: The key idea of the concept of "elemental sustainability" is to ensure that all extant elements in the periodic table are kept available for future generations [41]. Apart from iridium, the rare element indium, which is being used in indium tin oxide (ITO), could also be critical here.

15.1.3.1 Availability Issues

In 2009, 29% of computer displays, 17% of TVs, and 8% of mobile devices ready for end-of-life management were collected for recycling in the United States, according to a study of the US Environmental Protection Agency. The majority of devices in all of the three product categories were disposed, primarily in landfills [42]. Currently, thin film coatings for displays in smartphones and tablets account for more than half (80%) of the indium consumption [43]. The biggest competing industry for indium is highly efficient thin film solar cell photovoltaics. Indium reserves and resources are estimated to be 49 000 t [44], which makes indium one of the most scarce elements currently employed in OLEDs. Estimations on reserves for platinum group metals (PGMs) indicate that there are still between 91 000 and 338 000 t remaining [45].

Models for the future development of the annual production of PGMs predict that the annual production could grow until the second half of this century [45]. After that point we might start to run out of PGMs depending on the further exploration of new resources and the establishment of more extensive recycling methods. A striking approach that could open up new resources for PGMs is space mining. The abundance of iridium in certain asteroids is about five orders of magnitude higher than in the Earth's crust [46]. If the exploitation of these deposits was realized, a continuous supply of PGMs would be secured in the future [47]. Nevertheless, from a conservative point of view, scientists and engineers should not rely heavily on this possibility. In the last years, the annual production of iridium was about 3 t, according to the US Geological Survey [48]. Besides optoelectronics, PGMs are required for a variety of other industrial applications. For platinum the largest share is required for catalytic converters, whereas iridium is mainly used in alloys. Both iridium and platinum are used as heterogeneous catalysts in the chemical industry [49].

A former study [50] predicts for which raw materials the annual production would have to be increased significantly to cover the demand induced by emerging technologies in 2030. For indium, the increasing demand will mainly be caused by new display technologies, thin layer photovoltaics, and white LEDs, whereas platinum might play a key role for the fabrication of fuel cells for electric cars. These new fields of application for indium and platinum only add to the aforementioned demand by already existing industries, which is most likely to persist. Therefore, replacing scarce elements as indium, iridium, and platinum with more abundant alternatives, whenever possible, is most sustainable.

15.1.3.2 Recycling Considerations

It should be considered that currently, only a small amount of optoelectronic devices is collected for recycling. Incorporating those metals in OLEDs, where they are introduced in thin layers, with thicknesses in the order of several nanometers, also basically means diluting them to a degree where recycling is not feasible anymore. Even though there is not much reliable data available regarding recycling of rare metals from electronic devices [51], some conclusions can be drawn: As of today, the collection rates for smartphone and similar devices are not satisfactory [52]. Recycling is performed by pyrometallurgical routes, preferably after the battery has been removed [52]. While there are no accurate numbers available for iridium and platinum in OLEDs, estimations exist for LCD displays [52]. According to those, 700 mg m^{-2} of indium is included, which cannot be recycled with today's technology [53]. This seems to be related to the rather small concentration of rare metal per device, since other metals that are present in higher concentrations, such as gold, copper, silver, and palladium, can be recycled in satisfactory yields [53].

For metals that are used in the emitting layer of an OLED device, the situation is actually worse: The thickness is often between 30 and 100 nm, and the metal–organic compounds contain other elements (e.g. Irppy$_3$ contains 30 wt% of iridium). Often, the emitting molecules are diluted in host materials, further bringing down the concentration.

15.1.4 Realization of Efficient and Stable Blue OLEDs

15.1.4.1 The Blue Gap

Apart from sustainability concerns, there is also a technical issue that favors TADF over PHOS materials: the **blue gap**, which denotes the current trade-off between the efficiency and the stability of blue emitters. Figure 15.3 indicates that as of today, blue pixels of commercial products contain FLUO materials, even though they are less efficient. The reason for this is that – even after almost 20 years of industrial and academic research on the field of PHOS emitters – science failed to produce a blue PHOS material that combines efficiency, stability, and a proper color point. If a blue emitter were to show high efficiency and long lifetime, as TADF is promising to do, this would lead to great

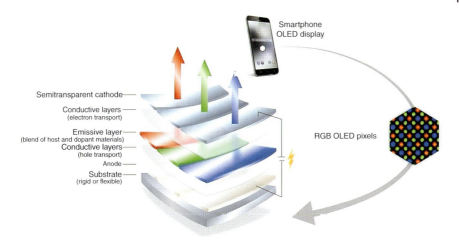

Figure 15.5 Simplified structure of an OLED as it is used in a smartphone display. The light of the blue, red, and green pixels is used directly to make up the displayed image. Currently, the surface area of blue pixels is in the order of 52% of the total display. *Source*: Picture courtesy of CYNORA.

opportunities to create better products with even lower power consumption and better resolution, something that is of great commercial interest at present.

Figure 15.5 indicates the impact of the blue gap in the current blue fluorescent emitters. All OLED displays currently require relatively large blue pixel areas to reach enough brightness in the display. In smartphones with red, green, and blue pixels, the blue pixel makes up to 52% of the total area. Having TADF or PHOS pixels with a much higher efficiency would enable the display manufacturers to make smaller blue pixels to yield the same amount of light, which would pave the way again to increase the display resolution. Additionally, customers would benefit from a longer battery life of their mobile device, which is closely connected to the power consumption of the display.

OLED TVs have a more complex stack architecture, essentially making white light from red-, green-, and blue-emitting layers and then using color filters to separate the colors again for the different pixels. Having a more efficient blue, which again makes up about 50% of the display area, would effectively reduce the power consumption of the TV by switching to a less complex stack design, which could potentially reduce manufacturing costs.

15.1.4.2 Key Performance Indicators

Knowing that, for the display quality, the color coordinates measured in the Commission Internationale de l'Éclairage (CIE) system [54] are most relevant and that aspects like doping concentration [55], outcoupling [56], microcavities [57], and emitter orientation [58] are affecting the color of a device, it is not feasible to use aspects like the peak emission, spectral shape, or spectral width to compare different devices.

As a status indicator, the so-called **blue index** is an often-used tool. Keeping in mind that a slight offset of the emission color may be corrected with color filters in a display, the CIE_y coordinate of the device's electroluminescence (EL) spectrum is divided by the efficiency and measured in candela per ampere ($cd\,A^{-1}$). With state-of-the-art blue fluorescent materials [59–62], it is possible to realize efficiency values around $10\,cd\,A^{-1}$ at CIE_y values around 0.1, yielding a blue index of 100. With phosphorescent or TADF emitters, this value can reach up to 300–400.

Another aspect to mind is the efficiency roll-off [63], the loss of efficiency at higher operating voltage or brightness, which should not be significant in an OLED display in the relevant brightness (for example, between 10 and $3000\,cd\,m^{-2}$ brightness regime). If the roll-off of one pixel color is stronger (or, as a matter of fact, weaker) than for the others, issues can arise when operating the display at different brightness levels.

15.2 Mastering Blue OLEDs with TADF Technology

15.2.1 Current Status of Blue TADF Technology: Academia

The general approach for emitter design is to combine electron-rich (donor) and electron-poor (acceptor) moieties in a molecule, so that the HOMO and LUMO orbitals are separated, thus decreasing the singlet–triplet gap (ΔE_{ST}). Typical examples for donor moieties are heterocyclic structures such as carbazole or dimethylacridine (DMAC) derivatives, while acceptor moieties vary from benzonitriles to triazines or sulfones. In this subsection, we collected a list with blue TADF materials published between 2012 and the beginning of 2017. Table 15.2 contains all relevant references together with information concerning the responsible principal investigator (PI), the used donor and acceptor types, the best OLED performance, and references to the molecular structure of key emitters, which are collected in Figures 15.6–15.10 (for a discussion of the different molecular structures, kindly refer to one of the other chapters in this book).

From the collected data, two main conclusions can be derived. First, academic research proves that it should be possible to achieve very high performance with blue TADF technology. Even at very low CIE_y coordinates between 0.1 and 0.2, efficiencies in the order of 20% EQE were realized.

Second, there is still a lot of room for improvement. In general, only few academic researchers in the TADF field deal with the degradation, and stability parameters are hardly ever reported. One noteworthy exception is a study by the group of Lian Duan, which reports promising performance values, displayed in Figure 15.11 [77]. These results are in themselves excellent for the device lifetime, compared with what academic researchers published, and among the best in the field looking at color point and efficiency. However, much like most references displayed in Table 15.2, they reveal that the overall stability is still too low for industrial application and that the efficiency roll-off is strong, reducing the excellent initial efficiency by more than a factor of two at a practical luminance of $500\,cd\,m^{-2}$.

Looking at the molecular building blocks, it is apparent that most researchers focus on a narrow set of donor unit that contains triarylamine cores, which are

Table 15.2 Status of sky- to deep-blue TADF technology as published in the scientific literature.

PI	Donor	Acceptor	Performance	Examples
Adachi and coworkers [64]	Arylamine, carbazole	Diphenyl sulfone	5% EQE max, *no CIE*	Group 1
Adachi and coworkers [65]	Carbazole derivatives	Benzophenone	14.3% EQE, CIE (0.17, 0.27)	Group 2
Adachi and coworkers [65]	DMAC	Diphenyl sulfone	20% EQE, CIE (0.16, 0.20)	Group 3
Adachi and coworkers [67]	Carbazole derivatives	Triazine	15% EQE, CIE (0.18, 0.28)	Group 4
Lee and coworkers [68]	Carbazole	Triazine	15% EQE, CIE (0.16, 0.17)	Group 5
Adachi and coworkers [69]	Carbazole derivatives	Triarylborane	14% EQE, CIE (0.17, 0.30)	Group 6
Kim and coworkers [70]	Azasiline	Triazine	22% EQE, CIE (0.15, 0.20)	Group 7
Lee and Lee [71]	DMAC	Thioxanthene dioxide	20% EQE, CIE (0.15, 0.13)	Group 8
Kido and coworkers [72]	Carbazole derivatives	Bis(phenylsulfonyl)benzene	11.7% EQE, CIE (0.18, 0.19)	Group 9
Kitamoto et al. [73]	Phenoxazine	Phenoxaborin	15% EQE, *no CIE*	Group 10
Zhang and coworkers [74]	Carbazole	Phenyl-cyanopyridine	9% EQE, CIE (0.18, 0.26)	Group 11
Adachi and coworkers [75]	Acridan derivatives	Phenoxaborin	20.1% EQE, CIE (0.14, 0.16)	Group 12
Duan and coworkers [76]	Carbazole derivatives	Benzonitrile	21.5% EQE, CIE (0.16, 0.26)	Group 13
Duan and coworkers [77]	Carbazole derivatives	Benzonitrile	21.2% EQE, CIE (0.21, 0.41)	Group 14
Yasuda and coworkers [78]	Diphenylacridan	Pyrimidine	12% EQE, CIE (0.16, 0.24)	Group 15
Adachi and coworkers [79]	Acridan	Phenazaborin	18% EQE, CIE (0.15, 0.23)	Group 16
Hatakeyama et al. [80]	Near-range charge transfer		20% EQE, CIE (0.12, 0.13)	Group 17
Adachi and coworkers [81]	Diphenylacridan	Benzonitrile	16% EQE, CIE (0.16, 0.16)	Group 18
Lee and coworkers [82]	Carbazole derivatives	Triazine	15.7% EQE, CIE (0.17, 0.24)	Group 19
Yasuda and coworkers [83]	Acridan derivatives	Pyrimidine	20% EQE, CIE (0.16, 0.23)	Group 20
Adachi and coworkers [84]	Carbazole derivatives	Triazine	18% EQE, CIE (0.15, 0.10)	Group 21
Lee and coworkers [85]	Carbazole	Benzonitrile	14% EQE, CIE (0.14, 0.12)	Group 22
Xu and coworkers [86]	Phenoxazine	Phosphine oxide	15% EQE, CIE (0.17, 0.20)	Group 23
Kim and coworkers [87]	Azasiline	Diphenylsulfone, triazine	5% EQE, CIE (0.15, 0.09)	Group 24
Kido and coworkers [88]	DMAC	Pyrimidine	18% EQE, CIE (0.16, 0.15)	Group 25
Gong and coworkers [89]	Carbazole	Triazine	15% EQE, CIE (0.15, 0.20)	Group 26

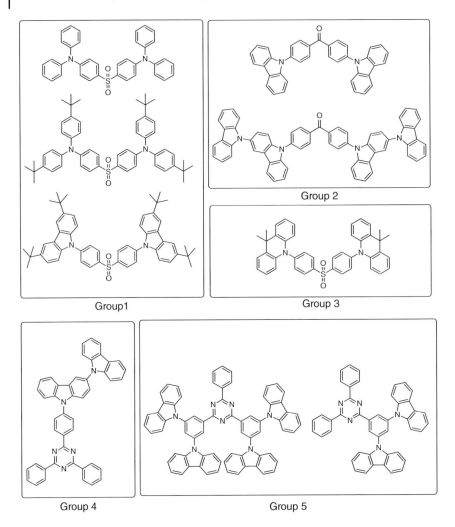

Figure 15.6 Selected structures of published blue TADF emitters (part 1).

often fused to five-membered carbazole-type donors or bridged to six-membered derivatives. The used bridges range from sp³-hybridized carbon (→ acridanes) to oxygen and sulfur (→ phenoxazine, phenothiazine) to silicon (→ azasiline). The acceptor units on the other hand cover more chemical space, ranging from aromatic systems substituted with electron-withdrawing groups to fused nitrogen-containing heterocycles. Often, other heteroatoms such as sulfur, phosphorous, and boron are found here.

15.2.2 Current Status of Blue TADF Technology: Industry

During the last 2 years, we have witnessed breathtaking improvement of TADF materials, which now can surpass conventional PHOS materials in terms of color point and efficiency, with similar stability. A main driving force behind this

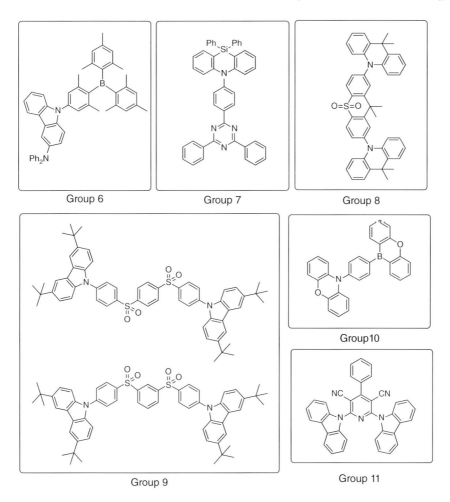

Figure 15.7 Selected structures of published blue TADF emitters (part 2).

progress was the development of suitable screening algorithms for molecular design, which make use of density functional theory (DFT) calculations. These computational tools drastically increased the efficiency of material development by allowing to only synthesize highly efficient TADF materials, thus leading to very short material development cycles with a very steep learning curve per cycle.

Results based on this development approach were recently published by the team at CYNORA [90]. We reported several blue materials, among them a blue emitter reaching 14% EQE at 500 cd m^{-2} and a lifetime LT$_{80}$ of 420 h at 500 cd m^{-2} starting luminance [90]. This material featured an emission maximum below 480 nm.

In here, we briefly discuss the properties of two other recently developed materials, TADF 1 and TADF 2, which are shown in Figures 15.12 and 15.13, as well as Table 15.3.

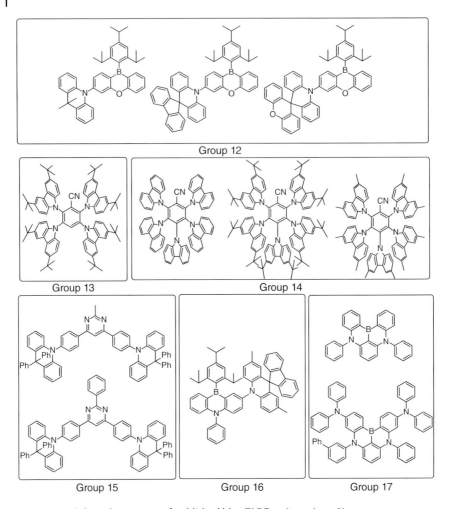

Figure 15.8 Selected structures of published blue TADF emitters (part 3).

The electrooptic JVL characteristics of this material in a simple device architecture, using the literature-known host mCBP [92] (short for 4,4′-bis (3-methylcarbazol-9-yl)-2,2′-biphenyl), show a voltage of about 4.2 V at 500 cd m^{-2}. The so-called blue index, which can be calculated by dividing the current efficiency in cd A^{-1} and the CIE$_y$ color coordinate, approaches 80 at 500 cd m^{-2} in a bottom-emitting device.

Ir(dmp)$_3$, short for iridium (III) tris[3-(2,6-dimethylphenyl)-7-methylimidazo [1,2-f] phenanthridine], is currently representing one of the most stable blue iridium emitters with a rather blue color point and decent efficiency. In a recent study by Forrest and coworkers [91], the basic performance was reported in the same host used to obtain the data shown in Table 15.3. The LT$_{80}$ at 500 cd m^{-2} as well as the blue index of Ir(dmp)$_3$ was estimated from the values given in the publication. Even using a simple, nonoptimized screening architecture and normal R&D-grade purity, the stability LT$_{80}$ at this starting luminance is in the order of

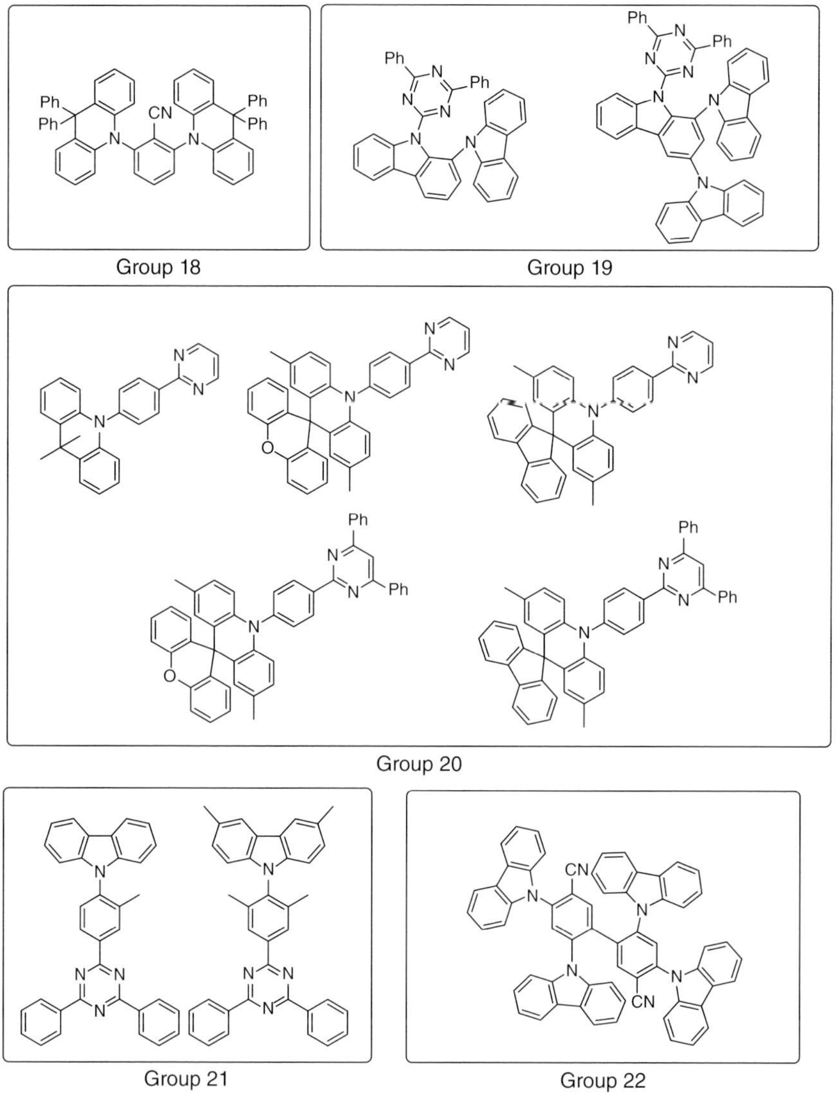

Figure 15.9 Selected structures of published blue TADF emitters (part 4).

100 h, which is in the range of the best phosphorescent materials when also considering the better color and the higher efficiency [91] (Table 15.3). Considering that phosphorescent materials are under investigation since 1997 [6], while TADF has just been studied since 2011 [8], these results demonstrate the breathtaking development curve for TADF.

Moreover, deep-blue materials with high performance have as well been developed. Emitter TADF 2, for example, displayed performances up to 20% in OLEDs with a CIE_y value of 0.17 (Figure 15.14). The efficiency of this emitter reached similarly high performance as the best materials displayed in Table 15.2.

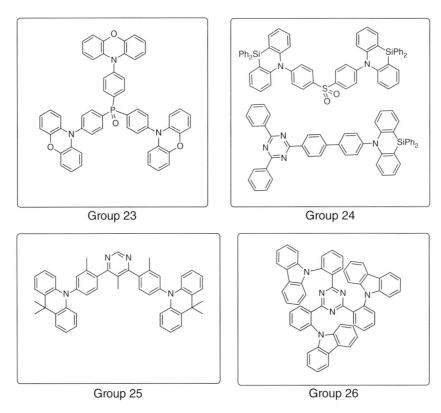

Figure 15.10 Selected structures of published blue TADF emitters (part 5).

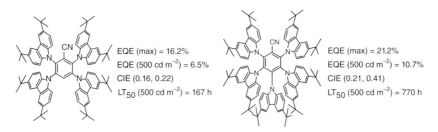

Figure 15.11 Performance of 4TCzBN and 5TCzBN. The lifetime was reported at a starting luminance of 500 cd m^{-2} [77].

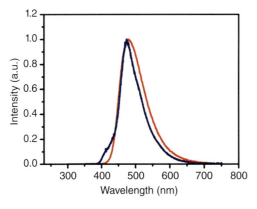

Figure 15.12 Comparison between photoluminescence (PL) (red curve) of a mid-blue TADF emitter and its electroluminescence (EL) (blue curve) performance in a bottom-emitting OLED.

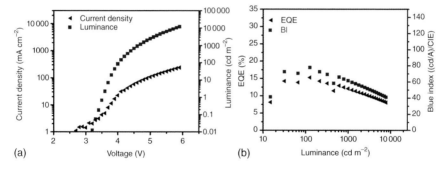

Figure 15.13 Performance of a recent deep-blue TADF 1 OLED material. (a) Current-density–voltage characteristics of the device. (b) Dependence of the EQE with luminance. Due to the used host system and stack architecture, a high maximum luminance and low turn-on voltage are achieved. At a practical luminance of 1000 nits, the EQE is above 12%. However, the device still shows a significant efficiency roll-off – at 150 nits, a maximum EQE of 15% is reached. This shows that even with the nonoptimized emitter TADF 1, better performance can be expected in an optimized stack architecture.

Table 15.3 Device performance of the CYNORA material (TADF 1) versus Ir(dmp)$_3$, the currently most stable and efficient blue phosphorescent emitter [91], in 13 wt% mCBP (PHOS).

	CIE (1000 nits)	EQE (1000 nits)	Blue index[a]*	LT$_{80}$ (500 nits)
TADF 1	(0.17, 0.27)	12%	78	94
PHOS	(0.16, 0.31)	8%	c. 53	c. 100

a) Blue index: calculated by dividing the efficiency in cd A^{-1} by the CIE$_y$ value.

15.3 An Alternative Approach: TADF Emitters as (Co) Hosts

15.3.1 General Remarks

Forrest proposed three main approaches to combine both device lifetime and efficiency in blue OLEDs [93]: (i) reducing the excited state lifetime, (ii) reducing the exciton density, and (iii) reducing the effectiveness of bimolecular recombination processes that can promote the excitation to very high energies.

Apart from that, general rules concerning emitter design do apply, too. For example, it is crucial to use robust, stable compounds that are not prone to decomposition or morphological changes such as crystallization, which need to be provided in excellent purity [94].

To understand various aspects of the use of TADF as hosts, it is critical to understand the fundamental differences between Förster-type and Dexter-type energy transfer, the two key mechanisms for the energy transfer between two species (Figure 15.15).

Because of the exchange of electrons during the so-called Dexter transfer [95], it is mainly important that the distance between two species that are involved in

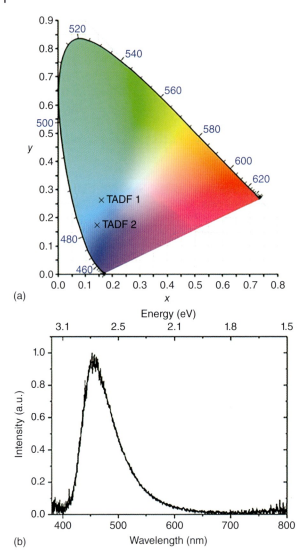

Figure 15.14 The color of OLEDs is measured in the CIE color coordinate system. The blue sector is defined by CIE_y values roughly below 0.30 (a). To satisfy the color expectations of the display industry, CIE_y values below 0.20, ideally even around CIE_y 0.10, will be needed. With current TADF materials, CIE_y values in the order of 0.15 can be reached (b). The shown EL spectrum corresponds to CIE_y 0.17, which is associated with a peak efficiency of 20% EQE.

a transfer step is in the order of 0.5–2 nm – for both singlet and triplet excitons. This short-range transfer is the only way for triplet excitons to migrate. During the transfer, the total spin of the system is conserved. It relies on orbital overlap. The rate for this process can be expressed with

$$k_{\text{Dexter}} \sim J \, \exp(-2r/L) \qquad (15.1)$$

where r is the distance between the species, L the sum of the van der Waals radii of the species, and J the spectral overlap, defined by

$$J = \int \text{emission}\,(\lambda)_{\text{donor}} \, \text{absorption}\,(\lambda)_{\text{acceptor}} \, \lambda^4 d\lambda \qquad (15.2)$$

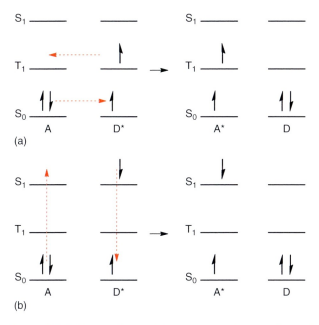

Figure 15.15 Elemental step of the energy transfer according to the mechanisms described by Dexter (a) [95] and Förster (b) [96]. In the Dexter process, electrons are directly exchanged, and the spin of the whole system is conserved, whereas the spin of each species (D and A) is conserved in the Förster process. Because of this, only the Dexter process allows triplet excitons to migrate.

The Förster transfer [96] does not rely on a direct exchange of electrons, but on dipole–dipole interaction. The spin of each participating species remains, which allows only singlet excitons to be transferred with this process. The rate for this process can be expressed with

$$k_{\text{Förster}} \sim \frac{J\phi}{k_r r^6} \tag{15.3}$$

where, in addition to the factors defined above, the radiative decay rate of the donating species, k_r, and the photoluminescence quantum yield (PLQY), ϕ, are of relevance.

The rate is not an exponential function of r, which allows for the Förster process also to occur when $r = 20$ nm.

15.3.2 First Attempts of Using TADF as Hosts

The fact that energy transfer from TADF molecules to fluorescent emitters is possible was established early: In 2008, Li and coworkers demonstrated that a green-emitting Cu(I)-TADF emitter is strongly quenched when introducing a neighboring layer with a low triplet energy. Thus, introducing a DPVBi layer between the TADF-containing emission layer (EML) and the ETL reduced the efficiency of the device from 16.7 to 5.7 lm W^{-1} [97]. In 2009, it was shown that the triplet energy of the ETL can affect the efficiency of a TADF OLED due to quenching of TADF triplet excitons by low-lying triplets of fluorescent molecules [98]. Because of the conceptual analogies between TADF-type emitters and ambipolar host materials [99–102], it is fair to speculate that the first uses of TADF molecules for host applications were realized unbeknownst to the researchers, which was already discussed elsewhere [10].

15.3.3 Discussion of Various Concepts

Often, the term *hyperfluorescence* is associated with the use of TADF materials as hosts in OLEDs [65, 103]. This is problematic, since *hyperfluorescence* was already established several decades ago [104] in medical imaging to describe an increased FLUO during fluorescein angiography, a technique used by ophthalmologists to study the retina.

Also, there are different ways of implementing TADF materials as hosts, which are often described with this term in a confusing manner. Because of this, we will treat the various concepts in separate sections in here.

15.3.3.1 TADF as Host for Other TADF Emitters

Kim and coworkers recently pointed out that the availability of suitable host materials for TADF is somewhat limited, because of the required high triplet energy, in combination with properly adjusted energy levels [105]. Duan and coworkers demonstrated an interesting potential solution to this: By using the same donor and acceptor motifs in emitter and host molecules, devices with high-efficiency and low driving voltage were realized (Figure 15.16) [76]. It was

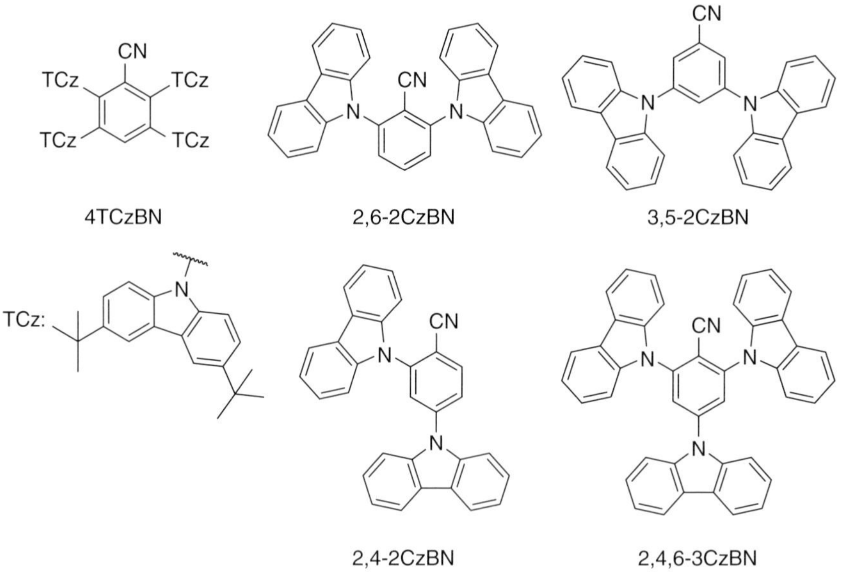

Figure 15.16 Materials used in a study by Duan [76]. 4TCzBN was used as the emitter, while 2,6-2CzBN, 3,5-2CzBN, 2,4-2CzBN, and 2,4,6-3CzBN were evaluated as host materials. All hosts showed a high triplet energy and a good alignment with the HOMO and LUMO energy of 4TCzBN. The best host based on necessary driving voltage was 3,5-2CzBN, which was attributed to a high PLQY indicated by PL studies, good alignment of frontier orbital energies, and a good ambipolar charge carrier mobility.

pointed out that the charge carrier mobility in the EML also needs to be high to bring down the driving voltage.

15.3.3.2 TADF as Host for Fluorescent Materials

Lee and coworkers pointed out a valid issue of this approach: the necessity of a three-component EML [106]. While the studies discussed in the previous section suggest that it should be possible to technically realize three-component EMLs where the relative amount of the components differs over three orders of magnitude, transferring them into mass production will be challenging and require optimization of the processing technology. To avoid these costly and elaborate expenses would require two-component setups of TADF hosts and fluorescent dyes.

In a pioneer work, Duan demonstrated the combination of PIC-TRZ and DIC-TRZ, which both emit around 500 nm, with DDAF, an orange-emitting fluorescent dye that emits at 580 nm (Figure 15.17) [107, 108]. Using Alq$_3$ as a conventional fluorescent host, DDAF gave a peak efficiency in the order of 3% EQE, while the TADF-type hosts gave a peak efficiency of 11% EQE. In both cases, doping concentrations of 1 wt% were used. In this work, it is proposed to use fluorescent molecules with sterically demanding substituents in future approaches in order to reduce charge trapping on the dopant.

The suitability of TADF materials for host applications seems to be connected to the absence of concentration quenching. While many emitting materials get less efficient at high doping concentrations [109], other materials may actually require high concentrations to emit efficiently (so-called aggregation-induced emission (AIE)) [110, 111].

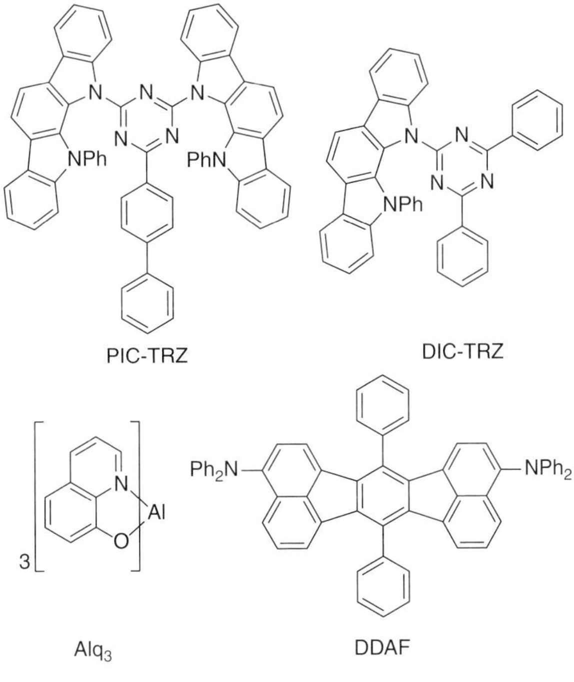

Figure 15.17 Materials used in a study by Duan [107, 108]. PIC-TRZ and DIC-TRZ served as TADF-type host materials, while Alq$_3$ is a fluorescent reference host. The dopant was DDAF, an orange-emitting fluorescent emitter.

An ideal TADF-type host material would actually not show any photophysical changes when modifying the concentration. Apart from Cu(I)-type TADF emitters that are usually devoid of concentration quenching even in neat films or powder [11, 39, 112–114], this issue has been investigated for a great number of metal-free TADF emitters as well [64, 115, 116].

Interestingly, the presence of DMAC (Figure 15.18) as a donor seems to reduce concentration quenching. While it has been speculated that steric hindrance (resulting in a reduced probability for Dexter transfer processes) could be responsible for this beneficial behavior [117]; this is yet to be investigated systematically.

Just recently, phenoxazine and phenothiazine (Figure 15.19) have been identified as alternatives for DMAC donors, yielding high efficiencies well over 15% EQE and even improved roll-off behavior in nondoped devices compared with conventional OLEDs with DPEPO as host materals [117].

15.3.3.3 TADF as Host for Phosphorescent Emitters

In addition, TADF-type hosts were used in combination with phosphorescent emitters [108]. The driving motivation is again the triplet energy: As stated earlier, phosphorescent emitters emit via a transition between the T_1 and S_0

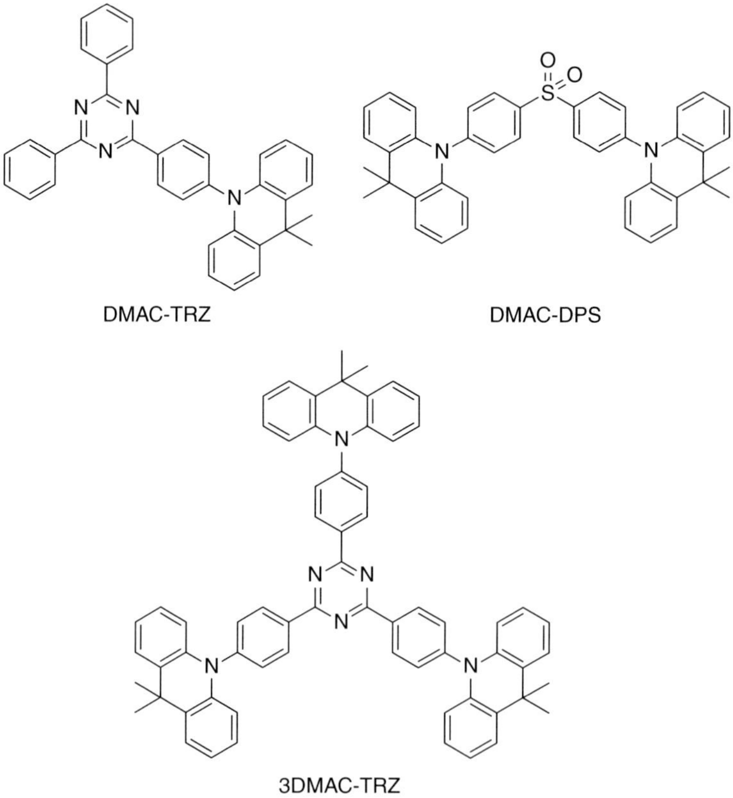

DMAC-TRZ

DMAC-DPS

3DMAC-TRZ

Figure 15.18 Numerous organic TADF emitters have been reported to not show concentration quenching [64, 115, 116]. Often this is connected with the presence of dimethylacridine units as donors.

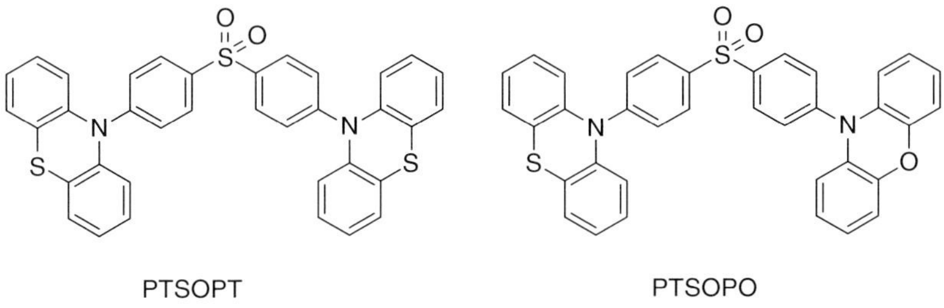

Figure 15.19 Lee et al. introduced PTSOPT and PTSOPO, two sky blue TADF emitters that were successfully used in efficient, nondoped OLED devices [117].

levels, and, again, the triplet energy of host materials needs to be larger than the triplet energy of the dopant. Especially when moving toward phosphorescent high-energy emission, TADF emitters are particularly attractive for such applications because of the small singlet–triplet splitting, ΔE_{S1-T1}.

A broader review on TADF and ambipolar host materials has been previously published [10]. Some examples are shown in Figure 15.20. D2ACN was published in 2008 in a study describing its use as a host material in red PHOLEDs [118]. After the realization of the beneficial properties of TADF materials as dopants in 2011 [8], the related molecule Spiro-CN was used in a TADF OLED [119].

Combining the orange emitter PO-01 in 10 wt% concentration with the TADF-type host BBIC-TRZ (Figure 15.21) gave excellent performance values with efficiencies in the order of 20% EQE, almost negligible roll-off up to brightness values of 2000 cd m^{-2} and a high operational stability of LT$_{50}$ = 1900 h at 1000 cd m^{-2} starting luminance under constant current conditions [108]. It is speculated that a potential reason for this surprisingly high stability is the possibility of energy transfer from the host to the dopant via both Förster and Dexter processes. Since TADF hosts are capable of transforming triplet into singlet excitons, a higher ratio of the favorable Förster-type transfer processes can occur, reducing the triplet exciton density. Interaction between triplet excitons and polarons – so-called triplet-polaron quenching (TPQ) – is assumed to be a major degradation pathway in phosphorescent OLEDs [93, 120, 121].

While it seems intriguing to fuse the properties of TADF-type emitters with the proven performance of phosphorescent materials, there are some drawbacks. For instance, part of the benefits of the metal-free TADF systems (sustainability,

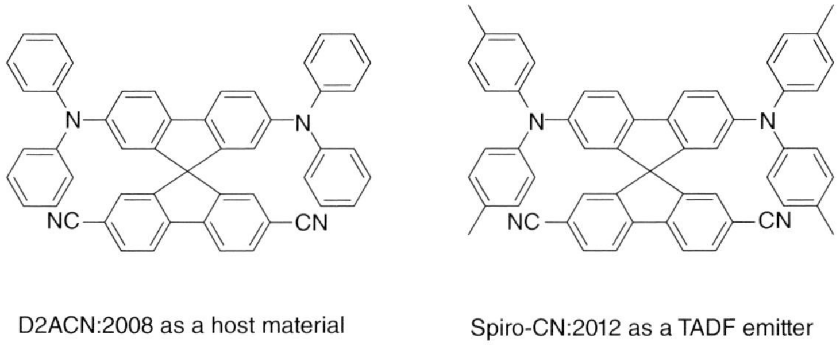

D2ACN:2008 as a host material

Spiro-CN:2012 as a TADF emitter

Figure 15.20 D2ACN is a greenish-yellow TADF-type host, which was used in combination with a red phosphorescent emitter (10% EQE$_{max}$) [118]. Spiro-CN, a derivative of this molecule, was used as a TADF emitter in combination with a fluorescent host (5% EQE$_{max}$) [119].

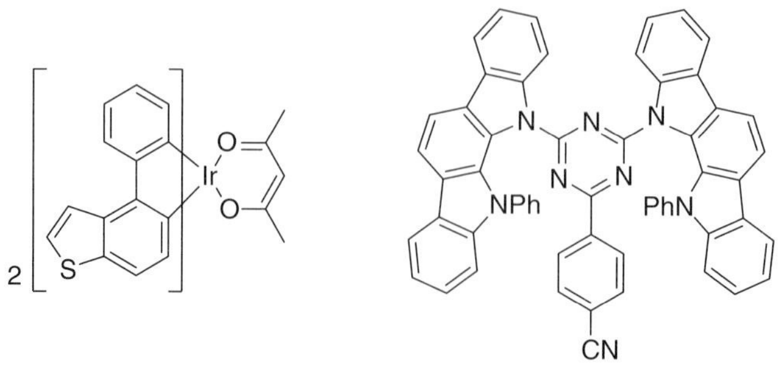

Figure 15.21 Structures of the phosphorescent emitter PO-01 and the TADF-type host BBIC-TRZ [108].

freedom of broad patents, potential cost issues) is lost when combining them with heavy-metal-containing systems. Most importantly, since it is not yet proven if the reason for the instability of blue PHOLEDs is tied to the phosphorescent dopant, the hosts, or both, using TADF-type hosts is no tool to solve any issues based on degradation of the phosphorescent dopants themselves.

15.4 Outlook: What to Expect from TADF Technology in the Future

In 2015, it was already established by CYNORA and others [9] that it is possible to realize a favorable deep-blue color point with TADF material technology. Recently, great progress toward more stable blue TADF devices was achieved, with emitters featuring emission below 480 nm, 14% external quantum efficiency, and lifetime values LT_{80} of 420 h (measured at 500 cd m^{-2} brightness) [90]. Within a relatively short period of R&D, TADF emitters have now reached a similar performance to PHOS emitters with a blue color point. The cornerstone of these successes was a fast translation of quantum chemical predictions from DFT calculations into material design and a continuous improvement of the underlying theory, which leads to improved materials in each learning cycle. Following this trend, blue TADF technology can reach market readiness in the very near future.

Further improvement can be expected through the realization of more sophisticated stack architectures: It is known that mCBP, a component used in the aforementioned early-stage devices, features several stability issues, potentially limiting the stability of these OLEDs [122]. Thus, the use of alternative, more stable hosts will lead to even longer lifetimes.

Nevertheless, it is also clear that the basic stability of both the materials and the stack architectures still needs to be improved fundamentally. Looking back at a very steep learning curve displayed by research-driven companies such as CYNORA, as well as the great academic progress [77, 105, 123], this necessary advancement seems achievable in a short amount of time. TADF will contribute to the next material-driven advancement of the OLED industry, making OLEDs ready for even more applications soon.

References

1 Tang, C.W. and VanSlyke, S.A. (1987). *Appl. Phys. Lett.* 51: 913.
2 Volz, D., Wallesch, M., Fléchon, C., Danz, M., Verma, A., Navarro, J.M., Zink, D.M., Bräse, S., and Baumann, T. (2015). *Green Chem.* 17: 1988–2011.
3 Shinar, J. (2003). *Organic Light-Emitting Devices: A Survey*. Springer-Verlag.
4 Yersin, H., Finkenzeller, W.J., Walter, M.J., Djurovich, P.I., Thompson, M.E., Tsuboyama, T., Akira, S.O., Ueno, K., Chi, Y., Chou, P.-T., Yang, X.-H., Jaiser, F., Neher, D., Xiang, H.-F., LAi, S.-W., Lai, P.T., Che, C.-M., Tanaka, I., Tokito, S., Van Dijken, A., Brunner, K., Börner, H., Langeveld, B.M.W., Mak, C.S.K., Nazeeruddin, M.K., Klein, C., Grätzel, M., Zuppiroli, L., Berner, D., Bian, Z.-Q., and Huang, C.-H. (2008). *Highly Efficient OLEDs with Phosphorescent Materials*, 1e. Wiley-VCH.
5 Van Slyke, S.A., Chen, C.H., and Tang, C.W. (1996). *Appl. Phys. Lett.* 69: 2160.
6 Baldo, M.A., O'Brian, Y., You, A., Shoustikov, A., Sibley, S., Thompson, M.E., and Forrest, S.R. (1998). *Nature* 395: 151–154.
7 Baldo, M.A., Lamansky, S., Burrows, P.E., Thompson, M.E., and Forrest, S.R. (1999). *Appl. Phys. Lett.* 75: 4.
8 Endo, A., Sato, K., Yoshimura, K., Kai, T., Kawada, A., Miyazaki, H., and Adachi, C. (2011). *Appl. Phys. Lett.* 98: 083302.
9 Bergmann, L., Zink, D.M., Bräse, S., Baumann, T., and Volz, D. (2016). *Top. Curr. Chem.* 374: 22.
10 Volz, D. (2016). *J. Photonics Energy* 6: 020901.
11 Yersin, H., Rausch, A.F., Czerwieniec, R., Hofbeck, T., and Fischer, T. (2011). *Coord. Chem. Rev.* 255: 2622–2652.
12 Yersin, H. and Monkiwius, U. (2008). DE 102008033563 A1.
13 McCarthy, J.T. (1995). *McCarthy's Desk Encyclopedia of Intellectual Property, Bureau of National Affairs*, 3e.
14 Krebs, F.C. (2009). *Sol. Energy Mater. Sol. Cells* 93: 394–412.
15 Smith, A.R.G., Lee, K.H., Nelson, A., James, M., Burn, P.L., and Gentle, I.R. (2012). *Adv. Mater.* 24: 822–826.
16 Zuniga, C.A., Barlow, S., and Marder, S.R. (2011). *Chem. Mater.* 23: 658–681.
17 Yoon, Y., Lee, H., Kim, T., Kim, K., Choi, S., Yoo, H.K., Friedman, B., and Lee, K. (2013). *Solid State Electron.* 79: 45–49.
18 Höfle, S., Pfaff, M., Do, H., Bernhard, C., Gerthsen, D., Lemmer, U., and Colsmann, A. (2014). *Org. Electron.* 15: 337–341.
19 Kaake, L.G., Barbara, P.F., and Zhu, X.-Y. (2010). *J. Phys. Chem. Lett.* 1: 628–635.
20 Meyers, A. and Weck, M. (2004). *Chem. Mater.* 16: 1183–1188.
21 Volz, D., Baumann, T., Flügge, H., Mydlak, M., Grab, T., Bächle, M., Barner-Kowollik, C., and Bräse, S. (2012). *J. Mater. Chem.* 22: 20786.
22 Moad, G., Chen, M., Häussler, M., Postma, A., Rizzardo, E., and Thang, S.H. (2011). *Polym. Chem.* 2: 492.
23 Furuta, P.T., Deng, L., Garon, S., Thompson, M.E., and Fréchet, J.M.J. (2004). *J. Am. Chem. Soc.* 126: 15388–15389.

24 Gao, H., Poulsen, D.A., Ma, B., Unruh, D.A., Zhao, X., Millstone, J.E., and Fréchet, J.M.J. (2010). *Nano Lett.* 10: 1440–1444.
25 Ma, B., Lauterwasser, F., Deng, L., Zonte, C.S., Kim, B.J., Fréchet, J.M.J., Borek, C., and Thompson, M.E. (2007). *Chem. Mater.* 19: 4827–4832.
26 Wang, X.-Y., Kimyonok, A., and Weck, M. (2006). *Chem. Commun. (Camb.)* 3933–3935.
27 Volz, D., Hirschbiel, A.F., Zink, D.M., Friedrichs, J., Nieger, M., Baumann, T., Bräse, S., and Barner-Kowollik, C. (2014). *J. Mater. Chem. C* 2: 1457.
28 Nuyken, O., Jungermann, S., Wiederhirn, V., Bacher, E., and Meerholz, K. (2006). *Monatsh. Chem. Chem. Mon.* 137: 811–824.
29 Nuyken, O., Bacher, E., Braig, T., Fáber, R., Mielke, F., Rojahn, M., Wiederhirn, V., Meerholz, K., and Müller, D. (2002). *Des. Monomers Polym.* 5: 195–210.
30 Gather, M.C., Köhnen, A., Falcou, A., Becker, H., and Meerholz, K. (2007). *Adv. Funct. Mater.* 17: 191–200.
31 Köhnen, A., Riegel, N., Kremer, J.H.-W.W.M., Lademann, H., Müller, D.C., and Meerholz, K. (2009). *Adv. Mater.* 21: 879–884.
32 Zhong, C., Liu, S., Huang, F., Wu, H., and Cao, Y. (2011). *Chem. Mater.* 23: 4870–4876.
33 Gu, C., Fei, T., Yao, L., Lv, Y., Lu, D., and Ma, Y. (2011). *Adv. Mater.* 23: 527–530.
34 Baba, A., Onishi, K., Knoll, W., and Advincula, R.C. (2004). *J. Phys. Chem. B* 108: 18949–18955.
35 Huyal, I.O., Koldemir, U., Ozel, T., Demir, H.V., and Tuncel, D. (2008). *J. Mater. Chem.* 18: 3568.
36 Liaptsis, G. and Meerholz, K. (2013). *Adv. Funct. Mater.* 23: 359–365.
37 Liaptsis, G., Hertel, D., and Meerholz, K. (2013). *Angew. Chem.* 52: 9563–9567.
38 Zink, D.M., Baumann, T., Nieger, M., and Bräse, S. (2011). *Eur. J. Org. Chem.* 2011: 1432–1437.
39 Czerwieniec, R., Yu, J., and Yersin, H. (2011). *Inorg. Chem.* 50: 8293–8301.
40 Zink, D.M., Grab, T., Baumann, T., Nieger, M., Barnes, E.C., Klopper, W., and Bräse, S. (2011). *Organometallics* 30: 3275–3283.
41 Anastas, P.T. and Warner, J.C. (1998). *Green Chemistry: Theory and Practice*. New York: Oxford University Press.
42 U. S. Environmental Protection Agency and U. S. Office of Resource Conservation and Recovery (2011). Electronics Waste Management in the United States Through 2009.
43 Peiró, L.T., Méndez, G.V., and Ayres, R.U. (2013). *Environ. Sci. Technol.* 47: 2939–2947.
44 Jackson, B. and Mikolajczak, C. (2011). *Availability of Indium and Gallium*. Indium Corporation.
45 Mudd, G.M. (2012). *Ore Geol. Rev.* 46: 106–117.
46 Henderson, P. and Henderson, G.M. (2009). *The Cambridge Handbook of Earth Science Data*. Cambridge University Press.
47 Libourel, G. and Corrigan, C.M. (2014). *Elements* 10: 11–17.

48 U.S. Geological Survey (2007). *Mineral Commodity Summaries.* U.S. Geological Survey.
49 Brenan, J.M. (2008). *Elements* 4: 227–232.
50 Angerer, G., Marscheider-Weidemann, F., Lüllmann, A., Erdmann, L., Scharp, M., Handke, V., and Marwede, M. (2009). *Rohstoffe für Zukunftstechnologien: Einfluss des branchenspezif ischen Rohstoffbedarfs in rohstoffintensiven Zukunftstechnologien auf die zukünftige Rohstoffnachfrage.* Institut für Zukunftsstudien und Technologiebewertung IZT gGmbH.
51 Chancerel, P., Rotter, V.S., Ueberschaar, M., Marwede, M., Nissen, N.F., and Lang, K.-D. (2013). *Waste Manage. Res.* 31: 3–16.
52 Binnemans, K., Jones, P.T., Blanpain, B., Van Gerven, T., Yang, Y., Walton, A., and Buchert, M. (2013). *J. Cleaner Prod.* 51: 1–22.
53 Manhart, A., Buchert, M., Bleher, D., and Pingel, D. (2012). Electronics Goes Green 2012+ (EGG), pp. 1–5.
54 Trezona, P.W. (2001). *Color Res. Appl.* 26: 67–75.
55 dos Santos, P.L., Ward, J.S., Bryce, M.R., and Monkman, A.P. (2016). *J. Phys. Chem. Lett.* 7: 3341–3346.
56 Bocksrocker, T., Eschenbaum, C., Preinfalk, J.B., Hoffmann, J., Asche-Tauscher, J., Maier-Flaig, F., and Lemmer, U. (2012). *Renewable Energy and the Environment Optics and Photonics Congress*, LM3A.4. Washington, DC: Optical Society of America.
57 Zhang, X.-W., Liu, L.-M., Li, J., Zhang, L., Jiang, X.-Y., Zhang, Z.-L., Wang, H., and Liu, X.-Y. (2011). *J. Disp. Technol.* 7: 515–519.
58 Komino, T., Tanaka, H., and Adachi, C. (2014). *Chem. Mater.* 26: 3665–3671.
59 Kuma, H. and Hosokawa, C. (2014). *Sci. Technol. Adv. Mater.* 15: 034201.
60 Zhu, M. and Yang, C. (2013). *Chem. Soc. Rev.* 42: 4963–4976.
61 Li, K., Liu, M., Yang, S., Chen, Y., He, Y., Murtaza, I., Goto, O., Shen, C., Meng, H., and He, G. (2017). *Dyes Pigm.* 139: 747–755.
62 Li, Y., Liu, J.-Y., Zhao, Y.-D., and Cao, Y.-C. (2017). *Mater. Today* 20 (5): 258–266.
63 Murawski, C., Leo, K., and Gather, M.C. (2013). *Adv. Mater.* 25: 6801–6827.
64 Zhang, Q., Li, J., Shizu, K., Huang, S., Hirata, S., Miyazaki, H., and Adachi, C. (2012). *J. Am. Chem. Soc.* 134: 14706–14709.
65 Lee, S.Y., Yasuda, T., Yang, Y.S., Zhang, Q., and Adachi, C. (2014). *Angew. Chem. Int. Ed.* 53: 6402–6406.
66 Reineke, S. (2014). *Nat. Photonics* 8: 269–270.Reference [66] is not cited in the text. Kindly check and provide the text citation or else delete the reference from the list.
67 Hirata, S., Sakai, Y., Masui, K., Tanaka, H., Lee, S.Y., Nomura, H., Nakamura, N., Yasumatsu, M., Nakanotani, H., Zhang, Q., Shizu, K., Miyazaki, H., and Adachi, C. (2014). *Nat. Mater.* 14: 1–37.
68 Kim, M., Jeon, S.K., Hwang, S., and Lee, J.Y. (2015). *Adv. Mater.* 27: 1–6.
69 Suzuki, K., Kubo, S., Shizu, K., Fukushima, T., Wakamiya, A., Murata, Y., Adachi, C., and Kaji, H. (2015). *Angew. Chem. Int. Ed.* 54: 15231–15235.
70 Sun, J.W., Baek, J.Y., Kim, K.-H., Moon, C.-K., Lee, J.-H., Kwon, S.-K., Kim, Y.-H., and Kim, J.-J. (2015). *Chem. Mater.* 27: 6675–6681.

71 Lee, I. and Lee, J.Y. (2016). *Org. Electron.* 29: 160–164.
72 Liu, M., Seino, Y., Chen, D., Inomata, S., Su, S., Sasabe, H., and Kido, J. (2015). *Chem. Commun.* 51: 16353–16356.
73 Kitamoto, Y., Namikawa, T., Ikemizu, D., Miyata, Y., Suzuki, T., Kita, H., Sato, T., and Oi, S. (2015). *J. Mater. Chem. C* 3: 9122–9130.
74 Liu, W., Zheng, C.-J., Wang, K., Chen, Z., Chen, D.-Y., Li, F., Ou, X.-M., Dong, Y.-P., and Zhang, X.-H. (2015). *ACS Appl. Mater. Interfaces* 7: 18930–18936.
75 Numata, M., Yasuda, T., and Adachi, C. (2015). *Chem. Commun.* 51: 9443–9446.
76 Zhang, D., Cai, M., Bin, Z., Zhang, Y., Zhang, D., and Duan, L. (2016). *Chem. Sci.* 7: 3355–3363.
77 Zhang, D.D., Cai, M., Zhang, Y., Zhang, D.D., and Duan, L. (2016). *Mater. Horiz.* 3: 145–151.
78 Park, I.S., Lee, J., and Yasuda, T. (2016). *J. Mater. Chem. C* 4: 7911–7916.
79 Park, I.S., Numata, M., Adachi, C., and Yasuda, T. (2016). *Bull. Chem. Soc. Jpn.* 89: 375–377.
80 Hatakeyama, T., Shiren, K., Nakajima, K., Nomura, S., Nakatsuka, S., Kinoshita, K., Ni, J., Ono, Y., and Ikuta, T. (2016). *Adv. Mater.* 28: 2777–2781.
81 Noda, H., Kabe, R., and Adachi, C. (2016). *Chem. Lett.* 45: 1463–1466. doi: 10.1246/cl.160814.
82 Kim, M., Choi, J.M., and Lee, J.Y. (2016). *Chem. Commun.* 52: 10032–10035.
83 Park, I.S., Komiyama, H., and Yasuda, T. (2017). *Chem. Sci.* 8: 953–960.
84 Cui, L.-S., Nomura, H., Geng, Y., Kim, J.U., Nakanotani, H., and Adachi, C. (2017). *Angew. Chem. Int. Ed.* 56: 1571–1575.
85 Cho, Y.J., Jeon, S.K., Lee, S.S., Yu, E., and Lee, J.Y. (2016). *Chem. Mater.* 28: 5400–5405.
86 Duan, C., Li, J., Han, C., Ding, D., Yang, H., Wei, Y., and Xu, H. (2016). *Chem. Mater.* 28: 5667–5679.
87 Sun, J.W., Baek, J.Y., Kim, K.-H., Huh, J.-S., Kwon, S.-K., Kim, Y.-H., and Kim, J.-J. (2017). *J. Mater. Chem. C* 5: 1027–1032.
88 Komatsu, R., Ohsawa, T., Sasabe, H., Nakao, K., Hayasaka, Y., and Kido, J. (2017). *ACS Appl. Mater. Interfaces* 9: 4742–4749. doi: 10.1021/acsami.6b13482.
89 Cha, J.-R., Lee, C.W., and Gong, M.-S. (2017). *Dyes Pigm.* 140: 399–406.
90 Press release (October 2016). www.cynora.com.
91 Zhang, Y., Lee, J., and Forrest, S.R. (2014). *Nat. Commun.* 5: 1–7.
92 Li, G., Fleetham, T., Turner, E., Hang, X.C., and Li, J. (2015). *Adv. Opt. Mater.* 3: 390–397.
93 Forrest, S.R. (2015). *Philos. Trans. R. Soc. A: Math. Phys. Eng. Sci.* 373: 20140320.
94 Fujimoto, H., Suekane, T., Imanishi, K., Yukiwaki, S., Wei, H., Nagayoshi, K., Yahiro, M., and Adachi, C. (2016). *Sci. Rep.* 6: 38482.
95 Dexter, D.L. (1953). *J. Chem. Phys.* 21: 836.
96 Förster, T. (1948). *Ann. Phys.* 437: 55–75.

97 Su, Z., Li, W., Chu, B., Xu, M., Che, G., Wang, D., Han, L., Li, X., Zhang, D., Bi, D., and Chen, Y. (2008). *J. Phys. D: Appl. Phys.* 41: 085103.

98 Zhang, Q., Komino, T., Huang, S., Matsunami, S., Goushi, K., and Adachi, C. (2012). *Adv. Funct. Mater.* 22: 2327–2336.

99 Leung, M., Hsieh, Y.-H., Kuo, T., Chou, P.-T., Lee, J.-H., Chiu, T.-L., and Chen, H.-J. (2013). *Org. Lett.* 15: 4694–4697.

100 Kim, D., Zhu, L., and Brédas, J.-L. (2012). *Chem. Mater.* 24: 2604–2610.

101 Wagner, D., Hoffmann, S.T., Heinemeyer, U., Münster, I., Köhler, A., Strohriegl, P., Ho, S.T., Mu, I., and Ko, A. (2013). *Chem. Mater.* 25: 3758–3765.

102 Takizawa, S., Montes, V.A., and Anzenbacher, P. (2009). *Synthesis (Stuttg)* 2452–2458.

103 Adachi, C. (2013). *SID Symp. Dig. Tech. Pap.* 44: 513–514.

104 Turner, R.G., James, D.G., Friedmann, A.I., Vijendram, M., and Davies, J.P. (1975). *Br. J. Ophthalmol.* 59: 657–663.

105 Sun, J.W., Kim, K.-H., Moon, C.-K., Lee, J.-H., and Kim, J.-J. (2016). *ACS Appl. Mater. Interfaces* 8: 9806–9810.

106 Song, W., Lee, I., and Lee, J.Y. (2015). *Adv. Mater.* 27: 4358–4363.

107 Zhang, D., Duan, L., Li, C., Li, Y., Li, H., Zhang, D., and Qiu, Y. (2014). *Adv. Mater.* 26: 5050–5055.

108 Duan, L. (2015). *SID Symp. Dig. Tech. Pap.* 2: 498–501.

109 Zhang, Y.Q., Zhong, G.Y., and Cao, X.A. (2010). *J. Appl. Phys.* 108: 083107.

110 Li, J., Jiang, Y., Cheng, J., Zhang, Y., Su, H., Lam, J.W.Y., Sung, H.H.Y., Wong, K.S., Kwok, H.S., and Tang, B.Z. (2014). *Phys. Chem. Chem. Phys.* 17: 1134–1141.

111 Hong, Y., Lam, J.W.Y., and Tang, B.Z. (2011). *Chem. Soc. Rev.* 40: 5361–5388.

112 Flügge, H., Rohr, A., Döring, S., Fléchon, C., Wallesch, M., Zink, D., Seeser, J., Leganés, J., Sauer, T., Rabe, T., Kowalsky, W., Baumann, T., and Volz, D. (2015). Reduced concentration quenching in a TADF-type copper(I)-emitter. Proc. SPIE 9566, Organic Light Emitting Materials and Devices XIX, 95661P (22 September 2015). doi: 10.1117/12.2185935.

113 Wallesch, M., Volz, D., Zink, D.M., Schepers, U., Nieger, M., Baumann, T., and Bräse, S. (2014). *Chem. Eur. J.* 20: 6578–6590.

114 Volz, D., Zink, D.M., Bocksrocker, T., Friedrichs, J., Nieger, M., Baumann, T., Lemmer, U., and Bräse, S. (2013). *Chem. Mater.* 25: 3414–3426.

115 Wada, Y., Shizu, K., Kubo, S., Fukushima, T., Miwa, T., Tanaka, H., Adachi, C., and Kaji, H. (2016). *Appl. Phys. Express* 9: 032102.

116 Zhang, Q., Tsang, D., Kuwabara, H., Hatae, Y., Li, B., Takahashi, T., Lee, S.Y., Yasuda, T., and Adachi, C. (2015). *Adv. Mater.* 27: 2096–2100.

117 Lee, I.H., Song, W., and Lee, J.Y. (2016). *Org. Electron.* 29: 22–26.

118 Hung, W.-Y., Tsai, T.-C., Ku, S.-Y., Chi, L.-C., and Wong, K.-T. (2008). *Phys. Chem. Chem. Phys.* 10: 5822.

119 Nakagawa, T., Ku, S.-Y., Wong, K.-T., and Adachi, C. (2012). *Chem. Commun.* 48: 9580.

120 van Eersel, H., Bobbert, P.A., Janssen, R.A.J., and Coehoorn, R. (2014). *Appl. Phys. Lett.* 105: 143303.

121 Coehoorn, R., Van Eersel, H., Bobbert, P.A., and Janssen, R.A.J. (2015). *Adv. Funct. Mater.* 25: 2024–2037.
122 Klubek, K.P., Tang, C.W., and Rothberg, L.J. (2014). *Org. Electron.* 15: 1312–1316.
123 Nakanotani, H., Masui, K., Nishide, J., Shibata, T., and Adachi, C. (2013). *Sci. Rep.* 3: 2127.

Index

a

absorption   20–22
  coefficient   64
  spectra of exciplexes   338
  spectrum   102
  strength analysis   21
acridine orange   277
ACRXTN and ACRSA   282–285
adiabatic electronic energies   299
$Ag_2Cl_2(dppb)_2$ 34
$Ag(dbp)(P_2\text{-nCB})$ 38, 44
  characterization of   40–45
  emission quenching   36–38
Ag(I)-based TADF compounds
  $Ag_2Cl_2(dppb)_2$ 34
  $Ag(phen)(P_2\text{-nCB})$ 34–38
  characterization of $Ag(dbp)(P_2\text{-nCB})$ 40–45
  emission quenching in $Ag(phen)(P_2\text{-nCB})$ 36–38
  energetically lower-lying 4d-orbitals   34
  ligand-centered (LC) orbitals   34
  nido-carborane-bis-(diphenylphosphine) $(P_2\text{-nCB})$ 34
  sterical hinderance   38–40
Ag(I) complexes   1–46, 84–85
$Ag(phen)(P_2\text{-nCB})$ 34–38, 40
aggregation-induced emission (AIE)   403, 472, 563
angular dependence_TADF   215
anionic bidentate ligand diphenyl-bis(pyrazol-1-yl)borate $(pz_2Bph_2^-)$ 142
anthracene crystal   199
atomic force microscope (AFM), ITO/PEDOT   484
Au(I) complexes   84, 150, 159
Avogadro number   21

b

B3LYP/def2-SVP level of theory   15, 20, 21, 23, 24
benzothiazole   403, 404
benzoxazole   403, 404
bidentate binap   84
bilayer-type exciplex   416
2,2′-bipyridyl based ligands   181–182
1,3-bis(carbazol-9-yl)benzene (mCP)   143, 170, 517
1,3-bis{3-[3-(9-carbazolyl)phenyl]-9-carbazolyl}benzene (CPCB)   514
2,7-bis(9,9-dimethyl-acridin-10-yl)-9,9-dimethylthioxanthene-S,S-dioxide (DDMA-TXO2)   449, 457–460
bisdiimine-type copper(I) complexes $([Cu(NN)_2]^+)$ 131
bis(diisobutylphenylphosphino)amido (PNP) ligands   157
bis(diphenyl-phosphino)methane (dppm)   181
bis(diphenylphosphino)ethane (dppe)   181
bis[2-(diphenylphosphino)phenyl]ether (POP)   131, 168, 181, 214, 449
6,6′-bis(2-4diphenylquinoline) (B1PPQ)   338

2,7-bis(phenoxazin-10-yl)-9,9-dimethylthioxanthene-*S,S*-dioxide (DPO-TXO2)   431, 432, 436, 441–443
bis-4,6-(3,5-di-3-pyridylphenyl)-2-methylpyrimidine (B3PYMPM)   332, 365, 414
1,3-bis(9H-carbazol-9-yl) benzene (mCP)   474, 475, 478, 484, 485, 487, 495
2-(bis(trimethylsilyl)methyl)-6-fluoropyridine (3 L)   100
2-(bis(trimethylsilyl)methyl)-6-methylpyridine (2 L)   100
2-(bis(trimethylsilyl)methyl)pyridine (1 L)   100
2-(bis(trimethylsilyl)methyl)quinoline (4 L)   100
blended single-layer exciplex   414–416
blue fluorescent emitter   486–490
blue OLEDs
  academia   552–554
  efficient, stable   550–552
blue PHOLED emitters   257
Boltzmann constant   2
Boltzmann distribution   4, 109, 121, 235, 263
Boltzmann statistics   274, 301, 318
Boltzmann thermal distribution   4
Born–Oppenheimer approximation   299, 304, 305
buffer hole-injection layer (Buf-HIL)   515
tert-butylcarbazole donor   526

C

canonical partition function   259
carbazolyl-modified 1-(2-pyridyl)-pyrazole based ligands   191–192
cationic Cu(I)-bis-phen complexes   275, 276
cationic Cu(I) complexes   62
cationic $[Cu(NN)_2]^+$ and $[Cu(NN)(PP)]^+$ complexes   178
charge-neutralized Cu(I) complex   140
charge-neutral three-coordinate Cu(I) complexes   146–155
charge-separated state   347
charge separation versus charge recombination   343–345
charge transfer (CT)
  emitter   528
  exciplex   332
chelating diphosphine POP ligand   180
chemical structures, of host   517
closed quantum system   312
combined steady-state and decay data analysis   250–252
Commission Internationale de l'Éclairage (CIE)   504, 551
comprehensive efficiency analysis of OLEDs   206–209
computation of ISC and RISC rate constants   264–265
Condon approximation   260
  electronic spin-orbit coupling matrix elements   261–262
  overlap of vibrational wavefunctions   262–263
confined exciplex systems   356
conformational rigidity   270
contact radical ion pair (CRIP)   342–343
copper (I) complexes
  four-coordinated
    Cu(I)-bis-phenanthroline complexes   275—277
  three-coordinated
    Cu(I)-NHC-phenanthroline complex   270–275
Coulomb energy   434, 435
crystalline Cu(I) complexes   128
$[Cu(CH_3CN)_4(POP)]BF_4$ 141
$Cu_2Cl_2(N^\wedge P)_2$ 27
$[Cu(dbp)P_2]BF_4$ 133
$[Cu(dbp)(POP)]BF_4$ 133
$[Cu(dmp)_2]^+$ (dmp = 2,9-dimethyl-1,10-phenanthroline) complex   123
$[Cu(dmp)(dtbp)]^+$ (dtbp = 2,9-di-*tert*-butyl-1,10-phenanthroline)   179

Cu(dmp)(phanephos)$^+$ 15–17, 20, 23, 28–30
[Cu(dmp)(phanephos)](PF$_6$)   18, 19, 134
[Cu(dmp)(POP)]BF$_4$ 131
[Cu(dppbz)$_2$]BF$_4$ (**48**) and [Cu(dppbz)(POP)]BF$_4$ 145
[Cu(dppbz)$_2$]$^+$ (dppbz = 1,2-bis(diphenylphosphino)benzene complex   123
Cu(dppb)(pz$_2$Bph$_2$)   9, 22–25, 29, 30
Cu$_2$(4 L)$_2$ 111
Cu(I) complexes
  with large $\Delta E(S_1–T_1)$
    absorption   20–22
    DFT and TD-DFT calculations   16
    flattening distortions and non-radiative decay   16–18
    radiative $S_1 \to S_0$ rate   20–22
    Strickler-Berg relation   20–22
    TADF properties   18–20
  molecular and electronic structure on emissive properties   178–181
  for OLEDs
    energy levels of molecular orbitals in tetrahedral geometries   122, 123
    ligand variation   123–125
  with small $\Delta E(S_1–T_1)$
    DFT and TD-DFT calculations   22–23
    emission spectra and quantum yields   23
    temperature dependence of the emission decay time and TADF   28–30
    triplet state T$_1$ and spin-orbit coupling   23
Cu(I)-NHC-phenanthroline complex   270–275
Cu(I)-phenanthroline complex, ultrafast dynamics of a   313–316
Cu(L$_{iPr}$)(SPh)   152
Cu(L$_{Me}$)(SPh)   152
[Cu(μ-C∧N)]$_2$ (C∧N = 2-(bis(trimethylsilyl)methyl)pyridine derivatives)

absorption spectrum   102
DFT calculation   103, 104, 112
emission decay kinetic analysis   105, 106, 108, 109
emission properties   104, 105
measurement   111
OLED device   110, 112
outline   100
planer 8-membered cyclic ring   112
synthesis   111
X-ray crystallographic study   101, 102
X-ray structure analysis   112
Cu$_4$(μ-I)$_4$ cubane unit   81
Cu$_2$(μ-X)$_2$ cores   155, 156
[Cu(NN)(PP)]$^+$ complexes   132, 133, 142
  with NN ligands   134–142
  with phen or bipy derivatives as ligands   130–134
[Cu(N^N)(POP)] complexes   78
[Cu(N^N)(P^P)]   61, 77–79
[Cu(N^N)(P^X)]   77–79
[Cu(phen)(POP)](BF$_4$)   189
[Cu(P^P)$_2$]   61, 78
cuprous emissive complexes   177
Cu(pympz)(POP)]$^+$ 78
[Cu(pytfmpz)(POP)]$^+$ 78
Cu$_2$(3L)$_2$ 111
[Cu(tmbpy)(POP)]BF$_4$ 133
[CuX(PPh$_3$)$_2$(4-Mepy)] (X = Cl, Br, I) complexes   79
cyano-based TADF molecules   391–396
cyclometalated iridium(III) complexes   93, 120, 122

d

DCz-DBTO2, temperature dependent emission of   434
decay data analysis   249–252
decay time τ 6
delayed fluorescence (DF)   199, 203–204, 300–301
  kinetics /of the ^{1}CT prompt state   449
  intensity   249

delayed fluorescence (DF) (contd.)
  lifetime 233
  TADF 467
  time resolved emission in solid state 440, 446
delayed $S_1 \rightarrow S_0$ fluorescence 5
$\Delta E(S_1\text{-}T_1)$ minimization 7
density functional theory (DFT) 16, 64, 267, 469, 555
density matrix formalism of MCTDH 311–312
Dexter energy transfer 383, 487, 495
DFT and TD-DFT calculations 16, 22–23
DFT-based multireference configuration interaction 267
diazafluorene 404–405
1,4-diazatriphenylenes (ATP) 402, 403
dibenzothiophene-S,S-dioxide acceptors 531
9,10-dicyanoanthracene (DCA) 336
1,4-dicyanobenzene (DCB) 343
diimine ligands ($[Cu(NN)_2]^+$) 177
dilute methyl cyclohexane (MCH) 431
dimeric Cu(I) chlorido NHC-picolyl complexes 82
dimethylacridine (DMAC) 552, 564
2,9-di-n-butyl phenanthroline ligand 38
dinuclear Cu(I) complex $[Cu_2(NN6)_2P_4](BF_4)_2$ 137
dinuclear Cu(I) complexes 80–83
  $[(dppbz)CuX]_2$ 146
  for OLEDs
    other dinuclear Cu(I) complexes 157
    possessing $Cu_2(\mu\text{-}X)_2$ cores 155–157
dinuclear Cu(I) halide complexes 83
dinuclear TADF-type Cu(I) emitters 122
diphenyl ketone 408–410
2-Di-phenylphosphinopyridine (PyrPHOS) 83
diphenyl sulfoxide 405–407
diphosphine ligands 147
dipole moment operator 64

dipyrido[3,2-a:2′,3′-c]phenazine (dppz) 137
donor-acceptor (D-A) molecular systems, TADF
  cyano group 391–396
  design principles of 384–386
  intramolecular or intermolecular architectures 391
  molecular designs and properties of 380
  nitrogen heterocycle 396
down-hill process 259, 260
DPEPO (bis[2-(diphenylphosphino) phenyl]ether) 214
$[(dppbz)Cu(\mu\text{-}X)]_2$ (X) 155
DPTZ-DBTO2
  emission spectra of 437
  phosphorescence spectra of 438
  steady-state fluorescence spectra of 439
Direct Singlet Harvesting 46
Duschinsky rotation effect (DRE) 308
DV-CDBP 511, 512, 518
DV-MOC-DPS 511

e
electrical excitation 240
  conditions for efficient electroluminescence 241–244
  steady-state 240–241
electric dipole radiation 201
electrochemical cyclo-voltammetry measurements 437
electroluminescence (EL) 476, 482, 501
  devices 490
  efficiency 242
  emitters 257–258
  4P-NPB 489
  properties, of emitters 507
  quantum efficiency 202–203
electroluminescence (or external) quantum efficiency (EQE) 202–203, 282, 382, 390
electron-accepting moieties 365–368, 391, 396
electron-donating 368
  moieties 360

electron donor molecules   360
electronic excitation   2, 267, 271, 277, 286, 331, 347, 471
electronic spin-orbit coupling (SOC)   259
    matrix elements   261–262
electronic spin-spin coupling (SSC)   259
electron paramagnetic resonance (EPR) spectroscopy   302
electron-rich pyridylpyrazole ligands NN20–NN22   141
electron transfer (ET)   333, 437, 471
electron transport layer (ETL)   202, 392, 393, 458, 481
El-Sayed rules   261, 262, 268, 345
El-Sayed's law   302
emission decay kinetic analysis   105, 106, 108, 109
emission decay time   5–7, 28–30, 41, 45, 46, 151, 158, 177, 182, 213, 274
emission layer (EML)   1, 2, 4, 201, 202, 481, 484, 486, 561
emission properties   22, 23, 29, 34, 40, 52, 81, 99, 104–105, 132, 133, 143, 148, 273, 277, 428–438
emission quenching in Ag(phen)(P2-nCB)   36–38
emission spectra and quantum yields   23
emissive charge-neutral three-coordinate Cu(I) complexes   154
emissive Cu(I) complexes   123, 160, 177, 178, 180–182, 184
emissive [Cu(NN)(PP)]$^+$ complexes   194
emitter host interactions   258, 457–459
emitters   2, 36, 61–86, 119–170, 353
energetically lower-lying 4d-orbitals   34
energy diagram for TADF   459
energy gap law   76, 131, 140, 262, 280, 436, 524
energy level diagram   426

energy separation $\Delta E(S_1-T_1)$ 30–34
eosin dyes   6, 19, 28, 29, 44, 321, 426, 485, 493
E-type delayed fluorescence   62, 121, 199, 258, 426
evaporation chamber   548
excimer   104, 232, 331, 386
exciplex   332
    bilayer-type   416
    blended single-layer   414
    decay processes of
        charge-separation vs charge recombination   343–345
        CRIP and SSRIP   342, 343
        fluorescence rate   340–341
        ISC   345–346
    electron acceptor molecule   360
    electron donor molecule   360
    electron-accepting moieties   365–368
    electron-donating moieties   360–365
    electronic structures of
        charge-transfer (CT)   332
        formation   332–335
        HOMO and LUMO   332
        local-excitation (LE)   331
        molecular orbitals   332
        stabilization   335, 337
    OLEDs. See organic light-emitting diodes (OLEDs)
    optical properties of
        absorption spectra   338–339
        photoluminescence   336–338
    organic solid films
        prompt vs. delayed fluorescence   347–350
        spectral shift   350–352
        triplet LE state   346
excitation energies and radiative rate constants
    DFT-based multireference configuration interaction   267
    fluorescence and phosphorescence rates   268–269
    time-dependent density functional theory   266

excited states 450
  cycles 235–238
  properties 452–455

f

fac-Ir(ppy)$_3$ derivatives 93
FASAF, chemical structure of 433
FDQPXZ 524
Fermi's Golden rule (FGR) approach 259, 278, 302, 303
first order Fermi's golden rule 320
first order mixing coefficient 297
first-order perturbation theory 260
flattening distortion dynamics of the MLCT excited state 76–77
flattening distortions and non-radiative decay 16–18
fluorescence (FLUO) 544
  blue 199
  efficiency 239
  materials, TADF 563–564
  OLED 119
  and phosphorescence decays 232
  and phosphorescence rates 268–269
  quantum yield 234
  rate constant 340
formed 335
Förster resonant energy transfer (FRET) 258
Förster energy transfer 383
Förster ENT 358
four-coordinated Cu(I)-bis-phenanthroline complexes 270, 275–277
Fourier transform photocurrent spectroscopy (FTPS) 338
Franck–Condon factors 2, 16, 36, 76, 414
Frank–Condon weighted density of states 446
full wave at half maximum (FWHM) 482
fullerenes 245, 379, 380

g

gated iCCD detection system 440
generic Jablowski diagram for TADF 460
green and blue emitters 77–79
grid based quantum nuclear wavepacket dynamics 305
ground and excited-state wavefunctions 298
gyromagnetic factor 261

h

halogen-bridged binuclear copper complex [(dppbz)Cu(μ-X)]$_2$ 147
halogen-bridged Cu(I) complexes 81
halogen-bridged dinuclear Cu(I) complexes 61, 95, 156
heavy atom effect 159, 313, 425
heptazine 401
heteroatom lone pair orbitals 427
heteroleptic Cu(I) complexes 77, 83, 133, 134, 180, 182–188, 191–194
heteroleptic [Cu(N^N)(P^P)] complexes 61, 78
heteroleptic [Cu(NN)(PP)]$^+$ complexes 177, 178, 181
heteroleptic diimine/diphosphine [Cu(NN)(PP)]$^+$ complexes
  with 2,2′-bipyridyl based ligands 181
  with carbazolyl-modified 1-(2-pyridyl)-pyrazole based ligands 191–192
  with 1-phenyl-3-(2-pyridyl)pyrazole based ligands 192–193
  with 3-phenyl-5-(2-pyridyl)-1H-1,2,4-triazole based ligands 193–194
  with 2-(2′-pyridyl)benzimidazole and 2-(2′-pyridyl)imidazole based ligands 182–185
  with 1-(2-pyridyl)-pyrazole based ligands 189–190
  with 2-(2-pyridyl)-pyrrolide based ligands 188
  with 5-(2-pyridyl)tetrazole based ligands 185–187
  with 3-(2′-pyridyl)-1,2,4-triazole based ligands 187–188

heteroleptic PyrPHOS dinuclear complexes 83
highest occupied molecular orbital (HOMO) 3, 63, 123, 178, 200, 270, 298, 332, 382, 469, 470, 491, 545
highly efficient thermally activated delayed fluorescence device 214–218
highly emissive d^{10} metal complexes
  advantages 61
  Ag(I) complexes 84–85
  Au(I) complexes 84–85
  Cu(I) complexes 61
  dinuclear Cu(I) complexes 80–83
  flattening distortion dynamics of the MLCT excited state 76, 77
  green and blue emitters 77, 79
  molecular structures of Cu(I) complexes 65
  molecular structures of d10 metal complxes 84
  Pd(0) complexes 84–85
  phosphorescence and TADF mechanisms 62, 64
  photophysical properties 72
  Pt(0) complexes 84–85
  structure-dependent photophysical properties 64–76
  three-coordinate Cu(I) complexes 79, 80
hole transport layer (HTL) 201, 202, 547
HOMO-LUMO excitation 8, 40, 45
homoleptic [Cu(NN)$_2$]$^+$ complexes 177–179
homoleptic [Cu(phen)$_2$]$^+$ complex 179
homoleptic [Cu(PP)$_2$]$^+$ complexes 134
homoleptic complex [Cu(POP)$_2$]$^+$ 134
homoleptic copper(I) bisdiimine complex [Cu(dtp)$_2$]BF$_4$ 3 (dtp = 2,9-di-*tert*-butyl-1,10-phenanthroline) 131

i

internal quantum efficiency (IQE) 93, 156, 199, 202, 218, 257, 258, 279, 286, 332, 353, 382, 466, 486, 487, 501
intersystem crossing (ISC) 1, 345–346, 425, 467, 501
  rate constants
    beyond the Condon approximation 263–264
    computation of ISC and RISC rate constants 264
    Condon approximation 260–263
    statical approaches 265
intraligand charge-transfer (ILCT) 125
intramolecular charge transfer (ICT) 426, 469
  emission 121
  processes 277
intramolecular D-A molecules
  bilayer-type exciplex 416
  blended single-layer exciplex 414–416
ionisation potential 434
Ir-based phosphorescent materials 120
isoemissive point 442, 445
isolated heteroleptic complexes [Cu(dmp)(PP)]$^+$ 134
i-Pr (propyl) substituted DPTZ-DBTO2 456

j

Jahn-Teller distortion 80, 150, 313

k

Kasha's rule 300
ketones 245
kinetics of the ^{1}CT prompt state 449

l

^{3}LE$_D$ phosphorescence (PH) 441, 444
ligand-centered (LC) orbitals 34
ligand-to-ligand charge transfer (LL'CT) 3, 139
light-emitting dendrimers 502, 528
light emitting electrochemical cells (LEEC) 1, 133, 177
light-outcoupling efficiency 203, 501

light-outcoupling factor 202, 205, 218, 222
linear vibronic coupling (LVC) model 307
liquid crystal displays (LCDs) 466, 501
lock-in technique 452
low efficiency roll-off triplet-triplet-annihilation device 218–222, 405, 407, 411, 416, 417, 490, 495
lowest unoccupied molecular orbital (LUMO) 3, 63, 123, 200, 270, 298, 332, 382, 469, 470, 491, 546
low order Taylor expansion 306
low-priced pure blue-phosphorescent OLEDs 120
luminescence decay curves of Ag(dbp)(P$_2$-nCB) 42
luminescent Cu(I) complexes 83, 122, 134, 156, 179, 191, 270
luminescent cyclometalated Ir(III) complexes 61, 62
luminescent dinuclear copper complexes
  [Cu(μ-C∧N)]$_2$ (C∧N = 2-(bis(trimethylsilyl)methyl)pyridine derivatives) 100
  cyclometalated iridium(III) complexes 93
  distortion 93
  fac-Ir(ppy)$_3$ derivatives 93
  luminescence properties 99, 100
  1,10-phenanthroline derivatives 93
  structure 94–99
luminescent neutral [Cu(N∧N)POP] complexes 78

m

mCP (1,3-bis(N-carbazolyl)benzene) 214
M06/def2-SVP level of theory 38
measured decay time 6, 16, 19, 29, 217
metal complexes exhibiting TADF 157–159
metal-free TADF emitters
  acridine orange 277
  delayed fluorescence 278
  dibenzothiophene-S-S-dioxide DBTO2 acceptor 278
  Duschinsky effects 278
  mechanism of the triplet-to-singlet upconversion 282–285
  phenothiazine (PTZ) donor 278
  photophysics 279
  reverse internal conversion 278
  second-order spin-orbit 260
  1,2,3,5-tetrakis(carbazol-9-yl)-4,6-dicyanobenzene (4CzIPN) 279
(metal + ligand L) to ligand L′ charge transfer (^{1,3}MLL′CT) 34
metal-to-ligand charge 262
metal-to-ligand charge transfer (MLCT) 3, 17, 62, 93, 100, 103, 120, 259
2-methyltetrahydrofuran (2-MeTHF) 125, 450
microscopic spin-orbit Hamiltonians 261
m-MTDATA:PBD exciplexes 357
molecular fluorescence 300
monodentate halide anions 146
mononuclear Cu(I) complexes for OLEDs
  bis(diimine) type 129, 131
  Boltzmann equation 127
  charge-neutral three-coordinate Cu(I) complexes 146–155
  [Cu(NN)(PP)]$^+$ complexes
    with NN ligands 134–142
    with phen or bipy derivatives as ligands 131–134
  decay rate of emission 127
  electroluminescence processes 126
  electron exchange 126
  emission mechanism 127
  emission processes of phosphorescent emitters 126
  equilibrium constant 127
  free energy change 127
  intramolecular exciplex 126
  non-radiative rate constant 127
  photophysical properties 128
  radiative constant 127
  radiative rate constant 126
  rate constants 128

simple excited state model   126
SOC interaction   126
temperature-dependent non-radiative processes   128
tetrahedral Cu(I) complexes with the LUMO on the PP ligand   129, 142–146
mononuclear Cu(I)-phenanthroline complexes   313
MTXSFCz   490–492
   absorbance PL   491
   molecular structure of   491
multi-configurational Ehrenfest (MCE)   310
multi-configurational time-dependent Hartree (MCTDH) approach   265, 310–311

n

National Television System Committee (NTSC)   504, 512, 523
natural transition orbital (NTO) analysis   124, 144, 388
neutral Cu(I) complexes   62, 81, 140, 142
neutral mononuclear Cu(I) complexes   139
N-heterocyclic carbene (NHC) Cu(I) complexes   153, 270
nido-carborane-bis-(diphenylphosphine) (P_2-nCB)   34
nitrogen heterocycle-based TADF molecules
   benzothiazole and benzoxazole derivatives   403–404
   diazafluorene derivatives   404–405
   1,4-diazatriphenylenes (ATP)   402–403
   heptazine derivatives   401–402
   oxadiazole   401
   pyrimidine derivatives   400–401
   quinoxaline derivatives   404
   thiadiazole   401
   triazine derivatives   396–400
   triazole   401
nonadiabatic coupling elements   304, 309

non-isotropic emitter orientation   204–205, 209, 214, 216, 218, 222
non-polar zeonex matrix   446
non-radiative (NR) quenching   450
non-radiative exciton quenching processes   202
non-radiative rate constant   107, 108, 121, 126–128, 474, 476
nonpolar solvents   139, 343, 430, 433–435, 450, 468
nonradiative decay processes   122, 258, 349
normal fluorescence (NF)   120, 467
NPB, TPBi exciplexes   357
N-phenylcarbazole (N-PhCz)   469
n–π*/π–π* mixed state transitions   431

o

optical modeling   205–206
optical properties of the DPTZ-DBTO2   451
optical transition dipole vectors (TDVs)   200, 204, 205, 208, 215–218, 220, 222
organic donor-acceptor systems   258, 262, 269, 286
organic light-emitting diode (OLED)   177, 501
   black box   209–214
   comprehensive efficiency analysis   206–209
   delayed fluorescence   199, 203–204
   device   110–112, 483
   electroluminescence quantum efficiency   202–203
   emitter materials   544–547
   highly efficient thermally activated delayed fluorescence device   214–218
   low efficiency roll-off triplet-triplet-annihilation device   218–222
   non-isotropic emitter orientation   204–205
   optical modeling   205–206
   optical transition dipole vectors   200
   TADF materials

organic light-emitting diode (OLED) (contd.)
    device structures and operation mechanisms   380–382
    emitters   382
    host materials and sensitizers   382–383
    host-free   383
    research of   380
    TX   465
    Using Exciplexes   353
    to vacuum-deposited counterparts   526–527
    working principle   200–202
organic photo-voltaics (OPVs)   331, 347
organoboron-based TADF molecules   411–412
oxadiazole   137, 401, 402

p

p-and e-type delayed fluorescence   199
Pd(0) complexes   84
perturbation theory   259, 260, 268, 269, 271, 303, 319
PF01 (4,4′-bis[phenyl(9,9′-dimethylfluorenyl)amino] biphenyl) for HTL (40 nm)   110
phase alternating line (PAL) systems   504
phenanthroline-based ligands   179, 180
1,10-phenanthroline (phen)   34
phenothiazine   446
    donors   531
phenothiazine donor and a dibenzothiophene-S,S-dioxide acceptor (PTZ-DBTO2)
    absorption of   431
    chemical structures of   430
phenothiazine donor unit (PTZ)   452
phenothiazine-dibenzothiophene-S,S-dioxide (PTZ-DBTO2)   429
3-phenyl-5-(2-pyridyl)-1H-1,2,4-triazole based ligands   193–194
1-phenyl-3-(2-pyridyl)pyrazole based ligands   192–193

1-phenyl-3-(2-pyridyl)pyrazole diimine ligands   192–193
phosphorescence (PHOS)   6, 544
    and TADF mechanisms   62–64
    and triplet state measurements   438–439
    decay time   6
    lifetime   232
phosphorescent emitters   61, 199
    TADF   564
phosphorescent OLEDs (PHOLEDs)   120, 209, 257, 297, 493
phosphorescent red and green emitting materials   199
photoexcitation
    conditions for efficient TADF   239–240
    excited-state cycles   235–238
    fluorescence and phosphorescence decays   232–233
    rate equations   232
    steady-state fluorescence and phosphorescence intensities   233–234
    TADF on-set temperature   238
photoluminescence of DPTZ-DBTO2   450
photoluminescence of exciplexes   336–338
photoluminescence quantum efficiency (PLQY)   545
photoluminescence quantum yield (PLQY)   177, 181, 348, 382
photophysical properties, of emitters   505
photophysics of TADF
    absorption of PTZ-DBTO2   431
    absorptivity   437
    2,7-bis(phenoxazin-10-yl)-9,9-dimethylthioxanthene-S,S-dioxide   431
    broad Gaussian profile   432
    chemical structure of FASAF   433
    Coulomb energy   434
    dibenzothiophene-S,S-dioxide   429
    dual fluorescence   434
    dynamical processes   455–457

electrochemical cyclo-voltammetry measurements   437
electron transfer   437
emission spectra of DPTZ-DBTO2   437
emission Stokes-shift   430
emitter host interactions   457–459
energy diagram   459
energy gap law   436
excitation power   437
excited state properties   452–455
excited states involved   450–451
fluorene units   429
Gaussian band shape   430
intensified CCD cameras   428
ionisation potential   434
kinetics of the ^{1}CT prompt state   449–450
Lippert-Mataga plot   433
next generation materials   428
nonpolar solvents   435
n–π*/π–π* mixed state transitions   431
optical absorption spectra   431
phenothiazine-dibenzothiophene-S,S-dioxide (PTZ-DBTO2)   429
phosphorescence and triplet state measurements   438–439
photoluminescence quantum yield   438
picosecond and nanosecond fluorescence lifetime measurement   428
polar solvents   435
PTZ-DBTO2 absorption spectrum   430
red edge absorption   436
solvatochromism   433, 435
solvent orientation polarizability parameter   433
solvent polarity   436
standard integrating sphere measurement   437
temperature dependent emission of DCz-DBTO2   434
time correlated singlet photon counting techniques   428
time resolved emission in solid state   446–448
time resolved emission in solution   440–446
twisting mechanism   434
X-ray crystal structure of DPO-TXO2   432
platinum group metals (PGM)   549
PMMA-doped emitter   36
4P-NPB   486, 487
polarized angular dependent emission spectroscopy   206, 207
polar polyethylene oxide (PEO)   322
polar solvents   280, 285, 435, 437, 471, 514
polycyclic aromatic hydrocarbons   245
polymer-based OLED   199
poly(para-phenylenevinylene)   297
polystyrene (PS) host matrix   533
poly(vinyl carbazole) (PVK)   133
potential energy surfaces   304–309
p-polarized angular dependent emission pattern   211
prompt and delayed fluorescence intensities   245–248
prompt fluorescence (PF)   234, 300
prompt versus delayed fluorescence   347–350
prototype OLEDs   142
pseudo-equilibrium constant   63
pseudo Jahn–Teller distortion   313
pseudo-Jahn-Teller (PJT) effect   68, 125, 313
[Pt(0)(binap)$_2$]   84
Pt(0) complexes   84
P-type delayed fluorescence   426
Purcell effect   202, 205, 208
Purcell factor   205, 208, 353, 358
2-(2′-pyridyl)benzimidazole ligands   137, 139
2-(2′-pyridyl)benzimidazole unit   139
2-(2′-pyridyl)benzimidazolylbenzene   183, 184
2-(2′-pyridyl)benzimidazolyl diimine ligands   183, 185
2-(2′-pyridyl)benzimidazolyl ligand   184

1-(2-pyridyl)-pyrazole based ligands 189–190
2-(2-pyridyl)-pyrrolide based ligands 188–189
1-(2-pyridyl)-pyrazole (czpypz) based diimine ligand 191
5-(2-pyridyl)tetrazolate (NN13) 140, 185–187
5-(2-pyridyl)tetrazole based ligands 185–187
3-(2′-pyridyl)-1,2,4-triazole based ligands 187–188
3-(2′-pyridyl)-1,2,4-triazole diimine ligands 187–188
pyrimidine 360, 365, 400
PyrPHOS dinuclear complexes 83

q

quadratic vibronic coupling (QVC) model 307
quantum mechanical considerations 6, 27, 32–34, 45
quantum-well (QW) 481–483, 485
quantum yield of triplet formation 230, 410
quinoxaline 360, 365, 366, 404, 525

r

radiative decay rates 18, 19, 21, 32, 33, 38, 45, 205, 274, 275, 298, 300, 323, 382, 452, 467, 468, 477, 524, 561
radiative exciton fraction 202–204, 206, 208–213, 218, 219, 221, 222
radiative fluorescence rate $k_r(S_1 \rightarrow S_0)$ 7
radiative quantum efficiency (RQE) 202, 210, 211, 213, 217
radiative rate constants 62–64, 69, 107, 108, 119, 121, 128, 266–269
radiative singlet-singlet rate $k_r(S_1 \rightarrow S_0)$ 30
radiative $S_1 \rightarrow S_0$ rate 20
rate equations 232, 240, 263, 347
Rayleigh-Schöodinger perturbation 269
refractive index 21, 33, 64, 208, 340, 341, 390

reverse intersystem crossing (rISC) 5, 63, 120, 258, 302, 426, 467, 502
PTZ-DBTO2 316–322
triplet-to-singlet transformation 377, 378
roll-off effect 3, 120
Rosenberg-Parker method 245, 249
[Ru(bpy)$_3$]$^{2+}$ (bpy = 2, 2′-bipyridine) 120

s

saturation/triplet-polaron quenching 3
sensitizers 331, 356–360, 377, 382–383, 399, 410, 416, 417
single particle density operators (SPDO's) 312
singlet and triplet excitons 2, 93, 117, 120, 121, 178, 298, 377, 381, 382, 466, 486, 487, 495, 502, 560
singlet exciton sensitizers 383, 416
singlet harvesting mechanism 2, 5
slow spin-lattice relaxation (SLR) 26
small energy separation $\Delta E(S_1-T_1)$ 2, 3, 32
small-molecule OLEDs 257
small splitting $\Delta E(S_1-T_1)$ 8
small zero-field splittings (ZFS) 26
solid state lighting (SSL) 501, 502
solution-processed blue TADF materials and devices 504–512
solution-processed green TADF materials and devices 512–523
solution-processed TADF polymers and dendrimers
  non-TADF monomer, TADF macromolecule 532
  phenothiazine donors 531
  TTA 528
solution-processed yellow-to-red TADF materials and devices 523–526
solvatochromism 341, 433, 435, 471
solvent orientation polarizability parameter 433
solvent-separated radical ion pair (SSRIP) 342–344

spectral shift   19, 64, 79, 150, 350–352, 414
spin angular momentum   63
spin-orbit coupling (SOC)   1, 23–27, 62, 63, 70, 81, 83, 85, 120, 177, 202, 259–262, 297, 302, 332, 381, 466
spin-orbit coupling matrix element (SOCME)   260–262, 264, 268, 269, 276, 277, 281, 286, 302, 315, 316, 318
spin-coating or blade-coating   528
spin-dependent interaction operator   259
spin-lattice relaxation (SLR)   4, 28
spin-orbit coupling (SOC)   1–4, 6, 23, 26, 27, 44, 46, 120, 131, 157, 177, 178, 259–263, 277, 284, 297, 302, 332, 345, 454, 466, 502
$S_1 \to S_0$ fluorescence   3, 30
steady-state data analysis   245–249
steady-state fluorescence and phosphorescence intensities   233
S–T energy gaps   121, 122, 214, 258, 259, 262, 270, 279, 280, 427, 495
sterically encumbered phenanthroline ligands   179
Stokes shift   69, 70, 80–82, 85, 140, 190, 263, 280, 430
Strickler-Berg equation   64
Strickler-Berg relation   20–22
strongly green–blue emitting [Cu(pypz)(POP)]$^+$ 78
substituted phen ligands   34
synthesized heteroleptic d^{10} metal complexes   79

t

TADF-type heteroleptic copper(I) emitters   142
TADF-type OLEDs   121, 144, 152, 157
Tamm–Dancoff approximation (TDA)   266, 316
TB-1PXZ   510, 511, 522
temperature dependence, emission decay time and TADF   28–30
temperature-dependent radiative constants   64
2,4,5,6-tetra(carbazol-9-yl)-1,3-dicyanobenzene (4CzIPN)   512
1,2,4,5-tetracyanobenzene (TCB)   342, 343
tetrahedral Cu(I) complexes   122, 123, 127
tetrahedral Cu(I) complexes with the LUMO on the PP ligand   129, 142–145
tetrahedral d^{10} copper (I) complexes   160
tetrahedral four-coordinate Cu(I) complexes   79–80
tetrahedral TADF-type Cu(I) complexes   129
1,2,3,5-tetrakis(carbazol-9-yl)-4,6-dicyanobenzene (4CzIPN)   279–282
tetra-ter-butylperylene (TBPe)   285
thermally activated delayed fluorescence (TADF)   199, 377, 502
  approaches   559–566
  blue OLEDs
    academia   552–554
    efficient, stable   550–552
  chromophore   529
  Cu-complex   379
  data analysis
    combined steady-state and decay data   250–252
    decay data   249–250
    steady-state data   245–249
  development of   377, 379
  diphenyl ketone   408–410
  diphenyl sulfoxide   405–407
  donor-acceptor (D-A) molecular systems   377
  emitter materials, OLED   544–547
  emitters   554, 555
    copper (I) complexes   269
    emitter-host interactions   258
    excitation energies and radiative rate constants   266
    intersystem crossing   259
    metal-free TADF emitters   277

thermally activated delayed fluorescence (TADF) (contd.)
  metal-to-ligand charge-transfer (MLCT)   259
  singlet and triplet-coupled open-shell configuration   258
  spin-orbit coupling (SOC)   259
  thermally stable dyes   258
 energy diagram of   467
 energy transfer   561
 eosin   379
 fluorescent materials   563–564
 intramolecular D-A molecules   413
 kinetics
  Arrhenius equation   229
  Boltzmann-weighted average   230
  complex schemes   244
  direct ISC   229
  electrical excitation   240–244
  excitation types   231
  photoexcitation   232
  quantum yield of triplet formation   230
  radiative rate constants   229
  rISC   229
  three-state scheme   229
 luminescent efficiency of   389–391
 material design
  Ag(I)-based TADF compounds   34–36
  Boltzmann constant   2
  Cu(I) complex with large $\Delta E(S_1\text{-}T_1)$ 15–22
  Cu(I) complex with small $\Delta E(S_1\text{-}T_1)$ 22–30
  energy separation $\Delta E(S_1\text{-}T_1)$ 30
  Franck-Condon factors   2
  intersystem crossing   1
  ligand-to-ligand charge transfer (LL'CT)   3
  metal-to-ligand charge transfer   3
  molecular parameters, and diversity of materials   4–15
  photocatalysis   1
  photoluminescence quantum yields   2
  photon generation   2
  photophysical principles   1
  relaxation properties   1
  saturation/triplet-polaron quenching   3
  $S_1 \rightarrow S_0$ fluorescence   3
  $S_1 \rightarrow S_0$ fluorescence rate   30–34
  singlet harvesting mechanism   2
  spin-orbit coupling   1
  triplet emitters   2
  triplet harvesting effect   2
  zero-field splitting   3
 OLEDs   379, 380, 467
 on-set temperature   238
 organoboron   411–412
 phosphorescent emitters   490–495, 564–566
 polymers   412
 processing aspects   547–549
 with quantum-well structure   481–486
 singlet-triplet energy splitting   386–389
 sustainability aspects   549–550
 TADF emitters   562–563
 x-bridged diphenyl ketone   410–411
 x-bridged diphenyl sulfoxide   407–408
thermally stable dyes   258
thiadiazole   401, 402
thiones   245
thioxanthone (TX)   465
 OLEDs   465–467
 phosphorescent emitters   490–495
 TADF emitters with QW   481–486
 TX-based TADF emitters   468–490
 WOLED based 4P-NPB   486–490
9-H-thioxanthen-9-one-10,10-dioxide (TXO)   469
third generation-emitter materials   199, 222
three-coordinate Cu(I)
 complexes   79, 80, 147, 155
 phenathroline complexes   80
three-coordinated Cu(I)-NHC-phenanthroline complex   270–275

time correlated singlet photon counting techniques   428
time-dependent density functional theory (TDDFT)   16, 64, 81, 266–270, 276–283, 286, 316, 318, 475
time-dependent Hartree (TDH) method   310
time-dependent Schrödinger equation   303
time-dependent self-consistent field (TDSCF)   310, 311
time-independent electronic Schrödinger equation   303
time-resolved EL and PL spectroscopy   207
time-resolved electroluminescence (EL)   207, 221
time resolved emission in solid state   446–448
time resolved emission in solution   440–446
time resolved fluorescence decay of DPO-TXO2   441, 442
time resolved gated emission measurements   428
time resolved normalized emission spectra of DDMA-TXO2   449
transition dipole moment   23, 32, 62, 63, 204, 205, 210, 219, 222, 269, 276, 285, 297, 340, 358, 384, 390, 525
transition dipole vector (TDV)   200, 201
transition electric dipole moment (M), 64
triazine derivatives   396–400
triazolylpyridine ligand N24–N27   141
4,4′,4″-tri(9-carbazoyl)triphenylamine (TCTA)   481
triphenylamine (TPA)   395, 401, 403, 412, 469
triphenylphosphine (PPh_3)   69, 79, 180–182
triplet emitters   2
triplet exciplex   127, 345–348, 358, 386, 414, 415
triplet exciton-polaron quenching (TPQ)   383
triplet fusion (TF)   278, 425
triplet harvesting effect   2
triplet lifetime   120, 313, 443, 444
triplet relaxation processes   230
triplet state T_1 and spin-orbit coupling   23–27
triplet–triplet annihilation (TTA)   8, 104, 120, 178, 199, 356, 383, 425, 502, 528
4,4′,4″-tris(N-carbazolyl)-triphenylamine (TCTA)   332, 414
1,3,5-tris(N-phenylbenzimidazole-2-yl)benzene (TPBI)   477, 478
tris-or bis(cyclometalated) iridium(III) complexes   120
$T_1 \rightarrow S_0$ and $S_1 \rightarrow S_0$ decay processes   4
Tully's Trajectory Surface Hopping (TSH)   309
twisted intramolecular charge transfer (TICT)   63, 471
twisting mechanism   434
two-emitting states model   121
TX-based TADF emitters
 aggregation-induced emission   472
 delayed component   477
 high radiative decay rate   469
 mCP   474
 OLED structure   475
 TCTA   481
 temperature dependence   474
 TICT   471
 TPBI   477
 TXO-TPA and TXO-PhCz   469, 471
TXO-PhCz   471
 emission characteristics   470
 mCP   485
 molecular structure   478
 TADF   482
 transient PL   478
TXO-PhCz:TPBI   477–480
TXO-TPA
 absorption spectra   471
 AIE   472
 blue fluorescent emitter   486

TXO-TPA (contd.)
  emitting layer   475
  mCP   474, 486
  molecular structures   468
  phosphorescent   471
  single crystal structure   473
  TCTA   481
  temperature dependence   474
  TPBI   477
  transient PL   475
  triplet energy   471
  UV-vis absorption   472
TXO-TPA:mCP   474, 476, 486

u

ultrafast spin-vibronic dynamics   315

v

vacuum thermal evaporation (VTE)   547
vapor-deposited OLEDs   81, 155, 157
variational multi-configurational Gaussian wavepacket (vMCG)   310
vibrational thermalisation   300
vibrational wavefunctions, overlap of   262
vibronic coupling
  adiabatic electronic energies   299
  Born–Oppenheimer approximation   299
  delayed fluorescence   300–301
  density matrix formalism of MCTDH   311–312
  election exchange interaction   298
  electronic ground state   298
  Fermi's Golden rule (FGR) approach   303
  first order mixing coefficient   297

  ground and excited-state wavefunctions   298
  model Hamiltonian   306–309
  multi-configurational time-dependent Hartree approach   310–311
  perturbation theory   303
  poly(para-phenylenevinylene)   297
  potential energy surfaces   304–306
  radiative rate   297
  rISC   302–303
  rISC of PTZ-DBTO2   316–322
  singlet and triplet excitons   298
  systematic material design   299
  time-dependent Schrödinger equation   303
  time-independent electronic Schrödinger equation   303
  Tully's Trajectory Surface Hopping (TSH)   309
  ultrafast dynamics of a Cu(I)-phenanthroline complex   313–316

w

white organic light emitting diodes (WOLED)   486, 488, 490
work function (WF)   200, 515

x

xanthene dyes   245
X-ray crystallographic study   101–102
X-ray crystal structure of DPO-TXO2   432

z

Zeonex   237, 238, 245–251, 438, 446, 448, 450, 452, 457, 458
zero-field splitting (ZFS)   3, 6, 7, 26, 70, 120, 230, 260